T 174·7 FUN.

Fundamentals of
Nanotechnology

Fundamentals of
Nanotechnology

Gabor L.   John J.   Harry F.   Joydeep
Hornyak • Moore • Tibbals • Dutta

CRC Press
Taylor & Francis Group
Boca Raton   London   New York

CRC Press is an imprint of the
Taylor & Francis Group, an **informa** business

Cover design by Jim Von Ehr, II and Michael Hamers, Lightspeed Commercial Arts, Boulder, CO.

The Zyvex nProber™ is designed to probe sub-100 nm features on semiconductor devices. The system is composed of a state-of-the-art Zyvex Nanomanipulator, an FEI Quanta™ 200 FEG SEM, a Keithley 4200 parametric analyzer, an advanced anti-contamination system and custom software that controls and integrates components. There are 8 encoded positioners. The XYZ fine resolution is 5 nm with a probe tip of 50 nm diameter. There is 200 nm resolution over a distance of 1 mm. The image is courtesy of Zyvex Instruments, LLC, Richardson, Texas. More images of this instrument are included in chapter 2 of the text.

CRC Press
Taylor & Francis Group
6000 Broken Sound Parkway NW, Suite 300
Boca Raton, FL 33487-2742

© 2009 by Taylor & Francis Group, LLC
CRC Press is an imprint of Taylor & Francis Group, an Informa business

No claim to original U.S. Government works
Printed in the United States of America on acid-free paper
10 9 8 7 6 5 4 3 2

International Standard Book Number-13: 978-1-4200-4803-2 (Hardcover)

This book contains information obtained from authentic and highly regarded sources. Reasonable efforts have been made to publish reliable data and information, but the author and publisher cannot assume responsibility for the validity of all materials or the consequences of their use. The authors and publishers have attempted to trace the copyright holders of all material reproduced in this publication and apologize to copyright holders if permission to publish in this form has not been obtained. If any copyright material has not been acknowledged please write and let us know so we may rectify in any future reprint.

Library of Congress Cataloging-in-Publication Data

Hornyak, Gábor Louis.
   Fundamentals of nanotechnology / Gabor L. Hornyak ... [et al.].
      p. cm.
   Includes bibliographical references and index.
   ISBN-13: 978-1-4200-4803-2 (alk. paper)
   ISBN-10: 1-4200-4803-1 (alk. paper)
   1. Nanotechnology. 2. Nanostructured materials industry. I. Title.

T174.7.H67 2008
620'.5--dc22
                                          2008028901

Visit the Taylor & Francis Web site at
http://www.taylorandfrancis.com

and the CRC Press Web site at
http://www.crcpress.com

$\mathscr{C}$ONTENTS

SECTION 5:   BIOLOGICAL AND ENVIRONMENTAL NANOENGINEERING   479

Chapter 11   NANOBIOTECHNOLOGY   481

REFACE

Fundamentals of Nanotechnology is the companion volume to *Introduction to Nanoscience*. Whereas *Introduction to Nanoscience* addresses the scientific aspects of nano (e.g., sans technological applications), *Fundamentals of Nanotechnology* explores the practical applications of nano by discussing new materials, chemicals, coatings, pharmaceuticals, components, mechanisms, devices and systems. *Fundamentals of Nanotechnology*, although focusing primarily on applications, is not devoid of science—it has plenty. Enough scientific background is presented in order to understand the principles behind the technology.

Teaching Nanoscience and Nanotechnology.   The purpose of this book is to serve as a one semester generalized text in nanotechnology at the upper division level in college. As with *Introduction to Nanoscience*, teaching a highly interdisciplinary topic to an audience with a varied background and mathematical experience is quite a challenge. In the preface to *Introduction to Nanoscience*, we describe the challenges that face instructors bent on teaching nanoscience (or nanotechnology). In short, the issues facing them include one or all of the following:

1. Teaching a subject that is growing exponentially with each year.
2. Teaching a subject that is highly interdisciplinary and that consists of contributions from the fundamental monolithic "liberal arts and sciences" of physics, chemistry, and biology and those within the realm of engineering (electrical, mechanical, chemical) and computer science.
3. Teaching a course to students that may have varied backgrounds and technical and mathematical expertise.

Ideally, teachings from *Introduction to Nanoscience* should precede teachings from *Fundamentals of Nanotechnology*. Although we recommend this order to present the curriculum, it is not absolutely necessary to do so.

Textbook Strategy.   The text is divided into five general divisions: (1) Perspectives, (2) Electromagnetic Nanoengineering, (3) Mechanical Nanoengineering, (4) Chemical Nanoengineering, and (5) Biological and Environmental Nanoengineering. Similarly to *Introduction to Nanoscience*, a generalized *physics* → *chemistry* → *biology* order is still in effect, but one that is overlain on an engineering scaffold:

> *Electromagnetics and Mechanics* → *Catalysis and Composites* → *Biotechnology, Biomimetics, Nanomedicine* → *Environment*

The "Perspectives" division consists of two chapters. In chapter 1, "Introduction," we review definitions, discuss the revolutionary (or evolutionary) impact of nano, and present a short business course. Traditionally, scientists and engineers are not business savvy and vice versa (our apologies upfront to those who have managed to span and mingle within each domain successfully!). In this way, this text is somewhat unusual. The world is driven by business and markets. Why not expose scientists and engineers to some of those aspects of our practical world early on in the text? A short essay on nanotechnology, education, and workforce development is also presented. Once again, we try to provide a real-world value to the text. We are all confronted with career choices at one time or another in our lives. Why not place seeds in the heads of students—seeds that may sprout into a career centered on nanotechnology directly or upon some field ancillary to the science, R&D, or manufacturing? There are many kinds of stimulating and lucrative careers involving nanoscience and nanotechnology that directly involve the science and the technology but also those that exist outside the laboratory and production line.

Then there are the buildings that house the science and technology—the fortresses of science and engineering, the nuclei from which new ideas and developments radiate, the stages where minds act out scenarios ranging from the believable to the unattainable. What about them? How have they evolved to support the study of the incredibly small? How have they evolved to accommodate the new demands placed on research that is highly interdisciplinary and must house equipment and instruments that measure dimensions on the order of nanometers and less? Lastly, we familiarize the student with the preexisting infrastructure (local, regional, national, and global) that centers on nanotechnology. As is the case with any other thing, nanotechnology has not developed in a vacuum. There is the human infrastructure from which the funding, policies, strategies, and levels or organization operate.

Chapter 2 is titled "Nanometrology: Standards and Nanomanufacturing." Intrinsically, we all understand the importance of metrology and standards. In this chapter, we present a conscientious and deliberate effort to do so. Without metrology, we have no standards and no manufacturing—at least in the way we are accustomed. The contribution of nanodevices in particular has opened portals to the quantum domain and the metrology that ensues. Today's metrology has evolved down to the level of the nanometer and beyond. The semiconductor industry has done much to establish micrometrology over the past 30 years or so. Terms such as "critical dimension" and others are now commonplace in that industry. With nanoscale devices already existing in computers, nanoscale metrology is proceeding at a great clip. There is understandable effort to adapt as quickly as possible to the changing paradigm of this new frontier.

Nanometrology dwells at the interface of macro- and microscale metrology and quantum measurement. Arguments concerning the degree of "quantum" and degree of "bulk" in nanometrology are not necessarily important although they do help in defining boundaries. Some nanometrologies simply do not involve formalized quantum considerations—others by necessity do. The metrology of the micron world is well defined and the usual embodiment of uncertainty is well understood. Certain aspects of the metrology of the nano-world involve a unique kind of uncertainty—that of the Heisenberg uncertainty. Just how important is Heisenberg uncertainty in nanometrology? We can draw our own conclusions after studying the text. Nanomanufacturing, some practices already in place, and molecular nanomanufacturing are also reviewed in this chapter.

The second division of the book, "Electromagnetic Nanoengineering," is composed of three chapters that address electronics, optics, and magnetics, respectively. Chapters 3 through 5 are centered on the electromagnetic character of components and nanodevices. Each chapter delves into the salient aspects of each topic and relates them to nano phenomena and ultimately to their applications. Chapter 3, "Nanoelectronics," provides a basic background for electronics, nanoelectronics, as well as for molecular electronics—the conduction of electrons through molecules. It is, after all, the electronic behavior of nanomaterials that makes them special. The threshold of the transfer of a single electron has been reached thanks to nanoscale devices. Nature has been transferring electrons one at a time for quite some time. We provide discussions about band structure, semiconductors, and transistors—all physical phenomena and devices that make our twenty-first century electronics go round and round.

Chapter 4, "Nano-optics," addresses topics centered on optical properties of nanomaterials—certainly a consequence of the behavior of electrons at the nanoscale. Topics include interactions of light with matter, the surface plasmon, scattering, and the electronic basis of color. Quantum dots are treated prominently in this chapter. We will discover how the emission of light is a function of particle size. Nanophotonics, of course, is offering new routes to solve technological challenges. Chapter 5, "Nanomagnetism," explores the magnetic properties of zero-, one-, and two-dimensional nanomaterials. Of course, applications of electromagnetic and optical phenomena via nanomaterials is presented in all the chapters.

The third division of the book is titled "Mechanical Nanoengineering." It consists of chapter 6, "Nanomechanics;" chapter 7, "Nanostructure and Nanocomposite Thin Films;" and chapter 8, "Applications of Nanostructure and Nanocomposite Thin Films." The mechanical engineer is a major player on the stage of nanotechnology. Many nanoscience and nanotechnology programs are grounded in materials science and engineering departments. This is by no accident. The mechanical engineer has shifted his or her focus from the macroscopic world to that of the micro and nano domains.

Chapter 6, "Nanomechanics," is intense, challenging, and provides a well-rounded approach to nanomechanics. It describes atomic forces and principles and develops insight into molecular dynamics. Two- and three-atom chains, interaction potentials, external forces, and dynamic motion comprise the introductory section and subject matter of this chapter. The latter parts of the chapter discuss a variety of topics ranging from carbon nanotube structure and mechanical properties, nanomechanical measurements, and nanoindentation to microelectromechanical systems (MEMS) and nanoelectromechanical system (NEMS). Chapter 7, "Nanostructure and Nanocomposite Thin Films," is the first part of a two-chapter series that covers nanostructure and thin film engineering. Part two of this miniseries (chapter 8, "Applications of Thin Films") focuses on the applications of the concepts presented in chapter 7. Chapters 7 and 8 focus on coatings technology.

The fourth division of the book centers on chemical topics and is appropriately titled "Chemical Nanoengineering." Chapter 9, "Nanocatalysis," provides a glimpse into the world of catalysis, and chapter 10, "Nanocomposites and Fibers," explores in more detail mechanical engineering principles as they apply to nanomaterials. The importance of catalysts in today's commercialized society cannot be understated. And, more than coincidentally, the amount of catalysts that are nanoscale simply add more importance to the previous statement as well as to the importance of nanotechnology.

Chapter 10, "Nanocomposites and Fibers," traces the evolution of composite materials from early times to the present in which carbon nanotubes are woven together to make the world's strongest fibers. Mechanical properties and testing are important aspects of any engineering material development. A general overview of mechanical (as well as thermal, electrical, and chemical) properties and testing is provided in this chapter. We then discuss carbon fibers, filaments, and nanotubes and review the basic kinds of composite material hosts that utilize nanocomponents: polymers, metals, and ceramics. With this chapter, we conclude the brief foray into chemical nanoengineering.

The fifth and final division of the book covers biological, medical, and environmental aspects of nanotechnology. The fifth division consists of four chapters. Chapter 11, "Bionanotechnology," introduces topics related to biotechnology with a nano spin. These include both natural and synthetic biotechnology systems. Chapter 12, "Biomimetics" represents an interface between the natural and synthetic worlds. Although not all of our engineering developments were inspired by biology, why not try and learn from the best. Chapter 13, "Medical Nanotechnology," deals with topics such as nano-encapsulated pharmaceuticals; medical imaging to prosthetics of various sorts are presented and discussed.

The book concludes with chapter 14, "Environmental Nanotechnology," which is an important chapter. The late professor and Nobel laureate Richard Smalley was a major proponent of applying nanotechnology to environmental problems. Types of environmental sensors, water and soil quality monitoring and mitigation, air quality monitoring and mitigation, and energy are some of the broad categories discussed. We leave the students with a consideration of the environment—our environment—and a key, perhaps, on how to keep it in good shape for future generations.

Thrust of the Text. In addition to its function as a pedagogical tool, *Fundamentals of Nanotechnology* should project a broad application of nanotechnology to students. It is our goal to provide interest and to stimulate; to forecast the future in some sense; and to open windows to potential careers and business paths. Much is at stake in our uncertain world. Geopolitical and economic forces drive our future. Nanotechnology will play a role. Let us make sure that role is productive and serves to enhance the human condition without neglecting (or damaging) our house.

Links to Other Books.
www.nanoscienceworks.org/textbookcommunity/introtonanoscience

$\mathscr{A}$CKNOWLEDGMENTS

I would first and foremost like to acknowledge the work of the coauthors **Professor John J. Moore**, director of the Interdisciplinary Materials Science Graduate Program and the Advanced Coatings and Surface Engineering Laboratory at the *Colorado School of Mines*; **Dr. Harry F. Tibbals**, director of Bioinstrumentation Resource Center at the University of Texas *Southwestern Medical Center* at Dallas; and **Dr. Joydeep Dutta**, director of the Center of Excellence in Nanotechnology at the *Asian Institute of Technology* in Bangkok, Thailand. Special acknowledgment is also extended to **Professor Scott W. Cowley** for his effort in putting together chapter 9. Special thanks are forwarded on behalf of **Dr. Jianliang Lin**, research faculty, Metallurgical and Materials Engineering, for his contribution to chapters 7 and 8 involving nanocomposites and thin films; and to graduate student **Masood Hasheminiasari**, Advanced Coatings and Surface Engineering Laboratory, for his contribution to chapter 6—both at the *Colorado School of Mines*.

We are also grateful to **Michael Burke**, a contributor to chapter 1 who described the business aspects of nanotechnology and took us by the hand from starting a nanobusiness to potential exit strategy scenarios; and to **Mihail Roco** of the National Nanotechnology Initiative whose coordination and inspiration have pushed nanotechnology to the forefront of federal government efforts.

We would like to specially acknowledge the contributions of **Jim R. Von Ehr, II**, the chairman and CEO of *Zyvex Instruments* and *Zyvex Labs* as well as the managing director of *Zyvex Asia*, not only to this text in the form of figures and quotations, but also to his overall efforts to develop and promote nanotechnology in the United States and the world. His contribution of $3.5 million to establish the University of Texas at Dallas *NanoTech Institute* and the founding of the *Texas Nanotechnology Initiative* serve as examples of his dedication to nanotechnology.

As one will certainly conclude upon reading chapter 2, much of the development of nanotechnology is taking place in the laboratories of the *National Institute for Standards and Technology* (NIST)—laboratories and scientists at Boulder, Colorado and Gaithersburg, Maryland. There is no doubt that NIST is by far the leader in paving the way for metrology, standards, and applications of nanotechnology. A common link exists between NIST and one of its state-of-the-art buildings, the Advanced Measurement Laboratory (AML), and that link is *HDR Architecture, Inc.*, the engineering and architecture firm that delivered its design. We specially thank **Ahmad Soueid** (and friends **Josh Rownd** and **Mark Jamison** of HDR), national director for HDR's Advanced Research Program, for providing information and resources about the AML. HDR is at the forefront of buildings for advanced technology and has sponsored several conferences concerning the design, engineering, and architecture of such structures that house the future of nanotechnology (Buildings for Advanced Technology, BAT series). HDR is responsible for the design of Purdue's *Birck Nanotechnology Center*, Brookhaven National Lab's *Center for Functional Nanomaterials*, Sandia-Los Alamos labs' core and gateway facilities at the *Center for Integrated Nanotechnologies* (CINT), *Nanotech 1* at the University of Southern Florida, and the *Quantum-Nano Centre* at the University of Waterloo in Canada.

There is no doubt that without the direction, creative insight, and expertise of the management and staff at CRC Press, this book would not have taken its final form. First and foremost, we all would extend our special thanks to **Nora Konopka**, publisher, Engineering and Environmental Sciences Division of CRC Press, Boca Raton, Florida, who has "gently" but "firmly" guided us through this process a second time and who generously donated her mesmerizing voice to the slidecast that features the books; **Jessica Vakili**, project coordinator, for her seemingly endless energy, patience, and support; **Ashley Gasque** and **Andrea Dale**, editorial assistants to Nora; **Kacey C. Williams**, content and community director (CRC Press, Nashville, TN), for her work developing the Web material with us; **Vance McCarty** who has done a magnificent job with Nanoscienceworks.org in featuring the text, its study guide, and author bios; the expertise and guidance of **Glenon C. Butler**, project editor who oversaw the nuts and bolts of the publishing process; **Randy Burling**, manager, production editing; the marketing efforts of **Arlene Kopeloff**, editorial assistant to **Catherine Giacari**; and **Anitha Johny**, account manager, *SPi*, Pondicherry, India who coordinated the copyediting and essentially pulled the book together. Also, special thanks to **Rick Beardsley**, VP of Production and Manufacturing, with whom I will be putting together a "First Author's Guide," once the contracted work is completed of course!

Special appreciation and thanks are extended to **Mike Hamers** of *Lightspeed Commercial Arts* in Niwot, Colorado for his expertise in designing the cover and rendering images production ready; and to **Professor Anil K. Rao**, a professor of biology at the Metropolitan State College of Denver, who helped develop many of the images seen throughout the text.

Gabor L. Hornyak

In addition to the many specific acknowledgments noted in the text and figures, I wish to thank **Professor Johann Deisenhofer** and **Dr. Richard Baxter** for discussions, images, and source materials in the sections on photosynthesis and biomimetic protein drug design. Also, I wish to thank **Drs. D. Wawro** and **R. Magnusson** of *Resonant Sensors Incorporated* for contributing material to the sections on photonic devices, and **Professor Anil Rao** for images and material on bone and shell composite nanomaterials. In addition to those cited in the text, I would like to thank faculty and researchers at UT Southwestern, UT Dallas, UT Arlington, and at Baylor, Rice, Colorado, and Durham universities for information, discussions, and opportunities to observe labs and procedures; and **Ely Silk** for information on diatoms and permission to use images from his Web site at http://www.viewsfromscience.com/documents/webpages/methods_p3.html.

For general discussions, inspiration, and support, I would like to thank **Professor Ruth Ann Word MD**, and the members of her team at UT Southwestern and Sandia Laboratories; special thanks also to **Professors Cole Giller MD, John Truelson MD, Shou Tang MD, Berj Gimi MD, J.-C. Chiao PhD, Austin Cunningham PhD, Roy Chaney PhD, Ingvar Sodal PhD, Bert Moore PhD**, and the staff of the *Callier Center*; special thanks for expert and helpful research assistance from **Jack W. Martin BBA**, senior information resource specialist and supervisor, UT Southwestern library.

I would like to acknowledge with gratitude the support of my parents, **Retha Davis Tibbals** and **Harry Fred Tibbals Jr.**, and **Samuel Thornton Tibbals** and **Faith Tibbals Iverson**; special thanks to teachers and **Professors Gerry Meisels, John Bear, Wayne Rabalais, Juan Oro**, and others from the University of Houston and Baylor University, as well as **Jerry Barrett, Hiroyasu Matsuda, Yasuharu Suematsu, Sachio Kawashima, Kazuhiro Fuchi, Ben Nishijima, Robert Dawes, David White, Gunter Gross, J. Franzen, Lubert Leger, Karen Buchanek Hector, Yuksel Inel, Robert E. Lyle, Bill Glaze, Gary Carmignani, Paul Kent, David Bellamy, Fred McLafferty, Rupert Smith** and **Mark Smith**, and in memoriam to Linford Medalist **Tomas C. Franklin, Jim McClary, Dr. W. E. Wentworth, Marie-Therese LeBlanc, Walter, Ruth**, and **Walter Emery Caughey, Joachim Burkhardt**, and **Dr. Ed Hayes**. To the many others to whom I am indebted and names that cannot be included here, I apologize in advance and hope to make appropriate acknowledgments in future books.

I wish to thank **Gabor Louis Hornyak** for initiating and carrying through this project, my coauthors for help and cooperation, **Mike Hamers** for excellent graphics support, and the staff of *Taylor & Francis*, especially **Nora Konopka, Jessica Vakili**, and **Glenon Butler**, and **Mariasusai Anithajohny** at *SPi*, Pondicherry, India.

Finally I wish to thank **Cindy** for her patience, intelligent and helpful discussions, inspiration, and her support in the effort toward completing this work.

Harry F. Tibbals

I would like to take this opportunity to thank several people for their suggestions, comments, and encouragement during the preparation of the manuscript—notably, **Heinrich Hofmann** of *Ecole Polytechnique Federale de Lausanne*, Switzerland; **Jons Hilborn** of *Uppsala University*, Sweden; **Mamoun Muhammed** of the *Royal Institute of Technology*, Sweden; **Sanjay Mathur**, Germany; **Frederico Rosei** of *INRS*, Canada; **Ioan Marinescu** of the *University of Toledo*, Ohio; and **Vivek Subramanian** of *University of California Berkeley*; **A. K. Raychaudhuri** and **Samir Pal** of the *S. N. Bose National Centre for Basic Sciences*, Kolkata, India; and **Dr. C. Thanachayanont** of *MTEC/NSTDA*, Thailand. It would not have been possible for me to contribute to this work without the active support of all the members of the *NANOTEC Center of Excellence in Nanotechnology* at AIT supported by the National Nanotechnology Center (NSTDA) of the Thai Ministry of Science and Technology (MOST). I would like to thank **Wiwut Tanthapanichakoon**, executive director of National Nanotechnology Center of Thailand (NANOTEC), for his kind support. Special thanks to **Said Irandoust** and **Worsak Nakulchai** for constant encouragement and to all my graduate students who supported me during the writing—notably, **Abhilash, Hemant, Sunandan, Jafri, Ruh-Ullah, Rungrot, Tanmay, Alfredo, Indrani, Cesar, Botay**, and **Ann**. I acknowledge the dedication of **Louis Hornyak**, who, with his ready suggestions and advice, was the driving force to complete this manuscript in such a short notice. Finally, I thank my loving wife, **Sonali**, and our two beautiful children, **Joya** and **Jojo**, for their patience and support and for not complaining much in spite of missing me most of the time during the period I have been working on this book.

Joydeep Dutta

I would like to thank **Dr. K. Yamamotoa, Professor Stan Veprek, Dr. Y. H. Lu, Dr. A. A. Voevodin, Dr. J. A. Picas, Dr. H. Okuno, Dr. J. Alami, Dr. T. Hata, Dr. C. R. M. Grovenor, Dr. J. A. Thornton, Dr. R. Messier, Professor D. L. Smith**, and **Professor I. Petrov** for allowing me to use their original figures in this work. I would also like to thank **Dr. J. A. Rees** of Hiden Analytical Ltd., United Kingdom, for many valuable discussions on the discharged plasma diagnostics and **Dr. William D. Sproul** of Reactive Sputtering, Inc., United States, for fruitful discussions on the reactive magnetron sputtering and pulsed magnetron sputtering. Support of the research programs from the Department of Energy, Advanced Technology Institute, Li Foundation, and the North American Die Casting Association are gratefully acknowledged.

I am especially grateful to **Professor Ivar Reimanis**, who has reviewed the chapters and provided valuable feedback. I am also particularly indebted to the following authors who provided some of the figures and plots: **M. Reith, S. M. Foiles, M. L. Baskes, M. S. Daw, W. Schommers, S. Baskouts, M. Meyyappan, Ali Yazdani, M. F. Yu, O. Lourie, M. J. Dyer, K. Moloni, T. F. Kelley, R. S. Ruoff, B. Bhushan, M. D. Ventra, S. Evoy, J. R. Heflin, G. Genta**, and **L. J. Hornbeck**.

John J. Moore

Authors

Gabor L. Hornyak has an interdisciplinary background in biology, chemistry, and physics. He received his PhD in chemistry from Colorado State University in 1997; BS and MS degrees in biology (human genetics) from the University of Colorado at Denver in 1976 and 1981, respectively; and a BS degree in chemistry from the University of California at San Diego in 1990. Dr. Hornyak worked in the aerospace industry as a senior scientist from 1978 to 1990 in San Diego at the Convair Division of General Dynamics in coatings, adhesives, and corrosion. From 1997 to 2002, he worked at the National Renewable Energy Laboratory on development of chemical vapor deposition synthesis of carbon nanotubes. He has over 15 years of experience in nanotechnology research and development. Dr. Hornyak played a major role in creating an awareness of the promise that nanotechnology brought to the citizens and institutions in Colorado and the surrounding region (2003–2005). He is the editor of the "Perspectives in Nanotechnology" series—a group of books dedicated to bringing topics to the general public that address issues about nanotechnology that are outside the laboratory and production line.

John J. Moore is the trustees' professor and head of department of metallurgical and materials engineering at the Colorado School of Mines (CSM), Golden, Colorado, where he is the director of the interdisciplinary graduate program in materials science, director of the Advanced Coatings and Surface Engineering Laboratory, and director of the Reaction Synthesis Laboratory.

Dr. Moore was awarded a BSc in materials science and engineering from the University of Surrey, Surrey, United Kingdom; a PhD in industrial metallurgy from the University of Birmingham, Birmingham, United Kingdom; and a DEng from the School of Materials of the University of Birmingham, Birmingham, United Kingdom.

Dr. Moore worked as a student apprentice at Stewarts and Lloyds Ltd., United Kingdom, from 1962 to 1966, and as manager of industrial engineering and production control at Birmid-Qualcast Industries Ltd., United Kingdom, from 1969 to 1974. He is currently serving on the board of directors of Hazen Research, Inc., Golden, Colorado (2002–present) and is chairman of the scientific advisory board, XSunX, Aliso Viejo, California (2005–present).

Prior to his appointment at CSM, Dr. Moore served as professor and head, Department of Chemical and Materials Engineering, University of Auckland, New Zealand, from 1986 to 1989; professor of metallurgical engineering at the University of Minnesota, Twin Cities, from 1979 to 1986, and senior lecturer of chemical metallurgy at Sandwell College, United Kingdom, from 1974 to 1979.

Dr. Moore's research interests and activities include physical and chemical vapor deposition of thin films and coatings; synthesis and processing of advanced ceramic, intermetallic, and composite materials using plasma and combustion synthesis techniques; synthesis, processing, and properties of biomaterials; and powder metallurgy processing of advanced materials. He has published more than 620 papers in materials science and engineering journals; holds 14 patents; and is the author, coauthor, or editor of 12 books not including *Fundamentals of Nanotechnology*. Dr. Moore is a fellow of the Institute of Materials, United Kingdom; a fellow of American Society for Metals (ASM) International; a fellow of the American Ceramic Society; a fellow of the Institute of Materials, Minerals, and Mining, United Kingdom; and a chartered engineer. He is also an honorary professor at the University of Salford, Salford, United Kingdom and at the University of Auckland, Auckland, New Zealand; he has also been awarded an honorary doctorate from the Moscow State Institute of Steel and Alloys, Russia.

Harry F. Tibbals has served as director of the Bioinstrumentation Resource Center for the University of Texas Southwestern Medical Center since 1997, where he is responsible for providing engineering support to clinical and basic biomedical life science researchers. His funded research includes development of pressure and electrochemical sensors for medical applications, testing and evaluation of life support systems for NASA use in space flight and extravehicular activity, and development of technology for Alcon Research Ltd. for diagnosis of diseases of the eye. Dr. Tibbals's work involves consultation and team leading on a wide variety of analytical, materials, and systems technology in support of medicine and biomedical research. He is frequently called upon to advise on risks and cost benefits for technology decisions.

Prior to joining UTSW, Dr. Tibbals was product line manager, Digital Cardiology Products for Jamieson Kodak, working in Dallas, Texas, and Rochester, New York; and at hospital cardiac catheterization laboratories throughout the United States and Europe. He was responsible for developing and obtaining FDA approval for the first fully digital imaging systems capable of showing the living human heart with medical radiology standards of precision and accuracy. He also consulted for the development of patient anesthesiology systems. As president of Biodigital Technologies, Inc. and board member and consultant to Martingale Research Corporation from 1989 to 1995, he led teams in the development of real-time systems for the identification of bacterial and viral disease agents, and systems for monitoring, analyzing, and reducing environmental hazards. From 1988 to 1991, Dr. Tibbals was product line manager on contract to Inmos and later SGS Thompson for digital signal processing products and applications. Clients for product and development management projects included NEC, Teledyne, Marathon Energy Systems, Coors, Bank One, Shelby Technologies, Innovative Systems SA, Optical Publishing Inc., Gentech, Colorado Medical Physics, and others. He was a consultant and project manager for the development of a production and distribution control system for the world's largest nitrogen fertilizer complex at BASF in Lugwigshafen, Germany.

For most of the 1980s, Dr. Tibbals worked for Rockwell International on trusted and secure systems in government and private areas, serving as principal investigator, project engineer, and systems engineer on projects in the United States and around the world. He was product planner and Rockwell representative on standards committees in the development of the first Rockwell Digital Facsimile Modem systems, which achieved more than 90% market share over a sustained period following its introduction. From 1983 to 1985, he was a principal design engineer for Mostek Systems Technology, heading work on standards, design, and introduction of the VMEbus product line and the development of applications that became dominant in telecommunications, process control, high-end workstations, and scientific and medical instrumentation.

During the 1970s, Dr. Tibbals served on the academic and research staff of Glasgow and Durham Universities in the United Kingdom, where he worked with the Edinburgh Regional Computing Centre, the Digital Cartography Unit, and the Glycoprotein Research Laboratory. He also taught for the Open University and Jordanhill College. He held visiting research and teaching positions at Bogazici University and the University of North Texas, working on instrumentation and systems for environmental monitoring and remediation.

Dr. Tibbals earned his BS degree in chemistry and mathematics at Baylor University in 1965, where he held scholarships in chemistry, English, and Old Testament studies, and was an undergraduate research fellow in electrochemistry. He was awarded his PhD from the University of Houston in 1970 for theoretical and experimental research in nonequilibrium statistical mechanics and kinetics of ion–molecule reactions. He won an SRC postdoctoral fellowship in physical silicon chemistry at the University of Leicester from 1970 to 1972. He has published a number of refereed scientific and technical papers; has received grants and study awards in computer systems architecture, man–machine interfaces, and applications of computers in chemistry; and holds two patents. In 1990, he was awarded a grant from the National Center for Manufacturing Science to visit key technology centers in Japan and study applications of advanced signal processing technology, including fuzzy logic and neural networks. He received a grant from Rockwell to organize a series of three symposia on networks in brain and computer architecture from 1986 to 1988. He served as adjunct professor in the University of Texas at Dallas School of Human Development, and he was a member of the advisory board for the University of Texas at Arlington Advanced Automation and Robotics Center and for the Rutgers Center for Advanced Information Processing.

Dr. Tibbals has served the IEEE Dallas CN group as treasurer, program chair, and board member. He is a member of the American Chemical Society, the Biophysical Society, the Royal Society of Chemistry (chartered chemist), The American Association for the Advancement of Science, The British Computer Society, and the Sigma Xi Society for Scientific Research. He has served on the boards of the Audubon Society Prairie and Cross Timbers Chapter and the Dallas Nature Center. He has served as an advisor to government bodies and companies of information systems and technology for environmental monitoring and improvements. He served on the U.K. Science and Religion Forum and the Commission on Caring for Creation, and he worked with George Dragus and Nicholas Madden on the generation of computerized concordances for early Greek Christian patriarchal writings.

Joydeep Dutta was born on May 5, 1964, and is currently director of the Center of Excellence in Nanotechnology and an associate professor in microelectronics at the Asian Institute of Technology (AIT), Bangkok, Thailand, whose faculty he joined in April 2003. He received his PhD in 1990 from the Indian Association for the Cultivation of Science, India. In 1991 and 1992, he did postdoctoral work at the electrotechnical laboratory (Japan) and at Ecole Polytechnique (France) before moving to Switzerland in 1993, where he was associated with the Swiss Federal Institute of Technology, Lausanne (EPFL), until 2003. From 1997 to 2001, Dr. Dutta worked in technical and managerial qualities in high-technology industries in Switzerland before returning to academia in 2002. He has been a member of the board of two companies working in high-technology electronics and environmental consulting, respectively.

Dr. Dutta has been teaching courses in microelectronics fabrication technology, nanomaterials and nanotechnology, optoelectronic materials and devices, failure analysis of devices, and emerging technologies at AIT. He has also taught nanomaterials (since 1997) at EPFL, and in Uppsala University (2003–2005) and Royal Institute of Technology (2005–present), both in Sweden.

Dr. Dutta is a fellow of the Institute of Nanotechnology (IoN) and the Society of Nanoscience and Nanotechnology (SNN) and a member of several professional bodies, including the Institute of Electrical and Electronics Engineers (IEEE), Materials Research Society (MRS), Society of Industry Leaders, Gerson Lehrman Group Council, the NanoTechnology Group Inc., and the Science Advisory Board—all in the United States; the Asia-Pacific Nanotechnology Forum (APNF), Australia; and the U.K. Futurists Network, United Kingdom, among others. He has reviewed projects of various scientific organizations of different countries (lately in Sweden and Ireland) and has organized a few international conferences and served as a member in several others.

Dr. Dutta's broad research interests include nanomaterials in nanotechnology, self-organization, microelectronic devices, and nanoparticles and their applications in electronics and biology. He recently completed a textbook on nanoscience, and he has over 100 research publications, over 350 citations (Scopus), five chapters in science and technology reference books, three patents (five ongoing applications), and has delivered more than 50 invited lectures.

CHAPTER OPENING CAPTIONS AND CREDITS

Section 1

Perspectives

INTRODUCTION

I never think of the future—it comes soon enough.

ALBERT EINSTEIN

Fundamentals of Nanotechnology is about technology—the manufacturing of high-tech (and not so high-tech) goods. It is about converting nanoscience into products. It is not so much about science although we do not refrain from relating to scientific principles, concepts. Our focus is on topics associated with materials, manufacturing (fabrication), devices, and applications. It is, in every way, a complement to *Introduction to Nanoscience*, a text that focuses primarily, as the title suggests, on the science and not on the technology. The format of the text is that of a cooperative, interdisciplinary effort that is typical of nano—albeit with emphasis on engineering—where biology, chemistry, and physics (and computer science) come together to contribute to this very broad subject.

The chapter opens with a short course on how to start, maintain, and exit a business. Although it applies to nearly any kind of business, it most certainly applies to nanobusiness. This is an unprecedented approach, to include such a topic in a technology textbook, but it has to be done in order to keep pace with our changing world. We accomplish our due diligence by offering this material in this chapter as well as a section on education and workforce development. It is one thing to understand the technology and pass the course with flying colors; it is quite another thing to know what to do with it or plan a career centered on its impact.

Nanometrology, nanomanufacturing, and research have to be conducted somewhere. Where they take place is actually quite important. As one might gather from the prefix, nano implies things that are very small, and being small, working with nanomaterials, measuring them, and making them all require special facilities and considerations. We discuss nanometrology and nanomanufacturing in *chapter 2* but we also turn our attention to the buildings that house them—the buildings for advanced technology. This is a rather unusual section in that material of this ilk is not usually found in science and engineering texts—but we have already broken that rule by adding a short course in business development and operation. However, by understanding what it takes to make nano happen, the student should acquire an even broader appreciation of the subject matter—and perhaps interest in a career path along the lines of architecture and engineering buildings for advanced technology may just be a beneficial byproduct of this chapter.

Lastly, we provide a catalog of nanotechnology institutions and products. We highly recommend that you visit some of the Web site links provided so that a broader perspective of nanotechnology can be acquired. Following this chapter, a special chapter on nanometrology is presented.

1.0   PERSPECTIVES OF NANOTECHNOLOGY

We start this adventure in 1959, at the onset of the modern age of nanotechnology, with Richard Feynman's lecture "There's Plenty of Room at the Bottom." It was in this lecture that Feynman alerted the consciousness of the scientific community about the untapped potential of nanomaterials. In *Introduction to Nanoscience*, we discussed the important historical contributions to nanotechnology and the remarkable observations of nature, wittingly or unwittingly, that were made throughout the ages and the scientific aspects of nano. We now strive to maintain focus on applications, integrated structures, and devices. However, before we embark, a small amount of reorientation is required. We start by recycling a few core definitions.

1.0.1 *Review of Definitions*

In 2003, Dr. Rachel Brazil of the Royal Society of Chemistry stated quite succinctly her definition of a powerful but relatively nonspecific term—*nanotechnology* [1].

> At present, the term is used to encompass a wide spectrum of nanoscience, from nanoparticles in sunscreen to the production of 'nanobots' for in vivo medical applications. In defining nanotechnology, distinctions need to be made between 'science' and 'technology'. A narrower definition of the type of 'technology' covered by the term may also be considered, limiting nanotechnology to technology producing functional devices fabricated and operating on the scale of nanometres.

The National Science and Technology Council, Committee on Technology, Subcommittee on Nanoscale Science, Engineering, and Technology (NSET) formally established the following boundaries of nanotechnology in the year 2000 [2].

> Research and technology development at the atomic, molecular or macromolecular levels, in the length scale of approximately 1–100 nanometer range, to provide a fundamental understanding of phenomena and materials at the nanoscale and to create and use structures, devices and systems that have novel properties and functions because of their small and/or intermediate size. The novel and differentiating properties and functions are developed at a critical length scale of matter typically under 100 nm. Nanotechnology research and development includes manipulation under control of the nanoscale structures and their integration into larger material components, systems and architectures. Within these larger scale assemblies, the control and construction of their structures and components remains at the nanometer scale. In some particular cases, the critical length scale for novel properties and phenomena may be under 1 nm (e.g., manipulation of atoms at ~0.1 nm) or be larger than 100 nm (e.g., nanoparticle reinforced polymers have the unique feature at ~200–300 nm as a function of the local bridges or bonds between the nano particles and the polymer).

It is always a difficult matter to draw a hard line but one has to be drawn so that we can obtain the proper perspective and orientation. It also helps us navigate in the real world, specifically if funding and contractual issues are at stake. Let's move on and introduce (or reintroduce) the fundamental and defining characteristics of words and phrases that include the prefix *nano*.

In order to provide the necessary perspective, the following working definitions of nanotechnology, and its distinction from nanoscience, are listed below [3].

Nanoscale
The nanoscale, based on the nanometer (nm) or one-billionth of a meter, exists specifically between 1 and 100 nm. In the general sense, materials with at least one dimension below one micron but greater than one nanometer can be considered as nanoscale materials.

Nanoscience
Nanoscience is the study of nanoscale materials—materials that exhibit remarkable properties, functionality, and phenomena due to the influence of small dimensions.

Nanoscience is similar to materials science in that it is an integrated convergence of academic disciplines. There exist a couple of major distinctions between the two: size and biology. The size we understand by now but we also understand that materials science traditionally does not include biological topics.

Nanotechnology

Nanotechnology, based on the manipulation, control, and integration of atoms and molecules to form materials, structures, components, devices, and systems at the nanoscale, is the application of nanoscience, especially to industrial and commercial objectives.

Nanotechnology is a horizontal enabling convergent technology that cuts across all vertical industrial sectors while nanoscience is a horizontal integrating interdisciplinary science that cuts across all vertical science and engineering disciplines.

Nanotechnology is a disruptive technology with a high barrier of entry that will impact the development of enhanced materials and devices. Nanotechnology will require that a new genre of partnerships be formed among and between business, academe, and government. It will focus study and effort on potential societal implications of a new and certainly disruptive technology. Nanotechnology is predicted to significantly impact the wealth and security of nations. Nanotechnology is the next industrial revolution.

Nanotechnology is considered to be, more so than ever, a technology that will have great impact on all aspects of culture and society.

Nanotechnology is the application of nanoscience—plain and simple. You will see many different forms of the above definitions in the media and scientific literature, but essentially all the definitions, after distillation and purification, crystallize into a few key forms—in particular that nanotechnology consists of materials with small dimensions, remarkable properties, and great potential.

1.0.2   *Technology Revolution or Evolution?*

Technology (from the Greek *teknikos* meaning "art, artifice; to weave, build, join" and *tekton* meaning "carpenter") has played a major role in the history of civilization. There is not much question that technology is one of the pillars (and drivers) of civilization. After all, isn't it new technology that offers the developers of that technology an advantage in the game of life (survival)? However, a revolution implies rapid and dramatic change. What exactly are those technological advances that changed our civilization and in such a so-called revolutionary manner?

From agriculture, the *practice* of growing and harvesting food, sprung civilization. Indeed, one must agree that this is true. Early human lifestyle was altered forever as nomadic ways adapted by the hunter–gatherer gave way to the sedentary form of the farmer—truly a revolutionary change. However the processes of agriculture and irrigation were deliberately developed over an extended period of time—perhaps several thousand years. Innovations in agriculture continue as we speak—genetically modified organisms (GMOs) provide an excellent example of our high-tech foray into that arena. Although change in agricultural technology progressed quite slowly, even the gradual development of agriculture must be considered to be a revolutionary one based on its impact on civilization.

Other technological breakthroughs such as the advent of metal tools, implements, and weapons and the discovery of fire may have occurred rather spontaneously, for example, by accidental discovery (as in the case of fullerenes). These revolutionary breakthroughs are based on the discovery of a specific *type of material*—stone, copper, bronze, or iron. We can only speculate that the spread of these technologies, however, still proceeded rather slowly. There was no rapid-fire means of disseminating information (perhaps only by conquest) due to limitations in population and communication several millennia ago. Today no such barriers exist. High population density is a worldwide phenomenon, and information is spread globally in the blink of an eye. Revolutions can now be very fast. The *Industrial Revolution* of the 1800s ushered in the modern technological era. The emerging availability of hydrocarbon fuels in particular launched mechanization, mass production, transportation, and communication to levels never seen before. The telephone, television, computer, cell phone, and Internet have all changed our lives forever.

How does one then superimpose a scale of measure on a technological revolution? And what kinds of scales are relevant to nanotechnology? Revolutionary developments are measured in terms of speed, population, and impact (economic and cultural) or by weighing the overall flow of resources (in and out). For example, what percentage of a population is engaged directly in a technological revolution? "Engaged" in this case indicates the number of people working in the field, so to speak. With regard to agriculture, the percentage of population engaged in the field early on was quite low. As the practice spread over several thousands of years, greater than 90% of a population may have been rooted in the agrarian lifestyle, for example, in the Middle Ages. The pure technology sectors (R&D and manufacturing) of today may employ fewer that 10% of the total workforce but the impact (in terms of sales, use, and lifestyle) is quite enormous.

And what of nanotechnology? Is it really a revolution or is it just a natural progression in miniaturization? The advent of the transistor, integrated circuits, and the computer age certainly changed the way we accomplish our daily tasks. The *Biotechnological Revolution* added biological materials into the mix. The *Nanotechnology Revolution* is right on the back of this "revolution" inspired by biology. Many consider biotechnology to be a component of nanotechnology. All these pseudo-issues boil down to semantics and boundaries. We already have stated our opinion on boundaries. Within a generalized field such as nanotechnology that is highly interdisciplinary in nature, boundaries are consequently trampled and blurred—to a great extent.

Nanotechnology is expected to change the way we live. In that sense, it can be considered to be revolutionary. Nanotechnology is also the product of the natural evolutionary process of miniaturization. It is actually with the perception of nanotechnology (hype or not) that many are concerned. However, we should not belabor what is in essence a pseudo-dilemma—whether or not nanotechnology is the *Next Industrial Revolution*. Let's leave that discussion to the pundits, the politicians, and the omnipresent media. Rather, we should simply try and understand why these materials have remarkable properties—and within the context of this text, how nanomaterials are converted and integrated into applications. In this volume of course, we stress the applications of nanotechnology products and those enhanced (enabled) by nanotechnology. It is somewhat artificial to designate a technology to be revolutionary *before* it has run its course. **Table 1.1** outlines some of the salient features of a revolution.

TABLE 1.1	*The* Next Industrial Revolution *Compliance Check List*		
Criterion	**Yes**	**No**	**Comments**
Basis	X		Nanotechnology is firmly grounded in reality—the physical properties and phenomena associated with nanotechnology are real and significant—they are the drivers behind the "revolution."
Speed	X		Although speed is not the only major component of a revolution, the faster it happens, the more revolutionary is its impact and the more profound its legacy. Nanotech has hit the mainstream since the mid-1990s—not that long ago. Changes nowadays are expected to occur quickly.
Track record		X	Outside of computers, pharmaceuticals, and nanoparticles, the glamorous and highly touted quantum-dots, carbon nanotubes, and drug delivery nanomaterials have not yet made a significant impact on the economy but are making headway as we speak.
Research papers, conferences	X		The scientific community has fully embraced nanoscience and nanotechnology. There is no single scientific conference in physics, chemistry, engineering, biology or medicine that does not have presentations, posters, sessions or the whole conference dedicated to nano.
Nanocompanies	X		There are more and more nanocompanies every year. Small Times Magazine lists 3500 nanocompanies just in the United States [4]. Nearly all Fortune 500 companies have some involvement in nanotechnology.
Patents	X		The number of patents (and the trend) is burgeoning—if anything, this trend indicates revolutionary proportions.
Stock market		X	Not big yet.
Institutions	X		About 500 institutions ranging from nonprofits, industry associations, research labs, economic development organizations, and university and educational institutions are registered with Small Times Magazine as nano groups [4].
% Workforce		X	Technology workforce contributes less than 10% of the total U.S. workforce in general. Nanotechnology, in its broadest sense (computers, pharmaceuticals, etc.), is already involved in many sectors. By 2015, 2 million more "nanotech" jobs are expected to emerge.
Education		X	Concerted efforts are underway to promote nanotechnology in K-12 and higher. Although a great proportion of the U.S. population knows about nanotechnology through movies, it is not considered to be a significant part of curricula at this time, but this too is changing.
Products	X	X	Nanosized transistors are already part of computer chips—although developed rather quietly without much fanfare via the natural evolutionary process of miniaturization. There is no doubt that more products will undergo enabling/enhancement from nanotechnology.
Evolutionary component		X	Evolutionary components are not considered to be revolutionary. In this case, nano is very evolutionary—emerging from micro- and biotech industries—a natural, nonrevolutionary process of miniaturization, ever since the first timepiece was developed.
Hype	X		Since the days of the "turbo," nothing has been hyped more than nanotechnology.
Upside	X		The reality-based upside and promise of nanotechnology is tremendous.

1.0.3　*Outlook*

What Is the Status Quo of Nanotechnology in the United States?　Zyvex Labs, LLC, has recently spun out nanotube composites that are integrated into baseball bats, golf clubs, sailboat masts, ballistic armor, and radiation shielding [4]. The energy industry in collaboration with universities has developed new solar cells and new high-performance batteries. The medical industry in collaboration with universities has developed dendrimer-based viricides and advanced diagnostic instrumentation [4]. According to Jim Von Ehr, II (CEO of Zyvex), societal impacts need to be studied and carefully weighed before the release of products containing nanomaterials and devices. Encounters with luddites, nano-pretenders, and obstructionists must be sifted from those with legitimate environmental understanding and legitimate health and safety concerns. The U.S. government in the form of the National Nanotechnology Initiative (NNI) has taken a leadership role in addressing issues confronting the development of nanotechnology. Positives include the leadership role of the NNI, interdisciplinary cooperation, support of university research, and interagency cooperation between and among the agencies and departments of the federal government. Negatives include insufficient focus to the NNI's nine "focus areas"—and that *science and discovery*, although a wonderful couple, are not enough.

Invention and innovation are required to transform those discoveries into products. There is not enough focus on product development. The National Institute of Standards and Technology's (NIST) ATP opportunity (Advanced Technology Program) is one of the few government agency programs that is able to "fund the gap" between a new idea and a fundable prototype. Why is this important—because nano is a global competition—a competition with foreign governments (and their complementary economies) that do stress and fund commercialization.

According to Jim Von Ehr, II of the Zyvex Corporation, university funding is doing well but is unfocused; commercialization is not working as well as it should; the (costly) national lab system and its relationship to the economy should be redefined; and patent reform should occur sooner than later to avoid creating "patent thickets" that impede innovation [4]. The government's role in technology is to regulate and tax when required. He believes that the U.S. government is not supporting industry very well. Tax codes, accounting rules, financial reporting, liability procedures, and regulations should not be changed every year. The acquisition of talent by limiting foreign visas forces jobs overseas and losses to foreign competition. The U.S. industrial policy has become "come to school here, create new technology, go back to your home country, and commercialize it there."

In 2005, London outpaced New York in the number of initial public offerings (IPOs) and the United States received none of the largest 25. The Sarbanes–Oxley (SOX) Act of 2002 had a devastating effect on foreign investment in the United States. Before SOX, 90% of foreign funds were raised in the United States. Post-SOX, 90% of foreign funds are raised outside the United States. This demonstrates clearly how the government can impact the economy. The combination of the SOX policy and the inability to import high-tech talent drives jobs to other countries and companies out of business here. In summary: (1) move towards engineering, applications, and commercialization, (2) intensify focus on energy, healthcare, and nanomanufacturing; fund social impact studies when nanotech has a larger impact; and study environmental issues in a science-driven manner, and lastly, (3) industry will respond to incentives [4].

1.0.4   *The Nano Perspective*

We can safely say that without manufacturing there would be no nanotechnology—at least according to our definition. Upon implementation of the technology, we ask ourselves is it for the greater good? Will society as a whole benefit from this technology? Societal implications and technology have always been linked. We therefore should make things for everyone—at least in theory. In the perfect world, everybody should win. Nanotechnology is no different. We develop products for the greater good, for defense, for profit, for barter, and for numerous other reasons. Where does nanotechnology fit into this age-old relationship? The diversity of products, a few mentioned earlier, shows no signs of slowing. While reading, and hopefully studying, this text, please keep in mind how this new technology impacts products, develops new ones, and enhances our security and quality of life. Think also how this new technology can impact our fears as well as our hopes.

1.1   THE BUSINESS OF NANOTECHNOLOGY

Although most will not admit it in public, and we do not indicate anything by stating this, most scientists are not particularly business savvy, and to be fair to scientists, most business folks do not know a hoot about science or technology—at least beyond the fundamentals. Although there certainly are exceptions to this general conundrum, do you all essentially agree? We therefore have decided to include a *short course in business* in this textbook. The age of the *scientist as entrepreneur* is upon us as more and more universities interact with the commercial sector. Biotechnology first and now nano have pushed the envelope of interaction between academia and industry. As a future scientist or engineer, one must start thinking in terms of a career in nanotechnology, and even better, start thinking in terms of starting a business.

There are companies and there are nanocompanies. A nanocompany is not necessarily a small company. It may actually be a big company, perhaps a Fortune 500 company. What kinds of companies are drawn to this new technology? What differences are there between creating a nanotechnology company from say a biotech one? Nanotechnology has a high barrier of entry—unlike that faced by the "dot-coms" of the 1980s and 1990s—where all that was needed were a couple of geniuses, a PC, and a garage. Nanotechnology requires expertise (e.g., PhD level), costly equipment (e.g., transmission electron microscopes, ultrahigh vacuum), clean rooms, and highly trained tech people. Development and manufacture of nanomaterials, more than ever, also requires partnerships between and among government, business, and academia. Because nanotechnology is a worldwide phenomenon, competition is intense on a global scale.

The people and entities that support nanotechnology—the intellectual property managers, the patent attorneys, the building designers, the consultants—are all faced with new challenges brought on by nanotechnology and the interdisciplinary nature of the subject matter. Hopefully we will be able to raise questions, pose challenges, and explain some of what surrounds this global phenomena in the following sections of the text.

1.1.1 Background

Throughout history, humans have sought improvement by innovative approaches to the various tasks of living. Although the invention of stone arrowheads and spear points, the wheel, and a host of other early innovations precede written record, they were in by some perspectives more impressive in their impact on society than the invention of the electric light and telephone in the late nineteenth century. One characteristic of each of those early inventions is that someone accomplished the equivalent of starting a business. With the wheel, for instance, the wheelwright developed his skills to be able to make a consistently round product of a standard size.* The specialists able to run with these innovations were the forerunners of modern businesses.

The modern concept of a business, (e.g., a company with employees, products, a board of directors, a CEO, and shareholders) dates back only a few hundred years. Throughout history, however, the purpose of business was to sell products and services to buyers—the customers. From the time that early humans graduated from a simple hunter–gatherer status to increased levels of stratification with diverse skill sets, the equivalent of business was established. After all, what is the difference between a flint-knapper making arrowheads and spear points in exchange for food and shelter and a modern multinational corporation making electronic components in exchange for money? In terms of the lowest common denominator, the answer is nothing—both are trading one item of value for another at an agreed rate of exchange.

Over the millennia, this simple form of proto-business, essentially a sole proprietorship, began to evolve and grow into defined organizations offering a range of goods—from a bread maker with a few employees and several different products to a modern multinational corporation with thousands of products and tens of thousands of employees. At the same time, various restrictions, requirements, and other hurdles were put into place to make certain types of businesses more difficult to enter than others. From early on, businesses began to need some sort of registration with civil authorities, if for no other reason than tax assessment. Today, a business must register with the appropriate governmental agency in the country or state that it operates and must have various tax account numbers for payment of a variety of taxes. It also is responsible for various reporting requirements, depending on its structure.

Nanoscience and nanotechnology have already and will continue to yield many new and wonderful discoveries and inventions. But, as has been the case throughout history, these inventions remain interesting curiosities unless they are applied to develop products and/or processes that make their way into the mainstream of commerce. Take the wheel, for example. It is easy to understand that many applications, from the simple wheel/axle wagon to more complex

---

* Imagine what it would be like to have a chariot with different-sized wheels. If it were somehow impossible to make wheels consistently, would it have ever developed?

applications like block-and-tackle, watches, and precision gearing were made possible by applying the basic principle of the wheel. These are merely direct applications of the wheel. Indirectly, the wheel has led to many support products and even whole industries. For example, because wheels require lubrication, someone started manufacturing, marketing, and distributing lubricants. Others developed new materials to make the wheel last longer and perform better. Today, it is difficult to imagine a product or process that doesn't have a wheel somewhere within its mechanism. Nanotechnology has the potential to do the same. Indeed, many observers have stated emphatically that nanotechnology-based companies will engender another Industrial Revolution that is expected to exert profound effects on the world economy and society.

The U.S. National Science Foundation (NSF) has estimated that the world nanotechnology industry will grow from approximately $35 billion in 2005 to $1 trillion in 2015, employing over 2 million workers.* Such rapid growth is unprecedented but reflects a conviction that nanotechnology will affect and become a part of nearly every segment of industry over the next decade.

1.1.2 Companies

What Makes a Company?   A company is not just a good idea, but is also a fine starting point. A company is a provider of goods and/or services that customers are willing to buy. It is also a person or a group of people able to supply those goods or services. This implies that the company is meeting some need, either existing or created.

The concept of an existing need is obvious, but what is a *created need*? The history of invention is the history of created needs. Nearly every adopted technology is either superior to an existing technology or it can outperform the previous technology. For example, conventional stoves and ovens are quite adequate for food preparation but the microwave oven provides an additional feature—convenience and speed in food preparation. Its development spawned an industry of prepackaged foods ready to "be popped into the microwave" and ready to eat in minutes. Later models were developed with oscillating or spinning cooking surfaces. A series of secondary products including "microwavable" cookware and special splatter covers appeared. Today, roughly a half century after the first microwave oven was introduced, most home kitchens and break-rooms in the United States have one.

The relevance of the example above is that not only can an inventor create a product and then a need for it, but also the acceptance of that product can spur the creation of a whole new industry. The entrepreneur needs to open his or her mind to the myriad possibilities for products, and therefore businesses, that can manifest themselves with the advent of new technologies, and their support. It will help the entrepreneur to consider exploitable market niches and products that will appear as the field of nanotechnology develops and expands its impact

---

*These numbers appear in Roco, M.C., *Journal of Nanoparticle Research*, 5, 181–189 (2003) and other sources from the same author. While this is a generally accepted figure, other sources place the market by the middle of the second decade of the twenty-first century as high as $2.6 trillion.

on society. Conversely, it will help the entrepreneur to understand what is displaced (and even disappears) as a result of a new technology.*

Successful entrepreneurs do not simply invent wondrous gadgets. They look at the gadgets they (or others) have invented and ask what could be needed to improve and to fully exploit and support them.

What Is a Nanotechnology Company? This seemingly simple question is actually a very complex one. One definition of a nanotechnology company is that it manufactures a product less than 100 nm in at least two dimensions, or that the manufacturing process itself is controlled at that size or smaller. The two-dimensional requirement would rule out companies making, for example, graphite lubricant because while sheets of graphene are one carbon atom thick (less than 1 nm), the other two dimensions are generally much, much larger than 100 nm. If a company were manufacturing coatings with particle sizes of consistently less than 100 nm, they would also qualify. However, we can safely call companies that manufacture thin films nanocompanies regardless of the lateral dimensions of their product. The process control issue would exclude bulk chemical manufacturers from joining the ranks of nanocompanies. Although chemical synthesis, as represented on paper, consists of reaction between atoms and molecules, the reality is that it occurs in 10,000 L reactors. The only real controls of these processes are physical in nature, for example, temperature, pressure, rate of mixing, rate and order of reactant addition, etc. By contrast, manufacturing an integrated circuit with components that are 65 nm in size do qualify as a nanotechnology process.

Nanotechnology companies also arise from the service sector. For our purposes, we define the service sector as comprised of companies that are not directly involved in developing or manufacturing a technology. Examples include:

- Consulting engineers specializing in the development of processes that fit the above definition
- Technology transfer professionals who seek out nanotechnology clients
- Patent law firms or agents having experience in filing and prosecuting nanotechnology-related patents
- Toxicology laboratories with the ability to study the effects of nanotechnology products on living organisms and the environment
- Contractors who were experienced in building the specialized manufacturing facilities needed or indeed a host of other service providers.

1.1.3 Sources of Nanotechnology Inventions

Although many discoveries and applications arise from work in laboratories of companies, a significant proportion, probably a majority of them, will originate

---

* For example, what happened to the typewriter industry when personal computers with word processors, such as the one used to write this book, became affordable and nearly universal? What about the typewriter repairman or ribbon and carbon paper manufacturers?

in academic and government laboratories, including both basic discovery and product development. However, these venues are not equipped to develop and commercialize products. The federal government mandates that research funded by the U.S. government be commercialized by a U.S. company. As a result, and over the years, institutions and laboratories have established bureaucratic structures and procedures for transferring newly developed technologies and products to U.S. companies. The general approach starts with the filing of a patent.* The university or national lab normally covers the cost of the patent filing. The invention is then licensed either exclusively or nonexclusively, usually for some fee plus royalties based on product sales, to a promising company. The terms for licenses vary considerably.[†‡]

Of course, many entrepreneurs who receive licenses from academic or government labs are in partnership (or will be soon) with the inventors employed by the institutions. Although the inventors understand the potential of the invention, not many have a clear idea of how to go about founding, registering, funding, and operating a company. The remainder of this section will provide a general overview of these subjects.

1.1.4 Founding a Company—What to Do First?

Most of the details you encounter below is nothing new—it applies to any and all businesses. Although nanotechnology is revolutionary in many ways, the structure of the business and the process required to move forward are relatively

---

* When examining the patent status of the invention, the entrepreneur should keep in mind that nanotechnology is a worldwide effort and the filings need to be prepared such that they can be filed as geographically broadly as possible. A U.S.-only patent, commonly filed by universities and government labs, limits the value since anyone outside the United States can simply copy the technology and potentially produce products for sale in worldwide markets. The reason for limited filings by these institutions is not lack of understanding but lack of budget. The subject of patents is covered elsewhere in this book.

† License terms and negotiations are beyond the scope of this book and the reader is referred to the Association of University Technology Managers (AUTM, www.autm.org) and the Licensing Executives Society (LES, www.les.org) for more information about license terms and negotiations. Both offer courses on these subjects, for a fee.

‡ In the last decades of the twentieth century, many companies began to actively seek inventions and technologies from academic, government, and even other industrial sources and established professional positions devoted to the process of in-licensing of technology of potential value and secondarily out-licensing (or outright selling) of those no longer of internal interest. As a result of their experiences, a number of these technology transfer professionals have offered their services as consultants to companies seeking to acquire technologies. They can be found either by Web searches or through listings with the AUTM and LES. The entrepreneur with limited (or no) experience in this arena should seriously consider having professional help in negotiating terms to acquire rights to technologies and inventions. The role these professionals play is to help establish terms, generally financial and business terms (due diligence requirements, termination rights, and a host of other issues), but most of them are not attorneys and while they will also participate in contract negotiations they are not responsible for the actual contract drafting and language. An experienced commercial attorney is essential for this important task, with preference for one who has written contracts for closely related technologies.

ubiquitous and require a modicum amount of common sense. Students who receive degrees in science become good scientists but not necessarily good business people.

Before embarking on your entrepreneurial journey (always a wild ride), several questions need to be addressed. In doing so, the entrepreneur defines the company in order to decide what structure to use when registering it. These questions include:

- What type of company?
  - Manufacturing
  - R&D—contract or grant-based
  - Consulting
  - Service provider
- How much financing is required to launch the company?
  - Fees, including legal, to set up the company and obtain operating licenses
  - Salaries for entrepreneur, partners, and any initial employees
  - Facilities cost
  - Licensing fees for technology
  - Operational setup costs: supplies, Web site, office and laboratory equipment, and furnishings—everything that may be needed from paper clips and business cards to atomic force microscopes
  - Running costs until other sources of finance are available
- How will it be financed?
  - Entrepreneur alone
  - Insiders and family/friends
  - Angel/venture funding
  - Eventually through IPO
  - Product or service sales (with no need for outside financing)
- How will ownership be distributed?
  - Entrepreneur alone
  - Entrepreneur and partners or family members
  - Entrepreneur and financial backers
  - Employees through stock incentives
  - Large shareholder base through public offering
- Where will the company be registered?
  - If in the United States, which state?
  - If outside the United States, where?

All of these factors plus many others, including personal biases, play a role in the choice of structure. A good way to make certain these issues are considered is to write a *business plan*.

The principle purpose of a business plan, at least in the first draft form, is to focus the entrepreneur on defining the company in terms of products and markets, hurdles to success, financial needs, revenue and cash flow, personnel, facilities, and equipment needs. It will later be useful in raising money, whether from venture or angel sources or from banks but its greatest value to the entrepreneur is to his own understanding. It should include:

- Executive summary (actually written last)
- Company description—when, where, and by whom founded, assets, etc.

- Market served, including a market analysis
- Products
- R&D
- Manufacturing issues
- Management team
- Other personnel needs
- Hurdles and key milestones, with estimated dates and costs
- Financial analysis
  - Revenues and timing
  - Expenses and their timing
  - Break-even
  - Pre-revenue financing needs

It should be detailed enough to show that the potential entrepreneur has thought out the issues well but should not be too long (or no one will read it). As a rule of thumb it should be under 50 pages including financial tables and executive summary—that by itself should be a maximum of two pages.*

1.1.5 Business Structures

There is a wide variety of business structures in the United States. Since businesses are incorporated at the state rather than federal level, some variation in available structures that the entrepreneur in nanotechnology may consider are as follows:

- Various type of corporations
- Various forms of partnerships
- Limited liability companies
- Sole proprietorships

There are a few major characteristics that differentiate them. These include:

- Taxation status
- Liability to the owners
- Governance requirements (i.e., management and reporting requirements)
- Number of shareholders permitted
- Ease of raising capital

For instance, if the company requires a large capital budget to build the capacity to manufacture products, the most likely source would be through venture financing followed by a public offering of stock. This is best done with a General C Corporation. Venture capital funds, angel investors, and banks are familiar

---

* There are many sources of information and hints on writing business plans. Entering "business planning" into any search engine will turn up a large number of sites that will, sometimes with no fee, offer to help an entrepreneur with preparing business and marketing plans. Business plan software is available from several sources and can be found in many software stores. Reviews of business planning software can be found at www.homeofficereports.com/Business%20Plan.htm. There are also numerous books available from any bookseller. One that provides a good overview is T. Berry, *Hurdle: the Book on Business Planning*, Palo Alto Software (2002).

and comfortable with this structure. Investment bankers, for example, have experience with IPOs. Governance and reporting requirements for C Corporations provide a degree of confidence to investors that the company will be managed responsibly, or at least with shareholder and board of directors oversight, so as to provide the best possible return on the investments. The C Corporation fits the general image of a "company" in the minds of most: it has a CEO who reports to a board of directors, other corporate officers responsible for finance, R&D, manufacturing, HR, etc. To investors, it has two other important characteristics. The C Corporation can have an unlimited number of shareholders and a variety of classes of stock,* making it attractive as an investment assuming it also has products or potential products that could provide those investors a good return. Those products do not have to be ready to go to market in order to secure funding, but the company must produce sufficient data to show that they can be producing revenue in a timeframe acceptable to the investors.

A C Corporation also provides, in general, the most flexibility in offering ownership incentives to employees or outside service providers like consultants. Their compensation can include share grants either through outright sales or, as is more often the case, through so-called option grants to these individuals. This subject is an important one in light of recent U.S. Federal laws.†

If the company is a consulting business, the entrepreneur could establish a *sole proprietorship* or, if working with a partner a *general partnership*. These do not require registration but may require business licenses in some jurisdictions.‡ No special tax forms are needed because the income from the business flows into the owner's personal tax returns. The individual is personally responsible for all debts and obligations and therefore all personal property is at risk unless otherwise legally protected. Partnerships need to be structured very carefully since each partner is liable for all business debts and obligations incurred by the other partners, and in the case of sole proprietorships, all personal property is at risk unless protected by some means. In setting up a general partnership, care should be taken in putting together a partnership agreement (PA), a legal contract between the partners that governs the relationship. In general, this PA cannot

---

* In general, the class of common stock is what is traded on the various stock exchanges. Special classes of stock are often used when raising money before a public offering, and can have different values and voting rights, can even be interest bearing and include rights to convert to common stock at some fixed rate. When considering special classes of stock, legal advice is essential.

† For the interested reader, the Public Company Accounting Reform and Investor Protection Act of 2002, also known as the Sarbanes–Oxley Act, was passed in the wake of several corporate scandals, ostensibly to restore public confidence in corporate accounting practices. It has been the source of much controversy, with some claiming it interferes inappropriately with the operation of corporations, others claiming it does not go far enough, and yet others claiming that the arguments on both sides are overblown. This important act has had one effect that could be considered positive, the creation of a new line of business in software and consultation for compliance. Visit http://fl1.findlaw.com/news.findlaw.com/hdocs/docs/gwbush/sarbanesoxley072302. pdf for a full copy of the Act or use a search engine to find general articles that will provide some background.

‡ If the founder does not want to operate under his own name, he should register a trade name with his state authorities, the so-called Doing Business As or dba name.

limit an individual owner's liability for the business activities of other owners.*

Even though there is paperwork involved, instead of a sole proprietorship or a partnership, an entrepreneur should consider a structure that has the same tax consequences overall to limit their personal liability. Either a *Subchapter S Corporation* or a *Limited Liability Company* (LLC) will do that. The advice of an experienced business attorney should always be obtained before deciding the best type of structure for your business.

1.1.6 *Registering a Company—Where?*

In the United States, companies are incorporated (or registered) under state laws, not at the federal level. Each state has its own requirements for paperwork, reporting and taxation. While a company can register in any state, it generally needs to have an office, or *Registered Agent*, in the state it chooses. Many U.S. companies, especially those expecting to operate in several states, choose to register in a state that has advantageous tax or other corporate policies, often Delaware or Nevada, though many states are changing business laws to make themselves more attractive. Information about these and other potentially attractive states can be found on various Web sites to help the entrepreneur understand the basics of registration.†

Companies must also register in the state in which they are physically located or in which they have offices, employees, or other facilities, even if their primary registration is in one of the so-called business friendly states. For instance, a company with offices and manufacturing facilities in North Carolina, may choose to register in Delaware but then must register as a "foreign" company in North Carolina. If it establishes a research site in South Carolina or a sales representative with an office in California it will then register in each of those states as a foreign company as well.

An entrepreneur may initially decide to register the company in the state in which offices and laboratories are located. This may be done to reduce paperwork, including the filing of multiple tax returns and annual reports. If the business expands to several states, the primary registration can be relocated to one of the business friendly states. However, the list of these states is growing as more and more are competing for businesses, so such a move should be carefully investigated before any action is taken.

The business of nanotechnology is worldwide and although the above discussion has centered on U.S. companies and registrations, many of the same issues face the entrepreneur who wants to start a business elsewhere in the world. Of course, registration and corporate structures, though similar, vary from country to country.

---

* Except that the limited liability partnership can shield each partner from malpractice by other partners.

† Two such sites are www.mycorporation.com and www.corporate.com, though a search engine will also provide other sources. These sites not only provide information but also will aid the entrepreneur in registering his company in one of those states. Many business attorneys will assist in preparing registration documents, ownership contracts, bylaws, etc. and will act as registered agents for the business.

The nanotechnology entrepreneur may decide to establish subsidiary companies in countries other than the one in which he initially establishes his organization. Intergovernmental commercial treaties are in place to make the establishment of whole- or majority-ownership of corporate subsidiaries in countries other than the residence of the parent relatively easy—although various rules and restrictions may apply. In general, due diligence must be exercised if an entrepreneur wishes to establish a presence in multiple countries, for example, consulting legal entities for advice both in the home country and the foreign country is recommended. Information about the requirements and costs of registration in a wide range of countries is readily available on the Internet.*

Registering a Company—The Process. The Internet has changed the way businesses are registered. As late as the 1990s, a person wishing to register a company in the United States would have to obtain a form, fill it out manually, and either mail or carry it to the business registration office in his state, along with a check, sometimes certified, to pay the fees. Today, most states encourage registration via online forms and even charge a premium for using a paper form. Online registration has become streamlined so that someone who has registered several companies can usually complete the process in 15 min or less, and then pay the fees by credit card, never leaving his office.†

Once the company has been formed and registered, the entrepreneur needs to prepare the paperwork: company bylaws, employment contracts, quarterly (or even more often) and annual reporting to tax authorities, and many other details. Regular meeting of the board of directors is required as is an annual meeting of all shareholders. There are many more requirements that differ with the size of the company. Many of these details are both time consuming and tedious, but they have to be done or the company could be liable for various penalties. The entrepreneur, especially the scientist, is wise to either be prepared for paperwork or consult a business-oriented professional.

---

* A Web search can be made for individual country registration requirements by using a search engine and a simple search string such as "business registration (country name)." A World Bank Web site: http://www.doingbusiness.org/ExploreTopics/Starting Business/ contains a broad compilation of estimated costs and registration requirements for all countries. U.S. embassies around the world have an officer known as the commercial attaché on their staffs. Many other countries have similar positions. The job of the attaché is to assist companies in understanding the business climate in these locations and helping businesses make appropriate contacts and negotiating the paperwork. These individuals and their staff can be a valuable resource.

† Once the company has been registered in the United States, the next step is to obtain an employer identification number from the Internal Revenue Service (IRS). This, too, is done online by going to the IRS site, www.irs.gov, and filling out and submitting Form SS-4, Application For Employer Identification Number. The number is assigned at the time of completion and a copy is mailed to the filing address. It may be necessary to obtain other permits and licenses from the state or states in which the company is registered, such as a sales tax license or a tax exemption number. In some jurisdictions a city or county tax or other permit may be required. Most of the applications for these are now available online and many, if not most, of them can be filled out and filed electronically.

1.1.7   Finances

Financing is the single-most important aspect of a new business. While this may seem obvious, one of the biggest problems for the entrepreneur is determining, and then acquiring, the amount of money needed to sustain operations until the company has sufficient product sales to pull the load. Unless the entrepreneur is already wealthy and can either fund the company from his own pocket or has partners, friends or family with money, time needs to be spent, perhaps a great deal of time, soliciting interest in his company from groups that do have money and who may be willing to invest it. After investing their own money and that of partners, family, and friends, most entrepreneurs turn to angel investors, venture capitalists, and then public equities markets. Angel investors are individuals or groups who invest their own money in companies that interest them. Usually the investment is small, under one million dollars (and often much less), and the investor receives equity in the company, sometimes a board position.*

Venture capitalists (VCs) are professionals who seek out investments for large funds they manage. These funds tend to be very large and investments they make are measured in millions of dollars. Venture capital fund managers have a reputation of being very exacting in choosing companies for investment. They often will want to participate in governance via board positions and may also demand preferred stock.† VCs have extensive contacts in the investment banking industry and can provide invaluable assistance when seeking additional finance via other venture funds, loans, or public stock offerings.

Capital markets, or stock exchanges, were established to provide an avenue for raising capital to finance major manufacturing businesses. In the United States, many business owners and investors are familiar with companies trading on the New York Stock Exchange and the NASDAQ system. However, there are public equities markets in every country and trading shares, via the Internet, that are able to facilitate investment. There is no reason to consider only the major U.S. markets (there are other stock exchanges in the United States and globally) as sources of capital. Increasingly, companies are choosing to register shares for sale on markets in Canada, various European countries, Singapore, Hong Kong, and others.

1.1.8   Managing the Company

Management, especially in a technical setting, is a complex and often delicate task. Entire racks of books about management can be found on the shelf and in

---

* Some entrepreneurs will balk at this. The author's advice is not to do so, since the angel investor usually has been successful in business and can offer valuable help in managing the company. Also, he is likely to know other potential angel investors or venture capitalists. When seeking more capital, such contacts are indispensable, and will provide more credibility than anything else, with the possible exception of a marketed product.

† Preferred stock can come in many different forms, but is "superior" to common stock. This means that if the company is sold or liquidated, the holders of these types of shares receive preference in any distribution.

the catalog of any bookseller. Numerous case studies have been published about successful and unsuccessful management practices. The subject is too large to address in this text but from the author's point of view, the most successful approach to management is respect. Employees at all levels who feel they are treated with respect will help make a company succeed by working harder and longer hours. On the other hand, an atmosphere of disrespect often results in underperformance and, sometimes, outright sabotage.

The most important task of the senior managers of any company is understanding the strategic goals and plans of the company and keeping themselves and their employees focused on them. This is especially important in a small, growing company.

1.1.9 *Developing and Manufacturing a Product*

This complex subject will vary over a wide range depending on the product. Product development and process development go hand-in-hand: the world's finest mousetrap will not be a success if it cannot be manufactured consistently and at a cost that makes it affordable to the consumer while providing a profit to the manufacturer.

For certain products, the development cycle is quite short because of facilitated production. A good example is the manufacture of carbon nanotubes formed by chemical vapor deposition. If purification and other properties are not considered, the cost of production is very low and fairly well understood. Thus, a new company entering this field could be selling product within a few months.

For others, the cycle may be very long. An example may be a nanoparticle drug delivery system. Besides the extensive testing any pharmaceutical product requires, that is, years to show both safety and efficacy, there may well be additional testing of any effects on the environment that may arise as a result of its manufacture or use. Another such product could be a nanoparticulate coating for glass to reduce glare or transmission of a certain wavelength. Such nanoparticles are expected to enter the environment as a result of the use of coated glass and data on the effects of that exposure could take a significant effort to produce and analyze. Such testing should be carried out or the manufacturer could be held liable for any problems.*

Building a manufacturing facility will also vary greatly with the product. Integrated circuit manufacturers working at the nanoscale will invest billions of dollars in manufacturing facilities, equipment and training, as will manufacturers of bio-nanotechnology products. A company manufacturing the window coating mentioned above may well be able to do so for a much smaller investment.

Product development and manufacturing are areas that the entrepreneur in nanotechnology should be very careful when considering. They can be very expensive in both time and finances and under- or overestimating the time and cost is very easy to do. With the advent of numerous and ingenious bottom-up fabrication processes, the cost of manufacturing can be driven to reasonable levels for start-ups.

---

* The asbestos litigation is an example.

1.1.10   Marketing

The old adage that "if you build a better mouse trap the world will beat a path to your door" is only partly true. The world needs to know the mousetrap exists and where your door is located; and then it needs to know why it is a better mousetrap. That is the function of marketing.

For every shelf of management books at a bookseller there are at least four shelves of books about marketing, advertising, and sales. These books when taken collectively contain conflicting messages and methods about how to go about developing a marketing and advertising campaign and then selling the product. An entrepreneur with little or no marketing experience may not want to indulge in marketing strategies. When the company has a product identified and has it under development, the best approach is to hire a professional marketing person—let them develop and execute the marketing plan.

1.1.11   Exits

Once a nanotechnology company is successful, generally by having a product either ready for market or actually producing revenue, the founders are often faced with a choice: to stay on and manage a commercial operation or move on, allowing professional managers with the appropriate experience to take over. This is actually a difficult subject for many entrepreneurs. They have seen their company succeed and they often want to continue to grow the enterprise. The problem is that while a start-up company often needs the single-minded drive, enthusiasm, direction and, even, charisma of the founder, when it comes down to actually producing products and making a profit, management needs to evolve—a condition more evident if the company has become publicly held through a stock offering. In this case, the company comes under pressure from stock markets to increase revenue and profits. The type of executive manager who focuses exclusively on that issue, while still supporting new ideas and growth of product lines, is needed to manage the company in that environment.

At this point, the founders can choose to remain, perhaps in different roles, or leave the company. If they remain, they need to understand the differing needs and be willing to put the appropriate management into place: finance, marketing, R&D, manufacturing, HR, legal/patent etc. They then should act as enablers, helping and supporting these managers to do their jobs and make the company even more successful.

Leaving, although traumatic perhaps for both the founders and the associated staff, may be the best choice. Very often, founders leave and use the assets they have from their company to start other companies or to act as sources of funding for other entrepreneurs.* The successful exit is defined by another mechanism. Successful small companies (those that actually develop products and make a profit selling them) are very likely to be bought out by another company. If that small company is publicly-traded, the stock market sets a value

---

* In fact, the author has known a number of "serial entrepreneurs" who, having been successful with their first company, have gone on to found others based on different technologies or applications. Since they have a track record of success they often find it easier to attract financial backers in subsequent companies.

based on share price. How markets value traded companies is generally based on a multiple of earnings or potential earnings per share per year. The market capitalization is simply the per-share trading price at the close of any day multiplied by the number of shares actually issued to shareholders (as opposed to being held in reserve by the company) and, therefore, varies with the trading price. Usually, when a company is being sold, the total price is more than the market-based valuation, with the premium resulting from the negotiating acumen of the company being purchased.

If the company is not a public company, the sale price is determined by negotiations. If an entrepreneur is involved in selling a company with a revenue and profit stream it should be possible to calculate a value, taking into account future product revenues. If there is no revenue stream, the value is based on a model of future cash flows. It would be advisable to engage the services of a business evaluator with experience in high-technology companies to calculate a market value when revenues are either very low or not yet realized. Some of the best sources of such evaluators are the large accounting firms, who all have consulting arms prepared to do this kind of evaluation.

The above description of the founder being successful and able to exit the company with significant assets, that is, money or its equivalent, is the dream of all entrepreneurs. However, the reality of business is that many do not succeed. The technology often times cannot be developed in a commercially feasible form due to the lack of money, or that a competing product appears first, or for a wide range of other reasons. This situation does not necessarily cause bankruptcy. In fact, most businesses in the United States that close their doors do not owe money; they just do not succeed.

When the decision is made to "wind up" a company, that is, close it down, there are often assets left in terms of cash, equipment, intellectual property, etc. Many of these assets are sold or, as in the case of a lease of space, transferred to another company. the resulting cash is usually divided among the shareholders pro rata, with holders of preferred shares receiving the distribution first.

Many successful entrepreneurs have been through the closing of a company. It is not a reflection of failure on their part. They learn from the experience and then go on to found other companies.

1.2 EDUCATION AND WORKFORCE DEVELOPMENT

Societal components collectively and integrally bend, mend, and mold our civilization. We now direct our attention to education and workforce development. With regard to the broader sense of societal implications, specifically that new technology brings along changes that are positive, negative, or indifferent, we will do our best to achieve a balanced presentation. We strive in this section to stay below and beyond the hype associated with nanotechnology, a technology and societal driver that pundits around the globe consider to be revolutionary.

As a student, you are standing at the threshold that opens into your future, your livelihood, and your potential contribution to society. The underlying purpose of this section is to inform, provoke thought, stimulate, and hopefully, to encourage direct action. One must seriously consider the following prediction by the NNI to understand the significance of what is to come. Specifically, that products impacted by nanotechnology are expected to contribute over $1 trillion to the global economy by 2015 [5]. The strategic plan released by the NNI in December 2004 opens with the following vision [2].

> The vision of the National Nanotechnology Initiative is a future in which the ability to understand and control matter on the nanoscale leads to a revolution in technology and industry.

This vision indicates a revolution (from the Latin *revolutus* to "turn, roll back," with the "general sense of great change in affairs") is in the making based on the promise of nanotechnology; a revolution that has the potential to radically transform both technology and society. We need to simply look back a few years to the twentieth century to understand the impact of new technology. The legacy of the automobile, the television, and the computer are well understood. Are nanoscience and nanotechnology capable of exerting such historical changes? The NNI strategic plan goes on to state [5]

> ... the NNI will expedite the discovery, development, and deployment of nano-technology in order to achieve responsible and sustainable economic benefits, to enhance the quality of life, and promote the national security.

The expectations for nanotechnology are quite lofty. **Goal 3** of the strategic plan is centered on the development of education and workforce development. The introductory paragraph of **Goal 3** emphasizes the important link between edu-cation and workforce development and the all-important tangible infrastructure that must be in place, seemingly a priori:

> A well-educated citizenry, a skilled workforce, and a supporting infrastructure of instrumentation, equipment, and facilities are essential foundations of the initiative. Nanoscale science, engineering and technology education can help to (1) produce the next generation of researchers and innovators, (2) provide the workforce of the future with math and science education and technological skills they will need to succeed, and (3) educate the citizenry capable of making well-informed decisions in an increasingly technology-driven society.

For any responsible strategy, the vision, goals, objectives, and the timetable are identified well before any reasonable action is to take place. Our hope is to convince you that a career in nano or related fields will offer challenge, accomplishment, and most importantly, responsible change in society. We repeat once again the inspirational statement attributed to Nobel Prize winner Dr. Richard Smalley, *"Be a scientist—Save the world"* [6].

1.2.1 *Technological Revolutions—The Workforce Point of View*

Education, workforce development, infrastructure, and industry are intimately intertwined and have collectively defined much of our civilization for thousands of years. Once again we are confronted with a familiar conundrum—what is the exact meaning and potential impact of nanotechnology and why would it be considered revolutionary? In order to gain a relevant perspective, we must review some basic history. Several thousands of years ago, developments in agriculture and irrigation transformed the nomad into the farmer. Prehistory ended with the birth of civilization. Agriculture is a *practice* that is defined as the production and distribution of food. With the creation of sessile agrarian societies, the livelihood of the majority of the population was based on farming and affiliated occupations. With the accumulation of assets, and eventually wealth, the need for armies, rulers, merchants, and clerics followed suit as a division in classes ensued.

On the other hand, the Industrial Revolution was based on a single-minded *manufacturing philosophy*, more specifically, the mass production of products. The invention of the printing press by Johannes Gütenberg in 1447 was an incipient form of mass production. The Industrial Revolution happened relatively quickly compared to the fundamental one of agriculture, but there was still plenty of time to adapt. Although all offspring of the Industrial Revolution, products such as the television, automobile, and computer have sprung minirevolutions of their own. Throughout history, we have had revolutions based on a *practice* (agriculture), based on a *manufacturing philosophy* (industrial), based on a *fuel* (oil), based on *products* (steam engine, railroad, automobile, telephone, and computer), based on *nature* (biotechnology), and based on *size* (micro and nano).

The currently burgeoning revolution in biotechnology is so because of our greater understanding of the *disciplines* of biology, biochemistry, and genetics. And what of the contributions of micro and later on, nano? What brand do we place on the micro- and nano-revolutions? Micro and nano are of course prefixes that relate to size. We now actually have revolutions, micro earlier and now nano, based on *SIZE* and size alone! The drive to manufacture smaller and smaller components materialized in the first portable watch, created by Peter Heinlen in 1524. The integrated circuit became the embodiment of miniaturization that has now morphed into nanocircuits.

Because nanotechnology has the capability to affect every type of product made (to the best of our knowledge, everything is still made of atoms and molecules), new jobs are certain to follow suit once a great idea leaves the lab—at least to countries that value education and innovation (is that us?). Will we have to overhaul our compartmentalized approach to education to one that is fundamentally interdisciplinary? According to an accepted definition of revolution, these changes are supposed to represent radical and far-reaching consequences, both of a technological and social nature. With each and every revolution throughout our history, the time scale of change has been diminished from that of its predecessor—overall, from several thousands of years to a few decades today. Is a revolution in education and workforce development needed or should we plod on with the status quo?

1.2.2 *The State of Education and Workforce Development*

It is widely viewed that the status of science and engineering education in the United States, and perhaps the Western world for that matter, is in a state of decline. According to Nobel laureate Richard E. Smalley [6], the United States in particular is falling behind developing nations such as China, India, and indus-trialized nations such as Japan with regard to the generation of new scientists and engineers. There is not enough space in this section to offer proper treat-ment of the multifold and complex factors responsible for the apparent decline. Without the addition of tedious detail, it is generally accepted that (1) the United States is producing fewer engineers and scientists; (2) although still the leader, more and more intellectual property is being produced outside the United States than ever before; (3) more and more high-technology jobs are being shipped overseas; (4) the structure of K-12 education is not optimized to promote science and engineering; (5) higher education is generally considered to be under funded with increasingly higher tuition; (6) the influx of foreign intellectual talent is limited due to current emphasis on security and increased prosperity in other nations; and (7) significant commitment in time and effort with diminished employment expectation results in fewer home-grown graduate students in science and engineering. According to Richard Florida, the author of *Flight of the Creative Class*, global competition for creative talent will be the defining issue of the twenty-first century [7]. Please acknowledge that this brief list is not apocalyptic in its message, just a statement of accepted trends. Isaac Asimov is noted for stating

> Science can be introduced to children well or poorly. If poorly, children can be turned away from science; they can develop a lifelong antipathy; they will be in a far worse condition than if they had never been introduced to science at all.

At the juxtaposition of any revolution, especially one that is suddenly upon us, opportunity abounds for those that are prepared. There are many means of navigating the impending maelstrom based on two generalized factors: (1) acknowledgement that the world is changing and (2) adapting by necessity to that change just like our predecessors did thousands of years ago. Economic development groups across the country have three goals: (1) jobs, (2) jobs, and (3) jobs. A civilization without jobs is incomprehensible, intangible, and actually unimaginable. Without employment there is no prosperity, at least within the existing paradigm—the paradigm that defines our quality of life. Along with each revolution, job descriptions have changed, oftentimes drastically.

In 1958, Congress passed the National Defense Education Act following the launch of Sputnik. In 2005, Louis V. Gerstner, Jr., a former chairman and CEO of IBM, stated in a *Newsweek* article [8]

> America is no longer winning the skills race. South Korea, with one sixth of the U.S. population, graduates as many engineers as the United States. China graduates four times as many; India, five times as many. Just as more than half of America's current science and engineering work force is approaching retirement, the flow of foreign talent is starting to dry up. For the first time in my memory, we're at the wrong end of a brain drain, as foreign-born grads in science, technology and engineering either return home after getting U.S. degrees or stay home in the first place.

According to RAND Corporation reports, alarms concerning potential skilled labor shortfalls oftentimes are not always justified [9]. However, they go on to say that

> …although previously anticipated STEM workforce shortages have not material-ized in the economic sense, the implications of a shortage of skills critical to U.S. growth, competitiveness and security justify continued examination.

1.2.3   *Current Workforce and Education Programs*

Awareness about the current state of education and the workforce is at a high level, especially with the onset of nano. New programs available to attract or create scientists and engineers are springing up in every nation. Factors such as the age of the existing workforce, dependence on foreign nationals and out-dated curriculum coupled with declining federal funding, and the accelerated pace of science and discovery bolster the case for urgency. There is a fierce global competition to commercialize nanotech-based products. The United States does not necessarily have the lead. The NNI has defined the needs for education and development of a workforce of the twenty-first century, and is backing up their proposal with funding initiatives. Emphasis on nanoscience and nanotechnology must begin in the K-12 school system and extend through community college, university, and vocational schools [10].

Programs developed by the Nanotechnology Institute (NTI) in Pennsylvania serve as excellent examples of advanced thinking and multilevel partnerships forged among business, academia, and government that promote nanotechnology education and workforce development. The organization is a collaboration of academic and research institutions with a state-funded economic development group, the Ben Franklin Technology Partners of Southeastern Pennsylvania. The following excerpt from their Web site summarizes the NTI workforce development mission (http://www.nanotechinstitute.org/nti/index.jsp).

> A highly-skilled, technically trained, nanotechnology workforce will be needed if the region is to gain full value from the commercialization opportunities that nanotechnology will generate. The NTI is partnering with regional community colleges to anticipate this need by building the partnerships and securing the tools and resources to develop and implement education, training and work-force development strategies.

Nanotechnology as we have learned has a high barrier of entry. Nanotechnology will require highly trained scientists and technicians, but those jobs need to be in place to attract these special graduates.

1.2.4   *The Workforce of the Future*

The *knowledge worker*, a phrase invented by Peter Drucker in 1959, identified the emergence of a changing workforce, one that relied more on brains than brawn [11]. Robert E. Kelley wrote an influential book in 1985 called *The Gold Collar Worker: Harnessing the Brainpower of the New Workforce* [12]. In it he writes that American business is suffering from a *brain drain* due to severe mismanagement of its most valuable resource—*brainpower*. If you recall, the 1980s was the decade of the infor-mation age (*a.k.a.* revolution). Kelley stated that the gold-collar worker should

> …engage in complex problem-solving, not bureaucratic drudgery or mechanical routine; they are imaginative and original, not docile or obedient. Their work is challenging, not repetitive, and their results are rarely predictable or quantifiable especially if they're scientists or researchers. Gold-collar workers are everywhere…

According to Dr. Mary Ann Roe, the author of *Cultivating the Gold Collar Worker*, nanotechnology along with the other emerging technologies, will require a special workforce. She states that [13]

> Complex powerful currents swirl through the nation and the world, altering the economic landscape, while offering extraordinary opportunities for well-prepared individuals… Technology-driven, these currents propel dynamic change, force innovation and create new types of work in the private sector.

According to Dr. Roe, this belongs to the techno-professional who wears a gold-collar—an individual with a white-collar mind but blue-collar hands; a worker who is able to create as well as operate and think as well as do.

We try to envision the future and the worker therein. Within the lexicon of social prognosticators of this day and age, we are told often that the average employee changes job description, or the job itself, every five years. We predict that the new workforce, whether of white, blue, or gold collar, will have flexibility in the job market, a high level of training and education, an interdisciplinary background, advanced communication and computer skills, and perhaps an understanding of the international education and economic community. We boldly predict that the face of K-12 must change in order to accommodate the new age of technology. Instead of teaching to the test in a compartmentalized fashion, classrooms must open curricula to new ideas and efficiently teach interdisciplinary programs in the sciences. Partnership and close communication between educational institutions, the universities and community colleges in particular, and industry is necessary to ensure that there is balance between the needs of industry and a supply of educated and trained students. We are all aware of the shortage of trained technicians in the biotechnology sector.

1.2.5 *Planning Ahead and Potential Career Paths*

There is no guarantee of success. That key ingredient is always up to the individual. Contemplate for a moment the vast complexity of today's economy. Surely there are numerous paths to a fulfilling career. Make a list of potential paths for your future career. Then, try and integrate those paths with a nanotechnology theme.

The best approach begins with *planning* now. It is never too early to make plans. Therefore, start a *journal* to keep track of any ideas, thoughts, or references that happen your way. *Flexibility* is always a survival advantage. Omnivorous species like crows tend to out compete counterparts that rely on a limited food sources. Devour knowledge. Consider also that many of us end up doing work that we never intended to do. Nonetheless, a good strategy is to place your focus on one academic discipline such as chemistry but learn enough about physics, engineering, and biology to be able to understand your coworkers of the future. *Networking*, a good way to achieve that good strategy, is perhaps one of the most powerful tools that an individual can acquire. It really is "who you know" much of the time. Attend conferences—as many as physically possible. Open your horizons and *partnerships* to include business, academia, and government, the big three that are currently working together to pave the road for the nanotech revolution. Traditionally, business folks do not mingle much with academics. That tradition is changing drastically as universities are acquiring the entrepreneurial spirit and commercializing intellectual property like never before.

Research all available *funding sources*. Conduct investigations to find federal, state, local government, and private sources. For example, funding for women scientists and underrepresented groups are out there just waiting to be tapped. Stay abreast of *societal implications*, understand social and global trends, and track investments into nano-based businesses. Take a serious look into *careers that are tangential to the science and technology*. For example, there is a great need for a new generation of patent lawyers who possess interdisciplinary skills and knowledge. Health and safety, job sourcing and staffing, education at all levels, economic development, public policy, regulation, international commerce, and investment are but a few examples of jobs outside the technology. Lastly, and certainly of some great importance, make plans now to *start a nanobusiness*. Take care of your education (perhaps an MBA in the plan), do the research, conceive a novel product or process, make the contacts, establish the necessary partnerships, locate and secure funding and, yes, then you are in business. Although things "do come to those that wait," it is a much better policy to get informed and go after it. We wish the best of luck to you all.

1.3 BUILDINGS FOR NANOTECH

We now jump from creating and operating a business and finding and securing employment to buildings that house them. Nanotechnology along with bio-technology are placing exceedingly stringent demands on laboratory design, manufacturing strategy, and construction. This section presents a short introduction into another world—the domain of architecture and construction—where our new facilities try to keep pace with the *Nano Age*.

Old buildings are constantly being replaced due to wear and tear, but a new paradigm of building design, construction, and operation is upon us nonetheless and on many fronts. New buildings need to be designed and constructed to accommodate the rapid changes brought about by new technologies in the fields of biology and nanotechnology, new technologies that rely on interdisciplinary cooperation in a big way. New designs also include more efficient means to improve indoor air quality, electricity production, energy efficiency (heating), waste management, water conservation, and daylight lighting designs—some already enhanced by the new technologies.

New buildings also include environmentally approved technologies and practices that include use of recycled materials and reduction of toxic material components. Although we can spend a significant amount of time discussing new building philosophies and practices, we must focus on the relevant thrust of this chapter—how building and facility design and construction conform to the needs of nanotechnology and biotechnology—and conversely, how nanotechnology can contribute to new buildings. Now more than ever, the goals of research centers, are to conduct world class research, attract researchers and students, attract money, attract industry, provide jobs and, at the same time, be flexible [14].

Several "monuments to science and technology" have been constructed worldwide to support nanotech and biotech R&D. A few notable buildings are mentioned in this text: the Center for Integrated Nanotechnologies (CINT) of Sandia and the Los Alamos National Laboratories in Albuquerque, New Mexico; the Birck Nanotechnology Center at the University of Purdue in West Lafayette, Indiana, and the state-of-the-art leader amongst all buildings that support advanced technology—the NIST's Advanced Measurement Laboratory (AML) in

Gaithersberg, Maryland. We shall focus primarily on NIST's AML facility. Please refer to www.HDRInc.com, aml.nist.gov and www.nanobuildings.com for complete information concerning all of their advanced technology facilities.

1.3.1 Nanotechnology in Buildings—Environmental Aspects

Although we will present the strategy and design elements required in constructing buildings to support nanotechnology research, development, and manufacturing, it is only fair that we reserve this section to discuss the impact of the nanotechnology on building materials and construction in general. The construction business is estimated to be on the order of $1 trillion per year worldwide [15]. Many kinds of building materials already take advantage of nanomaterials [15]:

The Now
- Flexible solar panels
- Self-cleaning windows
- Self-cleaning concrete using catalytic TiO_2
- Wi-Fi paint (an additive mixed with paint that reduces transmission of radio waves through walls)
- Selective absorbing–reflecting solar windows
- Scratch-resistant flooring
- Antimicrobial steel surfaces
- TiO_2 and AgO antimicrobial coatings (public places)
- Nano-cement (enhanced physical properties)
- Nansulate coating (corrosion prevention, insulation)
- Translucent concrete (enables transit of light)
- Nanotech-enabled gypsum drywall (water resistant, durable, stronger, lighter)
- Nanotech-enabled steel (corrosion resistant, higher strength, high plasticity)
- Aerogel glasses (high R-value thermal insulation, acoustic insulation)
- Nanogel (Cabot Corp.) traps air at the molecular level → thin insulating layers

According to the USDA Forest Service Research Laboratory, nearly 2 million housing units were constructed in the United States in 2004 [16]. Wood comprised, by volume, 80% of all the building materials used. Half of the wood products are engineered wood composites. Nanotechnology is expected to contribute to the next generation of wood-based products that exhibit enhanced strength, properties, and endurance similar to those of carbon-based composite materials [16]. The new materials are designed to be biodegradable [16]. And of course, nanotechnology will promote the development of "intelligent wood"—biocomposite products with built-in arrays of nanosensors. By the way, wood is a natural nanomaterial. According to the USDA [16]:

> Building functionality into lignocellulosic surfaces at the nanoscale could open new opportunities for such things as self-sterilizing surfaces, internal self-repair and electronic lignocellulosic devices. The high strength of nanofibrillar cellulose together with its potential economic advantages will offer the opportunity to make lighter weight, strong materials with greater durability.

The U.S. Department of Energy's Twenty Year Plan identifies strategies for environmental friendly buildings—try and select the ones you believe involve nanotechnology. The United States has 5% of the world's population but consumes 25% of the world's resources and discards 40% of the world's waste, of which 50% is due to construction activities [17]. If we can at least optimize our buildings, perhaps a dent can be made on some of these statistics. We listed a few current innovative products enhanced by nanotechnology earlier. Let us continue by painting a picture of the near and far futures.

The Near Future
- Cellular building materials
- Disaster-resistant materials
- Intelligent materials (e.g., self-repairing, self-adjusting)
- Superior moisture barrier materials
- Nontoxic materials
- Resource-efficient materials
- Superior insulating materials
- Superior weather-resistant capability (e.g., low maintenance)
- Smart materials with sensors able to detect loads, temperatures, decay, fire, etc.
- Fiber-cement siding (self-cleaning, thin layer of silica)
- Anti-bacterial house fittings (Ag particles or light-activated particles)
- Fire retardant materials that incorporate clays
- Easy-clean water repellant surfaces
- UV-resistant paint coatings
- Thin film coatings for roofing materials

Other aspects of construction include household devices. Home and laboratory products enabled by nanotechnology include [18]:

- Water treatment systems (nanofiltration, light-activated nanoparticles)
- Energy-efficient lighting, LED, and electroluminescent lighting
- Solar energy generation and advanced energy storage

Buildings in the Not-So-Far Future [19]
- Earthquake-proof buildings using carbon nanotube reinforcement
- Nano-reinforced glass structural and enclosure elements
- Quantum dot lighting
- Next generation nanosensors

The interior lighting paradigm will also be impacted by nanotechnology—by solid-state lighting devices made with nanocomposite materials. Carbon nanotube-based organic composites, known as "ultra-low energy high brightness" (ULEHB) lights, are expected to produce the same quality light with a fraction of the energy. Many other materials are under development: self-healing concrete, UV-IR radiation blockers, smog-mitigating coatings, and LEWs and LECs (light-emitting walls and ceilings). According to George Elvin of Nanowerk. com, the "smart home" is also on the horizon. Smart homes are decorated with nanotechnology-based sensors that monitor temperature, humidity, and toxins as well as transmit medical information to your doctor. From larger structures to appliances, sensors would be able to monitor vibration, stress, crack propagation—the foundation for "intelligent buildings" [15,19].

There is no doubt that nanotechnology will impact (and has already impacted) the way buildings are constructed, the materials selected for building structure and function, and the way people interact within buildings. It is not hype. It is due to the remarkable properties of nanomaterials that there is the excitement and anticipation. Yes, there is hype out there and the nano-pretenders abound, but atoms and molecules do not spin tall tales—they just spin.

Through nanotechnology, we will be able to develop the building materials of the future—those that absorb and radiate heat, offer earthquake-proof stability, the electrical wires that conduct over long distances efficiently, windows that reflect or absorb radiation depending on the current need, solar cells that power our instruments, and cooling systems and lighting that do not damage our environment. Yes, it is possible to accomplish all of the above. Yes, much of it will, and already has, come from our knowledge of nanotechnology. The materials and processes just have to be cost-effective and competitive—complete with an analysis of long-range recovery of costs and comfort with environmental friendly interactions.

In the United States, buildings consume 39% of the total energy, 12% of the total water, 66% of the total electricity, and contribute to 38% of the total carbon dioxide emissions [20]. What can nanotechnology (and our own policies) do to help mitigate these numbers?

Environmental Protection Agency's (EPA) Elements of Green Building
- Energy efficiency and renewable energy
- Water stewardship
- Environmentally preferable building materials and specifications
- Waste reduction
- Toxic material reduction
- Indoor environment mitigation
- Smart growth and sustainable development

In addition to renewable energy sources, equipment and appliances should conform to EPA's *Energy Star* program—the Building Design Guidance [21,22]. Designers, architects, and builders should consider the following:

- *Include statement of energy design intent* (SEDI)—describes the energy performance outcome of your building in the bid package
- *Specify design team participation*—during construction to ensure that energy performance features are incorporated
- *Include approval process for change orders*—accountability!
- *Document construction methods*—include manufacturer's literature, summary of energy efficient features, and explanation of anticipated functions to assist construction team
- *Select qualified manufacturers*—rejection of unapproved alternatives and coordination of manufacturers to enhance compatibility
- *Seek incentives*—local utility company incentives and government incentives to offset costs
- *Communication of superior design intent*—label with Energy Star mark

What is a green building material [23–26]? A green material is made of renewable resources. The use of green building materials reduces the environmental impact of extraction, transportation, processing, fabrication, installation, reuse, and recycling and disposal [26]. Green building materials reduce maintenance-replacement costs over the life of the material, provide for energy conversation, improve occupant health and productivity, lower costs associated with changing space configurations, and offer greater design flexibility [26]. Material selection criteria, in addition to its suitability for the intended function, include resource efficiency, indoor air quality, energy efficiency, water conservation, and affordability [25,26].

The Leadership in Environmental Design. The Leadership in Environmental Design (LEED) is a green building rating system developed by the U.S. Green Building Council. The LEED concept sprung from the Natural Resources Defense Council (NRDC) in 1994 and a consortium of nonprofits, government agencies, architects, engineers, builders, developers, product manufacturers, and others [27]. Buildings are rated according to the following six criteria: sustainable sites, water efficiency, energy and atmosphere, materials and resources, indoor environmental quality, and innovation and design processes. Certification categories start at the baseline with "Certified" and then progress from "Silver" to "Gold" to "Platinum" with Platinum being the best possible rating.

1.3.2 The Needs of Scientists and Engineers (And Equipment and Instrumentation)

Architects and designers need to know what engineers and scientists require. There are, for example, several challenges that arise from integrating technologies. The transfer of energy across multiple length scales; optical amplification of quantum dots in a two-dimensional crystal; combining top-down with bottom-up fabrication; interfacing biological and synthetic systems; and interfacing mechanical force and fluid transport (e.g., nanomechanics and nanofluidics) at the nanoscale. Such integrated nanotechnologies are expected to impact our world, according to N.D. Shinn of Sandia National Laboratory in New Mexico, who states "…connecting scientific disciplines and multiple length-scales is the key to success," and require special buildings to house processes and procedures [28].

The CINT model is a prime example of an integrated building in which integrated science is accomplished. CINT makes use of a "core-gateway model" that emphasizes interaction between two national labs (Sandia and Los Alamos) and universities. More about CINT is presented later in this chapter.

Nanotechnology research requires observations of reactions in real time (at femtosecond time scales or lower). Nanotechnology research requires increased resolution—the capability to see atoms and gauge the level of material behavior at the molecular level without interference from the outside world [29]. The underlying goal of every new advanced technology building is to, of course, have built-in flexibility [14]. Collaboration, interaction, and integration are key components of any new design—and these ingredients must all coexist within one building or complex. A case in point is the interdisciplinary collaboration among multiple disciplines and interaction between academia, business, and government. In addition, new buildings, need to accommodate technology transfer like never before, evolve into a technology transfer user facility [14].

Interactive Spaces.   A state-of-the-art research and manufacturing center is composed of many kinds of *spaces*—the result of forethought, adaptability, and an effective floor plan design. *Technical spaces* are places where characterization, laboratory work, and manufacturing are accomplished. These include clean rooms, imaging facilities, production (e.g., lithography), and metrology. *Nontechnical spaces* include offices, meeting rooms, libraries, conference rooms, and auditoriums. These spaces must be strategically placed around, within, or between technical spaces to enhance interaction between and among scientists, entrepreneurs, and managers.

A key element of advanced building design that is acquiring exponentially more importance with each passing year is the following—how do the architect and engineer accommodate the increasing level of interdisciplinary needs required for nanotechnology? We will need dry labs, wet labs, semiconductor clean rooms, bio-clean rooms, quiet labs, ultraquiet labs, labs for metrology and imaging, office spaces, interactive spaces, and spaces for support. Potential users may arise from numerous university departments—chemistry, physics, biology, chemical engineering, mechanical engineering, agronomy, mathematics, biophysics, computer sciences—and from industry and government.

Environmental Considerations [29].   Environmental stability is important in any laboratory or manufacturing facility. Factors that contribute to environmental stability include temperature, humidity, contamination, air velocity, and vibration. Environmental stability is defined over space (uniformity) and over time (drift) [29]. Uniformity is influenced by the amount of "in and out" of a room, the distribution scheme, and the room design. Drift is influenced by measurement and control systems, sensor locations, and equipment placement.

Temperature and Humidity.   Temperature control for a room that houses a TEM is a challenging prospect. The room temperature surrounding a high-powered (+300 kV) TEM should conform to 20 ± 0.01 to $0.25°C$. The drift of temperature should be no more than $0.5°C \cdot h^{-1}$ (fluctuation $0.05°C \cdot min^{-1}$) and air velocity less than $5 m \cdot min^{-1}$. The heat generated from a generic TEM column and support equipment is 500 W and 800–1200 W respectively. In order to achieve ca. control, internal sources of heat gain need to be minimized (e.g., lighting), a high air exchange rate needs to be maintained ($>300 AC \cdot h^{-1}$), and heat transfer through walls needs to be minimized [29]. Added features such as a heated floor pad aid in maintaining such a tight rein on temperature.

Humidity control, obviously, is vital for several highly sensitive measurement techniques. For example, the temperature and humidity requirement for JEOL's JXA-8200 electron probe microanalyzer is $\pm 1°C$ and 60% rh or less so that dew does not condense on the cooling water hose. At NIST's AML, humidity control is exerted to a level of $\pm 1\%$ in metrological areas and $\pm 5\%$ in other laboratories.

Clean Rooms.   A clean room (or cleanroom) is an isolated space within which a high level of particulate contaminant control is in effect. A clean room environment is designed to reduce environmental pollutants such as dust, microbes, aerosol particles, and chemical vapors [30]. According to ISO-14644 (The International Standards Organization)

TABLE 1.2	*United States Clean Room Standards*				
US FED STD 209E	**≥0.1 µm**	**≥0.2 µm**	**≥0.3 µm**	**≥0.5 µm**	**≥5 µm**
1	35	7	3	1	
10	350	75	30	10	
100		750	300	100	
1,000				1,000	7
10,000				10,000	70
100,000				100,000	700

Source: US FED-STD-209E Cleanroom Standards. With permission.

Cleanrooms and associated controlled environments provide for the control of airborne particulate contamination to levels appropriate for accomplishing contamination-sensitive activities. Products and processes that benefit from the control of airborne contamination include aerospace, microelectronics, pharmaceuticals, medical devices, healthcare, food, and others. Many factors besides airborne particulate cleanliness must be considered in the design, specification, operation, and control of cleanrooms and other controlled environments.

This is no easy feat to accomplish considering that the air outside in a typical city has on the order of 3.5×10^7 particles · m^{-3} 500 nm or larger. Microprocessors, for example, are assembled in a clean room. Other aspects of clean rooms such as temperature, humidity and pressure are also strictly regulated.

According to Federal Standard 209E, a *Class 10000* clean room should have no more than 10,000 particles larger than 0.5 µm in a cubic foot of air. A *Class 100* clean room should have no more than 100 particles larger than 0.5 µm in a cubic foot of air. A *Class 1* clean room should be essentially contaminant free (**Table 1.2**). Hard disk manufacturing requires a Class 100 clean room. The ISO, headquartered in Geneva, Switzerland, also recommends standards for clean rooms based on a logarithmic scale.

ISO-14644-1 Standards. Following normalization between cubic feet and square meters, clean room standards are converted accordingly: *Class 1* = ISO 3, *Class 10* = ISO 4, *Class 100* = ISO 5, *Class 1000* = ISO 6, and *Class 10,000* = ISO 7 (**Table 1.3**). Other standards are also in existence.

TABLE 1.3	*International Standards Organization Clean Room Standards*					
ISO-14644-1	**≥0.1 µm**	**≥0.2 µm**	**≥0.3 µm**	**≥0.5 µm**	**≥1 µm**	**≥5 µm**
ISO 1	10	2				
ISO 2	100	24	10	4		
ISO 3	1,000	237	102	35	8	
ISO 4	10,000	2,370	1,020	352	83	
ISO 5	100,000	23,700	10,200	3,520	832	29
ISO 6	1,000,000	237,000	102,000	35,2000	8,320	293
ISO 7				352,000	83,200	2,930
ISO 8				3,520,000	832,000	29,300
ISO 9				35,200,000	8,320,000	293,000

Source: International Organization for Standardization (ISO-14644). With permission.

Air entering a clean room is filtered for dust and the air inside is recirculated through high efficiency particulate air (HEPA) and ultra-low penetration air (ULPA) filters. Employees in a clean room facility must enter through airlocks that may contain an air shower and wear bunny suits that cover all of the person including hands and shoes. Any fibrous or particulate materials are prevented from entering the clean room (e.g., pencils, fabrics, soda pop). Clean rooms also maintain a positive pressure within to prevent egress of unfiltered air. The support equipment infrastructure and maintenance of a clean room (air conditioning, filter systems, etc.) may be enormous and complicated respectively.

Electricity. According to the IEEE (Institute of Electrical and Electronics Engineers) Std. 1100–1999

> Power quality is the concept of powering and grounding electronic equipment in a manner that is suitable to the operation of that equipment and compatible with the premise wiring system and other connected equipment.

Power disturbances arise from external or internal sources. External sources include lighting, faults, and utility switch surges such as voltage reduction and line maintenance. Internal sources arise from mechanical equipment (chillers, fans, pumps), elevators, shop equipment, and laboratory equipment. There are several means of reducing electrical disruptive effects by applying power conditioners, transient voltage surge suppressors reduction (dedicated circuits minimize line noise, transients), shielded-isolation transformers, uninterruptible power supplies (UPS), and standby generators [31].

Common and normal mode transients from line and load side sources, noise from and between lab equipment, stray ground currents, ELF (extra or extremely low frequency) and EMI (electromagnetic interference) sources, acoustic noise sources, vibration sources, irregular voltage and frequency, and sources of heat are some common problems facing architects and engineers who design any laboratory. With nano-capable equipment, such issues are exacerbated and special care must be taken to minimize electrical disruption so that they do not influence data acquisition and measurements. System wiring configurations also play an important role in reducing electrical disruptions. For example, grounding systems include lightning protection, safety wiring, communication system grounds, signal reference grounds, and instrument reference grounds. Isolated grounding of individual equipment is recommended [31]. For more information, please refer to www.HDRInc.com.

Vibration and Acoustics. Structural and mechanical design of advanced technology facilities must address vibration-sensitive issues [32,33]. Mechanical vibrations in general, depending on energy and their potential targets, can be detrimental to human health, comfort impairment, sensitive equipment, and structural components [34]. Many laboratory operations are sensitive to vibrations. These include metrology, high-end imaging (TEM, SEM), photolithography, and probe development [32,33]. Other methods and procedures are impacted by vibrations to some degree. These include optical microscopy, mass spectrometry, and other characterization methods. Low-end imaging, theory and modeling instruments, and nontechnical spaces, of course, are less sensitive to vibrations. The bottom-line, like for any requirement, must be balanced with cost-effectiveness, facility mission, aesthetics, facility operation costs, and future flexibility.

Sources of external vibration are numerous. Machinery, road traffic, and continuous construction activity are considered to be sources of *continuous vibration*. In this form, vibration continues uninterrupted for a defined period. *Impulsive vibrations* are infrequent and sporadic and are defined as "three distinct vibration events in an assessment period" (e.g., dropping heavy equipment and loading and unloading supplies). Rapid buildup to a maximum level followed by a damped decay of short duration is a characteristic of impulsive vibrations. *Intermittent vibrations* arise due to trains, passing heavy vehicles, and intermittent construction activity. Intermittent vibrations can be interrupted periods of continuous vibrations or repeated periods of impulsive vibrations that arise from continuous or repetitive sources.

Internal sources arise from periodic excitations (e.g. constant speed rotating equipment), footfall (e.g. from the scientists), AVAC, fluid transmission in pipes, and low-frequency airborne acoustic noise (e.g. chit-chatting) [34]. The effect of mechanical vibrations depends on the vibration amplitude, frequency range, duration, the predominant component, and time of day [34]. There are several standards that describe recommended vibration criteria (VC).

VC Curves.   Generic vibration criteria curves are provided to architects and engineers who design housing for vibration-sensitive instruments [35]. With the onset of the microelectronics, medical, and biopharma industries in the 1980s, *criterion curves* were developed to lay down generic standards for vibration for a wide range of instrumentation, equipment, and tools.

The VC curve is represented in the form of a set of *one-third octave band velocity spectra* [34,35]. The curves are defined in terms of constant velocity (RMS) within the 8–80 Hz frequency range. The energy-averaged RMS velocity is calculated within proportional bandwidths, for example, a one-third octave bandwidth spectrum at each frequency range that lies within 23% of the center frequency (or 71% of the peak value) and is considered in all three orthogonal directions [34]. RMS velocity (as opposed to the "peak" or "peak-to-peak" criteria) is measured in terms of $\mu m \cdot s^{-1}$. The RMS velocity is related to the product of the frequency f and the wavelength λ (displacement) of the vibration. Displacement, velocity, and acceleration are all interrelated.

$$v = 2\pi f \lambda \qquad (1.1)$$

VC curves extend from 4 to 100 Hz. Pneumatic isolation systems (e.g., for AFMs) may resonate with floor vibrations in the 1–3 Hz range. There is usually less concern for vibrations above 100 Hz. Relevant generic vibration criteria for advanced buildings are as follows: VC-A/B $\rightarrow$ 50 to 25 $\mu m \cdot s^{-1}$; VC-D/E = 6 to 3 $\mu m \cdot s^{-1}$; NIST-A $\rightarrow$ 0.025 μm displacement for $1 \leq f \leq 20$ Hz or 3 $\mu m \cdot s^{-1}$ for $20 \leq f \leq 100$ Hz; and NIST-A1 $\rightarrow$ 6 μm displacement for $f \leq 5$ Hz or 0.75 $\mu m \cdot s^{-1}$ for $5 \leq f \leq 100$ Hz [35]. Each instrument has a unique RMS v versus f profile. For example, *Omicron AFM* VC criteria lie within 0.5 and 1 $\mu m \cdot s^{-1}$ ($1 \leq f \leq 10$ Hz); a *Mann 3696 Stepper* is 0.8–80 $\mu m \cdot s^{-1}$ ($1 \leq f \leq 100$ Hz); a *Hitachi S-4700ii SEM* range is 10–20 $\mu m \cdot s^{-1}$ ($1 \leq f \leq 10$ Hz); and a *JEOL 2010 HRTEM and 6400 and 5800 SEMs* are 6–20 $\mu m \cdot s^{-1}$ ($1 \leq f \leq 10$ Hz).

For offices and theory and modeling facilities, RMS velocities between 400–800 $\mu m \cdot s^{-1}$ are tolerable. General labs fall into the range of 50–100 $\mu m \cdot s^{-1}$ (VC-A±). Clean room levels depend on the designation of the clean room: Class 1000 $\rightarrow$ 25 $\mu m \cdot s^{-1}$ (VC-B); Class 100 $\rightarrow$ 6 $\mu m \cdot s^{-1}$ (VC-D); Class 10 $\rightarrow$ 3 $\mu m \cdot s^{-1}$ (VC-E). Metrology lab VC criteria range from 3–6 $\mu m \cdot s^{-1}$. AFM and atom pushing

$\approx 3\,\mu m \cdot s^{-1}$ (VC-E or NIST-A) and nano-instrument development is defined under $1.25\,\mu m \cdot s^{-1}$ [32,33]. Vibration limits in general for nanolabs of $0.75–3\,\mu m \cdot s^{-1}$ are not unusual [36]. More information about vibrational concerns, criteria, and recommendations can be found on www.colingordon.com.

Electromagnetic Shielding.   Electromagnetic field (EMF) shielding is the process of reducing the intensity of EM radiation between two areas. Placing a barrier made of a conducting material is able to accomplish this task. In particular, sensitive equipment needs to be shielded from radio frequency-electromagnetic sources (RF-shielding). Such shielding (e.g., Faraday cages) diminishes the interaction between radio waves and electrostatic fields. Magnetic shielding may also be required. TEMs, SEMs, E-beam writers, and semiconductor inspection systems are some examples of sensitive equipment that require shielding.

Mechanical Noise.   Noise is measured in terms of decibels (dB). A *decibel* is a dimensionless unit of measure that gauges the intensity of sound. It is a measure that is used in electronics, signal transfer, and communications. The dB is the logarithm of a ratio—the ratio may be that of power, sound pressure (acoustics), voltage, or other indicators of intensity. The sound-intensity-level dB(SIL) decibel reference is standardized to the level of threshold intensity of hearing in humans in air—ca. 1 kHz with an intensity of $I_o = 10^{-12}\,W \cdot m^{-2}$. One decibel is

$$1\,dB = 10 \log\left(\frac{I}{I_o}\right) \qquad (1.2)$$

or in other words, 1 dB equals ten *bels*. The exponent of the power of ten in the final log term is known as the bel.

The dB(SPL) is the decibel (sound-pressure-level) relative to $20\,\mu Pa$ (2×10^{-5} Pa), the minimum sound a human can hear in air (e.g., a mosquito at 3 m distance). Other symbolism is used depending on the frequency scale used for calibration. The dB(A), dB(B), and dB(C) are based on different frequency weighting. For example, dB(A) is based on the A-scale range. Some nighttime zoning restrictions call out a maximum of 45 bB(A). Ambient noise is not to exceed 55 dB (10 Hz < f < 10 kHz) within 0.5 m surrounding an electron microscope as specified by the manufacturer [36].

The primary source of mechanical noise arises from HVAC (heating, ventilation, and air conditioning) noise and vibration. Equipment operation (e.g., pumps) and air movement through ducts (aerodynamic noise) cause significant noise if not constructed properly. Air turbulence is the cause of high vibration levels at low frequencies (59 dB @ 31.5 Hz) in air ducts (lined plenum). The problem disappears by installing silencers, changing the duct size (made smaller), and reducing the air duct velocity. Adding acoustical tiles to the walls and acoustical transparent curtains help to mitigate the effect of the reflective electron microscope suite doors. Fan deficiencies (static pressure, airflow, and fan speed) should also be addressed. Most advanced technology buildings place as much mechanical equipment as possible outside the building. These include cooling towers, exhaust fans, scrubbers, and pumps. What is known for certain is that conventional noise control solutions don't necessarily work for advanced technology buildings [36].

1.3.3 *Advanced Facilities That Support Nano and Biotech*

National Institute of Standards Advanced Measurement Laboratory.   The NIST has recently opened the doors to the most advanced measurement laboratory in the world—the NIST Advanced Measurement Laboratory. According to NSET (National Science and Technology Council, Executive Office of the President) in 2004

> NIST has recently constructed the most technologically advanced facilities in the world, the Advanced Measurement Laboratory, which will support industry in the conduct of this research with new ways to more accurately measure, quantify and calibrate important processes and properties.

The NIST's AML, costing \$235 million to construct, is a 49,843 m² (536,507 sq. ft) facility that is composed of five separate wings, two of which are buried 12 m (39 ft) underground [37,38]. The facility, under the auspices of the Department of Commerce, will provide

> …sophisticated measurements and standards needed by U.S. industry and the scientific community for key 21st century technologies such as nanotechnology, semiconductors, biotechnology, advanced materials, quantum computing and advanced manufacturing. NIST research efforts planned for the new facility range from improved calibrations and measurement of fundamental quantities such as mass, length and electrical resistance to the development of quantum computing technology, nanoscale measurement tools, integrated micro-chip-level technologies for measuring individual biological molecules, and experiments in nanoscale chemistry.

The minimum standard criteria for air quality, temperature control, vibration, and humidity control are cutting edge (**Fig. 1.1**). Air quality is controlled to 3.5 particles per liter (100 particles per cu. ft) as compared to 3,500 particles per liter (100,000 particles per cu. ft) in most modern labs. In standard cu. ft. terms, this converts to 100 particles per cu. ft. Temperature is controlled from ±0.1 to 0.01°C as compared to ±2°C in most labs [37].

The NIST's AML consists of five interconnected units. The five buildings are united with above ground walkways and underground tunnels. The AML is environmentally stable with regard to humidity, temperature, vibration, electromagnetic interference, and contamination. Metrology activities are conducted in two underground facilities.

There are two single-floor analytical characterization laboratories (housing 187 laboratory modules). Most of the analytical labs are controlled to 20 ± 0.25°C, a few to ±0.1°C. Two underground metrology laboratories (housing 151 laboratory modules) are located 9 m (40 ft) below ground level. There are two types of labs: "quiet" metrology labs are dedicated to measurement, and "rotating or dynamic" labs involve some kind of moving equipment. Selected labs are isolated on concrete slabs (equipped with air springs that are able to cancel out even the slightest of vibrations) to reduce any level of vibration. A multi-floor nanofabrication facility serves as an ultra-clean room.

The Nanofab building has a 18,000 sq. ft "raised-floor" *Class 100* clean room—adaptable for upgrade to *Class 10* if required. A sub-fabrication area is located beneath the clean room, and an interstitial space separates the clean room floor from the mechanical suite. NIST has invested in a completely new set of equipment dedicated to producing 150 MM wafers. The equipment includes furnaces, LPCVD (liquid phase chemical vapor deposition), rapid thermal annealers, three reactive

FIG. 1.1

The level of environmental control in NIST's new AML is compared to other laboratory facilities. The NIST facility has 338 reconfigurable modules. A 8.520 m² (91,700 sq. ft) nanofab facility Class 100 clean room is rated at <3.5 particles·L⁻¹. Enhanced air quality is achieved with HEPA (high efficiency particulate air) filters for general-purpose laboratories. Most laboratories have baseline temperature control within ±0.25°C and 48 laboratories have control to within ±0.1 or ±0.01°C. Vibration isolation is achieved to a level of 3 µm·s⁻¹ or less and down to 0.5 µm·s⁻¹ in 27 low-vibration modules. Humidity control is held at ±5% relative down to ±1% in special laboratory sections. Electrical power filtering provides institute-wide uninterruptible power and counter measures of voltage spikes, drop-outs and other dirty power problems that limit accuracy and precision, reduce analytical sensitivity and cause long-running experiments to crash. Green building features include natural daylighting, energy conservation, and recycling.

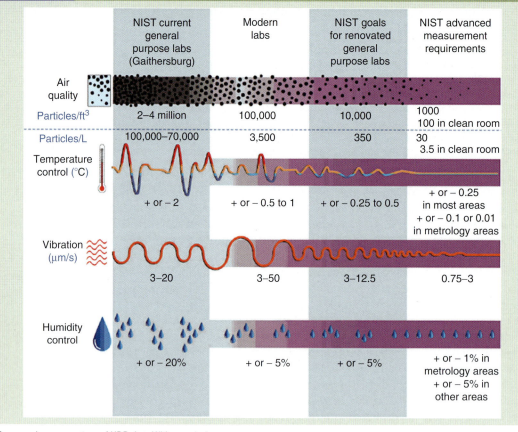

Source: Image courtesy of HDR, Inc. With permission.

ion etchers (RIEs), three metal deposition tools (thermal evaporator, electron beam evaporator and ion sputterer), contact lithography capability, electron beam lithography and focused ion-beam lithography. Field emission SEM, spectroscopic ellipsometers, contact profilometers, microscopes, and image capture instrumentation support product and research monitoring and metrology.

A selection of incredible images acquired at AML are displayed in **Figure 1.2**.

FIG. 1.2

(a) Silicon step rulers that range in height from 10s to 100s of nanometers to a single mono-layer measuring 0.3 nm. The microscope used to make the image sits on an isolated concrete slab equipped with air springs to cancel the smallest vibrations. (Image courtesy of J. Fu, NIST.) (b) 12 cobalt atoms in a circle on copper. The interference of electron waves produces a daisy pattern. An instrument autonomously picks up and places the atoms. (Image courtesy of J. Strocio and R. Celotta, NIST.) (c) A microheater is used to detect toxic gases. Variations in the thickness cause changes in color. NIST's ALM facility will produce arrays of these sensors. (Image courtesy of NIST.) (d) MgO cubes decorated with gold nanoparticles are imaged by a new 3-D chemical imaging method using a scanning TEM, a tilting stage, and sensitive detectors. (Image courtesy of J. Bonevich and, J.H. Scott, NIST.) (e) Magnetic domains of new generation logic devices are depicted. Changes in color correspond to changes in the direction of the magnetic field. This image was taken 12 m underground by the highest resolution magnetic imaging instrument in the United States. (Image courtesy of J. Unguris, NIST.) (f) Interference colors indicate the thickness of a clear organic film. Deep blue-brown indicates ~1 μm thickness that corresponds to the thickness required to embed a particle for analysis. The work was accomplished in NIST's ultraclean room. (Image courtesy of C. Zeissler, NIST.) (g) An infrared image illustrates temperature variations as a round swipe cloth is heated to detect the presence of explosives. Dark areas correspond to 40°C. Lightest areas are around 200°C. (Image courtesy of G. Gillen, NIST.)

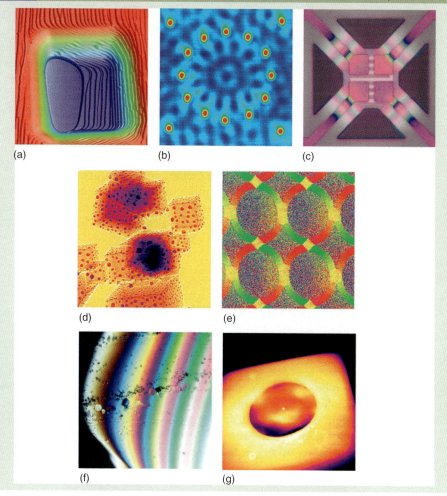

(a)   (b)   (c)

(d)   (e)

(f)   (g)

Center for Integrated Nanotechnologies (CINT).   The CINT (also designed by HDR Architecture, Inc.) is located in Albuquerque, New Mexico and was constructed in 2006.

> The vision of CINT is to become a world leader in nanoscale science by developing the scientific principles that govern the design, performance, and integration of nanoscale materials.

CINT emphasis is to take scientific discovery and the integration of nanostructures into the micro- and macroscopic worlds.

CINT's core facility is a $9.8 million 35,600 sq. ft gateway located near two national laboratories: Sandia and Los Alamos. CINT is called a gateway facility because it is able to link two national laboratories with collaborative users from universities and industry. Nanophotonics, nanoelectronics, complex nanomaterials, nanomechanics, and nanoscale bio-micro interfaces are some of the divisions within the CINT facility. From the description of the focus areas given above, one can conclude that integration is indeed a prime directive at CINT—and revisiting the challenges defined earlier in this chapter, attempts to accommodate energy across multiple length scales, combination of top-down and bottom-up fabrication strategies, and interfacing biological and synthetic systems was certainly accomplished in good faith.

CINT has allocated 20,000 sq. ft for office suites, visitor accommodations, computer bays, and communication links. The synthesis wing consists of 15,000 sq. ft and houses chemical benches, hoods, equipment, and bench-top characterization tools. A 12,000 sq. ft integration wing possesses a *Class 100* clean room equipped with flexible fabrication. The characterization wing is a vibration-isolated 15,000 sq. ft facility that houses scanning probe microscopes, nanomechanics tools, laser optics, and microelectronics. Utilities, storage, and service space account for 20,000 sq. ft of the total space. The CINT overall is over 90,000 sq. ft.

Purdue's Birck Nanotechnology Center.   We shall present one more facility, the recently constructed Birck Nanotechnology Center at Purdue University (by HDR Architecture, Inc.). The facility cost $45 million to build. It is 220,000 sq. ft and contains Class 10, 100, and 1000 clean rooms. There are over 100 labs and modules and nearly 100 faculty from 24 schools or departments participate in research at the facility.

1.4   NATIONAL AND INTERNATIONAL INFRASTRUCTURE

We have reviewed business development, education, careers, and buildings. One more cog in this mechanism that requires introduction is a topic about how all these buildings (and people) are organized. Nanotechnology has brought about, more than ever before, the need for cooperation between and among academia, industry, and government, and between and among nations of our world. In the next few paragraphs, we review a few organizations and infrastructure that support advanced technology research, development, and societal implication awareness.

The logical step in the hierarchy is to create an infrastructure of user facilities (buildings, equipment, and instrumentation). There are several statewide and few nationally recognized infrastructure networks dedicated to nanoscience research and development. The new generation facilities in general support a collaborative environment and incorporates sustainable design solutions with flexible, modular designs.

1.4.1 *Research and Development Organizations*

The National Nanotechnology Infrastructure Network (NNIN). The NSF released a solicitation in 2003 to invite universities to submit proposals for membership in the NNIN. On March 1, 2004, the NNIN was officially launched. The NNIN is an integrated partnership of 13 user facilities that is dedicated to support nanoscale fabrication, synthesis, characterization, modeling, design, computation, and hands-on training [39]. The NNIN supports academia, small and large industry, and government in the capacity of a research facilitator with open access by providing leading edge tools, instrumentation, and expertise.

The member institutions include Cornell Nanoscale Facility, Stanford Nanofabrication Facility, University of Michigan Solid State Electronics Laboratory, Georgia Institute of Technology Microelectronics Research Center, University of Washington Center for Nanotechnology, Penn State Nanofabrication Facility, Nanotech at the University of California at Santa Barbara, the Minnesota Nanotechnology Cluster (MINTEC), Nanoscience at the University of New Mexico, the Microelectronics Research Center at the University of Texas, the Center for Imaging and Mesoscale Structures at Harvard, the Howard Nanoscale Science and Engineering Facility, and the Triangle National Lithography Center at North Carolina State University [40].

The NNIN, according to its Web site, "provides unparalleled opportunities for nanoscience and nanotechnology research." The NNIN provides extensive support in nanoscale fabrication, synthesis, characterization, modeling, design, computation, and hands-on training [39]. The NNIN embraces and promotes educational programs to inform the public, educate K-12 students, employ and train undergraduates, develop curriculum, develop technological workforce, train teachers, and numerous other programs and objectives.

The Department of Energy's National Science Research Centers (NSRC). The U.S. Department of Energy (DOE) has five centers for nanoscience **(Fig. 1.3)**. (1) the *Molecular Foundry* (Lawrence Livermore National Laboratory) is involved in advanced light sources, electron microscopy, and the "nanowriter"; (2) advanced proton source, intense pulsed neutron source, and an electron microscope center for materials research are found at the *Center for Nanoscale Materials* (Argonne National Laboratory); (3) there is the *Center for Functional Nanomaterials* at Brookhaven National Laboratory; (4) the *Center for Nanophasic Materials* Sciences (Oak Ridge National Laboratory); and (5) Los Alamos and Sandia's Center for Integrated Nanotechnologies is involved with semiconductor research,

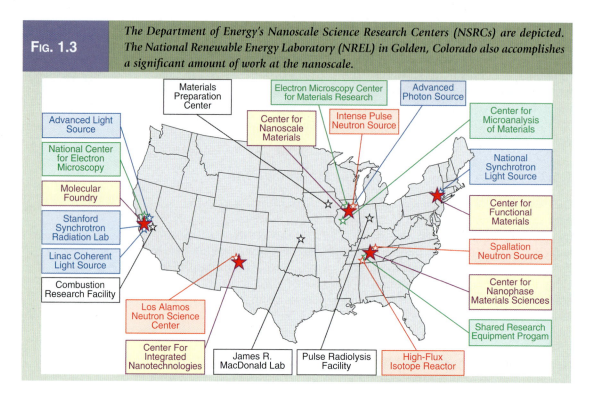

FIG. 1.3 *The Department of Energy's Nanoscale Science Research Centers (NSRCs) are depicted. The National Renewable Energy Laboratory (NREL) in Golden, Colorado also accomplishes a significant amount of work at the nanoscale.*

microelectronics, combustion research, and magnetic field laboratories. All of these facilities, and those of the NIST and other facilities in Europe and Asia, are concerned with integrating science and engineering across all scientific disciplines and across all size scales [28].

DOE's investment in nanotechnology is expected to exceed $500 million from 2007 to the present [40]. The Molecular Foundry, located near U.C. Berkeley, is a new center of high-powered nanotech R&D. The center contains clean room facilities, high-powered computing capability, highly sensitive equipment and instrumentation, and top-notch researchers [40].

1.4.2 Economic Development Organizations

The NanoBusiness Alliance (NBA). The NanoBusiness Alliance (www. nanobusiness.org), known as the "world's leading nanotechnology trade organization" made the national scene in nanotechnology business ca. 2002. The organization was founded by Mark Modzelewski and others who realized that nanotechnology was fast coming of age. The current executive director is Sean Murdock who recently testified before the House Science Committee in April 2008 in support of the National Nanotechnology Initiative Amendments Act of 2008.

1.4.3 Organizations Centered on Societal Implications

National Nanotechnology Initiative.   The NNI (at www.nano.gov) was established in 2001. Mihail C. Roco is its executive director. The goals of the NNI are quite comprehensive and far-reaching:

- Advance a world-class nanotechnology research and development program
- Foster the transfer of new technologies into products for commercial and public benefit
- Develop and sustain educational resources, a skilled workforce, and the supporting infrastructure and tools to advance nanotechnology
- Support responsible development of nanotechnology

Although the NNI is involved in many other areas (e.g., the NNIN and other programs), its efforts have spearheaded recognition of the importance of societal implications of nanotechnology. A number of excellent resources are available from the NNI Web site. Please consult this Web site often and in detail. Intergovernmental agency and department efforts are under the auspices of the National Nanotechnology Coordination Office (NNCO).

Center for Responsible Nanotechnology (CRN) [41]. The Center for Responsible Nanotechnology is a nonprofit research and advocacy group that is focused on the major societal and environmental implications of nanotechnology. The group was formed in 2002 by Mike Treder and Chris Phoenix. They also stress the development, evaluation, and implications of molecular manufacturing (e.g., fourth generation nanotechnology). Their mission statement reads as follows.

> The mission of CRN is to: 1) raise awareness of the benefits, the dangers, and the possibilities for responsible use of advanced nanotechnology; 2) expedite a thorough examination of the environmental, humanitarian, economic, military, political, social, medical, and ethical implications of molecular manufacturing; and 3) assist in the creation and implementation of wise, comprehensive, and balanced plans for responsible worldwide use of this transformative technology.

Foresight Institute. The mission of the Foresight Institute is to "ensure the beneficial development of nanotechnology." The foresight institute is essentially a think tank and public interest institute, this according to their Web site at www.foresight.org. The group was founded in 1986 and lists as the foremost challenges (with solutions through nanotechnology):

- Providing renewable clean energy
- Supplying clean water globally
- Improving health and longevity
- Healing and preserving the environment (and maximizing productivity of agriculture)
- Making information technology available to all
- Enabling space development

In January of 2008, the Foresight Institute released a roadmap for nanotechnology development [42,43].

1.4.4 Nanotechnology News Services

A great way to keep abreast of nanotech news is to subscribe to a variety of free news services. Several notable services are listed below.

- Nanoscienceworks Newsletter (www.nanoscienceworks.org)
- Nanotechnology Now (www.nanotech-now.com)
- Nanotechweb.org, IOP Publishing (www.nanotechweb.org)
- Nature Nanotechnology (www.nature.com/nano/index.html)
- Nano Vip Newsletter (www.nanovip.com/nanotechnology-newsletter)
- Nanotechnology.com, the international small technology network (www.nanotechnology.com)
- Small Times Magazine (www.smalltimes.com and www.nanotechnews.com)
- Nanowerk Nanotechnology News (www.nanowerk.com)
- The A to Z of Nanotechnology (www.azonano.com)
- Nanodot (www.foresight.org/nanodot/)
- Nanotechnology News Network (www.nanonewsnet.com)
- Chemical & Engineering News (www.pubs.acs.org/nanofocus/)
- Nano Techwire (www.nanotechwire.com)
- Nano World News (www.nsti.org/news/)

The list keeps on growing and growing.

For business, the Forbes/Wolfe Nanotech Reports (www.newsletters.forbes.com), Scott Livingston's Axiom Capital Management (www.axiomcapital.com), and Lux Capital (www.luxcapital.com) should keep you updated about the investment aspects of nanotechnology.

1.4.5 *International Organizations and Institutes*

There are numerous organizations overseas. Europe has several that are involved in every aspect of nanotechnology. The list is by no means complete. Please find out what is going on in your world.

North America
- Nanotechnology—National Research Council Canada (www.nrc-cnrc.gc.ca)
- NRC National Institute for Nanotechnology (www.nint-innt.nrc.gc.ca)
- National Institute for Nanotechnology—University of Alberta (www.uofaweb.ualberta.ca/nint)
- Canadian NanoBusiness Alliance (www.nanobusiness.ca)

Europe
- Nanoforum.org: European Nanotechnology Gateway (www.nanoforum.org)
- Nanotechnology (cordis.europa.eu/nanotechnology)
- ENTA—European Nano Trade Alliance (www.euronanotrade.com)
- Institute of Nanotechnology (www.nano.org.uk)
- Swiss Federal Institute of Technology (Ecole Polytechnique Fédérale Lausanne, EPFL) (www.epfl.ch)

Asia
- Asian Institute of Technology (www.ait.ac.th)
- National Nanotechnology Center (NANOTEC) (www.asia-nano.org)
- Asia Nano Forum (www.asia-nano.org)

- The Australian Research Council Nanotechnology Network (www. ausnano.org)
- Nanotechnology Research Institute (NRI) (unit.aist.go.jp/nanotech)
- Nanotechnology India (www.indiannanotechnology.com)

Africa
- South African Nanotechnology Initiative (SANi) (www.sani.org)
- Focus Nanotechnology Africa, Inc. (www.fonai.org)

South and Central America
- Laboratorio Nacional de Nanotecnologia (LANOTEC) in Costa Rica (www.cenat.ac.cr/cenat)
- Brazilian Nanotechnology Networks
- Development of Nanoscience and Nanotechnology (Brazil)
- Nanoscience Millennium Institutes (Brazil)
- Fundacion Argentina de Nanotecnologia (FAN) (www.fan.org.ar)

1.5 NANOTECHNOLOGY PRODUCTS

Fundamentals of Nanotechnology is about applications of nanoscience to commerce and industry. In other words, nanotechnology is about development and manufacturing of products enhanced by the remarkable properties of nanomaterials. We therefore present, to conclude the perspectives aspect of this text, a list of some products that you may or may not be aware that are associated with nanotechnology. Most of the products were found on the NNI's Web site, www. nano.gov. In addition, please consult www.nanotechproject.org/ for an extensive list of hundreds of products that have been enhanced with nanotechnology.

Automotive Industry
- Step assists, bumpers, paints, coatings, glare reduction, catalytic converters
- Cooling chips to replace compressors with no moving parts

Recreation
- Lighter stronger tennis racquets, long-lasting tennis balls, smart golf balls
- Nanotube reinforced masts for sailboats, new materials for hull and deck
- Golf shafts, skis, fog eliminators

Personal Use and Food
- Sunscreens, cosmetics, stain-free clothing
- Silver nanoparticle food storage containers, cutting boards, pans
- Nonstick bake ware
- Umbrella (based on lotus leaf)

Medicine, Therapeutics, and Hygiene
- Dental-bonding agents, burn and wound dressings
- Medical imaging with quantum dots
- Targeted drug delivery and gene therapy

- Water filters
- Lab-on-a-chip diagnosis
- Sanitized toilets

Structural Materials and Industrial Applications
- Stronger, lighter polymers; enhanced concrete, enhanced steel
- Wear-resistant nanoceramic coatings
- Catalysts
- Carbon nanotube-reinforced materials
- Various nano glues, nanoseal wood, nano-enhanced insulation
- Self-cleaning glass
- Exterior paint

Electronics and Computing
- Sub-100 nm transistors (old technology)
- Carbon nanotube triodes
- Organic LEDs and organic electroluminescent displays
- Cordless power tool batteries
- Carbon nanotube displays
- Protective self-assembling film layers for displays
- Cellular memory

Satisfy your own curiosity and research the amazing number of products out there that have already been enabled or enhanced with nanotechnology.

Acknowledgments

We would like to acknowledge Michael Burke, CEO & President of NanoThread, Inc. for his contribution to section 1.1. We are also indebted to the NNI for their central focused efforts to keep the United States abreast in the quest to commercialize nano.

References

1. R. Brazil, Nanotechnology—the issues: A response to the request for initial views on the Royal Society and the Royal Academy of engineering study on nanotechnology, Royal Society of Chemistry, London (2003).
2. M. C. Roco, J. Murday, C. Teague, S. Hays, and C. Merzbacher, The national nanotechnology initiative: Strategic plan, National Science and Technology Council Committee on technology, NSET (2004).
3. G. L. Hornyak, J. Dutta, H. F. Tibbals, and A. K. Rao, *Introduction to nanoscience*, CRC Press, Boca Raton, FL (2008).
4. J. Von Ehr, II, Comments at PCAST (President's Council of Advisors on Science and Technology) public meeting, Zyvex Labs, LLC (2007).
5. M. C. Roco and W. S. Bainbridge, eds., Societal implications of nanoscience and nanotechnology, Kluwer Academic Publishers, Boston, 3–4 (2001).
6. R. E. Smalley, Our energy challenge, Presentation, Columbia University, New York (2003).
7. R. Florida, *The flight of the creative class: The new global competition for talent*, Harper-Collins, New York (2005).

8. L. V. Gerstner, Jr., Sputnik was nothing, *Newsweek*, November 28 (2005).

9. T. K. Kelly, W. P. Butz, S. Carroll, D. M. Adamson, and G. Bloom, The U.S. scientific and technical workforce: Improving data for decision making, Conference Proceedings, RAND Science and Technology, RAND Corporation (2004).

10. M. C. Roco and W. S. Bainbridge, *Nanotechnology: Societal implications—maximizing benefits for humanity*, Report of the National Nanotechnology Workshop, December 2–3, 2003, Arlington, Virginia; National Science Foundation, Springer Science & Business Media (2005).

11. P. F. Drucker, *Landmarks of tomorrow: A report on the new "post-modern" world*, Harper, New York (1959).

12. R. E. Kelley, *The gold-collar worker: Harnessing the brainpower of the new work force*, Addison-Wesley, Reading, MA (1985).

13. M. A. Roe, Cultivating the gold collar worker, *Harvard Business Review*, 79, 32–33 (2001).

14. M. Jamison, HDR, Inc., Buildings for nanotechnology research, Presentation, Colorado Nano/Micro Summit, Boulder CO, May 24 (2004).

15. G. Elvin, Risks in architectural applications of nanotechnology, Nanowerk Spotlight, www.nanowerk.com/spotlight/spotid=1007.php (2006).

16. T. H. Wegner, J. E. Winandy, and M. A. Ritter, Nanotechnology opportunities in residential and non-residential construction, USDA Forest Products Laboratory, 2nd International Symposium on Nanotechnology in Construction, Bilbao, Spain (2005).

17. A. Marsh, Nanotechnology, green building, sustainable design musts for world survival, profit, *Commercial Property News*, www.cpnonline.com/cpn/specialties/article_display.jsp?vnu_content_id=1003382018 (2006).

18. Opportunities for nanotechnology, www.nanovic.com.au/?a=industry_focus.building (accessed 2008).

19. G. Elvin, The nano revolution, A science that works on the molecular scale is set to transform the way we build, *The Magazine*, Architect OnLine, www.architectmagazine.com/industry-news.asp?sectionID=1006&articleID=492836&artnum=1 (2007).

20. Green building, en.wikipedia.org/wiki/Green_building (2008).

21. Green buildings, U.S. Environmental Protection Agency, www.epa.gov/green-building/ (2007).

22. Building design guidance, U.S. Environmental Protection Agency, www.energystar.gov/index.cfm?c=new_bldg_design.new_bldg_design_guidance#construction (2007).

23. L. M. Froeschle, Environmental assessment and specification of green building materials, *The Construction Specifier*, October, 53 (1999).

24. D. M. Roodman and N. Lenssen, A building revolution: How ecology and health concerns are transforming construction, Worldwatch Paper 124, Worldwatch Institute, Washington, DC, March, 5 (1995).

25. R. Spiegel and D. Meadows, *Green building materials: A guide to product selection and specification*, John Wiley & Sons, Inc., New York (1999).

26. G. Dick, Green building materials, California Integrated Waste Management Board, www.ciwmb.ca.gov/GreenBuilding/Materials/ (2007).

27. Leadership in energy and environmental design, U.S. Green Building Council, LEED Rating Systems, www.usgbc.org/ (2008).

28. N. D. Shinn, CINT: A new model for a nanoscience research user facility, CINT-Sandia National Laboratories, Buildings for Advanced Technology II, Workshop, January 21–23, Mesa, AZ (2004).

29. N. Toussaint, Establishing and maintaining critical environments, HDR, Inc., Presentation, Buildings for Advanced Technology, Mesa, AZ, January 22 (2004).

30. Cleanroom, en.wikipedia.org/wiki/Clean_room#ISO_14644-1_cleanroom_standards (2007).

31. D. Bechtol, Electrical power and grounding, PE, HDR Architecture, Inc., Buildings for Advanced Technology II, Workshop, Mesa, AZ (2004).

32. H. Amick, Vibrations…. should i worry? What about? Colin Gordon & Associates, Buildings for Advanced Technology II, Workshop, Mesa, AZ (2004).

33. H. Amick, On generic vibration criteria for advanced technology facilities: With a tutorial on vibration data representation, *Journal of the Institute of Environmental Science*, 40, 35–44 (1997).

34. J. M. Proenca and F. Branco, Case studies of vibrations in structures, *Revue Européene de Génie Civil*, 9, 159–186 (2005).

35. C. G. Gordon, Generic vibration criteria for vibration sensitive equipment, International Society for Optical Engineering (SPIE), Conference on Current Developments in Vibration Control for Optomechanical Systems, Denver, CO (1999).

36. A. Yazdanniyaz, Mechanical system noise issues: Case studies, Arup, Presentation, Buildings for Advanced Technology II (BAT-II), Workshop, Mesa, AZ, January 22 (2004).

37. E. M. Vogel, NIST advanced measurement laboratory (AML) Nanofab, Presentation, National Institute of Standards and Technology (2004).

38. M. Baum, NIST launches advanced measurement laboratory: Research environment among the world's best for nanotech, leading-edge science, NIST News Release, www.nist.gov/public_affairs/releases/aml_dedication.htm (2004).

39. National Nanotechnology Infrastructure Network (NNIN), www.nnin.org (2004).

40. V. McCarthy, DOE's nano-centric research centers set to fuel new era in nano collaborative, Nano World News, Nano Science and Technology Institute (NSTI), www.nsti.org/news/item.html?id= 127 (2007).

41. Center for Responsible Nanotechnology, www.crnano.org/about_us.htm (2008).

42. K. E. Drexler, J. Randall, S. Corchnoy, A. Kawczak, and M. L. Steve, eds., Productive nanosystems: A technology roadmap, Battelle Memorial Institute and Foresight Institute (2007).

43. From here to there: Nanotechnology roadmap, Press Release, http://foresight.org/roadmaps, January 29 (2008).

Problems

1.1 What is the combined decibel level if you are talking in a restaurant at 70 dB and the adjacent noise from the kitchen contributes another 70 dB?

1.2 With regard to vibrations—acceleration, velocity, and displacement are all related by simple equations. How so?

1.3 Vibrations that influence equipment range from 1 to 100 Hz. Give examples of sources of 1-, 25-, 50-, and 100-Hz vibrations.

1.4 Clean rooms of Class 100 indicate that there are no more than 100 0.5-μm particles per cubic foot. What applications would require such tight control and why?

1.5 What products can you think of that cannot be enabled or enhanced by nanotechnology?

1.6 a. If someone is exposed to sound intensity of 1×10^{-12} W·m^{-2}, how does this translate into decibels? b. What is the intensity of sound in terms of W·m^{-2} of jet plane noise at takeoff if the recorded decibel level is 140?

1.7 What is your vision of a building (research center) of the future?

1.8 List all potential employers that involve nanotechnology but are not involved in science, technology, or manufacturing—for example, an intellectual property attorney.

1.9 List three reasons why it is not a good idea to place a TEM room next to a loading dock. (Hint: There is one subtle reason you might miss).

1.10 Is nanotechnology an industrial revolution in the making?

1.11 When starting a nano-based business, discuss the importance of partnerships.

1.12 Research the economic cluster model in Arizona. What do you think?

1.13 Do you agree that government should be heavily involved in funding nanotech? Why or why not?

NANOMETROLOGY: STANDARDS AND NANOMANUFACTURING

Whether you can observe a thing or not depends on the theory which you use. It is the theory which decides what can be observed.

I think that a particle must have a separate reality independent of the measurements. That is an electron has a spin, location and so forth even when it is not being measured. I like to think that the moon is there even if I am not looking at it.

ALBERT EINSTEIN

Chapter 2

THREADS

Nanomanufacturing is where the "rubber meets the road." It is one manifestation of the proverbial "moment of truth." It is the point of "put up or shut up." It is the reality check of nanoscience. It is the discriminator between hype and reality. Nanomanufacturing is the process of transforming nanoscience into a useful material, an application. Standardization of practice, product performance and packaging specifications accompany any manufacturing process, and products that are made or will be made by nanomanufacturing are no exception to this rule. The first chapter introduced nanotechnology and its relations to business development, education and workforce development and provided some

THREADS (CONTD.)

insight into where nanotechnology is housed and how it is trekking across our globe. This chapter, the second chapter in the *Perspectives* division of the text, provides relevant background in metrology, standards and manufacturing.

One of the links between and among science, technology and commerce is metrology. Nanometrology is no exception to this rule and is the embodiment of measurement at the nanoscale. Nanometrology and nanomanufacturing go hand in hand. Industry standards forge a necessary link between the two and are therefore an integral part of any commercial process. In this chapter, we introduce all three and dwell on their interrelationship. Metrology at the nanoscale presents new operational paradigms—ones that link the macroscopic and quantum world like never before. Quantum standards of all things! Our science and technology now exist in a world where quantum standards are required. How far we have come indeed! Measurement of long held standards such as the meter and kilogram are now defined from the bottom-up, from quantum mechanics and quantum phenomena.

Following the *"Perspectives"* division, the next three chapters delve into the electromagnetic manifestations of nanotechnology that comprise the *"Electromagnetic Nanoengineering"* division of the text.

Nanomanufacturing is the next great threshold of nanotechnology. As our capabilities move beyond creating simple aggregations of "nano-stuff", we find ourselves wanting to control quality and repeatability, just as we do in conventional manufacturing. It is not enough to simply make small things; we need to control the process that makes those things so the things perform the right functions. Ideally, we want to control nanotechnology to the point where we can make large quantities of identical things, using the discreteness of matter at the atomic scale to achieve our goal of adaptable, affordable, atomically precise manufacturing. Living cells make proteins in this manner, getting every atom in the right place, using molecular-scale manufacturing factories called ribosomes. The flexibility of living systems in manipulating molecules to transform the sun's energy into useful chemical forms, and to use those chemicals to transform inert minerals into structures as varied as diatom shells, oak trees, or elephants, gives us an existence proof that this technology is both feasible and valuable.

Metrology is important to manufacturing at any scale. We must be able to measure what we're building, if we hope to control our manufacturing processes. With nanotechnology, this is more challenging than normal, due to the size scales. We can't use light-based observation, since the structures we want to observe are usually smaller than the wavelength of light. We either need exotic techniques like high-energy accelerators, or must build probe-based tools that reach into the nanoworld in order to touch, manipulate, and measure objects. As an action-oriented company, Zyvex has chosen the "reach into the nanoworld and manipulate it" approach.

The challenge for tools companies in the nano space is to survive and grow on what are still tiny markets. We will advance the field quickly and all thrive if we can innovate new tools and new markets. Innovative companies can quickly become market leaders. As we move to a world of "digital matter", innovation and speed of execution become increasingly important to a company's success.

Opportunities abound, and as we get better at nanomanufacturing, we will come to find that "imagination is our only limit." Physics and chemistry provide constraints, but human ingenuity lets us work within those constraints to achieve our aims, whether those are flying hundreds of people across the ocean at nearly the speed of sound, videoconferencing to someone on the other side of the world, or my personal favorite, eating fresh red raspberries in the middle of winter. We accomplished all those because creative people said "I can…"

JAMES VON EHR, II,
CEO and Chairman, Zyvex Corporation

2.0   THE TRANSITION, THE NEED

Many great ideas sit on shelves in university labs, offices, and libraries—great ideas that never make it to the next level due to a variety of factors: (1) the idea was ahead of its time, e.g. the manufacturing technology required was not available, (2) gap funding to develop *definitive proof-of-concept* was not acquired in a timely manner and the idea melded into oblivion, (3) the transfer of technology between the university and a commercial venture was not facilitative or effective, (4) although characterized as a valid realistic technology, its shelf life expired, or (5), the road to commercialization is too long (**Fig. 2.1**). Unavoidably, some ideas turn out to be not so great, not so practical. There are numerous examples.

FIG. 2.1

An idea, a concept, a process, a device, or an invention is valued highly by the scientist or inventor (and the university) who developed it (red line). A businessperson, on the other hand, views that same idea, invention, etc., as a project that requires investment and development, complete with all the ancillary business trimmings. In other words, that same idea, concept, or invention is not worth much as much to the business developer as it is to the inventor—a disconnect—even though it may show great promise down the road (blue line). It is important that these two extremes meet somewhere at sometime—preferably early on. As time goes by and as the product is developed, it gains in value (blue line). If it sits on a university shelf, it loses value over time (red line).

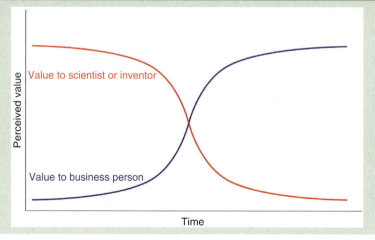

Intellectual property, technology transfer, and business development are just several facets of the process. There are also practical aspects of transferring a laboratory prototype into a real application that involve aspects such as measuring, standardizing, scale-up, manufacturing, reliability/testing, and packaging. For a nanoscience to become a nanotechnology, we need to understand metrology, and engineering and manufacturing practices as they apply to nanotechnology and how best to construct the equipment, instrumentation, facilities, and buildings that accommodate them. Numerous metrology and manufacturing challenges must be addressed before we are ready to place our stamp on the next industrial revolution.

2.0.1 *Background to Nanometrology*

Metrology has been with us forever. Our need to measure and standardize made civilization what it is today. Nanometrology is a natural extension of metrologies that sprung from microscale R&D and manufacturing. However, like never before, the face of metrology is changing dramatically—in a way that is driven by nanotechnology and the need to measure parameters that reside on the fringe of the limits of uncertainty and theory. Welcome to the *Nano Age*.

History.   In antiquity, the length of one's forearm was known as a *cubit*. Many cultures adopted unique versions of the cubit. Day-to-day things were measured and recorded by comparing them to various parts of the body. The human experience is rich with examples of such whimsical measurement. The Greeks popularized the *foot* as a means of placing physical dimensions on paper. Of course, the measure depended on the size of an individual's foot—hardly a standardizable commodity. Horse height (at the shoulder) is still measured to this day in terms of *hands*—the height of which depends, however, on whose hand is abstracted for the measurement. One of the best systems developed to measure length (down to ~1.7 mm resolution) is credited to cultures inhabiting the Indus Valley several millennia ago. With the onset of commerce that went beyond simple barter, the need for standardization assumed greater importance.

The Romans popularized the *natural cubit*, about 1.5 ft. The Roman foot was divided into twelve *unciae* (inches), each about 25 mm in length [1]. The *mille passus* (or 1000 paces, each pace equal to 5 ft) and the mile (5000 ft) were adopted by the early Britons during the Roman occupation [1]. The *mille passus* was changed later into the statute mile (5280 ft) by edict of Queen Elizabeth I. Mass was measured first in terms of the *grain* (of wheat?). The Romans introduced the *libra* (one pound) from which the symbolism *lb* is obtained. The Roman pound was divided into twelve *uncia* (ounces). Imperial units became official following passage of the *British Weights and Measures Act of 1824*. To this day, with small modifications here and there, the Imperial system survives. The sexagesimal system of numbers dates back to the ancient Babylonians who divided the circle into 360°. Hence, we refer to degrees, minutes, and seconds much like cultures did so in antiquity.

Ironically, the metric system was first defined by John Wilkinson of England in 1668. The French adopted this system, based on the number 10, in 1791, with some influence from Benjamin Franklin two years after the onset of the French Revolution. The rest is history. Mixing the Imperial system with the metric system, however, has caused disasters. In 1983, metric conversion errors resulted in a Boeing 767 jetliner to run out of fuel. The pilots, according to the *New York Times*, were able to make a dead-stick landing [2]. In another notorious incident, NASA's Mars $125 million orbiter landed (crashed) prematurely due to metric-to-Imperial units conversion errors [3].

Metrology.   Metrology is the science of weights and measures [4]. According to the U.S. Military Dictionary [5], metrology is

> ...the science of measurement, including the development of measurement standards and systems for absolute and relative measurements.

According to the *International Bureau of Weights and Measures, IBWM* (or *Bureau International des Poids et Mesures*)

> Metrology [from Greek metron 'measure', and logos 'study of'] is the science of measurement. Metrology includes all theoretical and practical aspects of measurement... the science of measurement, embracing both experiment and theoretical determinations at any level of uncertainty in any field of science and technology.

Metrology therefore has a scientific (fundamental) component as well as an industrial (applied) one. According to the IBWM [6]:

> Scientific or fundamental metrology concerns the establishment of measurement units, unit systems, the development of new measurement methods, realisation of measurement standards and the transfer of traceability from these standards to users in society. Applied or industrial metrology concerns the application of measurement science to manufacturing and other processes and their use in society, ensuring the suitability of measurement instruments, their calibration and quality control of measurements.

OK, we get the picture. We now define nanometrology. An excellent resource is available from Ref. [7], Nanometrology, IBWM, bipm.org/en/home/(2007) [7].

Nanometrology.   Nanometrology, of course, is metrology at the nanoscale [8].

> ...or particles to self-assemble at the nanoscale to form new materials with unusual properties. Nanometrology, i.e., the ability to conduct measurements at these dimensions, to characterize the materials—that, together, will form the metrological infrastructure essential to the success of nanotechnology. The MSEL Nanometrology Program incorporates basic measurement metrologies to determine material properties, process monitoring at the nanoscale.

MSEL is the acronym for the *Materials Science and Engineering Laboratory* at NIST, the *National Institute of Standards and Technology* (formerly the *National Bureau of Standards, NBS*).

According to MSEL [8],

> No previous materials technology has shown so prodigiously a potential for concurrent advances in research and industry as does the field of nanomaterials in mechanical devices, electronic, magnetic and optical components, quantum

computing, tissue engineering and other biotechnologies, and as-yet unanticipated exploitations of as-yet undiscovered novel properties of nanoscale assemblies and particles.

The MSEL Nanometrology Program to develop nanometrology infrastructure is straightforward—to incorporate preexisting basic measurement metrologies applied for material properties, process monitoring, nanomanufacturing, analytical techniques, advanced imaging, and multiscale modeling [8].

The MSEL effort, obviously, is intended to cover a wide range of nanometrological domains: (1) *mechanical property measurement* such as elastic properties, plastic deformation, adhesion, friction, stiction, and tribological behavior; nanoindentation (thin films and nanostructures); the use of AFM, acoustic microscopy, surface acoustic wave spectroscopy, and Brillouin light scattering; development of micro- and nanoscale structures and test methods to measure strength and fracture behavior of interfaces and materials having very small volumes, (2) *chemical and structural characterization and imaging* with neutron and x-ray beams, scattering and spectroscopy methods, chemical bond identification and orientation, polyelectrolyte dynamics, and equilibrium structures; and development of electron microscopic techniques for 3-D imaging, advanced scanning probe microscopies, and physical probing of cell membranes with carbon nanotubes, (3) *fabrication and monitoring of nanomanufacturing processes* such as electrochemical and microfluidic fabrication methods; nanocalorimetrics to study interfacial reactions, in situ observations of nanoparticle and nanotube dispersion, and alignment and advanced instrumentation for nanotribological experimentation, and (4) *theory modeling* to predict material behavior from the nanoscale to the macroscale; development of finite-element analysis, multiscale Green's function methods, classical atomistic simulations, ab initio and quantum mechanical density functional theory (DFT); and interfacing of models at different length scales to ensure accuracy of physics of components and systems [8].

Terminology. There are several terms associated within the domain of metrology that one should become familiar. We list a few important ones.

1. A *measurand* is the quantity that is subjected to measurement—electric current, length, density, etc.
2. *Accuracy* is the ability to obtain the true value of a measurement. It is the agreement between measurement and the true value. It does not reflect upon the quality of an instrument [9].
3. *Precision* is the ability to repeat measurements in the same way. Precision reveals information about the quality of an instrument.
4. *Uncertainty* is a parameter associated with the result of a measurement—a confidence level within which the measured value is thought to reside. It is the interval around the value that is reinforced by further measurement. The combined distribution of measurements is assumed to take normal form. Experimental uncertainty due to random measurement effects is different from systematic errors manifested by equipment or independent sources. *Uncertainties* and *probabilities* go hand in hand, for example, the measured value has the stated probability to lie within the confidence interval [10].

5. *Error* is the disagreement between a measurement and the true value of a measurand. In some ways, it is the opposite of accuracy. Errors arise from a number of scenarios: Abbe error, thermal expansion, repeatability, human error, vibrations, and surface contacts. These are not the same as uncertainties. Do not confuse *error* with *uncertainty*, for example, a measurement with a large uncertainty may have negligible error.

6. *Calibration* is the process of determining and documenting the deviation of a measured result from that of an accepted standardized "true" value [10].

7. *Traceability* is a process whereby the measured result can be compared to a standard for the specific measurand at one or subsequent stages. The tolerance on a product is usually looser than that of the primary standard, shown below [10].

Product → Manufacturer's Test Equipment → In-House Calibration Laboratory → Accredited Calibration Laboratory → National or International Standard

Measurements derived with a scanning probe microscope provide an example of a *traceability chain*. The chain, starting with the definition of the meter and ending with SPM (scanning probe microscopy) measurements, proceeds as follows [11]:

Definition of the SI unit, the Meter → Realization of the Meter → Calibration of the Laser Frequency → Interferometrically Traceable AFM → Calibration of Physical Transfer Standards (1-D pitch, 2-D pitch, flatness, and step-height standards) → SPM Calibration → Traceable SPM Measurements

8. *Accepted values* are values generally agreed upon as the standard. Accepted values may not be the true values. Question everything (within reason of course)!

We leave it up to the student to review other concepts and axioms of statistics such as *significant figures*, the *mean*, the *median*, *relative uncertainty*, and the *standard deviation*.

The number of tools involved in nanometrology is enormous and we will cover a few general kinds. Electron beam and scanning probe instruments, considered to be the workhorses of nanotechnology research and development, will be discussed in more detail of course. These instruments are in every way the purest methods of nanotechnology. Specialized spectroscopic techniques continue to serve nanotechnologists.

The field of nanometrology is enormous but of appropriate magnitude to support this apparently vaster field of nanotechnology. Although we are interested in measuring that which is very small, we are also interested in measuring things well beyond the smallness of nanoparticles—the single bond, the single molecule or atom, and the single electron transfer. We are also interested in how these collective atomic and molecular events come together to determine the behavior of bulk materials—from the bottom up. Metrology unconsciously takes place in all of nature. Metrology, for us humans, is required at ALL levels of complexity within the material continuum with which we deal on a daily basis.

2.0.2 Background to Nanomanufacturing

Nanofabrication and *nanomanufacturing* processes are, in many ways, the natural evolutionary extension to the nanoscale of preexisting manufacturing and micromanufacturing practices. The terms "synthesis" and perhaps "nanofabrication" are more appropriately applied in the laboratory setting, while the term "nanomanufacturing" resides as the exclusive domain of nanotechnology—in the factory and assembly line.

Several recently developed techniques are unique to nanoscale fabrication and manufacturing, and therefore must be considered to be innovative (if not revolutionary). Nanofabrication and nanomanufacturing are processes by which nanoscale materials or components are produced from the bottom up (from atoms and molecules), from the top down (from bulk materials) in small steps of high precision or by various mixtures of the two [12]. *Molecular manufacturing* (via the molecular assembler route) is a subset of the general field of bottom-up nanomanufacturing and is discussed in more detail later on in the text.

The National Science Foundation (NSF) established the *NanoManufacturing Program* in 2001 [13]. The NSF program objectives include (1) scale-up of nanotechnology for high rate production, reliability, robustness, yield, efficiency, and cost, (2) integration of nanostructures in functional microdevices and meso/macroscale systems and interfacing across dimensional scales, (3) interdisciplinary research, multi-functionality across energetic domains (mechanical, thermal, fluidic, chemical, biochemical, electromagnetic, and optical), (4) systems approach encompassing nanoscale materials, structures, fabrication and integration processes, production equipment and characterization, theory, modeling, simulation and control tools, biomimetic design, and (5), nanomanufacturing education and training of the workforce, involvement of socio-economic sciences, addressing the health, safety, and environmental implications, development of manufacturing infrastructure, as well as outreach and synergy of the academic, industrial, federal, and international communities. The program has a special interest in environmental, health, and safety aspects, as well as ethical, legal, and societal implications of nanomanufacturing [13].

Several centers for nanomanufacturing have cropped up around the globe (chapter 1). The mission of the NSF Center for High-Rate Nanomanufacturing (CHN) centered at Northeastern University in Boston, Massachusetts (plus the University of Massachusetts at Lowell and the University of New Hampshire) is to develop tools and processes that enable high-rate/high-volume bottom-up, precise, parallel assembly of nanoelements and polymer nanostructures [14,15]. Their fundamental purpose is to develop means of bridging the gap between nanoscience and nanotechnology and to provide education about nanomanufacturing and its surrounding concerns—environmental, economic, societal implications, and development of an emerging workforce [14,15]. The CHN has constructed a state-of-the-art facility within which to accomplish its twenty-first century manufacturing—the *George J. Kostas Facility for Micro and Nanofabrication* [16].

Metrology and manufacturing are bound tightly to one another—it would be impossible to manufacture any material or device without guidelines that are embedded in previous metrology and previous manufacturing. During the *Industrial Revolution* (ca. 1800 to 1920), a manufacturing process known as "mass production" was introduced. Clearly in order to mass produce a product, the product has to be the same size, shape, and function, all within some degree of uniformity. The low-limit tolerance specified by designers in those bygone days was in the neighborhood of 1 mil (25 μm) [17]. Clocks, firearms, automobiles, sewing machines, and numerous other machines of the industrial revolution were manufactured whole scale within such low tolerances. Manufacturing processes included machining, stamping, casting, forging, and other techniques that assembled interchangeable machine parts—all doable within a 1-mil tolerance [17]. The mechanical Vernier caliper was the measurement device of choice used to transfer design specification metrology to the stamp, gauge block, cast, and cutting tool [17].

The *Semiconductor Revolution* arrived on the scene in the early 1950s and its metrology was based on the micron, for example, the microcircuit. As we all know so well, the entire computer industry is based on semiconductor materials and integrated circuits. Manufacture of the integrated circuit requires planar multi-leveled lithographic processes that in turn require accurate patterning, transfer, and alignment [17]. Operational length scale down to 100 nm was made possible by optical laser interferometer technology. The optical-based laser early on and electron-based beam later were the vehicles of choice employed to transfer design specification metrology to the mask and the resist materials. Micromanufacturing has since employed more sophisticated pattern transfer technology that utilizes ion beams and higher energy optical sources. The evolution of these technologies helped prepare the way to the domain of the nanoworld.

The *Nanotechnology Revolution* has, in essence, just begun. Who can say with certainty that the nanotech revolution sprung into the scientific mainstream in 1990, 1995, or 2000 or whenever? We shall leave that debate to science and industrial historians. Regardless, we are on the verge of mass-scale production of nanosystems that consist of electronic, photonic, magnetic, mechanical, chemical, biological materials, and devices and all possible permutations and combinations thereof [17]. We have actually already breached that threshold with the sub-100-nm computer transistor since 2003 and as Intel's latest transistor is a remarkable 30 nm in size.

With nanotechnology, we no longer operate in the pure world of classical physics. The length scale is now that of the nanometer. Like with anything else, associated uncertainties exist at the nanometer level but also at that of the sub-nanometer. Accuracy is transferred to products by optical, electron, and atomic imaging; and microscopes capable of viewing the nanodomain. Our cutting tools have most certainly changed over the past hundred years or so—when exactly was the transition between one that was purely mechanical to one that is based on focused ion beams?

2.0.3 *The Nano Perspective*

No matter how one adheres to the evolutionary theme, traditional challenges assume special "revolutionary" significance as we attempt to interface more intimately with the nanoworld. According to the NNI (National Nanotechnology Initiative) report *Instrumentation and Metrology for Nanotechnology* [18], revolutionary advances are required in order to realize products and manufacturing processes based on nanotechnology. What form will these revolutionary developments take? Are there undiscovered analytical techniques beyond the current paradigms or will techniques and instrumentation evolve from preexisting technology with enhanced (revolutionary) capabilities? A combination of both is most likely.

Serious advances have contributed to nanocharacterization over the past few decades [19]. Atomic scale chemical sensitivity and mapping, structure and morphology determination (by tomography), electron holography (for electrical, magnetic, and thickness properties at high resolution), dual-characterization and fabrication tools, and higher data collection rates and sample analytical throughput are just a few of the numerous advances—above and beyond the development of the scanning and transmission electron microscopes and the scanning probe techniques.

There are a handful of key challenges facing nanomanufacturing and nanometrology [14,15,19]:

R&D Laboratory
- Development of communication between nano- and micro (and macro)-domains
- Assembly and connection between and among nano-elements and -devices
  - Made of similar or different materials
  - 2-D interfaces
  - 1-D electronic or mechanical connections
- Integration of top-down and bottom-up processes
- Development of higher-capacity computing systems and advanced modeling techniques
- Reproducibility and repeatability from lab to lab
- Blending of classical physics and quantum properties
  - Quantum entanglement
  - Heisenberg uncertainty issues
  - Scaling laws
- Characterization methods and instrumentation that require
  - Multifaceted instrumentation development
  - Enhanced 3-D characterization
  - Significantly enhanced characterization speed
  - Nondestructive interfacial characterization
  - In situ characterization
  - Quantitative measurement of dispersion of embedded nanomaterials
  - Complete in situ identification and tracking of cellular processes
  - Greater resolution, accuracy, and precision
  - Data recording without artifactual interference
  - Enhanced computer interfaces
- Physical properties
  - Ohmic contact
  - Heat transfer
  - Sticking and friction
  - Fluidics
  - Cooling
  - Surface area mitigation

Manufacturing
- Manufacture of nanocomponents and devices with high throughput
- Uniformity (bottom-up fabrication has high throughput but domain size is limited)
- Reliability (although progressive, relatively unknown at this time for many processes)
- High rate scale-up potential
- Contamination (do we need better than 100-class cleanrooms?)
- Efficient detection of defective elements and domains and their subsequent mitigation
- Packaging and stability
- Development of more nanometrology standards

Societal
- Technology transfer (systems need to be streamlined to keep up globally)
- Patent processing (new intellectual property paradigms with streamlined processing)
- Skilled workers? engineers and scientists? Where will they come from?
- Redefinition of partnerships and funding increases
- Occupational health and safety concerns
- Environmental safeguards
- Other societal elements

How does one make a device that is governed by quantum behavior, one that is susceptible to weak van der Waals forces, one that is influenced by thermal fluctuations in its immediate vicinity, or a device that may not comply with the second law of thermodynamics like we are accustomed? Does Nature possess inherent metrology that is able to provide quality inspection to all of its machinations? What happens to defective nanostructures in the biological domain? I think we are well aware of the answer to the last query. Nanotechnology has already made the scene in the academic realm and is contributing more and more to industrial applications. The precise control and transfer of materials and devices to nanodimensions is a key concern as sub-nanometer control is becoming more frequent.

The National Science and Technology Council in their report "Science for the 21st Century" stresses the need for nanometrology development and standards [20]. Along with this need, a parallel requirement to build laboratories and nanofabs is imperative. Nanometrology requires special facilities—facilities that have rooms without vibration, with enhanced electromagnetic shielding and thermal control to tolerances less than $0.1–0.01°C$ [21]. Nanotechnology is tagged with the moniker "revolutionary." This may or may not be true across the board, but it is most certainly setting the world of metrology on its head.

2.1 NANOMETROLOGY AND UNCERTAINTY

"The history of science is the history of measurement," so stated James M. Cattell, one of the founders of modern psychology [22,23].

If Professor Cattell had the opportunity to read Porter (1995, 2003)[24], then he might have extended his statement by saying that the history of measurement is also the history of commerce and government as well as the history of many other significant aspects of modern social and political life.

… so stated George Englehard, Jr., author of *Historical views of the concept of invariance in measurement theory* [23]. Without measurement we have no quantitative information. We only have information based on purely qualitative descriptions. This is, however, not to understate the importance of qualitative observation and recording. Standardized quantitative measures are extremely important and are usually the gateway to in depth analysis of a phenomenon or property.

Metrology is divided into four general groups: (1) theory, (2) techniques, (3) instrumentation, and (4) legislation (standards) [25]. The demands on the range of knowledge and metrological skills of engineers and scientists have increased dramatically. Ironically, this was stated in 1986, more than 20 years ago [26]. One can just imagine the importance placed on accuracy and precision in acquiring measurements at the nanoscale.

SI Units.   The seven SI base units are the meter (length, x, l, r, or m), the kilogram (mass, m or kg), the second (time, t or s), the ampere (electric current, I, i, or A), the Kelvin (temperature, T or K), the mole (amount of substance, n or mol), and the candela (luminous intensity, I_V or cd). These parameters, however, rely on one another in several ways and are therefore not entirely independent. The ohm (Ω), for example, is in terms of $m^2 \cdot kg \cdot s^{-3} \cdot A^{-2}$. All of these base SI units are directly relevant in nanometrology. We are now interested in how our perspective of them may become altered.

We are mostly familiar with the means and methods that such standards were set in the past, but new methods that radically differ from them are making the scene. These new methods, making an analogy with experimentation, represent proof-of-concept of measurement from an independent approach—a practice that is always good for good science. For example, the ohm can now be defined with extremely high accuracy from experiments involving the quantum Hall effect (QHE) and the value of the von Klitzing constant—that $R_{K\text{-}90}$ can be used to establish a reference standard of resistance with one standard deviation estimated to be 1×10^{-7} and a reproducibility that is significantly better than values obtained from traditional methods. The BIPM(Bureau International des Poidset Mesures) in 2000 has admitted as much. The QHE provides an "invariant reference" for resistance that is derived from physical constants with reproducibility better than two orders of magnitude than the best uncertainty derived from traditional SI units [27]—the power of the quantum domain!

There are several reasons for this kind of validation in addition to having roots in natural constants: (1) measurement results do not depend on external parameters (e.g., ambient conditions), (2) the values do not drift with time, and (3) the measures can be reproduced anywhere, a condition that "simplifies and improves traceability of any measurement to the primary standards" [27]. The discovery of the QHE and the Josephson effect (more about them later) made available to the scientific community two quantum standards from which the other basic seven standards can be derived [27]. Truly revolutionary! A bottom-up approach to metrology is making its mark.

2.1.1   Nanometrology

Norio Taniguchi, the person who coined the term nanotechnology, stated in 1974 that uncertainties between 100 and 1 nm would become the standard required for measurement and manufacturing [28]. According to J. B. Bryan in 1979, again somewhat prophetically, "sophisticated instrumentation and ultra-precise dimensional metrology" would be required [29]. From the days of mechanical machining to microlithography, the drive towards miniaturization, better precision, and better accuracy followed a steady drumbeat (**Fig. 2.2**). The need for accurate positioning along with precise carving tools led to the development of smaller and smaller actuators, motors, and sample stages that relied on the development of new materials and thin films and new methods of fabrication, but also now require the participation of disciplines such as chemistry and biology [30].

Traditional engineering materials (metals and alloys, ceramics, semiconductors, organics and polymers, and composites), all enhanced by nanotechnology, now include biological components. Nanoscale development has got us into intimate contact with natural biological processes—a domain where all materials and functions are based on nanoscale phenomena. Objects and devices have undergone miniaturization over the past several decades. Nanometrology evolved, was reinvented, then evolved further to keep up. Nanometrology is rapidly expanding its sphere of influence and will soon establish the rules of engagement at the nanoscale.

2.1.2   Uncertainty

A new age of metrology is upon us. Assignment of uncertainty to nanoscale materials and phenomena presents a new set of metrological challenges that has evolved from traditional challenges [31]. New models and procedures are needed to analyze and interpret new data.

Measurements are approximations of the true value of a physical quantity (the *measurand*), and therefore the result is reliable only when defined in terms of the uncertainty of the measurement [32]. Uncertainty can be statistical in nature (evaluated by statistical methods) or deterministic (evaluated by other means). Uncertainty arises from random effects. The student should review some of these terms by consulting any general statistics text. Standard uncertainty (standard deviation) μ_i is usually derived from the square root of the variance μ_i^2 [32]. A popular method to determine standard uncertainty is statistical analysis of independent observations that employ the method of least squares to fit a curve. Analysis of variance (ANOVA) can be applied to identify and quantify random effects.

The NIST relies on the *combined standard uncertainty* $\mu_c(y)$ in the definition of many measurement results (commercial, industrial, and regulatory). Combined uncertainties result from individual standard uncertainties (e.g., subject to the *law of propagation of uncertainty* and the method of *root sum squares*, RSS). Combined uncertainties are used in conjunction with the determination of fundamental constants, fundamental metrological research, and international comparisons and realizations of SI units [32]. However in cases of health and safety (or perhaps nanometrology?), *expanded uncertainty* is required—an extra confidence level U surrounding the measurement y of the *measurand Y* such that

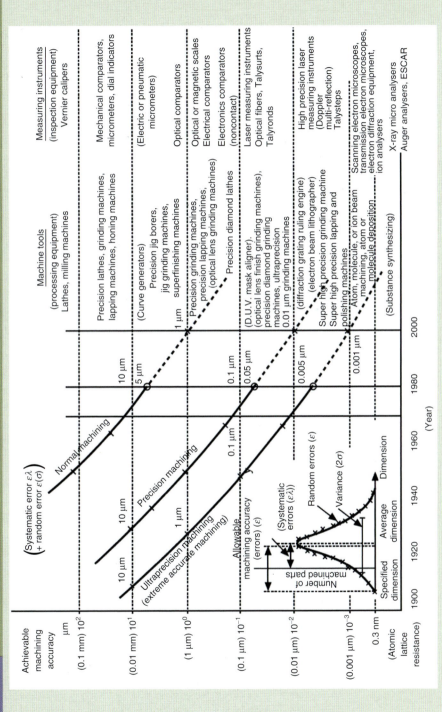

Fig. 2.2 The evolution of machining from the 1900s to the present day has changed dramatically. Achievable machining accuracy has been reduced from 100 μm to values well below that of the micron. The primary tools involved in high technology manufacture have evolved from those that are purely mechanical in nature, such as the lathe and milling machines, to tools that involve gratings and ion beams. The concomitant evolution of measurement devices occurred alongside the production tools—from the Vernier caliper to optical and electronic comparators to lasers and to the electron microscopes we use today.

Source: N. Taniguchi, Current status in, and future trends of, ultraprecision machining and ultrafine materials processing, Tokyo Science University, Annals of the CRIP, 32, 573–582 (1983). CRIP is the College International pour la Recherche en Produtique. With permission.

$$U = k\mu_c(\gamma) \tag{2.1}$$

$$\gamma - U \leq Y \leq \gamma + U \quad \text{or} \quad Y = \gamma \pm U \tag{2.2}$$

The factor k is the called the coverage factor. If there is a quantity z described by a normal distribution with expectation μ_z and standard deviation σ, the interval $\mu_z \pm k\sigma$ represents the interval of confidence. For example, $k = 1$, $k = 2$, and $k = 3$ for 68.27%, 95.5% and 99.73% of the distribution of values, respectively [32].

Uncertainty in a measurement depends on many factors. For example, NIST in conjunction with Mitutoyo America Corporation achieved "remarkable low uncertainty" in their quest to control roundness [33]. They were successful in measuring roundness with an uncertainty of less than 3 nm. According to Theodore Doiron of NIST's Engineering Metrology Group, the heralded development was due to "a special error separation measuring method, a tight temperature controlled environment and the stability of the *Mitutoyo RA-H500* roundness tester" [33].

2.1.3 *Heisenberg Uncertainty*

The *Heisenberg uncertainty principle* limits what we can do in metrology. A new brand of physics, bolstered by nanotechnology and quantum mechanics, explores in more depth what can be done with quantum principles within the sphere of influence of the Heisenberg limit [34]. The Heisenberg uncertainty principle presents a lower limit on the product of standard deviations (e.g., for position and momentum and other complementary couples). The Heisenberg uncertainty principle governs the theoretical limitation of the combined accuracy of pairs of simultaneous but related measurements. In other words, the combined uncertainty of such complementary measurements cannot be less than Planck's constant h (or more specifically, $h/4\pi$). Specific complementary measurement pairs include position–momentum and energy–time couples. The Heisenberg uncertainty principle does not place limitations on the accuracy of single measurements, nonsimultaneous measurements, or of simultaneous measurements of other types of related couples outside those restricted by the Heisenberg principle [35].

The Heisenberg Uncertainty Principle. The standard *quantum limit* scales as $1/\sqrt{N}$, where N can be the number of atoms, ions, electrons, or photons, and the *Heisenberg limit* is defined as scaling with $1/N$. One consequence of the Heisenberg uncertainty principle is that measurement of atoms and subatomic particles is governed by statistics and probability rather than by actual physical, observable measurements. The Heisenberg uncertainty principle, in other words, restricts what we can do. However, with the dawn of quantum information science, this state of affairs may undergo a drastic change and soon.

We have, for all practical purposes, the ability to measure the position of a macroscopic object with 100% certainty and that the uncertainty is for all practical purposes equal to zero ($\Delta x \rightarrow 0$). If we also know the velocity (and mass) of the macroscopic object, we can usually predict where it is going, where it has been, and vice versa. Such is the case in classical physics [36]. In the domain of classical physics, exact simultaneous values can be assigned to all physical quantities [37]. These are tools that allow us to make predictions about the physical characteristics of a classically physical object, for example, the flight of a jet plane. As objects assume smaller dimensions, however, the Heisenberg limit assumes more importance. The question to ponder is at what point does the Heisenberg limit, or the standard quantum limit for that matter, become a serious consideration in metrology.

In order to observe (measure) the position and/or the momentum of an electron, the electron must undergo a perturbation—perhaps one caused by an interaction with a photon that *is* used for the measurement [36]. Therefore, by probing the position of the electron (e.g., so that $\Delta x \to 0$), more than likely you have (or rather the photon has) altered the electron's momentum, perhaps drastically and bumped that parameter into the domain of the unknown. The momentum of the photon (or electron, atom, or other materials) is known from

$$p = \frac{h}{\lambda} \tag{2.3}$$

and the change in the electron's momentum following collision results in an uncertainty

$$\Delta p \approx \frac{h}{\Delta \lambda} \tag{2.4}$$

where $\Delta \lambda$ is the uncertainty in the electron's position. The product of the uncertainty in position and the uncertainty in momentum results in the Heisenberg uncertainty principle*

$$(\Delta x)(\Delta p) = \left(\frac{h}{\lambda}\right)\lambda = h \to \frac{h}{4\pi} = \frac{\hbar}{2} \tag{2.5}$$

or, since the value of the uncertainty represents a minimum

$$(\Delta x)(\Delta p) \geq \frac{h}{4\pi} = \frac{\hbar}{2} \tag{2.6}$$

The *h*-bar version of Planck's constant is a more accurate rendition of the uncertainty principle. The exact value (with uncertainty) is

$$\hbar = \frac{h}{2\pi} = 1.054\ 571\ 628(53) \times 10^{-34} \mathrm{J \cdot s} = 6.582\ 118\ 99(16) \times 10^{-16} eV \cdot \mathrm{s} \tag{2.7}$$

In simple terms, the measurement process itself limits the ability to measure position and momentum simultaneously [36]. If, at the subatomic scale we were to know position, say of an electron, exactly (e.g., $\Delta x \to 0$), then the uncertainty in its momentum must approach infinity ($\Delta p \to \infty$).

Energy and time also form a complementary pair of uncertainties [36].

$$(\Delta E)(\Delta t) \geq \frac{h}{4\pi} \tag{2.8}$$

---

* *x*-space with a characteristic linewidth of Δx is also localized in *k*-space (e.g., the particle's wave number) with a characteristic width of $1/(2\ \Delta x)$ and then $\Delta x \Delta k \geq 1/2$. Since measurement of a particle's wave number is equivalent to measuring its momentum *p* and that $p = hk$, the uncertainty in *k* of order Δk translates to an uncertainty in *p* of order $\Delta p = h\Delta k$. Therefore, $\Delta x \Delta p \geq \hbar/2$ (or $h/4\pi$).

EXAMPLE 2.1 *Measurement of an Electron's Position*

A free electron (such as one found in an electron beam) has kinetic energy equal to 25 eV. What is the velocity of the electron? If this velocity is known to within 0.5% accuracy, what is Δx of the position of the electron?

Solution:

Electron velocity

$$25\ e\text{V} = (25)1.6 \times 10^{-19}\,\text{C}\left(\frac{\text{J}}{\text{C}}\right) = 4.0 \times 10^{-18}\text{J}$$

$$\text{KE}_e = \tfrac{1}{2}m_e v_e^2;\quad 4.0 \times 10^{-18}\text{J} = \tfrac{1}{2}\left(9.1 \times 10^{-31}\ \text{kg}\right)v_e^2$$

$$v_e = \sqrt{2\left(\frac{4.0 \times 10^{-18}\text{J}}{9.1 \times 10^{-31}\ \text{kg}}\right)} = 3.0 \times 10^6\,\frac{m}{s}$$

The uncertainty in position is 0.5%.

Therefore, the uncertainty in momentum is

$$\Delta p = m_e\,\Delta v = 9.1 \times 10^{-31}\ \text{kg}\ [(0.005)\ (3.0 \times 106\text{m} \cdot \text{s}^{-1})] = 1.4 \times 10^{-26}\ \text{kg} \cdot \text{m} \cdot \text{s}^{-1}$$

The uncertainty in position follows from

$$\Delta x \approx \frac{h}{\Delta p} = \frac{6.6 \times 10^{-34}\ \text{J} \cdot \text{s}}{1.4 \times 10^{-26}\ \text{kg} \cdot \text{m} \cdot \text{s}^{-1}} = 47\ \text{nm}$$

Detection of an electron can be accomplished by bombarding the electron with a photon of velocity c, the speed of light. The position of an electron is known with an inherent uncertainty $\Delta x \approx \lambda$. The time it takes for this photon to traverse the distance equal to the uncertainty of the position of the electron is

$$\frac{\Delta x}{c} \approx \frac{\lambda}{c} = \Delta t \tag{2.9}$$

EXAMPLE 2.2 *The Position of a High-Speed Projectile*

A 22-caliber bullet with mass 2.6 g is able to achieve a velocity of 390 m · s⁻¹. If we are able to determine its velocity to ±0.02% ($v = 390 \pm 0.078$ m · s⁻¹), what is the minimum uncertainty in its position? Is this value consequential?

How does this value compare to that of the electron in Example 2.1 above?

Solution:

Calculation of Δv

$$v = 390 \pm 0.078\ \text{m} \cdot \text{s}^{-1},\quad \Delta v = 2 \times 0.078\ \text{m} \cdot \text{s}^{-1} = 0.16\ \text{m} \cdot \text{s}^{-1}$$

$$\Delta x = \frac{h}{2\pi\Delta p} = \frac{h}{2\pi(m_e\Delta v)} = \frac{6.6 \times 10^{-34}\text{J} \cdot \text{s}}{2\pi\left(2.6 \times 10^{-3}\ \text{kg}\right)\left(0.16\ \text{m} \cdot \text{s}^{-1}\right)} = 2.5 \times 10^{-31}\text{m}$$

The uncertainty in position of the electron is significant—on the order of several nuclear diameters. The uncertainty in position of the bullet is of little consequence to the measurement.

Also, since $E = h\nu = \frac{hc}{\lambda}$; $\Delta E = \frac{hc}{\lambda}$, the Heisenberg uncertainty with regard to energy and time reduces to

$$(\Delta E)(\Delta t) \approx \left(\frac{hc}{\lambda}\right)\left(\frac{\lambda}{c}\right) = h \qquad (2.10)$$

What is the physical meaning of this expression? It means that in order to measure the energy of a particle exactly, an infinite amount of time is required [36,38]. The measurement of the frequency ν of a photon is a measure of its energy. A suitable experiment, for example, is the measure of the energy of an excited state of an electron. The energy (frequency) can only be determined during the lifetime of the excited state, and the observed emission line, therefore has a characteristic energy width, $\Delta E = h\Delta\nu$. The diffraction limit in optics corresponds to this Heisenberg uncertainty limit.

2.1.4 Quantum Entanglement

We decided to include a section on this interesting phenomenon although its rigorous understanding is beyond the scope of this text. Nonetheless, please try and get a feel for what is presented.

An interesting quantum phenomenon that has potential effects on metrology, and one that has assumed greater importance over the past few years, is the phenomenon of *quantum entanglement*. Entanglement is a quantum phenomenon in which two or more quantum states are intimately linked—however separated in space—that leads to correlations between observable physical properties of systems—some that are remote with regard to one another [39]. Quantum entanglement, from the universal perspective, is actually the vehicle that serves to link the quantum world to that of our macroscopic classical world of physical observables. We espouse the concept of continuum throughout this text and in our sister book, *Introduction to Nanoscience*, whether it be in the form of an energy continuum, a material continuum, or the artificial "academic discipline" continuum. Here we have yet another continuum to consider, that of a metrological quantum mechanical–classical physics continuum. Traditionally, we have not needed to make direct links between the quantum domain and macroscopic physical properties. We do today and the phenomenon of quantum entanglement helps supply that link, and uncoupling that entangled couple may just allow for a whole new generation of applications.

Thomas Durt of Vrije University in Brussels believes that entanglement is omnipresent [40].

> When you see light coming from a faraway star, the photon is almost certainly entangled with the atoms of the star and the atoms encountered along the way … and the constant interactions between electrons in the atoms that make up our body are no exception. … We are a mass of entanglements.

It was Erwin Schrödinger who coined the word entanglement [40]:

> When two systems, of which we know the states by their respective representatives, enter into temporary physical interaction due to known forces between them, and when after a time of mutual influence the systems separate again, then

they can no longer be described in the same way as before, viz. by endowing each of them with a representative of its own. I would not call that one but rather the characteristic trait of quantum mechanics, the one that enforces its entire departure from classical lines of thought. By the interaction the two representatives [the quantum states] have become entangled.

Albert Einstein, although awarded a Nobel Prize for his work on the photoelectric effect, could not accept the probability-based nature of quantum mechanics, for example, "God does not play dice with the universe," so he said. Or, making his point with more aplomb, Einstein stated that

> I find the idea quite intolerable that an electron exposed to radiation should choose of its own free will, not only its moment to jump off, but also its direction. In that case, I would rather be a cobbler, or even an employee in a gaming house, than a physicist.

Einstein believed that something more deterministic existed underneath the probability, and hence, the concept of entanglement emerged in 1935 [41,42]. Einstein called quantum entanglement "spooky action at a distance." From a deterministic perspective, by evaluating the spin of one electron, only then does one know exactly the spin of the other electron. Before that point in time, electron spins exist, only within a range of probabilities. In other words, it is only after a direct observation that spins become fixed [41,42]. Entanglement does not minimize the principle of cause and effect, but it does underscore the importance of quantum particles, particles that are usually defined by probability values rather than fixed values [41,42].

The *Fano effect* is a characteristic of atomic spectra, bulk solids, and semiconductor heterostructures that offers clues about the relationship between the quantum and a bulk observable characteristic—the spectrum. The Fano effect describes the relationship between quantum interference of two competing optical pathways: the first pathway that couples the energy of a ground state with an excited state, and the second that connects the ground state with a continuum of energy states. It is actually a common phenomenon that occurs when a discrete quantum state (of an atom or molecule) interacts with a continuum state (of a surrounding host material) [43]. The effect is easily observed in spectra that display asymmetric lineshapes. In other words, the Fano effect influences the way an atom or molecule absorbs light and is expressed in its spectral response [43].

Experiments with small nanoscale systems, however, are made difficult by factors related to the Heisenberg uncertainty principle—in that the interaction of a nanoscale system and the continuum state that surrounds it may not be detected easily [43]. However, M. Kroner et al. have shown that photons scattered from a quantum dot with increasing laser intensity enabled them to observe weak interactions related to the Fano effect, for the first time, to a nanoscale phenomenon [43].

Two-Photon Correlation and Parametric Down Conversion.   A means of generating a two-photon entangled system is shown in **Figure 2.3**. Quantum correlation of paired photons is made by an incident photon of higher energy impinging on a nonlinear optical crystal. A nonlinear optical material induces dielectric polarization P response in a nonlinear way to the applied electric field E. Usually a

FIG. 2.3 *A schematic representation of parametric down-conversion is depicted. A single photon that impinges upon a nonlinear optical crystal is converted into two photons with lower energy. The energy and momentum of the first photon are preserved by the two output photons.*

function of intense laser exposure, an example of a nonlinear optical phenomenon is second harmonic generation from such a crystal.

According to Alan Migdall, a physicist in the Optical Technology Division of the NIST in Gaithersberg, Maryland [44],

> … such quantum correlation promises metrologists something of a free lunch: absolute measurements that don't require absolute standards.

As a matter of fact, quantum correlation techniques provide a method of determining the quantum efficiency of detectors without the need of calibration standards [44]. Migdall goes on to state that measurements of absolute infrared radiance without calibrated standards and without an infrared detector are also possible due to this phenomenon. This statement should cause same entanglement in your thinking.

Parametric down conversion is a process by which entangled photons are produced (**Fig. 2.3**). Input photons decay into entangled photons through the action of a nonlinear crystal. Therefore, if one knows the energy of the input laser and one of the output photons, then information about the entangled pair is known. And, even more interestingly, if measurements were made on the first photon of the entangled pair, it would be possible to predict the outcome of measurements on the second photon [44]. In other words, detection of one photon predicts the existence and properties of the second photon such as wavelength, emission time, direction, and polarization—all at a distance [44]. Measurement of the quantum efficiency of a detector is an important aspect of metrology (refer to chapter 14). By such methods, 150×10^{-18} s level of resolution has been achieved that approached the Heisenberg uncertainty limit on the simultaneous creation of photon pairs—and all this occurred in 1999, the Stone Age of nanotechnology.

We won't spend more time on this aspect but in general a quantum mechanical entangled state (photons) is referred to as a *NOON* state.

$$\left|\psi_{NOON}\right\rangle = \left|N\right\rangle_a \left|0\right\rangle_b + e^{i\theta}\left|0\right\rangle_a \left|N\right\rangle_b \qquad (2.11)$$

One can see immediately from where the acronym *NOON* originates. The equation represents a superposition of N particles in mode a with zero particles in mode b and vice versa. *NOON* states are exploited to measure precise phase shifts in optical interferometers. A Mach–Zehnder interferometer, for example, is used to measure phase shift when one beam of a pair of collimated beams is perturbed. Applications of such two (or more) entangled photon systems include lithography, quantum sensing, quantum imaging, "quantum Rosetta stones," [45] and quantum metrology.

In 2007, T. Nagata et al. demonstrated super-resolution and sensitivity with four-photon ($N = 4$) *NOON*-states [46]. Optical phase measurements are routinely employed to determine distance, position, displacement, orientation, acceleration, and optical path [46]. The standard quantum limit, however, exists without consideration of quantum entanglement. Nagata et al. achieved higher precision by the exploitation of entanglement [46].

Recently, J.L. O'Brien et al. of the Centre for Quantum Photonics of the H.H. Wills Physics Laboratory and Department of Electrical and Electronic Engineering, University of Bristol in the United Kingdom reported control of single photons via an optically controlled NOT gate. A NOT gate is a gate based on digital logic that operates on the principle of a truth table, for example, input "A," output "NOT-A" [47,48]. Most photonic quantum technology relies on large-scale assemblies fixed to optical tables and complex interferometers (for subwavelength control). O'Brien and his team devised a milliscale device from silica waveguides on a silicon chip (an entire optical table). The characteristics of their apparatus are as follows, according to the authors [47,48]:

> ... high visibility (98.5 ± 0.4%) classical interference, high-visibility (94.8 ± 0.5%) two-photon quantum interference, high-fidelity controlled-NOT entangled logic gates (logical basis fidelity, F = 94.3 ± 0.2%) and on-chip quantum coherence confirmed by high fidelity (>92%) generation of two-photon path entangled states.... We fabricated 100s of devices on a single wafer.

We add this information to demonstrate the relative simplicity of the setup. Quantum metrology measurements utilizing four-photon communication have already demonstrated by O'Brien and his colleague Takeuchi in 2003 (Hokkaido University, Japan) [47,48]. The team also claims that exotic quantum processes are not necessary to reach the ultimate limits of measurement [49].

Without nonlocalized entanglement, standard interferometers are limited by statistical shot-noise that scales with $1\sqrt{N}$ where N is the number of particles (photons) passing through the interferometer over time. With quantum correlations (entanglement), the interferometer sensitivity is improved by a factor of $\sqrt{N}$ to scale as $1/N$—the limit imposed by the Heisenberg uncertainty principle. Exploitation of quantum entanglement is used to overcome quantum projection noise (QPN) (noise due to fluctuations in population of a "fixed number" of atoms or, alternatively, QPN arising from statistical nature of projecting a superposition of two states into one state when a measurement is made). The use of cold atoms and precise spectroscopy measurement has enhanced the frequency measurement accuracy of atomic clocks, but the measurement is limited by QPN of ensembles of uncorrelated particles [50].

For example, noise can be reduced by a process called "squeezing" (amplitude or phase squeezing). Photons are the best choice for quantum metrologies, because they deliver a small amount of noise (decoherence). Photons are good candidates for quantum bits in quantum communication, quantum lithography, quantum computing, and of course, quantum metrology [46]. Other topics not discussed in detail can be elaborated upon in excellent references.

Unentangled Photon Metrology? According to B.L. Higgins et al. at the Center for Quantum Dynamics at the Griffith University in Brisbane, Australia, advances in precision measurement have always been the gateway to scientific discoveries [51]. He also states that entangled states are not necessary for precise and accurate measurements. For example, he replaces entangled input states with "multiple applications of the phase shift on entangled single-photon states." Measurement precision at the fundamental level, of the optical phases used in length metrology, for example, are difficult to achieve due to limitations of the number of quantum sources available, for example, photons [51]. He goes on to say that standard measurement techniques that apply each resource independently are subjected to phase uncertainty that scale as $1/\sqrt{N}$, also know as the standard quantum limit. The Heisenberg limit, on the other hand, scales to $1/N$, the true quantum limit and a $\sqrt{N}$ factor enhancement over the standard quantum limit. By interferometry methods alone, this limit cannot be beaten [52].

Conventional classical measurement relies on N independent physical systems that are separately prepared and detected. The experimental values (e.g., measurement) derived by such classical techniques rely on averaging the outcomes of N independent measurements. For quantum measurement, N physical systems are prepared in an entangled state. Measurement occurs over all N states simultaneously with a single delocalized measurement (of probability) that collectively considers all N components [52].

Nanomechanical Oscillators and the Uncertainty Principle. Nanoelectromechanical systems (NEMS) are the miniaturized version of microelectromechanical systems (MEMS) except with higher frequency ranges, higher quality factors Q, higher sensitivity, and lower power demands [53]. Q is a measure of the damping of oscillations in a mechanical system as a function of frequency—higher Q indicates more stable vibrations. For example, high Q gallium nitride (GaN) wires 30–500 nm in diameter and 5–20 μm in length grown at NIST were stimulated to vibrate between 400,000 to 2.8 million times per second [54,55]. Another feature of nano-oscillators is that they can be cooled close to their ground states under cryogenic temperatures.

Such nanomechanical oscillators are perfect tools to study quantum effects like back-action, coherent states, and superposition [53]. Quantum back-action is a phenomenon that reflects the Heisenberg uncertainty principle—for every measurement there is always a perturbation in the object subjected to measurement. In addition to gigahertz-level vibrations and resolution with atomic spacing, MEMS and NEMS systems are capable of detecting force in the zeptonewton range (10^{-21} N) and mass in the zeptogram range (10^{-21} g) [56]. We are now prepared to observe quantum behavior in a mechanical device.

Now consider this. If a superconducting single-electron transistor (SSET) were coupled to a resonating nanomechanical structure (e.g., a silicon nitride beam) and a voltage applied between them that corresponds to a quantized energy state of electrons passing through the SSET, what do you think would happen? According to Keith C. Schwab of Cornell University, Ithaca, New York, and his research team, "counter-intuitive" developments unfold [53,57]. Positions of the beam during oscillation alter SSET conductivity. Therefore, from the measurement of current in the SSET (from the observer's point of view), determination of the resonator position was made possible. Schwab et al. also determined that random charge fluctuations within the SSET influenced the frequency, position, and damping rate of the resonator [53,57]. The authors stated that damping in the resonator mode due to a back-action force caused the temperature to drop from 550 to 300 mK. In essence, the SSET acted more like an absorber of energy rather than a generator of heat. This phenomenon will no doubt generate new ways to cool down components of future NEMS mechanisms [53,57].

Keith Schwab et al. essentially investigated the relationship between a nanomechanical oscillator device and the uncertainty principle [58]. The purpose of the experiment was to probe the limitations placed by quantum mechanics on the resolution of position. In this case, the research team coupled a single-electron transistor (SET) and an ultrasensitive microwave detector network to a nanomechanical device (a 19.7-MHz resonator of high Q). The SET served as an appropriate amplifier for a radio frequency nanomechanical system [58]. The goal for the research was to conduct position detection approaching that set by the Heisenberg uncertainty principle, at mK temperatures [53]. Schwab et al. claim that "ultimate sensitivity should be limited by both the quantum fluctuations of the mechanical oscillator and the quantum fluctuations of the linear amplifying system" [58].

The nanomechanical device consisted of an aluminum layer on a silicon nitride resonator that was 8.7 μm in length and 200 nm wide. They were able to detect movement due to the stimulus of observation (e.g., back-action). He states:

> We made measurements of position that are so intense - so strongly coupled - that by looking at it we can make it move.... Quantum mechanics requires that you cannot make a measurement of something and not perturb it. We're doing measurements that are very close to the uncertainty principle; and we can couple so strongly that by measuring the position we can see the thing move.

Schwab is working on building a nanodevice that exhibits superposition quantum behavior—a nanodevice that can be in two places at the same time [53].

2.1.5   Applications

Please note that any application of quantum entanglement implies an integral relationship with metrology (nano or otherwise). Quantum entanglement phenomena are expected to contribute to metrology—some fairly far out. Certain government agencies, for example, believed that communication faster than the speed of light was possible. This was based on entanglement operating instantly at some distance [42]. However, theoretically, this is not possible because the message would in effect be read in the past. A more practical application of entanglement is that of encryption. If one half of a set of entangled pairs is sent to a complementary communications link, interception is rendered impossible due to disintegration of the entanglement.

Nanotechnology And?   How does this all relate to applications and perhaps nanotechnology? According to Professor Ali Mansoori of the Departments of Bioengineering and Chemical Engineering and Physics at the University of Illinois at Chicago, the Heisenberg principle applies strictly to subatomic particles like electrons, positrons, and photons [59]. Nanotechnology, on the other hand, involves the position and momentum of atoms, molecules, for example, larger particles that do not come under the umbrella of the strict confines of the Heisenberg principle [59]. This does not imply that nanosystems are not affected by the Heisenberg uncertainty principle. As a matter of fact, it is through the application of nanomaterials that we are able to probe the interface between the quantum limit and the classical world of physics.

With regard to information technology, nanotechnology is expected to have great impact—in actuality, it already has. Future trends however indicate that smaller devices are inevitable. With their development, certain fundamental limits will be breached [60]. As fewer electrons, perhaps even single electrons, are switched to execute operational commands, at what point does this become physically feasible? Or, when the size of magnetic domains (e.g., bits) are reduced, at what point is stability compromised due to particle interactions with thermal fluctuations?

Quantum lithography and Two-Photon Microscopy.   Quantum lithography and two-photon microscopy provide a fertile ground for applications of entangled quantum states. The resolution of objects smaller than the wavelength of the applied beam is difficult due to light scattering around objects (e.g., the Rayleigh limitation). The reduction of the wavelength, while keeping the wavelength of the radiation field constant, although paradoxical, is what needs to be done [52]. According to Vittorio Giovannetti et al. at the *National Enterprise for Nanoscience and Nanotechnology* of Italy's Istituto Nazionale per la Fisica della Materia (NEST-INFM) and Scuola Normale Superiore in Pisa, Italy, quantum effects can help. By using equipment that is sensitive to the de Broglie wavelength, using two or more photons, the wavelength can be reduced from one wavelength to half that wavelength without changing the source frequency

$$\lambda = \frac{2\pi\hbar}{p} = \frac{2\pi\hbar c}{E} = \frac{2\pi c}{\omega} \rightarrow \frac{2\pi\hbar c}{2E} = \frac{\lambda}{2} \qquad (2.12)$$

In essence, two entangled photons (as a single entity!) enter an apparatus consisting of mirrors and beamsplitters. The entangled photons originate from a light wave that is split and then recombined on a surface. The recombined pattern is the same pattern that would have been generated from the original wave but at half the wavelength. For example, packets of 10 entangled photons originating from infrared radiation could end up as x-rays. More photons reduce the practical λ even further [52]. Beating the Rayleigh limit (e.g., best features limited by half the wavelength) is a piece of cake.

Quantum Computing.   Although computers rely on quantum mechanics today, information is encoded (transmitted and processed) in accordance to classical principles (as a "0" or a "1"), e.g., two well-defined states. Another application is that of *quantum computing*, an application that has received significant attention over the past few years. In this case, the quantum information can exist as a superposition of different states (not so well-defined)—both in *0ñ* and *1ñ* states, for example. Interference in a double-slit experiment is an example of superposition of states. Quantum interference in principle is the same except that only one particle is required due to the potential of entanglement. Single-particle quantum interference has been demonstrated with photons, electrons, neutrons, and atoms [61].

In quantum computing, information is stored with quantum particles (photons, atoms, or electrons) called quantum bits (or *qubits*). The problem is, according to the Heisenberg uncertainty principle, that accessing the particles should destroy the information carried by the qubits. With quantum entanglement, interaction with qubits is possible without "frying your quantum memory," this according to Brian Clegg, author of the *God Effect: Quantum Entanglement, Science's Strangest Phenomenon* [41,42]. Clegg also believes that quantum computers are not likely to make the scene anytime soon.

The weirdest one—are you ready for this—is the prospect of *teleportation*. Within the bounds of the Heisenberg uncertainty principle, interacting with a quantum particle changes that particle, for example, one cannot input a particle (scan it) and send it somewhere and have it reassembled into an exact copy. However, if only half of an entangled pair is "scanned" and the other half of the pair is placed through a logic gate, it is possible to make an identical particle at some distance. The downside is that the original particle (or human) loses its identity. According to Brian Clegg, this has actually been accomplished with large molecules [41]. Nobel Prize winner Brian Josephson (of Josephson junction fame) stated that *telepathy* can be explained by quantum entanglement. Brian Clegg concludes

> What entanglement (and quantum theory in general) does do is remind us is that the real world is much stranger than we imagine. That's because the way things are in the world of the very small is totally different to large scale objects like desks and pens. We can't rely on experience and common sense to guide us on how things are going to work at this level. And that can make some of the effects of quantum physics seem mystical. In the end, this is something similar to science fiction writer Arthur C. Clarke's observation that "any sufficiently advanced technology is indistinguishable from magic."

Welcome to the *Nano Age*! On that note, we move onto the next section, *Quantum Metrology*.

2.2   QUANTUM METROLOGY

Quantum metrology is about time, frequency, and photonics; and fundamental constants and the path towards quantum detection. There is a concerted effort to standardize measurements at the quantum level, for example, to improve measurement accuracy, traceability, new nanoscale metrology solutions, and the development of quantum standards [62]. The measurement continuum extends from the classical domain of physical objects down to particles like atoms and single molecules. The electromagnetic domain, over increasing ranges of frequency, must also address quantum fields within which individual photons (*quanta*) are of interest to nanotechnologists and others—energy of single atoms, nanoparticles, photons, and phonons. The "continuum of measurement accuracy" is given by the following scheme:

Classical Metrology → Classical-to-Nanometrology → Quantum Metrological Transition → Quantum Metrology

Determination of the length scale of the metrological transition between the classical to the quantum is challenging and depends on which physical property is under consideration and the lower limit of the energy scales [62]. For example, electric current is no longer a continuous property at the nanoscale (e.g., the Coulomb blockade). The same is true for ferromagnetism (at the superparamagnetic limit). Add to this that nanoparticles are susceptible to thermal fluctuations, and require cooling to ~100 nK for measurement acquisition. The challenge is quite formidable. The ultimate goal of quantum metrology is to provide measurements, based in quantum theory, with better accuracy than those provided by classical means. Quantum metrology is not as new as one might believe. For example, the redefinition of the meter has been based on the wavelength of light for quite some time. Ultimately, all SI unit based mechanical standards will be redefined by electrical units derived from quantum standards [63].

We discussed some of the obstacles, issues, and progress in development methods based on quantum entanglement and the relationship between measurements and the Heisenberg uncertainty principle. A new age of measurement is upon us; about that there is no uncertainty.

2.2.1 Atomic Clocks, the Meter, and Time

Time and Frequency. Measurement of time and frequency is a high priority in metrology. An oscillating clock is a device that is capable of measuring these critical parameters to high accuracy and precision. If one starts with N number of cold ions in the ground state that are excited by an electromagnetic pulse, then a superposition of the ground and excited states is created. A second identical pulse is applied to measure a phase factor φ between the two states, typically by measuring the intensity. From N independent ions, ω from the phase factor φ has an error of $\Delta\varphi = 1/\sqrt{N}$ and $\Delta\omega = 1/(\sqrt{N}\cdot t)$. For entangled states, ions not acting independently, $\Delta\varphi = 1/N$ and $\Delta\omega = 1/(N\cdot t)$, an enhancement of $\sqrt{N}$.

Much of nanometrology depends on time—reaction rates, stability functions, resonant frequencies, and other natural phenomena that exist at the nanoscale. The second, the nanosecond and smaller increments of time, therefore must also be standardized. A cesium (Cs) fountain atomic clock housed at NIST, Boulder, Colorado contributes to the international group of atomic clocks (Coordinated Universal Time, UTC) [64]. The uncertainty of this clock is 5×10^{-16} s. **Figure 2.4** shows the progress made with regard to uncertainty in standardized clocks since 1940.

Frequency and time intervals are measured during the up-down movement of a cesium cluster—hence the term fountain (**Fig. 2.5d**). With the assistance of several lasers, cesium atoms are pushed together and cooled (to near absolute zero). Another laser launches the cluster vertically ca. 1 m high as the others are shutoff. Gravity then pulls the cluster down into a microwave cavity that acts to

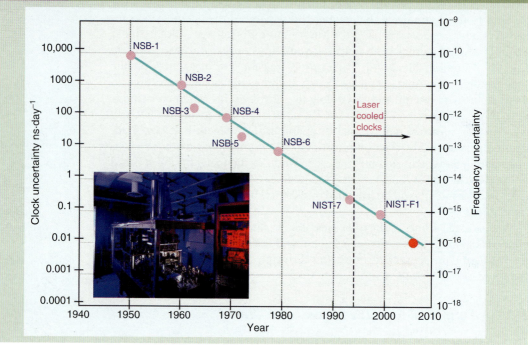

Source: Tom O'Brian, NIST Time and Frequency Division, Boulder, Colorado. With permission.

stimulate fluorescence, which is measured by a detector. After tuning to stimulate maximum fluorescence, the natural resonance frequency of Cs is attained, and it is that frequency that is used to define the second—9,192,631,770 Hz [64,65]. The Cs clock oscillator is a source of microwaves. For more information, please refer to http://tf.nist.gov/cesium/fountain.htm. The cooling and the mechanical motion allow for longer duration of observation of the Cs cluster, and it is from this extended observation that such accurate time intervals are determined [64,65].

Recently, advances have been made to improve the cesium fountain—better cooling techniques in particular that drive the temperature down to ca. 100 nK [62]. Enhanced cooling leads to reduced "Cs cold-collisional shifts" that contribute to the fountain systematic uncertainty [62]. Quieter laser sources have also improved the accuracy of the clock [62]. Newer atomic clocks based on optical (rather than microwave) measurements are expected to be 20 times more stable than the current atomic clocks and have the potential to be 100x to 1000x more stable, such as the mercury ion clock [62] An optical clock has been developed that oscillates at ~10^{15} Hz—arising from a single atom of mercury. The detector (also a vital factor without which no such clock is possible) is provided by a femtosecond (10^{-15} s) laser counter, called a comb [66]. Stephen Webster of the National Physical Laboratory (UK) states that

> The detector laser acts as a reduction gear that is capable of downshifting the optical frequency to a countable radio frequency without impairing the purity of the optical frequency.

In other words, the detector (the large gear) is able to reduce the oscillation of the mercury emission (the small gear) by a factor of 500,000 times—enough to reduce the signal to the microwave domain and be able to be counted electronically—with exceptional accuracy [66]. **Figure 2.5** displays various renditions of NIST atomic clocks.

Enhancing signal-to-noise ratio improves accuracy in frequency measurements. We mentioned quantum entanglement earlier in this chapter. *Entangled atomic clocks*, in quantum-entangled states (an instantaneous interaction between two

FIG. 2.5 *(a) NIST and the University of Colorado at Boulder teamed up to set the record for the most accurate operational atomic clock, based on thousands of strontium atoms trapped in grids formed by lasers. (b) An ultracold plasma of 26,000 Be ions fluoresce when excited by a laser. (c) The newest and most accurate atomic clock uses blue laser light to cool and trap strontium atoms in a vacuum chamber.*

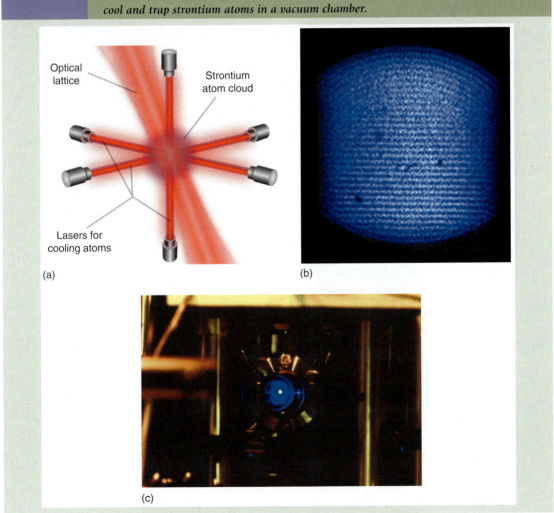

(a)

(b)

(c)

(continued)

**FIG. 2.5
(CONTD.)**

(d) The cesium atomic fountain clock mechanism is displayed. (e) Aluminum and beryllium ions are trapped in this very small device—an extremely small atomic clock that could become the most accurate clock yet. (f) The NIST-chip-scale atomic clock includes a laser, a lens, an optical attenuator (that reduces laser power), a waveplate (that alters polarization), a cesium vapor cell, and a photodiode. (g) Ytterbium atoms are trapped within pancake-shaped wells. A yellow laser provides excitation between lower and higher energy levels.

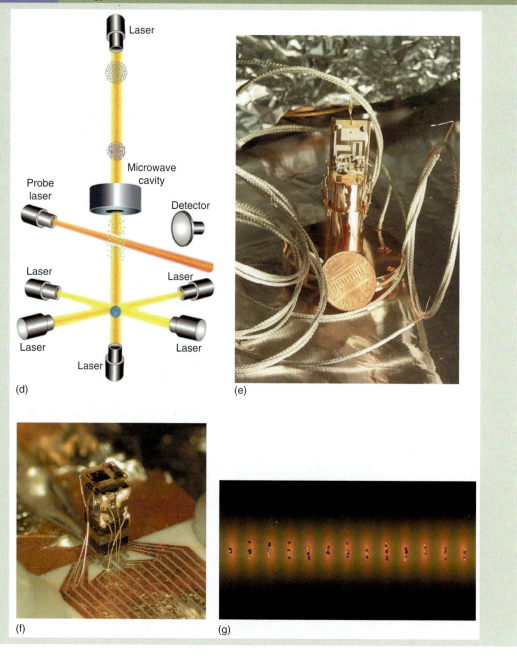

(d)

(e)

(f)

(g)

FIG. 2.5 (CONTD.) *(h) An electromagnetic trap encloses one mercury ion. The optical frequency of the ion is the basis for this small atomic clock. Mercury ion clocks are the most accurate—losing 1 s every 400 million years. (i) Two magnetic coils and an optical lattice (the red laser beam) serve to trap ytterbium atoms. The blue-violet laser beams cool the clock's atomic timepiece and thereby slow its motion.*

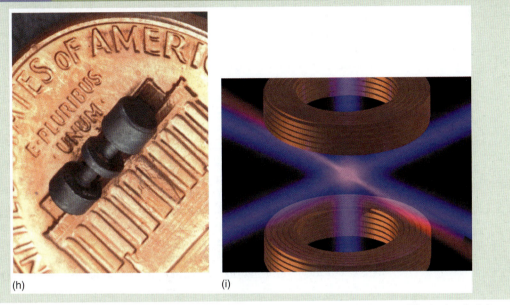

(h)   (i)

Source: (a), (b), (d–g) and (i): National Institute of Standards and Technology; (c) M. Boyd and T. Ido, JILA (Joint Institute for Laboratory Astrophysics, NIST and University of Colorado at Boulder); and (h), J. Berquist and D. Wineland, NIST. With permission.

particles regardless of distance) from a suitably designed pair of atoms, should be able to enhance clock measurements in the presence of background noise [67].

The development of an entangled atomic clock would be accurate and precise enough to test the value of the 'Fine Structure Constant', one of the fundamental constants of physics.

The Sommerfeld fine-structure constant is a measure of the strength of electromagnetic forces that govern the interaction between electrons and photons and was first used to explain the splitting or "fine structure" of the energy levels of the hydrogen atom [64,65]. It is a dimensionless ratio that involves four physical constants: Planck's constant, the elementary charge (the square of), vacuum permittivity (or vacuum permeability), and the speed of light.

$$\alpha = \frac{e^2}{\hbar c 4\pi\varepsilon_o} = \frac{\mu_o c e^2}{2h} = \frac{1}{137.035\,999\,07(09)} \tag{2.13}$$

The measurement has a precision of 0.70 ppb [68] and can be determined directly from quantum Hall effect measurements, the anomalous magnetic moment of the electron, or by direct substitution of constants [69].

Most recently, a Sr (strontium) lattice clock at 1×10^{-16} fractional uncertainty was reported by A. D. Ludlow et al., using a calcium (Ca) clock for remote optical evaluation. (*Science*, 319, 1805–1808 (2008)). The researchers evaluated clock performance comparisons over km-scale distances. They found that such lattice-confined neutral atom clocks outperformed the best efforts of Cs-based atomic clock primary standards [70]. The clock is accurate to within one second in every 200 million years. The current Cs standard clock is accurate to 1 s in every 80 million years or so [71]. The prototype mercury single-ion clock shown in **Figure 2.5h** is accurate to 1 s in 400 million years.

Optical frequency combs are accurate and precise tools that measure frequencies. Beginning with lasers that emit a continuous tandem of short pulses (fs) that consist of millions of colors, the characteristics of light are then converted into frequency numbers (lines) that resemble a comb in appearance. The timing between the pulses ordains the spacing between the teeth of the comb. According to NIST, the teeth of the comb can be used to measure light emitted by lasers, atoms, stars, and other sources with high precision. Why do we bring up optical combs in our discussion of atomic clocks? Such combs greatly simplify frequency metrology while simultaneously improving accuracy. New kinds of atomic clocks are expected with 100x the accuracy of Cs clocks ($f = 9 \times 10^9$ Hz, in the microwave domain) as a result of this technology. Obviously "ticks of a clock" that correspond to optical frequencies are expected to be more accurate—oscillations at frequencies approaching 500×10^9 Hz! Nothing exists at the time that is capable of counting such a high number of oscillations directly; therefore, optical frequencies must be stepped down via gears to the microwave region. For more information concerning this latest development of nanometrology, please consult www.nist.gov/public_affairs/newsfromnist_frequency_combs.htm.

2.2.2   *The Quantum Triangle*

The age of atomic resolution and manipulation is well developed, but another age is emerging rapidly and impacting metrology and other fields of science and technology—single-electron detection and manipulation. Are we able to manipulate a single electron to do our bidding? Are we able to see a single electron? Are we able to measure the charge of a single electron?

In 1879, Edwin Hall, at the age of 24, discovered that a transverse electrical field is formed in a solid if placed in a magnetic field that is perpendicular to the flowing current. He placed a thin film of gold connected to a battery in a strong magnetic field. He found that a small but measurable voltage, V_H, proportional to the magnetic field times the current was generated across the film, orthogonal to the magnetic field and the applied current (**Fig. 2.6**). It's as if the electrons behaved like a "cloud of mosquitoes in a cross-wind, in this case a magnetic wind that pushes the electrons towards one edge" [72]. The Hall potential arises in the direction of the wind [72].

The QHE is obviously the quantized version of the continuous bulk form of the Hall effect that relates Planck's constant h and the elementary charge e. Nobel laureate Klaus von Klitzing discovered in 1980 that two-dimensional electronic systems placed in a strong magnetic field at low temperatures showed plateau (rather than continuous) behavior as a function of the number of electrons. He was awarded the Nobel Prize in Physics in 1985 for his contribution. He showed that the quantized steps of the Hall effect can be expressed with high

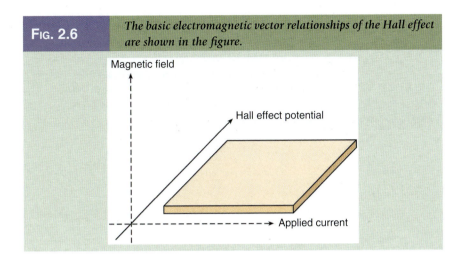

FIG. 2.6 *The basic electromagnetic vector relationships of the Hall effect are shown in the figure.*

precision as integral steps proportional to the ratio between two fundamental physical constants: h and e. The von Klitzing constant R_K has become the new standard unit for electrical resistance.

Metrology, the Volt, and the Ohm. Impedance, temperature, strain, electrical, and power standards are mostly based on resistance standards, for example, the ohm [73,74]. Efforts are underway at NIST's AML (Advanced Measurement Laboratory) to redefine the legal ohm with quantum Hall resistance (QHR) measurements. With the use of cryogenic current comparators (CCC), direct scaling of the QHR up to $1\,M\Omega$ was possible in 2002. This level has since been amplified in 2004. High-ratio scaling is complemented by efforts to develop thin-film resistance standards for the *quantum metrology triangle* (QMT) (**Fig. 2.7**) [73,74]. The QMT relates voltage (V), current (I), and frequency (f) via a triangle consisting of the Josephson junction, quantum Hall, and single-electron tunneling effects (SET) that are important for nanometrologically precise electrical measurements [75]. Verification of the value of the elementary charge e and the fine structure constant α is possible by QMT experiments and calculations.

The challenge facing QMT experiments is to test the aforementioned Josephson voltage standard, the QHR, and the SET. Testing needs to be done in an extremely quiet high-ratio CCC combined with orders-of-magnitude larger SET sources. The ultimate goal is to produce useful levels of quantum current.

Single-electron metrology in the form of the single-electron electrometer has already become a viable technology [63]. The definition of single-electronics implies the controlled movement, positioning, and measurement of a single electron [63]. The potential effect of a single electron must not be understated, for example, a metallic nanosphere with radius 1 nm charged with one electron (charge, q or e = the elementary charge, 1.6022×10^{-19} C) produces an electric field on its surface in vacuum of $1.4\,GV \cdot m^{-1}$ [76]. This is enough to repel other electric charges. Please note that isolation of a single electron has not been accomplished—only that a single electron has been added to a preexisting electronic system. The "single-electron box" is a device that is capable of exhibiting single-electron charging effects. Consisting basically of a nanometal grain, a

FIG. 2.7

The QMT: the relationships between frequency f, voltage V, and current I are functions of physical constants h and e, Planck's constant, and the value of the elementary charge, respectively. In this way, the current measure can be derived from measurement of frequency and not necessarily the kilogram (SI units) [63]. The current standard of the ampere is derived from the Josephson array voltage, the QHR, and Ohm's law [63].

tunnel junction (a thin oxide coating) and a bias voltage energy source. Electrons can be added over a threshold voltage ($V_{\text{threshold}} = e/C$) to the particle according to the Coulombic energy relation

$$E_C = \frac{e^2}{2C} > k_B T \tag{2.14}$$

where C is the capacitance of the junction. Obviously the Coulombic energy must be greater than the ambient thermal energy $k_B T$ in order for quantum effects to be noticed [63].

As the bias voltage is raised, the electronic box can be filled with a predetermined number of electrons—resulting in a Coulomb blockade current (or charge) versus voltage behavior. Electrons can only be added or subtracted from a conducting island by tunneling—each time modifying the island potential by $\pm E_C$. Current is therefore inhibited (blocked) and is a classical case of Coulombic repulsion of charge carriers. Stepwise episodes exceeding the Coulombic blockade potential results in Coulomb staircase behavior. The current, therefore, through a tunneling junction is quantized. If a tunnel junction is under conditions of constant current I, charge accumulates on the tunnel junction similar to a capacitor and then single-electron tunneling oscillations appear with frequency

$$f = \frac{I}{e} \tag{2.15}$$

In superconductors, these are Cooper pairs that form Bloch oscillations.

$$f = \frac{I}{2e} \tag{2.16}$$

B. Josephson of Cambridge predicted that electrons should be able to tunnel in pairs, for example, Cooper pairs form between two superconductors separated by an insulating but very thin tunneling junction. A Cooper pair is a pair of electrons that can be coupled via lattice vibrations (phonons) over distances of hundreds of nanometers in superconductors (also condensation of electrons into a state that forms a bandgap). Are Cooper pairs an example of entangled electrons?

If DC voltage is applied to this junction, an AC current of frequency f is generated equal to $f = 2eV/h$. If an applied current of frequency f is applied, then a DC voltage equal to V_n would be generated: $V_n = nhf/2e$. The voltage measured across a Josephson junction V_J that is irradiated by a microwave EM field with frequency f_J is

$$V_J = n \frac{h}{2e} f_J \tag{2.17}$$

where n is the integral voltage step number [76].

The current generated by an electron pump is proportional to the charge q (an integral multiple of e)

$$I = qf \tag{2.18}$$

which is driven by an external frequency source (e.g., the clock frequency of the pump). Experiments have demonstrated accuracy of 1.5×10^{-8} at pA current levels [27]. From QHR and that $I = ef$, the Hall voltage is

$$V_H(i) = \left(\frac{R_K}{i} \right) ef \tag{2.19}$$

where R_K is the QHR (the von Klitzing constant).

The accuracy and precision of single-electron boxes depend on several factors [63]: (1) the Coulomb energy must exceed the ambient thermal energy as stated above, (2) capacitance C must be smaller than 12 aF to observe charging effects at liquid nitrogen temperatures and 3 aF at 300 K, (3) this requires that grain size be smaller than 15 and 5 nm, respectively, (4) for logic operations, the operating temperature has to be a factor of 50x lower and granules smaller than 1 nm, (5) quantum fluctuations must be negligible (number of electrons localized on any single island with no electron sharing between islands), and (6) all tunnel junctions must be opaque to electrons, for example, confined at a level of tunnel resistance [63]

$$R_K = \frac{h}{e^2} = 25812.807 \, \Omega \tag{2.20}$$

where R_K is the integer QHE standard resistance that links voltage to current. Notice that the form of the QHE is similar to Ohm's law: $V = \Omega I$. The voltage standard (Josephson constant) is defined as

$$K_J = 483597.9 \, \text{GHz} \cdot \text{V}^{-1} \tag{2.21}$$

The Josephson "superconductor–insulator–superconductor" junction when irradiated with microwave energy generates a DC potential determined by the frequency of the microwave radiation, the value of the elementary electronic charge, and Planck's constant. It is essentially a frequency-to-DC voltage converter [77,78].

$$e\left(R_K K_J\right) = in\left(\frac{f_J}{f}\right) \tag{2.22}$$

$$e\left(R_K K_J\right) \rightarrow a\left(\frac{f_J}{f}\right) = 2 \tag{2.23}$$

where f is the frequency of the SET pump cycle, f_J is the frequency of the microwave, a is a constant that represents a ratio of experimental integers such as the Josephson step number n, the Hall plateau number i, and the number of electrons pumped per cycle [77]. The dimensionless value is known to a combined relative uncertainty of 7.8×10^{-8} [77].

Hall resistance of semiconductors is measured in a magnetic field in which current flows in the x direction and the Hall voltage is measured in the y direction. The ratio of the two is the Hall resistance [63]. The AC Josephson effect implies that voltage applied across a Josephson junction (a superconducting tunnel junction separated by a thin barrier) produces an AC current with frequency proportional to $f = 2eV/h$. Conversely, irradiation of a Josephson junction with electromagnetic sources produces a voltage $V = hf/2e$ across the junction [63].

The accuracy of the QHR has been achieved to a few parts per 10^{10} and is established as a universal quantity DC standard for resistance that is independent of host material or device [76]. The reproducibility of the QHR is two orders of magnitude better than that of the ohm based on SI standards [76]. The von Klitzing constant R_K and the Josephson constant K_J have contributed significantly to a new age of electrical measurement, one quite well suited for nanoscale research and manufacturing. Work is underway to establish a primary AC resistance standard based on QHE [76].

2.2.3 The Single-Electron Transistor

A device that is able to combine atomic-scale resolution with ultrasensitive levels of charge detection is certainly desirable. Quantum mechanical tunneling and the Coulomb blockade are two fundamental phenomena responsible for the action of a SET [79]. The device is sensitive to nearby sources of electric fields, for example, electrons. N.B. Zhitenev and T.A. Fulton of Lucent Technologies in New Jersey have developed a scanning SET microscope that is capable of atomic resolution and detecting charge smaller than 0.001 e [79]. The limit of 0.001 e is the state-of-the-art value obtained at temperatures less than 4 K. The SETs of Zhitnenev et al. are based on the metal–metal tunnel junction developed by I. Giaever in 1960—two metal superconductors separated by a thin tunneling junction a few nanometers in thickness [79,80]. The device is shown in **Figure 2.8** below.

Recently, physicists in Finland and at Stony Brook University, New York, Jukka Pekola and Dmitri Averin, respectively, constructed a SET nanodevice that

FIG. 2.8

N.B. Zhitenev and T.A. Fulton explain the operation of the SET [79]. The input electrode is at a preset bias voltage. The output electrode is connected to the ground potential. The conducting island is insulated with two separate layers—thin films that serve as tunneling junctions. Current I flows from the input electrode to the output electrode. The gate with bias voltage V_G has capacitance C_G and is connected to the conducting island. Current is dependent on V, V_G, resistance R_J, and capacitance C_J of the junctions. On the voltage scale of e/C ($C = 2C_J + C_G$), the dependence of I is nonlinear. The current is a periodic function of V_G oscillation (as e/C_G) at constant V. A Coulomb blockade suppresses current over the range of V and VG at low V. Higher V demonstrates a dependence of current on $2R(V − e/2C)$.

Source: N. B. Zhitenev and T. A. Fulton, Scanning single-electron transistor microscopy, In Encyclopedia of nanoscience and nanotechnology, Vol. 5, J. A. Schwarz, C. I. Contescu, and K. Putyera, eds., Marcel–Dekker (2004), Taylor & Francis (2005). With permission.

is able to measure current one electron at a time [81]. The device, cooled to 0.1 K, consists of a conducting island connected to two tunnel junctions—electrons flow in through one tunnel junction and out the other tunnel junction. A constant voltage was applied across the island junctions and an oscillating voltage applied to the gate electrode, similar to the scenario described above, to control the flow of electrons [82]. One electron is able to tunnel through the thin insulating junctions. Repulsion between electrons prevents more than one electron from tunneling through the junction at any one time [82]. Measurement of the amplitude and the mean value of the gate voltage allowed investigators to determine the current that flowed through the transistor per every cycle of oscillation

$$I = eN_e f \tag{2.24}$$

or the electron charge e times the number of electrons N_e times the frequency f of the gate voltage. The device has proven to be extremely precise but the accuracy needs to be improved—uncertainty is on the order of 10^{-2}. Accuracy could be improved, according to Pekola et al., if 10 such devices were placed in parallel. Such a configuration would produce 100 pA of current—enough to measure with greater accuracy [82]. This component would contribute to a complete "QMT" in which the current, voltage, and resistance are all related via the fundamental physical constants as described above, for example, Planck's constant h and the elementary charge e.

2.3   Nanometrology Tools

Pure nano analytical probe methods consist of electron beams and scanning microscopes. Both are capable of sub-nanometer resolution. Resolution of electron beam probes depends on the wavelength (or energy), the spot size of the electron beam and the sensitivity of the detector(s). Resolution of scanning probe microscope methods depends on the sharpness of the tip, the mechanical properties of the cantilever, the capability of the piezo-scanner, and the sensitivity of the input electronics and detector. Both methods offered the first glimpses into the nanoworld as they helped launch the *Nano Age*.

A new generation of single-atom/single-molecule metrology is upon us. The AFM, STM, SEM, TEM, and SERS are just some of the analytical (and hence, metrological) tools available for us to probe the nanoworld.

2.3.1   *Electron Beam and Atomic Force Tools*

We restrict our discussion in this section to electron beam and atomic force methods.

Electron Microscope Metrology.   Scanning (SEM) and transmission (TEM) and scanning transmission (STM) electron microscopy are utilized in particle (or facet, hole) size determination and morphological characterization. Both techniques rely of the wavelength of the electron beam as the critical factor responsible for ultimate resolution. The wavelength of the beam for either technique is dependent on the de Broglie equation that we know so well by now

$$p = \frac{h\nu}{c} = \frac{h}{\lambda} \qquad (2.25)$$

where

  p is momentum
  h is Planck's constant
  λ is the wavelength
  c is the speed of light
  ν is the frequency

The classical form of the wavelength from energy is given by

$$\lambda = \frac{h}{\sqrt{2m_e e\mathrm{V}}} \qquad (2.26)$$

and for electron diffraction

$$n\lambda = 2d\sin\theta \qquad (2.27)$$

where d is the interplanar distance.

Nanometrological inspection of nanocrystallites, nanotubes, quantum wells, and other sub-micrometer devices is often accomplished by SEM [83]. It is generally known that the size of nanometer structures appears larger and the distance between nanometer structures appears to be less than its real dimensions—when compared to TEM results [83]. The cause of the enlargement is linked to the radial distribution of secondary signal electrons, the effective resolution of the SEM, and the probe diameter d. Critical dimension (CD) SEM precision has evolved to 0.3 nm for the 45 nm technology node [84]. Recall that SEM images are two-dimensional projections of three-dimensional surfaces; therefore, it is difficult to provide accurate 3-D profile data, especially over large areas [85,86]. For this reason, SPMs (scanning proble methods) have greater potential than optical microscopes, SEMs, and surface roughness measure instruments for nanometrology of 3-D nanostructured surfaces [85,86].

SEM is one of the workhorses of the $200 billion semiconductor industry. The ITRS (International Technology Roadmap for Semiconductors) states that SEM will continue to provide at-line and inline imaging for characterization of cross-sectional samples, particle and defect analysis, and CD measurements [87]. Customers require effective CD (critical dimension) and defect review to transcend the 45-nm threshold. The semiconductor industry requires that SEM standard artifacts related to magnification, calibration, performance, and measurement of linewidth be addressed [87].

Scanning Probe Microscope Metrology. Nanoscale metrology with AFM is correlated with a unique set of issues. Errors such as overestimation of the size of positive features (e.g., facets) and underestimation of the size of negative features (e.g., pores) are a direct byproduct of the sharpness of the tip, contact angle with the surface, and the sensitivity and control of the force application system. There are four underlying sources of image artifacts in AFM methods: probes, scanners, image processing, and vibrations. AFM images, for example, are the result of convolution of the object of interest and the geometry of the probe [88]. A general rule of thumb is that if the probe is much smaller than the feature on the surface, probe-induced artifacts are minimized and the image is reproduced with good accuracy. For example, if a 100-nm feature is imaged with a 10-nm diameter probe, a good image should be acquired with a minimal number of artifacts [88]. On the other hand, if the probe tip is bigger than a protruding feature, images appear larger than the object are recorded (**Fig. 2.9**). On the flip side, if imaging of pores (holes) is desired, the pore diameter appears smaller if a larger probe tip is applied. Regardless, in most cases, although the depth profile may be inaccurate with a large probe, the pore diameter and the order in a pattern may still be accurately imaged. The shape of objects may be distorted as well with large probes. With some larger silicon probe tips, artifactual triangular patterns represent the geometry of the tip rather than the geometry of the feature [88].

Artifacts can be induced by other factors associated with AFM image acquisition and processing. Nonlinear motion, hysteresis, and the geometry of the ceramic piezoelectric scanner may distort images. The major sources of error in large-area measurements are due to the angular motions of the scanning actuator/stages [85,86]. Probe-sample angles less than nonperpendicular (or close to 12° off of normal) can be corrected in some AFMs. The artifact produced in this case is an image that may appear smaller on one side than the other. "Edge overshoot," in which one side of the feature is distorted, can be caused by hysteresis. Scanner drift, in which distortion is observed at initial contact with a feature, is caused

FIG. 2.9

Line profiles, top, are determined for the most part by the geometry of the sample, but the overall image quality relies heavily on the geometry of the probe tip as well. Top right: Large probe tips with relatively low tip curvature exaggerate the size of protruding images (positive features). Top left: Conversely, a sharper tip (smaller the diameter or more extensive tip curvature) renders a better and more accurate image. Nanotubes or quantum dots often look larger than expected; however, the height of the nanomaterials is correct when considering the line profile. Bottom right: With pores (negative features), the size (diameter) is often smaller than it is actually. The line profile is determined by the geometry of the probe and not as much by the geometry of the sample [88]. Bottom left: Although sharp-tipped thin probes are better for depth profiling, all types are able to give decent diameter data and an outlay of repeating patterns [88].

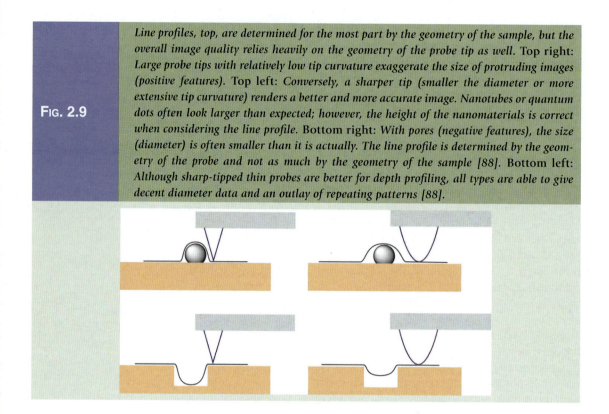

by thermal effects induced by external sources. Most of these kinds of distortions can be detected by imaging a calibration standard.

Image-processing artifacts are more rare. Too much application of low pass filters may cause steps in features to be distorted (too much smoothing)—too little may result in sharpness at steps. Synthetic periodic structures that appear to be the atomic structure may also be introduced by the use of Fourier filtering. Vibrations absolutely cannot be tolerated during image acquisition of atomic structure (or in actuality, any microscopic structure). A building floor acts as a "transducer" for many sources of vibrations, both external and internal. Acoustic vibrations (e.g., a voice) are not desired, especially during image capture. Other sources of AFM artifacts arise from surface contamination, faulty electronics, and vacuum leaks. An excellent article by Paul West and Natalia Starostina [88] of Pacific Nanotechnologies, Inc. (LOT-Oriel Gruppe Europa) can be found at www.lot-oriel.com/site/site_down/pn_artifacts_deen.pdf.

Analysis by scanning probe (SPM) techniques such as atomic force (AFM), chemical force (CFM), friction force (FFM), and atomic force acoustic (AFAM) microscopies rely on substrate surface chemistry (surface energy) for contrast. A new nanometrology, developed by the NIST in Gaithersburg, Maryland, is based on gradient micropatterned ($\nabla\mu p$) substrates intended to serve as reference substrates for SPM calibration [89]. The microscale chemical pattern-surface energy gradients provide a means of calibrating image contrast against traditional chemical measures such as contact-angle determination and surface spectroscopy [89]. The patterns were formed from a combined microcontact printing/vapor deposition of monochlorosilane self-assembled monolayer on silicon substrates [89].

Veeco Instruments, Inc. has recently introduced a new high-throughput AFM designed for quality assurance testing of semiconductor wafers [90]. The *InSight 3D Atomic Force Microscope* platform is capable of accuracy and precision for nondestructive, high-resolution three-dimensional measurements of 45- and 32-nm semiconductor features. Approximately 30 wafers can be processed per hour. According to Veeco, the new instrument is able to "*address critical dimension (mid-CD precision = 1.5 nm or 1% @ 3σ), depth (1 nm or 1% 3σ) and chemical mechanical planarization (CMP) metrology in a production environment.*" The *InSight 3D AFM* is designed to compete with other on-line nanometrology techniques. "*CD-SEM and scatterometry are precise methods but not accurate enough, causing significant measurement issues. Veeco's InSight provides the lowest measurement uncertainty for CD metrology, which leads to improved process control.*" The new instrument provides twice the accuracy of previous AFMs. Improvements in the X–Y stage, auto-focus laser pattern recognition system, new probe designs, and system reliability are responsible for the enhanced capability [90].

Atomic resolution is required in scanning probe devices. The ability to "see" single electrons and even smaller levels of charge is also important. The combination of a sharply refined scanning probe and a SET should be a powerful combination that enables us to achieve atomic resolution and minimal levels of charge [79].

2.3.2 Spectroscopic Tools

Single-Molecule Measurement and Imaging. Single-molecule measurement and imaging has acquired significant importance in the area of biomolecular study. In particular, the study of enzyme action offers a gateway into understanding the mechanisms of nanomachines [91]. Since the early 2000s, new methods have evolved to detect single biomolecules. For instance, study of biomolecules by using a dielectric bead with an attached biomolecule trapped by optical tweezers has been demonstrated recently [92–94].

Two new methods have gained prominence: super-resolution imaging using single-molecule fluorescence and single-molecule trapping by Brownian motion suppression [91]. Insights into analysis of single-molecule spectra have also made progress in the past few years [95].

Single-molecule spectra are easy to obtain in solid samples. However, to study biological molecules in situ is challenging due to Brownian motion effects exhibited at room temperature, for example, the diffusion of a 10-nm nanoparticle through a 0.3-µm laser spot area (or 100 fL) occurs in less than 1 ms [91]. Various means have been devised to circumvent this difficulty: Immobilization on a transparent glass surface, encapsulation within the aqueous environment of a gel pore or a vesicle, or application of the tethered bead-laser tweezer method mentioned earlier [91]. The ABEL (anti-Brownian electrokinetic) trap is able to constrain single molecules with nanoscale resolution [96]. The trap is able to monitor the Brownian motion of the particle by fluorescence microscopy, and then by applying a feedback voltage to control the electrical drift the Brownian motion is cancelled. The trap works well on nanoscale materials (molecules) that are able to provide an optical image and an electric charge [96]. A.E. Cohen et al. in 2005 have shown that the ABEL trap is capable of trapping fluorescent polystyrene nanospheres with diameter as little as 20 nm [96]. According to the authors, the ABEL trap can be applied to single-molecule spectroscopy, nano-manufacturing, and the detection of biological particles [96].

Single-molecule detection procedures have been applied to the analysis of bacterial neurotoxin (a protease) translocation across membranes [97]. In their 2007 paper, A. Fischer and M. Montal reported the analysis of the dynamics of *botulinum* neurotoxins by single-molecule assay with millisecond resolution. From this they were able to determine the mechanism of transport of the toxin across cell membranes. Botulin neurotoxins are composed of two protein chains linked by a disulfide bond: a 50-kDa light chain (LC, the cargo) and a 100-kDa heavy chain (HC, the channel/chaperone). They concluded that the neurotoxin transit occurred according to the following scheme: an insertion step (with LC unfolding), entry into the HC channel, conduction of the LC through the HC channel, release of the LC cargo (reduction of the disulfide bridge), and LC refolding into the cytosol [97]. The processes were monitored by way of single-channel conduction currents.

Short Wavelength Sources.   Short wavelength sources include x-rays and free-electron lasers (FEL) that are able to produce short coherent photon pulses. The underlying micro- and nanostructure of materials plays an important role in the nano-, micro-, and macroscopic properties of novel materials that can be probed by short-wavelength techniques [98]. Characterization of such nanostructures is therefore a prerequisite to engineering design of devices, for example, that consider the degree of micro- and nanostructural properties like strain and stress, crystalline defects, and texture [98]. X-rays are sensitive to structural features on the scale of the chemical bond. In addition to standard static x-ray diffraction, structural changes triggered by infrared radiation can be measured by short x-ray pulses [99].

Synchrotrons produce stable and intense x-ray beams with low divergence. Picosecond pulses (10–400 ps) make this source suitable for the study of dynamic behavior with nanosecond time resolution [99]. Time-resolved x-ray diffraction is an excellent technique with which to measure phase transitions in nanomaterials. However analysis of living systems by synchrotron-based x-ray microscopes is limited to 20 nm due to the onset of radiation damage. Application of higher energy pulses is a means to overcome this limitation. Single-particle diffraction and flash imaging using wavelengths from 20 nm down to 0.2 nm allows for single-molecule resolution by x-rays [99]. To image noncrystalline samples, the photon pulses need to be shorter than the time it takes to damage the molecule (e.g., <100 fs and 10^{11} photons per 100-nm spot) [99]. Interest in solving the structure of noncrystallizable proteins is one of the driving forces behind this technique.

Larger structures require longer wavelength sources. Free-electron lasers produce photons that overlap with the soft-x-ray domain. The wavelength of such devices has been decreased from 32 nm down to 13 nm. By means of <6-nm wavelength sources, single-particle imaging by x-rays are on the horizon.

Scanning Nano-Raman Spectroscopy (SNRS).   SNRS, also called tip-enhanced Raman spectroscopy (TERS), has been shown to be an effective means of analyzing nanoscale silicon-based structures [100]. SNRS or TERS with side-illumination optics have been shown to optimize near-field and far-field (background) Raman signals by optimizing the beam polarization [100]. The breakthrough yielded signals with an order of magnitude improvement in contrast with a lateral resolution of Raman images of ca. 20 nm [100].

Surface-Enhanced Raman Spectroscopy (SERS). An advanced form of Raman spectroscopy (SERS), (or SERRS—surface enhanced resonance Raman spectroscopy or scattering) is well known for attomole (10^{-18}) or better levels of detection with improvements since the discovery of the surface enhancement phenomomen in 1974 [102]. SERRS relies on the coincident overlap of the chromophore transition with the excitation frequency. The SERRS system includes silver nanoparticles upon which the analyte is able to adsorb the appropriate excitation energy [103]. From configurations like these, single-molecule detection has been achieved using SERRS [104,105]. According to A. Macaskill et al., SERRS is a more sensitive and specific detection method for the direct analysis of DNA compared to fluorescence measurements because SERRS is able to produce fingerprint spectra. This renders the technique better suited for multiplexed identification of target molecules without the need for separation. [103]. With fluorescence techniques, broad and overlapping spectra often limit its usefulness in that regard [103]. DNA by itself does not function as a classical chromophore, but attachment of an appropriate dye molecule (a fluorescent label) renders it so.

Single-Molecule (Atom) Spectroscopy and Detection. Single-molecule (atom) detection is easily obtainable via SPM methods. On the surface, it would seem that this goal would be more difficult to accomplish via spectroscopic methods, but, nonetheless, this too has also been achieved. Single-molecule spectroscopy and single-molecule detection provide important data especially when coupled with other scanning, imaging, and spectroscopic analysis [106]. Researchers in Genoa, Italy have managed to combine a commercial confocal (or multi-photon) scanning head, a mode-locked titanium–sapphire laser, and an inverted microscope to detect single-molecule fluorescence of indo-1, rhodamine-6G, fluorescein, and pyrene—the result of the ability to selectively collect signals from highly confined volumes [106]. Dilute solutions of the dyes were applied to glass slides by the process of spin coating.

Scatterometry. CD and other profile properties can be obtained from periodic structures by a procedure known as scatterometry. Scatterometry analyzes the light (usually extreme ultraviolet, $\lambda = 13.5$ nm) diffracted from such structures. According to the 2004 ITRS, 3-σ metrology of at least 0.2 nm is required for the 32-nm nodes by 2013. Currently, 0.7-nm precision is available for printed and physical isolated lines of 90-nm dimension [84].

Scatterometry is based on the analysis of collimated light reflected from a grating (or structured) surface. The resulting diffraction pattern relies on the wavelength of the incident light and the period (structure) of the features on a surface. For example, a perfectly flat surface would not produce a diffraction pattern of any order. Reflection depends on

$$R = \frac{I_{out}}{I_{in}} \qquad (2.28)$$

where I_{out} and I_{in} are the intensities before and after reflectance. The quantity R depends on the grating pitch (the period), geometry, refractive index, number of multilayers, and base substrate properties [84]. The factors that influence R from the perspective of the incident beam include wavelength, angle of incidence, polarization, and azimuth angle [84]. Obviously the reflectance profile will be complicated. Analysis is therefore accomplished by comparing data to a reference profile derived from Maxwell's equations [84]. For example, if the method is required to deliver a measurement resolution of 0.1 nm, then the method must be able to *differentiate between the fits of two-signature curves 0.1 nm apart in the line or CD—within the constraints of the signal-to-noise ratio* [84]. Counting bin numbers over all pairs of spectra generates uniqueness scores. Hopefully, in all cases, apples are being compared to apples.

Manufacturing process control factors include (1) targets fitting within a prescribed area that are a function of the effective spot size of the incident probe, (2) scribe CD structures are employed to track processing shifts, and (3) characterization is required of two-dimensional and three-dimensional structures by providing grating average values for CD, sidewall profile, thickness, and optical properties of the stack [84]. The industry uses both angular- and wavelength-dependent scatterometry. Issues include (1) that the process works best with periodic structures, (2) relies on accurate values of optical properties of materials, (3) standards are needed (as indicated above) to validate the process, and (4) may become limited when analyzing structures with ever smaller dimensions [107].

2.3.3 *Nanomechanical Tools*

Challenges of Nanomechanical Metrology. Nanomechanical methods face, according to the NNI report on instrumentation and metrology [18], numerous challenges and barriers. It does seem like everything is a challenge in nanometrology, but the status quo represents, without question, the dynamic state of flux that nanometrology is undergoing and the importance of nanotechnology. **Table 2.1** lists an edited review of challenges and barriers facing nanomechanical metrology [18].

Mechanical behavior for many small, synthetic and biological systems is not well known. In order to get the most out of mechanical devices (e.g., performance and reliability), mechanical property measurements and testing need to be conducted. Properties such as localized strength and fracture toughness of materials and interfaces and the effect of surface phenomena need to be studied [108]. Depth-sensing nanoindenters have application as a universal testing machine. The Manufacturing Engineering Laboratory of NIST has devised a compression load test configuration using a nanoindenter that yields information on applied load and load-point displacement [108]. Such testing coupled with finite element analysis (FEM) should be able to mitigate the need for mechanical information about nanomaterials.

Nanoindentation. Quantitative nanoindentation methods are the most versatile mechanical testing systems applied to nanomaterials and nanodevices, along with AFMs. There are over 1000 commercial nanoindentation instruments available [109]. The nanoindentation technique yields information about hardness

TABLE 2.1	*Challenges Facing Nanomechanical Metrology*
Metrology	**Challenges**
Standards and calibration	Traceability of force and displacement, international collaboration on standards, nanoscale forces, and contact mechanics
Modeling	Computational power, data storage, and mining
Experimentation	Testing fixtures, sample development, positioning, and manipulation
Multiple technique integration	Atomic/molecular-scale resolution, single-event spatial resolution, time and position information synchronization
Automated measurements with high throughput	Speed, automation, yield, quality, size, and conditioning Robust probes, periodic reference standards Lack of wide range of testing environments (T, *f*), lack of models, high-speed methodologies, and well-characterized nanoscale probes
New instruments	Tip wear, control, cm to nm positioning, decoupled lateral and vertical force sensors and need for standards, quantitative mechanical property mapping
Real-time measurements	Contact area, surface treatment, robustness, real-time measurement, interfacial testing, thermal drift

and elastic properties from force, area, and displacement data acquired from indentations on the order of 5–10 nm [109]. However, worldwide, there exists a dearth of force calibration, standardized methods, and standard reference materials. Displacement is easily measured from interferometry but force measurements pose more problems due to reliance on artifactual SI standards (e.g., the kg mass). NIST is working on ways to trace force measurements to electronic and length SI units in the range between 1 mN and 10 nN [109].

Nanotribology and Surface Properties.　There is a need to establish standards and methods to measure nanoscale adhesion, friction, and surface forces and characterization of surface properties (e.g., texture) at the nanoscale [110]. Normal and lateral forces of nanocontacts in particular need to be characterized at the nanonewton level. Understanding how meniscus and electrostatic forces affect friction measurements is an ongoing concern. Thin-film stress measurements are important to determine the robustness of electrodeposited thin films [111]. Mechanical stresses develop during nucleation and growth in such films and lead to loss of adhesion and the generation of bulk surface defects. Cantilever-based devices have been devised that are able to measure and resolve forces on the order of $0.03\,\text{N}\cdot\text{m}^{-1}$ of films as they form. For example, profiles of stress formation during the deposition of 250 nm of Au on borosilicate glass revealed that there is a rapid rise in tensile stress during the first 20 nm of deposition [111]. It was also determined that the highest tensile stresses correlate with high nucleation densities obtained at more negative electrodeposition potentials.

Adhesion.　Adhesion and mechanical properties characterization from axis-symmetric testing is a promising method to quantify adhesive performance while studying debonding mechanisms simultaneously [112]. Traditional "one-per" tests (serial testing) do not offer enough throughput, and as a result, combinatorial methods have become more attractive. The Johnson–Kendall–Roberts method (JKR) utilizes a hemispherical lens that is pressed into a sample. By monitoring the contact area between the lens and the sample, it is possible to describe the work of adhesion [112]. The Multilens Combinatorial Adhesion Tester (MCAT), equipped with an array of microlenses, is able to track the contact radii of 400 specimens simultaneously.

Peel tests, one of the primary methods applied by industry to evaluate adhesives, is a method that also suffers from the limitations of serial testing. Combinatorial and other high-throughput methods have also been applied to accelerate the rate of adhesive performance testing [113]. A thin polymer applied to a substrate is characterized by a surface energy gradient and simultaneously, an orthogonal temperature gradient. Failure maps can be constructed that reveal the transition from adhesion to failure as a function of surface energy, annealing time, and annealing temperature [113].

Computer Modeling of Nanomechanical Behavior.   Nanoscale mechanical failure is difficult to measure [114]. Finite element method (FEM), classical atomistic simulations, and density functional theory (DFT) offer means of addressing this dilemma. FEM is an excellent procedure with which to model elastic behavior of bulk materials but comes up short with regard to failure analysis because failure mechanisms depend on atomic scale behavior. Classical atomistic simulations are able to handle thousands to millions of atoms but fall short at the chemistry end of the spectrum. DFT is rather accurate and describes the chemistry accurately. Unfortunately, even in this day and age, central processing unit (CPU) time is at a premium and DFT models are not able to correlate more than a few hundred atoms simultaneously. Use of all three models is often required to understand mechanical behavior from the atomic to the macroscopic level [114]. The results of computer modeling techniques are correlated with experimental nanoindentation and AFM data to provide quantitative predictions of material behavior.

Roundness.   Roundness measurements, mentioned briefly earlier, determine the circularity of the contour of a 3-D object. Roundness is the quality or state of being round. Eccentricity is the deviation from roundness, but in general, out-of-roundness includes the presence of surface defects as well. Roundness has been traditionally evaluated according to NBSIR 79-1758, a document created but the National Bureau of Standards (now NIST). The world's highest accuracy for roundness measurements are obtained on a Mitutoyo RA-H5000 located in the AML of NIST in Gaithersberg, Maryland. Measuring uncertainty of 3 nm and long-term repeatability within 1 nm have been achieved with this instrument [115].

2.4 NANOMETROLOGY AND NANOMANUFACTURING STANDARDS

According Sarah Gale of *Small Times*, the magazine of the MEMS and nanoworlds, *"…the rush to develop standards for nanotechnology that the world will embrace is in full throttle"* as metrology and standards organizations worldwide are converging to establish standards for nanotechnology [116]. There are several key ingredients required to realize this ambitious goal including generalized components such as defining a common language, material dimensions and quality, and the foundation for nanomanufacturing practices [116].

2.4.1 Standards for Nanotechnology

Just like the NNI and the National Institutes of Health, other organizations have established plans of attack concerning the development of nanotechnology including several metrology organizations that are involved in the process of doing exactly that—creating a roadmap. Some of the most important standards organizations are listed in **Table 2.2**.

All of these organizations realize that global cooperation is imperative if universal standards regulating nanometrology, practices, and standards (and nanomanufacturing) are to play a viable role in the commercialization of nanotechnology. An excellent source of nanotechnology standards news can be found online at Nanotechnology Standards News (IHS Engineering), (http://engineers. ihs.com/news/topics/nanotechnology-standards-news.htm) [117].

The Institute of Electrical and Electronics Engineers (IEEE) has already released a standards document for nanotechnology and an overview of the roadmap to guide standardization efforts [118,119]. The IEEE roadmap reveals the complexity of developing standards for new materials, especially materials that exist at the nanoscale. The complexity and challenge of developing the IEEE roadmap can be seen by reference to the draft which was made available for the public in April, 2007 [120]. The roadmap identifies 25 key technologies, 74 standards opportunities, and 13 "high priority" technologies.

TABLE 2.2	Standards and Testing Organizations of Note		
Organization		**Acronym**	**Web Address**
American Institute of Chemical Engineers		AIChE	www.alche.org
American National Standards Institute		ANSI	www.ansi.org
American Society of Mechanical Engineers		ASME	www.asme.org
American Society for Testing and Materials International		ASTM	www.astm.org
British Standards Institute		BSI	www.bsi-global.com
Institute for Electrical and Electronics Engineers		IEEE	www.ieee.org
International Organization for Standards		ISO	www.iso.org
International Union of Pure and Applied Chemists		IUPAC	www.iupac.org
National Institute of Advanced Industrial Science and Technology		AIST	www.aist.go.jp
National Institute of Standards and Testing		NIST	www.nist.gov
National Sanitation Foundation International		NSF	www.nsf.org
Semiconductor Equipment and Materials International		SEMI	www.semi.org

TABLE 2.3	*Terminology for Nanotechnology (ASTM E 2456-06)*		
Term	**Definition**	**Term**	**Definition**
Agglomerate	A group of particles held together by relatively weak forces (e.g., van der Waals or capillary) that may break apart into smaller particles upon processing.	Nanotechnology	A term referring to a wide range of technologies that measure, manipulate, or incorporate materials, and/or features with at least one dimension between approximately 1–100 nm. Such applications exploit the properties, distinct from the bulk/macroscopic systems, of nanocomponents.
Aggregate	A discrete group of particles in which the various individual components are not easily broken apart, such as in the case of primary particles that are strongly bonded together (e.g., fused, sintered, or metallically bonded particles).	Nanostructured	Containing physically or chemically distinguishable components, at least one of which is nanoscale in one or more dimensions.
Nanoscale	Having one or more dimensions from approximately 1–100 nm	Non-transitive particle	A nanoparticle that does not exhibit size-related intensive properties
Nanoscience	The study of nanoscale materials, processes and phenomena, or devices.	Transitive nanoparticle	A nanoparticle exhibiting size-related intensive properties that differ significantly from that observed in fine particles (100 nm < fine particle < 2.5 μm) or bulk materials.

IEEE.   At the tail of 2005, the IEEE released a standards document that defines test methods for measurement of electrical properties of carbon nanotubes—*IEEE 1650–2005* [118]. It was the first formal effort to identify the minimal information required for reporting results for a nanomaterial [116]. The purpose of the document was to ensure that results can be reproduced in other labs and thereby proven—and on a large scale [121].

ASTM.   ASTM (American Society for Testing and Materials) formed in 1898 by C.B. Dudley specifically to support the railroad industry develops and publishes voluntary consensus technical standards for technology and materials. It is considered by many to be the world's largest organization dedicated to standards. ASTM standards are developed from numerous volunteer consensus technical committees. ASTM developed one of the first nanotechnology standards in 2006 [122]. Logically, the first standard should revolve around "standard terminology related to nanotechnology." We have mentioned time and time again about the importance of language in science. Selected definitions originating from ASTM E 2456-06 are provided in **Table 2.3**.

2.4.2 NIST Efforts

The National Institute of Standards and Testing has undertaken efforts to develop metrology and standards for nanomaterials and processes. Part of the commitment is fully visible in the form of NIST's new AML in Gaithersburg, Maryland. NIST has also recently opened the Center for Nanoscale Science and Technology in 2006 that contains a nanofabrication facility and focuses on instrumentation research, metrology, and standards development. Standard reference materials are also being developed for the characterization of nanomaterials, for example, standards for gold nanoparticles ranging in diameter from 1 to 100 nm for instrument calibration, thin films, single-phase nanoscale particles for electron beam, and ion beam analytical imaging instrumentation. The Nanotechnology Characterization Laboratory (NCL) at the National Cancer Institute (NCI) is conducting tests on the efficiency and toxicity of nanoscale particles—in vitro and in vivo. The Department of Energy (DOE) laid the groundwork for five Nanoscale Science Research Centers (NSRC) across the country. Developing new tools and methods for nanomaterials include the transmission electron aberration corrected microscope (TEAM), dedicated to developing the next generation TEMs.

ANSI. Standards are in the process being developed to accommodate nanotechnology. In June 2004, the American National Standards Institute (ANSI) established a panel to help coordinate national efforts in standards development. The International Organization for Standardization (ISO) followed suit shortly thereafter. The international arm of ASTM, known as ASTM-International, is working with ANSI to develop and standardize nanotechnology terminology, characterization methods, health and safety protocols, intellectual property management, and international liaison and cooperation. ASTM has also formed a partnership with IEEE, the American Society for Mechanical Engineers, Semiconductor Equipment Manufacturers International, the International Institute for Chemical Engineers, and the National Sanitation Foundation to develop global terminology and standards.

Work is underway globally to develop (1) methods to detect nanomaterials within biological environments, (2) methods to understand heterogeneity in nanomaterials, (3) methods for measuring purity and metrology of particle size, (4) methods and standardized tools to assess nanomaterial shape, structure and surface area, (5) to detect and measure nanomaterial spatio-chemical composition, and (6) an inventory of nanomaterials and their use.

2.4.3 IEEE Roadmap for Nanoelectronics

The purpose of the IEEE Standards Association Nanoelectronics Standards Roadmap is to frame the big picture view of nanoelectronics standards, both in the near and far terms [120]. This includes creating standards for materials (nanoparticles such as nanopores, quantum dots, quantum-confined atoms, nanospheres, nanorods, nanoshells, complex nanoparticles, nanocrystals, and magnetic nanoparticles and ferrofluids); and nanoscale conductive interconnects like nanotubes and nanowires; and devices like sensors, storage devices, emitting, and switching.

All categories need to be identified and provided with standards—whether manufacturing or material—if commercialization is to proceed efficiently worldwide. Many of the standards will be extensions of preexisting standards that already support micron-level industrial practices developed over the past 20 years or so. Others are brand new and will have to be developed as we go along.

Equally impressive is another list provided in the IEEE roadmap that addresses "behaviors to exploit." The list is a "who's who" or more appropriately, a "what's what" of nanotechnology. It is a summary of the many properties and phenomena that are exhibited by nanomaterials and by itself is a valuable pedagogical tool.

2.5   NANOMANUFACTURING AND MOLECULAR ASSEMBLY

Nanomanufacturing is already a reality. Components for our computers already are stocked with nanosized transistors (<100 nm). However, bottom-up techniques are making headway into nanomanufacturing. Because nano-technology is amenable to bottom-up synthesis, one would expect that chemical and biological methods of synthesis be applied on a larger scale. There are however, growing pains associated with whole-scale uses of bottom-up synthesis methods. The development of micro- and nanomanipulators offers another route to devices and products. And lastly, biomolecular assembly, a process that is based on natural assembly lines characteristic of living things, offers us a unique look at bottom-up assembly adapted to a synthetic world.

2.5.1   *Lithographies*

The accurate placement of patterns is a critical element in future nanomanu-facturing practices. Current nanoelectronics, nanophotonics, and nanomag-netics all require the application of patterns during some part of their assembly. More accurate pattern gratings are needed for 2-D planar nano-manufacturing to become a reality. Solutions to pattern placement problems include eventual elimination of laser interferometers. M.L. Schattenburg of MIT, who states that current metrology tools are inadequate for nanomanu-facturing, provides an intriguing overview of nanometrology in nanomanu-facturing [17]. In general and historically, two basic criteria are required for viable manufacturing of 2-D systems: (1) CD metrology that regulates feature size and (2) accurate pattern placement (location and/or overlay of patterns on top of complementary structures). Rapid, reliable, and accurate inspection of products during assembly is another need. Currently, SEMs for CD evalua-tion have limited accuracy, AFMs equipped with nanotips are too slow, and laser interferometers are inadequate for metrology at this small scale [17]. The critical metrology component is the "metrology frame." The stable metrol-ogy frame consists of a length scale and a way to compare a manufactured component with that length scale. Microscopes, for example, traditionally are used to accomplish the latter.

According to Schattenburg, metrology frames "underpin all accurate metrol-ogy tools" that include lithography mask (reticle) writers, lithography scanners and steppers, CD metrology tools, pattern placement and overlay metrology tools, circuit and mask repair, and coordinate measuring tools [17]. Just to provide a perspective, according to the ITRS [17,123], metrology milestones indicate the direction the semiconductor industry is taking over a 15-year horizon [123]. From 2001 through 2016, CD is expected to decrease from 130 nm down to 22 nm; overlays from 45 nm down to 9 nm; mask image placement from 27 nm down to 6 nm. The corresponding frame errors, once again from 2001 through 2016, are expected to decrease from 11 nm down to 2.3 nm; and the length scale error from 2.8 nm down to 0.6 nm (based on MIT research efforts) [17]! From 2005 on or so, solutions are in the works and some issues, the overlay parameters in particular, have no working solutions at this time.

Now, let's discuss an amazing (little) device called the MIT Nanoruler. C.G. Chen and M.L. Shattenburg et al. of MIT in 2004 reported the development of a ruler based on a new form of interference lithography capable of large-scale work that is fast, accurate, and precise. The nanoruler was designed to pattern large gratings. For example, a 400-nm period grating was etched onto a 300-mm wafer in ca. 20 min [124]! The grating pattern lower limit of the 2004 device was 150 nm. The tool is housed in a cleanroom (housed in a larger cleanroom) that holds temperature fluctuations to ±0.005°C. Scanning beam interference lithography (SBIL), a concept conceived by Schattenburg, possesses precision "better than 1 nm" across a 300-mm wafer. SBIL consists of two narrow (2 mm diameter) ultraviolet laser beams that generate interference pattern fringes. A spot formed by the fringes is scanned linearly across the wafer's surface. The SBIL writes the grating with thousands of parallel fringe spots.

2.5.2 *Nanomanipulators and Grippers*

Manufacturing that is accomplished at the macroscale, like automobile assembly lines, is driven to a great part by automated components, for example, robots. We now explore their micro- and nano-counterparts. Manipulators with high positioning accuracy and stages with nanometer sensitivity are well represented in today's marketplace.

Nanomanipulators. Nanomanipulators are often coupled with scanning electron microscopes to provide a view of field operations for research, development, and production applications. Some models, like the *S100 nProber* (sub-100-nm) manufactured by the Zyvex Corporation, have four positioners that grasp, move, test, and position nanocomponents to accommodate four axes of movement. Transfer of samples and interfaces with preexisting laboratory equipment are possible with such devices. These nanomanipulators are equipped with a joystick and keypad to provide tight control over manipulation. **Figure 2.10** displays images of selected nanomanipulators and grippers, courtesy of the Zyvex Corporation. Zyvex also manufactures high aspect ratio *NanoEffector Probes* (tip radius <50 nm) that are able to probe small features with 50-nm contact and small geometries (four 50-nm contacts within 100 nm of each other). Eight probe tips can be utilized in a 500-nm workspace and is utilized by Zyvex's *nProber*. For more information, please consult Zyvex's website at www.zyvex.com.

FIG. 2.10

(a) A Zyvex nProber is capable of sub-100-nm manipulation. The semiautomated system consists of 8 encoded positioners, an xyz-encoded center stage (for step and repeat capability), and vision feedback for point and click positioning. (b) A field-emission gun scanning electron microscope (FEI Quanta 200 FEG SEM), a parametric analyzer, and advanced contamination system is shown. The FEI Quanta 200 provides optimized resolution, video rates, beam shift, vacuum technology and user control for nanoprobing integrated circuit and failure analysis. (c) A close up of the eight encoded positioners that enable advanced probing and increased throughput. (d) A Zyvex microgripper is shown.

Source: Images courtesy of Jim Von Ehr, II; Zyvex Instruments, Richardson, Texas. With permission.

Nanorobotics?   The field of nanorobotics, although infringing on the realm of science fiction, is pursued quite vigorously by several groups. A recent call for papers (expected publication in fall 2008) by the *International Journal of Robotics Research* is serious about expanding the world of robotics down to the nanoscale. Please refer to www.ijrr.org. The current state of the art and future challenges in nanorobotics are scheduled to be discussed. Their participation in nanomanufacturing is inevitable. Nanorobots are defined as

... intelligent systems with overall dimensions at or below the micrometer range that are made of assemblies of nanoscale components with individual dimensions ranging between 1 and 100-nm. Nanorobots will be able to perform at least one of the following actions: actuation, sensing, signaling, information processing, intelligence, swarm behavior at the nanoscale.

Such nanobots face critical design, fabrication, and control challenges, according to the journal. In particular, nanorobots would be exposed to unusual microenvironments that are vastly altered from macroscopic scenarios. Solicited paper topics include design and manufacture of nanobots, kinematic and dynamic modeling, robotic control, AFM/SPM-based assembly and manipulation, molecular self-assembly and swarm behavior, nanosensors, bio-nano robots, and applications (medical, biological, and industrial manufacturing) [125]. We shall wait and see. Crossing this threshold would indeed be an amazing milestone.

2.5.3 Bottom-Up Manufacturing

Bottom-up chemical methods of nanomanufacturing are expected to deliver a major contribution to manufacturing at the nanoscale. Although far from being realized, bottom-up techniques in general will provide inexpensive means to form nanostructures (dots, wires, and thin films) with high throughput. We shall review a few major kinds of bottom-up manufacturing techniques. Detailed descriptions of these techniques are found in our sister book, *Introduction to Nanoscience* (CRC Press, 2008) [101]. Challenges facing self-assembly on a larger scale include attainment of long-range order, overall robustness, development of methods to mitigate contaminants, errors and defects; and control of microenvironments.

Microfabrication. Microfabrication techniques are essentially top-down "machining" processes. By subtraction from a bulk substrate material, the following repeating process takes place to form circuits, components, and devices:

Substrate → Application of Thin Resist Film → Mask and Lithography → Etch → Liftoff of Mask → Characterization and Release

Lithography involves cleaning (following each step), application of the mask, exposure to pattern, development, curing, and inspection. Lithography is usually an optically driven process but direct writing is also accomplished by electron or ion beams and AFM techniques.

Nanofabrication with Soft Lithography. In the soft lithography process, polydimethylsiloxane (PDMS) is poured into a predetermined pattern (formed by traditional lithography). Following curing, the PDMS is peeled out as a negative replica of the primary mask. This rubbery material is then attached to a substrate and becomes a stamp for the pattern. The stamp's surface is coated with thiols and then pressed against a surface consisting of a thin film of gold. The thiols form a self-assembled monolayer on the gold. George Whitesides and his group at Harvard University developed this process known as micro-contact printing (μCp).

Nanofabrication with Direct Manipulation. STM, AFM, spin-polarized STM, and dip-pen nanolithography are a few methods that create 2-D hierarchies by direct writing. STM, for example, utilizes a sharp, metal-conducting tip that is brought into close contact (within 1 nm) of a sensitized surface. The surface can then be "etched" by the STM tip that is rastered to produce a pattern. Dip-pen lithography is capable of depositing a wide variety of component species ranging from carbon nanotubes, colloidal particles, self-assembled monolayers, reactive species, and many more. An AFM tip immersed in a liquid meniscus (that is a reservoir containing the targeted delivery materials) is rastered across a surface according to a predetermined pattern.

Nanosphere Lithography. Nanosphere lithography employs spherical latex particles (or other suitable materials) that are assembled into a close-packed pattern. By methods such as evaporation or sputtering, materials are allowed to fill the interstices of the spherical matrix to form a negative pattern based on the original template. The process was developed by Richard P. van Duyne of Northwestern University.

Other Template Methods. Templates include porous templates made of metal oxides and various silicas. Aluminum oxide templates, formed via an anodized process, contain nearly perfect arrays of parallel pore channels with diameters a few nanometers to larger than 250 nm. The anodic process to form the films is extremely "low tech" and requires some highly polished aluminum, one of four kinds of mineral acids (sulfuric, oxalic, chromic, or phosphoric), a cooling apparatus, and a DC-power supply. The diameter of the pores formed during anodizing is directly proportional to the applied voltage. Highly ordered arrays are formed by performing the anodizing twice. Removal of the first oxide layer after a 24 h period leaves a templated surface of perfectly ordered scallops. The array of hexagonally packed scallops have attained the steady-state equilibrium with regard to diameter and distribution. Anodizing over this layer yields an equivalent distribution of pore channels, already in an ideal, ordered state. Pore channels can be filled with whatever material required. For example, carbon nanotubes can be grown in the pore channels to form substrates for flat panel displays. Through-hole anodic membranes have utility as maskants for patterning surfaces.

Self-Assembly. Supramolecular host–guest chemistry, molecular self-assembly, and inorganic–organic hybrid materials are just some processes and/or materials that rely on self-assembly in order to form nanostructured materials. Self-assembly is the spontaneous formation of a material from preexisting components that is reversible (to some degree—of course, in manufacturing, we want things to be nonreversible), controllable, and is based on stochastic chemical processes. There is static self-assembly, dynamic self-assembly, and programmed self-assembly. More information about self-assembly can be gleaned from *Introduction to Nanoscience*.

Micelles, the meniscus effect, crystallization, and polymerization occur in nature unabated. The action of protein folding, the self-assembly of the tobacco mosaic virus, and the molecular engine called the ribosome are all based on intermolecular binding forces and substrates held together with strong covalent bonds. The environment is very important in self-assembly. Other drivers such as energy minimization, nucleation, and the effect of templates also influence self-assembly. Nature is a good tutor. Applications of self-assembly to synthetic systems is dependent on the degree of control, for example, controlling the local environment, external fields, particle dispersion, and timing.

2.5.4 *Molecular Scale Assembly Lines*

Richard Feynman first proposed in 1959 the concept of "working with atomic precision" to fabricate nanosystems [126]. Most of the discussion, however, in this section is gleaned from Eric Drexler's essay *Productive nanosystems: The physics of molecular fabrication* (*Physics Education*, 40, 339–346 (2005)). Drexler is one of the leading proponents of molecular manufacturing, and he states the following [126]:

> Fabrication techniques are the foundation of physical technology and are thus of fundamental interest. Physical principles indicate that nanoscale systems will be able to fabricate a wide range of structures, operating with high productivity and precise molecular control. Advanced systems of this kind will require immediate generations of system development, but their components can be designed and modeled today.

Nature, after all, does precisely this—fabricate materials bottom up from molecular assembly systems that deliver precise control over structure—with a few glitches here and there. Ribosomes are biological molecular machines that are directed by digital codes to fabricate proteins—a testament to a functional nanosystem that has demonstrated incredible success (and reliability) over eons that use inexpensive materials that are "biodegradable." The yearly output of such molecular machines is on the order of billions of tons of exactly the same product(s). Drexler goes on to state that synthetic systems should be able to outperform biological counterparts. An excellent example that supports this claim is the contrast between our avian friends and modern day aircraft. To provide balance, however, we add that synthetic systems are far inferior to their biological counterparts in many other ways.

Drexler envisions the assembly of precise, intricate structures by enjoining molecules with direct mechanical control. He claims that this is possible if tight control over molecular motion, bonding transformations that exclude unwanted molecular encounters while simultaneously ensuring positional accuracy down to 1 Å. Enzymes, after all, are able to do just that—precise and accurate control down to the atomic level. Reliability would be a beneficial outcome of this manufacturing methodology. The secret to achieving the successes achieved by biological systems is dependent on the nature of scaling laws and their applicability to nanosystems [126].

Productive Nanosystems.   Nature perhaps will be our greatest tutor—via biomimetics—copying nature should certainly give us a head start over developing such systems from scratch. An interesting juxtaposition has been reached. We emulate natural systems to learn how to build nanosystem molecular assemblies in the long term yet our technology is mostly "very synthetic." For example, most of our technology does not emulate nature at all. In accordance, Drexler's plan, actually, intends to move away from assembly in fluid media (as in the case of biological materials). Due to "unnecessary drag" caused by fluid and gaseous environments to media and restrictions imposed by biomolecules that are essentially large monomers with low bond density and stiffness, performance can be

Fig. 2.11

A molecular planetary gear is shown. The structure is held together with strong covalent bonds and is able to rotate at a frequency of 1 GHz while delivering power density of 10^{15} $W \cdot m^{-3}$. A molecular machine such as this resembles a conventional machine; but do these molecular analogues really perform like conventional machines? And, what factors exist at the nanoscale that serve to disrupt this notion? According to molecular dynamic models, at temperatures much less that $k_B T$, molecular models have shown that "bearing interfaces" of strong covalent solid molecular machines show no static friction. Dynamic friction, however, is likely through phonon interactions. Another aspect is thermal fluctuations. Thermal fluctuations could provide a real barrier to accelerated use of nanomaterials. Thermal fluctuations, as one might imagine, contribute to increasing uncertainty in position and reduce energy barriers. One solution to reduce thermal fluctuations is to operate at extremely low temperatures.

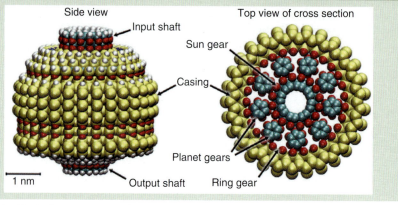

Source: K. E. Drexler, Productive nanosystems: The physics of molecular fabrication, *Physics Education*, 40, 339–346 (2005). With permission.

enhanced by moving out of the liquid/gaseous environment and employing materials that are stronger, harder, and stiffer [126]. In other words, nanosystem productivity should be enhanced by extrapolation of our macroscopic machines down to the nanoscale in the form of a modern assembly line. This type of philosophy has already been accomplished at the microscale. An example of a futuristic molecular planetary gear is shown in **Figure 2.11**. A molecular mill mechanism is shown in **Figure 2.12**. The figure represents a repetitive mechanism that is able to add one atom to a cluster.

Nanotribology. Friction is another parameter that requires more study at the nanoscale. Friction (from the Latin *frictionem* "a rubbing down", from *fricare* "to rub") is an age-old engineering issue. It was from a frictional process after all that some of our ancestors were able to make fire. The ancient Egyptians figured out that moving monumental masses on wooden rollers were easier pulled on wet sand than dry [127]. Pork fat was used in the Middle Ages to lubricate axels of wagons [127]. Leonardo Da Vinci in the late 1400s and early 1500s, by way of his inclined plane experiments, introduced friction in the context of modern engineering by sliding a rectangular block down a planar surface [128,129]. He found that the magnitude of friction had a linear dependence on load but was independent of geometrical area of contact and the type of material (the universal coefficient of friction ≈0.25) [127]. Guillaume Amontons in the late 1600s

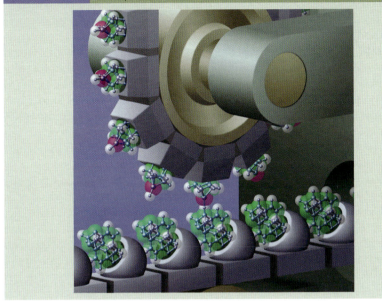

FIG. 2.12

In this artist's rendition, the concept of atomic precision is depicted by a mechanical system that forces components to react, for example, guiding molecular reactions. In the image, a molecular machine attaches a hydrogen atom to a cluster. The machine components are shown undersized and without detail.

Source: Image courtesy of Eric Drexler. With permission.

reaffirmed Da Vinci's results and found that frictional force is directly proportional to the perpendicular force and independent of the type of material (universal coefficient of friction ≈0.33). Leonhard Euler related frictional force with gravitational force and was the first to distinguish between static and kinetic friction. Later on, Charles Augustin Coulomb in the mid-1700s studied the independence of friction from velocity.

With perhaps a peek into the nanodomain, Coulomb suggested that micrometer roughness was responsible for friction. It was later shown that highly polished surfaces demonstrated more friction than roughened surfaces. John Theophilus Desaguliers in the early 1700s implied that a molecular component to friction was at work—adhesion that was dependent on geometrical contact area [127]. Interestingly, it is rather difficult to derive classical friction laws from fundamental atomic principles [128]. Philip Bowden and David Tabor of Cambridge University proposed an adhesion bonding mechanism between atoms to friction [128]. G.A. Tomlinson in 1929 proposed that phonons were responsible for the mechanism of friction. With the advent of the quartz crystal microbalance (QCM) and the lateral force microscope (a type of AFM), the mechanisms of friction were explored with greater frequency.

The phononic mechanism was proven by J. Krim et al. by showing how krypton monolayers demonstrated measurable sliding friction over a gold surface [129]. Phonon effects were measured by a QCM. In another study, Abdelmaksoud et al. showed that trace levels of interfacial slippage between tricresylphosphate (TCP) and metal or metal oxide yielded low macroscopic coefficients of friction while rigidly attached TCP showed high macroscopic coefficients of friction [130]. Static friction, as opposed to sliding friction, requires more force to overcome. At the nanoscale, when considering monolayers sliding over atomically smooth substrates, there is no static friction—ca. 10^5 times less than for macroscopic counterparts [128,129]. Coefficients of friction derived from nanotribological experiments tend to disagree by several orders of magnitude with their macroscopic counterparts [128,129].

Ernst Meyer at the University of Basel in 2004 studied friction with a lateral force microscope sliding over the surface of an atomically smooth sodium chloride surface. He showed that there is a transition from stick slip to sliding due to atomically modulated friction phenomena. Once overcoming the stick-slip regime, an ultra-low domain of friction is encountered, for example, superlubricity [127]. MEMS friction testers (tens of contact regions) capable of detecting forces as small a $5\,\mu N$ have been developed to bridge the gap between macroscopic (numerous contact regions) and nanoscale friction (single contact region).

At the macroscale, over $100 billion is lost in the United States annually due to friction [128]. Much of this kind of friction depends on the interaction with gravity. Gravity has no effect at the nanoscale, rather electromagnetic forces are dominant. Material shear at the macroscale is relatively insignificant. This shear, however, of a single atomic layer of a nanomachine can be terminal to its operation [128]. Superlubricity, nonetheless, is expected to help solve some of the frictional issues encountered today. In order for this to occur, according to J. Krim, understanding of friction at the atomic-, nano-, and mesoscale is imperative. The future of friction, she goes to say, will be part of the system design process and reduced or eliminated "before the fact" rather than application of lubricants "after the fact."

Self-Assembly.   The competition to such mechanized "mechanically guided systems" is the natural process of Brownian motion-driven self-assembly—a process that takes place usually in aqueous media. Biology exploits self-assembly via the action of programmable machines to make precursor building blocks [126]. However, Drexler believes that directed mechanical assembly will replace biologically based self-assembly that produces interfaces that are "soft and weak." Although Brownian motion-based self-assembly is prevalent today, in the long run, it will not survive the nanomanufacturing demands required by future technology, according to Drexler.

Scaling Laws.   Classical scaling laws are mathematical laws that predict how variation in one quantity affects variations in other quantities. Scaling laws are commonly referred to as power laws. The Stefan–Boltzmann law, for example, is a power law. Scaling laws are able to relate one function to that function at different size regimes. For example, ants are able to lift ca. 20× their own weight in mass while humans are not quite able to achieve that level of perfection. Put another way, if there were no scaling laws, then we should be able to lift 2 tons of weight (assuming a 200-lb person) in proportion to what an ant is able to lift (e.g., a linear transposition).

The simplest way to get a feel for scaling laws is to study mechanical systems. In mechanical systems, mass is proportional to volume. However, if a linear dimension of an object is reduced by a factor of x, then the volume of that object is reduced by a factor of x^3. At the nanoscale (ca. 10^{-8} m), several classical scaling laws retain a sense of relevance—especially with regard to mechanical functions [126]. These laws do break down at smaller scales where atomic structure and quantum effects exert more influence. Please refer to **Table 2.4**. Classical

TABLE 2.4	*Physical Properties: Scaling and Magnitude*		
Quantity	**Scaling**	**Magnitude**	
Magnetic force	L^4	10^{-19}	N
Volume	L^3	10^{-24}	m^3
Mass	L^3	10^{-21}	kg
Electrostatic energy	L^3	10^{-19}	J
Torque	L^3	10^{-15}	m N
Gravitational force (weight)	L^3	10^{-20}	N
Area	L^2	10^{-16}	m^2
Force (at working stress)	L^2	10^{-7}	N
Mechanical power	L^2	10^{-7}	W
Electrostatic force (constant field)	L^2	10^{-11}	N
Electric current	L^2	10^{-6}	A
Length	L^1	10^{-8}	m
Deformation (constant stress)	L^1	10^{-11}	m
Motion time	L^1	10^{-8}	s
Stiffness	L^1	10^4	N m^{-1}
Voltage (constant field)	L^1	10^0	V
Gravitational stress	L^1	10^{-5}	N m^{-2}
Mechanical working stress	L^0	10^9	N m^{-2}
Modulus of elasticity	L^0	10^{12}	N m^{-2}
Electrostatic stress (constant field)	L^0	10^5	N m^{-2}
Adhesive strength (dispersion forces)	L^0	10^9	N m^{-2}
Strain	L^0	10^{-3}	—
Density	L^0	10^3	kg m^{-3}
Speed	L^0	10^0	m s^{-1}
Current density	L^0	10^{10}	A m^{-2}
Electric field	L^0	10^8	V m^{-1}
Amplitude of thermal vibrations	$L^{-1/2}$	10^{-12}	m
Acceleration	L^{-1}	10^8	m s^{-2}
Spring stiffness	L^{-1}	10^4	N m^{-1}
Deformation (constant force)	L^{-1}	10^{-11}	m
Mechanical power density	L^{-1}	10^{17}	W m^{-3}
Electrical resistance	L^{-1}	10^0	Ω
Motion frequency	L^{-1}	10^8	s^{-1}
Relative productivity (scaled parts)	L^{-1}	10^5	s^{-1}
Relative productivity (atomic parts)	L^{-4}	10^3	s^{-1}

Source: K. E. Drexler, Productive nanosystems: The physics of molecular fabrication, *Physics Education*, 40, 339–346 (2005). With permission.

continuum scaling laws as they apply to magnitudes down to 10^{-8} m break down at smaller dimensions where atomic structure and, if smaller, quantum effects need to be considered [126].

For a more detailed discussion, please refer to Drexler's paper mentioned earlier in this text.

Impediments to Mechanical Nanosystem Fabrication.   Scale-down of a macroscopic machine to the nanoscale is not just a matter of making it smaller. Other factors need to be considered. For example, forces that are irrelevant at the macroscale such as van der Waals forces can be significant at the nanoscale. There are several aspects that could contribute to the failure of a molecular system: (1) thermal fluctuations that allow energy barriers to be breached along a path to an undesirable state, (2) structures rearrange or become fragmented, and (3) displacement of a nanotool during a molecular operation causes unwanted bonding transformations (threshold displacement of $\Delta x_i \approx 0.1$ nm). Drexler believes that many of these challenges can be overcome by designing molecular tools using stiff materials positioned by 100-nm scale devices [126].

2.6   Concluding Remarks

We come to the end of the *Perspectives* division of the text. We have experienced a brief foray into the wondrous world of nanometrology and have briefly touched upon nanomanufacturing. Great efforts are expended on roadmaps describing just how to go about tackling nanometrology [131]. Many believe that indeed nanometrology is the next big thing in measurement [132]. Are we really interested in quantifying the surface coverage of conjugated molecules on functionalized nanoparticles [133] or understanding the transition between stick-slip friction and continuous sliding in atomic friction [134]? Based on what you have soaked up so far, I hope that your answer is in the affirmative. The federal government and the international community do understand the importance of nanometrology because it is metrology, after all is said and done, that allows us to practice commerce, conduct meaningful research, and make decisions—some of the fundamental bases of our civilization [135]. Will we ever be able to quantify and standardize self-assembly processes [136]? And, how has nature gone about metrology and manufacturing all these years?

We hope that the student has gained a sense of importance of quantum metrology and quantum phenomena and how nanomaterials are capable of playing a major role in the modulation of quantum measurements and properties.

References

1. History of measurement, en.wikipedia.org/wiki/History_of_measurement, modified 12 (2007).
2. Jet's fuel ran out after metric conversion errors, *New York Times*, July 30 (1983).
3. NASA's metric confusion caused mars orbiter loss, www.cnn.com/TECH/space/9909/30/mars.metric/ (1999).

4. Metrology, Computer Encyclopedia, computerlanguage.com/ (2007).
5. Metrology, *The Oxford essential dictionary of the U.S. military*, Oxford University Press, Oxford (2007).
6. Metrology, International Bureau of Weights and Measures, IBWM, bipm.org/en/home/ (2007).
7. Nanometrology, International Bureau of Weights and Measures, IBWM, bipm.org/en/home/ (2007).
8. Nanometrology, Materials Science and Engineering Laboratory, www.msel.nist.gov/Nanometrology.pdf (2007).
9. Accuracy, precision, and uncertainty, Science Division, Bellevue Community College, scidiv.bcc.ctc.edu/Physics/Measure&sigfigs/Measure&sigfigsintro.html (accessed 2007).
10. Introduction to metrology, MIKES, Mittatekniikan Keskus Centre for Metrology and Accreditation, Finnish National Standards Laboratories, cs.utu.fi/rlahdelma/ADMSSIM/L01_Metro.pdf (2005).
11. A. Lassila, Traceable dimensional measurements on nanometer range, Annual Report, *MIKES*, Mittatekniikan Keskus Centre for Metrology and Accreditation, Finnish National Standards Laboratories (2004).
12. Nanomanufacturing, en.wikipedia.org/wiki/Nanomanufacturing (2007).
13. Nanomanufacturing (NM), Solicitation, Civil, Mechanical and Manufacturing Innovation, National Science Foundation, nsf.gov/funding/pgm_summ.jsp?pims_id=13347 (2008).
14. Center for high-rate nanomanufacturing, nano.neu.edu/industry/industry_showcase/industry_day/ (2007).
15. A. Busnaina, Center for high-rate nanomanufacturing (CHN), Presentation, nsec.neu.edu (viewed 2007).
16. The George J. Kostas facility for micro and nanofabrication, kostas.neu.edu/ (2007).
17. M. L. Schattenburg, Nanometrology in nanomanufacturing, Presentation, Massachusetts Institute of Technology, NASA Tech Briefs, Cambridge, MA snl.mit.edu/papers/presentations/2003/Schattenburg/MLS-NASA-Tech-2003.pdf (2003).
18. M. T. Postek and R. J. Hocken, *Instrumentation and metrology for nanotechnology*, Report of the National Nanotechnology Initiative Workshop, Gaithersburg, MD (2004).
19. R. Cavanaugh, Instrumentation and metrology for nanocharacterization, chap. 2 In *Instrumentation and metrology for nanotechnology*, M. T. Postek and R. J. Hocken, eds., Report of the National Nanotechnology Initiative Workshop, Gaithersburg, MD (2004).
20. J. Marburger and R. R. Colwell, Science for the 21st century, National Science and Technology Council, Executive Office of the President; Washington, DC, July (2004).
21. E. M. Vogel, NIST advanced measurement laboratory (AML) Nanofab, Presentation, National Institute of Standards and Technology (2004).
22. J. M. Cattell, Mental measurement, *Philosophical Review*, 2, 316–332 (1893).
23. G. Englehard, Historical views of the concept of invariance in measurement theory, In *Objective measurement: Theory into practice*, M. Wilson, ed., Vol. 2, pp. 73–79, Ablex, Norwood, NJ (1994).
24. T. M. Porter, *Trust in numbers: The pursuit of objectivity in science and public life*, Princeton University Press, Princeton, NJ (1995).
25. A. J. Fiok, J. M. Jaworski, R. Z. Morawski, J. S. Oledzki, and A. C. Urban, Theory of measurement in teaching metrology in engineering faculties, Warsaw University of Technology, sierra.iem.pw.edu.pl/~jsol/pedeefy/art34.pdf (accessed 2007).
26. A. J. Fiok, J. M. Jaworski, R. Z. Morawski, J. S. Oledzki, and A. C. Urban, Theory of measurement in teaching metrology in engineering faculties, *Measurement*, 6, 63–67 (1988).
27. B. Jeckelmann and B. Jeanneret, The quantum hall effect as an electrical resistance standard, *Measurement Science and Technology*, 14, 1229–1236 (2003).

28. N. Taniguchi, Current status in, and future trends of, ultra-precision machining and ultrafine materials processing, *Annals of the CIRP*, 32, 573–582 (1983).

29. J. B. Bryan, The Abbe principle revisited, *Precision Engineering*, 1, 129–132 (1979).

30. D. J. Whitehouse, Nanometrology, chap. 8, In *Handbook of surface and nanometrology*, Institute of Physics (IOP) Publishing, a subsidary of Taylor & Francis, Bristol (2003).

31. A. Weckenmann, T. Wiedenhöfer, and K.-D. Sommer, Determining measurement uncertainty in nanometrology with the example of nanoclusters, *tm-Technisches Messen*, 71, Part 2, 93–100 (2004).

32. J. W. Lyons, Guidelines for evaluating and expressing the uncertainty of NIST measurement results, National Institute of Standards and Technology, physics.nist.gov/Pubs/guidelines/sec6.html (1993).

33. J. Salsbury, Nanometrology—NIST and Mitutoyo achieve world's highest accuracy roundness measurements, *Mitutoyo News*, mitutoyo.com/pressrelease/NISTPR%20.pdf (2004).

34. C. M. Caves, Quantum-limited measurements: One physicists crooked path from quantum optics to quantum information, Department of Physics presentation, University of New Mexico (viewed 2008).

35. Uncertainty Principle, *The Columbia Electronic Encyclopedia*, www.cc.columbia.edu/cu/cup/ through encyclopedia2.thefreedictionary.com/Heisenberg+limit (2007).

36. J. D. Wilson and A. J. Buffa, *College physics*, Prentice-Hall, Upper Saddle River, NJ (2000).

37. J. Hilgevoord, The Uncertainty Principle, *Stanford Encyclopedia of Philosophy*, plato.stanford.edu/entries/qt-uncertainty/ (2006).

38. D. Halliday, R. Resnick, and J. Walker, *Fundamentals of physics*, 7th ed., John Wiley & Sons, Inc., New York (2005).

39. Quantum entanglement, en.wikipedia.org/wiki/Quantum_entanglement (2008).

40. M. Brooks, Entanglement: The weirdest link, *New Scientist*, 181, 2440–2427 (2004).

41. P. Comstock, The strange world of quantum entanglement, Interview of Brian Clegg, *California Literary Review*, calitreview.com/topics/science/51/ (2007).

42. B. Clegg, *The god effect: Quantum entanglement, science's strangest phenomenon*, St. Martin's Press, New York (2006).

43. M. Kroner, A. O. Govorov, S. Remi, B. Biedermann, S. Seidl, A. Badolato, P. M. Petroff, W. Zhang, R. Barbour, B. D. Gerardot, R. J. Warburton, and K. Karrai, The nonlinear Fano effect, *Nature*, 451, 311–314 (2008).

44. A. Migdall, Correlated photon metrology without absolute standards, *Physics Today*, 52, 41–46 (1999).

45. H. Lee, P. Kok, and J. P. Dowling, A quantum Rosetta stone for interferometry, *Journal of Modern Optics*, 49, 2325–2338 (2002).

46. T. Nagata, R. Okamoto, J. L. O'Brien, K. Sasaki, and S. Takeuch, Beating the standard quantum limit with four-entangled photons, *Science*, 316, 726–7294 (2007).

47. A. Politi, M. J. Cryan, J. G. Rarity, S. Yu, and J. L. O'Brien, Silica-on-Silicon Waveguide Quantum Circuits, *Science Express*, www.sciencemag.org/sciencexpress/recent.dtl (2008).

48. Tiny photonics logic circuit achieves quantum entanglement. *R&D Magazine*, March 28 (2008).

49. J. L. O'Brien, Precision without entanglement, Perspectives, *Science*, 318, 1393–1394 (2007).

50. G. Santarelli, P. Laurent, P. Lemonde, A. Clairon, A. G. Mann, S. Chang, A. N. Luiten, and C. Salomon, Quantum projection noise in an atomic fountain: A high stability cesium frequency standard, *Physics Review Letters*, 82, 4619 (1999).

51. B. L. Higgins, D. W. Berry, S. D. Bartlett, H. M. Wiseman, and G. J. Pryde, Entanglement-free Heisenberg-limited phase estimation, *Nature*, 450, 393–396 (2007).
52. V. Giovannetti, S. Lloyd, and L. Maccone, Quantum-enhanced measurements: Beating the standard quantum limit, *Science*, 306, 1330 (2004).
53. A. Naik, O. Buu, M. D. LaHaye, A. D. Armour, A. A. Clerk, M. P. Blencowe, and K. C. Schwab, Cooling a nanomechanical resonator with quantum back-action, *Nature*, 443, 193–196 (2006).
54. S. M. Tanner, J. M. Gray, C. T. Rogers, K. A. Bertness, and N. A. Sanford, High-Q GaN nanowire resonators and oscillators. *Applied Physics Letters*, 91, 203117 (2007).
55. High-Q NIST nanowires may be practical oscillators, *Nanotech-Now*, www.nanotech-now.com/news.cgi?story_id=26600 (2007).
56. K. C. Schwab and M. L. Roukes, Putting mechanics into quantum mechanics, *Physics Today*, 58, 36–42 (2005).
57. D. Nikbin, Quantum back-action has cooling effect, *Physics World*, physicsworld.com/cws/article/news/25905 (2006).
58. K. C. Schwab, Nanomechanics coupled to SET, Homepage, Cornell University, www.kschwabresearch.com/articles/detail/7 (2008).
59. G. A. Mansoori, *Principles of nanotechnology*, World Scientific Publishing Co. Pte. Ltd., Singapore (2005).
60. G. Aeppli, P. Warburton, and C. Renner, Will nanotechnology change IT paradigms? *BT Technology Journal*, 24, 163–169 (2006).
61. A. Zellinger, Fundamentals of quantum information, *Physics World*, physicsworld.com/cws/article/print/1658 (1998).
62. Science & technology programme in quantum metrology, Department of Trade and Industry (DTI), National Measurement System Directorate (NMSD), April 2004 to March 2007, London (2007).
63. C. Wasshuber, Computational single-electronics, In *Computational Electronics*, S. Selberherr, ed., Springer-Wein, New York (2001).
64. NIST-F1 Cesium fountain atomic clock: Primary and frequency standard for the United States, National Institute of Standards and Testing (NIST), Boulder, CO, http://tf.nist.gov/cesium/fountain.htm (2008).
65. S. Jefferts, D. Meekhof, T. Heaver, and E. Donley, NIST-F1 Cesium fountain atomic clock, National Institute of Standards and Technology, Boulder, CO, tf.nist.gov/cesium/fountain.htm (2007).
66. J. Roach, New clock will lead to more accurate measure of time, National Geographic News (2001).
67. C. F. Roos, M. Chwalla, K. Kim, M. Riebe, and R. Blatt, "Designer atoms" for quantum metrology, *Nature*, 443, 316–319 (2006).
68. Fine structure constant, en.wikipedia.org/wiki/Fine-structure_constant (2007).
69. Current advances: The fine structure constant and Quantum Hall effect, NIST Reference on Constants, Units and Uncertainty, physics.nist.gov/cuu/Constants/alpha.html (2000).
70. A. D. Ludlow, T. Zelevinsky, G. K. Campbell, S. Blatt, M. M. Boyd, M. H. G. de Miranda, M. J. Martin, J. W. Thomsen, S. M. Foreman, J. Ye, T. M. Fortier, J. E. Stalnaker, S. A. Diddams, Y. Le Coq, Z. W. Barber, N. Poli, N. D. Lemke, K. M. Beck, and C. W. Oates, Sr lattice clock at 1×10^{-16} fractional uncertainty by remote optical evaluation with a Ca clock, *Science*, 319, 1805–1808 (2008).
71. New atomic clock super-accurate, Denver and the West, *The Denver Post*, February 15 (2008).
72. J. Kinaret, M. Jonson, A. Bárány, P. Sylwan, and S. Björn-Rasmussen, Mosquitoes in a cross-wind, Poster, Nobel Prize in Physics 1998, The Royal Swedish Academy

of Sciences; nobelprize.org/nobel_prizes/physics/laureates/1998/illpres/practice. html (1998).

73. Physics of Quantum Hall resistance standards, National Institute of Standards and Testing, Boulder CO, www.boulder.nist.gov/div814/div817b/whatwedo/qhall/ qhall.htm (2002).

74. Metrology of the Ohm, Electronics and Engineering Laboratory (EEEL), National Institute of Standards and Testing, www.eeel.nist.gov/817/whatwedo-817g/ohm/ ohm.html (2004).

75. S. Liou, W. Kuo, Y. W. Suen, W. H. Hsieh, C. S. Wu, and C. D. Chen, Shapiro steps observed in a superconducting single electron transistor, *Chinese Journal of Physics*, 45, 230 (2007).

76. B. Doucot, B. Duplantier, V. Pasquier, and V. Rivasseau, eds., *The Quantum Hall effect: Poincaré seminar 2004*, Birkhäuser Verlag, Basel (2005).

77. T. Yang, N. B. Belecki, and J. F. Mayo-Wells, A practical Josephson voltage standard at 1 V, NVL-NIST, nvl.nist.gov/pub/nistpubs/sp958-lide/html/315-318.html

78. R. E. Elmquist, M. E. Cage, Y. Tang, A-M. Jeffery, J. R. Kinard, R. F. Dziuba, N. M. Oldham, and E. R. Williams, The ampere and electrical standards, *Journal of Research of the National Institute of Standards and Technology*, January-February (2001).

79. N. B. Zhitenev and T. A. Fulton, Scanning single-electron transistor microscopy, In *Encyclopedia of nanoscience and nanotechnology*, Vol. 5, J. A. Schwarz, C. I. Contescu, and K. Putyera, eds., Marcel-Dekker, New York (2004); Taylor & Francis (2005).

80. I. Giaever, Energy gap in superconductors measured by electron tunneling, *Physics Review Letters*, 5, 147–148 (1960).

81. J. P. Pekola, J. J. Vartianen, M. Möttönen, O.-P. Saira, M. Meschke, and D. V. Averin, Hybrid single-electron transistor as a source of quantized current, *Nature Physics*, Published Online, www.nature.com/nphys/journal/vaop/ncurrent/abs/nphys808. html, 9 December (2007).

82. B. Dumé, Ampere could be defined one electron at a time, Nanotechweb.org, www. nanotechweb.org/cws/article/tech/32295 (2008).

83. J. Cazaux, Errors in nanometrology by SEM, *Nanotechnology*, 15, 1195–1199 (2004).

84. J. Allgair and B. Bunday, A review of scatterometry for three-dimensional semiconductor feature analysis, *Future Fab International*, 19, future-fab.com/documents. asp?grID=216&d_ID=3390 (2005).

85. J. Aoki, W. Gao, and S. Kiyono, A high-precision AFM for nanometrology of large area micro-structured surfaces, *EUSPEN Glasgow*, 254–255 (2004).

86. J. Aoki, W. Gao, S. Kiyono, and T. Ono, A high-precision AFM for nanometrology of large area micro-structured surfaces, *Key Engineering Materials*, 295–296, 65–70 (2005).

87. Scanning electron microscope-based metrology, In Office of microelectronics programs: Programs, activities and accomplishments, NISTIR-7367, Electronics and Electrical Engineering Laboratory (EEEL), National Institute of Standards and Technology, January (2007).

88. P. West and N. Starostina, *AFM image artifacts*, Pacific Nanotechnologies, LOT-Oriel Gruppe Europa, www.lot-oriel.com/site/site_down/pn_artifacts_deen.pdf (viewed 2008).

89. D. Julthongpiput, M. J. Fasolka, W. Zhang, T. Nguyen, and E. J. Amis, Gradient chemical micropatterns: A reference substrate for surface nanometrology, *Nano Letters*, 5, 1535–1540 (2005).

90. Veeco introduces new insight 3D atomic force microscope: Unparalleled accuracy for in-line CD, depth and CMP metrology applications, Veeco.com (2007).

91. W. E. Moerner, New directions in single-molecule imaging and analysis, *Proceedings National Academy of Sciences*, 104, 12596–12602 (2007).

92. C. Bustamante, Z. Bryant, and S. B. Smith, Ten years of tension: Single-molecule DNA mechanics, *Nature*, 421, 423–427 (2003).

93. K. C. Neuman and S. M. Block, Optical trapping, *Review of Scientific Instruments*, 75, 2787–2809 (2004).

94. G. Hummer and A. Szabo, Free energy surfaces from single-molecule force spectroscopy, *Accounts of Chemical Research*, 38, 504–513 (2005).

95. I. Gopich and A. Szabo, Fluorophore-Quencher distance correlation functions from single-molecule photon arrival trajectories, *Journal of Physical Chemistry B*, 109, 6845–6848 (2005).

96. A. E. Cohen and W. E. Moerner, Method for trapping and manipulating nanoscale objects in solution, *Applied Physics Letters*, 86, 93–109 (2005).

97. A. Fischer and M. Montal, Single molecule detection of intermediates during Botulinum neurotoxin translocation across membranes, *Proceedings of National Academy of Science*, 104, 10447–10452 (2007).

98. T. A. Siewert, and D. Balzar, Metrology for nanoscale properties: X-Ray methods, In *Nanometrology*, p. 8, National Institute of Standards and Testing, www.msel.nist.gov/Nanometrology.pdf (viewed 2008).

99. C. Caleman, Towards single-molecule imaging: Understanding structural transitions using ultrafast x-ray sources and computer simulations, Acta Universitasis Upsaleinsis, Uppsala (2007).

100. N. Lee, R. D. Hartchuh, D. Mehtani, A. Kisliuk, J. F. Maguire, M. Green, M. D. Foster, and A. P. Sokolov, High contrast scanning Nano-Raman spectroscopy of silicon, *Journal of Raman Spectroscopy*, 38, 789–796 (2007).

101. G. L. Hornyak, J. Dutta, H. F. Tibbals, and A.K. Rao, *Introduction to nanoscience*, CRC Press, Boca Raton, FL (2008).

102. C. L. Haynes, C. R. Yonzon, X. Zhang, and R. P. Van Duyne, Surface-enhanced Raman sensors: Early history and the development of sensors for quantitative biowarfare agent and glucose detection, *Journal of Raman Spectroscopy*, 36, 471–484 (2005).

103. A. Macaskill, A. A. Chemonosov, V. V. Koval, E. A. Lukyanets, O. S. Federova, W. E. Smith, K. Faulds, and D. Graham, Quantitative surface-enhanced resonance Raman scattering of phthalocyanine-labeled oligonucleotides, *Nucleic Acid Research Advance Access*, 1–6 (2007).

104. K. Kneipp, Y. Wang, H. Kneipp, L. T. Perelmean, I. Itzkan, R. R. Dasari, and M. S. Filed, Single molecule detection using surface-enhanced Raman scattering, *Physical Review Letters*, 78, 1667–1670 (1997).

105. S. Nie and S. Emory, Probing single molecules and single nanoparticles by surface-enhanced Raman scattering, *Science*, 275, 1102–1106 (1997).

106. F. Cannone, G. Chirico, and A. Diaspro, Two-Photon interactions a single fluorescent molecule level, *Journal of Biomedical Optics*, 8, 391–395 (2003).

107. Optical grating scatterometry, Optical Technology Division, Physics Laboratory, National Institute of Standards and Technology; physics.nist.gov/Divisions/Div844/facilities/scatterometry/scatterometry.html (2005).

108. E. R. Fuller and G. D. Quinn, Mechanical metrology for small-scale structures, In *Nanometrology*, p. 4, National Institute of Standards and Testing, www.msel.nist.gov/Nanometrology.pdf (viewed 2008).

109. D. T. Smith, Nanoindentation methods and standards, In *Nanometrology*, p. 5, National Institute of Standards and Testing, www.msel.nist.gov/Nanometrology.pdf (viewed 2008).

110. S. M. Hsu, Nanotribology and surface properties, In *Nanometrology*, p. 6, National Institute of Standards and Testing, www.msel.nist.gov/Nanometrology.pdf (viewed 2008).

111. G.R. Stafford and O. Kongstein, Nanostructure fabrication processes: Thin film stress measurements, In *Nanometrology*, p. 17, National Institute of Standards and Testing, www.msel.nist.gov/Nanometrology.pdf (viewed 2008).

112. A. M. Forster and S.-H. Moon, Combinatorial adhesion and mechanical properties: Axisymmetric adhesion testing, In *Nanometrology*, p. 19, National Institute of Standards and Testing, www.msel.nist.gov/Nanometrology.pdf (viewed 2008).

113. C. M. Stafford and Y. M. Chiang, Combinatorial adhesion and mechanical properties: Innovative approaches to peel tests, In *Nanometrology*, p. 20, National Institute of Standards and Testing, www.msel.nist.gov/Nanometrology.pdf (viewed 2008).

114. L. E. Levine and A. M. Chaka, Nanomechanics: Coupling modeling with experiments, In *Nanometrology*, p. 18, National Institute of Standards and Testing, www.msel.nist.gov/Nanometrology.pdf (viewed 2008).

115. J. Salsbury, Nanometrology: NIST and Mitutoyo achieve world's highest accuracy roundness measurements, *Mitutoyo News*, NISTPR.pdf (viewed 2008).

116. S. F. Gale, The state of standards: Nano, *Small Times*, 8, 21–23 (2008).

117. IEEE Completes Standards Roadmap for Emerging Nanoelectronic Applications, *Nanotechnology Standards News*, electronics.ihs.com/news/ieee-nanoelectric-roadmap.htm (2007).

118. IEEE 1650–2005, IEEE Standard test methods for measurement of electrical properties of carbon nanotubes, grouper.ieee.org/groups/1650, December (2005).

119. P. W. Brazis, IEEE nanoelectronics standards roadmap (NESR) overview, Institute of Electrical and Electronics Engineers, andards.ieee.org/getieee/nano/nanoelectronics_roadmap_v1.pdf (2005).

120. Nanoelectronics standards roadmap, IEEE nanotechnology standards roadmap initiative, IEEE Standards Association (IEEE-SA), April 17 (2007).

121. D. Gamota, in The state of standards: Nano, article by, S.F. Gale, *Small Times*, 8, 21 (2008).

122. Standard terminology relating to nanotechnology, *ASTM E 2456-06*, P. Picariello, Staff Manager of Committee E56 on Nanotechnology, Subcommittee E56.01 on Terminology & Nomenclature, ASTM International, West Conshocken, PA, www.astm.org (2006).

123. A. Allan, D. Edenfeld, W. H. Joyner, A. B. Khang, M. Rodgers, and Y. Zorian, 2001 Technology roadmap for semiconductors, *Computer*, 35, 42–53 (2002).

124. P. Konkola, C. Chen, R. K. Heilmann, C. Joo, J. Montoya, C.-H. Chang, and M. L. Schattenburg, Nanometer-level repeatable metrology using the Nanoruler, *Journal of Vacuum Science and Technology B*, 21, 3097–3101 (2003).

125. C. Mavroidis and A. Ferriera, eds., *Current state of art and future challenges in nanorobotics*, International Journal of Robotics Research (2008).

126. K. E. Drexler, Productive nanosystems: The physics of molecular fabrication, *Physics Education*, 40, 339–346 (2005).

127. E. Meyer, R. M. Overnay, K. Dransfeld, and T. Gyalog, *Nanoscience: Friction and rheology at the nanoscale*, World Scientific Publishing Co., Pte. Ltd., Singapore (2002).

128. J. Krim, Friction at the nanoscale, *Physics World*, physicsworld.com/cws/article/print/21309, IOP Publishing (2005).

129. J. Krim, D. H. Solina, and R. Chiarello, Nanotribology of a Kr monolayer: A quartz-crystal microbalance study of atomic-scale friction, *Physical Review Letters*, 66, 181–184 (1991).

130. M. Abdelmaksoud, J.W. Bender, and J. Krim, Bridging the gap between macro- and nanotribology: A quartz crystal microbalance study of tricresylphosphate uptake on metal and oxide surfaces, *Physical Review Letters*, 92, 176101–176103 (2004).

131. M. Désenfant and M. Priel, Road map for measurement uncertainty evaluation, *Measurement*, 39, 841–848 (2006).

132. D. Graham, Nanometrology—Is it the next big thing in measurement? *The Analyst*, 132, 95–96 (2007).

133. L. F. Pease, III., D.-H. Tsai, R. A. Zangmeister, M. R. Zachariah, and M. J. Tarlov, Quantifying the surface coverage of conjugate molecules on functionalized nanoparticles, *Journal of Physical Chemistry C*, 111, 17155–17157 (2007).

134. Instrumentation, metrology, and analytical methods, chap. 2 In *Environmental, health, and safety research needs for engineered nanoscale materials*, The National Nanotechnology Initiative, NSET Subcommittee, Office of Science and Technology Policy, Washington DC (2006).

135. A. Socoliuc, R. Bennewitz, E. Gnecco, and E. Meyer, Transition from stick-slip to continuous sliding in atomic friction: Entering a new regime of ultralow friction, *Physical Review Letters*, 92, 134301–134303 (2004).

136. J. A. Pelesko, *Self assembly: The science of things that put themselves together*, Chapman & Hall/CRC, Taylor & Francis Group, Boca Raton, FL (2007).

Problems

2.1 Explain, in your own words, the meaning of the Heisenberg uncertainty principle.

2.2 Calculate the uncertainty in the position of the moon if its velocity is $1.01 \, \mathrm{km \cdot s^{-1}}$ and if we can measure its velocity to 10^{-8} km. Use $\Delta x \Delta p \approx h$. Is the value of the uncertainty consequential?

2.3 What is the uncertainty in position of an electron of an atom if there is $2.0 \times 10^7 \, \mathrm{m \cdot s^{-1}}$ uncertainty in its velocity?

2.4 The modern meter is defined as the length of the path traveled by light in vacuum during the interval of 1/299 792.458th of a second. From this data, calculate the speed of light?

2.5 The uncertainty in calculating the length of the meter has dropped five orders of magnitude since the early 1900s. What is the wavelength of the sources and why were they chosen?

Source	Uncertainty	Wavelength
Pt-Ir Bar	2×10^{-6}	N/A
λ Cd Lamp	7×10^{-8}	?
λ^{86}Kr Lamp	4×10^{-9}	?
λ HeNe Laser	2.5×10^{-11}	?

2.6 What source was used to provide the timing element in the HeNe laser measurement of the meter given above?

2.7 From 1904–1960, the reproducibility of measurements against the prototype meter bar (Pt-Ir) was on the order of $\pm 0.25 \, \mu$m. What is this in terms of uncertainty?

2.8 Does the Heisenberg uncertainty principle place an ultimate limit on the measurement of position?

2.9 Devise a scenario for molecular assembly based on a natural model.

2.10 What is the difference between accuracy and precision?

2.11 Which excited state has a broader spectral signature—one with a lifetime of 10^{-10} s or one with a lifetime of 10^{-8} s? How much broader is the peak?

2.12 Three groups of four students each are asked to make measurements of the diameter of spherical nanoparticles (in nm) by TEM. The nanoparticles were purchased from a vendor. The students compared the diameter to a scale bar placed in the corner of the image.

Student group	#1	#2	#3	#4	#5	#6
Nano	20.1	20.935	17.0	22.23	16	18.11
Micro	19.011	19.113	19.201	19.166	19.205	19.022
Milli	21	20	21	20	21	20
Pico	18.23	18.26	18.20	18.23	18.24	18.25
Centi	22.1	20.5	18.5	21.0	19.0	19.5

Rank the groups according to the following criteria:

a. Uncertainty (highest to lowest)
b. Accuracy (best to worst)
c. Precision (best to worst)
d. Error (most to least)
e. Significant Figures (most to least)

The professor managed to find the specification sheet provided by the vendor for the nanoparticles and announced to the groups that the diameter is 20 ± 1.0 nm. Rerank accuracy and error as stipulated below.

a. Accuracy (best to worst)
b. Error (most to least)

Do more significant figures help you improve accuracy?

2.13 If an instrument is capable of attomole level detection, how close is that to single-molecule detection? How many molecules are there in an attomole? At what mole-level is a single molecule?

2.14 In a laser-tweezer system, the object trapped by the optical laser tweezer is the large dielectric polystyrene bead ($d \approx 1$ mm). A biomolecule is tethered chemically to the surface of the latex bead. Can the laser tweezer trap a single molecule without the assistance of the polystyrene sphere? Estimate the laser power required if 10 mW is required to trap a 1 μm diameter polystyrene sphere and that the gradient optical forces in the tweezer system are proportional to the polarizability α of the trapped object that is in turn proportional to the volume d^3 of the potential trapped molecule. Hint: consult W.E. Moerner, New directions in single-molecule imaging and analysis, *Proceedings of National Academy of Science*, 104, 12596–12602 (2007).

2.15 If the uncertainty of the cesium fountain atomic clock at NIST in Boulder, Colorado is 5×10^{-16} s, how many years would it take for the clock to gain or lose 1 s?

Section 2

Electromagnetic Nanoengineering

NANOELECTRONICS

... there is plenty of room to make [computers] smaller .... nothing that I can see in the physical laws ... says the computer elements cannot be made enormously smaller than they are now. In fact, there may be certain advantages.

RICHARD FEYNMAN, 1959

Chapter 2 introduced some of the basics of quantum metrology while also introducing, albeit not in a direct fashion, some electronics. *Chapter 3*, a primer on nanoelectronics, is the first chapter in the *Electromagnetic Nanoengineering division* of the text. The division consists of three chapters that among them address three aspects of electromagnetics: electronics, optics, and magnetism. Following this division, things become quite mechanical as we move on to the three chapters of the *Mechanical Nanoengineering division* of the book.

Nanoelectronics is at the heart of any nanosensor or nanodevice. As with other properties, we expect electronic behavior to behave differently at the nanoscale than it does at the bulk scale. It is a domain in which a metal conductor, if reduced enough in size, now becomes a semiconductor. It is a domain in which electrical conduction, if forced through a small enough device, now exhibits staircase behavior rather than a continuous smooth curve.

3.0 ELECTRONICS AND NANOELECTRONICS

Electronics has a rich history. Most of us are familiar with the great events and discoveries and we therefore will not dwell upon them for too long. By the early nineteenth century, most physicists and chemists, including Michael Faraday, were convinced that the flow of charged particles were responsible for electrolysis and other electrical phenomena. Nanoelectronics is an extension of electronics down to the nanoscale. Macroscopic phenomena are all altered at this scale as new properties emerge. Distinctions between nanoscale and quantum phenomena all begin to emerge.

3.0.1 Basic Electronic Terminology and Symbols

Not all of us had the "blessed good fortune" of taking electrical engineering courses while in college. Therefore, for the benefit of those of us who missed out on that special opportunity, a brief review of electronics is given below for the purpose of perspective and orientation (**Table 3.1**).

3.0.2 Fundamental Types of Electronic Materials (and Nanomaterials)

Electronics, as the name implies, is a field of engineering that involves electrons—moving and storage. Electronics is an applied engineering discipline and industrial sector that involves the shuffling of electrons back and forth in order to perform designated tasks, to perform work. Electronics is a branch of physics that is concerned with the behavior of electrons in devices—whether in the vacuum, gas, liquid or solid phases.

There are several kinds of electronic materials—dielectric, optical, ferroelectric, magnetic, semiconductor, metal, superconductor, organic, polymer, and most recently, biomolecular. **Table 3.2** summarizes the different kinds of electronic materials and provides a representative example of each.

TABLE 3.1	*Basic Electronic Definitions and Relations*	
Name	**Form**	**Units, comments**
Ampere	A (3.1)	The "old" SI definition of the ampere: A force (generated between two parallel, straight wires of infinite length with negligible cross section in vacuum and carrying current in the same direction) equals to 2×10^{-7} N·m^{-1}
Coulomb	C (3.2)	
Elementary charge	e (3.3)	Elementary charge $e = 1.6022 \times 10^{-19}$ C. 1 C = 6.2415×10^{18} charge carriers. 1 A = 1 C·s^{-1}
Current	I or i (3.4)	Current is in terms of Amperes and is symbolized by I or i.
Current density	$J = \dfrac{A}{m^2}$ (3.5)	Current density is in terms of amperes per square meter
Volt	$V = \dfrac{J}{C}$ (3.6)	1 V equals 1 J·C^{-1}
Power	W (3.7)	Electric power is given in terms of watts: J·s^{-1}. The kilowatt·hour is a measure of total energy used. Also $W = i \cdot V = V \cdot A$, the rate of electrical energy transfer
Charge	$q = it$ (3.8) $i = \dfrac{dq}{dt}$ (3.9)	Total charge in coulombs, (C·s^{-1}, current) × (t, time). Charge is quantized: $q = N_e$, where N_e is the number of electron charges
Coulomb's law	$F = \dfrac{1}{4\pi\varepsilon_o}\left(\dfrac{q_1 q_2}{r^2}\right)$ (3.10)	Force in newtons, N, where ε_o is the permittivity of free space and r is the radial distance between the two charges in m ($\varepsilon_o = 8.854 \times 10^{-12}$ C^2·N^{-1}·m^{-2} or C·V^{-1}·m^{-1}). $1/4\pi\varepsilon_o = 8.99 \times 10^9$ N·m^2·C^{-2}
Electrostatic energy	$E = \dfrac{1}{4\pi\varepsilon_o}\left(\dfrac{q}{r^2}\right)$ (3.11)	Taking a test charge from a distance r from a point charge, the magnitude of the electric field E is F/q_1
Electrostatic potential	$V = \dfrac{1}{4\pi\varepsilon_o}\left(\dfrac{q}{r}\right)$ (3.12)	The electric potential V due to a test charge q relative to a particle the zero potential taken to infinity. r is the radial distance
Charge density	$D = \varepsilon_o E$ (3.13) $D = \varepsilon_o \kappa E$ (3.14) $n_{p,n,e,h} = \dfrac{N_{p,n,e,h}}{m^3}$ (3.15)	κ is the relative permittivity—a dimensionless number commonly known as the *dielectric constant*. E is the strength of the electric field in V·m^{-1} n is the charge carrier density in m^{-3}. N is the number of charge carriers where p, n, e, h are positive, negative, electrons, holes, or other forms of charge carriers, respectively ($n_{Cu} = 8.49 \times 10^{28}$·m^{-3} and $\rho_{Cu} = 1.69 \times 10^{-8}$ Ω·m; $n_{Si} = 1 \times 10^{16}$·m^{-3} and $\rho_{Si} = 2.5 \times 10^3$ Ω·m)
Ohm's law and current density	$V = IR$ (3.16) $J = \sigma E$ (3.17)	Voltage (in volts, V) equals current (I, in amperes A or C·s^{-1}) times resistance (R or Ω, in Ohms). Current density J is in terms of A·m^{-2} or C·s^{-1}·m^{-2}. Conductivity σ is in terms of reciprocal ohm·cm (usually as Ω^{-1}·cm^{-1}). E is the strength of the electric field: $E = $ V·L^{-1} (usually as V·cm^{-1})
Current density	$J = n_e \upsilon e$ (3.18)	Current density as a function of electron carriers in terms of charge per unit area. n_e is the number of electrons per unit volume; υ is the velocity of the electrons in m·s^{-1}. J is in terms of A·m^{-2} or C·m^{-2}·s^{-1}

(*continued*)

Name	Form		Units, comments
TABLE 3.1 (CONTD.)	**Basic Electronic Definitions and Relations**		
Farad	$$F = \frac{A \cdot s}{V} = \frac{C}{V} = \frac{C^2}{J} = \frac{C^2}{N \cdot m}$$ $$= \frac{s^2 \cdot C^2}{m^2 \cdot kg} = \frac{s^2 \cdot A}{m^2 \cdot kg}$$	(3.19)	The SI unit for capacitance. Derivation of the SI units of F is shown at the left [1]
Resistivity and conductivity	$$\rho = \frac{\Omega A}{m}$$ $$\sigma = \frac{1}{\rho}$$	(3.20) (3.21)	*Resistivity* is resistance times area divided by length in terms of $\Omega \cdot m$ *Conductivity* is the reciprocal of resistivity in terms of $\Omega^{-1} \cdot m^{-1}$. Also called siemens (or *mhos* or reciprocal ohms, $A \cdot V^{-1}$)
Carrier mobility	$$\mu = \frac{\bar{v}}{E}$$ $$\bar{v} = \frac{J}{eN_e}$$	(3.22) (3.23)	Where $\bar{v}$ is the drift velocity (in $m \cdot s^{-1}$), the average velocity of electrons due to an external electric field. It is assumed to be constant in a crystal (or changes vary slowly) [3]
Conductivity/ carriers	$$\sigma = nqu$$ $$\sigma = n_e q_e u_e$$ $$\sigma = n_n q_n \mu_n + n_p q_p \mu_p$$	(3.24) (3.25) (3.26)	Conductivity is dependent on the charge volume density of carriers n in m^{-3} (negative and positive) × the total charge q × the mobility. For multivalent ions, the magnitude of q depends on the valency Z_i of the ion. For monovalent ions and electrons, $q = e$. The equation for a pure electronic condition is shown by equation (3.26) [2]

3.0.3 *Fundamental Kinds of Electronic Devices*

During your day-to-day experience, you will most likely run across an array of electronic devices—from your personal devices like cell phones, digital cameras, GPS trackers, computers, and DVDs to sophisticated devices found outside your home such as ATMs, cash registers, security equipment, sophisticated laboratory equipment, and automobile gadgets. All of these devices share similar components—transistors, capacitors, voltage regulators, rectifiers, converters, and many more. Many of these already are marketed as nano-versions—many are totally new kinds of devices (can you think of any?). There are many kinds of electronics that are unique to the nano and quantum domains. Some of them were already described in chapter 2.

The *transistor* is the heart of modern day electronic devices and regulates the flow of current or voltage. Therefore, it is a solid-state device capable of amplification, switching, and signal generation. Semiconductor transistors consist of p-type silicon and n-type silicon to form the p/n-junction, commonly known as the depletion zone—where positive and negative charges cancel each other (**Fig. 3.1**). Applications of transistors include their use in rectifying junctions, amplifiers, and switches. A *rectifying junction* is a junction that allows current to flow in just one direction. Most transistors in today's devices consist of *complementary metal oxide semiconductors* (CMOS), devices that employ two complementary transistors per gate in microprocessors. *Gates* control output that is dependent on the combination of inputs (e.g., logic gates). *Field effect transistors* (FET) are used to amplify weak signals. MOSFETs are metal oxide semiconductor FETs.

Batteries are centers of electric potential derived from chemical energy that provide static charge potential, or electromagnetic force, required to power devices. Batteries provide the necessary static charge potential to drive reactions,

TABLE 3.2	*Electronic Materials*		
Material	**Description**	**Examples**	**Applications**
Conductors	Materials that have a surplus of "free" charge carriers. Electrons in the conduction band behave like an electron gas. Materials with relatively low electrical resistance and no bandgap. The measure of electrical conductivity is σ, the ratio of the current density J to the electric field strength E in terms of siemens per meter ($S \cdot m^{-1}$ or $\Omega^{-1} \cdot m^{-1}$). Conductivity between $10^4 < \sigma < 10^8 \ \Omega^{-1} \cdot m^{-1}$	Metals like Ag, Cu, Au, Pt, Al, metal alloys, ionic SLNs, tap water, graphite, and some kinds of carbon nanotubes	Charge carriers in integrated circuit devices, transmission wires, magnetic coils, single-electron devices, low resistance connectors between devices
Super-conductors	Materials with incredibly low electrical resistivity with exclusion of an interior magnetic field (Meissner effect). The resistance is essentially zero (ca. 10^{16} times less than room temperature values) when below some critical temperature T_c (usually between 0.01 and 125 K) [3]. There are two types of superconductors: Type I can be destroyed by a magnetic field and Type II only gradually so. Type I is characterized by the presence of Cooper pairs	27 elements, many alloys, Cu_xO_y, ceramics and Se- or S-based organic materials [3]	Super-efficient electronic circuits that generate no heat—lossless power transmission lines, high-speed levitated trains, faster computers and switching devices called cryotrons [3], also NMR, MRI, and Josephson junctions. Superconducting digital electronics, quantum computing, microwave communication systems
Ballistic conductors	Crystalline character of highly pure materials allows electrons to transit without scattering—different from superconductivity due to lack of Meissner effect. Conductivity given by Landauer's equation. Current density 100 to 1000x better than Al	Multiwalled nanotubes are ballistic conductors at room temperature [4]	Energy-efficient electron transport and information processing
Semi-conductors	Materials that have a bandgap with energy in the IR and UV-visible range. Conductivity: $10^{-4} < \sigma < 10^4 \ \Omega^{-1} \cdot m^{-1}$. The bandgap energy E_g is <2 eV [2]. Electrons (e) and holes (h) are charge carriers	*Intrinsic semiconductors* are pure elemental materials (Si) where $N_e = N_h$. *Extrinsic semiconductors* are doped with impurities	Transistors, solar cells, rectifier junctions
Insulators	Conductivity is very low: $10^{-10} < \sigma < 10^{-16} \ \Omega^{-1} \cdot m^{-1}$. Characteristic bandgap energy >2 eV [2]. Low density of electron charge carriers (N_e)	Industrial ceramic materials—Al_2O_3 ($\kappa = 10.1$), BeO ($\kappa = 6.7$), polyester ($\kappa = 3.6$), borosilicate glass, polyethylene	Insulators, capacitors, ferroelectric materials, piezoelectric materials

(*continued*)

TABLE 3.2 (CONTD.)	*Electronic Materials*		
Material	**Description**	**Examples**	**Applications**
Dielectric materials	Insulating materials (see above) that are able to store charge via process of polarization by external field. Electrons are bound to atoms. There is very limited or no mobility of charge carriers. Insulators have completely filled valence bands and empty conduction bands for example, no intraband transitions. Bandgaps are typically >5 eV, and therefore, limited IR or visible transitions (except phonons)	NaCl and other alkali halides, fused quartz, polymeric materials	Capacitors, insulators
Ferroelectric materials	Capacitors below T_c with the ability to instantaneously switch polarity by the application of an external electric field—a condition called spontaneous polarization	Ceramic materials like $BaTiO_3$ due to its tetragonal microstructure	Capacitors, filters, resonant tunneling devices, ferro-electric RAM's, holographic 3-D storage, photonic networks
Piezoelectric materials	Piezoelectric materials are able to convert electricity into mechanical energy. Conversely, when pressure is applied, piezos generate electricity. The fraction of mechanical energy converted to electrical energy is k. Piezoelectric materials are considered ferroelectrics	Quartz or SiO_2, ($k = 0.1$); $BaTiO_3$, ($k = 0.1$) and combinations of $PbTiO_3$ and $PbZrO_3$ (PZT) [Shack]	Transducers convert one kind of energy into another kind of energy. AFM and STM raster mechanics, quartz crystal microbalances, radios, electronic devices
Organics and polymers	Electronic materials made of organic (carbon-based) materials rely on charge transfer complexes and conjugated π-systems. Mechanisms include: mobility gaps, tunneling, phon-assisted hopping; resonance stabilization, delocalization of π-electrons. Organic photoelectric complexes often involve a chromophore and a semiconductor	Conductive polymers: polyaniline (PANI), polypyrrole (PPY), polyacetylene (PA) Chromophores: ruthenium bipryridine (Rubpy), chlorophyll	Molecular electronics, solar cells, quantum devices Photosynthesis
Biomolecular materials	Use of biological materials in electronic circuits. Proteins, lipids, etc., are able to transfer charge, transfer charged molecules, undergo color changes	DNA Bacteriorhodopsin Self-assembled monolayers Proteins/lipids	Neuroelectronic interfaces Biological motors Storage elements Electronic switches, gates, biosensors, and biological transistors [5]
Nano-electronic materials	Analogous to macroscopic counterparts but possess nanoscale dimensions. Biomaterials can produce cascades. Electronic properties of nanomaterials may be quantized	Carbon nanotubes Quantum dots Organic thin films Inorganic nanowires Small molecules (e.g., rotaxanes and catenanes) Charge transfer complexes	Single-electron transistors Nanowires NEM systems Unimolecular rectifiers Computer transistors Field-effect transistors Nanosensors

Fig. 3.1

Schematic rendition of a prototype silicon electronic device. When forward bias is applied to the p-type semiconductor terminal (as depicted in the figure), the holes are pushed away from the positive contact and electrons away from the negative contact towards the p/n junction and electric current is made to flow. When no potential is applied, the depletion zone consists of canceled hole–electron pairs, a process known as recombination. When increasing bias potential, the depletion zone is thinned so that charge carriers are able to tunnel through the junction as electrical resistance is decreased.

devices, or processes. Nanotechnology has already created a new generation of batteries that outperform anything on the market today. Battery components include the *anode*, the *cathode*, and the *electrolyte* (solid or liquid).

A *capacitor* is a passive electronic component that is able to store energy. Its mode of energy storage is not like that of a battery. Capacitors, consisting of a dielectric material sandwiched between two conducting plates, store energy in the form of charge. *Capacitance* is directly proportional to the surface area of the electrical contact with the dielectric and measured in terms of farads (**Table 3.1**). Capacitors are used in conjunction with transistors and form two core elements in integrated circuits. A *resistor* is another passive component that provides a source of resistance in an electronic circuit, for example, resisting the flow of electrons. Its function is to produce a voltage drop across its terminals. Resistors also produce heat when activated and are usually made of carbon particulates with a binder material. A *converter* is a device that processes AC power into DC power or vice versa. *Transducers* are devices that convert a physical signal into one that is electronic. The stimulus may be physical (e.g., pressure), chemical, or biological. These devices all have nanoscale analogs that operate in similar ways. Nanoscale sensors, transducers, transistors, rectifying junctions, resistors, and capacitors all group together in nano-integrated circuits to perform predetermined functions.

3.0.4   *The Nano Perspective*

The first level of function in the biological world is at the nanoscale—as atoms and molecules react stochastically without any perceived function or objective except to find the lowest energy state. Nature has optimized the nanoworld—optimized its *special nanoelectronics* and has shown an incredible track record for several hundreds of millions of years—one that humbles us for sure. We are, however, at the stage where atoms and molecules can be manipulated by us with due intent—to do our bidding as it were. Yes, we have controlled atoms, molecules, and electrons before in our chemical reactions, but we have never controlled them singly. Nanotechnology is the application of nanoscience to industry and commerce. We take nanoscience and turn it into devices, materials, and applications. We take an atom, a molecule, or an electron and tell it to "stay" or "go."

Carbon nanotube supercapacitors, triodes, and diodes; the use of biological and complex organic molecules in circuits, nanowire electrical conduits, quantum dot relays, gigantomagnetic capability, nanoresonators, single-electron transfer, single-photon optics, zeptomole level analytical capability, atomic level and better resolution, advanced solar cells and components, quantum computing and spintronics and nanoscale field emitters—the list goes on and the possibilities are endless.

3.1 MICROELECTRONICS

In 1904, Sir Ambrose Fleming invented the thermionic valve (diode or two electrode rectifier). The age of microelectronics officially began with the invention of the germanium transistor, used initially to amplify electrical currents, in the late 1940s by J. Bardeen. W. Brittain, and W. Shockley, all of Bell Labs. It was smaller, cheaper, and more efficient than the vacuum tube it replaced. Use of silicon grew popular due to its superior properties and the ability to grow it as a single crystal with high purity. Silicon also readily forms a thin oxide layer (better than germanium)—a layer that serves as a gate oxide and demonstrates facility during fabrication and function. Then William Shockley left Bell Labs to form Shockley Semiconductor after he developed the junction (sandwich) transistor.

In 1957, eight colleagues, known as the "Fairchild Eight," left Shockley's company to form Fairchild Semiconductor, and two of the eight, R.N. Noyce and G. Moore, left to form Intel [6]. The planar process to form integrated circuits was underway [7]. The process makes use of silicon's thin oxide layer through which impurities are able to diffuse. The concept of the integrated circuit was developed in 1958 when R.N. Noyce (Fairchild and the Intel) and J. St. Clair Kilby (Texas Instruments) independently placed transistors and circuits on a silicon wafer or chip [6]. Noyce and Kilby did not know each other at the time of their breakthrough, but their two respective companies engaged in a lawsuit that lasted well into the 1960s [8]. According to Gordon Moore of Intel (the author of Moore's Law), 10^{17} transistors are manufactured every year—an amount equivalent to the number of ants in the world [6]. That figure breaks down to about three billion transistors manufactured per second!

3.1.1 *Introduction to Band Structure*

Charge carriers and their mobility are responsible for electrical conduction. Charge carriers can be electrons, holes, or positive and negative ionic species. Semiconductors differ from metals in that electrical conductivity increases as a function of temperature—whereas in metals conductivity decreases with increasing temperature due to increasing resistance. The average velocity of the charge carriers is the product of the carrier mobility μ times the strength of the electric field E.

The Electronic Structure of Atoms and Molecules. According to the *Pauli exclusion principle*, no two electrons can be in the same exact state, for example, two electrons in the same orbital must have opposite spins to form a closed shell. Each electron of an atom is defined by four quantum numbers that form a unique set: n, l, m, and s (principal, two orbital angular momentum, and spin, respectively). The electrons fill consecutive levels of atomic orbitals. The s orbital can accommodate 2 electrons, the p 6, the d 10, and the f 14, and so on and so forth.

Types of Solid Materials. When metal atoms come together to form molecules or clusters, a reshuffling of electrons and energy levels among the valence electrons takes place (core shell electrons that exist in filled shells remain unchanged). If these atoms have unfilled valence shells, electrons become mobile and contribute to a phenomenon known as an electron cloud. In other words, valence electrons donated by atoms become delocalized and spread over the entire structure—an extension of molecular orbital theory. The luster, high electrical and thermal conductivity, and malleability of metals are due to this electron cloud. Why is a metal conducting and why is an insulator non-conducting?

Ionic solids are formed by strong nondirectional Coulombic attractions between ions and have large bond energies on the order of 2–4 eV per atom. Low electrical conductivity (no free electrons), water solubility, and visible light transparency (visible photons do not have enough energy to excite electrons) are properties associated with ionic solids. Covalent solids (Group IV elements like carbon and silicon and Group III–V molecules like gallium arsenide) have strong localized, directional bonds with high bond energies on the order of 4–7 eV per atom. Covalent solids have stable closed shell configurations. Higher melting points, low electrical conductivity, and solubility in nonpolar media (hydrocarbons) are characteristic of covalent solids. Metallic solids are characterized by Coulombic attractions between the positive atom cores and the negatively charged and delocalized valence electron gas (or electron sea). The cohesive bond energy of metals is less, between 1–4 eV and metals therefore are able to absorb visible light. The bonds in metals are nondirectional and allow for free electron movement throughout the crystal structure of the metal.

Band Theory. Sodium serves as a good example for the formation of electronic bands in metals. Sodium has one electron in its valence shell, the $3s^1$ electron. If two sodium atoms form a bond, two molecular orbitals are formed: A σ bonding orbital and a σ^* antibonding orbital. The energy of an occupied bonding orbital is lower due to constructive orbital overlap (constructive interference) between the two atomic orbitals and the resulting enhanced amplitude in the internuclear region [9]. This gives rise to bond strength between the two atoms. The antibonding orbital has higher energy due to destructive interference between two electrons that cancels their respective amplitude and generates a node. These electrons do not reside in the internuclear region [9].

FIG. 3.2

Band structure of sodium (Na) is depicted (only even numbered clusters are shown for ease of illustration—for Na_3, there would be an orbital midway in energy between the bonding and antibonding orbitals—a nonbonding orbital). The black dots represent electrons. The electron configuration of sodium is $1s^2 2s^2 2p^6 3s^1$. There is one valence electron. When a molecular orbital is formed between two Na atoms, splitting occurs to form HOMO and LUMO molecular orbitals. As more atoms are added to the cluster, a band is formed. In metals, the bandgap is small and can be overcome by thermal energy and electrons can be easily promoted into the conduction band. The progression is described by the following: atom → molecule → cluster → nanoparticle → bulk material. In each case, going from atoms to the bulk, the bandgap energy decreases as more states become populated.

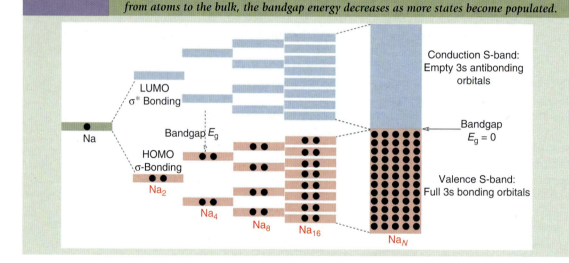

Electrons in a metal (or crystal) become free when excited and move into unoccupied molecular orbitals (LUMO) that lie just above the occupied molecular orbitals of highest energy (HOMO). In a metal, this requires little energy, which is provided by thermal energy $k_B T$ (~25 meV). From molecular orbital theory, electronic bands are formed when many atoms come together to form a solid. **Figure 3.2** depicts the change in the electronic orbital structure of sodium as more atoms are added. Compare the electronic structures of sodium (11 electrons, $1s^2 2s^2 2p^6 3s^1$) with neon (10 electrons, $1s^2 2s^2 2p^6$). Sodium is conducting and neon has a band gap of ~20 eV—the energy between the closest filled and empty bands. The determination of electrical conduction relies on two parameters: whether the band is full or not and the degree of separation between the bands.

Bands made of just s-orbitals are called s-bands. Bands that include p-orbitals also form if the energy gap between them is not too great. Many metals that have the capability to form d-bands are constructed from overlap of many d-orbitals [9]. Bandgap energies exist between orbital energies (e.g., s, p, and/or d) or between HOMO and LUMO bands. Magnesium (Mg), for example, has two electrons in the 3s shell but three low-lying 3p orbitals. In this case, the valence band (the 3s band) is full and the conduction band, the 3p band, is empty. Due to the overlap between the 3s and 3p bands, called an sp-band, there is no bandgap and therefore, Mg is a conductor.

The Fermi Energy.   A Fermi–Dirac distribution describes the population of electrons (or Fermions, particles with 1/2 spins, that obey the Pauli exclusion principle) as a function of the Fermi energy, the Boltzmann constant, and temperature. At absolute zero, electrons populate all available energy states below the Fermi energy E_F (**Fig. 3.3**). Due to thermal excitation at higher temperatures, electrons start to occupy the conduction band and deplete the valence band. Electrons that occupy states at the higher energy states of the valence band (the HOMO) become mobile with thermal stimulation. The Fermi–Dirac behavior is described by the following relation.

$$f(E) = \frac{1}{e^{\left[(E - E_F)/k_B T\right]} + 1}$$

(3.27)

where
  $f(E)$ is the Fermi function
  E_F is the Fermi level
  k_B is the Boltzmann constant

FIG. 3.3   *The Fermi function at T = 0 K (left) and T > 0 K (right). The probability (between 0 and 1) that an electron will populate an energy state is a function of the band structure and the temperature. As temperature increases, thermal excitation of electrons promotes high-lying electrons from the valence band into the conduction band. The Fermi level is defined at 0 K and represents the energy of the highest filled state in the valence band. At f(E_F), the value is exactly 0.5. When E > E_F, enough electrons are promoted that conduction of electricity is possible.*

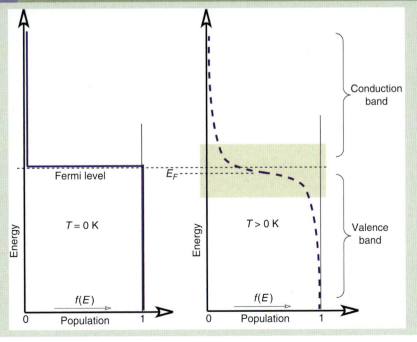

The distribution describes the population of electrons in the band structure as a function of energy and temperature—the probability that an energy level is occupied by an electron.

There are two terms that you will encounter from time to time: *intraband transitions* and *interband transitions*. When a photon excites an electron to a higher level within the same band, it is called an intraband transition—a transition from occupied to empty states within the same band with the participation of a phonon [3]. Transitions within the conduction band for example are called

EXAMPLE 3.1	*Probability of Electron Promotion*

What is the room temperature (300 K) probability of electron promotion from the valence band into the conduction band of four common materials: diamond, graphite, germanium, and silicon? (Hint: Assume that the Fermi level lies halfway between the conduction and valence bands).

Solution for Diamond:
The bandgap for diamond is 5.5 eV.

$$E_F = \frac{E_g}{2} = \frac{5.5\,eV}{2} = 2.75\ eV; \text{ and that } k_B = 1.381 \times 10^{-23}\,J \cdot K^{-1} = 8.617 \times 10^{-5}\,eV \cdot K^{-1}$$

$$f(E)_{Diamond} = \frac{1}{e^{\left[(2.75\,eV)/(8.617 \times 10^{-5}\,eV \cdot K^{-1} \cdot 300\,K)\right]} + 1} = \frac{1}{e^{106.4} + 1} = \frac{1}{1.58 \times 10^{46}} = 6.31 \times 10^{-47}$$

Solution for Graphite:
The bandgap for graphite is 0.0 eV. It is a good conductor of electricity. The band is a p-band that consists of abutting π and π^* molecular orbitals in the solid.

$$f(E)_{Graphite} = \frac{1}{e^{\left[(0\,eV)/(8.617 \times 10^{-5}\,eV \cdot K^{-1} \cdot 300\,K)\right]} + 1} = \frac{1}{e^0 + 1} = \frac{1}{2} = 0.50$$

Solution for Pure Silicon (no doping):
The bandgap for silicon is 1.09 eV.

$$E_F = \frac{E_g}{2} = \frac{1.09\,eV}{2} = 0.545 eV$$

$$f(E)_{Silicon} = \frac{1}{e^{\left[(0.545\,eV)/(8.617 \times 10^{-5}\,eV \cdot K^{-1} \cdot 300\,K)\right]} + 1} = \frac{1}{e^{21.43} + 1} = \frac{1}{1.43 \times 10^9 + 1} = 6.98 \times 10^{-10}$$

Solution for Pure Germanium (no doping):
The bandgap for silicon is 0.72 eV.

$$E_F = \frac{E_g}{2} = \frac{0.72\,eV}{2} = 0.36 eV$$

$$f(E)_{Germanium} = \frac{1}{e^{\left[(0.36\,eV)/(8.617 \times 10^{-5}\,eV \cdot K^{-1} \cdot 300\,K)\right]} + 1} = \frac{1}{e^{13.9} + 1} = \frac{1}{1.12 \times 10^6 + 1} = 8.95 \times 10^{-7}$$

Both silicon and germanium offer a much better chance at conduction than diamond. Although not in doped states, the two materials are genuine intrinsic semiconductors.

FIG. 3.4 *Electrical conductivities of some common materials.*

intraband transitions. Intraband transitions are not quantized and occur only in metals with maximum energy correlating to the minimum and maximum band edge [3]. If more energetic radiation is applied, interband transitions become stimulated. The mechanism of absorption depends on the energy difference between the bands. Interband transitions occur in both metals and semiconductors and are similar to optical resonance excitation of bound electrons [3]. A transition from the valence band to the conduction band is an example of an interband transition.

3.1.2 Basic Conductor and Semiconductor Physics

Conductors. **Figure** 3.4 lists materials that traverse the electrical conductivity continuum.

In metals, valence and conduction bands overlap. There are then no forbidden energy gaps. Metals, therefore, are excellent conductors. At the temperature of absolute zero, electrons are in the lowest energy levels with the highest filled level known as the Fermi level E_F. At ambient temperatures, some electrons from these highest energy levels are excited by thermal stimuli to empty levels above. The E_F represents the energy at which the energy levels are half filled or half empty.

Conductors or insulators are characterized by the degree of separation of valence and conduction bands and whether the bands are filled or empty. If N atoms make up a solid, each band may consist of $2N$ electrons, in the sense relating to filling of electronic orbitals. Let us consider an example consisting of two elements, Ne and Na. The number of electrons in sodium is odd, therefore the last band up is only half filled. Ne on the other hand has an even number of electrons and the last "band" is therefore full (**Table** 3.3).

TABLE 3.3	*Electronic Configurations of Neon and Sodium*	
Name	**No. of electrons**	**Configuration**
Neon	10	$1s^2\,2s^2\,2p^6$
Sodium	11	$1s^2\,2s^2\,2p^6\,3s^1$ or [Neon] $3s^1$

For a solid material to conduct electricity, conduction electrons must be induced to flow by addition of some energy from an external applied voltage. The energy level of conduction electrons is approximated by the thermal energy k_BT, equal to ca. 0.025 eV, at room temperature. The relative separation of bands therefore is a consideration, and we should also consider the energy of the next unfilled band. Thus in this example, we have:

Neon
- The energy separation between the nearest empty band and the closest filled band is around 20 eV.
- For conduction to take place in neon, the electrons in the upper most band, completely filled, would have to "jump" across the band gap of 20 eV, very unlikely with only 25 meV of energy k_BT (thermal energy)!

Sodium
- The band is partially filled, thus a "little bit" of energy can be added to the electrons—no quantum jump is involved.
- The final electrons "added" to make sodium, the $3s^1$ electrons, are very easy to strip from the "neon" ion core. They become the conduction electrons, often known as the *free electrons* in the metal.

The electronic configurations of the following elements that happen to be excellent conductors as solids are listed in **Table 3.4**:

This is not to say that all conductors have one electron in an unfilled band, but certainly the best conductors do!

Insulators. An insulator has a wide forbidden energy gap as we have noted earlier. The valance band is completely filled, and the conduction band is completely empty. Since there are no free electrons present, an insulator should not be able to conduct electricity. An electrical insulator, therefore, resists the flow of electricity. Application of a voltage difference across a good insulator results in negligible electrical current. In comparison, a conductor allows current to flow readily. Electrical wiring and electronic circuits require both insulators and conductors. For example, wires typically consist of a current-carrying metallic core sheathed in an insulating coating.

Resistivity. Resistivity is the measure of a material's effectiveness in resisting current flow. Materials with resistivities higher than 10^8 $\Omega \cdot$m are considered as

TABLE 3.4	*Electronic Configurations of Conductors*	
Name	**No. of electrons**	**Configuration**
Aluminum	13	[Neon] $3s^2$ $3p^1$
Copper	29	[Argon] $3d^{10}$ $4s^1$
Silver	47	[Krypton] $4d^{10}$ $5s^1$
Gold	79	[Xenon] $4f^{14}$ $5d^{10}$ $6s^1$

good insulators. Glass, rubber, and many plastics are well-known examples of good conductors. Resistivities as high as 10^{16} $\Omega \cdot m$ can are attained in some exceptional insulating materials. Conductors on the other hand have resistivity values as low as 10^{-8} $\Omega \cdot m$.

Charge carriers are responsible for transporting current. One kind of charge carrier is the electron. In most conductors, electrons move relatively freely, while in insulators electrons do not move freely. When atoms of simple metals combine to form a solid, the outer valence electrons become free for conduction. In an ideal insulator, all electrons stay tightly bound to the atoms, so there are no electrons that can be readily moved through the material for conduction.

In order to understand better the mechanism of conduction (and hence, insulation), consideration of electronic band structure is required. The electrons in an isolated atom possess discrete energies, a consequence of quantum mechanics. These discrete levels evolve into bands of allowed energies when the atoms condense into a solid. Forbidden regions separate the allowed bands, as schematically displayed in **Figure 3.2**. The electrons in a solid fill in the bands, from lower to higher energy.

The distinction between an insulator and a conductor is how the electrons fill in the allowed bands. For a simple metal, the highest band containing electrons will be only half full. The thermal energy (at ordinary temperatures) will be sufficient to generate conduction electrons—electron states of slightly higher energy are available in the incompletely filled band.

In comparison, the highest energy band containing electrons is completely full in a good insulator. The thermal energy of the electrons is not sufficient for promotion from this band, known as the *valence band* (VB), to the next band with available energy states, known as the *conduction band* (CB). The gap between the valence and conduction band, known as the band gap, is at least several electron volts (*eV*) wide in an insulator—thermal electron energies are 100 times smaller.

Semiconductors. The energy band model of a semiconductor is similar to that of an insulator except that the forbidden energy gap is relatively narrow in comparison. At very low temperatures, semiconductors behave like insulators. However, at higher temperatures, a few electrons are promoted from the VB to the CB by simple thermal excitation. Electrons promoted to the CB now are able to conduct electricity in the presence of an applied electric field. Corresponding electron vacancies, or holes, created in the VB contribute to over all conductivity. Two commonly used semiconductor materials are silicon (bandgap = 1.1 *eV*) and germanium (bandgap = 0.7 *eV*).

The topic of semiconductors is enormous and we will provide only a cursory overview. Since the invention of the transistor ca. 40 years ago, three single-component, ten binary, and a few tertiary semiconductor materials (e.g., mercury cadmium telluride [HgCdTe] and aluminum gallium arsenide [AlGaAs]) have been developed. These single and binary semiconductors, ordered by the magnitude of their bandgap, are listed in **Table 3.5**. Others not listed include ones that have significant potential like mercury telluride (HgTe), manganese selenide (MnSe), gallium antimonide (GaSb), indium nitride (InN), scandium nitride (ScN), aluminum nitride (AlN), zinc selenide (ZnSe), and boron nitride (BN).

TABLE 3.5	*Common Semiconductor Materials*		
Material	**Bandgap (eV)**	**Bandgap type**	
InSb	0.230	Direct	
InAs	0.354	Direct	
Ge	0.664	Indirect	
Si	1.124	Indirect	
InP	1.344	Direct	
GaAs	1.424	Direct	
CdTe	1.475	Direct	
AlAs	2.153	Indirect	
GaP	2.272	Indirect	
ZnTe	2.394	Direct	
SiC	2.416	Indirect	
GaN	3.503	Direct	
C	5.5	Indirect	

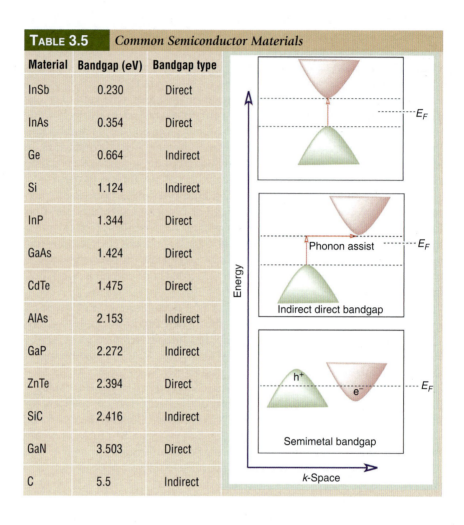

Table 3.5 also indicates the nature of the bandgap, that is, whether it is direct or indirect gap material, as this determines to a large extent just what kinds of applications the semiconductor may be suited for. Direct bandgap semiconductors are those in which the minimum energy of the CB lies directly on top of the maximum energy of the VB (in momentum or *k*-space). Therefore, in direct bandgap semiconductors, electrons in the conduction band have a high probability to combine directly with holes that exist in the valence band without any loss of momentum. When recombination does occur, a photon is released that corresponds to the bandgap energy (radiative or spontaneous recombination) [10]. Gallium arsenide and copper indium diselenide (CuInSe$_2$) are good examples of direct bandgap semiconducting materials. Direct bandgap materials find applications in light-emitting diodes and lasers [10].

In crystalline silicon, recombination does not occur as directly. In "momentum space," the VB minimum and CB maximum are not aligned perfectly. Silicon is an example of an indirect bandgap semiconducting material. In this case, direct transition is forbidden (e.g., does not conserve momentum). Recombination does occur however but it is in a form that is not mediated by the emission of a photon. Energy from excitation is released in the form of phonons. *Semimetals*, another class of semiconductors, are those in which the lowest energy of the conduction band is actually lower in energy than the highest state of the VB. In this scenario, however, the lowest state of the VB is shifted in momentum space from the highest state of the VB. Arsenic, antimony, bismuth, HgTe, and graphite are examples of semimetal materials. Boron nitride (white), although having the same structure (isoelectronic) as graphite (black), is a semiconductor. Graphical representations of direct, indirect semiconductors, and semimetals are shown in the image in **Table 3.5**.

Intrinsic Semiconductor: Group IV Elements (Si and Ge).   An intrinsic semiconductor is a pure elemental semiconducting material (ideal) that has no chemical impurities. No atoms are displaced from their proper sites in the structure of the material. This state is highly unlikely however. For such an ideal material at absolute zero, the valence band is filled completely and the conduction band is absolutely empty. At temperature, $T > 0\,K$, two processes occur: (1) Due to thermal vibration of the lattice, electron–hole pairs (EHPs) are generated due to break up of some covalent bonds and (2) recombination of EHPs to produce covalent bonds also occurs in thermal equilibrium. The number of electrons per unit volume (n_i) in CB is equal to the number of holes per unit volume (p_i) in VB. This means that the Fermi level, E_F, is located in the middle of the forbidden energy gap. Hence, at $T > 0$ charge carriers are obtained by breakdown of some covalent bonds.

Extrinsic Semiconductors.   The electrical properties of a semiconductor are drastically altered when doped by "impurity atoms." Such a solid is called an extrinsic semiconductor. There are two types of extrinsic semiconductors: *n-type semiconductors* and *p-type semiconductors*.

n-Type Semiconductors.   When a group IV element (Si/Ge) is doped with a group V element (P/As/Sb), an n-type semiconductor is formed (**Fig. 3.5**). At temperatures

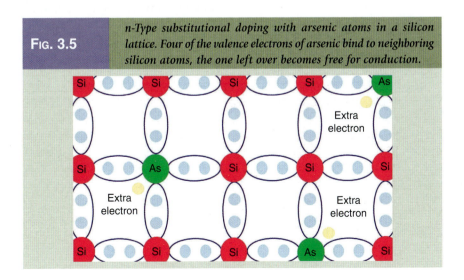

FIG. 3.5 *n-Type substitutional doping with arsenic atoms in a silicon lattice. Four of the valence electrons of arsenic bind to neighboring silicon atoms, the one left over becomes free for conduction.*

FIG. 3.6	*Energy level diagram of a typical n-type semiconductor. Here E_{Fi} is the Fermi level of an intrinsic semiconductor.*

above absolute zero, four of the impurity atom electrons play the same role as the four valence electrons of the VB. The fifth valence electron is unbonded and is therefore considered to be free. Thus, addition of donor atoms provides for additional allowed energy states to the semiconductor (**Fig. 3.6**).

At very low temperatures, $T < 100\,K$, electrons are attached to the donor atoms, that is, they occupy the E_d state (see **Fig. 3.6**). At around $100\,K$, the thermal energy enables the extra impurity electron to shift into one of the many empty states of the nearby conduction band, where it has effective mass m_e^*, mobility μ_e, and can carry current. Note that in an n-type material, there are two types of charge carriers from three processes:

> Doping—produces electron as charge carriers (majorities carriers)
> Thermal vibration—produces electrons and holes
> Recombination—takes away electron and holes
> > $n \equiv$ defined as the electron density
> > $p \equiv$ defined as the hole density

In an extrinsic semiconductor in thermal equilibrium, N_d is the donor concentration. The product of concentrations of majority and minority carriers in thermal equilibrium is independent of the doping impurity concentration and is a function of only of the temperature and the semiconductor material, that is, $n_p = n_i2$ (constant) for every T.

Here, $n \gg n_i \gg p$.

Arsenic and phosphorous have one more valence electron than silicon (Si). When it is placed substitutionally in the lattice of a Si atom, four of the valence electrons bind to neighboring Si atoms, the one left over becomes free for conduction. In the band structure picture, impurity states are created much closer to the conduction band. These states are known as donors, since they "donate" electrons to the conduction band. It is much easier to "liberate" the electron for conduction at lower thermal energy. A semiconductor doped with donors is known as an "n-type" semiconductor (n for negative charge carrier).

p-Type Semiconductors.   When a group IV element (Si/Ge) is doped with a group III element (In/Ga), a p-type semiconductor is formed. In this case, there is a vacancy in the covalent bond structure.

FIG. 3.7 *Energy level diagram of a typical p-type semiconductor. Here E_{Fi} is the Fermi level of an intrinsic semiconductor.*

Addition of the impurity adds allowed energy states (E_a) that are empty at absolute zero. At higher temperature ($T > 100\,\text{K}$) electrons from VB can jump to E_a and leave a vacancy behind (**Fig. 3.7**). In a p-type material, the following processes occur:

Doping—produces holes as charge carriers (majority carriers)
Thermal vibration—produces electrons and holes
Recombination—takes away electrons and holes
Here, again $n_p = n_i 2$. Here, $p \gg n_i \gg n$.

Gallium or boron has one fewer valence electron than silicon—when it is placed substitutionally in the lattice for a Si atom only three of the neighboring Si atoms are successfully bound, a hole remains, which also aids conduction since electrons may now move more freely through the lattice, "jumping" into a hole and leaving a hole behind to be filled (**Fig. 3.8**).

FIG. 3.8 *p-Type substitutional doping with gallium (Ga) atoms in a silicon (Si) lattice. Three of the valence electrons of Ga bind to neighboring Si atoms leaving a hole for conduction.*

TABLE 3.6	Resistivity of Antimony-Doped Germanium Compounds	
Donor concentration (μ %)	**Resistivity (Ω · cm)**	
5	109	
50	105	
150	10	
300	0.1	

In the band structure picture, impurity states are created much closer to the valence band. These states are known as acceptors, since they can accept electrons from the valence band. Again, it is much easier to "liberate" the electron for conduction at lower thermal energy. A semiconductor doped with acceptors is known as a "p-type" semiconductor (p for positive charge carrier).

Doping changes the resistivity of semiconductors drastically. A typical example of antimony-doped germanium with different antimony concentrations and its relative resistivity is shown in **Table 3.6** above.

A diode is created when a p-type material is joined to an n-type material to create a p/n-junction (**Fig. 3.1**). The p-type material will have a high concentration of holes and few electrons. The n-type material will have a high concentration of electrons and few holes. It is obvious that there will be a large concentration gradient of both free electrons and free holes. Therefore there must be a large electric field in the p/n-junction region, which will oppose the tendency of the electron and hole concentrations to even themselves out. As previously mentioned this electric field arises when the carriers diffuse down their concentration gradients. As they do so they uncover (remove the screening carriers from) the fixed doping atoms, which have positive and negative charges for donors (n-type) and acceptors (p-type), respectively.

The formation of a p/n-junction is schematically shown in **Figure 3.9**, in which two pieces of uniformly and oppositely doped semiconductors are brought together. The p/n junction is not actually created in this way, but it is useful to imagine such a process in order to illustrate the basic physics. Initially the two pieces of semiconductors consist of fixed-charge doping atoms surrounded by an equal number of oppositely charged free carriers. When they are joined together, the electrons and holes (which are moving around at random) will begin to recombine with each other at the interface between the two regions. As they do so, the screening free carriers (electrons or holes) are removed from the junction region leaving the fixed positively and negatively charged doping atoms exposed. A powerful electric field will be established which will have a direction that opposes further movement of holes and electrons towards each other (and hence inhibit subsequent recombination).

The p/n-junction, consisting of a p-type and n-type silicon layer in contact, is the site of free electrons (n-type) and holes (p-type), canceling each other to form a rectifying junction, for example, the depletion zone does not contain mobile charge carriers. Doping Si with P, which has one more electron than Si, creates an n-type silicon. Conversely, doping silicon with B, which has one fewer electron in its valence shell, causes the formation of a p-type semiconductor.

Formation of a p/n-junction (schematic only). The doping atoms are shown with a circle around the charge sign, while the free electrons do not have a circle. When the n and p regions contact each other there is rapid recombination of electrons and holes at the interface. A depletion region forms, containing a strong electric field. Very soon, the field strength is sufficient to prevent further migration of electrons and holes towards each other, and equilibrium is reached. The depletion region width is greatly exaggerated in the figure.

When a positive bias voltage (positive pole, forward bias) is applied to the p-type side, charge flows across the interface (**Fig. 3.9**). If polarity is reversed (reverse bias), very little charge can flow.

When equilibrium is established (after a few microseconds) the silicon wafer will consist of three regions: a neutral n-type region; a neutral p-type region; and a narrow-charged depletion region. The depletion region is typically $0.1–1\,\mu m$ wide and has a very low concentration of free electrons and holes (hence the name). The electric field is confined to the depletion region, and is nearly zero in the uniformly doped n-type and p-type regions that comprise the bulk of the wafer (which is typically $500\,\mu m$ thick). The electric field strength will rise from near zero at the two edges of the depletion region to a maximum in the center, and can reach a strength of $106\,V\cdot cm^{-1}$.

Of course, if the n-type or p-type regions are not uniformly doped (e.g., from a solid-state diffusion of dopants) then the electric field will be nonzero outside the depletion region as well. **Figure 3.10** shows in cross section the main regions of a particular diode together with the electric field strength. The electric field strength is nonzero to the left of the depletion region because that region has been diffused, and has a nonuniform doping concentration. The electric field

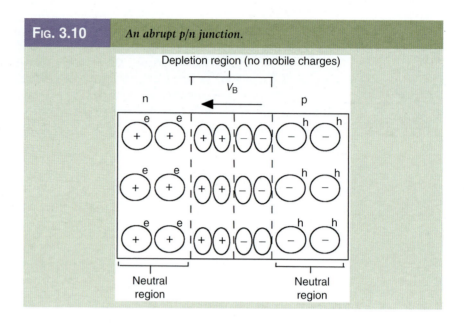

FIG. 3.10   *An abrupt p/n junction.*

strength is zero in the right-hand p-type region because the doping concentration is uniform. The electric field strength in the depletion region has a triangular shape and rises to a maximum in the center of the depletion region.

The direction of the electric field (conventional field, + to −) is from left to right because the n-type region has fixed positive charges (the phosphorus atoms that donated an electron to the crystal) and the p-type region has fixed negative charges (the boron atoms that accepted an electron from the crystal). The electric field opposes further movement of holes down the concentration gradient from the p-type silicon to the n-type silicon and similarly for electrons. At equilibrium, the electric field exactly balances the concentration gradient.

Because of the concentration gradient, electrons diffuse from n to p and holes from p to n. Hence a potential barrier VB builds up across the junction. VB causes drift currents which exactly balance the diffusion currents (in equilibrium), that is, there is no net flow of electrons or holes across the junction.

Barrier height VB in absence of external bias

Consider **Figure 3.11**.

(a) Fermi level is shifted due to doping.
     In general,

$$E_F = E_{Fi} + \frac{KT}{2} \ln \frac{\overline{n}}{\overline{p}} \qquad (3.28)$$

E_{Fi} is the Fermi level in an intrinsic semiconductor

$\overline{n}$ is the electron concentration $\Big\}$ at temperature T in thermal equilibrium
$\overline{p}$ is the hole concentration

(i) n-type:

$$\frac{\bar{n}}{p} \simeq N_d$$

$$\bar{p} = \frac{n_i^2}{N_d} \tag{3.29}$$

N_d is the donor concentration

Substituting equation (3.29) in equation (3.28)

$$E_F = E_{Fi} + KT \ln \frac{N_d}{n_i}$$

i.e., $E_F > E_{Fi}$ $\hspace{2cm}$ (3.30)

$$E_F - E_{Fi} = q_N^V = KT \ln \frac{N_d}{n_i}$$

(ii) p-type:

$$\bar{p} \simeq N_A$$

$$\bar{n} = \frac{n_i^2}{N_A} \tag{3.31}$$

N_A is the acceptor concentration

$$\therefore E_F = E_{F_i} - KT \ln \frac{N_A}{n_i}$$

i.e., $E_F < E_{F_i}$ $\hspace{2cm}$ (3.32)

$$E_F - E_{F_i} = qVp = -KT \ln \frac{N_A}{n_i}$$

From equations (3.30) and (3.32) we see that Fermi level in n-type is shifted up as compared to intrinsic Fermi level and is shifted down in p-type due to doping.

(b) At the junction, Fermi level must be the same on both sides. Thus, the electron energy levels are higher on the p-side as compared to the n-side (**Fig. 3.11**).

$$V_B = V_N + |V_P|$$

$$= \frac{KT}{q} \ln\left(\frac{N_d N_A}{ni^2}\right) \tag{3.33}$$

Forward bias (**Fig. 3.12**):
V_A is the applied voltage and V_t is the total voltage.

$$V_t = V_B - |V_A| \tag{3.34}$$

FIG. 3.11 *The Fermi level is the same for both sides of the junction, therefore, the electron energies are higher on the p-side compared to the n-side.*

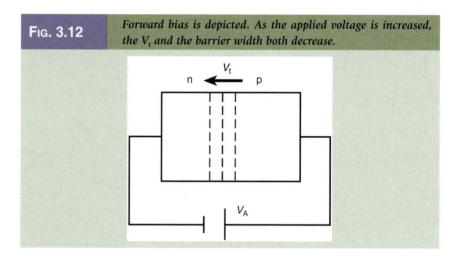

FIG. 3.12 *Forward bias is depicted. As the applied voltage is increased, the V_t and the barrier width both decrease.*

As V_A is increased, V_t and barrier width both decrease.
Reverse bias (**Fig. 3.13**):

$$V_t = V_B + |V_A| \tag{3.35}$$

As V_A increases, V_t decreases.
The Poisson equation in one dimension is given by

$$\frac{d^2V(x)}{dx^2} = -\frac{dE(x)}{dx} = -\frac{Q}{\varepsilon_r \varepsilon_o} \tag{3.36}$$

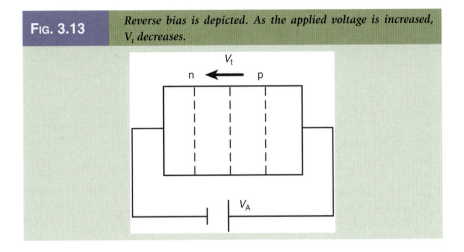

FIG. 3.13 *Reverse bias is depicted. As the applied voltage is increased, V_t decreases.*

where
  ε_o is the permittivity of vacuum
  ε_r is the dielectric constant of the material
  Q is the charge density
  E is the electric field
  V is the potential.

Solving under the abrupt junction approximately, we get

$$V_t = V(-d_n) - V(d_p)$$

$$= \frac{q}{2\varepsilon_r\varepsilon_o}\left(N_d d_n^2 + N_A d_p^2\right) \tag{3.37}$$

where
  $Q = qN_d$ on the n-side, and $Q = -qN_A$ on the p-side.
  d_n is the barrier width on the n-side
  d_p is the barrier width on the p-side

Now,

$$N_d d_n = N_A d_p \tag{3.38}$$

$$\therefore\ d_n^2 = \frac{2\varepsilon_r\varepsilon_o}{q}\frac{N_A}{N_d(N_A+N_d)}V_t$$

Similarly,

$$d_p^2 = \frac{2\varepsilon_r\varepsilon_o}{q}\frac{N_A}{N_d(N_A+N_d)}V_t$$

$\therefore$ Total barrier width

$$d = d_n + d_p$$

$$= \left[\frac{2\varepsilon_r\varepsilon_o}{q}V_t\frac{(N_A+N_d)}{N_A N_d}\right]^{1/2} \tag{3.39}$$

FIG. 3.14 *(a) Plot of the I–V characteristics of a diode (V_r is the cut-in voltage, V_{br} the breakdown voltage; (b) I–V characteristics of a pure conductor.*

I–V Characteristics.   Current through the p/n junction (**Fig. 3.14**) is given by

$$I = I_o \left[\exp\left(\frac{qV}{KT} \right) - 1 \right]$$

where
  V is the applied voltage
  K is the Boltzman constant
  T is the absolute temperature

1. When the junction is forward biased, V is positive
   If

$$V \gg \frac{4KT}{q} \quad \text{i.e.,} \quad V \gg 100 \text{ mV}$$

   then

$$\exp\left(\frac{qV}{KT} \right) \gg 1$$

   then

$$I = I_o \exp\left(\frac{qV}{KT} \right)$$

2. When the junction is reverse biased, V is negative
   Then, if

$$\exp\left(\frac{qV}{KT} \right) \ll 1, \quad I = -I_o \tag{3.40}$$

I_o is thus called the saturation value for a reverse bias condition.

FIG. 3.15	*(a) NPN-type; (b) NPN-transistor symbol; (c) PNP-type; (d) PNP-transistor symbol.*

3.1.3 Transistors

The transistor is a three-layer semiconductor device consisting of either two n-type and one p-type layers of material (referred to as npn transistors) or two p-type and n-type layers (referred to as pnp transistors) (**Fig. 3.15**). The middle range is called the base and the two outer regions are called the emitter and the collector. In most transistors, the collector region is made larger than the emitter region since it is required to dissipate more heat. The emitter is heavily doped, the base is lightly doped and is thin, while the doping of the collector is in between the other two. The emitter is so called since it emits electrons (holes in case of a pnp transistor) into the base. The base passes most of these electrons (hole for pnp) onto the collector. The collector gathers these electrons (holes for pnp) from the base.

A transistor has two p/n junctions, the one between the emitter and base is called the emitter junction while the one between the base and collectors is called the collector junction. There are four possible ways of biasing these two junctions (**Table 3.7**).

TABLE 3.7	*Junction Bias*		
Condition	**Emitter junction**	**Collector junction**	**Region of operation**
FR	Forward biased	Reverse biased	Active
FF	Forward biased	Forward biased	Saturation
RR	Reverse biased	Reverse biased	Cut-off
RF	Reverse biased	Forward biased	Inverted

3.2   NANOSCALE ELECTRONICS

3.2.1   *Background*

Fabrication of future nanotechnology devices is expected to take place from the bottom up. This is in stark contrast to most integrated circuit device manufacturing that is done today from the top down. Top-down manufacturing is a reductionist manufacturing philosophy. In top-down fabrication, electronic features are created by subtracting material from a bulk source. The top-down philosophy has demonstrated an incredible track record over the past several decades. Sadly, Si-based microelectronics cannot travel down the path of Moore's law forever. In 1997, Gordon Moore pointed out that there were probably five more generations (at 18 months per generation) left of improvements down the trackfollowing his own "Moore's law" before processes run into the wall of physical limitations. This implies that the ability of the industry to double the computing power of a microchip within that period could conservatively begin a noticeable slowdown in the mid-2000s. [11]. Fabrication of computer chips, and in no uncertain terms, will come up against the wall of economic unfeasibility as further scaling down of microelectronics continues.

Higher and higher energy lithography sources and support equipment are required to achieve the resolution required for nanometer-scale features. Because of the different functions they perform, logic chips are a more complicated design than memory chips are. As a result, the design complexity is expected to increase exponentially with the number of transistors to be integrated, and this in turn impacts the design-cycle time, causing a tendency for the actual integration level to fall short of the scales potentially attainable with current technology. This expenditure in investment is expected to be economically impractical as some juxtaposition between economics, smaller features, and complexity is attained along somewhere soon down the line.

3.2.2   *The Current State of Microelectronics and Extensions to the Nanoscale*

There are many kinds of physical limitations that would hinder our "staying the course" of traditional manufacturing practice. Stray signals and heat-dissipation limitations will be encountered as more transistors are packed onto a chip, and the increasing magnitude of difficulty of fabricating ever-smaller devices will severely impact future progressive steps. Experts expect such challenging issues to dramatically express themselves as transitions approach the 100-nm landmark. Because of this and other factors, the explosive increase in transistor densities and processing rates of improved integrated circuits projected by Moore's law is countered by mounting costs required for facilities to manufacture transistors, chips, and wafers. Inevitably, market equilibrium will be reached when the continually decreasing physical scale of the microelectronics reaches the rapidly increasing economic factors that impact production. This impasse could be reached as early as 2015. The cost of a facility, for example, that is required to fabricate high technology devices and components is projected to be nearly

$200 billion. That event will signal the end of the long and remarkable advances in CMOS computer-chip processing power.

We all are aware how storage capacity and computer speed increased significantly over the years. We too are well aware how the concomitant cost of manufacturing integrated microcircuits, transistors, chips, and devices has been reduced. The primary bearer of the load of the electronic industry has been, and still is, photolithography—in particular, the use of visible and ultraviolet sources. Photolithography, however, has a fundamental resolution limit—70 nm [11]. Even with the advent of shorter wavelength sources such as extreme ultraviolet (EUV), x-rays, and electron beams, obstacles in the form of resolution limits, radiation damage, and increasing costs to develop the next generation of high-vacuum, high-energy lithography equipment have inspired investigators to develop other means.

Much of modern IC R&D and manufacturing processes is devoted to optimization of chip size, wafer size, and detection of defectivity and interconnects—all of this just to keep pace with Moore's law. We have to, however, ask this question. Can we fabricate circuits <1 μm by simply extending practices associated with current device production technologies? The answer is an unequivocal no! What then is required for us to get down to that remarkable level of properties and phenomena in components and devices that are manufactured whole scale? One of the first steps to take is to develop an interdisciplinary "off the beaten path" approach because the limitations of current manufacturing options are on the horizon. For example, ArF lasers with a wavelength 193 nm, a current technology, is able to achieve critical dimensions down to 100 nm. To resolve smaller structures would require light of smaller wavelength (higher energy). If we go beyond the extreme ultraviolet, yes Martha, we are in the domain of the x-ray. Consequences of taking such an "energetic step" is to search for new materials that are able to stand up against x-ray bombardment. This will only add to the already astronomical R&D expenses involved in modem IC manufacture.

3.2.3 Nanotechnology-Based Strategies: Single-Electron Tunneling

Bottom-up strategies are nanotechnology-based methods that offer real solutions to the dilemma described in the previous section. Strategies using molecules, biological entities (proteins/DNA), particle deposition, etc., are straightforward techniques to synthesize, fabricate, and explore new component and device hierarchies. Self-organization is a fabrication strategy that allows for building patterns, processes, and structures at a higher level(s) through multiple and diverse interactions among components at lower level(s), for example, molecules. The self-organization process is inspired from nature. As we state over and over in this text, nature has perfected these processes over millions of years, evolved many complex structures that function as sensors, actuators, memory, and logic units. Synthetic sensors have a linear response because they are easily connected to the next stage of processing. In comparison, biological sensors usually have nonlinear characteristics, for example, they are "compensated by feedback systems" within which they play an integral role.

The dynamic range and overload capability, as a result, are often spectacular. As examples of memory devices, nature stores biological information in DNA where approximately 50 atoms are used for one bit of information about the cell—currently our level of "one bit" far exceeds this modest number. Modern hard disks can have an areal density of $150\,GB \cdot in.^{-2}$, a level that corresponds to more than a million atoms. However, all these changes will take more than a decade from making an impact in the marketplace at the same level as today's IC technology!

Electron Tunneling and the Single-Electron Transistor.   Once again we delve into the fundamentals of single-electron transfer. In 1985, Dmitri Averin and Konstantin Likharev of the University of Moscow proposed the idea of a new three-terminal device called a single-electron tunneling (SET) transistor that was eventually fabricated by Theodore Fulton and Gerald Dolan at Bell Labs in the United States in the following two years. Unlike FETs, single-electron devices are based on the tunneling effect that has its fundamentals in quantum mechanical phenomenon. The tunneling effect of electrons is observed if two metallic electrodes are separated by an insulating barrier about 1-nm thick (the length of 10 hydrogen atoms). Electrons at the Fermi energy level are able to "tunnel" through the insulator, even though in classical terms their energy would be too low to overcome the potential barrier.

The electrical behavior across a tunnel junction depends on how effectively the barrier transmits electron waves (that decrease exponentially as a function of barrier thickness) and on the number of electron-wave modes that impinge on the barrier (given by the area of the tunnel junction divided by the square of the electron wavelength). A single-electron transistor takes advantage of the quantum transfer of charge through the barrier. Such charge transfer becomes quantized when the junction is made sufficiently resistive.

If a tunnel junction interrupts an ordinary conductor, electric charge will move through the system by both a continuous and a discrete process. Since only discrete electrons can tunnel through junctions, charge will accumulate at the surface of the electrode against the isolating layer, until a high enough bias has built up across the tunnel junction (see right side of **Fig. 3.1**). Then one electron will be transferred. Likharev coined the term "dripping tap" as an analogy of this process. In other words, if a single-tunnel junction is biased with a constant current I, the so-called Coulomb oscillations will appear with frequency

$$f = \frac{I}{e} \tag{3.41}$$

where e is the charge of an electron (**Fig. 3.16**). Notice that equation (3.41) is identical to equation (2.15) presented in chapter 2 and describes SET oscillations.

The charge continuously accumulates on the tunneling junction barrier until it is energetically favorable for an electron to tunnel. This discharges the tunnel junction by an elementary charge e. Similar effects are observed in superconductors.

FIG. 3.16   *Current-biased tunnel junction showing Coulomb oscillations.*

There, charge carriers are Cooper pairs, and the characteristic frequency becomes $f = I/2e$; (equation (2.16)) is related to what are known as the *Bloch oscillations* (**Fig. 3.17**).

The current-biased tunnel junction is one very simple circuit that is able to control the transfer of electrons. Another one is the electron-box (see **Fig. 3.2**). A particle is only on one side connected by a tunnel junction. On this side electrons can tunnel in and out. Imagine for instance a metal particle embedded in oxide, as shown in **Figure 3.18**.

The top oxide layer is thin enough for electrons to tunnel. To transfer one electron onto the particle, the Coulombic energy $E_c = e^2/2C$, where C is the capacitance of the particle, is required. Neglecting thermal and other forms of energy, the primary energy source is the bias voltage V_b. So long as the bias voltage is small, smaller than a threshold $V_{th} = e/C$, no electron is able to tunnel because

FIG. 3.17   *The electron-box can be filled with a precise number of electrons.*

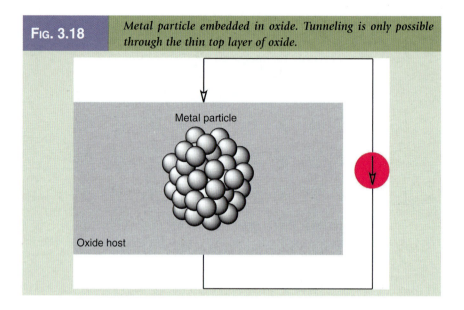

FIG. 3.18 *Metal particle embedded in oxide. Tunneling is only possible through the thin top layer of oxide.*

Metal particle

Oxide host

not enough energy is available to charge the island. This behavior is known as the *Coulomb blockade* behavior. Raising the bias voltage will populate the particle with one, then two, and three electron, and so on and so forth, leading to a staircase-like characteristic.

It is easily understandable that these single-electron phenomena, such as *Coulomb oscillations* and *Coulomb blockade*, only matter if the Coulomb energy exceeds that of the thermal energy. Otherwise thermal fluctuations will disturb the motion of electrons and will wash out the quantization effects. The necessary condition is

$$E_c = \frac{e^2}{2C} > k_B T \qquad (3.42)$$

where

k_B is Boltzmann's constant
T is the absolute temperature

What does this expression imply? It means that the capacitance C has to be smaller than 12 nF for the observation of charging effects at the temperature of liquid nitrogen and smaller than 3 nF for charging effects to appear at room temperature. F is the unit for capacitance, the farad. *A second condition* for the observation of charging effects is that quantum fluctuations of the number of electrons on an island must be negligible. Electrons need to be well localized on the islands. If electrons are not localized on islands, charging effects would not be observed since islands would not be separate particles but rather one larger agglomerated uniform space.

The charging of one island with an integer number of the elementary charges would be impossible, because one electron is shared by more than one island. In this case, the Coulomb blockade would vanish because no longer would a lower limit of charge exist. This leads to the requirement that all tunnel junctions must be opaque to electrons in order to confine electrons on islands. This tunnel junction "transparency" is described by the tunnel resistance R_T and must fulfill the following condition for discrete charging effects to be observable. This should be understood as an order-of-magnitude measure, rather than an exact threshold.

$$R_T > \frac{h}{e^2} = 25,813 \ \Omega \tag{3.43}$$

where h is Planck's constant.

Therefore, these effects are experimentally verifiable for very small high-resistance tunnel junctions like those associated with small particles with small capacitances and/or at very low temperatures. Advanced fabrication techniques such as the production of granular films with particle sizes down to 1 nm and a deeper physical understanding allow today the study of many charging effects at room temperature.

In summary, if a voltage source charges a capacitor, through an ordinary resistor, the charge on the capacitor is strictly proportional to the voltage and shows no sign of charge quantization. But if the resistance is provided by a tunneling junction, the metallic area between the capacitor plate and one side of the junction forms a conducting "island" surrounded by insulating materials. In this case the transfer of charge onto the island becomes quantized as the voltage increases, leading once again to the so-called Coulomb staircase (**Fig. 3.17**). The Coulomb staircase is seen only under certain conditions. The energy of the electrons due to thermal fluctuations must be significantly smaller than the Coulomb energy, which is the energy needed to transfer a single electron onto the island when the applied voltage is zero. This Coulomb energy is given by $e^2/2C$, where e is the charge of an electron and C is the total capacitance of the gate capacitor and the tunnel junctions. Secondly, the tunnel effect itself should be weak enough to prevent the charge of the tunneling electrons from becoming delocalized over the two electrodes of the junction, as happens in chemical bonds.

SET Function.   So how does a SET transistor work? The key point is that charge passes through an island in quantized units. For an electron to jump onto an island, its energy must equal the Coulomb energy $e^2/2C$. When both the gate and bias voltages are zero, electrons do not possess enough energy to enter an island, and as a result current does not flow. As the bias voltage between the source and drain is increased, an electron passes through the island when the energy in the system reaches the Coulomb energy, for example, the Coulomb blockade. The critical voltage needed to transfer an electron onto the island, equal to e/C, is called the Coulomb gap voltage.

Now imagine that the bias voltage is set under the Coulomb gap voltage. If the gate voltage increases, the energy of the initial system (with no electrons on the island) gradually increases concomitantly while the energy of the system with one excess electron on the island gradually decreases. At the gate voltage corresponding to the point of maximum slope on the Coulomb staircase, both of these configurations equally qualify as the lowest energy states of the system. This lifts the Coulomb blockade, thereby generating electrons to tunnel into and out of an island.

The Coulomb blockade is lifted when the gate capacitance is charged with exactly minus half an electron, which is not as surprising as it may seem. The island is surrounded by insulators, which means that the charge on it must be quantized in units of e, but the gate is a metallic electrode connected to a plentiful supply of electrons. The charge on the gate capacitor merely represents a displacement of electrons relative to a background of positive ions.

Researchers have long considered whether SET transistors could be used for digital electronics. Although the current varies periodically with gate voltage (in contrast to the threshold behavior of the FET), the SET would still form a compact and efficient memory device. However, even the latest SET transistors suffer from "offset charges." That means that the gate voltage needed to achieve maximum current varies randomly from device to device. Such fluctuations make it impossible to build complex circuits.

One way to overcome this problem might be to combine the island, two tunnel junctions, and the gate capacitor that comprise a single-electron transistor in a single molecule. An intrinsically quantum behavior of SET transistors should not be affected at the molecular scale. In principle, the reproducibility of such futuristic transistors would be determined by chemistry, and not by the accuracy of the fabrication process per se.

It is not yet clear whether electronics based on individual molecules and single-electron effects will replace conventional circuits based on scaled-down versions of FETs. Only one thing is certain: That if the pace of miniaturization continues unabated, the quantum properties of electrons will become crucial in determining the design of electronic devices before too long—most likely before the end of the next decade.

Single-electron transistors are not so different from traditional MOSFETs (**Fig. 3.19**).

FIG. 3.19   *Comparison of SET and MOSFET.*

1. In the SET, electron conduction takes place one electron at a time while in a MOSFET, many electrons simultaneously take part in conduction.
2. In the SET, the drain gate controls Coulomb blockade behavior while in MOSFETs the gate controls the channel.
3. The SET requires an opaque junction where $R_T > R_Q \sim 26 \, k\Omega$ while MOSFETs require highly transparent junctions.

3.2.4 *Nanotechnology-Based Strategies: Molecular Wires*

The concept of fabricating nanoscale devices with molecules was initiated in the early 1970s by chemists Ari Aviram of the IBM Thomas J. Watson Research Center in Yorktown Heights, New York, and Mark Ratner, a professor at Northwestern University, Chicago, Illinois. The two started working on the idea of synthesizing molecular electronic elements and devices and proposed the concept of a rectifier consisting of a single molecule. [12]. Excellent reviews on molecular electronics research is available in the literature [13–15]. In 1997, Professor Robert Metzger and his colleagues at the University of Alabama produced the molecule named *hexadecylquinolinium tricyanoquinodimethanide* [16]. They used this molecule to measure the bi-directional current flow through it between two aluminum electrodes. The rectification ratio of these two currents, the ratio indicating the preferential flow direction, ranged from 2.4 to 26.4 for different room temperature samples. These results demonstrated a definite preference for the direction of the electron flow. The chemical configuration of this molecule is shown in **Figure 3.20**.

FIG. 3.20

The DC electrical conductivity of g-(n-hexadecyl)quinolium tricyanoquinodimethanide (or $C_{16}H_{33}$-Q-3CNQ) assembled in Langmuir–Blodgett multilayers demonstrated unimolecular rectification behavior. In other words, the conductivity of electrons from one of this impressive molecule is not the same as conductivity originating from the other end. Monolayers with the negative end of the zwitterion on the substrate and the tail (the hydrocarbon chain) up and away from the substrate conducted electrons away from the substrate.

Zwitterionic $C_{16}H_{33}Q$-3CNQ

Source: R. M. Metzger, B. Chen, U. Höpfner, M. V. Lakshmikantham, D. Vuillaume, T. Kawai, X. Wu, H. Tachibana, T. V. Hughes, H. Sakurai, J. W. Baldwin, C. Hosch, M. P. Cava, L. Brehmer, and G. J. Ashwell, *Journal of American Chemical Society*, 119, 10455–10466 (1997). With permission.

There are several compelling reasons to consider that molecular electronics will play a major role in the future of nanoelectronics. Essentially all electronic processes in nature, from photosynthesis to signal transduction, are mediated by molecular structures. There are four major advantages for molecular circuits:

1. *Size*. The size scale of molecules is between 1 and 100 nm, which allows to make functional nanostructures with accompanying advantages in cost, efficiency, and power dissipation.
2. *Assembly and Recognition*. Intermolecular interactions can be utilized to form structures by nanoscale self-assembly. Molecular recognition can be used to modify electronic behavior, to obtain both switching and sensing capabilities on a single-molecule scale.
3. *Dynamical Stereochemistry*. Many molecules have multiple distinct stable geometric structures or isomers (an example is the rotaxane molecule in **Figure 3.21d**, in which a rectangular slider has two stable binding sites along a linear track). Such fixed geometry isomers have very distinct optical and electronic properties. For example, the retinal molecule (in eyes) switches between two stable structures, a process that transduces light into a chemoelectrical pulse that allows vision.

FIG. 3.21

Examples of molecular transport junctions. The top panel depicts molecules with various localized, low-energy molecular orbitals (colored dots) bridging two electrodes L (left) and R (right). In the middle panel, the black lines are unperturbed electronic energy levels; the red lines indicate energy levels under an applied field. The bottom panel depicts representative molecular structures. (a) A linear chain, or alkane. (b) A donor–bridge–acceptor (DBA) molecule, with a distance l between the donor and acceptor and an energy difference EB between the acceptor and the bridge. (c) A molecular quantum dot system. The transport is dominated by the single metal atom contained in the molecule. (d) An organic molecule with several different functional groups (distinct subunits) bridging the electrode gap. The molecule shown is a rotaxane [2], which displays a diverse set of localized molecular sites along the extended chain. Two of those sites (red and green) provide positions on which the sliding rectangular unit (blue) can stably sit. A second example of a complex molecule bridging the electrodes might be a short DNA chain.

Source: Adapted from R. E. Hummel, *Electronic properties of materials*, Springer-Verlag, Berlin (1985).

| **FIG. 3.22** | *Schematic representation of the self-assembly of molecular devices.* |

4. *Synthetic Tailorability.* The molecules can be tailor made to the expected composition and geometry to control a molecule's transport, binding, optical, and structural properties. The tools for such molecular synthesis are highly developed (**Fig. 3.22**)

But what are these devices specifically? To begin with, there are two generalized classes of molecules that have demonstrated possession of the characteristics acceptable for molecular-scale electronic devices: (1) carbon nanotubes and (2) polyphenylene-based chains.

Polyphenylene Molecular Wires. The polyphenylene-based chains, which are made of chains of aromatic benzene rings, are much smaller molecules than carbon nanotubes are. It then seems more likely to provide an immediate source of molecular-scale rectifiers and switches. As a result, these materials have established a base for much of the recent research into molecular electronic wires and devices. An example of such a device is shown in **Figure 3.23**, conducting molecular wire assembly, demonstrated by Tour [17]. The basic structure of this molecule is shown in the figure.

To understand the basics of what is involved in producing a current carrier of such a small magnitude, let us first look at how one would produce such a "simple" conductor or wire.

If we remove two hydrogen atoms from a benzene ring (C_6H_6), then we obtain a phenylene structure (C_6H_4), which is a ring with free sites or what we call two bonding sites. Now if we have many phenylene groups together, the missing hydrogen atoms will act as attraction sites; a chain of phenylene groups can be formed as depicted in **Figure 3.23 (Top)**. The chain is terminated with a phenyl-group molecule, which is a benzene ring with only one hydrogen removed (C_6H_5). Since this single strand of nanofilament can be of arbitrary length, it is called a "polyphenylene" chain, where "poly" means "more than one." Other molecular groups can be adapted to this chain to specialize it for specific properties or characteristics that may be desired.

For example, triply bonded ethynyl or acetylenic links are often inserted as spacers to eliminate interference from adjacent hydrogen atoms in order to

FIG. 3.23

(Top): "Tour wires." A conductive polyphenylene molecular chain comprised of a series of modified benzene rings (Middle): A wire with an insulating island is depicted. (Bottom): Molecular rectifying diode, originally proposed by Aviram and Ratner, uses chemically doped polyphenylene molecular wires for the conduction of electrons. The dopants are intramolecular modifiers. Two intramolecular doping agents, X and Y, are applied to the polyphenylene backbone structure [18]. "X" is an electron donating group (e.g., "n-type") and "Y" is an electron accepting group (e.g., "p-type"). The donor–acceptor complex is separated within the molecular diode by a semi-insulating bridge structure "R" (e.g., dimethylene groups) that provides a potential energy barrier to preserve the potential drop between the Tour wires and X and Y coplanar assemblies to allow for tunneling. The thiol linkages, "S," also provide potential barriers to the Au-electrodes that serve to maintain isolation of the molecular wire structure. None of the barriers have potential that is insurmountable when a proper bias is applied [18].

enhance the conductivity of the molecular wire. A three-ring polyphenylene chain demonstrated approximately 30×10^{-9} A of current passing through the molecule. The technique for synthesizing conductive polyphenylene-based chains has been refined by James Tour so well with great repeatability over the past few years that they have come to be known as "Tour wires." [17]. Adding aliphatic methylene groups to the chain makes the molecule an insulator, and depending on where the polyphenylene-based molecular wire is added, aliphatic methylene could function either as a "resistor" or as an "insulator."

Both rectifying diodes and resonant-tunneling diodes have been recently demonstrated. A molecular-rectifying diode, based on the work by Aviram and Ratner, has been successfully demonstrated by teams led by Metzger [16] at the University of Alabama and Reed at Yale University.[17]. Although other configurations of polyphenylene-based molecular rectifiers are being designed, molecular-resonant tunneling diodes, which are quantum devices that employ quantum effects in their simplest form by utilizing energy quantization for the "on"-"off" switching of an electric current, have been found to integrate rather nicely. The energy quantization "regulates" the variation of voltage bias across the source and drain contacts of the diode. However, current can pass equally well in either direction for the resonant-tunneling diode, unlike that of the rectifier.

In order to control the desired direction of the electron flow, a set of "insulators" acting as tunnel barriers are placed very close together to create a potential-energy "well" or "island" similar to the case explained in single-electron transistors. The probability that electrons can tunnel depends upon the difference between the energy level of the incoming electrons and that of the electrons within the potential well. If the energy quantum state of the electrons arriving in the device is different from the energy level allowed inside the potential well, current flow is not allowed and the device is switched "off." But if the energy quantum state of the arriving electrons matches one of the energy levels of the island barrier, a condition said to be "in resonance," then the state of the device is "on" and current flows.

Tour and Reed produced the first working "prototype," as shown schematically in **Figure 3.24**. By inserting two aliphatic methylene groups into the polyphenylene-based chain "wire" on either side of a single aliphatic ring, they were able to demonstrate a resonant-tunneling diode. **Figures 3.23 (Middle)** and 3.24 show that the "island" is formed by the CH_2 or methylene group located on either side of the benzene ring while the insulating property of the aliphatic

| FIG. 3.24 | *Polyphenylene molecule (Fig. 3.23 Middle) with resonant-tunneling diode configuration.* |

group, as mentioned earlier, acts as potential-energy barriers to the flow of electrons. The aromatic ring between the aliphatic groups provides a very narrow gap of approximately 0.5 nm, which becomes a region of lower potential energy that the electrons must pass through.

Because of the possibility of multiple quantum levels implicit with resonant-tunneling devices, the additional advantage of multistate switching behavior is permitted. A continuously increasing bias voltage can produce multiple energy levels that come into resonance with the incident electrons. This multiple switching behavior represents additional logic states for each device.

Carbon Nanotubes.   Similar progress has been experienced in the research of carbon nanotubes, also known as tubular fullerenes. A carbon nanotube, as depicted in **Figure 2.9**, is a cylinder of carbon atoms comprising a single molecule that measures from 1 to 20 nm (nominally 10 atoms) in diameter. Carbon nanotubes can be either conductors or semiconductors—depending upon their "chiral" or twist angles and diameters. In 1997, Cees Dekker and his associates at the Delft University of Technology in the Netherlands were the first to demonstrate the capability of nanotubes to act as wires [19,20]. The following year this same team demonstrated a transistor using a nanotube as one of its components. They then expanded their research into finding ways to adapt nanotubes as electronic devices and discovered that a kink in a nanotube caused it to act like a rectifier diode. Through manipulation of individual molecules, researchers at IBM Corporation then successfully produced an array of transistors out of carbon nanotubes. [20]. While a technique for the self-ordering of both the conducting and semiconducting nanotubes like in the case of organic molecules has so far eluded researchers, the IBM team has discovered a technique for producing only semiconducting nanotubes through the destruction of the conductors. This is a significant step towards the practical application of nanotubes in computer chips.

Application of molecular electronics requires flexible reaction chemistry capable of producing long strings with desired branch structures, and a high electrical conductivity for these large structures is a necessity. While polyphenylenes provide the more flexible reaction chemistry, and can be self-assembled more easily, carbon nanotubes are more conductive. Technical challenges yet remain in the development of nanoelectronic devices using the molecular approach [21]. A practical integration of molecules or nanoparticles or nanotubes into scalable, functional electronic devices will still take some more time to be achieved. But the total impact of nanotechnology on the future of electronics seems likely to be much greater than the influence of the silicon integrated circuit. The benefits of "molectronics" will ultimately evolve from learning how to tailor fundamental properties and phenomena at the nanoscale and the capability of controlling the features of a desired component or structure during its basic molecular assembly. The technological future with nanoelectronics will be revolutionary, spawning major new industries neither envisioned nor conjectured today, involving the design and chemical synthesis of molecules that are quantum electronic devices, self-assembly into desired circuits by following encoded instructions based on simple principles yet to be discovered. Nanoelectronics may change the perspective of nearly every human endeavor for the remainder of the twenty-first century.

References

1. Farad, en.wikipedia.org/wiki/Farad (2008).
2. J. F. Shackelford, *Introduction to materials science for engineers*, 4th ed., Prentice-Hall, Upper Saddle River, NJ (1996).
3. R. E. Hummel, *Electronic properties of materials*, Springer-Verlag, Berlin (1985).
4. C. Berger, Y. Yi, Z. L. Wang, and W. A. van de Heer, Multiwalled carbon nanotubes are ballistic conductors at room temperature, *Applied Physics A-Materials Science & Processing*, 74, 363–365 (2004).
5. S. Kornguth, Nanotechnology and biomolecular electronics, National Institute of Standards and Technology, www.atp.nist.gov/clso/nano_tech.htm (2005).
6. Transistors and chips, IEEE Virtual Museum, www.ieee-virtual-museum.org (2008).
7. M. Di Ventra, S. Evoy, and J. R. Heflin, *Introduction to nanoscale science and technology*, Kluwer Academic Publishers, Boston (2004).
8. J. Redin, A tale of two brains, www.xnumber.com/xnumber/kilby.htm (2007).
9. D. F. Shriver, P. W. Atkins, and C. H. Langford, *Inorganic chemistry*, W.H. Freeman & Co., New York (1990).
10. Direct bandgap, en.wikipedia.org/wiki/Direct_bandgap (2009).
11. C. G. Huang, S. I. Lee, and Y. D. Hong, Driving forces of future semiconductor technology, In *Future trends in microelectronics*, S. Luryi, J. Xu, and A. Zaslavsky, eds., p. 13, Wiley-Interscience publication, New York (1999).
12. A. Aviram and M. A. Ratner, Molecular rectifiers, *Chemical Physics Letters*, **29**, 277 (1974); R. M. Metzger et al., Electrical rectification in a Langmuir–Blodgett monolayer of dimethyanilinoazafullerene sandwiched between gold electrodes, *Journal of Physical Chemistry B*, **107**, 1021 (2003).
13. A. Nitzan, Electron transmission through molecules and molecular interfaces, *Annual Review of Physical Chemistry*, **52**, 681 (2001).
14. V. Mujica and M. A. Ratner, Molecular conductance junctions: A theory and modeling, In *Handbook of nanoscience, engineering, and technology*, W. A. Goddard III et al., eds., CRC Press, Boca Raton, FL (2002); C. Joachim, J. K. Gimzewski, and A. Aviram, Electronics using hybrid-molecular and mono-molecular devices, *Nature*, **408**, 541 (2000).
15. J. C. Ellenbogen, A brief overview of nanoelectronic devices. presented at the Proceedings of the 1998 Government Microelectronics Applications Conference (GOMAC98), Arlington, VA, 13–16 March; www.mitre.org/technology/nbanotech/GOMAC98_article.html (1998).
16. R. M. Metzger, B. Chen, U. Höpfner, M. V. Lakshmikantham, D. Vuillaume, T. Kawai, X. Wu, H. Tachibana, T. V. Hughes, H. Sakurai, J. W. Baldwin, C. Hosch, M. P. Cava, L. Brehmer, and G. J. Ashwell, Unimolecular electrical rectification in hexadecylquinolinium tricyanoquinodimethanide, *Journal of the American Chemical Society*, 119, 10455–10466 (1997).
17. M. A. Reed, C. Zhou, C. J. Muller, T. P. Burgin, and J. M. Tour, Conductance of a molecular junction, *Science*, 278, 252–254 (1997).
18. J. C. Ellenbogen and J. C. Love, Architecture for molecular electronic computers, chap. 7 In *Handbook of nanoscience, engineering and technology*, W. A. Goddard III, D. W. Brenner, S. E. Lyshevski, and G. J. Iafrate, eds., pp. 7-1–7-64, CRC Press, Boca Raton, FL (2003).
19. P. Avouris, P. Collins, and M. Arnold, Engineering carbon nanotubes and nanotube circuits using electrical breakdown. *Science*, 292(5517), (2001).
20. S. Tans, A. Verschueren, and C. Dekker, Single nanotube-molecule transistor at room temperature. *Nature*, 393, 49–51 (1998); Yao, et al. Carbon nanotube intramolecular junctions, *Nature*, 402, 273 (1999).
21. J. Tour, M. Kozaki, and J. Seminario, Molecular scale electronics: A synthetic/computational approach to digital computing. *Journal of American Chemical Society*, 120, 8486–8493 (1998).

Problems

3.1 Why are semiconductor quantum dots not very good in classical microelectronic applications? Give at least two reasons.

3.2 Discuss your understanding of tunneling of electrons in single-electron nano-structures?

3.3 Define Fermi energy. Describe the Fermi energy in a metal. Describe the Fermi energy for a semiconductor. What is its importance? How does Fermi energy relate to nanoparticles? Do semiconductors have a true Fermi energy?

3.4 What is the basic difference between classical microelectronics and single "electron" electronics?

3.5 What are compound semiconductors? Give some examples. What are some of their uses? Are there nanoparticulate compound semiconductors? If yes, give some examples.

3.6 What are charge carriers? Identify the charge carriers in metallic substances, semiconducting substances, and conductive liquids. Can nanoparticles be used as charge carriers?

3.7 Explain from the figure below the carrier recombination mechanisms in semiconductors?

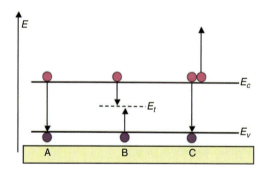

3.8 Explain the Coulomb blockade in your own words.

3.9 Calculate the intrinsic carrier density in silicon at 300 K.

3.10 When $E_C - E_F > 3k_BT$

$$n = N_C e^{(E_C - E_F)/k_BT} \quad \text{and} \quad N_C = 2\left(\frac{2\pi m_p^* k_B T}{h^2}\right)^{3/2}$$

Find the concentration of electrons and holes.

3.11 Show that for intrinsic semiconductors, the correct alternative expression for n and p are

$$n = n_i e^{(E_f - E_i)/k_BT} \quad \text{and} \quad p = n_i e^{(E_i - E_f)/k_BT}$$

3.12 Use classical Coulomb potential energy expressions in finding the lowest potential energy of two electrons that occupy four quantum dots in a quantum automata cell. Show the energy of the electrons is lower when the diagonal configuration is occupied than when the electrons occupy dots that are adjacent. The dimensions of the 2-D box is a × a × a × a.

3.13 Reconsider Problems 3.8–3.11. Is there anything that we need to modify to account for diminished size (as in the case of nanoparticles)?

3.14 Consider a molecular wire. How would you expect its electrical conductivity to behave? Does an electron in a conjugated linear molecule behave like a "particle in a box"? Plot what you would expect in terms of its *I–V* behavior.

3.15 What is an ohmic junction? Is there an issue with such junctions at the nano-scale? Why or why not?

3.16 Is overheating a problem in nanoscale circuitry?

3.17 Explain the role of electron tunneling in a single-electron transfer device. Have we been able to transfer a single electron?

3.18 What is ballistic conduction? How does it relate to the "mean free path" of an electron? What nanoscale material has potential to serve as a ballistic conductor? Is ballistic conduction related to super-conduction?

3.19 What is a unimolecular rectifier? Nanoionics?

3.20 Describe the action of a carbon nano-tube field emission device.

3.21 Why is it difficult to dope a nanoscale semiconductor?

NANO-OPTICS

Some physicists may be happy to have a set of working rules leading to results in agreement with observation. They may think that this is the goal of physics. But it is not enough. One wants to understand how Nature works.

PAUL DIRAC

Chapter 4 represents the second part of the *Electromagnetic Nanoengineering* section, following *chapter 3* on *Nanoelectronics*. It is interesting to note that electronics, optics, and magnetism are essentially inseparable. They are all part of the electromagnetic continuum. All involve electrons and photons. Nonetheless, we draw the boundary for our convenience. In *chapter 4*, we study the basics of light and its interaction with matter. From there, we go on to investigate light's interaction with nanomaterials. Once again, we find that the size of matter has a great influence on the optical response of the material.

Chapter 4 is sandwiched in between electronics (*chapter 3*) and magnetism (*chapter 5*). The order of the chapters should make no difference with regard to the subject matter at hand.

4.0 INTRODUCTION TO OPTICS

Optics (from the Greek *opticus* "of sight or seeing"; *optos* "seen visible," related to *ops* "eye") is the study of the interactions of light with matter and other optical phenomena. Infrared, visible, and ultraviolet are ranges of EM (electromagnetic) radiation that we are most familiar. X-rays, microwave, and radar are included in the extended family of EM phenomena. The field of quantum physics provided us a fundamental understanding of optics from the "bottom up." Classical optics (geometric optics) described the propagation of light through materials like prisms. Concepts such the light ray, refractive index, diffraction, and polarization were developed through the work of early scientists in the field. **Figure 4.1** depicts a page taken from the *Cyclopaedia*, compiled by Ephraim Chambers, published in 1728 London [1].

There are many pioneers in this field, too numerous to list and discuss, but we shall recount a few notable contributors [1–3]. The first lenses were derived from quartz that was polished dating to eighth century B.C. Democritus attempted to explain color and claimed it was due to the "roughness of constituent atoms." Euclid in the 4th–3rd century B.C. was one of the first to study optics from a geometrical sense. In his seven axioms, he defines visual rays, cones, and "seeing as a result of lines falling on objects"; angles and their relation to size and clarity; and left, right, and high rays. Archimedes (5th–4th century B.C.) studied *catoptrics*, reflections from surfaces and refraction. The Roman Seneca during the reigns of Caligula, Claudius, and Nero studied the properties of water with regard to magnification. Hero of Alexandria (first century) considered the

FIG. 4.1

Table of Opticks, *published in* Cyclopaedia *in 1728 London. The* Cyclopaedia *(or Universal Dictionary of Arts and Sciences) was compiled by Ephraim Chambers, an apprentice to a globe craftsman. His epitaph read* "Heard of by many, Known to few, Who led a Life between Fame and Obscurity, Neither abounding nor deficient in Learning, Devoted to Study, but as a Man, Who thinks himself bound to all Offices of Humanity, Having finished his Life and Labours together, Here desires to rest."

FIG. 4.1   *See caption on page 166.*

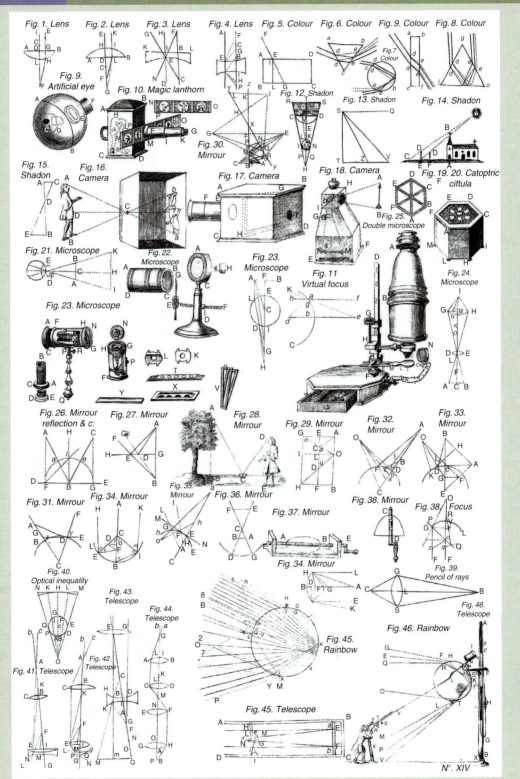

Fig. 1. Lens   Fig. 2. Lens   Fig. 3. Lens   Fig. 4. Lens   Fig. 5. Colour   Fig. 6. Colour   Fig. 9. Colour   Fig. 8. Colour

Fig. 9. Artificial eye   Fig. 10. Magic lanthorn   Fig. 7. Colour   Fig. 12. Shadon   Fig. 13. Shadon   Fig. 14. Shadon

Fig. 15. Shadon   Fig. 16. Camera   Fig. 17. Camera   Fig. 18. Camera   Fig. 19. 20. Catoptric ciftula

Fig. 30. Mirrour   Fig. 25. Double microscope

Fig. 21. Microscope   Fig. 22. Microscope   Fig. 23. Microscope   Fig. 11 Virtual focus   Fig. 24. Microscope

Fig. 23. Microscope

Fig. 26. Mirrour reflection & c: Fig. 27. Mirrour   Fig. 28. Mirrour   Fig. 29. Mirrour   Fig. 32. Mirrour   Fig. 33. Mirrour

Fig. 31. Mirrour   Fig. 34. Mirrour   Fig. 35. Mirrour   Fig. 36. Mirrour   Fig. 37. Mirrour   Fig. 38. Mirrour   Fig. 38. Focus

Fig. 34. Mirrour   Fig. 39. Pencil of rays

Fig. 40. Optical inequality   Fig. 43. Telescope   Fig. 44. Telescope   Fig. 45. Rainbow   Fig. 46. Telescope

Fig. 41. Telescope   Fig. 42. Telescope   Fig. 46. Rainbow

Fig. 45. Telescope

N°. XIV

phenomena of reflection, smooth versus porous surfaces (the beginning of nanophotonics?), and that light must travel at high speeds. The ninth century Middle Eastern scientist Al-Kindi stated that "all things in the world emit light rays in all directions." The tenth century Arab scholar Alhazan researched reflection from spherical and parabolic mirrors and explained why the sun and moon appear larger when near the horizon. Roger Bacon, in thirteenth century Europe, influenced greatly by Arabic scholars, conducted some of the first mathematical analyses on light [2,3].

Ancient Greeks and Romans filled glass spheres with water and used them to magnify objects. Spectacles made the scene ca. thirteenth century when Roger Bacon (1220–1292) used broken shards of a glass sphere for a lens. Modern optics took form, beginning in the sixteenth century, from the efforts of Kepler, Descartes, Huygens, and Newton. Johannes Kepler (1571–1630) explained the workings of the eye and eyeglasses. Willebrod Snell (1580–1626) discovered the law of refraction. Rene Descartes promoted a corpuscular theory of light and that light consists of "tiny globes that travel and bounce according to the laws of optics—color was due to different spins of the globes." Isaac Newton claimed that light consisted of discrete particles of different sizes with "immutable refracting powers" [3]. Interestingly, such corpuscular theories of light preceded those based on wave mechanics.

Thomas Young in 1801 proposed that the nature of light was wavelike and particulate per se. He found that interference patterns were produced by light emerging from adjacent pinholes. Interestingly, his idea was ridiculed in his homeland England [3]. Augustin-Jean Fresnel developed a mathematical treatment of wave optics. Laplace, Fourier, and Poisson supported the corpuscular theory of light. Young, Fresnel, and Francois Arago supported the wave theory.

James Clerk Maxwell explained EM theory and laid the foundations for the classical behavior of light. Rudolf Luneburg used Maxwell's equations to systematically treat ray and diffraction optics. Heinrich Hertz, discoverer of the quantum photoelectric effect, also used Maxwell's equations to prove that light waves are EM in nature.

Classical optics, however, was found to break down at the high frequency limit. The major tenet of "ultraviolet catastrophe" (Rayleigh-Jeans) predicted that an ideal black body radiator (a 3-D cavity) emits radiation with infinite power as the energy of the radiation increases (shorter and shorter wavelength)—that radiated power per unit frequency is proportional to the square of the frequency. However Max Planck, Albert Einstein, and others soon realized that this was not possible. Hence, through nonclassical physics, a solution to the dilemma was found based on the concept of the quantum. In essence, quantum theory predicts that radiated energy goes to zero at infinite frequencies within a cavity and that the total power is finite. Classical physics describes light in terms of rays and waves. Quantum physics describes light in terms of waves, particles, and specific energy.

The quantum mechanical treatment of light began when Joesph von Fraunhofer (1787–1826) measured the positions of hundreds of absorption lines in elemental spectra. Planck, Einstein, Niels Bohr, Erwin Schrodinger, Werner Heisenberg, and many more brilliant scientists established the foundation for modern quantum mechanics.

Modern Developments. Vacuum tubes and transistors in the 1960s evolved into integrated circuits in the 1980s and on into VLSI (very large-scale integration) in which ICs were created by combining thousands of transistor-based circuits into a single chip. In the 2000s, we are on the verge of molecular electronics. In an analogous way, crossing over from vacuum tubes and transistors, discrete fiber optic components were developed in the 1970s that evolved into planar optical waveguides in the 1980s and on into integrated optical circuits, until finally we are in the developmental period of optical photonic crystals. The frequency ranges and velocities of the two primary components, the electron and the photon, differ significantly: $f_{\text{electron}} \sim 10^{10}$ Hz versus $f_{\text{photon}} \sim 10^{15}$ Hz, $v_{\text{electron}} \sim 10^{5}$ m·s^{-1} versus $v_{\text{photon}} \sim 10^{8}$ m·s^{-1}.

Optical techniques have several advantages: sensitivity (single photons, single molecules), selectivity (fluorescence, filters), spectroscopy, (chemical analysis) and speed (femtosecond timescales).

History of Photonics. Photonics is the interaction of light with matter. It is the technology of light. Optical fibers and waveguides, semiconductor physics, light-emitting diodes (LEDs), laser diodes, p/i/n-photodiodes, avalanche photodiodes (APD), organic photonic devices, photonic crystal fibers, and many more devices are based on the interactions between light and matter. Photonics as a field began in the 1960s with the invention of the laser, and subsequently the laser diode and optical fibers and the Erbium-doped fiber amplifier in the 1970s. Some of the first applications of photonics were devoted to telecommunications. Whereas the optics industry focused on cameras and light sources, photonics pursued the development of optical fibers, laser LEDs, compact discs, digital videos, and LEDs.

The first visible spectrum LED was invented by N. Holonyak of GE in 1962 and he is considered to be the "father of the LED." In the 1990s, Shuji Nakamura of Nichia Corporation developed the first high-brightness blue LED based on InGaN (p-doped GaN). The first white LED used a Ce:$Y_3Al_5O_{12}$ or YAG (yttrium–aluminum–garnet) phosphor coating that mixed yellow (down-converted) light with blue to produce white light.

4.0.1 *Interactions of Light with Matter*

The propagation of EM radiation is mediated by the materials that it encounters. In vacuum, EM radiation travels at ca. 10^8 m·s^{-1}. The absorption of light makes an object dark or opaque to the wavelengths or colors of the incoming wave. Some materials are opaque to some wavelengths of light but transparent to others. Glass and water are opaque to ultraviolet light, but transparent to visible light. The wavelengths of light absorbed by a material depend on the material composition, and hence properties derived from them. If a material or matter absorbs light of certain wavelengths or colors of the spectrum, an observer will not see these colors in the reflected light. On the other hand if certain wavelengths of colors are reflected from the material, an observer will see them and see the material in those colors. For example, the leaves of green plants contain chlorophyll, a pigment that absorbs the blue and red colors of the spectrum and reflects the green. Leaves therefore appear green. The fundamental interactions of matter with light are reviewed in **Figure 4.2**.

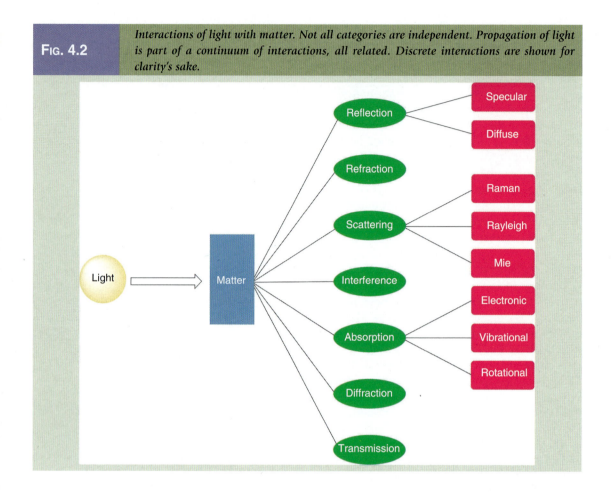

FIG. 4.2 *Interactions of light with matter. Not all categories are independent. Propagation of light is part of a continuum of interactions, all related. Discrete interactions are shown for clarity's sake.*

Light (EM radiation) can be viewed classically or from the quantum mechanical point of view. Since light is an EM phenomenon that consists of oscillating electric and magnetic fields and that electrons and protons are charged particles with motion that generate oscillating electric and magnetic fields, interaction between them is inevitable. The energy of electrons is quantized and the energy of EM radiation is quantized as well. We review some basic terminology and relate them to materials without going into too much detail (**Table 4.1**). Keep in mind that all interactions between light and matter can be described by the sum of all absorption and emission of photons and the most probable paths that photons take. All EM radiation exists within a continuum with matter related through $E = mc^2$. All photon and matter interactions are the same; it is just their expression that we are able to distinguish, catalog, and measure. Additionally, it is also with the size of the matter that makes a difference in those properties that we call observables.

Optical properties of nanomaterials are interesting to study due to the variety of new possible applications, some of which have already been demonstrated or have been predicted theoretically. These interesting properties arise either due to the intrinsic absorption or scattering processes. When light is incident on a material, the intensity of the incoming light onto the material reduces by either

TABLE 4.1	*Interactions between Radiation and Matter*	
Phenomena	**Description**	**Examples**
Reflection	Change in the wavefront at an interface between two material media in which the incident wave is directed back into the medium of its origin. A reflected wave that preserves the geometrical structure of the incident wave to produce a mirror image is called *specular reflection*. The angle of incidence is equal to the angle of reflection: $\theta_i = \theta_r$. *Diffuse reflection* occurs when light strikes a roughened surface and is the macroscopic embodiment of scattering. Smooth surfaces reflect and rough surfaces scatter. In specular reflection, momentum of the light (an elastic process) is considered to be preserved.	Shiny mirror surfaces reflect light. Metals are able to reflect light due to the presence of mobile electrons on their surface. The electrons are able to generate an opposing field that deflects that of the impinging radiation.
Scattering	When radiation is not reflected, other mechanisms account for its distribution. Light impinging on a roughened surface may be deflected in all directions with loss of coherence and phase. This is called scattering. The degree and kind of scattering is dependent on the size of the particles or surface facet and the wavelength of the impinging light. Scattering is the microscopic form of reflection—in the general sense.	Molecules, small and somewhat larger particles in the nano- and micro-domains are responsible for scattering light. Scattering is highly size-dependent— thereby, it is very relevant to nano-optics. SERS is based on Rayleigh scattering— both elastic and inelastic ramifications thereof.
Refraction	Refraction is related to reflection. Light incident on a surface is either reflected (or scattered) or refracted— usually both. It depends, like all EM phenomena, on the wavelength and the nature of the material (e.g., the refractive index). Refraction is described by Snell's law: $$\frac{n_1}{n_2} = \frac{\sin\theta_2}{\sin\theta_1} \qquad (4.1)$$	Transparent materials and liquids are able to refract (bend) light. The refractive index of air (or vacuum) is equal to 1.
Absorption	Absorption is a molecular phenomenon with mechanisms of electronic transitions, vibrations, and rotations.	Chromophores and fluorophores are examples of organic materials that have specific electronic transitions. Raman and IR spectroscopy are based on the absorption of wavelengths that induce vibrations, both symmetrical (Raman) and asymmetrical (IR).
Transmission	Transmission is the ability of light to pass through a material. Transmission is, in a general way, complementary to absorption. Transmission is a wavelength and material-dependent process.	Transmission of light is what is left over after reflection, scattering, and absorption have occurred.
Diffraction	Diffraction is the change in direction and intensity of waves after interaction with an obstacle (e.g., an aperture) that is of size on the order of the wavelength. It is the bending of waves around a feature or spreading of them through an aperture.	Diffraction is important in optical imaging techniques. The ultimate resolution of EM wavelength based optics is limited by the Heisenberg principle. The practical resolution of optical methods is limited by $1/2\lambda$. Diffraction gratings make use of this phenomenon in spectrometers.

the absorption in the material, transmission through the material, or reflection and scattering from the material as shown in **Figure 4.2**.

Intrinsic absorption of an optical material occurs by three fundamental processes: electronic absorption, lattice or phonon absorption, and free-carrier absorption. *Electronic absorption* occurs due to the interaction between the incident radiation and the motions of charged particles in a material (observed towards the higher frequency end of the infrared spectrum). Only EM radiation with sufficient energy to cause an electron to transfer between the valence band (VB) and conduction band (CB) will be absorbed by this mechanism. *Lattice absorption* characteristics are observed at the lower frequency regions, in the middle to far-infrared wavelength range, define the long wavelength transparency limit of the material, and are the result of the interactive coupling between the motions of thermally induced vibrations of the constituent atoms of the substrate crystal lattice and the incident radiation. Hence, all materials are bound by limiting regions of absorption caused by atomic vibrations in the far-infrared and motions of electrons and/or holes in the short-wave visible regions.

In the interband region, the frequency of the incident radiation has insufficient energy ($E = h\nu$) to transfer electrons to the CB and cause absorption; here the material is essentially loss free. In addition to the fundamental electronic and lattice absorption process, *free-carrier absorption* in semiconductors can be present. This involves electronic transitions between initial and final states within the same energy band (*intraband transitions*).

The absorbance of an object is quantified by the *Beer–Lambert law*. In absorption, the frequency of the incoming light wave is at or near the energy levels of the electrons in the matter. The electrons will absorb the energy of the light wave and change their energy state. After that, the electron returns to the ground state emitting a photon of light or the energy is dissipated internally. Beer–Lambert's law is usually expressed as follows:

$$A = -\log\left(\frac{I_T}{I_o}\right) \tag{4.2}$$

where
I_T is the intensity of the transmitted light through material
I_o is the intensity of the impinging light.

4.0.2 *The Nano Perspective*

Nano-optics is the study of optical phenomena and techniques near or beyond the diffraction limit [4]. Phenomena associated with nano-optics were some of the first nanotechnology to be controlled by humans. Nature also abounded with examples of nano-optical phenomena albeit unbeknownst to us in both synthetic and natural cases. Optical nano persisted throughout our history in the form of stained glass, photography, and quantum dots now. With the advent of the scanning electron microscope (SEM) and its application, we were able to decipher the colors of the Lycurgus cup. With transmission electron microscope (TEM) and atomic force microscope (AFM), we were able to see quantum dots for the first time. Nano-optics is evolving alongside nanotechnology and new instrumentation designed to see them better.

4.1   THE SURFACE PLASMON

There are several challenges facing optics in general as miniaturization continues unabated to the nanoscale. For example, spatial resolution is limited by diffraction. The diffraction limit is described by the Heisenberg principle:

$$\Delta x \cdot \Delta p_x \geq \frac{\hbar}{2} \tag{4.3}$$

and for photons

$$\Delta x \cdot \Delta k_x \geq \frac{1}{2} \tag{4.4}$$

where k is the wave vector. In order to shrink or overcome this limit, confocal microscopy and near-field microscopy (NSOM) have been developed recently. In addition, Raman spectroscopy and multiphoton fluorescence are capable of spatial resolution less than 200 nm, which is the highest optical resolution. Quantum dots of course take center stage in this section. Their versatility, stability, and overall expectations are true representatives of things to come and of the future of nano-optics.

4.1.1   *The Surface Plasmon Resonance*

Since clusters are an ensemble of atoms, we can consider two types of effects that are related to the size of these clusters: *intrinsic effects* are related to the changes due to the surface-to-volume ratio of the cluster, and *extrinsic effects* that are related to the size of the cluster which varies according to the external constraints that are generated due to an applied field. The field of technology that involves the surface plasmon is called *nanoplasmonics*.

The interface between the cluster and the *surrounding media* (usually a dielectric material) or host plays an extremely important role in the electrical and optical properties. Numerous theories have been developed to explain the properties of clusters as a function of their nature, size, shape, and the surrounding matrix. No single theory however can explain all the effects in entirety.

One of the notions to be developed in this section is the definition of the surface plasmon. Since a large number of atoms of the metallic nanoparticle are actually on the surface, the neighboring electrons form a sort of an electron gas since they are in continuous interaction with their neighbors. [5]. The surface plasmons are thus collective excitation of free electrons on the surface of the clusters. In other words, plasma oscillations on metal surfaces are collective longitudinal excitations of the conduction electron gas. The plasmon is a quantum of a plasma oscillation. When the light is incident on nanoparticles, its electric field perturbs this electron cloud as electrons are excited into the CB. This creates surface charge separation in a small particle ($r \ll \lambda$) and is called dipolar resonance (**Fig. 4.3**). Higher-order resonances exist for larger particles that are best described by Mie scattering theory, a topic beyond the scope of this chapter.

The dipole moment per unit volume is known as the polarization. The dipolar oscillation of all electrons will have the same phase and when the frequency

FIG. 4.3

The surface plasmon of two types of metal particles are depicted. The direction of propagation of the light is perpendicular to the horizontal axis (as depicted) of the elongated shape. Therefore, the electric field of the light runs parallel to the horizontal axis of the elongated shape. In both the sphere and wire, the plasmon generates an opposing electric field to counter that of the light.

of the EM field resonates with the coherent electron motion, there is a strong absorption in the optical spectrum. It is this plasmon oscillation that is responsible for the ruby color imparted to gold colloids with small diameter (<20 nm). The frequency and width of the surface plasmon absorption depend on the size and shape of metal nanoparticles, the dielectric constant of the metal itself (e.g., the composition of the particle), and the dielectric constant of the surrounding medium.

The classical Drude model of electrical conduction, developed in the 1900s by Paul Drude, explains the transport properties of electrons in materials (especially metals). The Drude model is the application of the "kinetic theory of gases" to model the "electron gas" of a metal. Assumptions include that the material contains immobile positive ions and an "electron gas" of classical, noninteracting electrons of density n, with motion damped by a frictional force due to collisions of the electrons with the core nuclei themselves, and importantly with the surface boundary. Relaxation time is symbolized by τ. From the Drude model, we can see that if a metal nanoparticle has the size and shape in which its CB electrons have a higher possibility to polarize, it is easier for the surface plasmon resonance to occur with a lower frequency and sharp bandwidth. We also can use the Drude model to explain the surface plasmon resonance observed in noble metals, such as gold, silver, and copper, and that size and shape play a role in the development and maintenance of the surface plasmon (e.g., higher polarizability at lower frequency and sharper bandwidth than that exhibited in the bulk). Therefore, the plasmon resonance of the metal nanoparticles (noble metal particles in particular) are strongest and shifted towards the visible part of the EM spectrum. A metal colloid and clusters surrounded by dielectric media are shown in **Figure 4.4**.

The electric field strength of spherical metal nanoparticles (dielectric constant ε_m) surrounded by an insulating medium (dielectric constant ε_o) exposed to light is described by

$$E_{\text{sphere}} \propto \left(\frac{\varepsilon_m - \varepsilon_o}{\varepsilon_m + 2\varepsilon_o} \right)_\lambda \qquad (4.5)$$

FIG. 4.4

Top: *Metal colloids are embedded collectively within a monolithic dielectric surrounding medium.* Bottom: *Ligand-stabilized clusters are embedded within a localized dielectric. In each case, the surrounding medium affects the position of* λ_{max} *of the surface plasmon dipolar resonance. Bulk material complex dielectric functions adequately describe the optical response of colloidal metals but caution must be used when applying them to smaller entities such as clusters. Metal clusters with diameter <10 nm start to exhibit quantum properties.*

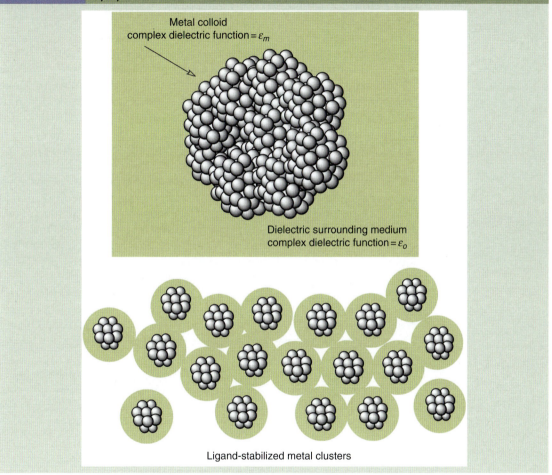

Metal colloid
complex dielectric function = ε_m

Dielectric surrounding medium
complex dielectric function = ε_o

Ligand-stabilized metal clusters

The factor "2" in the denominator is indicative of spherical shapes. From this equation, we can readily see that the localized electric field strength is maximized when

$$\left[\varepsilon_m = -2\varepsilon_o\right]_\lambda \tag{4.6}$$

the condition for the dipolar resonance. For nanoparticles that are greater than ca. 20 nm, other factors such as higher-order resonance absorption modes and scattering phenomena need to be considered. For example, minor differences in the dimensions of lithographically produced Ag particles shown by Jensen et al. [6] showed a significant effect on the resonance of the plasmon.

At the resonance, the extinction of Ag particles can be very large and increases linearly with particle volume.

We can use this characteristic of surface plasmon resonance (depending on the size and shape of nanoparticles) for sensor applications described in Ref. [6]. Moreover, for larger particles, the light cannot polarize the nanoparticles homogenously, and retardation effects lead to the excitation of higher-order modes or multipoles. Therefore, several resonances are generated leading to a broad extinction profile, or a few other low-energy peaks in the extinction spectra.

For nonspherical particles, the localized electric field induced by a surface plasmon is not evenly distributed around the surface [7] like nanorods and nanowires. The distribution of the local electric field is more complex, and there are some effects resulting in the significant increase of the local electric field. Thus, surface plasmon resonance in this case is different from that of spherical nanoparticles. For example, cylindrical or oblate nanoparticles are often described as nanorods with reference to their aspect ratio.

The plasmon resonance of a nanorod is divided into two bands. The first corresponds to the oscillation of the electrons being perpendicular to the major rod axis, often referred as the transverse surface plasmon absorption. The other corresponding to the oscillation of the electrons along the major rod axis is referred as the longitudinal surface plasmon absorption. The separation of two peak absorption frequencies depends on the aspect ratio; when this ratio increases, the separation increases. In addition, the λ_{max} of the nanorods is shifted to lower energies. **Figure 4.5** shows results from the report of El Sayed et al. [8]. The inset shows the peak absorption of transverse surface plasmon (squares) and the peak absorption of longitudinal surface plasmon (spheres) corresponding to the different aspect ratios. From this inset, we can observe that the transverse surface plasmon resonance is nearly unchanged for different aspect ratios, whereas the longitudinal surface plasmon resonance changes linearly with the aspect ratio.

The oscillation between the surface and the center of the nanoparticle (x_o) is generated by the incident light. The movement of the electrons generates an induced dipole that is periodic (**Fig. 4.6**). The equation that describes the movement of electrons (e.g., the induced electric field) is given by

$$E_{induced} = E_o \exp^{-iwt} \tag{4.7}$$

If the field is created by an incident EM wave of frequency ω, then the field experience by the electrons is represented by this time-dependent expression

$$m_e \left(\frac{\partial^2(x)}{\partial(t)^2} \right) + m_e \gamma \frac{\partial(x)}{\partial(t)} = e \cdot E_o \exp^{-iwt} \tag{4.8}$$

where m_e is the electron's rest mass, and γ is an absorption factor that takes into account the damping forces arising from friction discussed earlier.

The dipole oscillation for a nanoparticle between the center and the surface x_o is expressed as

$$p = ex_o \tag{4.9}$$

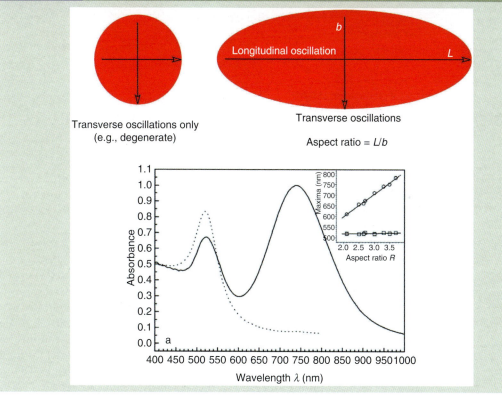

FIG. 4.5 Top: *The transverse and longitudinal plasmon resonance modes are depicted for two Au nanoparticle shapes. The sphere, of course, is isotropic with regard to the plasmon resonance and is therefore expected to display just one absorption peak.* Bottom: *Absorption spectrum of a nanorod (solid line) with aspect ratio = 3. The two peaks represent the transverse mode (left in the spectrum) and the longitudinal mode (at the right in the spectrum). The dotted line represents the absorption profile for spherical particles. The box in the upper corner plots the behavior of λ_{max} as a function of aspect ratio. As the particles become longer, the λ_{max} of the longitudinal mode (circles) shifts to lower energies. The λ_{max} of the transverse mode (squares) remains relatively unchanged* [8].

where p is the polarization due to one electron with charge e (similar to the more recognized expression for the dipole moment for a molecule: $\mu = \delta d$). With respect to the total number of electrons n, this becomes

$$P = np \tag{4.10}$$

The polarization is also defined from the perspective of the electric field:

$$P = n\alpha \cdot E \tag{4.11}$$

where α is the polarizability (remember your basic Raman spectroscopy).

Colloidal dispersions of metals behave differently from the bulk. If we consider a single representative spherical particle of diameter d with complex dielectric constant ε_m [32] embedded in a medium of dielectric constant ε_o in an oscillating

FIG. 4.6

Periodic oscillations are induced by the EM field of the impinging radiation. Electrons on the metal particle surface, represented as gold, oscillate in beat with the EM field. The electron cloud (dotted circles) oscillates on the surface of the metal. At the right, the respective polarization of the metal electrons is shown.

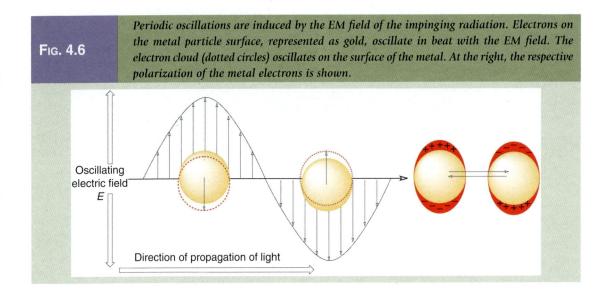

Oscillating electric field E

Direction of propagation of light

electric field, a density oscillation is built up. This is known as the surface plasma mode. The dielectric function under this mode is found proportional to $1/\tau^2$ where τ is the average size-limited scattering time—a main factor for the optical properties for small particles. So much for classical EM treatment.

A new effect, that is, the quantum size effect, plays a considerable role for tiny particles when the classical scattering concept breaks down. In this case the surface plasmon mode is treated quantum compared to the classically derived. Considering all such effects, silver is found to be the most effective metal with sharp response to surface plasmon resonance. Subsequent research in this field reveals that plasmon peak shift sensitivity depends on the interaction of the cluster with surrounding atoms or molecules and the medium. The more electrons that are spilled out of the geometrical surface of the metal cluster the more a red shift is favored. On the other hand the more electrons that are injected into the metal cluster through the surface, the more a blue shift is observed.

Nanoparticle EM response is adequately described by the quasi-static approximation. The major assumption is that the nanoparticle must be much smaller than the wavelength of light. If this is the case, then the electrodynamic condition becomes one of electrostatics. Under these conditions, the oscillations of the time-dependent external field occur slowly relative to the motions of the free electrons comprising the plasmon, for example, the opposing field is easily set up. As a result, only the dipolar resonance of the plasmon contributes to the optical response. There are no phase shifts (retardation effects), multipole phenomena (e.g., quadrupole scattering and extinction), radiation damping, and dynamic depolarization. This condition lends itself to treatment by *electrostatic* principles, a far simpler method to predict absorption than if an electrodynamic treatment is required. The physical particle size regime that qualifies for a quasi-static domain is $r < 10\,\text{nm}$ or so. The Maxwell–Garnett, equation 4.12, and Bruggeman (not shown) effective medium theories are used successfully to describe the optical response of small metal clusters

$$f_m \left(\frac{\tilde{\varepsilon}_m - \tilde{\varepsilon}_o}{\tilde{\varepsilon}_m + \kappa \tilde{\varepsilon}_o} \right) = \left(\frac{\tilde{\varepsilon}_c - \tilde{\varepsilon}_o}{\tilde{\varepsilon}_c + \kappa \tilde{\varepsilon}_o} \right) \tag{4.12}$$

where

f_m is the inclusion fraction (usually a metal)

ε_m, ε_o, and ε_c are the dielectric constants of the metal, host oxide, and composite, respectively.

The factor κ in the denominator is indicative of spheres if equal to 2.

4.1.2 Scattering

Mie theory offers a means of describing the optical response of particles larger than those that reside within the quasi-static limit—particles 20–50 nm and larger in diameter—or when the ratio of particle circumference to wavelength is larger than 10 [9]. As depicted in **Figure 4.2**, there are several kinds of scattering phenomena. Apart from absorption, scattering phenomena gives rise to novel applications of nanomaterials.

When we discuss about extinction in optical properties, for small nanoparticles (diameter much smaller than the wavelength of the incident light), only the dipole absorption contributes to the extinction of the nanoparticles and scattering is negligible. However, when the size of the nanoparticles increases (for large nanoparticles), the extinction is not only dependent on the higher-order multipole modes but also the light scattering of light can be thought of as the redirection of light that takes place when an EM wave (i.e., an incident light ray) encounters an obstacle or nonhomogeneity, let us say a scattering particle.

As the EM wave interacts with the discrete particle, the electron orbits are perturbed periodically with the same frequency as the electric field of the incident wave—similar to dipolar plasmon resonance we discussed earlier. The oscillation or perturbation of the electron cloud results in a periodic separation of charge within the molecule called an *induced dipole moment*. The oscillating induced dipole moment becomes a source of EM radiation. Scattered light is the result. The majority of light scattered by the particle is emitted at the identical frequency of the incident light, a process referred to as *elastic scattering*. In summary, the above comments describe the process of light scattering as a complex interaction between the incident EM wave and the molecular/atomic structure of the scattering object; hence, light scattering is not simply a matter of incident photons or EM waves "bouncing" off the surface of an encountered object [9].

Rayleigh Scattering. Rayleigh scattering is a molecular phenomena and the limit in size can be extended to ca. $d = 0.10\lambda$, for example, where photons do not lose energy. The intensity of Rayleigh scattering is proportional to $1/\lambda^4$. This means that Rayleigh scattering is more intense for the blue wavelength light—hence the blue color of the sky. Rayleigh scattering is elastic. SERS (surface enhanced Raman spectroscopy) relies on some of the equations we presented earlier in this section: the one describing the polarization and the ones describing the resonance condition. SERS utilizes scattered photons induced by a laser source to detect changes in vibrational states of analyte molecules. For example, Campion et al. state that for an analyte signal

$$G = \chi^{12} \left(\frac{\varepsilon_m - \varepsilon_o}{\varepsilon_m + 2\varepsilon_o} \right)_\lambda^2 \left(\frac{\varepsilon_m - \varepsilon_o}{\varepsilon_m + 2\varepsilon_o} \right)_s^2 \tag{4.13}$$

where

$\chi = r/(r + d)$ with r equal to the radius of the spherical metal nanoparticle
d is the distance of the analyte molecule from the surface of the metal particle.

The λ and s represent the laser and Stokes fields, respectively. Obviously the SERS signal is enhanced significantly as the relative denominators approach zero [10]. SERS is a true nanoscale phenomenon.

Mie Scattering. Scattering from molecules and very tiny particles ($<0.10\lambda$) is predominantly Rayleigh type of scattering. For particle sizes larger than this limit, Mie scattering predominates. This scattering produces a pattern like an antenna lobe, with a sharper and more intense forward lobe for larger particles. Mie scattering is not as strongly wavelength dependent and produces an almost white glare around the sun when a lot of particulate material is present in the air. It also gives us the white light we see from mist and fog. Mie scattering is compared to Rayleigh scattering in **Figure 4.7**.

In Mie theory, if nanoparticle size is less than the incident wavelength, the Mie size parameter is less than unity.

$$x = \frac{2\pi n_o a}{\lambda} \tag{4.14}$$

In this case, scattering is defined by Rayleigh scattering. The Mie scattering cross section is reduced to the Rayleigh scattering cross section.

$$C_{\text{scattering}} = \frac{8\pi}{3} \left(\frac{2\pi n_o a}{\lambda} \right)^4 a^6 \left(\frac{m^2 - 1}{m^2 + 2} \right)^2 \tag{4.15}$$

Mie scattering encompasses the general spherical scattering problem (absorbing or nonabsorbing) and places no limitations on particle size. $C_{\text{scattering}}$ is inversely

FIG. 4.7

Different kinds of scattering are depicted. Rayleigh scattering is the most intense due to short wavelengths (e.g., the blue sky). Mie scattering involves larger particles. Notice how here is a significant portion of forward scattering of photons for larger particles. These are truly nanophenomena.

Rayleigh scattering | Mie scattering | Mie scattering: Larger particles proportional to λ^4. Therefore, for a same-sized particle, the scattering efficiency $Q_{scattering}$ decreases as the wavelength becomes longer where

$$Q_{scattering} = \frac{C_{scattering}}{\pi a^2} \qquad (4.16)$$

Conversely, as the particle diameter a becomes larger, scattering efficiency is expected to improve.

Mie scattering theory therefore converges to the limit of geometric optics for large particles and may be used for describing most spherical particle scattering systems, including Rayleigh scattering. As you can see, we can create a continuum between quantum optics, quasi-static optics, and Rayleigh scattering, Mie scattering, and geometric optics of the classical world.

4.1.3 *Color Generation from Nanoparticles and Nanostructures*

Color Due to Interference. This mode of color is quite familiar to us—from the vivid colors displayed by bubbles made of water and surfactants. The color is based on interference (constructive) of light wavelengths as they traverse the thickness of the bubble film. Depending on the thickness of the bubble, different colors are made visible. Relative thickness of microtomed sections for TEM analysis is determined by interference colors. Bubble films are on the order of several hundred nanometers in thickness.

Jewelry made of anodized titanium often display brilliant colors due to the thin refractive oxide coating. Colors such as bronze ($\lambda \approx 300\,nm$), blue ($\lambda \approx 400\,nm$), yellow ($\lambda \approx 600\,nm$), and purple ($\lambda \approx 700\,nm$) are some colors available. Optimal oxide thickness is equal to ca. half the wavelength of the color desired. Of course, incidence angle and refractive index also play a role in the resultant color.

Color Due to Diffraction. The best example of diffraction colors is a CD (**Fig. 4.8**).

Color Due to Scattering. We have discussed above the phenomenon of scattering. Different colors are generated by different kinds of scattering, different particle sizes, and different wavelengths. The sky is blue because short wavelengths are scattered by molecules. The sky is red because long wavelengths (e.g., reds) are scattered by larger particles. The deficiency in scattering intensity is compensated by longer path lengths.

Color Due to the Surface Plasmon. Perhaps the most famous example of metal plasmon color is provided by the Lycurgus cup, housed in the British Museum. The cup, made in fifth century Rome, consists of Au and Ag nanoparticles housed in a soda lime NaO glass matrix. SEM analysis in the 1990s showed that noble metal nanoparticles were responsible for the color. A remarkable dichroism is also exhibited by the cup. Upon reflection, the cup looks green. Upon transmission, the cup is a crimson red. Gold-55 quantum dots occupy the smallest limit

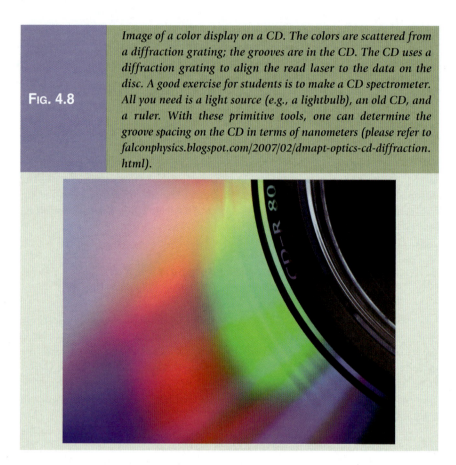

FIG. 4.8 *Image of a color display on a CD. The colors are scattered from a diffraction grating; the grooves are in the CD. The CD uses a diffraction grating to align the read laser to the data on the disc. A good exercise for students is to make a CD spectrometer. All you need is a light source (e.g., a lightbulb), an old CD, and a ruler. With these primitive tools, one can determine the groove spacing on the CD in terms of nanometers (please refer to falconphysics.blogspot.com/2007/02/dmapt-optics-cd-diffraction. html).*

for the plasmon resonance. They are, with their protecting ligand shell, around 4 nm in diameter. The color is a ruby red with λ_{max} at ca. 520 nm.

Color Due to Quantum Fluorescence. Semiconductor quantum dots are known for their intense fluorescent colors. Although made of exactly the same material, different colors are generated due simply to the difference in size of the quantum dots (QDs) (**Fig. 4.9**).

4.1.4 Applications of Nanoplasmonics

In nanophotonics and information technology, nanoscale metal structures embedded in organic devices could also potentially be beneficial in improving the light absorption, as well as the charge separation and charge collection processes. In biomedical applications, by linking specific antibodies to the metal surface, tumor cells can be targeted and imaged before pathologic changes occur at the anatomic level. HIV-I virus can be detected with Ag nanoparticles, and super paramagnetic nanoparticles used in MRI can detect disease in soft tissues (e.g., liver). Au nanoparticles applied on electrode surfaces improve electrode chemical analysis. In sensing, detecting low concentrations of molecules or

FIG. 4.9 *Fluorescence of CdSe quantum dots decreasing in size from left to right are depicted. More will be said about QDs in the next section. The wavelength of fluorescence emission of CdSe quantum dots is related to the size of the nanoparticle. The emission of longer wavelengths is indicative of larger size.*

Source: Image reprinted with permission from Professor Zhiqun Lin, Department of Materials Science and Engineering, Iowa State University.

biological species exploits the dependence of the surface plasmon resonance frequency on the dielectric constant of the surrounding medium. The sensitivity of single, small metal particles could approach the single-molecule detection limit for large biomolecules.

Waveguides.   Waveguides based on surface plasmon resonance are also possible. Arrays of closely spaced metal nanoparticles are able to interact by establishing coupled plasmon modes. The resonance creates a dipole field due to the oscillations of the plasmons as we discussed earlier. If a periodic array (a line or plane or other structure) of metal nanoparticles is created, electrodynamic interactions between the adjacent metal nanoparticles can combine to form a waveguide. The phenomena is based on the near-field coupling of adjacent metal particles. The propagation of light along the array can be forced inside such a pipe on a scale well below the diffraction limit and be made to take 90° turns that are significantly lower than the wavelength of the light [11].

The screening parameter utilized by effective medium theories (e.g., Maxwell–Garnett) incorporates a term known as the screening parameter κ (described briefly above). If parallel Au nanorods are placed within the pore channels of anodically formed porous alumina, they too have the ability to "screen" light into the transparent host dielectric medium. This concept forms the basis for development of transparent metal nanostructured materials [12]. The screening effect of a spherical particle (e.g., $\kappa = 2$) is not as strong as screening by higher aspect ratio parallel nanorods (e.g., $\kappa = 1$) [12].

SERS. Surface enhanced Raman spectroscopy (SERS) has exploited the phenomenon of surface plasmon for several decades. SERS techniques have developed to the point of single-molecule detection. The surface plasmon of the nano–noble–metal particle is able to enhance analytical signals several orders of magnitude. Illumination with a fiber optic source of a SERS substrate at the tip of the probe is all that is required to detect analyte samples at incredibly low levels.

4.2 QUANTUM DOTS

A quantum dot is a semiconductor nanostructure that is very small along all three spatial dimensions. What is very small? Small in this case implies that the motion of CB electrons, VB holes, and excitons are restricted in the three spatial directions. More clarification is required. Quantum dots (a.k.a. zero-dimensional structures) are nanostructures where all dimensions are comparable to the exciton Bohr radius. QDs generally have a diameter that is less than 10 nm. An electron–hole pair created when an electron leaves the VB and enters the CB is called an exciton.

4.2.1 *The Bohr Exciton Radius*

Excitons have a natural physical separation between the electron and the hole which is dependant on the material. This average distance is called the *exciton Bohr radius*. In large semiconductor crystals, the exciton Bohr radius is small compared to the size of the crystal, and the exciton is then relatively free to move around in the crystal. However, in nanocrystals, *quantum confinement* occurs that serves to restrict the motion of electrons, holes, and excitons in 1, 2, or 3 dimensions, respectively. Thus, the exciton Bohr radius may additionally be defined as the natural physical separation in a crystal between an electron in the CB and the hole it leaves behind in the VB. The size of this radius controls how large a crystal must be before its energy bands can be treated as continuous, which distinguishes a semiconductor bulk crystal from a quantum dot. Mathematically, the exciton Bohr Radii is described by

$$a_B = \frac{4\pi\varepsilon_o\varepsilon_r\hbar^2}{m_o e^2}\left[\frac{1}{m_e^*} + \frac{1}{m_h^*}\right] \tag{4.17}$$

where

ε_o and ε_r are the absolute and relative dielectric constants of the medium respectively

m_o is the rest mass of an electron

e is the fundamental electronic charge

m_e^* and m_h^* are the effective masses of an electron and hole exciton pair, respectively

In your spare time, provide a physical interpretation of this equation.

The difference in energy between the lowest energy state of the CB and the highest energy state of the VB of a semiconductor is the energy bandgap. In bulk semiconductors, CB and VB have continuous energy states separated by the bandgap that allows us to easily control by doping and other methods.

In semiconducting quantum dots, on the other hand, motion of charge carriers is confined due to the particle dimensions approaching the exciton Bohr radius. The CB and VB energy states split up and become *discretized*. Charge transfer occurs within these discrete levels, resulting in photoemission. Mathematically, these discrete energy levels can be expressed by the formula in equation

$$E_n = \left[\frac{h^2}{8ma^2}\right]n^2 \qquad (4.18)$$

where

E_n is the energy of the quantum dot at some discrete energy level n
m is the mass of the QD
a is the diameter of the QD

Such discretization of the bandgap shifts the emission of the QD to a higher energy, for example, a blue shift. Comparison of the bandgap of bulk semiconductors, QDs, and molecules are shown in **Figure 4.10**.

4.2.2 Tuning the Gap

The discrete energy states of QDs are similar to those of organic molecules. Electrons excited from VB to CB may lose energy radiatively or nonradiatively by discrete jumps within these levels. The optical properties of quantum dots have strong size dependence. Excitation as well as emission bands can be tailored by

FIG. 4.10

As the material assumes smaller dimensions, the VB and CB lose their continuity and become discretized (e.g., split into discrete energy levels). Note also how the energy gap increases with smaller size. This phenomena is responsible for the blue-shifted fluorescence emitted by smaller particles compared to larger particles made of the same material.

FIG. 4.11   *Graphical representation of the increase in "bandgap" with decreasing size.*

varying the size of QDs. In general, the smaller the diameter of the QD, the greater is its exciton binding energy and the higher its emission energy (**Fig 4.11**). The emission spectrum of a QD can be tuned from ultraviolet past into the infrared. The gap of ZnS, for example, is 3.6 *eV* for bulk material but it can be increased to about 4.5 *eV* if in the form of a QD of diameter between 1 and 4 nm [13]. The bandgap energy E_g is related to the diameter *a* of the QD from

$$E_g = E_{g(bulk)} + \frac{3\varepsilon_o \varepsilon_r \hbar^2 \pi^2}{2a^2} \left[\frac{1}{m_e^*} + \frac{1}{m_h^*} \right] \tag{4.19}$$

and

$$\Delta E_g = \frac{3\varepsilon_o \varepsilon_r \hbar^2 \pi^2}{2a^2} \left[\frac{1}{m_e^*} + \frac{1}{m_h^*} \right] \tag{4.20}$$

4.2.3   Luminescence

Quantum dot semiconductors, just like in bulk semiconductors, are amenable to doping. Concomitantly, QDs, in addition to fluorescence, also are able to undergo luminescent radiative pathways. Luminescence is the generation of light emission that is not fluorescence. This property is characteristic of semiconductors and the cause is excitation and subsequent recombination of electron–hole pairs. Luminescence in semiconductors is initiated by the absorption of photons. The former is called photoluminescence while the latter is called electroluminescence or sometimes cathodoluminescence (e.g., a cathode is used for the generation of electrons which cause the excitation). Regardless of the form of excitation energy, the result is excitation of electrons in the VB of the semiconductor and subsequent jumping of excited electrons from the VB to the CB. These excited electrons however do not stay in the excited state for too long and must release the excess energy and get to their normal state by recombining

| FIG. 4.12 | *Different luminescent recombination processes for electrons and holes in doped semiconducting QD (not to scale).* |

with holes. This happens of course by emission of energy either in the form of heat (via phonons) or light.

When a recombination results in emission of light it is called radiative recombination, otherwise it is nonradiative recombination. Radiative recombination is favored if donor/acceptor energy levels are introduced in the energy band of a semiconductor through doping. Dopants that reside within the crystal in between lattice sites are called interstitial dopants. Interstitial dopants do not give rise to luminescent recombinations. Dopant ions that reside within a semiconductor QD (e.g., by substituting lattice atoms) are called substitutional dopants, and this form results in luminescent recombinations that occur in three different ways (**Fig. 4.12**).

4.2.4 Applications

Crystalline semiconducting QDs have many applications: (1) as components in nonlinear optical devices, (2) in photoluminescent and electroluminescent materials, and most importantly (3) as fluorophores in biolabeling, medical imaging, and other types of image applications. ZnS QDs in particular have received a significant amount of attention over the past decade or so. ZnS has excellent optical properties and is chemically quite stable. ZnS has potential for use in optoelectronic devices such as high-field electroluminescence, cathodoluminescence, and field emission displays (FED).

Fluorophores. Fluorophores are molecules or functional organic groups that can emit light or fluoresce after absorbing light at a particular wavelength. In other words, the component of a molecule which makes it fluorescent is named a fluorophore. Fluorophores absorb energy at specific wavelengths and emit at other specific wavelengths. This emitted energy depends on the type of the chemical composition, the immediate environment (a ligand stabilized shell, for example), and functional groups. Fluorophores are particularly useful for bacterial, protein, and cell studies and in other biologically related investigations.

For example they can be incorporated into other biological proteins to produce fluorescence. This process of fusion is known as biological labeling or biolabeling in short. Biological labeling of living bacteria or cells, etc., has helped immensely in getting better understanding of the inner details of cells and their functions. This is relatively easy to accomplish because of the advanced development of fluorescence microscopies such as NSOM and confocal instruments.

Use of traditional organic dyes has proven to be rather difficult. First of all, it is difficult to label proteins with organic dye fluorophores because sufficient technical expertise and experience is required to inject the biolabels at the microscopic level with minimal damage to the object of the tag. Targeting specific proteins in vivo is also difficult with organic dyes. Biological materials can also be highly fluorescent and when organic dyes are attached, there is difficulty in resolving them from background fluorescence. Most organic dyes are not chemically stable (to light and heat), and this results in a decay of fluorescent properties. Therefore, organic dyes are usually stored in the dark at refrigerator temperatures (ca. 4°C). Since large amounts of chemical dyes are often required, poisoning of the cell may result.

In QDs, the absorption and emission lines are much farther apart than that of organic dyes. This is illustrated in **Figure 4.13** where the fluorescent isothiocyanate (FITC or fluorescein) absorption–emission profile is compared to that of a generic QD. In **Figure 4.13 (Middle)**, the degradation of luminescence efficiency between hybridized EUB338-conjugated QDs and 4′-6-diamidino-2-phenylindole (DAPI) organic fluorescent labels on *E. coli* is depicted.

Most organic dyes used for biolabeling have a narrow excitation band and often the emission band overlaps with it (**Fig. 4.13 Top**). Semiconducting QDs on the other hand can be excited by any wavelength above its bandgap with built-in tunability. The emission wavelength can be tuned by manipulation of the size, use of dopant materials, and their specific placement in the energy band structure of the semiconductor QD. Semiconductor fluorescent QDs can be synthesized for multicolor luminescence.

QDs have a high number of photons and can give high luminescent intensities [14]. QDs therefore are capable of a broad range of emission with narrow emission spectral bandwidth [14]. For example, ZnS-capped CdSe QDs demonstrated 20x more intensity and 100x better stability than the common organic dye rhodamine 6G [15].

QDs have been used in combination with bacteriophages, and subsequently tag infectious bacteria [16]. There is a need to identify infectious bacteria quickly and with great amplification of signal to increase sensitivity. A "quick and easy" method to do just that has been developed by research groups at the National Cancer Institute (NCI) and the National Institute of Standards and Testing (NIST). The team developed a phage-based method that utilized QDs for tagging (**Fig. 4.14**). Organic fluorophores like GFP and luciferase have disadvantages as discussed earlier. Specifically, they suffer from low signal-to-noise ratio (due to background autofluorescence) and low photostability (e.g., susceptible to photobleaching).

New fluorescent materials that incorporate CdSe QDs exhibit broadband absorption with narrow emission and with size-dependent local λ_{max} of absorption. ZnS QDs, due to an outer shell of a few atomic layers, further enhance the photostability of the QD fluorophore [16]. The method developed by the

FIG. 4.13

Top: *Generic spectra emphasizing the difference between organic dye (FITC) and a generic QD fluorescence. The organic fluorescence (top left) is within range of its excitation (absorption) wavelength whereas in the case of the QD, the absorption and emission regions are quite separated. Middle: Comparison between organic and QD fluorescence lifetime is depicted. Green relates to EUB338 QD fluorophores, blue to an organic dye DAPI. Bottom: Biolabeled E. coli with chitosan-capped Mn-doped ZnS QDs [31].*

Fig. 4.14 The process of tagging microorganisms with nanoengineered QD complex to detect infectious bacteria is shown. (a) First, the phage is genetically engineered to assemble a small surface peptide (a biotinylation peptide) that is fused to a capsid protein. Biotin (vitamin H) is then attached to the lysine residue of the tagged peptide. Streptavidin-functionalized QDs are then attached. (b) Western blot analysis of control (T7-myc phage) versus experimental (T7 biophage) results with streptavidin–horseradish peroxidase is shown. (c) Fluorescence micrograph of E. coli with 100-fold excess of biotinylated phage is depicted. (d) Bright field TEM of the same region is shown. Scale bars are 1 and 2 μm, respectively.

Source:　R. Edgar, M. McKinstry, J. Hqang, A. B. Oppenheim, R. A. Fekete, G. Giulan, C. Merril, K. Nagashima, and S. Adhya, *Proceedings of the National Academy of Science*, 103, 4841–4845 (2006). With permission.

NCI–NIST team utilizes in vivo biotinylation of genetically engineered T7 phage linked to streptavidin-modified QDs. Amazingly, this genetically engineered, nanoengineered phage–QD complex reduces amplification time to 20–45 min because each infected bacterium is able to produce 10–1000 phages that are detected by the QD complex.

Field Emission Displays. Fluorescent quantum dots have the potential to be used as cathodoluminescent phosphors, and hence can be used in FEDs. Most flat screen displays make use of LCDs that are electronically switched between a transparent and opaque state. However, viewing angles for these are very narrow, and faster switching speed is also a problem. Hence, instead of this light "on/off technology," light emitters as the picture elements should be used, equivalent to a miniature version of old CRT technology. This configuration gives much better viewing angles with full color display and faster switching speeds. In this display technology, like of that in a CRT, emission of electrons and corresponding emission of color is due to electrons impinging the picture elements. This technology is known as a *field emission display* (FED). In FEDs, electrons are emitted from cold cathodes unlike in CRTs and the most challenging task is to develop low-voltage phosphors with cathodoluminescent

| FIG. 4.15 | *Schematic representation of a generic nanophosphor-based FED is shown. Television tubes use ZnS with Cu+ doping to give green, ZnS doped with Ag gives blue, and Y-VO₄ gives a red color.* |

efficiency to get good resolution, stability, and brightness [17]. Good FEDs should have low current saturation and the nanophosphors showed less current saturation than the microphosphors [18].

There are more advantages to this nanoscale FED technology. The processing temperature of the nanophosphors are hundreds of degrees lower than the commercial microphosphors that are fabricated by mechanical milling with sizes of the order of $2\,\mu m$. These micron-sized particles pose a problem for screen efficiency of very high-resolution displays [19]. Nanophosphors, on the other hand, are fabricated at lower temperatures and are very small: 2–100 nm in diameter. These small particles, not surprisingly, can be made to luminesce at different wavelengths for the reasons we described before. This gives QD-based FEDs great flexibility. What do you think that the resolution of such a device could be, potentially? A schematic of a typical nano-emitter is shown in **Figure 4.15**.

4.3 NEAR-FIELD MICROSCOPIES

4.3.1 The Diffraction Limit

Measurement and standards need to be extended to the nanoscale (e.g., nanometrology). One of the areas where this is required is nanoscale optics. NSOM has made incredible progress in this area—to the point of single-molecule spectroscopy. The ability to detect, measure, and manipulate large, single biological molecules like proteins and RNA is leading to a greater understanding of folding and other mechanisms. Efforts are underway to accomplish all of this—and within living cells [20].

Far-field optics is limited by diffraction (**Fig. 4.16**).

$$\Delta x = 0.61\left(\frac{\lambda}{\text{NA}}\right) \quad \text{and} \quad \frac{\lambda}{\text{NA}} \to \frac{4\pi}{\Delta k} \to \frac{\Delta p}{\hbar} \qquad (4.21)$$

FIG. 4.16   *Resolution limited by the Heisenberg principle. In NSOM, an evanescent wave is created that when close to the surface is able to beat the diffraction limit.*

where
    NA is the numerical aperture.

$$(\Delta x \cdot \Delta p) = 1.22 \cdot h > \left(\frac{\hbar}{2}\right) \tag{4.22}$$

$$\text{Resolution limit} \approx \frac{\lambda}{2} \tag{4.23}$$

Optical resolution is therefore restricted by the diffraction limit defined by the uncertainty principle. Depending on the wavelength, for most standard optical (and UV) microscopes this limit is ca. 150–300 nm. Resolution limit of 30 nm with far-field microscopes was attained with nonlinear materials and solid immersions [4]. This is not possible using conventional UV-visible light microscopy because it has a wavelength range of 300–700 nm. Thus conventional optical microscopy fails because the resolution is restricted to half the wavelength used.

4.3.2 *Near-Field Microscopy*

The diffraction limit can be overcome by adding a nanoscale object in the near-field and resolution is improved significantly. The invention of NSOM overcomes the major problem of resolution limits. In order to obtain an illumination in the nano range, we need optical apparatus that is able to produce a small spot size with the beam in the near-field of the probe (with diameter between 10 and 100 nm). To overcome this problem a small aperture that illuminates a region on the sample that is a few nanometers in diameter is needed. We should keep in mind that the character of light changes as it passes through small apertures. Thus the localization of the light waves result in the formation of *evanescent waves*. As the distance from the aperture increases, the intensity of the evanescent waves decreases rapidly. Therefore the aperture has to be close to the object, only a fraction of wavelength away. That is why it is called near-field microscopy [21].

Much more research and progress is going on to design electronics devices with feature size on the nanometer scale, in which the electron is confined in the nanometer scale structure called quantum dot. In these structures the spatial confinement of electron approaches the de Broglie wavelength, so the matter wave properties of electron changes rigorously. Therefore, along with the electronic properties their optical properties also vary from that of bulk materials. With current techniques it is a challenging task to observe optical transition between the electronic states in quantum dots. But by using the nanometer scale light source (obtained from the conventional laser light from an aluminum-coated fiber tip), Guest et al. have now succeeded in exciting and detecting single optical transition on the nanometer scale in a solid material [21].

In NSOM, a subwavelength aperture (e.g., an aperture that is smaller than the wavelength of the illuminating source) is used as the scanning probe. The probe is scanned a few nanometers above the surface of the sample. The confined light transmits through the sample and is collected by traditional optical apparatus [22]. The probe tip is a small aperture at the end of a tapered optical fiber that is coated with aluminum. With such a probe, even though the original light source may be 500-nm wavelength, it is possible to beat the diffraction limit. Resolution down to 50 nm and better is achievable.

4.3.3 *Applications*

Near-field polarimetry was used to investigate the structure of isotactic polystyrene with subdiffraction limit resolution [22]. The crystallites of ultrathin polymer films (<100 nm) crystallites were imaged by this technique. The objective was to investigate the radial strain, local tilt of the crystal axis, and strain in the amorphous layer above and below the growth planes of the crystallites of the polymer by optically characterizing the birefringence profiles of the material [22]. Characterization of chain conformation near the growth front, amorphous layers (positioned over and under crystallites), and the orientation of folded chains was not possible by traditional diffraction limited methods. Use of the NSOM technique was applied to reveal the structure during early growth stages of the crystallites (**Fig. 4.17**) [22].

In anisotropic crystals, two rays of light produced by a double refraction have different velocities in the crystal (e.g., it takes longer time for the slow ray to traverse the crystal). The faster ray travels a distance ΔR beyond the surface of the crystal before the slow ray reaches the surface. ΔR is known as the retardation. Birefringence is the difference in refractive index within a material. Without retardation, for example, light emerging from a crystal can recombine by interference. If there is no retardation, the resultant ray is identical to the incident ray. If there is retardation, then the resultant ray is altered. If the light source is monochromatic, then the crystal will appear light or dark depending on the degree of retardation. Retardation is a function of crystal orientation and thickness.

FIG. 4.17 *Retardance and topographical images of PS crystallites acquired with NSOM polarimetric technique. (a) Retardance, (b) topography, (c) retardance with overlaid fast axis orientation, and (d) topography with overlaid fast axis orientation, and (e and f), high resolution scans. For a more detailed interpretation, see Ref. [22].*

4.4   NANOPHOTONICS

4.4.1   *Photonics*

According to the Photonics Dictionary [23]

> The technology of generating and harnessing light and other forms of radiant energy whose quantum unit is the photon. The science includes light emission, transmission, deflection, amplification and detection by optical components and instruments, lasers and other light sources, fiber optics, electro-optical instrumentation, related hardware and electronics, and sophisticated systems. The range of applications of photonics extends from energy generation to detection to communications and information processing.

Photonics is the study of the interactions of light with matter. Photonics is not just the study of light, it is more appropriately a technology that happens to be based on the interactions of light with matter. The goal of photonics is to develop components and devices. Photonics first began to emerge with the invention of the laser in 1960 and the laser diode a decade later. The invention of the optical fiber as a means of transmitting information via light, and therefore information, formed the basis for optical telecommunications. The field is now quite enormous and consists of a variety of subdisciplines and applications that includes laser technology, biological and chemical sensing, medical diagnostics and therapeutics, display technology, optical computing, fiber optics, metrology, holography, photodetection, and photonic crystals.

In 1987, Eli Yablonovitch at Bell Communications Research Center in New Jersey and Sajeev John of the University of Toronto created an array of 1-mm holes in a material with refractive index equal to 3.6 (a.k.a. Yablonovite) [24,25]. They found that the array prevented microwave radiation from propagating in any direction. It took nearly a decade to fabricate photonic crystals that do the same in the near-IR and visible range. They based their work on the premise that "photonic bandgap" (PBG) behavior would be similar to the behavior of electron waves in natural crystals, for example, that phenomena such as reciprocal space, Brillouin zones, dispersion relations, Block wave functions, and Van Hove singularities must also be applicable for photonic waves [26].

4.4.2   *Photonic Structures in Living Systems*

There are numerous examples of photonic systems that are displayed by living things. The nanostructure of the wing of the butterfly genus *Morpho* is composed of intricate photonic structures on the order of 500–600 nm that interact with light to produce a beautiful iridescent blue color. Other butterflies also have what are known as structural colors—interference colors that do not depend on chemical means or a pigment to become visible. Professor Peter Vukusic, a member of the Thin Film Photonics Group in the School of Physics at the University of Exeter, United Kingdom discussed the "optical tricks that have given butterflies, beetles and other creatures an evolutionary advantage" [27]. It was Isaac Newton and Robert Hooke who first connected photonic structures to the iridescent colors of a peacock's feather, the Abalone's shell (mother of pearl), and natural opals.

There exist two generalized mechanisms that produce color in butterfly wings. The first is due to the presence of pigment molecules that are able to absorb certain wavelength light and selectively transmit or reflect other wavelengths. The second mechanism is more interesting. It leads to remarkable iridescent colors and is based on structure, for example, a structural color. Interference colors are due to the thin-film interference or diffraction phenomenon. Interference in thin film as discussed above is due to the difference in the refractive indices of the two mediums, which form the interface for the light to pass through [26]. Wave mechanics has shown that the color of butterfly scales is due to interference phenomena. When light is incident on thin films some of the light reflects from the outer surface while the rest reflects from the inner surface of the thin film. One sees interference colors when the film thickness is on the order of the wavelength of visible light. These "reflected" wavelengths interfere with each other either constructively or destructively and that causes increase or decrease of amplitude of the original wavelength. Constructive interference is responsible for the color production.

The structural hierarchy of the blue *Morpho* butterfly wing is shown in **Figure 4.18**. Starting with scales (not shown), there exist veins (or ridges) that consist of photonic structures. If viewed from the side (**Fig. 4.18 Left**), one can see that the tiered structures correspond to light that interferes with the structure.

Source: Images are courtesy of Professor Pete Vukusic, School of Physics, University of Exeter, UK. With permission.

FIG. 4.18 — The ultrastructure of the blue Morpho butterfly is depicted. On the left, the veins of the structure are depicted. On the right, the transverse section of the veins is shown. The tiered structure is ca. 600 nm in height.

4.4.3 Photonic Crystals

Photonic crystals are periodic structures made of dielectric materials that interact with resonance with radiation at wavelengths consistent with the periodicity of their lattice network. One of the goals of nanophotonic crystals is to serve as waveguides and to confine light without experience losses. Photonic crystals, called "semiconductors of light" by Marion Florescu of NASA's Jet Propulsion Laboratory and Cal Tech, can be compared to semiconductors [28]. Semiconductors consist of periodic arrays on the atomic scale that control the flow of electrons. Photonic crystals consist of periodic arrays on the scale of wavelength of dielectric materials that control the propagation of light waves. Both material regimes have a characteristic bandgap. We understand the nature of the bandgap in semiconductors. We shall discuss PBGs later in this section. Photonic crystals are designed to confine, manipulate, and control the propagation of photons in three dimensions. Photonic crystals can be one-, two-, or three-dimensional (**Fig. 4.19**).

Applications of photonic crystals include their use in devices to control the flow of radiation, dielectric mirrors for antennas, microresonators, low-threshold nonlinear devices, microlasers and amplifiers, controlled miniaturization, and pulse sculpting (Florescu). PCs can act as a filter regardless of polarization or direction of motion of photons to localize photons, to inhibit the spontaneous emission of chromophores, to modulate stimulated emission, and to serve as waveguides [29]. Control of these phenomena will enable us to make LEDs, zero-threshold semiconductor diode lasers, and enhance performance of quantum optical devices as well as all other kinds of optical devices [29]. PCs and applications of photonics reach across the EM spectrum—from UV to radio waves.

The major characteristics of synthetic photonic crystals is that dielectric networks retain connectivity, the active elements have the same average optical path, and that there be a high ratio of dielectric indices. Even if the crystal is heterogeneous, the average optical path should be the same regardless of the media. There should be an overlap of frequency gaps along different directions. The beauty of photonics is that PCs can be fabricated to specifically overlap with the radiation range of interest.

Source: C. P. Poole and F. J. Owens, *Introduction to Nanotechnology*, John Wiley & Sons, Inc., New Jersey (2003). With permission.

The Photonic Bandgap.   The photonic bandgap (PBG) is a range in wavelength within which there is no absorption or propagation of light by a material. In other words, the PBG is a wavelength range in 3-D dielectric structures in which EM waves are forbidden regardless of the direction of propagation of the incident beam [26]. The PBG is the key to the way light might be controlled by a material structure. For example, by introducing defects in the structure or by doping, certain pathways can be closed or diverted. Creating periodic structures out of materials with different dielectric properties (high and low refractive index materials) serve as waveguides—similar to the way electrons are "guided" through a semiconductor via doped domains. The goal of photonic crystals is to exclude light transmission in all directions for specific wavelengths—once again, similar to semiconductors that exclude the flow of electrons for specific energy bands. It was actually the de Broglie model for electronic wave propagation in a crystalline solid that inspired research on the PBG.

The PBG is determined by the radius of the holes (or other features like dielectric rods), the periodicity of the holes (or rods), the lattice structure, the thickness of the material, and the refractive index. In essence, optical properties can be absolutely under control by designing the photonic device with proper attributes. Periodic structures along the direction of propagation of light restrict or allow modes or limit the density of states of photons. For example, the structures shown in **Figure 4.20** are two different kinds of photonic structures.

The behavior of light in a photonic crystal is described by Maxwell's equations for periodic dielectrics. Dispersion relations (dependence of energy on the wavelength or k-vector) are accurate because there is little interaction between photons in such a crystal (**Fig. 4.21**) [30].

If some features were removed from the lattice, for example, a couple of rows of rods or holes removed in a nonrandom way, that region becomes devoid of structure and would act as a waveguide. In such a configuration, an "allowed frequency" is now allowed to traverse the bandgap—similar in effect to doping a semiconductor material crystal [30].

In semiconductors, p or n-doping creates energy levels in the bandgap. The same is true in photonic crystals except that the PBG is activated to accommodate the passage of light.

FIG. 4.20 | Two types of 2-D photonic structures are depicted (e.g., an array of cylindrical holes or cylindrical rods). On the left, a "cavity" structure is shown. On the right, a structure suitable as a waveguide is shown. The black dots represent a periodic array of holes (or rods) in the photonic material. Photonic crystals are based on the symmetry and periodicity (distribution) of the dielectric functions of the materials [30].

Source: C. P. Poole and F. J. Owens, *Introduction to Nanotechnology*, John Wiley & Sons, Inc., New Jersey (2003). With permission.

Light is actually able to turn sharp corners in such a structure. Because the light is traveling in what was once forbidden territory, it has no way to escape the forbidden zone back and make its way back into the crystal's periodic array crystal [30].

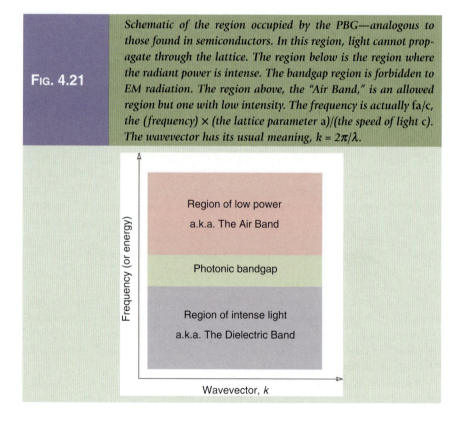

FIG. 4.21 | Schematic of the region occupied by the PBG—analogous to those found in semiconductors. In this region, light cannot propagate through the lattice. The region below is the region where the radiant power is intense. The bandgap region is forbidden to EM radiation. The region above, the "Air Band," is an allowed region but one with low intensity. The frequency is actually fa/c, the (frequency) × (the lattice parameter a)/(the speed of light c). The wavevector has its usual meaning, $k = 2\pi/\lambda$.

By creating such defects or altering the geometry of its members, resonant cavities can be produced in the crystal. This allows characteristics of photonic crystals to be tunable [30].

4.4.4 *Fabrication of Nanophotonic Crystals*

The ability to tune the properties of a photonic crystal (e.g., material selection, component geometry, size and orientation, and spacing) is the key to photonic applications. Use of photonic crystals as filters and laser components. For example, in spontaneous emission, coupling a photonic crystal to a laser allows for independent control of the rate of decay and the coupling between atoms and photons in the crystal [30].

Fabrication and Application of 1-D and 2-D Photonic Materials. 1-D and 2-D photonic crystals are quite easy to fabricate [29].

Physical or chemical deposition techniques layer by layer are able to generate 1-D structures (see **Fig. 4.10**). 2-D systems are a little more difficult but overall are still relatively easy to obtain with today's lithographic technology. Pattern transfer and selective etching allow for the creation of 2-D structures. Applications of these materials have already achieved commercial level status in the form of optical notch filters, dielectric mirrors, optical resonance cavities, and Bragg gratings on optical crystal [29].

Fabrication and Application of 3-D Photonic Materials. Three-dimensional photonic crystals are more problematic to fabricate. In 1994, K.M. Ho conceptualized a simple cubic lattice by stacking dielectric rods (**Fig. 4.22**), similar to

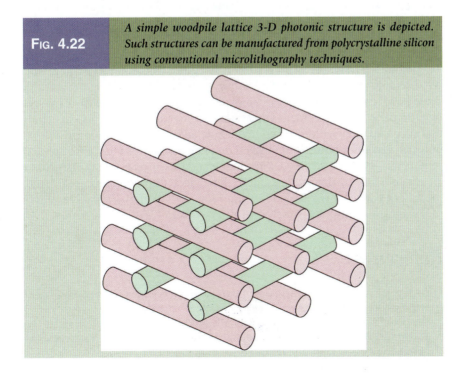

FIG. 4.22 *A simple woodpile lattice 3-D photonic structure is depicted. Such structures can be manufactured from polycrystalline silicon using conventional microlithography techniques.*

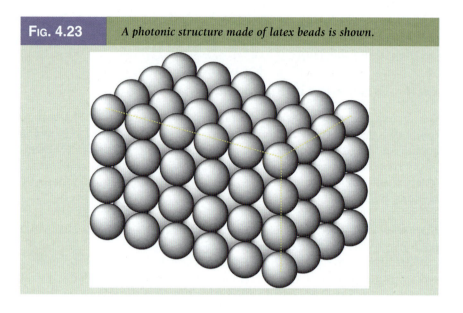

FIG. 4.23   *A photonic structure made of latex beads is shown.*

stacking popsicle sticks that we made structures out of in elementary school. A thin film of poly-Si (polycrystalline silicon) was first deposited on a silicon wafer and patterned by a standard photolithography technique. Features (the wood of the woodpiles) formed were accomplished by reactive ion etching. The gaps between the individual lumber were filled with SiO_2 and the process was repeated after another layer of poly-SI was deposited on top of the first layer. This time the pattern was placed 90° to the longitudinal direction of the first layer. This process was repeated until the desired configuration was achieved. Selective etching by hydrofluoric acid (HF) removed the interstitial SiO_2.

Photonic crystals can also be fabricated by holographic, two-photon, and self-assembly processes. Crystalline lattices of spherical colloids are varied and diverse due to the flexibility of the template starting materials. Particle diameters on the order of a few nanometers to several microns are available today. Materials that are compatible with this method include dielectric materials made of latex polymer and silica beads (**Fig. 4.23**). Long-range order on the scale of several centimeters has been achieved with these materials. Inverse opaline structures are also possible by this method but in these cases, the templating agents (e.g., the spheres) are dissolved leaving an array of porous cavities. Crystalline lattices of nonspherical colloidal particles are also possible.

References

1. Cyclopaedia or Universal dictionary of arts and sciences, /en.wikipedia.org/wiki/Cyclopaedia (2008).
2. History of optics, en.wikipedia.org/wiki/History_of_optics (2008).
3. T. Ohm, Optics highlights: An anecdotal history of optics from Aristophanes to Zernike, www.ece.umd.edu/~taylor/optics3.htm (2008).

4. B. Goldberg, Nanophotonics and nanoplasmonics: Control of light at one-hundreth of the wavelength, Presentation, NANOMAT 2007, Bergen, Norway (2007).

5. Optics, en.wikipedia.org/wiki/Optics (2008).

6. T. R. Jensen, M. D. Malinsky, C. L. Haynes, and R. P. Van Duyne, Nanosphere lithography: Tunable localized surface plasmon resonance spectra of silver nano-particles, *Journal of Physical Chemistry*, 104, 10549–10556 (2000).

7. P. C. Andersen and K. L. Rowlen, Brilliant optical properties of nanometric noble metal spheres, rods, and aperture arrays, *Applied Spectroscopy*, 56 (2002).

8. S. Link and M. A. El-Sayed, Shape and size dependence of radiative, non-radiative and photothermal properties of gold nanocrystals, *International Reviews in Physical Chemistry*, 19, 409–453 (2000).

9. C. N. Davies, Survey of scattering and absorption of light by particles, *British Journal of Applied Physics*, 5, S64-S65 (1954).

10. A. Campion, Raman spectroscopy of molecules adsorbed on solid surfaces, *Annual Reviews of Physical Chemistry*, 36, 549–572 (1985).

11. G. Cao, *Nanostructures & nanomaterials: Synthesis, properties and applications*, Imperial College Press, London (2004).

12. G. L. Hornyak, Optical characterization and optical theory of nanometal/porous alumina composite membranes, Dissertation, Colorado State University, Fort Collins, CO (1997).

13. A. A. Khosravi, M. Kundu, B. A. Kuruvilla, G. S. Shekhawat, R. P. Gupta, A. K. Sharma, P. D. Vyas, and A. K. Kulkarni, *Applied Physics Letters*, 67, 2506 (1995).

14. W. Chan, D. Maxwell, X. Gao, R. Bailey, M. Han, and S. Nie, Luminescent quantum dots for multiplexed biological detection and imaging, *Current Opinion Biotechnology*, 13, 40 (2002).

15. W. C. W. Chan and S. M. Nie, Quantum dot bioconjugates for ultrasensitive nonisotopic detection, *Science*, 281, 2016 (1998).

16. R. Edgar, M. McKinstry, J. Hqang, A. B. Oppenheim, R. A. Fekete, G. Giulan, C. Merril, K. Nagashima, and S. Adhya, High-sensitivity bacterial detection using biotin-tagged phage and quantum-dot nanocomplexes, *Proceedings of the National Academy of Science*, 103, 4841–4845 (2006).

17. H. Weller, U. Koch, M. Gutierrez, and A. Henglein, Photochemistry of colloidal metal sulfides. 7. absorption and fluorescence of extremely small ZnS particles (the world of neglected dimensions), *Berichte Der Bunsen-Gesellschaft-Physical Chemistry Chemical Physics*, 88, 649 (1984).

18. A. D. Dinsmore, D. S. Hsu, H. F. Gray, S. B. Qadri, Y. Tian, and B. R. Ratna, Mn-doped ZnS nanoparticles as efficient low-voltage cathodoluminescent phosphors, *Applied Physics Letters*, 75(6), 802 (1999).

19. R. O. Peterson, FED phosphors: Low or high voltage, *Information Display-Journal of the Society for Information Display*, 3, 22 (1997).

20. L. S. Goldner, Single molecule spectroscopy, National Institute of Standards and Testing, physics.nist.gov/Divisions/Div844/facilities/smspec/sm.html (2008).

21. ftp://203.159.21.137/dutta/Ruhullah-Optics%20in%the%20Nano-World.pdf

22. L. S. Goldner, G. N. Goldie, M. J. Fasolka, F. Ranaldo, J. Hwang, and J. F. Douglas, Near-field polarimetric characterization of polymer crystallite, *Applied Physics Letters*, 85, 1338–1340 (2004).

23. Photonics, *Photonics Dictionary*, www.photonics.com/directory/(2007).

24. E. Yablonovitch, Inhibited spontaneous emission in solid state physics and electronics, *Physical Review Letters*, 58, 2059–2062 (1987).

25. S. John, Strong localization of photons in certain disordered dielectric superlattice, *Physical Review Letters*, 58, 2486–2489 (1987).

26. E. Yablonovitch, Photonic bang-gap structures, *Journal of the Optical Society of America B*, 10, 283–295 (1993).

27. Natural photonics, *Physics World*, physicsworld.com/cws/article/print/18931 (2004).
28. M. Florescu, Photonics crystals: A new frontier in modern optics, Presentation, NASA Jet Propulsion Laboratory & California Institute of Technology, www.mcc. uiuc.edu/nsf/ciw_2006/talks/Florescu.ppt (2006).
29. M. Di ventra, S. Evoy, and J. R. Hefliln Jr., *Introduction to nanoscale science and technology*, Kluwer Academic Press, Boston (2004).
30. C. P. Poole and F. J. Owens, *Introduction to nanotechnology*, Wiley-Interscience, John Wiley & Sons, Inc., Hoboken, NJ (2003).
31. J. Dutta et al., Unpublished results from the AIT Center of Excellence in Nanotechnology, Bangkok, Thailand (2008).
32. G. Mie, Beiträge zur optik trüber Medien, speziell kolloidaler Metallösungen, *Ann Physics (Leipzig)*, 25, 377 (1908).

Problems

4.1 Why is the color of 20-nm gold particles red? What happens when two or three gold particles agglomerate together?

4.2 What is the Lycurgus cup and how does it relate to nanotechnology?

4.3 What is near-field scanning optical microscopy (NSOM)?

4.4 What is a Bohr radius? What does it have to do with electronic transitions in semiconductor materials?

4.5 What are excitons?

4.6 What led Bohr to his radical proposal of "quantum leaps" as an alternative to Rutherford's "planetary" model?

4.7 In solitary atoms, electrons are free to inhabit only certain, discrete energy states. However, in solid materials where there are many atoms in close proximity to each other, bands of energy states form. Where do these "bands" come from?

4.8 What is Fermi energy? What is its importance?

4.9 What is a photonic bandgap material? For visible bandgap, what particle sizes would you choose?

4.10 Niels Bohr in 1913 hypothesized that electrons in hydrogen were restricted to certain discrete levels. This comes about because the electron waves can have only certain wavelengths, that is, $n\lambda = 2\pi r$,

where r is the orbit radius. Based on this, one can show that the energy of hydrogen atom for $n = 2$ is 3.4 eV

4.11 How much energy is required for taking out an electron from a Si atom?

4.12 Calculate the energy difference between the ground state and the first excited state for an electron in a quantum dot of size 10^{-8} cm (mass of electron = 9.1×10^{-31} kg and Planck's constant = 6.626×10^{-34} J·s)

4.13 The probability that an electron will occupy a state at the energy E_C is the same as the probability that a hole will occupy a state at the energy E_V. What is the energy E_F of the Fermi level?

4.14 Assuming that $n = 1$ states for both electrons and holes participating in an optical transition, estimate the emitted photon energy from a GaAs (Bandgap = 1.42 eV) single-quantum well laser with a well length of 10 nm. Assume the hole mass to be 0.62 times the free space mass and the hole mass to be 0.067 times the free space mass of electrons. (Given: Free space electron mass = 0.511×10^6 eV/C^2; $\hbar = 6.58 \times 10^{-16}$ $eV\cdot s$; $c = 3.0 \times 10^{10}$ cm·s^{-1}.)

4.15 Suppose an electron accelerating at 5 kV strikes a copper target. What type of x-rays will be emitted?

NANOMAGNETISM

Here again it was the quantum theory which came to the rescue.

WALTHER NERNST

Chapter 5

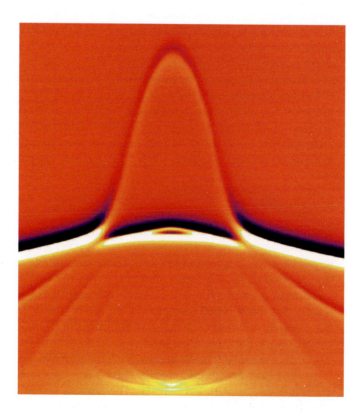

We conclude the *Electromagnetic Nanoengineering* section with *chapter 5* that deals with magnetic properties of nanomaterials. Once again, size is a critical parameter with regard to the magnetic response of nanomaterials. For several decades now, information has been stored via the action of magnetic materials.

Mechanical Nanoengineering is the next section of the book. In this section, mechanical properties and the properties of thin films and nanocomposites are discussed in *chapters 6* through *8*. This is followed by a section on *Chemical Nanoengineering* and applications of nanomatrials and the final section deals with *Biological and Environmental Nanoengineering* aspects of nanotechnology.

5.0   INTRODUCTION

Magnetism is one of the phenomena by which materials can exert attractive or repulsive forces on other materials. Some well-known materials that exhibit easily detectable magnetic properties are nickel, iron, cobalt, and their alloys. All materials are influenced to a greater or lesser extent by the presence of a magnetic field. Magnetism also does have other manifestations, particularly as one of the two components of electromagnetic waves such as light or the cell phone signals often known as electromagnetic waves.

What Is a Magnet?   A magnet is normally known as a piece of iron, steel, or magnetite that has the property of attracting iron or steel. The most well-known magnet is probably the lodestone, also often called magnetite, which is a naturally occurring rock that was first discovered in a region known as magnesia and was hence named after this region. Magnetism may be naturally present in a material or the material may be artificially magnetized and magnets can either be permanent or temporary. After being magnetized, a permanent magnet retains the properties of magnetism indefinitely while a temporary magnet (e.g., a magnet made of soft iron) is usually easy to magnetize but it loses most of its magnetic properties when the magnetizing cause is discontinued. Permanent magnets are usually more difficult to magnetize, but they remain magnetized practically forever. These materials which can be magnetized are called *ferromagnetic materials*.

5.0.1   History

The term "magnetism" was introduced by a shepherd by the name of Magnés (from the city of Magnesia in modern day Turkey) who found that his iron-tipped cane was attracted to magnetic rock deposits. The first scientific discussion on magnetism was enumerated by Thales (625 B.C. to about 545 B.C.) according to Aristotle. The earliest literary reference to magnetism lies in a fourth century B.C. book called *Book of the Devil Valley Master*, "The lodestone makes iron come or it attracts it," that was published in China, where also the first compass was reported by Shen Kuo (1031–1095), which is reported to have been first used by Alexander Neckham, by 1187, in Europe for navigation.

In 1269 Peter Peregrinus wrote the *Epistola de Magnete*, the first extant treatise describing the properties of magnets. The modern understanding of the interrelation between electricity and magnetism was triggered in 1819 by Hans

Christian Oersted, from Denmark, who accidentally discovered that an electric current could influence a compass needle. This landmark experiment is popularly known as Oersted's Experiment. Andre-Marie Ampere, Carl Friedrich Gauss, Michael Faraday, and others carried out separately experiments to enumerate the links between magnetism and electricity in greater detail.

Possibly one of the best contribution to the understanding of the electromagnetic nature of waves was clarified by James Clerk Maxwell who presented the now famous "*Maxwell's equations,*" unifying electricity, magnetism, and optics into a new field called *electromagnetism*. 1921–1928 was probably the best-known years where active minds joined to explain the origin of magnetism that changed physics and magnetism. Notable amongst them were the introduction of spins in atoms by Pauli, Uhlenbeck, Goudsmidt, and Dirac; the quantum theory of magnetism propounded by Heisenberg, Schrödinger, and Dirac; and the magnetic exchange interaction in materials proposed by Heisenberg.

5.0.2 *Magnetic Phenomena and Their Classical Interpretation*

Magnetic Flux. A group of magnetic field lines emitted outward from the north pole of a magnet is called *magnetic flux*. The symbol for magnetic flux is Φ (phi). The SI unit of magnetic flux is the weber (Wb). One *weber* is equal to 1×10^8 magnetic field lines.

Magnetic Field (H). A Current through a coil produces a magnetic field (H). If N be the number of turns of the coil of turn length L and I be the current passed through the coil, the relation for magnetic field (H) is given by

$$H = NI/L \text{ ampere-turns/m}$$

The above equation can be related as $B = (1 + \chi)\mu_o H$.

We stated earlier that different types of magnetism exist and that they are characterized by the magnitude and the sign of the susceptibility. Since various materials respond so differently in a magnetic field, we suspect that several fundamentally different mechanisms must be responsible for the magnetic properties. In the first part of this chapter we shall attempt to unfold the multiplicity of the magnetic behavior of materials by describing some pertinent experimental findings and giving some brief interpretations (**Fig. 5.1**).

Diamagnetism. In a diamagnetic material, there are no unpaired electrons, so the intrinsic electron magnetic moments cannot produce any bulk effect and the magnetization arises solely from the electrons' orbital motions. When a material is put in a magnetic field, the electrons circling the nucleus will experience a Lorentz force from the magnetic field, in addition to their Coulomb attraction to the nucleus. Depending on which direction the electron is orbiting, this force may increase the centripetal force on the electrons, pulling them in towards the nucleus, or it may decrease the force, pulling them away from the nucleus. This effect systematically increases the orbital magnetic moments that were aligned opposite the field, and decreases the ones aligned parallel to the field (in accordance with Lenz's law). This results in a small, bulk magnetic moment, with an opposite direction to the applied field.

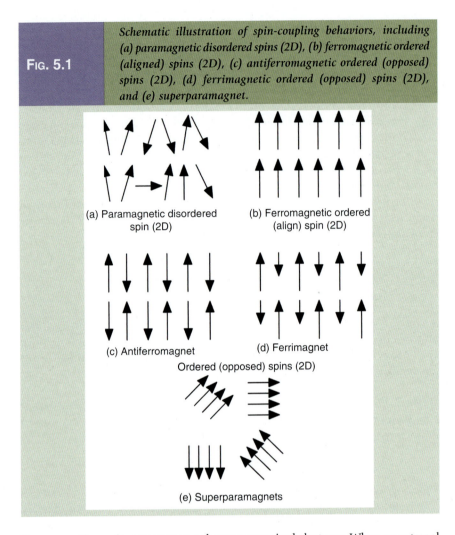

FIG. 5.1   *Schematic illustration of spin-coupling behaviors, including (a) paramagnetic disordered spins (2D), (b) ferromagnetic ordered (aligned) spins (2D), (c) antiferromagnetic ordered (opposed) spins (2D), (d) ferrimagnetic ordered (opposed) spins (2D), and (e) superparamagnet.*

(a) Paramagnetic disordered spin (2D)

(b) Ferromagnetic ordered (align) spin (2D)

(c) Antiferromagnet

(d) Ferrimagnet

Ordered (opposed) spins (2D)

(e) Superparamagnets

Paramagnetism.   In a paramagnet there are unpaired electrons. When an external magnetic field is applied, these magnetic moments will tend to align themselves in the same direction as the applied field, thus reinforcing it.

Ferromagnetism.   A ferromagnet has unpaired electrons. In addition to the electrons' intrinsic magnetic moments wanting to be parallel to an applied field, there is also in these materials a tendency for these magnetic moments wanting to be parallel to each other due to exchange interaction between electrons. Thus, even when the applied field is removed, the electrons in the material can keep each other continually pointed in the same direction. Every ferromagnet has its own individual temperature, called the Curie temperature, or Curie point, above which it loses its ferromagnetic properties. This is because the thermal tendency to disorder overwhelms the energy lowering due to ferromagnetic order. In ferromagnetic rare earth metals (e.g., Gd, Dy, Sm), the magnetic moments (e.g., the Si) are strongly localized at the atomic positions.

Antiferromagnetism.   In an antiferromagnet, there is a tendency for the intrinsic magnetic moments of neighboring valence electrons to point in opposite directions. When all atoms are arranged in a substance so that each neighbor is

FIG. 5.2   *Magnetic susceptibility of various kinds of magnetic material.*

"anti-aligned," the substance is antiferromagnetic. Antiferromagnets have a zero net magnetic moment. In varying temperatures, antiferromagnets can be seen to exhibit diamagnetic and ferrimagnetic properties. In antiferromagnetic materials, the local spin densities are nonzero, yet the overall spin (and magnetic moment) vanishes. Examples of antiferromagnets are MnO, FeO, CoO, and NiO. Chromium is a peculiar ferromagnet with nonvanishing local spin densities but with a periodicity which is incommensurate with the Cr lattice.

Ferrimagnetism.   Ferrimagnets retain their magnetization in the absence of a field. However, like antiferromagnets, neighboring pairs of electron spins like to point in opposite directions but there is more magnetic moment from the sublattice of electrons which point in one direction, than from the sublattice which points in the opposite direction. We can distinguish various kinds of magnetic properties using magnetic susceptibility, ratio of magnetization, and applied magnetic field (**Fig. 5.2**).

The interaction of localized moments can also be described using the molecular field approach where it is assumed that the interaction of spin S with all the other spins can be approximated with an effective magnetic field B_{mf} due to those neighboring spins. In a metal which is constructed with paramagnetic atoms, the bonding is mediated by the valence electrons, which are often also responsible for the atomic paramagnetism. In Na for example, the single 2s valence electron ($S = 1/2$) gives rise to the conduction band, which is know to be almost free electron like. This means that these electrons hardly feel the atomic potentials but instead an almost featureless average potential. In this state, the local spin density vanishes everywhere. In other words the bulk is nonmagnetic. This is not true for itinerant ferromagnetic metals such as the 3d transition metals (Fe, Co, Ni). The bulk is ferromagnetic where each atom contributes $0.54\,\mu_B$. This is due to an imbalance in the up-down spin densities. The local spin densities at the Fermi level of ferromagnetic metals do not vanish. This is important in the characterization of magnetic materials.

Several exchange-type interactions are responsible for magnetism in metals. One is the intra-atomic exchange, which causes atoms to attain a net spin that underlies Hund's rules. Exchange is also responsible for spin order. That interaction is the interatomic exchange coupling can be either indirect or direct.

5.0.3   The Nano Perspective

Why the Interest in Nanoscale Magnetic Materials?   There is a dramatic change in magnetic properties when the critical length governing some phenomenon (magnetic, structural, etc.) is comparable to the nanoparticle or nanocrystal size. Since a large amount of surface is exposed, effects due to surfaces or interfaces are stronger.

5.1   CHARACTERISTICS OF NANOMAGNETIC SYSTEMS

5.1.1   *Introduction to Nanomagnetism*

In magnetic materials, a magnetic field is produced because of the movement of electrons within the material, which produces a field around the material and a magnetization effect exists within an atom, as shown in **Figure 5.3**. Because of the charge of an electron, magnetic moment appears just like magnetic field is generated, when current flows through solenoid coils. A magnetic moment can appear even due to the orientation of the spins of an electron. The magnetic moment generated due to the orbital motion or spin motion of a single electron is called *Bohr magneton*, which is the smallest unit of magnetic moment of solids. This magnetic moment can interact with the magnetic field as that in the current loop.

$$\text{Bohr magneton} = \frac{qh}{4\pi m_e} = 9.27 \times 10^{24} \ [\text{A} \cdot \text{m}^2]$$

where
    q is the charge of an electron
    h is the Planck's constant
    m_e is the rest mass of an electron

In most materials, the electron spin will have its pair, up and down spin, and the spin will cancel each other resulting in a zero net spin, hence rendering these materials immune to any effects of an applied magnetic field (no magnetic moment). Following Hund's rule, certain materials such as the transition metals have unbalanced spins that lead to a nonzero net spin rendering a magnetic moment to the atoms (**Table 5.1**).

FIG. 5.3	*Electron orbit in an atom.*

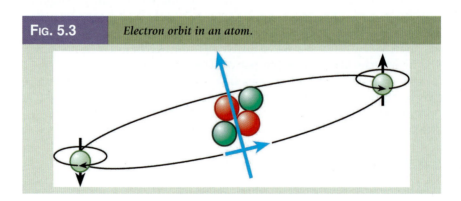

TABLE 5.1	Electron Configuration of Some Transition Metals					
Metal	**3d**					**4s**
Sc	↑					↑↓
Ti	↑	↑				↑↓
V	↑	↑	↑			↑↓
Cr	↑	↑	↑	↑	↑	↑
Mn	↑	↑	↑	↑	↑	↑↓
Fe	↑↓	↑	↑	↑	↑	↑↓
Co	↑↓	↑↓	↑	↑	↑	↑↓
Ni	↑↓	↑↓	↑↓	↑	↑	↑↓
Cu	↑↓	↑↓	↑↓	↑↓	↑↓	↑

Magnetism at nanometric sizes deals more with the newly discovered phenomena attributed to nano-sized materials such as spin-polarized tunneling, oscillatory exchange coupling between magnetic–nonmagnetic multilayers, magnetoresistivity etc., which take place mostly in low dimensional materials. When an object becomes so small that the number of surface atoms is a sizable fraction of the total number of atoms, then surface effects tends to be important. A typical size of a classically expected magnetic domain is of the order of $1\,\mu m$. As the dimensions of magnetic materials decrease down to nanometer scales, these materials start to exhibit new and very interesting physical properties mainly due to quantum size effects. Magnetic behavior is changed, the magnetization of noble metals is considerably induced, electron interference patterns are observed, oscillatory exchange interactions occur between adjacent layers separated by a nonmagnetic spacer, and magnetoresistivity is enhanced many orders of magnitude. All these new phenomena are affected by the imperfections of nanostructures. At the nanoscale the intrinsic properties of the materials become extrinsic and can be adjusted by size in the nanoscale, and thus devices virtually having any characteristic can be made realistic. The homogeneity and purities in chemical composition, crystalline structures, external morphology, etc., determine the physical properties. These new materials are important for magnetic sensors; giant magnetoresistance (GMR) reading heads are used to read magnetically stored data. The recording industry desires for ever denser and more reliable recording media and hence nanosize properties are of distinct interest.

Thus, in magnetic materials, electrons' orbital angular motion around the nucleus and the electrons' intrinsic magnetic moment are the most important sources of magnetization. The magnetic behavior of a material depends on the structure of the materials as well as the temperature (at high temperatures, random thermal motion makes it more difficult for the electrons to maintain alignment).

A qualitative, as well as a quantitative distinction between different types of magnetism, can be achieved in a relatively simple way following a method proposed by Faraday. The magnetic material to be investigated is suspended from

FIG. 5.4

Type of magnetic material investigation by suspension from one of the arms of a sensitive balance placed in an inhomogeneous magnetic field.

one of the arms of a sensitive balance and is allowed to reach into an inhomogeneous magnetic field (**Fig. 5.4**).

Diamagnetic materials are expelled from this field, whereas para-, ferro-, antiferro-, and ferrimagnetic materials are attracted in different degrees. It has been empirically found that the apparent loss or gain in mass, that is, the force F on the sample exerted by the magnetic field is

$$F = V \chi H \frac{dH}{dx}$$

where

dH/dx is the change of the magnetic field strength $|H|$ in the direction of the field

V is the volume of the sample

The magnetic material can be characterized by a material constant χ, called the susceptibility which expresses how responsive this material is to an applied magnetic field. Frequently, a second material constant, permeability μ, is used. This constant is related to the susceptibility by

$$\mu = 1 + 4\pi\chi$$

For empty space, and for all practical purposes, also for air, χ is 0 and thus $\mu = 1$. For diamagnetic materials one finds χ to be small and negative, and thus μ slightly less than 1. For para- and antiferromagnetic materials χ is again small, but positive. Thus, μ is slightly larger than 1. Finally, χ and μ are large and positive for ferro- and ferrimagnetic materials. The magnetic constants are temperature dependent, except for diamagnetic materials.

The magnetic field parameters at a given point in space are defined to be the magnetic field strength **H**, which we introduced above, and the magnetic flux density or magnetic induction **B**. In free space **B** and **H** are identical. Inside a magnetic material, the induction **B** consists of the free-space component (**H**), plus a contribution to the magnetic field which is due to the presence of matter.

$$\mathbf{B} = \mathbf{H} + 4\pi\mathbf{M}$$

where **M** is called the magnetization of the material. For a material in which the magnetization is thought to be proportional to the applied field strength we define

$$\mathbf{M} = \chi\mathbf{H}$$

Combining the above two equations we get

$$B = H(1 + 4\pi\chi) = \mu H$$

Finally, we need to define the magnetic moment, μ_m, through the following equation

$$M = \frac{\mu_m}{v}$$

that is the magnetization is the magnetic moment per unit volume.

It needs to be noted that in magnetic theory several unit systems are commonly in use. Scientific and technical literature on magnetism is still widely written in electromagnetic cgs (emu) units. The magnetic field strength in cgs units is measured in oersted (Oe) and the magnetic induction in gauss.

5.1.2　Characteristics of Nanomagnetic Materials

Size Dependence.　In general, the magnetic moment of transition metals decreases with the increase of number of atoms (size) in the clusters. However, the moment of different transition metal clusters are found to depend differently on the size of clusters. In the lower size limit, the clusters have a high-spin majority configuration and the behavior is more like an atom, but with the increase of size, the moment moves towards that of the bulk with slow oscillations.

Surface Magnetism.　The magnetic properties at the surfaces of ferromagnets are significantly altered from the bulk for several reasons. For example, since the number of nearest neighbors is reduced, a valence electron tends to spend more time at each ionic site, due to reduced coordination compared with the bulk. Weaker bonding to neighboring sites causes the ions to have a more isolated atomic character.

Magnetic Anisotropy and Domains in Small Particles.　Small magnetic particles therefore are typically monodomain, since the energy cost to form a domain wall outweighs the reduction in magnetic energy. Typically, critical sizes for monodomain particles are in the range of 20–2000 nm and depend on the ferromagnetic material under consideration.

5.1.3　Magnetization and Nanostructures

Changes in magnetization of a material occur via activation over an energy barrier. Each physical mechanism responsible for an energy barrier has an associated length scale. The fundamental magnetic lengths are the crystalline anisotropy length l_k, the applied field length l_H, and the magnetostatic length l_S, which are defined as follows:

$$l_k = \sqrt{J/k}$$

$$l_H = \sqrt{2J/HM_S}$$

$$l_S = \sqrt{J/2\pi M_S^2}$$

Here, k is an anisotropy constant of the bulk material due to the dominant anisotropy and J is the exchange within a grain. If more than one type of barrier is present, then magnetic properties are dominated by the shortest characteristic length. For most common magnetic materials, these lengths are of the order of 1–100 nm. For example, nickel at 1000 Oe and room temperature has lengths $l_S \cong 8$ nm, $l_k \cong 45$ nm, and $l_H \cong 19$ nm.

When a sufficiently large magnetic field is applied, the spines within the material align with the field. The maximum value of magnetization achieved in this state is called the saturation magnetization, M_S. As the magnitude of the field decreases, spins cease to be aligned with the field and the total magnetization decreases. In ferromagnets, a residual magnetic moment remains at zero field. The value of the magnetization at zero field is called the remanent magnetization, M_R. The ratio of the remanence magnetization to the saturation magnetization M_R/M_S is called the remanence ratio and varies from 0 to 1. The magnitude of the field that must be applied in the negative direction to bring the magnetization of the sample back to zero is called the coercive field. H_C magnetic recording applications require a large remnant magnetization, moderate coercivity, and (ideally) a square hysteresis loop. The different regimes of nanomaterials in the realm of nanomagnetism is summarized in **Figure 5.5**.

Classification of Magnetic Nanomaterials. The magnetic behavior of most experimental realizable systems is a result of contributions from both interaction and size effects. The correlation between nanostructure and magnetic properties suggests a classification of nanostructure morphologies. The following classification is designed to emphasize the physical mechanisms responsible for the magnetic behavior that **Figure 5.6** schematically represents. At one extreme are systems of isolated particles with nanoscale diameters, which are denoted by type A. These noninteracting systems derive their unique magnetic properties strictly from the reduced size from the components, with no contribution from interparticle interactions. At the other extreme are bulk materials with nanoscale structure denoted by type D, in which a significant fraction (up to 50%) of the sample volume is composed of grain boundaries and interfaces. In contrast to type A systems, magnetic properties here are dominated by interactions. The length of

Fig. 5.5 *Different regimes of magnetism with respect to the number of atoms in the material.*

Fig. 5.6 *Schematic representation of the different types of magnetic nanostructures. Type A materials include the ideal ultrafine particle system, with interparticle spacing large enough to approximate the particles as noninteracting. Ferrofluids, in which magnetic particles are surrounded by a surfactant, preventing interactions, are a subgroup of type A materials. Type B materials are ultrafine particles with a core-shell morphology. Type C nanocomposites are composed of small magnetic particles embedded in a chemically dissimilar matrix. The matrix may or may not be magnetoactive. Type D materials consists of small crystallites dispersed in a noncrystalline matrix. The nanostructure may be two phase, in which nanocrystallites are a distinct phase from the matrix, or the ideal case in which both the crystallites and the matrix are made of the same material.*

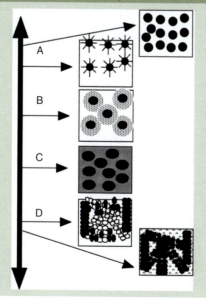

the interactions can span many grains and is critically dependent on the character of interface. Due to this dominance of interaction and grain boundaries, the magnetic behavior of type D nanostructures cannot be predicted simply by applying theories of polycrystalline materials with reduced length scale. In type B particles, the presence of a shell can help prevent particle interactions, but often at the cost of interaction between the core and the shell. In many cases the shells are formed via oxidation and may themselves be magnetic. In type C materials, the magnetic interactions are determined by the volume fraction of the magnetic particles and the character of the matrix they are embedded in.

Ferrofluids. A ferrofluid is a synthetic liquid that holds small magnetic particles in a colloidal suspension, with particles held aloft due to thermal energy. The particles are sufficiently small that the ferrofluid retains its liquid characteristics even in the pressure of a magnetic field, and substantial magnetic forces can be induced, which results in fluid motion.

A ferrofluid has three primary components. The carrier is the liquid element in which the magnetic particles are suspended. Most ferrofluids are either water based or oil based. The suspended materials are small ferromagnetic particles such as iron oxide, on the order of 10–20 nm in diameter. The small size is necessary to maintain stability of the colloidal suspension, as particles significantly larger than this will precipitate. A surfactant coats the ferrofluid particles to help maintain the consistency of the colloidal suspension.

The magnetic properties of the ferrofluid is strongly dependent on particle concentration and on the properties of the applied magnetic field. With an applied field, the particles align in the direction of the field, magnetizing the fluid. The tendency for the particles to agglomerate due to magnetic interaction between particles is opposed by the thermal energy of the particles. Although particles vary in shape and size distribution, an insight into fluid dynamics can be gained by considering a simple spherical model of the suspended particles (**Fig. 5.7** and **Table 5.2**).

| FIG. 5.7 | *Representative model of a typical ferrofluid.* |

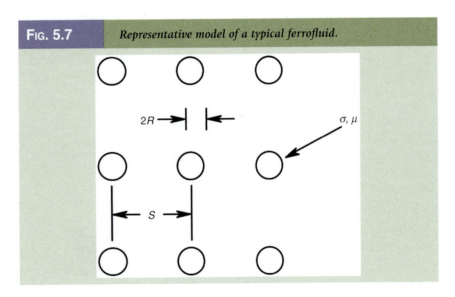

TABLE 5.2	*Material Parameters of a Typical Ferrofluid [1]*
Sample volume	1.7×10^{-6} m^3
Electrical conductivity of particles	$\sigma_f = 3 \times 10^6$ ($\Omega \cdot$ m)$^{-1}$
Electrical conductivity of ferrofluid	$\sigma_{\text{fluid}} < 10^{-7}$ ($\Omega \cdot$ m)$^{-1}$
Volume percentage of particles	3% by volume
Initial magnetic permeability of particles	$\mu \approx 100 \mu_0$
Particle mean radius	$R \approx 10^{-8}$ m
Density of magnetic particles	$\rho_{\text{FE}} = 7.8$ g $\cdot$ cm^{-3}
Fluid density	$\rho_f = 1$ g $\cdot$ cm^{-3}
Fluid viscosity of carrier fluid	$\eta = 1$ cp $= 0.01$ g $\cdot$ cm$^{-1} \cdot$ s^{-1}

The particles are free to move in the carrier fluid under the influence of an applied magnetic field, but on average the particles maintain a spacing S to nearest neighbor. In a low density fluid, the spacing S is much larger than the mean particle radius $2R$, and magnetic dipole–dipole interactions are minimal.

Applications for ferrofluids exploit the ability to position and shape the fluid magnetically. Some applications are

- Rotary shaft seals
- Magnetic liquid seals, to form a seal between regions of different pressures
- Cooling and resonance damping for loudspeaker coils
- Printing with magnetic inks
- Inertial damping, by adjusting the mixture of the ferrofluid the fluid viscosity may be changed to critically damp resonances accelerometers
- Level and attribute sensors
- Electromagnetically triggered drug delivery

Single-Domain Particles. Groups of spins all pointing in the same direction and acting cooperatively are separated by domain walls, which have a characteristic width and energy associated with their formation and existence. The motion of a domain wall is a primary means of reversing magnetization. The dependence of *coercivity* on particle size is similar to that schematically illustrated in **Figure 5.8**.

In large particles, energetic considerations favor the formation of domain walls. Magnetization reversal thus occurs through the nucleation and motion of these walls. As the particle size decreases towards some critical particle diameter D_{critical}, the formation of domain walls becomes energetically unfavorable and the particles are called single domain. Changes in the magnetization can no longer occur through domain wall motion and instead require the coherent rotation of spins, resulting in larger coercivity. As the particle size continues to decrease below the single domain value, the spins get increasingly affected by thermal

FIG. 5.8 *Qualitative illustration of the behavior of the coercivity in ultrafine particle systems as the particle size changes.*

fluctuations and the system becomes superparamagnetic. Particles with sufficient shape anisotropy can remain single domain to much larger dimensions than their spherical counterparts.

Single-Domain Characteristics. In a granular magnetic solid with a low volume fraction, one has a collection of single-domain particles, each with a magnetic axis along which all the moments are aligned. In **Figure 5.9**, the typical nanoparticle sizes for some common ferromagnetic materials (single-domain size, $D_{critical}$) and the stability of the magnetic properties or the superparamagnetic limit at room temperature ($D_{superparamagnetic}$) is shown in **Figure 5.9**.

In the absence of a magnetic field, parallel and antiparallel orientations along the magnetic axis are energetically equivalent but separated by an energy barrier

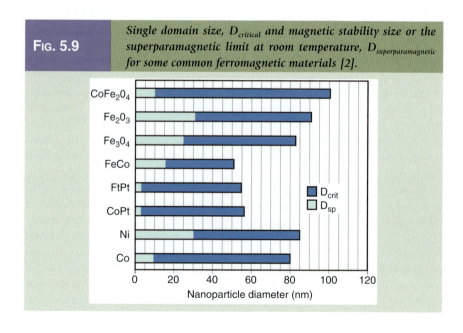

FIG. 5.9 *Single domain size, $D_{critical}$ and magnetic stability size or the superparamagnetic limit at room temperature, $D_{superparamagnetic}$ for some common ferromagnetic materials [2].*

of CV, where C is the total anisotropy per volume, and V is the particle volume. Since the size of each single domain remains fixed, under an external field, only the magnetic axes rotate. Thus the measured magnetization (M) of a granular magnetic field solid with a collection of single-domain particles is the global magnetization

$$M = \frac{\langle \vec{M}\vec{H} \rangle}{H} = M_s \langle \cos \theta \rangle$$

where θ is the angle between the magnetic axes of a particle. M_s is the saturation magnetization, $\vec{H}$ is the external field, and the average $\langle \cos \theta \rangle$ is taken over many ferromagnetic particles. The hysteresis loop of a granular solid is thus a signature of the rotation of the magnetic area of the single-domain particles. This should be contrasted with the hysteresis loop of a bulk ferromagnet, in which the sizes and the direction of the domains are altered drastically under an external field.

Superparamagnetic Materials.   Superparamagnetic materials consist of individual domains of elements that have ferromagnetic properties in bulk. Their magnetic susceptibility is between that of ferromagnetic and paramagnetic materials. **Figure 5.10** illustrates the effect of a superparamagnetic material (gray circle) on the magnetic field flux lines.

At sufficient high temperatures, the energy barrier (CV) is overcome by the thermal energy. Consequently, the magnetic moments within a particle rotate rapidly in unison, exhibiting the superparamagnetic relaxation phenomenon. The simplest form of the relaxation time can be described by the Arrhenius relation

$$\tau = \tau_0 \exp\left(\frac{CV}{k_B T} \right)$$

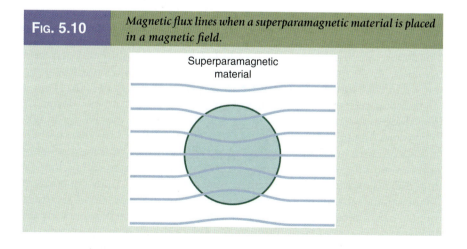

| FIG. 5.10 | *Magnetic flux lines when a superparamagnetic material is placed in a magnetic field.* |

Superparamagnetic material

where

τ is the relaxation time

τ_0 is the characteristic time

CV is the total anisotropy energy

The behavior of τ is dominated by the exponential argument. Assuming representative valves ($\tau_0 = 10^{-9}$ s, $k_B = 10^6$ erg/cm^3, and $T = 300$ K), a particle of diameter 11.4 nm will have a relaxation time of 0.1 s. Increasing the particle diameter to 14.6 nm increases τ to 10^8 s. For an instrument that can measure certain magnetic characteristics (e.g., magnetometry, AC susceptibility, Mössbauer spectroscopy) with a measurement time of τ_i, one can define a blocking temperature.

$$T_B = \frac{CV}{k_B \ln(\tau_i/\tau_0)}$$

At $T < T_B$, τ_i is less than τ and the instrument detects the ferromagnetic nature (e.g., a hysteresis loop) of the system. However, at $T > T_B$, because τ_i is longer than τ, the time-averaged value of the ferromagnetic characteristics (e.g., magnetization and coercivity) vanish within the measuring time of τ_i. Then the system is in an apparent paramagnetic or superparamagnetic state, even though within each particle the magnetic moments remain ferromagnetically aligned. Because of the superparamagnetic relaxation, the value of remnant magnetization (M_R) and coercivity (H_c) decrease with increasing temperature and vanish at the blocking temperature (T_B). Above T_B, all apparent ferromagnetic characteristics disappear. The blocking temperature in a superparamagnetic system decreases with increasing measuring fields, being proportional to $H^{2/3}$ at large magnetic fields and proportional to H^2 at lower fields.

5.2   Magnetism in Reduced Dimensional Systems

5.2.1   *Two-Dimensional Systems*

Thin Films. There is always a change in behavior as film thickness decreases beyond a certain limit. The magnetization of the sample is obtained by using usual spin wave theory. The spin wave vector perpendicular to the film plane is quantized for such thin films. The degrees of freedom for spin waves is decreased from three to two and only low energy spin waves with a wave vector in two dimensions (film plane) is considered. This effect is expected to reflect itself in critical exponents during the magnetic phase transition near critical temperature T_c. As a result magnetization in two dimensions falls off more rapidly than in three dimension with increasing temperature around T_c. The Curie temperature T_c is determined mainly by the number and coordination symmetry of exchange coupled neighboring magnetic atoms and strength of J. All these parameters are different in the case of very thin films. And the Curie temperature drastically changes in the thin-film case. T_c is found to depend on the number of layers.

T_c even vanishes when we go to a one-dimensional system; the magnetization can be enhanced for ultrathin films, especially for monolayers. The atoms try to maximize their spins as per Hund's rule and also obey Pauli's exclusion principle, which favors the increase of individual atomic moments at the surfaces. Also the electrons want to be far apart from each other to minimize the Coulomb energy.

Monolayers. For monolayers, magnetic moments are very large. Even a non-magnetic metal in its bulk form can become spontaneously ferromagnetic in its monolayer form though there are difficulties in growing ideal monolayers. For a high quality Cr monolayer on Fe(100) the atomic magnetic moment is observed to rise up to $3\,\mu_B$ instead of $0.4\,\mu_B$ for antiferromagnetic Cr in its bulk form.

Quantum Wells. With the decrease of the film thickness, the film can behave like a quantum well for spin carriers. These carriers are reflected by the walls of the wells and the wave functions interfere to form a standing wave with discrete energy levels, and the reflection coefficient is spin dependent. This effect influences almost all the physical properties, such as inverse photoemission and photoemission, magnetic anisotropy, magneto-optic response, electrical conductivity, Hall effect, and superconductivity.

5.2.2   One-Dimensional Systems

A one-dimensional magnetic system on an atomic level occurs as a chain of magnetic atoms even in the bulk form of some crystalline materials. These magnetic chains are magnetically decoupled from neighboring chains by non-magnetic intermediated atoms. Artificial one-dimensional magnetic structures are becoming more attractive for their magnetoresistance. Though the geometry is experimentally difficult for a continuous thin-film case however, it can be realized by growing parallel stripes separated by nonmagnetic very narrow spacer wires on a substrate. The current in the substrate plane can be applied to get high magnetoresistance. One dimensional system may contain many parallel magnetic stripes and many properties of the interface are changed. For instance, continuous spectra split into discrete energy levels; magnetization direction in the sample plane may be switched between perpendicular and parallel to the stripes.

5.2.3   Zero-Dimensional Systems

The most remarkable changes are observed in ultrafine magnetic particles since the surface-to-volume ratio is highest for "zero-dimensional" systems. The surface magnetism increases or decreases, or even can be disordered and/or dead for some surface-treated (e.g., Ni) particles at low temperatures. The smallest magnetic particles are magnetic molecules and clusters. The size of artificial magnetic particles may be reduced down to even 1 nm. The fine particles behave as a monodomain magnet because the particle diameter would be even smaller than domain wall thickness. When the particle size shrinks further the super-paramagnetic limit is reached. When the size of the spin in the case of a single domain ferromagnetic particle without crystalline anisotropy can be many orders of the magnitude greater than for a single atom it is known as *superparamagnetism*. The fine particle system on a substrate can be in a distinctly different phase from the bulk, depending on the particle density and size in the nanometer range. The ultimate goal is to achieve particle array for magnetic recording. Thus both signal-to-noise ratio and storage capacities would be increased a few orders of magnitude. If an assembly of identical uniaxial superparamagnetic particles is initially polarized along the easy axis, then the magnetization will reduce with increasing time. An important property of superparamagnetic particle systems is that they are nonhysteretic and their magnetization curves scale with H/T.

5.3 PHYSICAL PROPERTIES OF MAGNETIC NANOSTRUCTURES

5.3.1 Substrate Effects on Structures and Related Properties

At nanoscale the magnetic phases are dramatically changed and new magnetic phases arise in ultrathin films of epitaxially grown metallic Fe, for instance, on a single-crystal Cu(100) substrate. The magnetic properties, phase boundaries, and critical temperatures depend on the film thickness and on different crystal structures of the film (fcc) that is imposed by the substrate through adhesion at the interface. The most important parameter determining the magnetic structure is the exchange overlap integral of electronic wave functions on neighboring atoms. This overlap depends on neighboring atomic distances determined by crystal lattice parameters.

The major contribution to the magnetism comes from electronic spins while the magnetic anisotropies originate from the interaction between spins and orbit; exchange interactions are changed because of lack of neighbors. A significant magneto-elastic energy due to lattice mismatch is induced as well. Anisotropies can overcome the magnetostatic demagnetizing energy to give a perpendicular magnetization for ultrathin-film cases.

5.3.2 Oscillatory Exchange Coupling

Oscillatory exchange coupling is an important phenomena observed in ferromagnetic thin films separated by a nonmagnetic spacer. In this phenomena one of the layers polarizes the conduction electrons of the nonmagnetic metallic spacer. If the lifetime of the polarization is long enough, these polarized electrons carry this information to the layer across the spacer. Thus the second layer is coupled with the first one through polarized conduction electrons, which is also known as RKKY.

5.3.3 Spin-Polarized Tunneling

Quantum mechanical effects may be important in particular for very small particles at low temperatures, when quantum mechanical reorientation of the magnetization may occur. This process proceeds through a tunneling mechanism. When two ferromagnetic materials are separated by a very thin insulator the tunneling current from one layer to the other depends on the relative orientation of magnetization of the layers and the potential barrier, which in turn, depends on the insulator thickness and the type of the layer materials. Parallel orientation of the adjacent layer corresponds to lower resistivity. Indeed tunneling rate depends on the junction quality. The tunneling current is also affected by the temperature and voltage.

The magnetocrystalline anisotropy in the particle define a preferred direction along which the magnetization will be oriented at very low temperatures. Equal and opposite directions are energetically degenerate. If there is only one anisotropy axis, then the ground state will be degenerate, and the magnetization will be either parallel or antiparallel to the magnetization axis. However, in the more general case, if there are two or more anisotropy axes, then the ground state will be a superposition of states where the magnetization is oppositely aligned (compare this with the NH_3 molecule). Hence, if the magnetization is initially aligned along the easy axis (for example, by cooling it in a strong magnetic field), then it will oscillate at the tunneling frequency between the spin-up and spin-down state.

5.3.4 Magnetoresistivity

Magnetoresistivity is an important phenomena observed at the nanoscale, which is a change of electrical resistivity when an external magnetic field is applied to the substance. The magnetoresistivity can be varied by the relative orientation of magnetization in neighboring ferromagnetic layers separated by a nonmagnetic metallic thin layer. The prototype system is layers of Co separated by spacer layers of Cu. The in-plane resistance of this system depends sensitively on the magnetic field applied to the layers. The origin of the effect is a strong exchange coupling between the ferromagnetic layers. The magnetoresistivity also depends on the direction of the current with respect to the film normal. The parameters affecting the magnetoresistivity for multilayer films are spin-scattering parameters for outer surfaces and interfaces, effective masses, inner potentials, and relaxation times for magnetic and nonmagnetic films. All of these parameters are strongly influenced by imperfections of the samples. This property is ideal for a magnetic sensor as it directly relates the applied magnetic field and the electrical resistivity. The resistance in normal metals is increased in the presence of an external magnetic field and depends on the relative orientation of the current with respect to the field. However, the resistivity decreases with the applied field for magnetic–nonmagnetic composite systems when the size of the nonmagnetic metallic component separating magnetic components becomes smaller than the mean free path of charge carriers. This is called GMR and can be much larger than AMR.

5.3.5 Magnetic Moments of 3d Transition Metal Clusters

An overall decreasing trend of the magnetic moment with increasing cluster size is observed in all three cases, with some oscillating fine structure features. The

FIG. 5.11	Magnetic moment per atom in units of Bohr magneton for iron, cobalt, nickel, and rhodium clusters as a function of cluster size.

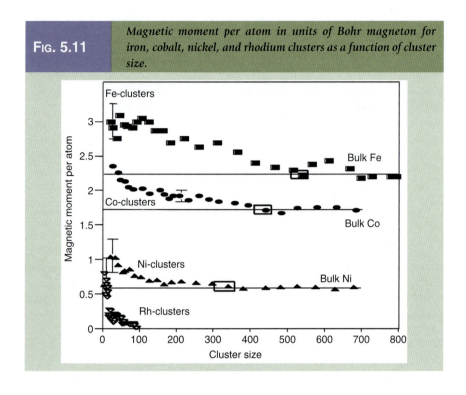

magnetic moment of an atom at the surface is generally larger than in the bulk for small sizes where the magnetic moments per atom for Fe, Co, and Ni are approximately $3\mu_B$, $2\mu_B$, and $1\mu_B$ respectively (Co is slightly higher, probably due to orbital effects.) These values correspond to the maximum spin that can be obtained with 7, 8, and 9 electrons in a d orbital. For example a nearly free iron atom has 8 valence electrons of which 1 is in the 4s orbital and 7 in the 3d orbital. Of those 7 and 5 are in spin-up states (forming the majority spin band) and 2 in spin-down states (the minority spin band). Hence the net magnetic moment per atom due to the 3d orbitals is $3\mu_B$. In fact for Fe, $\mu_{bulk} = 2.2\mu_B$; for Co, $\mu_{bulk} = 1.6\mu_B$, for Ni, $\mu_{bulk} = 0.6\mu_B$. Clusters with as few as 600 atoms already appear to have bulk-like magnetic moments (**Fig. 5.11**). It is interesting to note that measurements of Cr clusters ($N > 10$) have demonstrated that these clusters do not deflect. This indicates that they are nonmagnetic or more likely that they are antiferromagnetic.

5.3.6 The Temperature Dependence of Magnetic Moments

Loss of ferromagnetic order occurs at the Curie temperature where thermal motion overcomes the order imposed by the interatomic exchange interaction. In itinerant magnetism there are two distinct pictures: in the band picture, the magnetic moment reduction is caused by thermally induced electronic excitations from the top of the majority spin band to the Fermi level of the minority spin band, which reduces the total moment. In the localized moment picture, the global moment is reduced through local misalignments. The molecular beam method favors measurements of magnetic moments as a function of temperature over a wide range of temperature, ranging from 80 to 1000 K. In this way the ferromagnetic to paramagnetic phase transition can be probed. The magnetic moments have been measured as a function of temperature for several sizes and it is clear that they decrease with increasing temperature. It is also observed that for the magnetic moments of Ni and Co clusters the magnetization curves appear to converge to their respective bulk behaviors. Fe is anomalous and no obvious trend can be discerned. Small clusters usually prefer icosahedral structures.

5.4 RECENT PROGRESS IN NANOSCALE SAMPLE PREPARATION

5.4.1 *Epitaxial Methods*

Epitaxial layer-by-layer growths in ultrahigh vacuum systems are the most useful and common methods of preparation of nanosized samples, since lattice symmetry and size of the crystal strongly influence the magnetic properties. However, due to their relatively higher surface free energy, the magnetic materials are difficult to grow on any substrate. Also any particular ferromagnetic material can be deposited in more than one crystal structure. Moreover some artificial solids are possible to grow, exhibiting new physical properties.

The most critical parameters in magnetic multilayer growth are lattice and relative surface free energies mismatch between the substrate and overlayer film. The interdiffusion, chemical reactions, and alloying between the substrate and overlayer atoms give additional problems. For epitaxial multilayer film preparations, noble metals, semiconductors, and some oxides are used as single-crystal substrates. Many tricks have been developed to overcome problems such as low temperature and high growth rate. Each molecule has to stick at a nearest place with an energy minimum of the previous layer, and thus, growth symmetry follows the substrate structure. Electrochemical deposition techniques and self-organization are other useful and promising methods.

Magnetic materials are deposited on cylindrical pores to get nanowires perpendicular to the surface of the substrate. The beam of magnetic materials can be focused by interference field of an intense laser onto the substrate to form parallel wires.

5.5 NANOMAGNETISM APPLICATIONS

5.5.1 *Overview*

Nanosized magnetic iron oxide particles have been studied extensively due to their wide range of applications in ferrofluids, high-density information storage, magnetic resonance imaging (MRI), biological cell labeling, sorting and separation of biochemicals, targeting, and drug delivery [3]. For many of these applications surface modification of nanosized magnetic particles were accomplished by physical/chemical adsorption or surface coating of desired molecules, depending on the specific applications. A silica coating on the surface of nanosized iron oxide particles, for example, have also been shown to prevent their aggregation in liquids and improve their chemical stability.

For example, poly(1-vinylimidazole) polymer-grafted nanosized magnetic particles have been used in applications as magnetic carriers in a wide range of disciplines. Poly(1-vinylimidazole) is chosen to graft on nanosized magnetic particles, as the resultant organic–inorganic hybrid magnetic materials are anticipated to expand the sorbent-based separation technology to a multiphase complex system, ranging from biological cell sorting to industrial effluent detoxification and recovery of valuables. Metal ion binding properties of imidazole and poly(1-vinylimidazole) have been reported by many researchers for the removal of various metal ions from aqueous solutions.

MRI detectable and targeted quantum dots have also been developed. Quantum dots were coated with paramagnetic and pegylated lipids. The quantum dots are usually functionalized by covalently linking RGD peptides. With recent developments in chemistry and the synthesis of powerful, innovative, specific, and multimodal contrast agents, for example, by introducing fluorescent properties as well, MRI is becoming increasingly important for molecular imaging. Quantum dots, semiconductor nanocrystals in a size range of 2–6 nm, have gained much interest in the past few years for biological imaging purposes, especially because of their bright fluorescence, their photostability, and their narrow and tunable emission spectrum. The in vivo use requires the quantum dots to be water soluble and biocompatible. Efforts have been undertaken to achieve these goals, and quantum dots have been used successfully for imaging studies of live cells and animal models, mainly in combination with two-photon fluorescence microscopy.

The semiconductor currently used in integrated circuits, transistors, and lasers such as silicon and gallium arsenide (GaAs) are nonmagnetic. The carrier is almost independent of the spin direction. The miniaturization becomes more difficult in nanostructures. Magnetic semiconductors, alternative semiconductors which have both properties of magnetic materials and semiconductors, exchange interactions that can give rise to pronounced spin-related phenomena not just in nanostructures but also in more conventionally sized devices.

Semiconductor spin electronics can be divided into two fields such as semiconductor magneto-electronics and semiconductor quantum spin electronics. Semiconductor magneto-electronics are used to implement new functions by using semiconductor materials that are also magnetic or combinations of semiconductors and magnetic materials. For example, semiconductors such as optical isolators, magnetic sensors, and nonvolatile memory are possible to be developed if the magnetism can be controlled by light and electric field.

The other field that is referred as semiconductor quantum spin electronics is mainly focused on using the quantum mechanical nature of spin in semiconductors. For example, since the various types of spins in nonmagnetic semiconductors

FIG. 5.12	*The basic elements of spintronics devices are depicted. The quantum mechanical nature of spin is exploited in semiconductors. Such efforts lead to the development of solid-state quantum mechanical information processing devices.*

Mass + electric charge + spin
(intrinsic quantity of angular momentum)

Like that of a tiny bar magnet

West to east: spin up

In the magnetic field, spin up and -down have different energies

In an ordinary electric circuit, the spins are oriented at random and have no effect on current flow

Spintronic devices create spin-polarized currents and use the spin to control current flow

have a much longer coherence time than electrical polarization and can be controlled by light or electric fields, it is easy to manipulate spin as a quantum mechanical entity. Such properties lend themselves to the development of solid-state quantum information processing devices; in this way, spin in semiconductors is heralding a new era both in classical and quantum physics and technology (**Fig. 5.12**).

The spin of the electron was ignored in mainstream charge-based electronics. A technology has emerged called spintronics (spin transport electronics or spin-based electronics) [4]. It is not the electron charge but the electron spin that carries information. Spintronics offers opportunities for a new generation of devices. This technology combined standard microelectronics with spin-dependent effects that arise from the interaction between spin of the carrier and the magnetic properties of the material. In an ordinary electric circuit the spins are oriented at random and have no effect on current flow. Spintronic devices create spin-polarized currents and use the spin to control current flow. Devices that rely on an electron's spin to perform their functions form the foundations of *spintronics*.

The advantages of these alternative techniques would be nonvolatility, increased data processing speed, decreased electric power consumption, and increased integration densities compared with current semiconductor devices. The field of spintronics is addressed by the experiment and theory of the optimization of electron spin, the detection of spin coherence in nanoscale structures, the transport of spin-polarized carriers across relevant length scales and

TABLE 5.3	Comparison of Memory Technologies for the Year 2011 [5]			
	CMOS			
Technology	**DRAM**	**Flash**	**SRAM**	**MRAM**
Reference	SIA 1999	SIA 1999	SIA 1999	
Generation at introduction	64 Gb	64 Gb	180 Mb/cm^2	64 Gb
Circuit speed (MHz)	150	150	913	>500
Feature size (nm)	50	50	35	<50
Access time (ns)	10	10	1.1	<2
Write time	10 ns	10 μs	1.1 ns	<10 ns
Erase time	<1 ns	10 μs	1.1 ns	N/A
Retention time	2–4 s	10 years	N/A	Infinite
Endurance cycles	Infinite	10^5	Infinite	Infinite
Operating voltage (V)	0.5–0.6	5	0.5–0.6	<1
Voltage to switch state	0.2 V	5 V	0.5–0.6 V	<50 mV
Cell size	2.5 F^{2a}/bit 0.0005 μm^2	2F^2/bit	12F^2/bit	2F^2/bit

a F = minimal lithographic feature size.

heterointerfaces, and the manipulation of both electron and nuclear spins on sufficiently fast timescales.

Merging of electronics, photonics, and magnetics will ultimately lead to new spin-based multifunctional devices such as spin–FET (field effect transistor), spin–LED (light-emitting diode), spin RTD (resonant tunneling device), optical switches operating at very high frequency in the terahertz range, modulators, encoders, decoders, and quantum bits for quantum computation and communication. A summary of the future memory technologies involving nanomagnetism is shown in **Table 5.3**.

5.5.2 *Current Status of Spin-Based Electronics Devices*

There are a number of effects that couple magnetization to electrical resistance. These include

- Ordinary magnetoresistance (OMR)
- Anisotropic magnetoresistance (AMR)
- Giant magnetoresistance (GMR)
- Tunneling magnetoresistance (TMR)
- Ballistic magnetoresistance (BMR)
- Colossal magnetoresistance (CMR)

The GMR is perhaps the major innovation in spin-based electronics. GMR is observed in very thin-film materials constructed of alternate ferromagnetic and nonmagnetic layers. The resistance of the material is lowest when the magnetic domains in ferromagnetic layers are aligned and highest when they are anti-aligned (**Fig. 5.13**).

A spin valve (**Fig. 5.14a**) is a GMR-based device that comprises of two ferromagnetic layers (e.g., alloys of nickel, iron, and cobalt) isolated within a

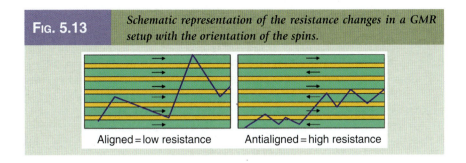

FIG. 5.13 *Schematic representation of the resistance changes in a GMR setup with the orientation of the spins.*

Aligned = low resistance   Antialigned = high resistance

thin nonmagnetic metal (usually copper), with one of the two magnetic layers being aligned (called *pinned*) whereby the magnetization in the pinned layer is relatively insensitive to any external magnetic fields. The other magnetic layer is called the *free* layer, and its magnetization can be changed by the application of a relatively small magnetic field. As the magnetizations in the two layers change from a parallel to a antiparallel alignment, the resistance of the spin valve increases typically from 5% to 10%. Antiferromagnetic layer (called *"pinning"* layer) is usually constructed by using an antiferromagnetic material that is grown on top of the pinned layer so that they can exchange their spins. The two films form an interface that acts to resist changes to the pinned magnetic layer's magnetization.

In alternative structures, the pinned layer was replaced with a synthetic antiferromagnet: Two magnetic layers separated by a very thin (~10 Å) nonmagnetic conductor (usually ruthenium metal is used). The magnetizations in the two magnetic layers are strongly antiparallely coupled and are thus effectively immune to outside magnetic fields. This structure improves both stand-off magnetic fields and the temperature of operation of the spin valve. Another way of making these structures is by growing a nano-oxide layer (NOL) at the outside surface of the soft magnetic film. This layer reduces resistance due to surface

FIG. 5.14 *Spin-dependent transport structures. (a) Spin valve and (b) magnetic tunnel junction.*

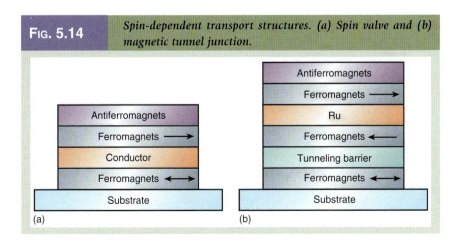

scattering, thus reducing background resistance and thereby increasing the percentage change in magnetoresistance of the structure. A magnetic tunnel junction (MTJ) (**Fig. 5.14b**) is such a device in which a pinned layer and a free layer are separated by a very thin insulating layer (commonly aluminum oxide). The tunneling resistance is modulated by an external magnetic field in the same way as the resistance of a spin valve is, and usually 20% to 40% change in the magnetoresistance can be observed. Also for its operation a saturating magnetic field is required which is equal to or somewhat less than that required for spin valves. Because the tunneling current density is usually small, MTJ devices tend to have very high resistances.

Applications of Spin-Based Electronics Devices.   The GMR-based galvanic isolator which is a combination of an integrated coil and a GMR sensor on an integrated circuit chip is perhaps a good example of the current application. GMR isolators introduced in 2000 eliminate ground noise in communications between electronics blocks, thus performing a function similar to that of optoisolators providing electrical isolation of grounds between electronic circuits. The GMR isolator is ideally suited for integration with other communications circuits and the packaging of a large number of isolation channels on a single chip.

Magnetic Data Storage.   MRAM (magnetic random access memory) that uses magnetic hysteresis to store data and magnetoresistance to read data is another application of the GMR effect. GMR-based MTJ or pseudospin valve memory cells are integrated on an integrated circuit chip and function like a static semiconductor RAM chip with the added feature that data are retained with the power off. Potential advantages of the MRAM compared with silicon electrically erasable programmable read-only memory (EEPROM) and flash memory are 1000 times faster write times, no wear out with write cycling (EEPROM and flash wear out with about 1 million write cycles), and lower energy for writing. MRAM data access times are about 1/10,000 that of hard disk drives.

Magnetic nonstructures are starting to play a role in technology, particularly in "non-volatile magnetic data storage." New combinations with semiconductor technology are developing such as MRAMs where the storage capacitor of a traditional semiconductor in replaced by a nonvolatile magnetic dot. The storage media to today's magnetic hard disk drives may be viewed as an array of magnetic nanoparticles. The magnetic coating of the disk consists of a ternary Co–Pt–Cr mixture, which segregates into magnetic Co–Pt grains. These grains are magnetically separated by Cr at the grain boundaries. Typical grain sizes are 10–20 nm using about 10^3 grains/bit at a recording density of 1 Gbit/in.^{2} for commercial devices. The grains segregate randomly, which introduces statistical noise into the read out signal due to the variation in grain size, coercivity, and domain structure. This explains the large number of grains that are required to reduce these fluctuations in a device.

There is a fundamental limit on the improvement of magnetic storage density. The limit is controlled by

1. Thermal flipping of the bits as the grains become smaller.
2. Energy barrier between the two stable magnetizations along the easy axis becomes comparable with kT. Eventually the superparamagnetic limit is reached where individual grains stay magnetized, but their orientation fluctuates thermally.

For typical magnetic storage media, the superparamagnetic limit imposes a minimum particle size of about 10 nm that is a maximum recording density of several terabits per square inch. This is almost four orders of magnitude higher than the density of 1 Gbit/in.^{2} found in top-of-the-line disc drives today. While the current particle size is already close to the superparamagnetic limit the number of particle/bits is still more than 10^3. There are many signal-to-noise issues on the way towards reducing this number and reaching the theoretical limit. Inconsistent switching of different particles and an irregular domain structure require averaging over many particles.

Controlling coercivity, size, orientation, and position of magnetic nanoparticles will be essential for reducing the number of particles needed to store a bit. For example, a large crystalline anisotropy can produce a higher switching barrier than the shape anisotropy above. Single-domain nanoparticles with high saturation magnetization and coercivity are being optimized for this purpose. The orientation of segregated grains can be controlled using multilayered structures where the first layer acts as a seed for small grains and subsequent layers shape the crystalline orientation for the desired anisotropy. A further improvement would be the move from longitudinal to perpendicular recording where the demagnetizing field does not destabilize the written domains.

The ultimate goal in magnetic storage is single-particle-per-bit on quantized recording. It is aimed at producing single-domain particles close to the superparamagnetic limit with uniform switching properties. Lithography is currently the method of choice for producing regular arrays of uniform magnetic dots. Dot arrays with a density of 65–250 Gbit/in.^{2} have been produced by electron beam lithography. The performance and parameters for various types of volatile and nonvolatile memory chip alternatives are shown in **Table 5.4**.

5.5.3 Sensors

Nanostructured material has also been used in the development of some reading head devices. The traditional inductive pick up of the magnetic signal is replaced by a magnetoresistive sensor in state-of-the-art devices.

Permalloy/(Cu/Co) multilayers are utilized in reading heads. Currently, reading heads in high-end disc drives are based on the 2% AMR of permalloy. The resistance is highest for the current parallel to the magnetization and lowest perpendicular to it, producing a sinusoidal orientational dependence. The magnetic stray field between adjacent bits with opposite orientation rotates the magnetization in the permalloy film with respect to the current and thus induces a resistance change. That is directly connectible into a read-out voltage.

Present activities with GMR are directed towards lowering the switching field while keeping a large magneto-resistance. To obtain the best of both characteristics one obtains soft permalloy layer for easy switching with a high-spin co-layer that enhances the magnetoresistance.

TABLE 5.4	Comparison of Performance and Parameters for Various Types of Volatile and Nonvolatile Memory Chip Alternatives [6]					
Parameter	**DRAM**	**SRAM**	**NOR flash**	**NAND flash**	**FeRAM**	**MRAM**
Read cycles	$>10^{15}$	$>10^{15}$	$>10^{15}$	$>10^{15}$ before cycling	10^{12}–10^{15}	$>10^{15}$
Write cycles	$>10^{15}$	$>10^{15}$	10^4–10^5	10^6	10^{12}–10^{15}	$>10^{15}$
Write voltage (V)	2.5–5	3.3–5	10–10	18	0.8–5	0.8–5
Cell write time (ns)	10–100	1–50	6×10^3	2×10^5	10–50	10
Write energy (pJ)	Few 10^{-2}		9000	1	1	10–100
Random access time (ns)	40–70	6–70	150	~10,000	40–70	40–70
Cell size (F^2)	8	~100	12	4.6	9–13	6–10
Retention (years)	None	None	10	10	10	10
Scaling issues	Charge		Tunnel oxide → read current → access time	Erase voltage tunnel oxide scaling.	3D + material texture	Switching field increases with scaling and uniformity
Status/forecast	256 Mb/1 Gb	4–16 Mb	32 Mb/128 Mb	256 Mb/1 Gb	1 Mb/4 Mb	Few kb/1 Mb (?)
Applications	PC memory	Cache memory	Program code and data	Data files (camera, MP3)	Contactless smartcard	Envisaged: embedded (SOC) and mass storage

Another type of a nonvolatile magnetic storage device avoids moving parts altogether at the expense of having to pattern the storage medium. This is a combination of magnetic memory elements with semiconductor circuits that sense and amplify the magnetic state (MRAM).

Further into the future are logic devices based on magnetic nanostructures. A bipolar spin switch has been demonstrated that acts like a transistor.

5.5.4   *Nanomagnetism for Biomedical Applications*

The use of magnetic nanoparticles for biological and medical applications has been developed recently. The size of the particles can range from a few nanometers to several micrometers and thus is compatible with biological entities ranging from proteins (a few nm) to cells and bacteria (several μm). The combination of biology and magnetism is useful, because the biochemistry enables a selective binding of the particles, while the magnetism enables easy manipulation and detection. Using magnetic field gradients, the magnetic particles can be subjected to significant forces even when they are embedded in a biological environment. The absence of ferromagnetism in most biological systems, which typically have only weak dia- or paramagnetism, means that the magnetic moment from the ferromagnetic particles can be detected with little noise in a biological environment.

FIG. 5.15 *Magnetic virus. The DNA is removed from the virus interior, which results in a rigid empty protein shell. This shell can be used as a template for ferromagnetic nanoparticles.*

Based on these ideas, a variety of nanomagnetism applications in the bio-medical field have emerged. A simple application is to bind magnetic particles to the interested biological systems, which then allows manipulating the biological material via the magnetic field gradient. This has been already applied to several problems, such as separating red blood cells from blood, cancer cells for bone marrow, drug delivery, and hypothermal treatment.

Applications in biology and medicine normally require that the particles are stable in aqueous solution at neutral pH values. For this reason, magnetite (Fe_3O_4) and maghemite (γ-Fe_2O_3) are most commonly utilized for biofunctionalized magnetic nanoparticles.

The ability to tag biological molecules with functionalized magnetic particles has been already exploited for biological sensors. Traditionally many biological sensors, like DNA microarray, use fluorescent markers for the detection of specific biological molecules. For example, the DNA can be bound to a fluorescent molecule, then DNA is exposed to an array with many different well-defined DNA strands and it will only stick to complementary matching DNA. The position of the light signal from the fluorescent marker then indicates which are the right DNA strands. Similarly it has been demonstrated that magnetic particles can be used as tags and the binding can be identified by detecting the stray magnetic field of the particles (**Fig. 5.15**). Key advantages of using magnetic particles versus fluorescent molecules have been argued to be the following:

- Magnetic particles typically have an unlimited shelf life compared to fluorescent markers which deteriorate with time (note: quantum dots circumvent this problem).
- The use of magnetic particles together with magnetoelectronic sensors allows for a complete electronic readout of the sensors.
- The sensitivity of the sensors based on magnetic versus fluorescent tagging are comparable.
- Magnetic tags allow for manipulation of the target molecules such that they can be moved towards the magnetic field sensor using magnetic field gradient

FIG. 5.16 *Scheme of localized magnetic hyperthermia applied to a breast carcinoma. The tumor region contains injected magnetic particles.*

Hyperthermia. There are several reports regarding the efficacy of hyperthermia for the treatment of musculoskeletal tumors [7–10]. Currently, the principal methods of hyperthermia for clinical use are microwaves or radiofrequency waves to generate heat in the tumor (**Fig. 5.16**). Using these methods, however, it is difficult to heat deeply seated tumors effectively and selectively [11]. New types of ferromagnetic thermoseeds are being developed experimentally to solve this problem, but they have never been used clinically because of their unreliable rise in temperature. Application of magnetic materials for hyperthermia of biological tissue with the goal of tumour therapy is known in principle for more than four decades. Much empirical work was done in order to manifest a therapeutic effect on several types of tumors by performing experiments with animals or using cancerous cell cultures. However, routine medical applications are not known till now and there is a demand for a more profound understanding of the related material properties to render that method reliably for tumour therapy of human beings. The heating of oxide magnetic materials with low electrical conductivity in an external alternating magnetic field is due to loss processes during the reorientation of the magnetization. If thermal energy $k_B T$ is too low to facilitate reorientation, hysteresis losses dominate which depend on the type of the remagnetization process (wall displacement or several types of rotational processes). With decreasing particle size, thermal activation of reorientation processes leads to a dependence on temperature and measurement frequency to superparamagnetic behavior of the particle ensemble and the occurrence of the so-called Néel losses [12]. In the case of ferrofluids, losses related to the rotational Brownian motion of magnetic particles may arise too, and hence care should be taken to analyze the results, depending on the temperature and measurement frequency.

Ocular. The anterior segment of the eye is bounded by the cornea and the lens–iris diaphragm, and contains the aqueous humor. The posterior segment begins behind the lens–iris diaphragm and includes, from inside outwards the vitreous (gel/fluid), retina (neurosensory tissue), and choroid (heavily muscular).

FIG. 5.17 *Internal tamponade using silicone magnetic fluids—magnetic fluid for use in eye surgery [13].*

Magnetized scleral buckle

Magnetic fluid

The retinal photoreceptors are supported by the choroid (**Fig. 5.17**). The retina and the choroid stay attached to each other with a suction pump, which keeps the subretinal space dry.

Retinal detachment is a major cause of vision loss in adults. It occurs when the retina separates from the choroid, resulting in eventual death of the retina and subsequent loss of vision. As one ages, the vitreous gel normally undergoes liquefaction, collapses, and separates from the retina. Separation of the vitreous gel may result in the formation of a tear in the retina at a site of vitreoretinal adhesion. The tear provides a pathway for the vitreous fluid to pass through and underneath the retina, overcoming the suction attraction of the retina to the choroid, thus detaching the retina from the underlying choroid.

The goal of surgery is to close the holes in the retina, preventing further fluid flow into the subretinal space, allowing for reattachment of the retina. Efforts are going on to develop an internal tamponade from modified silicone fluid containing spherically stabilized 4–10 nm sized metal particles, which could be held in place with an external magnetized scleral buckle. With an appropriate magnetic fluid inside the vitreous cavity, a stable, 360° internal tamponade might be achieved. The enriching magnetized scleral buckle and magnetic fluid would produce a ring of silicone oil in opposition to the retinal periphery. The central vitreous cavity would be free of the magnetic fluid and the lens, anterior structures, or macula, thus avoiding the complications of currently available treatment modalities.

There are still other applications on magnetic nanoparticles, such as; drug delivery, hyperthermia treatment. Key challenges for nanoparticles for biological application include the modification of nanoparticles for enhanced aqueous solubility, biocompatible or bio-recognition, optimization of their magnetic properties. Looking further ahead into the future there are still many possibilities for the magnetic nanoparticles in biomedical field. Many possible applications of magnetic nanoparticles in biology and medicine clearly offer plenty of opportunities for future research and development.

References

1. Z. Jibin and L. Yongping, *IEEE Transactions on Magnetics* 28, 3367 (1992).
2. K. M. Krishnan, A. B. Pakhomov, Y. Bao, P. Blomqvist, Y. Chun, M. Gonzales, K. Griffin, X. Ji, and B. K. Roberts, *Journal of Materials Science* 41, 793–815 (2006).
3. S. A. Wolf, D. D. Awschalom, R. A. Buhrman, J. M. Daughton, S. von Molnar, M. L. Roukes, A. Y. Chtchelkanova, and D. M. Treger, *Science*, 294, 1488–1494 (2001).
4. J. F. Gregg, I. Petej, E. Jouguelet, and C. Dennis, *Journal of Physics D: Applied Physics*, 35, R121–R155 (2002).
5. *Handbook of nanoscience, engineering and technology*, CRC Press, Boca Raton, FL (2003).
6. J. De Boeck, W. V. Roy, J. Das, V. Motsnyi, Z. Liu, L. Lagae, H. Boeve, K. Dessein, and G. Borghs, *Semiconductor Science and Technology*, 17, 342–354 (2002).
7. R. K. Gilchrist et al., *Annals of Surgery*, 146, 596 (1957).
8. C. Streffer and D. van Beuningen, in: J. Streffer (Ed.), *Hyperthermia and the therapy of malignant tumors*, Springer, Berlin (1987).
9. A. Jordan, R. Scholz, J. SchuKler et al., *International Journal of Hyperthermia*, 13, 83 (1997).
10. P. Burgman, A. Nussenzweig, and G. C. Li, in: M. H. Seegen-Schmiedt, P. Fessenden, and C. C. Vernon (Eds.), *Thermoradiotherapy and thermochemotherapy, Vol. 1: Biology, Physiology, Physics*, Springer, Berlin, 1995.
11. D. C. F. Chan, D. B. Kirpotin, and P. A. Bunn Jr., *Journal of Magnetism and Magnetic Materials*, 122, 374 (1993).
12. L. Néel and C. R. Hebd, *Seances Acad. Sci.*, 228, 664 (1949).
13. J. P. Dailey et al., *Journal of Magnetism and Magnetic Materials*, 194, 140 (1999).

Problems

5.1 What is the origin of magnetic moment in magnetic materials?

5.2 What is a Bohr magneton?

5.3 Which type of magnets have higher coercivity? Soft or hard?

5.4 What are the three different types of material response to an applied electromagnetic field called?

5.5 Hund's rule states that the lowest energy configuration for an atom is the one having the maximum number of unpaired electrons allowed by the Pauli principle in a particular set of degenerate orbitals. Show the spins in the various levels in the table below:

	Configuration	1s	2s	$2p_x$	$2p_y$	$2p_z$
C	$1s^2\,2s^2\,2p^2$					
N	$1s^2\,2s^2\,2p^3$					
O	$1s^2\,2s^2\,2p^4$					

5.6 We've discussed how M affects J, and the ability to transport charge, as manifested through magnetoresistive effects. One can also consider the converse: can a current J of carriers with a net spin polarization affect M?

5.7 Consider that the molar volume of magnetite is $4.4 \times 10^{-5}\,m^{-3}$ and magnetic moment of Fe_3O_4 is 9.27×10^{-24}, then calculate the magnetization of magnetite (Fe_3O_4) assuming that the magnetization is only due to the six 3d electrons of the Fe^{2+} ions and that only the spin angular momentum of the electrons contributes to the magnetic moment.

5.8 Determine the values of $\hat{S}$, L, and $\hat{J}$ for Cr^{3+} which has three electrons in the 3d sub-shell. All lower energy shells are filled.

5.9 Calculate the magnetization of Fe_3O_4 assuming that only the six 3d electrons of the Fe^{2+} ions contribute to the magnetization and that only the spin angular momentum of the electrons contribute to the magnetic moment. Consider the molar volume of magnetite as 4.4×10^{-5} m^{-3}.

5.10 What is giant magnetoresistance? What are magnetic multilayered monolayers?

5.11 How would you fabricate a magnetic nanowire?

5.12 What future applications do you see arising from such giant magnetoresistance?

Section 3

Mechanical Nanoengineering

NANOMECHANICS

Masood Hasheminiasari and John J. Moore

I am not afraid to consider the final question as to whether, ultimately in the great future we can arrange atoms the way we want; the very atoms, all the way down!… The principles of physics, as far as I can see, do not speak against the possibility of maneuvering things atom by atom. It is not an attempt to violate any laws… but in practice, it has not been done because we are too big… At the atomic level, we have new kind of forces and new kind of possibilities, new kind of effects.

RICHARD FEYNMAN, 1960

Chapter 6

FIG. 6.0	*Richard P. Feynman (1918–1988), scientist, physicist, and teacher who was a joint recipient of the Nobel Prize in Physics in 1965.*

Source: http://www.feynmangroup.com/company/whos_feynman.cfm, 2008.

THREADS

Chapter 6, Nanomechanics, is the first chapter in the *Mechanical Nanoengineering* section of the text. We are now ready to define, or attempt to embrace, the mechanics of nanomaterials. This chapter is about how small size can influence the mechanical behavior of materials and its impact on the bulk properties. There are two chapters in the *Mechanical Nanoengineering* section. In *chapters 7* and 8, thin films and nanocomposites are discussed. The sections all provide examples of applications of this nanotechology.

Following the mechanical-oriented chapters, we embark on the chemical engineering aspects of nanotechnology—specifically, the domain of the catalyst and the composite.

6.0 INTRODUCTION

This chapter is designed to develop an understandable description of what is called "nanomechanics," the mechanical behavior of an object in the nanometer level, that is, objects for which at least one dimension is in the nanometer range. Our main concern is not to focus on the rigid body dynamics or on the quantum mechanics in detail. Instead we will focus on the static deformation of solid objects, both with and without external forces in one, two, and three dimensions. Then, we will discuss about nanomechanical devices and their applications in the nanotechnology area.

In the first part of this chapter, the main goal is to build the foundations required to understand the concept of nanomechanics and provide some examples in order to show how the small size scale can impact and vary the bulk properties. In the second part of the chapter, we will describe some aspects of nano-microelectrome-chanical systems (NEMS/MEMS) and illustrate other nanomechanical devices.

This chapter begins by discussing a very basic mechanical problem, in which two atoms of a molecule bond together by their mutual interactions. The motion of these atoms is restricted to one dimension. Next we move to three atoms and then expand our previous discussion of a two-atom chain to a more general problem.

6.0.1 Two-Atom Chain Mechanics

Let's consider a molecule consisting of only two atoms. We assume that the atoms can only move horizontally towards or away from one another. We presume that there is an attractive force between the atoms; due to an electrostatic attrac-tion, if the atoms have opposite electric charges, to a covalent bonding or due to an attraction known as the van der Waals force, generated by induced dipole moments in each atom. So, we assume that there is a net force $f(r)$ between these two atoms, which is only a function of distance between two atoms, r. If the force is attractive, then the atoms will accelerate and combine with each other. This does not happen, because when the two atoms get too close to one another, the electron clouds of each atom repel one another through their electro-static repulsion, and are furthermore limited to the law of quantum mechanics from occupying the same volume of space. Thus, the attractive force becomes repulsive as the atoms approach each other.

It is usually more useful to work with the interaction potential energy $\phi(r)$ rather than force $f(r)$, which is defined by the relation

$$f(r) \equiv -\frac{d\phi}{dr} \tag{6.1}$$

The potential energy is also described as the negative of the work done by the force for a displacement $r - r_0$ from the point of zero potential energy r_0

$$\phi(r) = -W = -\int_{r_0}^{r} f(r)\,dr \tag{6.2}$$

which is equivalent to equation (6.1).

6.0.2 Interaction Potentials

One basic characteristic of atoms and molecules is the electric charge, e. Electric charge is related to the property of particles to exert forces on each other by means of electric field. Pair of particles with electric charges e_1 and e_2 exert repulsive (at $e_1 e_2 > 0$) or attractive (at $e_1 e_2 < 0$) forces on one another which are described by

$$F_1 = -F_2 = -\nabla_i V_c(r_1, r_2), \quad i = 1, 2 \tag{6.3}$$

where
    r_1 and r_2 are radii of the particles
    $V_c(r_1, r_2)$ is the electrostatic Coulomb potential

$$V_c(r_1, r_2) = \frac{1}{4\pi\varepsilon_0}\frac{e_1 e_2}{r}, \quad r = |r_1 - r_2| \tag{6.4}$$

where $\varepsilon_0 = 8.854188 \times 10^{-12}$ C/Vm is the permittivity constant in vacuum. The Coulomb potential is a long-range interaction and would be noticeable even at large separation distances, because the potential decays slowly with distance. But as atoms get closer the electrostatic field of the positively charged atomic nuclei or ion is neutralized by the negatively charged electron clouds surrounding the nuclei, which is called a short-range interaction potential. In general, the interaction potential can be based on purely theoretical calculations or phenomenological considerations.

Phenomenological interaction potential functions are often a more realistic view of atomic interaction than potentials which are derived exclusively from theoretical calculations. Phenomenological atomic interactions are in most cases based on a simple analytical expression which may or may not be justified from theory and contains one or more parameters adjusted to the experimental results. There are many interaction potentials developed for a two-atom system, pair-wise potentials, of which some are briefly mentioned in the next section.

Lennard–Jones Potentials. The general form of the Lennard–Jones potential is [1]:

$$\phi(r) = \frac{\lambda_n}{r^n} - \frac{\lambda_m}{r^m} \tag{6.5}$$

This potential was developed to treat inert gases, but it is often used to describe metals and other materials. The most common form is called Lennard–Jones (6-12) potential, where $n = 12$ and $m = 6$ and has the following form:

$$\phi(r) = 4\varepsilon \left[\left(\frac{\sigma}{r} \right)^{12} - \left(\frac{\sigma}{r} \right)^{6} \right] \tag{6.6}$$

Due to its simple form, this potential is often used to treat the cross-interaction of two different materials, as indicated in **Table 6.1**. The properties of the noble gases are estimated by 10% accuracy. Since this potential is designed for noble gases, we cannot expect to gain adequate results from the Lennard–Jones potential in metallic systems [2–4].

Buckingham Potentials. The Buckingham potential has both inverse 6th and inverse 8th power dependency, which make this potential complicated [5]. A simpler form of this potential which has eliminated the inverse 8th power functionality is called the modified version of the Buckingham potential and has the form:

TABLE 6.1	Lennard–Jones Potential Parameters for Different Materials		
Symbol	Mass ($\times 10^{-7}$ kg)	ε ($\times 10^{-21}$ J)	σ ($\times 10^{-10}$ m)
Ne	33.51	0.5315	2.786
Ar	66.34	1.6539	3.405
Kr	139.16	2.2075	3.639
Xe	218.02	3.0497	3.962
Cu	105.52	65.626	2.338
Ag	179.13	55.276	2.644

Symbol	Type	Mass (× 10⁻⁷ kg)	r_m (× 10⁻¹⁰ m)	ε (× 10⁻²¹ J)	α

TABLE 6.2 Modified Buckingham Potential Parameters for Different Nonbonded Materials

Symbol	Type	Mass ($\times 10^{-7}$ kg)	r_m ($\times 10^{-10}$ m)	ε ($\times 10^{-21}$ J)	α
C	sp,sp²	19.925	3.88	0.357	12.5
H	Hydrocarbon	1.674	3.00	0.382	12.5
O	Carbonyl	26.565	3.48	0.536	12.5
N	sp³	23.251	3.64	0.447	12.5
F	Flouride	31.545	3.30	0.634	12.5
Cl	Chloride	58.064	4.06	1.950	12.5
Br	Bromide	131.038	4.36	2.599	12.5
I	Iodide	210.709	4.64	3.444	12.5
S	Sulfide	53.087	4.22	1.641	12.5
Si	Silane	46.454	4.50	1.137	12.5
P	Phosphine	51.464	4.36	1.365	12.5

$$\phi(r) = \frac{\varepsilon}{1 - \frac{6}{\alpha}} \left[\frac{6}{\alpha} e^{\alpha\left(1 - \frac{r}{r_m}\right)} - \left(\frac{r}{r_m}\right)^{-6} \right] \tag{6.7}$$

where there are three independent parameters $(\varepsilon, r_m, \alpha)$ with ε as the depth of the energy minimum and r_m as the corresponding value of the distance r between two atoms. The steepness of the exponential is measured by α.

This potential is often used to describe the attractive and repulsive forces experienced by the pairs of uncharged, nonbonded atoms. **Table 6.2** illustrates a set of parameters for different materials [6].

Moreover, we can treat the interaction of atoms from different materials by the introduction of the mean value procedure:

$$\varepsilon_{12} = \frac{\varepsilon_1 + \varepsilon_2}{2}, \quad r_{m12} = \frac{r_{m1} + r_{m2}}{2} \tag{6.8}$$

Barker Potentials. Barker determined potentials for ground-state krypton–krypton and xenon–xenon interactions [7], which are in good agreement with a wide range of experimental data including second virial coefficients, gas transport properties, solid state data, long-range interactions, and measurements of differential scattering cross sections.

While the many-body interactions have been neglected, the third-order triple dipole three-body interactions have been considered. Therefore, Barker potentials can be considered as effective pair potentials

$$\phi(r) = \varepsilon \left[\phi_0(r) + \phi_1(r) + \phi_2(r) \right] \tag{6.9}$$

where

$$\phi_0(r) = e^{\alpha\left(1 - \frac{r}{r_m}\right)} \sum_{i=0}^{5} A_i \left(\frac{r}{r_m} - 1\right)^i - \sum_{i=0}^{2} \frac{C_{2i+6}}{\left(\frac{r}{r_m}\right)^{2i+6} + \delta},$$

$$\phi_1(r) = \begin{cases} e^{\beta\left(1-\frac{r}{r_m}\right)}\left[P\left(\frac{r}{r_m}-1\right)^4 + Q\left(\frac{r}{r_m}-1\right)^5\right] & r \geq r_m \\ 0 & r < r_m \end{cases} \tag{6.10}$$

$$\phi_2(r) = \begin{cases} e^{\gamma\left(1-\frac{r}{r_m}\right)^2}\left[R\left(\frac{r}{r_m}-1\right)^2 + S\left(\frac{r}{r_m}-1\right)^3\right] & r \geq r_m \\ 0 & r < r_m \end{cases}$$

Again, ε is the depth of the potential at its minimum, where the value of the interatomic distance is $r = r_m$. Parameters for krypton and xenon are listed in **Table 6.3**.

In the next figure, **Figure 6.1**, the Barker potential for krypton is compared with the corresponding Lennard–Jones and Buckingham potentials. The Lennard–Jones potential matches nearly exactly with the Buckingham potential, but the Barker potential shows a deeper minimum [8].

In conclusion, of all the pair potentials reviewed in this section, the *Barker* potentials are the most accurate with the broadest range of validity. Especially the independence of temperature predestines these functions as model potentials within molecular dynamics (MD) calculations for nanosystems. There are many

TABLE 6.3	Barker Potential Parameters for Krypton and Xenon	
Parameters	**Krypton**	**Xenon**
ε	2.787	3.898
r_m	4.0067	4.3623
α	12.5	12.5
A_0	0.23526	0.2402
A_1	−4.78686	−4.8169
A_2	−9.2	−10.9
A_3	−8	−25
A_4	−30	−50.7
A_5	−205.8	−200
C_6	1.0632	1.0544
C_8	0.1701	0.1660
C_{10}	0.0143	0.0323
δ	0.01	0.01
β	12.5	12.5
P	−9	59.3
Q	68.67	71.1
γ	0	−50
R	0	2.08
S	0	−6.24

FIG. 6.1	*A comparison between the Barker, modified Buckingham, and Lennard–Jones potentials for krypton.*

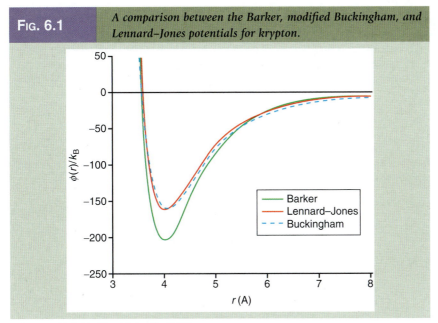

Source: M. Reith, World Scientific (2003).

other kinds of pair-wise interaction potentials such as the Morse potential, the Schommers potential, and many more potentials that are beyond the scope of this section. Among the different many-body potentials, the embedded atom potential (EAM) is one of the most used. Now, we briefly mention the EAM, which is usually applied to metallic systems.

Embedded Atom Potential. This multibody potential is specifically designed to treat metallic systems. One appealing aspect of the EAM potential is its physical picture of metallic bonding, where each atom is embedded in a host electron gas created by all neighboring atoms [9]. The atom–host interaction is more complicated than the simple pair-wise models. This interaction is described in terms of an empirical embedding energy function. The embedding energy function incorporates some important many-atom effects by providing the amount of energy required to insert an atom into the electron gas of a given density. The total potential energy, ϕ, includes the embedding energies, G, of all the atoms in the system, and an electrostatic Coulomb interaction V_c:

$$\phi = \sum_i G_i \left(\sum_{i \neq j} \rho_j^a(r_{ij}) \right) + \sum_{i,j>i} V_c(r_{ij}) \qquad (6.11)$$

Here, ρ_j^a is the averaged electron density for a host atom j, viewed as a function of the distance between this atom and the embedded atom i. Therefore, the host electron density is employed as a linear superposition which is assumed to be spherically symmetric. More information about the shape of the functions G, ρ, and V_c can be gathered from Clementi and Roetti [10] and Foiles et al. [11]. For a comprehensive review of the EAM, readers are referred to Daw et al. [12].

EXAMPLE 6.1 *Lennard–Jones Potential for Two Argon Atoms in an Ar_2 Molecule*

The Lennard–Jones potential is one of the easiest pair potentials available, which applies to atoms interacting through the van der Waals interaction. Therefore, this pair potential gives an excellent description of the interactions between inert gas atoms, such as argon, krypton, and xenon. A simple form of Lennard–Jones (6-12) potential is assumed:

$$\phi(r) = -\frac{A}{r^6} + \frac{B}{r^{12}}$$

with the parameters A determining the strength of the attractive interaction, and B the repulsive interaction. The potential energy has a minimum at equilibrium spacing ($r = r_0$), so we have:

$$\left. \frac{d\phi}{dr} \right|_{r=r_0} = 0$$

$$\therefore 6\frac{A}{r_0^7} - 12\frac{B}{r_0^{13}} = 0 \rightarrow (r_0)^6 = \frac{2B}{A}$$

$$r_0 = \left(\frac{2B}{A} \right)^{\frac{1}{6}}$$

The binding energy E_b, the difference between the potential energy minimum and that when the two atoms are infinitely far apart, is given by:

$$E_b = \phi(\infty) - \phi(r_0)$$

$$\phi(r_0) = -\frac{A^2}{2B} + \frac{A^2}{4B} = -\frac{A^2}{4B}$$

$$\therefore E_b = \frac{A^2}{4B}$$

In argon, the equilibrium spacing is found to be $r_0 = 0.38\,nm$, and the binding energy is $E_b = 10.4\,meV = 1.7 \times 10^{-21}\,J$ [13]. So, we can calculate the constants to be

$$A = 2r_0^6 E_b = 63 \,\overset{\circ}{A}^6 \, eV$$

and

$$B = r_0^{12} E_b = 9.4 \times 10^4 \,\overset{\circ}{A}^{12} \, eV$$

So the final form of pair-wise potential can be written as

$$\phi(r) = \frac{9.4 \times 10^4}{r^{12}} - \frac{63}{r^6} \, (eV)$$

The binding energy of an argon molecule is less than the thermal energy at ambient temperature, $k_BT = 26\,meV$. So, solid argon forms only at quite low temperatures, below 100 K. But, the binding energy for the much stronger ionic, metallic, and covalent interactions in typical solids is in the range of several tens of electron volts, rather than a few meV. Therefore, these types of bonds cannot simply be modeled with Lennard–Jones potentials.

6.0.3 *External Forces*

We assume that equal and opposite external forces f_{ext} are applied to each atom. The atoms will move apart until they reach a new equilibrium point. The potential energy associated with the external force is given by $\phi_{ext}(r) = -f_{ext}r$. So, the total potential energy is then $U_{tot} = \phi(r) + \phi_{ext}(r)$. For small f_{ext} the minimum for the total potential $U_{tot}(r)$ will shift to the new equilibrium position; for f_{ext} that is too large, no minimum occurs and there will be no equilibrium point, then the atoms will unbind.

It is often useful to understand how two atoms in a solid will respond to very weak forces, such that the atoms only displaced a very small amount from their equilibrium. We can use the Lennard–Jones interaction potential to model this phenomenon. For a very weak force, the very small shift in the equilibrium allows us to expand the interaction potential by using a Taylor series expansion.

$$\phi(r) = \phi(r_0) + \frac{d\phi}{dr}\bigg|_{r_0} (r-r_0) + \frac{1}{2!}\frac{\partial^2\phi}{\partial r^2}\bigg|_{r_0} (r-r_0)^2$$

$$+ \frac{1}{3!}\frac{\partial^3\phi}{\partial r^3}\bigg|_{r_0} (r-r_0)^3 + \cdots \tag{6.12}$$

$$\approx \phi(r_0) + \frac{1}{2}\frac{\partial^2\phi}{\partial r^2}\bigg|_{r_0} (r-r_0)^2$$

We have used the fact that $d\phi/dr(r_0) = 0$, and we have neglected the higher-order terms in the Taylor expansion. So, we are just left with a *harmonic potential approximation* for the interaction which depends on the square of displacement $u = r - r_0$ from equilibrium. For the Lennard–Jones potential, the curvature is given by equilibrium spacing and binding energy:

$$\frac{\partial^2\phi}{\partial r^2}\bigg|_{r_0} = 72\frac{E_b}{r_0^2} \tag{6.13}$$

The approximation works very well for very small displacements from equilibrium, but it fails as one moves far from the equilibrium. Furthermore, in the presence of a weak external force, the equilibrium point shifts to where $dU_{tot}/dr = 0$; using the expansion (equation 6.12) for interaction potential, this leads to

$$-f_{ext} + \frac{\partial^2\phi}{\partial r^2}\bigg|_{r_0} (r-r_0) = 0 \tag{6.14}$$

or

$$u \equiv r - r_0 = \frac{1}{\dfrac{\partial^2\phi}{\partial r^2}} f_{ext} = \frac{1}{k} f_{ext} \tag{6.15}$$

Thus, we find that the displacement u from equilibrium for small forces f_{ext} is linear with respect to an external force. The linear response for small displacement u is a generic property of most of the materials.

6.0.4   *Dynamic Motion*

In this section, we allow the atoms to move and have a kinetic energy in addition to their potential energy. We assume that the center of mass for our system remains at rest. And we introduce some parameters such as $r = r_2 - r_1$ and center of mass location:

$$r_{cm} = \frac{M_1 r_1 + M_2 r_2}{M_1 + M_2} \tag{6.16}$$

The positions of atoms can be rearranged in terms of r_{cm} and r as

$$\left.\begin{array}{l} r_1 = r_{cm} - \dfrac{M_2}{M_1 + M_2} r \\[2mm] r_2 = r_{cm} + \dfrac{M_1}{M_1 + M_2} r \end{array}\right\} \tag{6.17}$$

If the center of mass is at rest, so $dr/dt = 0$, then the velocities satisfy

$$M_1 \dot{r}_1 = -M_2 \dot{r}_2 \tag{6.18}$$

So, the kinetic energy can be written as

$$\begin{aligned} K &= \frac{1}{2} M_1 \dot{r}_1^2 + \frac{1}{2} M_2 \dot{r}_2^2 \\ &= \frac{1}{2} \mu \dot{r}^2 \end{aligned} \tag{6.19}$$

Using the reduced mass $\mu = M_1 M_2 / (M_1 + M_2)$ and with the momentum $p = \mu \dot{r}$, the kinetic energy is $K = p^2/2\mu$ and the Hamiltonian for the system, $H = K + U$ is then

$$H = \frac{1}{2\mu} p^2 + \phi(r) \tag{6.20}$$

and Hamilton's equations of motion yield

$$\mu \ddot{r} = -\frac{d\phi}{dr}(r) = f(r) \tag{6.21}$$

If we consider very small displacement, $u = r - r_0$, from the equilibrium spacing, then using the Taylor expansion of interaction potential (equation 6.12) reveals

$$\mu \ddot{u} = -\frac{d^2\phi}{dr^2}\bigg|_{r_0} u \tag{6.22}$$

which is the equation of motion for a *simple harmonic oscillator*, and has the general solution of the form

$$u(t) = u_0 \cos(\omega_0 t + \varphi) \tag{6.23}$$

EXAMPLE 6.2	*Natural Resonance Frequency of Argon Atoms Calculated by Lennard–Jones Potential*

For the Lennard–Jones interaction potential with two argon atoms, and with masses $M_1 = M_2 = 2\mu = 6.6 \times 10^{-23}$ g

$$k = \left.\frac{\partial^2 \phi}{\partial r^2}\right|_{r_0} = 52 \text{ meV/Å}^2 = 0.83 \text{ N/m}$$

We know find the natural frequency $\omega_0/2\pi = 0.8$ THz. This is low for a mechanical atomic resonance frequency and is due to the weak van der Waals forces in the argon molecule. The typical resonance frequency for covalently or ionically bonded atoms is on the order of 10 THz [14].

where the resonance frequency ω_0 is given by

$$\omega_0 = \sqrt{\frac{1}{\mu}\frac{\partial^2 \phi}{\partial r^2}} \qquad (6.24)$$

and amplitude u_0 and phase φ are defined by the initial conditions.

6.1   THREE-ATOM CHAIN

Now we add a third atom to the system, and restrict the motion to one dimension, and also assume that all the atoms are identical, with mass M. Then we consider equal and opposite external forces applied at two ends as shown in **Figure 6.2**. The two end atoms 1 and 3 will be displaced symmetrically and the middle atom does not move

$$u_1 = r_2 - r_1 - r_0 = f_{ext}/k$$

$$= \frac{1}{\frac{\partial^2 \phi}{\partial r^2}} f_{ext} \qquad (6.25)$$

with an identical expression for u_3.

Moreover, we now briefly describe the dynamic behavior of this three-atom chain as we have done for the two-atom system. Again, we assume that the center of mass remains at rest, so

FIG. 6.2	*Schematic representation of three-atom chain connected with springs.*

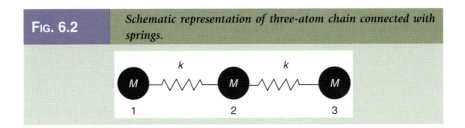

$$\dot{u}_1 + \dot{u}_2 + \dot{u}_3 = 0 \qquad (6.26)$$

The Hamiltonian for the system is defined by

$$H = K + U = \frac{1}{2M}(p_1^2 + p_2^2 + p_3^2) + \frac{k}{2}(u_1 - u_2)^2 + \frac{k}{2}(u_2 - u_3)^2 \qquad (6.27)$$

The corresponding equations of motion are

$$\left. \begin{array}{l} M\ddot{u}_1 = k(u_2 - u_1) \\ M\ddot{u}_2 = k(u_1 - 2u_2 + u_3) \\ M\ddot{u}_3 = k(u_2 - u_3) \end{array} \right\} \qquad (6.28)$$

The solutions where all degrees of freedom have the same harmonic and time dependence possess the form

$$u_n = A_n e^{-i\omega t} \quad \text{where } n = 1, 2, 3 \qquad (6.29)$$

Inserting these solutions to equation (6.28), we find the linear system of equations

$$\left. \begin{array}{l} -M\omega^2 A_1 = k(A_2 - A_1) \\ -M\omega^2 A_2 = k(A_1 - 2A_2 + A_3) \\ -M\omega^2 A_3 = k(A_2 - A_3) \end{array} \right\} \qquad (6.30)$$

We can rearrange and write the above system of equations in a matrix form, where $\omega_0 = \left(\frac{k}{M}\right)^{\frac{1}{2}}$ is the frequency of the system. So, the system of equations can be written as an eigenvalue–eigenvector equation

$$\left(\frac{\omega}{\omega_0}\right)^2 \begin{bmatrix} A_1 \\ A_2 \\ A_3 \end{bmatrix} = \begin{bmatrix} 1 & -1 & 0 \\ -1 & 2 & -1 \\ 0 & -1 & 1 \end{bmatrix} \begin{bmatrix} A_1 \\ A_2 \\ A_3 \end{bmatrix} \qquad (6.31)$$

This approach for the three-atom chain can be expanded to the *n*-atom chain in a similar way and readers are advised to go to Ref. [14] for a more detailed understanding.

6.2 LATTICE MECHANICS

A dynamic solution for the system of particles that comprise a stable lattice structure can be described by the Lagrangian formalism, which is briefly mentioned as follows. An arbitrary system of *n* particles can be described by the Lagrange equations of motion [15]

$$\frac{d}{dt}\left(\frac{\partial T}{\partial \dot{q}_j}\right) - \frac{\partial T}{\partial q_j} = Q_j, \quad j = 1, 2, \ldots, s \qquad (6.32)$$

where

    s is the number of independent degrees of freedom

    Q_j are the generalized forces.

In order to write the lattice equation of motion, we require the Lagrangian in terms of the particle displacement vectors, u_n. The kinetic energy of lattice particles can be written in a matrix form

$$T = \frac{1}{2}\sum_n \dot{u}_n^T M \dot{u}_n \tag{6.33}$$

where M is a diagonal matrix of particle masses written for one unit cell. This gives the lattice Lagrangian in the general form of

$$L = \frac{1}{2}\sum_n \dot{u}_n^T M \dot{u}_n - U(u) \tag{6.34}$$

Here, U is the lattice potential energy, and u is a formal notation for all the displacement vectors, u_n in a given lattice. Substituting equation (6.34) into the Lagrange equation of motion (6.32), written for one unit cell with s degrees of freedom

$$\frac{d}{dt}\frac{\partial L}{\partial \dot{u}_{n,s}} - \frac{\partial L}{\partial u_{n,s}} = f_{n,s}^{\text{ext}}, \quad s = 1, 2, ..., S \tag{6.35}$$

We obtain the lattice equation of motion

$$M\ddot{u}_n + \frac{\partial U}{\partial u_n} = f_n^{\text{ext}}, \quad n = (n_1, n_2, n_3) \tag{6.36}$$

where f_n^{ext} is the vector of external forces exerted on the current unit cell n. In equation (6.35) $u_{n,s}$ and $f_{n,s}^{\text{ext}}$ are individual components of the vector u_n and f_n^{ext}, respectively. One special form of the lattice equation of motion can be obtained within the *harmonic approximation*, which consists of expanding the potential energy in Taylor series about the equilibrium

$$U(u) = U(0) + \sum_{n,s} \frac{\partial U}{\partial u_{n,s}}\bigg|_0 u_{n,s} + \frac{1}{2}\sum_{n,n',s,s'} \frac{\partial^2 U}{\partial u_{n,s}\partial u_{n',s'}}\bigg|_0 u_{n,s}u_{n',s'} + \cdots \tag{6.37}$$

and ignoring the second-order and higher terms. The zero-order term can be ignored in equation (6.37), since a constant shift of the Lagrangian does not alter the equations of motion. Therefore, the harmonic approximation leads to the lattice Lagrangian

$$L = \frac{1}{2}\sum_n \dot{u}_n^T M \dot{u}_n + \frac{1}{2}\sum_{n,n'} u_n^T K_{n-n'} u_n \tag{6.38}$$

where the superscript T indicates a transposed vector and the equation of motion reaches the final matrix form

$$M\ddot{u}_n(t) - \sum_n K_{n-n'} u_{n'}(t) = f_n^{\text{ext}}(t) \tag{6.39}$$

Here, K are the lattice stiffness matrices, which represent linear elastic properties of the lattice structure. These matrices are composed of the atomic force constants, according to

$$K_{n-n'} = \left.\frac{\partial^2 U(u)}{\partial u_n \partial u_{n'}}\right|_{u=0} \tag{6.40}$$

For one-, two-, or three-dimensional lattice with nearest unit cell interactions, there are up to 3, 9, or 27 nontrivial K-matrices, respectively. The lattice equation of motion (6.39) is an ordinary differential equation with a finite difference. The effective solution of these equations involve transform methods, such as Fourier and Laplace transforms, which convert the operation of convolution into an ordinary matrix multiplication in the transform domain.

In order to derive the K-matrices, the potential energy of the atomic interactions needs to be written for one associate cell only; this gives

$$U = V(u_n - u_{n-1} + \rho) + V(u_{n+1} - u_n + \rho) \tag{6.41}$$

where V is a pair-wise potential such as the Lennard–Jones with defined equilibrium distance ρ. According to equation (6.40), the K-matrices yield

$$K_{-1} = \frac{-\partial^2 U}{\partial u_n \partial u_{n+1}} = k, \quad K_0 = \frac{-\partial^2 U}{\partial u_n^2} = -2k, \quad K_1 = \frac{-\partial^2 U}{\partial u_n \partial u_{n-1}} = k \tag{6.42}$$

Here, the linear force constant k depends on the parameters of the potential

$$k = 36 \cdot 2^{\frac{2}{3}} \varepsilon / \sigma^2 \,(\text{Lennard–Jones}) \tag{6.43}$$

EXAMPLE 6.3   *Monatomic Chain Lattice (Equation of Motion)*

The monatomic chain lattice, **Figure 6.3**, is the simplest example that can be described with previous understandings. We assume that the interaction exists only between the nearest lattice atoms; this can be viewed so that the cutoff radius of the potential function is close to 1.5 ρ, where ρ is the equilibrium distance for the potential.

FIG. 6.3   *Monatomic chain lattice interacting with its nearest neighbors.*

Substituting the *K*-matrices into the general equation of motion (6.39) yields

$$M\ddot{u}_n - k(u_{n-1} - 2u_n + u_{n+1}) = f_n^{\text{ext}} \tag{6.44}$$

6.3 STRESS AND STRAIN

In the first part of this chapter, we have introduced a brief description of solids from the atomic point of view and then broadened our knowledge to two-dimensional objects which can be expanded to three-dimensional objects. Classical dynamics assumes that all objects are infinitely rigid. But, the bonds between the atoms that make up the solid are not infinitely rigid, and many thermodynamic properties of insulators are based on the flexibility of these bonds. Therefore, rigid body dynamics is an approximation to the actual motion of solids.

Now, we consider a displacement vector $u(r)$ at a given position and then add a small Δr to the previous position; therefore, the new position vector is at $u(r + \Delta r)$. Then the relative displacement vector is defined as $\Delta u = u(r + \Delta r) - u(r)$. We can also expand each component u_i of displacement $u(r + \Delta r)$ in a vector Taylor series about the point r

$$u_i(r + \Delta r) = u_i(r) + \sum_{j=1}^{3} \frac{\partial u_i}{\partial x_j} \Delta r_j + \text{Higher-order terms} \tag{6.45}$$

Dropping the higher-order terms, we can write the relative displacement, Δu, with respect to the initial position in component form as

$$\Delta u_i = \sum_{j=1}^{3} \frac{\partial u_i}{\partial x_j} \Delta r_j \tag{6.46}$$

These derivatives can be assembled into a tensor D, where components are given by $D_{ij} = \partial u_i / \partial x_j$

$$D = \begin{pmatrix} \dfrac{\partial u_1}{\partial x_1} & \dfrac{\partial u_1}{\partial x_2} & \dfrac{\partial u_1}{\partial x_3} \\[2mm] \dfrac{\partial u_2}{\partial x_1} & \dfrac{\partial u_2}{\partial x_2} & \dfrac{\partial u_2}{\partial x_3} \\[2mm] \dfrac{\partial u_3}{\partial x_1} & \dfrac{\partial u_3}{\partial x_2} & \dfrac{\partial u_3}{\partial x_3} \end{pmatrix} \tag{6.47}$$

The tensor D is the basis for the definition of the strain tensor, which can be split into two species: symmetric and antisymmetric

$$D = S + \Omega$$

$$S = \frac{1}{2}(\nabla u + (\nabla u)^T) \tag{6.48}$$

$$\Omega = \frac{1}{2}(\nabla u - (\nabla u)^T)$$

where T here indicates the transpose. The tensor S is the strain tensor, and Ω is the rotation tensor. Written out in component form, the strain tensor is

$$S_{ij} = \frac{1}{2}\left(\frac{\partial u_i}{\partial x_j} + \frac{\partial u_j}{\partial x_i}\right) \tag{6.49}$$

and in tabular form

$$S = \begin{pmatrix} \dfrac{\partial u_1}{\partial x_1} & \dfrac{1}{2}\left(\dfrac{\partial u_1}{\partial x_2} + \dfrac{\partial u_2}{\partial x_1}\right) & \dfrac{1}{2}\left(\dfrac{\partial u_1}{\partial x_3} + \dfrac{\partial u_3}{\partial x_1}\right) \\[2ex] \dfrac{1}{2}\left(\dfrac{\partial u_1}{\partial x_2} + \dfrac{\partial u_2}{\partial x_1}\right) & \dfrac{\partial u_2}{\partial x_2} & \dfrac{1}{2}\left(\dfrac{\partial u_2}{\partial x_3} + \dfrac{\partial u_3}{\partial x_2}\right) \\[2ex] \dfrac{1}{2}\left(\dfrac{\partial u_1}{\partial x_3} + \dfrac{\partial u_3}{\partial x_1}\right) & \dfrac{1}{2}\left(\dfrac{\partial u_2}{\partial x_3} + \dfrac{\partial u_3}{\partial x_2}\right) & \dfrac{\partial u_3}{\partial x_3} \end{pmatrix} \tag{6.50}$$

The rotation tensor is

$$\Omega_{ij} = \frac{1}{2}\left(\frac{\partial u_i}{\partial x_j} - \frac{\partial u_j}{\partial x_i}\right) \tag{6.51}$$

and in tabular form

$$\Omega = \begin{pmatrix} 0 & \dfrac{1}{2}\left(\dfrac{\partial u_1}{\partial x_2} - \dfrac{\partial u_2}{\partial x_1}\right) & \dfrac{1}{2}\left(\dfrac{\partial u_1}{\partial x_3} - \dfrac{\partial u_3}{\partial x_1}\right) \\[2ex] \dfrac{1}{2}\left(\dfrac{\partial u_2}{\partial x_1} - \dfrac{\partial u_1}{\partial x_2}\right) & 0 & \dfrac{1}{2}\left(\dfrac{\partial u_2}{\partial x_3} - \dfrac{\partial u_3}{\partial x_2}\right) \\[2ex] \dfrac{1}{2}\left(\dfrac{\partial u_3}{\partial x_1} - \dfrac{\partial u_1}{\partial x_3}\right) & \dfrac{1}{2}\left(\dfrac{\partial u_3}{\partial x_2} - \dfrac{\partial u_2}{\partial x_3}\right) & 0 \end{pmatrix} \tag{6.52}$$

The tensor Ω gives the local rotation of volume element at the point r. Note that Ω does not represent the body rotation of the solid as a whole. Both S and Ω behave as second-order tensors, which under a rotation of the coordinate system given by transformation R, the strain tensor S', in the new coordinate system, have components which are given in terms of the old coordinate system

$$S'_{ij} = \sum_{m=1}^{3}\sum_{n=1}^{3} R_{im}S_{mn}R_{nj} \tag{6.53}$$

The displacement vector that gives the strain shown in figure is

$$u = (0,\, sx_1,\, 0) \tag{6.54}$$

The strain tensor is then given by

Pure Shear Strain

We now consider the case of pure shear strain, where the solid is strained as depicted in **Figure 6.4**. The two plane surfaces perpendicular to x_1 are shifted by Δl along x_2.

FIG. 6.4   *Schematic representation of pure shear strain applied to a cube.*

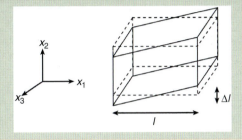

$$S = \begin{pmatrix} 0 & \frac{s}{2} & 0 \\ \frac{s}{2} & 0 & 0 \\ 0 & 0 & 0 \end{pmatrix} \tag{6.55}$$

Thus, the rotation tensor is given by

$$\Omega = \begin{pmatrix} 0 & -\frac{s}{2} & 0 \\ -\frac{s}{2} & 0 & 0 \\ 0 & 0 & 0 \end{pmatrix} \tag{6.56}$$

The motion depicted in **Figure 6.4** consists of both shear strain and rotation tensor. A slight change in displacement vector $u = (-\frac{1}{2}sx_2, \frac{1}{2}sx_1, 0)$ can make the rotation tensor to be zero while the strain tensor remains unchanged. The readers are encouraged to try to sketch what shape the cube would be under this displacement vector.

6.4   LINEAR ELASTICITY RELATIONS

We now focus on linear elastic materials and their responses to any external forces. The most general linear elastic relation that relates stress T to strain S is defined as below

$$T_{ij} = \sum_{k=1}^{3} \sum_{l=1}^{3} \alpha_{ijkl} S_{kl} \tag{6.57}$$

where the elastic moduli α_{ijkl} are constants. We assume that the material is homogeneous and the elastic moduli are independent of position within the solid. There are $3^4 = 81$ distinct components in the elastic moduli tensor. But, due to the symmetry of strain and stress tensors, this reduces to 36 components. Using the definition above, we can rewrite the elasticity relation in a 6×6 elasticity matrix

$$\tau_i = \sum_{j=1}^{6} c_{ij}\varepsilon_j \tag{6.58}$$

The constants $c_{ij}\varepsilon_j$ are called the elastic stiffness coefficients, which have the dimensions of N/m^2 in SI units.

6.4.1 Orthotropic and Isotropic Materials

A material that has a mirror symmetry about all three planes, $x_1 - x_2, x_2 - x_3$, and $x_1 - x_3$, is known as an orthotropic material, and these special types of symmetry can reduce the number of independent components of elastic moduli to 12 distinct constants as shown below:

$$c = \begin{bmatrix} c_{11} & c_{12} & c_{13} & 0 & 0 & 0 \\ c_{21} & c_{22} & c_{23} & 0 & 0 & 0 \\ c_{31} & c_{32} & c_{33} & 0 & 0 & 0 \\ 0 & 0 & 0 & c_{44} & 0 & 0 \\ 0 & 0 & 0 & 0 & c_{55} & 0 \\ 0 & 0 & 0 & 0 & 0 & c_{66} \end{bmatrix} \tag{6.59}$$

Moreover, in the case of isotropic materials, three mirror symmetries for independent 90° rotations about the axes are added and also an additional 45° rotation about one axis is found, which can reduce the number of independent components to two distinct elastic constants

$$c = \begin{bmatrix} c_{11} & c_{12} & c_{12} & 0 & 0 & 0 \\ c_{12} & c_{11} & c_{12} & 0 & 0 & 0 \\ c_{12} & c_{12} & c_{11} & 0 & 0 & 0 \\ 0 & 0 & 0 & c_{44} & 0 & 0 \\ 0 & 0 & 0 & 0 & c_{44} & 0 \\ 0 & 0 & 0 & 0 & 0 & c_{44} \end{bmatrix} \tag{6.60}$$

where $c_{44} = (c_{11} - c_{12})/2$, the two elastic constants required to describe isotropic materials are traditionally referred to as Lame constants (1798–1870) λ and μ.

$$\begin{aligned} \lambda &= c_{12} \\ \mu &= c_{44} = (c_{11} - c_{12})/2 \end{aligned} \tag{6.61}$$

6.4.2 Crystalline Materials

Many materials are neither isotropic nor orthotropic, but display more restrictive symmetries. Here, we cite the symmetries and the corresponding form of elastic

Material	a_0 (Å)	ρ (g/cm³)	c_{11} (GPa)	c_{12} (GPa)	c_{44} (GPa)
Silicon (Si)	5.4307	2.330	165	64	79.2
Germanium (Ge)	5.6200	5.323	129	48	67.1
Diamond (C)	3.5670	3.515	1040	170	550

TABLE 6.4 *Lattice Constants, Density, and Elastic Moduli of Diamond Structure Materials*

stiffness matrix for a few common semiconductor materials. A more comprehensive and detailed discussion can be found in Landolt-Bbornstein [16].

Diamond Structure. Crystals formed from elements in group IV of the periodic table typically have diamond structures. The diamond structure consists of two interpenetrating face-centered cubic lattices with origins offset by ¼ of the cubic diagonal. The form of elastic stiffness coefficients is that given for isotropic materials (equation 6.60) except that there are three independent coefficients. The lattice constants, density, and elastic moduli for silicon, germanium, and diamond are tabulated in **Table 6.4** [16].

Wurtzite Structure. Gallium nitride (GaN), aluminum nitride (AlN), indium nitride (InN), and zinc oxide (ZnO) are binary compounds having what is commonly called a wurtzite structure. The wurtzite structure is formed from two interpenetrating hexagonal close-packed lattices, each lattice filled with one type of atom. The form for the elastic constants matrix is given by equation (6.62).

$$c = \begin{bmatrix} c_{11} & c_{12} & c_{13} & 0 & 0 & 0 \\ c_{12} & c_{11} & c_{13} & 0 & 0 & 0 \\ c_{13} & c_{13} & c_{33} & 0 & 0 & 0 \\ 0 & 0 & 0 & c_{44} & 0 & 0 \\ 0 & 0 & 0 & 0 & c_{44} & 0 \\ 0 & 0 & 0 & 0 & 0 & \frac{c_{11}-c_{12}}{2} \end{bmatrix} \tag{6.62}$$

The lattice constants, density, and elastic stiffness for GaN, AlN, InN, and ZnO are shown in **Table 6.5** [16,17].

Material	a_0 (Å)	c_0 (Å)	ρ (g/cm³)	c_{11} (GPa)	c_{33} (GPa)	c_{44} (GPa)	c_{12} (GPa)	c_{13} (GPa)
GaN	3.189	5.185	6.095	374	379	101	106	70
AlN	3.112	4.982	3.255	345	395	118	125	120
InN	3.540	5.705	6.880	190	182	10	104	121
ZnO	3.249	5.207	5.675	209	218	44.1	120	104

TABLE 6.5 *Materials Characteristics of Crystals with Wurtzite Structure*

6.5   MOLECULAR DYNAMICS

Molecular dynamics is a simulation technique in which the time evolution of a set of interacting particles is obtained by integrating the equations of motion, which are derived from Newton's equations of motion

$$m_i \ddot{r}_i = m_i a_i = -\nabla_i U + F_i, \quad i = 1, 2, ..., n \tag{6.63}$$

applied to each atom i in a system containing n atoms. Here, m_i is the atomic mass, $\nabla_i U$ the first derivative (gradient) of potential energy, and F_i the force acting on atom i due to the interaction with other atoms. In order to solve the second-order differential equations of motion (6.63), several algorithms have been introduced. Here, we just mention these which are mostly applied in nanosystems.

6.5.1   *Verlet Algorithms*

As a direct solution to the second-order differential equations (6.63), the Verlet algorithm, a time-integration method, is widely used to solve the equations of motion. This method uses the current position r_i and acceleration a_i as well as the previous position r_{i-1} of an atom to derive the position r_{i+1} for the next time step in the following manner

$$r_{i+1} = 2r_i - r_{i-1} + a_i \Delta t^2 \tag{6.64}$$

with

$$r_i = r(t_i), \quad a_i = \frac{F(t_i)}{m_i}, \quad t_i = i\Delta t, \qquad i = 0, 1, 2, ..., N \tag{6.65}$$

where the interaction forces F_i and acceleration a_i have to be calculated for each particle according to the following equations

$$F_i = -\nabla_{r_i} U(r_1, ..., r_N)$$

$$a_i = -\frac{1}{m_i} \nabla_{r_i} U(r_1, ..., r_N) \tag{6.66}$$

Since the velocities do not appear directly they can be obtained by applying the central difference method

$$v_i = \frac{r_{i+1} - r_{i-1}}{2\Delta t} \tag{6.67}$$

The time step here is denoted by Δt. Since the atomic vibrations are on the order of approximately 100 fs, a time step smaller than that is required, typically of 2 fs. Several modifications have been proposed to improve the numerical precision of the basic Verlet algorithm [18]. One of these is the so-called half-step leap-frog scheme [19]:

$$v_{i+1/2} = v_{i-1/2} + \Delta t\, a_i$$
$$r_{i+1} = r_i + \Delta t\, v_{i+1/2} \tag{6.68}$$

Here the current velocities have to be calculated from the mid-step values

$$v_i = \frac{1}{2}[v_{i-1/2} + v_{i+1/2}] \tag{6.69}$$

Another derivative, the velocity Verlet algorithm [20], works without a mid-step at the cost of additional storage for a_i

$$r_{i+1} = r_i + v_i\,\Delta t + \frac{1}{2}a_i\,\Delta t^2$$
$$v_{i+1} = v_i + \frac{1}{2}[a_i + a_{i+1}]\Delta t \tag{6.70}$$

There are further derivatives [18], but basically all Verlet methods produce the same error and generate identical position trajectories. So, there seems to be no need to implement a more complicated Verlet algorithm than is given by equation (6.70).

6.5.2   *Nordsieck/Gear Predictor–Corrector Methods*

Nordsieck [21] and Gear [22] developed an integration technique on the basis of Taylor expansions of the positions, velocities, accelerations, and further derivatives

$$r(t + \Delta t) = r(t) + v(t)\Delta t + \frac{1}{2}a(t)\Delta t^2 + \cdots + \frac{1}{k!}q_k(t)\Delta t^k$$
$$v(t + \Delta t) = v(t) + a(t)\Delta t + \frac{1}{2}q_3(t)\Delta t^2 + \cdots + \frac{1}{(k-1)!}q_k(t)\Delta t^{k-1}$$
$$a(t + \Delta t) = a(t) + q_3(t)\Delta t + \frac{1}{2}q_4(t)\Delta t^2 + \cdots + \frac{1}{(k-2)!}q_k(t)\Delta t^{k-2} \tag{6.71}$$
$$q_i(t + \Delta t) = q_i(t) + q_{i+1}(t)\Delta t + \frac{1}{2}q_{i+2}(t)\Delta t^2 + \cdots + \frac{1}{(k-i)!}q_k(t)\Delta t^{k-i}, \quad i = 3, 4, \ldots,$$

where

$$q_k(t) = \frac{\partial^k}{\partial t^k}r(t) \tag{6.72}$$

Now for the position $r^{(0)}$ and its scaled derivatives $r^{(k)}$ with

$$r^{(0)} = r, \quad r^{(1)} = v\Delta t, \quad r^{(2)} = \frac{1}{2}a\,\Delta t^2, \quad r^{(k)} = \frac{1}{k!}q_k\Delta t^k \tag{6.73}$$

a simple Taylor series predictor becomes

$$\begin{bmatrix} r_{i+1}^{\sim(0)} \\ r_{i+1}^{\sim(1)} \\ r_{i+1}^{\sim(2)} \\ \vdots \end{bmatrix} = P \begin{bmatrix} r_i^{(0)} \\ r_i^{(1)} \\ r_i^{(2)} \\ \vdots \end{bmatrix} \tag{6.74}$$

where P is the Pascal triangle matrix with the binomial coefficients in its columns

$$P = \begin{bmatrix} 1 & 1 & 1 & 1 & 1 & 1 & \cdots \\ 0 & 1 & 2 & 3 & 4 & 5 & \cdots \\ 0 & 0 & 1 & 3 & 6 & 10 & \cdots \\ 0 & 0 & 0 & 1 & 4 & 10 & \cdots \\ \vdots & \vdots & \vdots & \vdots & \vdots & \vdots & \ddots \end{bmatrix} \tag{6.75}$$

The predictor does not generate the exact values for the position and its derivatives. But, with help of the predicted position $r_{i+1}^{\sim(0)}$ the forces of the time step $i+1$ can be calculated, and therefore the correct accelerations, a_{i+1}. The comparison with the predicted accelerations $r_{i+1}^{\sim(2)}$ from equation (6.74) gives a measure of the error corresponding to the predictor step:

$$\varepsilon_{i+1} = \frac{1}{2} a_{i+1} \Delta t^2 - r_{i+1}^{\sim(2)} \tag{6.76}$$

Then this error is used to improve the predicted values in a corrector step which has this form

$$\begin{bmatrix} r_{i+1}^{(0)} \\ r_{i+1}^{(1)} \\ r_{i+1}^{(2)} \\ \vdots \end{bmatrix} = \begin{bmatrix} r_{i+1}^{\sim(0)} \\ r_{i+1}^{\sim(1)} \\ r_{i+1}^{\sim(2)} \\ \vdots \end{bmatrix} + \begin{bmatrix} c_0 \\ c_1 \\ c_2 \\ \vdots \end{bmatrix} \varepsilon_{i+1} \tag{6.77}$$

Usually the Nordsieck/Gear algorithm works with $i = 3, ..., 8$ values for which the corrector vector can be found. If accuracy and long periods are not important for the simulations, then the Verlet algorithms have to be preferred. But, for high accuracy problems or long-time simulations the 6-value Nordsieck/Gear predictor–corrector yields better results, though at the cost of decreased step sizes. From equation (6.66), it becomes obvious that the problem of modeling a material is essentially that of finding the potential U, which reproduces the behavior of the material under the simulation conditions. Depending on the origin of the potential, there are three different MD techniques: empirical, tight-binding, and first principles.

The empirical methods employ classical potentials, which can be given by different techniques, for example, the dependence of the energy on the nuclei position can be extracted from the first principle description. Another choice is to fit the potential to experimental data. The simplest form of the many-body potential is in the form of a sum of pairwise terms, with the energy of a pair only depending on their relative distance, r_{ij}

$$U(r_1, ..., r_N) = \frac{1}{2} \sum_{i \neq j} \varphi(r_{ij}) \tag{6.78}$$

Unfortunately, the types of materials that can be realistically modeled using this approach are limited to the noble gases, where electronic bonding is absent and atoms are interacting through the weak van der Waals forces.

The potential for metals and other materials must incorporate the quantum mechanical effect of bond weakening, a consequence of the Pauli principle [23]. Several schemes were developed based on the analytical form

$$U = \frac{1}{2}\sum_{i \neq j} \varphi(r_{ij}) + \sum_{i} \phi(n_i) \qquad (6.79)$$

As before, φ is a two-body interaction potential part whereas ϕ is a function giving the energy of atom as a function of its coordination n_i. In general, the classical potentials for metals and semiconductors are designed from the start with a cutoff radius, which limits the interaction to only the nearest neighbor atoms. If an abrupt truncation in the term of a step function is employed, then the energy and its derivatives are not continuous functions of atomic coordinates, which can disrupt a minimization process or lead to the unwanted effects in a dynamic simulation. In order to resolve this issue, a smoothing function can be introduced which tapers the interaction to zero at a given distance. For example, the Brenner potential for carbon has incorporated a switch type function as described below [24]

$$U(r) = \begin{cases} 1 & r < r_1 \\ \dfrac{1}{2}\left\{1 + \cos\left[\dfrac{\pi(r - r_1)}{r_2 - r_1}\right]\right\}, & r_1 \leq r \leq r_2 \\ 0 & r > r_2 \end{cases} \qquad (6.80)$$

This has the property of leaving the interaction unchanged for distances less than the inner cutoff distance, $r_1 = 0.17\,\text{nm}$, and decreases to zero at the second cutoff, $r_2 = 0.2\,\text{nm}$. Additionally, the first derivative is continuous on the full range, which avoids problems in minimization and dynamic simulations.

All methods discussed above assume that the electronic system is in the ground state and follows the nuclear motion. This approximation is valid in most cases, but there are some physical situations, where this approximation is no longer appropriate. For instance, response of matter under intense laser pulses or behavior of materials in the plasma state where some of the species are in the excited or ionized states cannot be described using basic MD. In these cases, more complicated MD has to be considered. However, these physical phenomena are beyond the scope of nanomechanics discussed in this chapter.

6.5.3 *Molecular Dynamics Applications*

In this section, we review applications of the Newtonian equations of motion to some typical MD simulations in the field of nanomechanics and nanomaterials.

Inelasticity and Failure of Gold Nanowires. Nanowires are found to have great potential as structural reinforcements, as elements in electronic circuitry, and in many other applications [25,26]. The examples depicted here are MD simulations of the tensile failure of gold nanowires [27]. The wire size was initially 16 nm in length with a square cross section of length 2.588 nm. The wire was first relaxed to a minimum energy configuration with free boundaries everywhere, and then

FIG. 6.5   *Snapshots of the deformation of the 2.588 nm wire at a strain rate of 3×10^9 s^{-1} using the EAM potential. Atoms are colored according to their value of potential energy.*

Source: S. M. Foiles, M. L. Baskes, and M. S. Daw, *Physical Review B,* 33(12), 7893–7991 (1986).

thermally equilibrated at a fixed length to 300 K. Finally, a ramp velocity was applied to the nanowire ranging from zero at one end to a maximum value at the loading end. So, one end of the nanowire was fixed while the other end was elongated at a constant velocity at each time step corresponding to an applied strain rate of 3.82×10^9 s^{-1}.

The gold nanowire depicted in **Figure 6.5** resembles the same failure mechanisms as a macroscopic tensile specimen, such as necking and yielding. However, one very interesting quality of the gold nanowire is its incredible ductility, which is manifested in the elongation of extremely thin nanobridges, as seen in the last snapshots in **Figure 6.5**.

Thermal Stability of Nanosystems.   Mechanical stability of nanosystems due to temperature change is an example of material properties that can be altered extremely when we go from the macroscopic to the microscopic realm. Thermal stability and melting temperature of macroscopic systems are well defined and usually are stable up to the melting point. But in contrast to macroscopic systems, the melting point of nanosystems depends on the number of particles and is also a function of the shape of the system. For example, the melting temperature of macroscopic aluminum (Al) is about 933 K and the structure is stable close to the melting point. But, this is not the case for Al systems of nanometer size. Reith and Schommers [28] have performed a MD simulation on the basis of the Schommers pair potential for an Al nanosystem. The thermal stability of a three-dimensional object (F-shape) standing on the surface is studied. The structures are usually unstable far below the melting point and even dissolve. In **Figure 6.6**, a three-dimensional F-shaped structure consisting of aluminum atoms is shown. **Figure 6.6a** corresponds to the initial configuration. It corresponds to the crystalline structure of aluminum at zero absolute temperature.

FIG. 6.6	*A MD simulation of a nanosystem with the three-dimensional shape resting on a substrate. Both the nanostructure and substrate consist of aluminum.*

(a)   (b)

(c)   (d)

5 nm

Source: M. Reith, W. Schommers, and S. Baskouts, *Mod. Phys. Lett. B*, 14, 621 (2000).

The time step used in this calculation is 5×10^{-15} s; the particle number n of the nanosystem is $n = 1660$ (without substrate). After 2000 time steps the system has reached a temperature of 250 K (**Fig. 6.6b**), also after 6000 time steps (**Fig. 6.6c**) the temperature is 270 K, and the nanosystem remains at this temperature for a further step, **Figure 6.6d**. Though the melting point of Al is 933 K, it can be clearly seen from **Figure 6.6d** that this configuration is structurally disturbed already at 270 K, which is significantly below the melting temperature. Furthermore, it is typical for the behavior of nanostructures that a tiny change in the initial condition can lead to different final shapes as depicted in **Figure 6.7**. In the case of **Figure 6.7a**, the temperature reaches 400 K after 10^4 time steps. In **Figure 6.7b**, the temperature is 500 K after 5000 time steps. So, as we can see

FIG. 6.7	*The effect of initial condition variations on the final shape of the nanostructures. (a) The temperature reaches 400 K after 10^4 time steps and (b) the temperature is 500 K after 5000 time steps.*

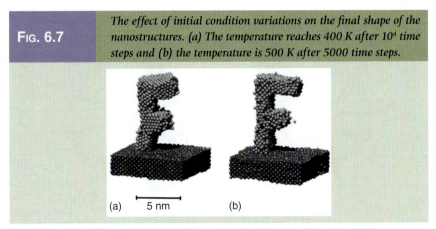

(a)   5 nm   (b)

Source: M. Reith, W. Schommers, and S. Baskouts, *Mod. Phys. Lett. B*, 14, 621 (2000).

from the figure, the time steps have an important effect on the final shape of the nanosystem. The system in **Figure 6.7b** is at a higher temperature but still is more stable than that in **Figure 6.7a** because it has a smaller number of time steps.

In conclusion, specific material properties of nanosystems may differ essentially from the corresponding properties of macroscopic systems. The thermal behavior of nanosystems is a complex function of the particle number, outer shape, and many other parameters.

6.5.4 *Nanomachines*

Nanomachines are defined as systems of at least two different materials with movable parts. In the macroscopic world, the design of a simple bearing and axle is not difficult. Both bearing and axle may be made of the same material and if the diameters are nearly the same, the axle is still movable due to the presence of a thin film of lubricant [8]. But, if we consider the same situation at the nanometer level, both parts will stick together forever. Furthermore, there are still other difficulties. Due to the atomic structure, smooth and sharp surfaces are not possible at the nanometer size.

Rotating parts are of particular focus in connection with nanomachines. The first design setup is to create a static structure and then velocity vectors, according to revolution speed, which are added to each atom before calculating the MD. **Figure 6.8** illustrates an example of a nanowheel of krypton. Here, the structure disintegrates because the centrifugal forces are too strong [8].

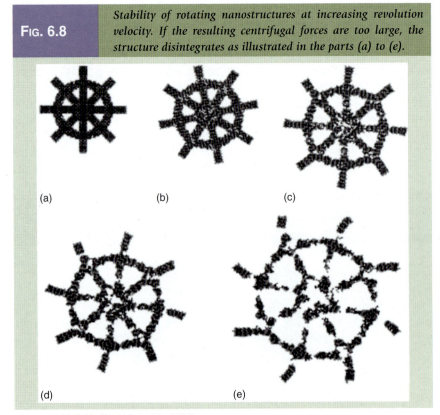

Fig. 6.8	*Stability of rotating nanostructures at increasing revolution velocity. If the resulting centrifugal forces are too large, the structure disintegrates as illustrated in the parts (a) to (e).*

(a)　　　　　　(b)　　　　　　(c)

(d)　　　　　　　　　(e)

Source: M. Reith, World Scientific (2003).

FIG. 6.9 — *Nanoturbine MD simulations: velocities of (a) 5×10^{10} revolutions per second and (b) 10^{11} revolutions per second.*

(a)

(b)

10 nm

Source: M. Reith, World Scientific (2003).

Moreover, a more complicated example of a nanomachine and its integrity is depicted in **Figure 6.9** [8]. This is a model of a nanoturbine that consists of two bearings and an axle standing on a substrate. As you can see from the figure, the fit of axle and bearing is rather loose. However, MD studies have shown that such a nanoturbine can remain stable up to 5×10^{10} revolutions per second, but it ruptures at 10^{11} revolutions per second [8].

6.5.5   *Wear at the Nanometer Level*

Within the frame of MD systems, friction in the macroscopic domain is not defined. At the microscopic MD level, the forces are derived as quantities which are dependent on the structural configuration (particle position) but not on the particle velocities. Therefore, a force which is proportional to the velocity cannot be introduced at the microscopic level and thus a friction constant in the macroscopic sense is not definable as a constant in the microscopic world. At this level, wear is described by complex processes. To understand this phenomenon, an example is provided here. **Figure 6.10** shows a MD model for a spinning wheel moving towards a thin film [8]. The wheel rotates at about 10^{12} revolutions per second and is kept at a constant temperature of 300 K and its diameter is approximately 10 nm.

When the wheel approaches the surface and contacts the surface, friction effects emerge and, depending on the magnitude of the vertical force, the wheel may be destroyed as depicted in the figure. Therefore, the friction at the microscopic level is a complex process and may not be described by one constant only. The wear at the microscale level depends on the specific structure of the surface and additionally on the shape and motion of the wheel.

Fig. 6.10	*A MD simulation for friction of spinning wheel moving towards the surface of a thin film. Both the film and the wheel consist of aluminum atoms. It can be seen that friction effects emerge in the form of a complex process when the wheel contacts the surface.*

Source:　M. Reith, World Scientific (2003).

6.6　Structure and Mechanical Properties of Carbon Nanotubes

Carbon nanotubes (CNTs) have been synthesized for a long time from the action of a catalyst over the gaseous environment originating from the thermal decomposition of hydrocarbons. There are several other techniques such as sputtering or evaporation that can be used to produce CNTs, but thermal decomposition is the most widely used method to synthesize CNTs. The accidental discovery of single-wall carbon nanotubes (SWCNTs) by Iijima et al. [29] and Bethune et al. [30] at NEC in 1990 made a huge impact on science and technology. CNTs consist of honeycomb lattices of carbon rolled into cylinders nanometer in diameter and micrometer in length. CNTs have incredible properties.

- 1/6 the weight of steel; 5 times its Young's modulus; 100 times its tensile strength; 6 orders of magnitude higher in electrical conductivity than copper and can be strained up to 15% without fracture
- Metallic or semiconductor depending on its chirality
- 10 times smaller than the smallest silicon tip in a scanning tunneling microscope (STM)

6.6.1　Structure of Carbon Nanotubes

To simplify the understanding of a carbon nanotube's shape, a perfect graphene sheet (single atomic layer made of a hexagonal display of sp^2 hybridized carbon

| FIG. 6.11 | *Schematic representation of graphite sheet and CNTs that are formed by rolling.* |

atoms) is depicted in **Figure 6.11**, and is then rolled into three different cylinders [31].

Though carbon atoms are involved in aromatic rings, the $C = C$ bond angles are no longer planar as they should ideally be. This means that the hybridization of carbon atoms is no longer pure sp^2 but involves some percentage of the sp^3 character. This angle change results in a more active surface of the carbon nanotube compared to its graphitic sheet. As illustrated in **Figure 6.11**, there are three different ways to roll a graphitic sheet into a SWCNT; some of the resulting nanotubes enable symmetry mirrors both parallel and perpendicular to the nanotube axis (armchair and zigzag configurations). The other way of forming CNTs is shown in **Figure 6.11** (chiral nanotube), which does not have any symmetry mirrors that were mentioned above. The various ways to roll graphene into tubes are therefore mathematically defined as the vector of helicity C_h and the angle of helicity θ as defined below:

$$C_h = na_1 + ma_2 \tag{6.81}$$

with

$$a_1 = \frac{a\sqrt{3}}{2}x + \frac{a}{2}y \quad \text{and} \quad a_2 = \frac{a\sqrt{3}}{2}x - \frac{a}{2}y \tag{6.82}$$

where $a = 0.246\,\text{nm}$

| **FIG. 6.12** | *Image of two neighboring chiral single-wall carbon nanotubes as produced by high-resolution tunneling microscopy.* |

Source: Image courtesy of Prof. Yazdani, University of Illinois at Urbana. With permission.

and

$$\cos\theta = \frac{2n + m}{2\sqrt{n^2 + m^2 + nm}} \qquad (6.83)$$

where n and m are integers of the vector of helicity considering the unit vectors a_1 and a_2. For example the zigzag type nanotubes have an angle of helicity of 0° (as shown in **Figure 6.11**), whereas armchair type CNTs have an angle of helicity of 30°. It is clear from **Figure 6.11** that having the vector of helicity perpendicular to any of the three overall C=C bond directions will produce the zigzag type nanotube $(n, 0)$, while having the vector of helicity parallel to one of the three C=C bond directions will provide armchair type structures (n, n). Because of the six-fold symmetry of the graphene sheet, the angle of helicity for the chiral (n, m) nanotubes is such that $0 < \theta < 30°$ [32]. **Figure 6.12** indicates two examples of what chiral single-wall carbon nanotubes look like, as seen by means of a high-resolution STM.

6.6.2 *Mechanical Properties of Carbon Nanotubes*

Considerable progress has been made in investigating the mechanical properties of SWCNTs and multiwall carbon nanotubes (MWCNTs). The theoretical predictions and experimental measurements are very promising and motivate further studies of future applications for lightweight and high-strength products. CNTs are very strong through the three-folded bonding of the curved graphene sheet, which is stronger than in diamond due to their differences in C–C bond length (0.142 versus 0.154 nm for graphene and diamond, respectively) [32].

The tensile strength of SWCNTs can be 20 times that of steel [33] and has actually been measured equal to about 45 GPa [34]. One important application of CNTs is in composite materials, which are reinforced by the introduction of SWCNTs or MWCNTs. Mechanical properties of CNTs can be measured by atomic force microscopy (AFM); the tests involve measurements of deformations under controlled forces in lateral [35] or normal [36] directions, and a tensile test can be done by incorporating two AFM tips at both ends of CNTs [37]. In order to apply continuum mechanics, we need to define the thickness of the nanotube or a graphene sheet for the continuum beam approximation.

Most researchers working in this field use a value of 0.34 nm close to the interlayer separation in graphite sheets. One important point in the measurements of elastic properties, such as Young's modulus and shear modulus, is the huge amount of precision that must be done in order to minimize any further errors. Even a very small error in the measurement of deflection enters into equations of beam deflection as d^4, which leads to a large uncertainty in the results [32]. Several techniques have been employed to measure the mechanical properties of nanotubes including bar, beam, and shell models. The bar model has been used in the experiment by Lourie and Wagner [38], in which the compressive response was calculated using micro-Raman spectroscopy. The reported values for Young's modulus are 2.3–3.6 TPa for SWCNTs, and 1.7–2.4 TPa for MWCNTs. Another test was performed by Yu et al. [33,37] using tensile loading of SWCNTs and MWCNTs. The Young's modulus obtained ranged from 320 to 1470 GPa for SWCNTs and from 270 to 950 GPa for MWCNTs.

A cantilever beam model was used in an experiment conducted by Wong et al. [35] in which individual MWCNTs were bent using an AFM tip. After fitting the measured data points to the analytical solution, a Young's modulus of 1.28 ± 0.59 TPa was obtained. In another model simulated by Salvetat et al. [39,40], the deflection of a simple-supported beam was modeled and a Young's modulus of ~1 TPa for MWCNTs, grown by arc discharge, was reported, whereas CNTs grown by the catalytic decomposition of a hydrocarbon gas showed a modulus of 1–2 orders of magnitude smaller.

There are numerous experimental measurements [41–45] and theoretical calculations [46–54] for mechanical properties of CNTs that due to the limited space in this chapter are just mentioned in the reference section. Readers are strongly recommended to review these references to get a more comprehensive understanding of how the mechanical properties of CNTs are obtained.

6.7 NANOMECHANICAL MEASUREMENT TECHNIQUES AND APPLICATIONS

Experimental tools are used to measure the mechanical properties of materials at the nanometer level. The most prominent measurement technique that evaluates the mechanical properties is scanning probe microscopy, which is a broad term for different techniques that can be performed in both contact and noncontact modes. Scanning probe microscopes (SPM) are tools that scan a sharp probe tip across the specimen and provide nanoscale information about the sample. One of the very common SPM and STM was invented by Binnig and Rohrer in the 1980s. A sharp tip is positioned very close to the surface of the specimen in an STM instrument and the tip is scanned through the sample.

One major drawback of the STM is that it is unable to image nonconducting surfaces, since electronic current is involved in the measurements. Another measurement technique, AFM, was proposed by Binnig et al. [55], not long after the invention of STM. The probe was allowed to contact the surface directly and the resulting probe deflection can be measured by optical methods to generate an image. In this technique, the interaction force between the tip and the sample surface is recorded. AFM provides information about the mechanical properties of the surface, but not of electronic properties.

In some AFM instruments, the static mode is used to obtain force–distance curves, which then can be analyzed to produce the local, reduced Young's modulus. In practice, these values are not measured directly and the raw data must be calibrated and converted to produce real displacement and force data. Although absolute values of mechanical properties are difficult to measure, relative measurements are still promising. Another difficulty is that the movement of the tip on the sample can occur not only vertically but also along the cantilever axis, which limits the validity of applied contact mechanics [32]. Because of these limitations, the static AFM method is not widely used especially in the case of stiff surfaces. Using acoustic vibrations of the AFM sample surface, we can access local elastic and inelastic properties of the specimen.

6.7.1 AFM Measurements: Mechanical Properties of CNTs

Mechanical properties of solids such as elasticity, inelasticity, and plasticity are usually measured on a macroscopic scale for materials. But recently new techniques such as AFM and STM have been developed to study these properties on the nanoscale region. CNTs are one of the promising nanomaterials that are widely used nowadays. Mechanical measurements on CNTs performed with the AFM have confirmed theoretical expectations [51] of their superior mechanical properties. As we mentioned previously, there are several ways to measure the mechanical properties of CNTs. One method is to bend the CNT by the AFM tip and calculate the Young's modulus from the force–displacement curve. Another method, which is a more direct measurement of the elastic properties of CNTs, has been performed on MWCNTs [33] and SWCNT ropes [37] under axial strain. Two AFM tips were used to hold MWCNTs or SWCNT ropes and then tensile strain was applied to them portioned inside the scanning electron microscope (SEM). The AFM tips were integrated with different cantilevers, one rigid with a spring constant of 20 N/m and the other compliant with a spring constant of 0.1 N/m, as depicted in **Figure 6.13**.

FIG. 6.13   *Tensile loading of individual MWCNTs. (a) SEM image of a MWCNT attached between two AFM tips and (b) higher magnification of image in (a).*

(a) — 10 μm   (b) — 2 μm

Source: M. F. Yu, O. Lourie, M. J. Dyer, K. Moloni, T. F. Kelley, and R. S. Ruoff, *Science*, 287, 637–640 (2000). With permission.

FIG. 6.14	(a) Schematic representation of two AFM tips straining the MWCNTs and (b) Plot of stress versus strain for different individual MWCNTs.

Source: M. F. Yu, O. Lourie, M. J. Dyer, K. Moloni, T. F. Kelley, and R. S. Ruoff, *Science*, 287, 637–640 (2000). With permission.

The rigid cantilever was driven using a linear piezomotor. On the other end, the compliant lever was bent due to the applied tensile force. The force is calculated as $F = kd$ where k is the spring constant of the flexible AFM cantilever and d its displacement in the vertical direction. The strain of the nanotube is $\delta L/L$, which is shown in **Figure 6.14**. The stress–strain curve was derived from the force–displacement data points and then the Young's modulus was obtained. For this setup configuration, Young's modulus values ranging from 270 to 950 GPa were found.

All these measurements for the mechanical properties of CNTs that were mentioned in this and previous sections are summarized in **Table 6.6**.

TABLE 6.6	*Summary of the Mechanical Properties of CNTs Measured Using SPM Methods*				
Young's modulus E (GPa)	**Tensile strength σ (GPa)**	**Shear modulus G (GPa)**	**Nanotube type**	**Method of testing**	**Ref.**
1300 ± 600	—	—	MWCNTs arc grown	Lateral bending	[35]
1000 ± 600	—	—	SWCNTs	Normal bending	[36]
1000 ± 600	—	~1	SWCNT ropes	Normal bending	[40]
1020	30	—	SWCNT ropes	Tensile loading	[37]
870 ± 400	—	—	MWCNT arc grown	Normal bending	[36]
270–950	11–63	—	MWCNTs arc grown	Tensile loading	[33]
400	—	—	SWCNT rope	Normal bending	[56]
12 ± 6	—	—	MWCNT catalytic	Normal bending	[36]
—	45 ± 7	—	SWCNT rope	Lateral bending	[34]

It should be noted that the absolute values of mechanical properties have large uncertainties due to the huge influence of the precision of tube diameters and lengths used in the experiments.

6.7.2 Nanoindentation

Nanoindentation is defined as a tool to measure mechanical properties of materials at the nanoscale. This method was first developed in the early 1980s evolving from traditional Vicker hardness testing. Nanoindenter tips with various shapes are used in this technique to analyze the resistance of materials to an external force. A schematic representation of continuous load–displacement data is shown in **Figure 6.15**, which consists of three different regions called loading, holding, and unloading regions. Some important quantities that are indicated in this figure are P_{max} (peak load), h_{max} (maximum displacement), and $S = dP/dh$ (the slope of unloading curve at maximum indentation depth).

It is noticeable that S has the dimensions of force per unit length, which is known as elastic contact stiffness. The elastic modulus of materials can be derived using the following equations:

$$E_r = \frac{(\sqrt{\pi} \cdot S)}{2\beta\sqrt{A}} \tag{6.84}$$

where
  β is a constant that depends only on the geometry of the indenter
  A is the projected contact area

FIG. 6.15	*Schematic representation of the nanoindentation load–displacement curve.*

E_r is called the reduced modulus, a parameter that considers the effect of a non-rigid indenter on the load–displacement behavior. The elastic modulus E of the test material is calculated using the following expression:

$$\frac{1}{E_r} = \frac{1-v^2}{E} + \frac{1-v_i^2}{E_i} \tag{6.85}$$

where
  v is the Poisson's ratio of the test material
  E_i and v_i are the elastic modulus and Poisson's ratio of the indenter, respectively.

It may seem inappropriate that we have to know the material's Poisson's ratio in order to calculate its modulus. But, even a rough number, say $v = 0.25 \pm 0.1$, produces only about a 5% error in the calculation of elastic properties of most materials. For different types of nanoindenter tips, a specific β constant is used for calculations. For indenters with square cross sections such as the Vickers pyramid, $\beta = 1.012$; for triangular cross sections such as the Berkovich and the cubic corner indenters, $\beta = 1.034$.

An additional mechanical property that is usually obtained from nanoindentation is hardness H as defined below:

$$H = \frac{P}{A} \tag{6.86}$$

where P is the applied load and A is the projected contact area of indentation at load P as a function of contact depth. Therefore, we must obtain the elastic contact stiffness (S) and projected contact area (A) in order to be able to derive the elastic modulus and hardness. There are two distinct methods that measure the stiffness and projected contact area, that is, the continuous stiffness measurement (CSM) method and the unloading stiffness measurement (USM) method.

In the CSM method, a small oscillating force is applied either to the sample or indenter during the indentation period and the contact stiffness value is calculated from the displacement response against the depth of indentation. Once the stiffness of contact S is defined, the elastic modulus and hardness can be obtained using equations (6.84)–(6.86).

In the USM method, there are several ways to calculate the contact stiffness. The method of Oliver and Pharr [57] is the most widely used. According to the method, data analysis procedure begins by fitting the load–displacement data acquired during unloading to the power–law relation:

$$P = B(h - h_f)^m \tag{6.87}$$

where
  P is the applied load to the test surface
  h is the resulting penetration
  B and m are empirically determined fitting parameters
  h_f is the final displacement after complete unloading

The contact stiffness S is then obtained using

$$S = \frac{dP}{dh} = Bm(h_{max} - h_f)^{m-1} \tag{6.88}$$

It is worth noting that in nanoindentation the projected contact area is not obtained by optical imaging. Rather it is computed as a function of the contact depth h_c, applying an empirical relationship given by

$$A = f(h_c) = C_0 h_c^2 + C_1 h_c + C_2 \sqrt{h_c} \tag{6.89}$$

The constants C_0, C_1, and C_2 are determined prior to the experiment by indenting a sample of known properties such as fused silica. The contact depth h_c is different from the total penetration depth h, and is estimated using

$$h_c = h - \varepsilon \frac{P}{S} \tag{6.90}$$

where ε is a constant that depends only on the geometry of the indenter. For cones, $\varepsilon = 0.72$ and for spheres, $\varepsilon = 0.75$. There is empirical justification for using $\varepsilon = 0.75$ for Berkovich and Vickers tips as well.

6.8　NANO-MICROELECTROMECHANICAL SYSTEMS (NEMS/MEMS)

MEMS are based on devices that have a characteristic length of 1 mm or less but more than 100 nm and consist of electrical and mechanical parts. The similar term commonly used in Europe is microsystem technology (MST) and in Japan it is called micromachines.

NEMS refer to nanoscopic devices that have a characteristic length of 100 nm or less and combine the electrical and mechanical parts. **Figure 6.16** compares MEMS and NEMS in the size range and their applications. These devices (NEMS/MEMS) are referred to as intelligent miniaturized systems that consist of sensing, processing, and/or actuating parts and combine mechanical and electrical components and operations.

6.8.1　MEMS Fabrication Techniques

In this section, we will discuss various MEMS fabrication techniques that are commonly used to fabricate different microdevices such as sensors and actuators. Micromachining is one of the most important steps in fabricating MEMS, and is described in more detail in the following sections.

Bulk Micromachining.　Bulk micromachining is the oldest MEMS technology available, which is currently by far the most successful in manufacturing MEMS devices, such as pressure sensors and ink-jet printer heads. The basic concept behind micromachining is to remove materials selectively. This can lead to the

FIG. 6.16   *Comparison of NEMS and MEMS dimensionality.*

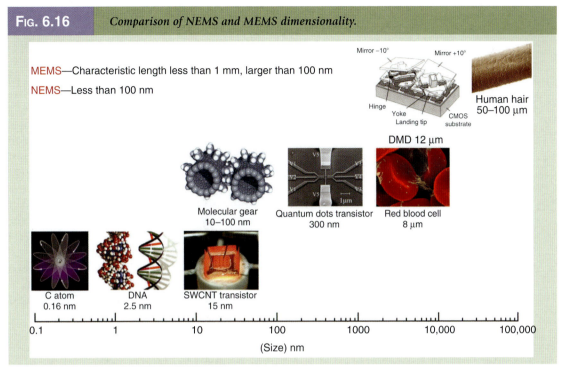

MEMS—Characteristic length less than 1 mm, larger than 100 nm

NEMS—Less than 100 nm

Mirror –10°   Mirror +10°

Hinge   Yoke   CMOS substrate   Landing tip

Human hair 50–100 µm

DMD 12 µm

Molecular gear 10–100 nm   Quantum dots transistor 300 nm   Red blood cell 8 µm

C atom 0.16 nm   DNA 2.5 nm   SWCNT transistor 15 nm

0.1   1   10   100   1000   10,000   100,000

(Size) nm

Source: A. P. Graham, G. S. Duesberg, R. Seidel, M. Liebau, E. Unger, F. Kreupl, and W. Hoenlein, *Diamond and Related Materials*, 13, 1296–1300 (2004); W. G. van der Wiel, S. De Franceschi, J. M. Elzerman, T. Fujisawa, S. Tarucha, and L. P. Kouwenhoven, *Reviews of Modern Physics*, 75, 1–22 (2003); Texas Instruments DLP Products, Plano, TX, http://www.dlp.com. With permission.

creation of different micromechanical components such as beams, plates, or membranes that are used to fabricate movable parts in MEMS devices.

The use of anisotropic wet etchants to remove silicon is the basic micromachining technique used in the semiconductor industry. Backside wet etching is used in order to create movable parts, as depicted in **Figure 6.17**.

One of the basic ways to control the etching process is to remove the etchant from the substrate at a specified time or thickness. But this technique is unable to produce features that are thinner than 20 µm. Therefore, new etch stop techniques

FIG. 6.17   *Wet anisotropic silicon back side etching.*

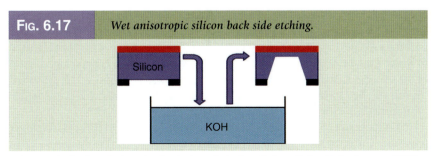

Silicon

KOH

Source: B. Bhushan, *Springer handbook of nanotechnology*, Springer-Verlag, Berlin, Germany (2004). With permission.

were introduced to have more precise results. Doped regions and electrochemical bias are used to slow down or stop the etch process, and hence create more controllable structures. An alternative method to wet etching is dry etching, which was developed more recently in order to fabricate high-aspect-ratio structures and to design processes for anisotropic dry etching. The most basic dry bulk micromachining method is based on the front side undercut of microstructures using XeF_2 vapor phase etching [61]. This technique has its limitations due to the fact that it is an isotropic etching. Another technique that uses both isotropic and anisotropic dry etch is single-crystal-reactive etching and metallization (SCREAM) [62], which can create structures with suspended components. Most of the dry etching techniques are plasma based and have several advantages compared to wet etching.

The main advantage of dry etching is the ability to produce smaller undercuts and also facilitates the creation of high-aspect-ratio vertical structures. One of the most important dry etching techniques is reactive ion etching (RIE), which combines both physical and chemical processes. In this method, the surface of the material is activated by the incident ions from the plasma and then reactive species react with the material producing faster etching in the vertical direction. Another technique, deep reactive ion etching (DRIE), which is based on ion etching, was recently introduced. In this method, the passivation and etching steps are performed in a two-step cycle sequence. The DRIE technique is capable of achieving an aspect ratio of 30:1 and silicon etching rates of 2–3 μm/min [32].

Surface Micromachining.   Another important new fabrication technique that is widely used to create movable parts on top of a silicon substrate is surface micromachining [63]. This technique is based on the deposition of thin films of polysilicon and other structural films on top of a sacrificial layer that is subsequently removed by etching. This process creates a movable micromechanical structure that can be integrated with on-chip electronics to produce MEMS devices. One of the major advantages of this technique is that extremely small sizes can be obtained.

The basic surface micromachining process is depicted in **Figure 6.18** [32]. As can be seen from the figure, the process starts with depositing a sacrificial layer on top of the silicon substrate and then the sacrificial layer is patterned. Furthermore, the structural layer is deposited and then patterned, which would be anchored to the substrate through the opening created during the previous step. Finally, the sacrificial layer is removed by etching and a movable part is obtained.

One of the most widely used techniques in industry is called multiuser MEMS processes (MUMPs). **Figures 6.19–6.26** show the sequential steps that are used to make a micromotor [64,65]. A thin film of silicon nitride is deposited on top of the silicon substrate and then a blanket layer of polysilicon (poly 0) is deposited on top of the silicon nitride layer; the wafer is then coated by UV-sensitive photoresist, which is shown in **Figure 6.19**.

In the next step, the photoresist is patterned using UV light through the mask level one. The photoresist in the exposed area is removed and the patterned area is left behind, as shown in **Figure 6.20**.

FIG. 6.18 *Surface micromachining fabrication process.*

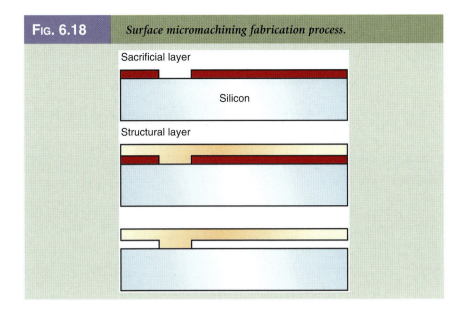

The unwanted part of (poly 0) is etched using RIE and, after this step, the photoresist layer is totally etched away by putting the structure in a solvent bath (**Fig. 6.21**). This method of patterning the wafers by photoresist and then etching the remaining photoresist is a commonly used method in MUMPs.

The first sacrificial layer is deposited on top of the previous layers and then is patterned lithographically using its photo mask. The unwanted oxide layer is removed applying RIE and the remaining photoresist is stripped, see **Figure 6.22**.

FIG. 6.19 *Schematic of micromotor deposition steps, step 1.*

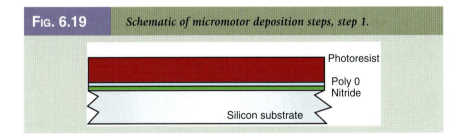

FIG. 6.20 *Schematic of micromotor deposition steps, step 2.*

FIG. 6.21 *Schematic of micromotor deposition steps, step 3.*

FIG. 6.22 *Schematic of micromotor deposition steps, step 4.*

Another layer of polysilicon (poly 1) is coated on top of the existing layers and then lithographically patterned using its specific mask. In the next step, another oxide layer (sacrificial layer) is deposited on top of the layers as depicted in **Figure 6.23**.

The second oxide layer is patterned twice to obtain a contact through the poly 1 and also the substrate layers. The wafer is patterned again and the unwanted oxide layer is removed using an appropriate etchant, as shown in **Figure 6.24**.

Next, the last polysilicon layer (poly 2) is deposited on top of the structure and is patterned through its photoresist mask and the metal layer is deposited on top of poly 2, as shown in **Figure 6.25**.

FIG. 6.23 *Schematic of micromotor deposition steps, step 5.*

FIG. 6.24 *Schematic of micromotor deposition steps, step 6.*

FIG. 6.25   *Schematic of micromotor deposition steps, step 7.*

FIG. 6.26   *Schematic of micromotor deposition steps, final step.*

Finally, the structures are released by immersing the chip into 49% HF solution. The (poly 1) "rotor" can be seen around the fixed (poly 2) hub, which can be moved electrostatically as illustrated in **Figure 6.26**.

6.8.2 NEMS Fabrication Techniques

Recent developments in nanotechnology have enabled the production of nanoscale devices such as NEMS and other instruments at the nanometer level. These improvements create new possibilities for next-generation communication and electromechanical devices. For example, as the dimensions shrink to the nanometer level in sensor technology, the sensitivity and accuracy of devices are improved.

NEMS are produced by nanomachining in a top-down approach (from large to small sizes) and bottom-up approach (from small to large scale) [66–70]. The top-down approach is based on fabrication methods that produce nanostructures (similar to micromachining used in MEMS technology) including electron beam lithography and STM writing which removes an atom at a time from the surface. The bottom-up approach includes chemical synthesis, thin film deposition techniques, molecular beam epitaxy (MBE), and various plasma techniques that could be physical or chemical.

As discussed in the previous section, UV and x-rays are two major lithographic techniques that are commonly used in MEMS fabrications but UV lithography does not provide nanometer resolution, and x-ray masks are difficult to make, while lithography by x-rays has its own safety problems. Therefore, more precise techniques such as *e*-beam lithography are mostly used in patterning the

FIG. 6.27   *Schematic illustration of nanoimprint fabrication.*

Source: B. Bhushan, *Springer handbook of nanotechnology*, Springer-Verlag, Berlin, Germany (2004). With permission.

layers [71]. This method uses an electron beam to expose an electron sensitive resist such as polymethylmethacrylate (PMMA) dissolved in trichlorobenzene (positive) or polychloromethylstyrene (negative) [72]. The *e*-beam gun is usually a component part in SEM and TEM instruments, but can obtain resolutions of only 10 nm. Another interesting technique which is being developed is called nano-imprint [73]. This technique uses an *e*-beam to fabricate hard material (mold) to stamp and deform polymeric resist and then these steps are followed by a RIE step to transfer the pattern into the substrate, as depicted in **Figure 6.27**. This technique is very economical, since a large number of nanostructures can be fabricated by a single stamp.

Another useful technique that can be applied in the fabrication of nanostructures is SPM. Electrons emitted from a biased SPM tip can be used in order to expose a resist (the same way as *e*-beam) [74]. Different instruments such as constant current STM, noncontact AFM, and AFM with constant current can be used to obtain the lithographic patterns.

6.8.3   NEMS/MEMS Motion Dynamics

In this section, the theory of beams and cantilevers and their motion is briefly described. Here, we assume that the beam is straight, untwisted, and has a constant cross section, as shown in **Figure 6.28** [75]. Moreover, the beam thickness (d) and width (w) are small compared to its length (l), which reduces the system to a one-dimensional problem. Furthermore, the normal stresses in lateral directions are negligible [76]. So with these assumptions, the only remaining normal stress σ_z can be written as

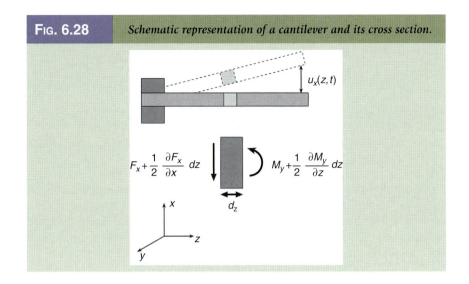

FIG. 6.28   *Schematic representation of a cantilever and its cross section.*

$$\sigma_z = kx \tag{6.91}$$

where k is constant and $x = 0$ is in the center of the beam. With no external momentum applied, the total bending momentum is defined as

$$M = M_y = \int_A x\sigma_z dA = k\int_A x^2 dA, \quad I_y = \int_A x^2 dA \tag{6.92}$$

From previous equations, one can see that $k = M_y/I_y$ and the cross section stress is derived as

$$\sigma_z = E\varepsilon_z = \frac{M_y x}{I_y} \tag{6.93}$$

where E is Young's modulus. If the deflection in the x direction is small, the second derivative of the deflection is the inverse of radius of the curvature r:

$$\frac{\partial^2 u_x(z,t)}{\partial z^2} \approx r^{-1} \tag{6.94}$$

Also the strain can be obtained by $\varepsilon = -x/r$. Therefore, we obtain the Euler–Bernoulli beam theory:

$$M_y = -EI_y \frac{\partial^2 u_x(z,t)}{\partial z^2} \tag{6.95}$$

As we know, the total momentum has to be zero and the equation of motion becomes

$$m\frac{d^2 u_x(z,t)}{dt^2} = \sum F_{int} = \frac{\partial f_x}{\partial z} dz \tag{6.96}$$

With the mass of the beam given by $m = \rho A dz$ (where ρ is the density of the beam and A is its cross section), the equation of motion becomes

$$\rho A \frac{d^2 u_x(z, t)}{dt^2} = -\frac{\partial^2 M_y}{\partial z^2} \tag{6.97}$$

And the final equation of motion can be written as

$$\rho A \frac{d^2 u_x(z, t)}{dt^2} + EI_y \frac{d^4 u_x(z, t)}{dz^4} = 0 \tag{6.98}$$

This linear fourth-order differential equation can be solved using separation of variables [76–78]. Here, we are not interested in the complete solution; instead we want to find the natural resonant frequency of the beam. This can be obtained by using a Fourier transformation [75]. After solving the Fourier transformation, one can obtain the natural resonant frequency as follows:

$$\omega_i = \frac{\beta_i^2}{l^2} \sqrt{\frac{EI_y}{\rho A}} \tag{6.99}$$

Here, β_i is constant, which depends on the boundary conditions used to solve the differential equation of the cantilever (clamped–free cantilever) or the beam (clamped–clamped beam). The moment of inertia of a beam with different cross sections can be found in the appropriate tables.

6.8.4 MEMS Devices and Applications

MEMS devices are inherently small and thus can offer such attractive characteristics as reduced size, weight, and less power dissipation compared to macroscopic systems. Moreover, these miniaturized systems can obtain better precision and also work at higher speeds than their macroscopic-sized components. In this section, we will review some of the devices that are commercially available nowadays and demonstrate their applications.

Pressure Sensor.   Pressure sensors are commercial devices that are widely used in various industrial and biomedical applications. These sensors can be based on four different mechanisms such as piezoelectric, piezoresistive, capacitive, and resonant sensing [32]. In this section two different types of piezoresistive and capacitive pressure sensors are mentioned.

A pressure that is applied to the sensor will deform the silicon band structure, thus altering the resistivity of the material. The device consists of a silicon diaphragm suspended over a vacuum cavity to form a pressure sensor. An external pressure applied to the diaphragm would introduce stress on the sensing resistors, resulting in a resistance change based on the external pressure. These resistors are temperature dependent and consume direct current (DC), which make them less attractive compared to capacitive pressure sensors.

Capacitive pressure sensors are more attractive than piezoresistive pressure sensors since they are temperature independent and do not consume any DC power and are very stable over time. Moreover, complementary metal oxide semiconductor

FIG. 6.29   *Cross section of a capacitive pressure sensor.*

(CMOS) microelectronic circuits can be easily interfaced with these sensors to improve the overall performance of the system. A schematic representation of a capacitive pressure sensor is depicted in **Figure 6.29**.

This device consists of a suspended silicon diaphragm over a vacuum cavity. As the pressure is applied to the outer surface of the device, the diaphragm deflects towards the cavity, resulting in an increase in the capacitance value. This change in capacitance is monitored with electronic devices, and the sensor capacitance values are converted to an output voltage corresponding to the diaphragm position. This voltage is used to generate a feedback signal to the top electrode to maintain the diaphragm at its nominal position.

Digital Displays.   The digital micromirror device (DMD) was introduced by Texas Instruments (TI) in 1987. The DMD is an integral part of TIs digital light processing (DLP) technology. This DMD technology can achieve higher resolution and brightness, and produce lightweight projection displays that generate images with higher fidelity and stability [79,80]. This technology has various applications such as computer projectors, high definition television (HDTV), and movie projectors (DLP cinema) [80]. A DMD consists of up to 2.07 million aluminum micromirrors with a typical area of 16×16 μm as illustrated in **Figure 6.30**.

FIG. 6.30   *Two pixels (two micromirrors) of a DMD chip.*

Source: L. J. Hornbeck, *MRS Bulletin*, 26, 325 (2001). With permission.

These micromirrors switch forward and backward thousands of times per second by electrostatic attraction.

Each micromirror is allowed to rotate by ±10°, corresponding to "on" or "off" position due to the electrostatic force. Light is reflected from any mirror that is on and passes through a projection lens and creates an image on a screen. The remaining light that is coming from off-mirrors would be reflected away from the projection lens to an absorber. The three DMD chips are used for projecting red, green, and blue colors in color displays. The DMD is fabricated using surface micromachining technology. Three layers of aluminum thin films are deposited and patterned to form the mirror and its suspension structure.

Polymeric material is used as a sacrificial layer and is removed by plasma etching to produce the suspended micromirror structures. This fabrication process is also compatible with CMOS fabrication, which enables the manufacturer to achieve a higher yield and lower cost compared to the fabrication of these DMD devices without any underlying circuit technology.

6.8.5 NEMS Devices and Applications

NEMS devices are fabricated using either top-down or bottom-up approaches. The top-down approach incorporates more precise lithographic techniques, such as electron beam lithography, which give better accuracy compared to the MEMS lithographic techniques, and the device can be fabricated by employing etching and lithography steps on the bulk material. This top-down approach is widely used to manufacture NEMS devices because it is based on existing technology that is used to produce integrated circuits (ICs) and MEMS structures. The only significant difference between the nanoscale and microscale processing steps is the patterning method used for various features. In contrast, the bottom-up approach follows the same path that nature constructs objects, by assembling atoms and molecules on top of one another.

In this section, the goal is to introduce some of the early developments in the field on NEMS devices and their applications. The first generation of NEMS devices are based on freestanding nanomechanical beams, oscillators, and tethered plates, which are fabricated using bulk and surface nanomachining processes [32]. Here, the processing steps of a nanomechanical beam of silicon are briefly described. Carr et al. [81] fabricated a submicron clamped–clamped mechanical beam and suspended plates with a nanometer tether. The processing steps are shown in **Figure 6.31**.

In the first step, PMMA is deposited on top of the silicon-on-insulator (SOI) substrate, and then the PMMA layer is patterned using electron beam lithography. An aluminum film is then deposited and patterned into the silicon etch mask. The nanomechanical beam is then patterned by RIE and the underlying SiO_2 layer is removed by immersing the device in a hydrofluoric acid solution. Applying these processing steps, nanomechanical beams of 7–16 μm in length, 120–200 nm in width, and 50–200 nm in thickness were successfully fabricated [81]. NEMS technology is still in the developmental stages with very limited commercial success. Nevertheless, NEMS devices have been used for precision measurements [82] and for probing the material's properties at the nanometer level [83,84].

FIG. 6.31 — *Cross-sectional schematics of a process to fabricate nanomechanical structures using silicon on insulator substrates.*

Legend: Si, SiO$_2$, Al, PMMA

Source: B. Bhushan, *Springer handbook of nanotechnology*, Springer-Verlag, Berlin, Germany (2004). With permission.

6.9 SUMMARY

In the first part of the chapter, the foundations of nanomechanics were introduced. A simple two-atom chain of a molecule was briefly discussed in order to derive the interaction potentials associated with a specific material and being able to calculate the force generated due to this potential. Also this approach was expanded to three and *n*-atom chains to study the static and dynamic behavior of materials at the nanoscale.

MD, a simulation technique, was reviewed and its applications on different systems reveal that failure mechanisms of materials at the nanoscale can be different from macroscopic failures. The tensile failure of a gold nanowire specimen shows an incredible ductility during its elongation, which is manifested in the elongation of extremely thin atom chains. Furthermore, the thermal stability of a nanosystem was modeled which indicated that these nanosystems are very sensitive to their initial conditions and number of particles used for the simulation.

CNTs exhibit fascinating mechanical properties which, compared to steel, have 1/6 the weight of steel, 5 times its Young's modulus, 100 times its tensile strength, and can be strained up to 15% without fracture. Nanomechanical measurement devices such as the AFM, STM, and nanoindenters can be used to obtain the mechanical properties of material experimentally at the nanometer level.

NEMS/MEMS devices and their applications are briefly mentioned in the last section of this chapter. Their different fabrication techniques were introduced and some commercial products that are based on this technology were reviewed.

Acknowledgments

We are especially grateful to Professor Ivar Reimanis, who has reviewed this chapter and provided valuable feedback. We are also particularly indebted to the following authors who provided some of the figures and plots: M. Reith,

S. M. Foiles, M. L. Baskes, M. S. Daw, W. Schommers, S. Baskouts, M. Meyyappan, Ali Yazdani, M. F. Yu, O. Lourie, M. J. Dyer, K. Moloni, T. F. Kelley, R. S. Ruoff, B. Bhushan, M. D. Ventra, S. Evoy, J. R. Heflin, G. Genta, and L. J. Hornbeck.

References

1. I. M. Torrens, *Interatomic potentials*, Academic Press, New York, London (1972).
2. L. Verlet, Computer experiments on classical fluids. I. thermodynamical properties of Lennard-Jones molecules, *Physical Review*, 159, 98 (1967).
3. R. G. Della Valle and E. Venuti, Quasi harmonic lattice dynamics and molecular dynamics calculations for the Lennard-Jones solids, *Physical Review B*, 58, 1 (1998).
4. T. Halicioglu and G. M. Pound, Calculation of potential energy parameters form crystalline state properties, *Physica Status Solidi A-Applied Research*, 30, 619 (1975).
5. R. A. Buckingham, The classical equation of state of gaseous helium, neon and argon, *Proceedings of the Royal Society of London Series A-Mathematical and Physical Sciences*, 168, 264 (1938).
6. K. E. Drexler, *Nanosystems: Molecular machinery, manufacturing, and computation*, John Wiley & Sons, Inc., New York (1992).
7. J. A. Barker, R. O. Watts, J. K. Lee, T. P. Schafer, and Y. T. Lee, Interatomic potentials for krypton and xenon, *Journal of Chemical Physics*, 61, 8 (1974).
8. M. Reith, *Nano-engineering in science and technology*, World Scientific Publishing Co. Pte. Ltd., Singapore (2003).
9. M. S. Daw and M. I. Baskes, Semiempirical, quantum mechanical calculation of hydrogen embrittlement in metals, *Physical Review Letters*, 50, 1285 (1983).
10. E. Clementi and C. Roetti, Roothaan-hartree-fock atomic wavefunctions, *Atomic Data and Nuclear Data Tables*, 14(3–4), 177–478 (1974).
11. S. M. Foiles, M. L. Baskes, and M. S. Daw, Embedded atom method functions for the fcc metals Cu, Ag, Au, Ni, Pd, Pt, and their alloys, *Physical Review B*, 33(12), 7893–7991 (1986).
12. M. S. Daw, S. M. Foiles, and M. L. Baskes, The embedded atom method: A review of theory and applications, *Materials Science Reports*, 9, 251–310 (1993).
13. N. W. Ashcroft and N. D. Mermin, *Solid state physics*, Saunders College, Philadelphia, PA (1976).
14. A. N. Cleland, *Foundations of nanomechanics*, Springer, Berlin, Germany (2003).
15. H. Goldstein, *Classical mechanics*, Addison-Wesley, Reading, MA (1980).
16. Landolt-Bornstein, *Numerical data and functional relationships in science and technology*, Vol. 11, Springer, Berlin (1979).
17. O. Ambacher, Growth and applications of group-III nitrides, *Journal of Physics D: Applied Physics*, 31, 2653–2710 (1998).
18. M. P. Allen and D. J. Tildesley, *Computer simulation of liquids*, Oxford Science Publications, New York (1990).
19. R. W. Hockney, The potential calculation and some applications, *Methods in Computational Physics*, 9, 136 (1970).
20. W. C. Swope, H. C. Andersen, P. H. Berens, and K. R. Wilson, WA computer simulation method for the calculation of equilibrium constants for the formation of physical clusters of molecules: Application to small water clusters, *Journal of Chemical Physics*, 76, 637 (1982).
21. A. Nordseick, On numerical integration of ordinary differential equations, *Mathematics of Computation*, 16, 22 (1962).
22. C. W. Gear, *Numerical initial value problems in ordinary differential equations*, Englewood Cliffs, NJ, Prentice Hall, NJ (1971).
23. V. M. Harik and M. D. Salas, *Trends in nanoscale mechanics*, Kluwer Academic Publishers, the Netherlands (2003).

24. D. W. Brenner, Empirical potential for hydrocarbons for use in simulating the chemical vapor deposition of diamond films, *Physical Review B*, 42, 9458–9471 (1990).

25. C. M. Lieber, Nanoscale science and technology: Building a big future from small things, *MRS Bulletin*, 28(7), 486–491 (2003).

26. P. Yang, The chemistry and physics of semiconductor nanowires, *MRS Bulletin*, 30(2), 85–91 (2005).

27. H. S. Park and J. A. Zimmerman, Modeling inelasticity and failure in gold nanowires, *Physical Review B*, 72, 054106 (2005).

28. M. Reith, W. Schommers, and S. Baskouts, Thermal stability and specific material properties of nanosystems, *Modern Physics Letters B*, 14, 621 (2000).

29. S. Iijima and T. Ichihashi, Single-shell carbon nanotubes of 1 nm diameter, *Nature*, 363, 603–605 (1993).

30. D. S. Bethune, C. H. Kiang, M. S. de Vries, G. Gorman, R. Savoy, J. Vazquez, and R. Bayers, Cobalt-catalysed growth of carbon nanotubes with single-atomiclayer walls, *Nature*, 363, 605–607 (1993).

31. M. Meyyappan, *Carbon nanotubes: Science and applications*, CRC Press, Boca Raton, FL (2004).

32. B. Bhushan, *Springer handbook of nanotechnology*, Springer-Verlag, Berlin, Germany (2004).

33. M. F. Yu, O. Lourie, M. J. Dyer, K. Moloni, T. F. Kelley, and R. S. Ruoff, Strength and breaking mechanism of multiwalled carbon nanotubes under tensile load, *Science*, 287, 637–640 (2000).

34. D. A. Walters, L. M. Ericson, M. J. Casavant, J. Liu, D. T. Colbert, K. A. Smith, and R. E. Smalley, Elastic strain of freely suspended single-wall carbon nanotube ropes, *Applied Physics Letters*, 74, 3803–3805 (1999).

35. E. W. Wong, P. E. Sheehan, and C. M. Lieber, Nanobeam mechanics: Elasticity, strength and toughness of nanorods and nanotubes, *Science* 277, 1971–1975 (1997).

36. J. P. Salvetat, J. M. Bonard, N. H. Thomson, A. J. Kulik, L. Forró, W. Benoit, and L. Zuppiroli, Mechanical properties of carbon nanotubes, *Applied Physics A-Materials Science and Processing*, 69, 255–260 (1999).

37. M. F. Yu, B. S. Files, S. Arepalli, and R. S. Ruoff, Tensile loading of ropes of single wall carbon nanotubes and their mechanical properties, *Physical Review Letters*, 84, 5552–5555 (2000).

38. O. Lourie and H. D. Wagner, Evaluation of Young's modulus of carbon nanotubes by micro-Raman spectroscopy, *Journal of Materials Research*, 13 (9), 2418–2422 (1998).

39. J. P. Salvetat, A. J. Kulik, J. M. Bonard, G. A. D. Briggs, T. Stockli, K. Metenier, S. Bonnamy, F. Beguin, N. A. Burnham, and L. Forro, Elastic modulus of ordered and disordered multiwalled carbon nanotubes, *Advanced Materials*, 11(2), 161–165 (1999).

40. J. P. Salvetat, G. A. D. Briggs, J. M. Bonard, R. R. Bacsa, A. J. Kulik, T. Stockli, N. A. Burnham, and L. Forro, Elastic and shear moduli of single-walled carbon nanotube ropes, *Physical Review Letters*, 82(5), 944–947 (1999).

41. M. M. J. Treacy, T. W. Ebbesen, and J. M. Gibson, Exceptionally high Young's modulus observed for individual carbon nanotubes, *Nature*, 381(6584), 678–680 (1996).

42. A. Krishnan, E. Dujardin, T. W. Ebbesen, P. N. Yianilos, and M. M. J. Treacy, Young's modulus of single-walled nanotubes, *Physical Review B*, 58(20), 14013–14019 (1998).

43. P. Poncharal, Z. L. Wang, D. Ugarte, and W. A. de Heer, Electrostatic deflections and electromechanical resonances of carbon nanotubes, *Science*, 283(5407), 1513–1516 (1999).

44. M. F. Yu, M. J. Dyer, J. Chen, and K. Bray, Multiprobe nanomanipulation and functional assembly of nanomaterials inside a scanning electron microscope, in: International Conference IEEE-NANO2001, Maui (2001).

45. D. A. Dikin, X. Chen, W. Ding, G. J. Wagner, and R. S. Ruoff, Resonance vibration of amorphous SiO_2 nanowires driven by mechanical or electrical field excitation, *Journal of Applied Physics*, 93, 226 (2003).

46. G. Overney, W. Zhong, and D. Tomanek, Structural rigidity and low-frequency vibrational-modes of long carbon tubules, *Zeitschrift fur Physik D-Atoms Molecules and Clusters*, 27(1), 93–96 (1993).

47. G. G. Tibbetts, Why are carbon filaments tubular, *Journal of Crystal Growth*, 66(3), 632–638 (1984).

48. G. H. Gao, T. Cagin, and W. A. Goddard, Energetics, structure, mechanical and vibrational properties of single-walled carbon nanotubes, *Nanotechnology*, 9(3), 184–191 (1998).

49. B. I. Yakobson, C. J. Brabec, and J. Bernholc, Nanomechanics of carbon tubes: Instabilities beyond linear response, *Physical Review Letters*, 76(14), 2511–2514 (1996).

50. S. Timoshenko and J. Gere, *Theory of elastic stability*, McGraw-Hill, New York (1988).

51. J. P. Lu, Elastic properties of carbon nanotubes and nanoropes, *Physical Review Letters*, 79(7), 1297–1300 (1997).

52. N. Yao and V. Lordi, Young's modulus of single-walled carbon nanotubes, *Journal Applied Physics*, 84(4), 1939–1943 (1998).

53. E. Hernandez, C. Goze, P. Bernier, and A. Rubio, Elastic properties of C and BxCyNz composite nanotubes, *Physical Review Letters*, 80(20), 4502–4505 (1998).

54. X. Zhou, J. J. Zhou, and Z. C. Ou-Yang, Strain energy and Young's modulus of single-wall carbon nanotubes calculated from electronic energy-band theory, *Physical Review B*, 62(20), 13692–13696 (2000).

55. G. Binnig and Heinrich, Scanning tunneling microscopy—from birth to adolescence, *Reviews Modern Physics*, 59, 615 (1987).

56. G.-T. Kim, G. Gu, U. Waizmann, and S. Roth, Simple method to prepare individual suspended nanofibers, *Applied Physics Letters*, 80, 1815–1817 (2002).

57. W. C. Oliver and G. M. Pharr, An improved technique to determining hardness and elastic modulus using load and displacement sensing indentation experiments, *Journal of Materials Research*, 7, 1564–1583 (1992).

58. A. P. Graham, G. S. Duesberg, R. Seidel, M. Liebau, E. Unger, F. Kreupl, and W. Hoenlein, Towards the integration of carbon nanotubes in microelectronics, *Diamond and Related Materials*, 13, 1296–1300 (2004).

59. W. G. van der Wiel, S. De Franceschi, J. M. Elzerman, T. Fujisawa, S. Tarucha, and L. P. Kouwenhoven, Electron transport through double quantum dots, *Reviews of Modern Physics*, 75, 1–22 (2003).

60. Texas Instruments DLP Products, Plano, TX, http://www.dlp.com.

61. B. Eyre, K. S. J. Pister, and W. Gekelman, Multi-axis microcoil sensors in standard CMOS, *Proceedings of the SPIE Conference on Micromachined Devices and Components*, Austin, pp. 183–191 (1995).

62. K. A. Shaw, Z. L. Zhang, and N. C. MacDonnald, SCREAM: A single mask single-crystal silicon process for microelectromechanical structures, *Proceedings of the IEEE Workshop Micro Electro Mechanical Systems*, Fort Lauderdale, pp. 155–160 (1993).

63. J. M. Bustillo and R. S. Muller, Surface micromachining for microelectromechanical systems, *Proceedings of the IEEE*, 86(8), 1552–1574 (1998).

64. MCNC, Make a micromotor, http://mems.mcnc.org/smumps/Mumps.html.

65. http://www.ece.wis.edu/~priasmor/mem.html.

66. B. Bhushan, *Handbook of micro/nanotribology*, 2nd ed., CRC press, Boca Raton, FL (1999).

67. G. Timp, ed., *Nanotechnology*, Springer, New York (1999).

68. E. A. Rietman, *Molecular engineering of nanosystems*, Springer, New York (2001).

69. H. S. Nalwa, ed., *Nanostructured materials and nanotechnology*, Academic press, San Diego, CA (2002).

70. W. A. Goddard, D. W. Brenner, S. E. Lyshevski, and G. J. Iafrate, *Handbook of nanoscience, engineering, and technology*, CRC press, Boca Raton, FL (2003).

71. P. Rai-Choudhury, ed., *Handbook of microlithography, micromachining and microfabrication*, SPIE, Bellingham, WA (1997).

72. L. Ming, C. Bao-qin, Y. Tian-Chun, Q. He, and X. Qiuxia, The sub-micron fabrication technology, *Proceedings of 6th International Conference on Solid-State and Integrated-Circuit Technology*, IEEE, pp. 452–455 (2001).

73. S. Y. Chou, Nano-imprint lithography and lithographically induced self-assembly, *MRS Bulletin*, 26, 512–517 (2001).

74. H. T. Soh, K. W. Guarini, and C. F. Quate, *Scanning probe lithography*, Kluwer, Boston, MA (2001).

75. M. D. Ventra, S. Evoy, and J. R. Heflin, *Introduction to nanoscale science and technology*, Springer, New York (2004).

76. G. Genta, *Vibration of structure and mechanics*, Springer (1999).

77. H. A. C. Tilmans, M. Elwenspoek, and J. H. J. Fluitman, Micro resonant force gauges, *Sensors and Actuators A-Physical*, 30, 35 (1992).

78. A. A. Shabana, *Vibration of discrete and continuous systems*, Springer-Verlag, New York (1997).

79. L. J. Hornbeck, Digital light processing update: Status and future applications, *Proceedings of the Society for Imaging and Photooptical Engineering*, 3634, 158 (Projection Displays V) (1999).

80. L. J. Hornbeck, The DMD™ projection display chip: A MEMS-based technology, *MRS Bulletin*, 26, 325 (2001).

81. D. W. Carr and H. G. Craighead, Fabrication of Nanoelectromechanical systems in single crystal silicon using silicon on insulator substrates and electron beam lithography, *Journal of Vacuum Science and Technology B*, 15, 2760–2763 (1997).

82. A. N. Cleland and M. L. Roukes, A nanometre-scale mechanical electrometer, *Nature*, 392, 160–162 (1998).

83. K. Schwab, E. A. Henriksen, J. M. Worlock, and M. L. Roukes, Measurement of the quantum of thermal conductance, *Nature*, 404, 974–977 (2000).

84. S. Evoy, A. Olkhovets, L. Sekaric, J. M. Parpia, H. G. Craighead, and D. W. Carr, Temperature-dependent internal friction in silicon Nanoelectromechanical systems, *Applied Physics Letters*, 77, 2397–2399 (2000).

85. L. Anand and Y. Wei, Mesoscopic modeling of the deformation and fracture of nanocrystalline metals, *IUTAM Symposium*, Beijing, China, 3–10 (2005).

86. L. A. Girifalco, R. A. Lad, Energy of cohesion, compressibility and the potential energy functions of the graphite system, *Journal of Chemical Physics*, 25(4) 693–697 (1956).

Problems

6.1 Calculate the electrostatic coulomb potentials of Li^+ ions in vacuum at the distances of 10, 5, and 1 nm. Also, explain why this potential fails as the particles get closer to one another. (Permittivity of vacuum, $\varepsilon_o = 8.854 \times 10^{-12}$ $C^2/(J \cdot m)$ and $e = 1.602 \times 10^{-19}$ C).

6.2 Determine the equilibrium spacing and binding energy of two Ne atoms using the Lennard–Jones (6-12) potential.

6.3 Plot a graph comparing Barker and Lennard–Jones (6-12) potentials for Xe atoms. Then, calculate the maximum relative error between these two potentials for xenon atoms.

6.4 Calculate the potential energy of an HCl molecule using modified Buckingham potentials. Assume that the spacing is $(r = 2r_m)$. Furthermore, evaluate the amount of force that exists between these two atoms at the above spacing.

6.5 Define the natural resonance frequency of the system depicted in **Figure 6.32** using Lennard–Jones (6-12) potential. Assume binding energy of 1.5×10^{-21} J, $r_0 = 0.35$ nm, and $m = 3 \times 10^{-23}$ g.

6.6 Find the eigenvalue $(\omega/\omega_0)^2$ and eigenvector $[A_1 \ A_2 \ A_3]$ for the system of three argon atoms using the system of equations (equation 6.31).

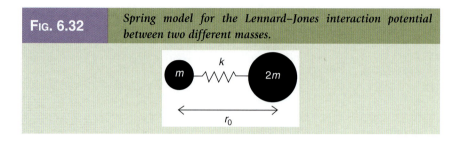

FIG. 6.32   *Spring model for the Lennard–Jones interaction potential between two different masses.*

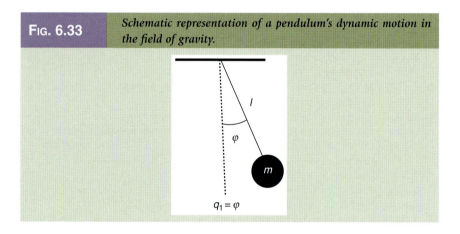

FIG. 6.33   *Schematic representation of a pendulum's dynamic motion in the field of gravity.*

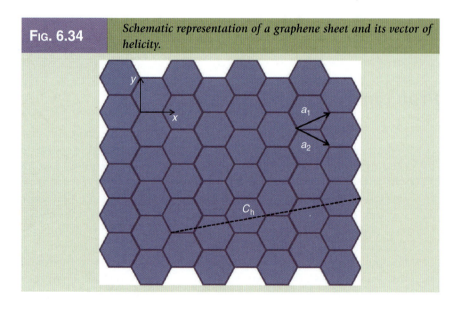

FIG. 6.34   *Schematic representation of a graphene sheet and its vector of helicity.*

6.7 Derive the Lagrangian and equation of motion for a pendulum depicted in **Figure 6.33** using the generalized coordinate q_i.

6.8 Given the displacement vector $u = (x_1 x_2, x_3^2, x_1 x_3)^T$ find the expression for the strain tensor S and rotation tensor Ω.

6.9 Elastic stiffness coefficients of an isotropic nanocrystalline nickel are found to be $c_{11} = 247$ GPa, $c_{12} = 147$ GPa [85]. Define the Lame constants for this material and then calculate its Young's modulus E, and shear modulus G. (Hint: The Young's modulus and shear modulus in terms of the Lame constants are given as follows: $E = \frac{\mu(3\lambda + 2\mu)}{\lambda + \mu}$ and $G = \frac{1}{\mu}$.)

6.10 Compare the level of accuracy between Verlet algorithms and Nordsieck/Gear predictor–corrector methods.

6.11 Explain the failure mechanism of a nanowire shown in **Figure 6.5** and then contrast its differences from macroscopic failure mechanisms.

6.12 The melting point of bulk aluminum is about 933 K. However, as mentioned in this chapter, the nanostructured Al can melt at about 270 K. Explain why this is the case and what other parameters influence the melting point of materials at nanometer scales.

6.13 Describe the common difficulties associated with constructing nanomachines which are absent at the macroscopic level.

6.14 Compare the wear mechanisms in macroscopic and microscopic domains by contrasting their differences.

6.15 Compute the angle of helicity for the zigzag and armchair type nanotubes. Also, explain why the angle of helicity ranges between 0° and 30° for all chiralities.

6.16 Define the vector of helicity C_h and the angle of helicity θ for **Figure 6.34**.

6.17 A carbon nanotube's diameter is defined as $d = |C_h|/\pi$. Prove that the nanotube's diameter has the form of $d = (a/\pi)\sqrt{n^2 + m^2 + nm}$, where $a = 0.246$ nm. Then, calculate the nanotube diameter for the picture shown in problem 6.16.

6.18 The Lennard–Jones potential energy of a carbon–carbon system is given by:

$$\phi = \frac{A}{\sigma^6}\left[\frac{1}{2}\gamma_0^6\left(\frac{\sigma}{r}\right)^{12} - \left(\frac{\sigma}{r}\right)^6\right], \quad [86].$$ Show that the equilibrium spacing r_0 is defined as $r_0 = \sigma\gamma_0$, and then calculate the binding energy of this carbon–carbon system assuming that $A = 24.3 \times 10^{-79}$ (J·m^6), $\sigma = 0.142$ nm, and $\gamma_0 = 2.7$.

6.19 Describe two different SPM methods that are used for nanomechanical measurements and indicate their advantages/disadvantages compared to each other.

6.20 A silica sample was tested by a CSM nanoindentation instrument and a reduced modulus of 71 GPa was obtained. Assume that the Berkovich diamond tip has a modulus of 1140 GPa, and Poisson ratios of the diamond tip and silica sample are 0.07 and 0.2, respectively. Calculate the modulus of this silica specimen applying these data points.

6.21 Explain the main differences between wet and dry etching and contrast their advantages/disadvantages upon one another.

6.22 Describe the functionality of a capacitive pressure sensor and its advantages compared to piezoresistive pressure sensors.

Nanostructure and Nanocomposite Thin Films

John J. Moore, Jianliang Lin, and In-Wook Park

What this tells us is that if you're building nanostructures, the surface is what's really important.

Paul Evans

*C*hapter 7

THREADS

Chapter 7 brings on some substance and detail about nanostructured and nanocomposite thin films. It is the second chapter in the *Mechanical Nanoengineering* division of the text. Following *chapter 7*, we provide applications of these thin films and with this chapter, we round out the division. In *chapter 10*, although part of the *Chemical Nanoengineering* division of the text, mechanical properties as they apply to nanocomposites are reviewed.

7.0 INTRODUCTION

Nanostructured coatings have recently attracted increasing interest because of the possibilities of synthesizing materials with unique physical–chemical properties [1,2]. A number of sophisticated surface-related properties such as optical, magnetic, electronic, catalytic, mechanical, chemical, and tribological can be obtained by advanced nanostructured coatings [3,4]. There are many types of design models for nanostructured coatings, such as three-dimensional nanocomposite coatings [2,5], nanoscale multilayer coatings [6,7], functionally graded coatings [1,4], etc. The optimized design of nanostructured coatings needs to consider many factors, for example, ion energy and ion flux of the depositing species, interface volume, crystallite size, single-layer thickness, surface and interfacial energy, texture, epitaxial stress and strain, and overall coating architecture, all of which depend significantly on materials selection, deposition methods, and process parameters [2,8].

In particular, pulsed reactive magnetron deposition techniques have been investigated, more recently, since it is possible to conduct reactive sputtering without arcing during deposition. Pulsed reactive sputtering can also change and control the plasma constituents, increase the ion energy and ion flux, and control microstructural growth of the thin film through ion bombardment [8]. The applications of pulsing in reactive magnetron sputtering opens up considerable opportunities for the control of ion energy and ion flux to optimize the deposition process and tailor the as-deposited coating structure and properties.

The focus of this chapter is to introduce the relationships between processing, structure, properties, and functionality of nanostructured coatings using various deposition processes, such as unbalanced magnetron sputtering (UBMS), hybrid coating system of cathodic arc evaporation (CAE) and magnetron sputtering (MS), pulsed closed-field unbalanced magnetron sputtering (P-CFUBMS), and high-power pulsed magnetron sputtering, as shown in **Figure 7.1**.

7.1 CLASSIFICATION OF NANOSTRUCTURED, NANOCOMPOSITE TRIBOLOGICAL COATINGS

7.1.1 *Nanoscale Multilayer Coatings*

Research on using nanoscale multilayers (i.e., "superlattices") to increase the hardness and toughness of coatings has provided significant advancements in

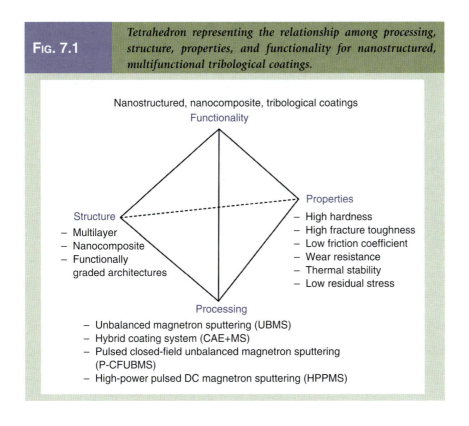

FIG. 7.1 *Tetrahedron representing the relationship among processing, structure, properties, and functionality for nanostructured, multifunctional tribological coatings.*

understanding the advantages of employing this type of coating architecture. Early research by Palatnik with multilayers of metals showed that significant improvements in strength were achieved when layer thickness was decreased below 500 nm [6,9]. In early modeling, Koehler [7] predicted that high shear strength coatings could be produced by alternating layers of high and low elastic modulus. Key elements of the concept are that very thin layers (≤10 nm) inhibit dislocation formation, while differences in elastic modulus between layers inhibit dislocation mobility. Lehoczky demonstrated [47] these concepts on metallic Al/Cu and Al/Ag multilayers and showed that a Hall–Petch type equation could be used to relate hardness to $1/(\text{periodicity})^{1/2}$, where periodicity is a minimum periodic length between layers in the multilayer coating architecture. Springer and Catlett [10], and Movchan et al. [11] reported on mechanical enhancements in metal/ceramic (e.g., Ti/TiN, Hf/HfN, W/WN, etc.)[12] and ceramic/ceramic (e.g., TiN/VN [13], TiN/NbN [14,15], $TiN/V_xNb_{1-x}N$ [16,17], etc.) laminate structures that followed a Hall–Petch relationship. These pioneering works were followed by intensive research in multilayers [18,19], which has produced coatings significantly harder than the individual components making up the layers. To achieve increased hardness, the layers must have sharp interfaces and periodicity in the 5–10 nm range.

The multilayer architectures, as shown in **Figure 7.2**, exhibiting high hardness are frequently called superlattices [20]. The different design architectures have been classified and some reports have formalized the multilayer design [4,21].

| FIG. 7.2 | *Cross-sectional TEM images and selected area diffraction patterns (SADP) of nanoscale multilayer coatings: (a) (Ti, Al)N/SiN, (b) (Ti, Al)N/WN, and (c) CrN/BCN.* |

(a) (Ti, Al)N/SiN   50 nm (b) (Ti, Al)N/WN   50 nm (c) CrN/BCN   50 nm

Source: K. Yamamotoa, S. Kujimeb, and K. Takahara, *Surface and Coatings Technology*, 200, 435–439 (2005). With permission.

Multilayer architectures clearly increase coating hardness and have commercial applications, especially in the tool industry. However, they can be difficult to apply with uniform thickness on three-dimensional components and rough surfaces. If the layers are not of the correct periodicity, the superlattice effect is lost. Another relatively new technology, nanocomposites, offers the same advantages as multilayers (plus other benefits) and their properties are not critically dependent on thickness or substrate geometry.

7.1.2   *Nanocomposite Coatings*

Nanostructured composite (i.e., "nanocomposite") coatings are usually formed from ternary or higher-order systems and comprise at least two immiscible phases: two nanocrystalline phases and, more commonly, an amorphous phase surrounding nanocrystallites of a secondary phase. The most interesting and extensively investigated nanocomposite coatings are ternary, quaternary, or even more complex systems, with nanocrystalline (nc-) grains of hard transition metal nitrides (e.g., TiN, CrN, AlN, BN, ZrN, etc.), carbides (e.g., TiC, VC, WC, ZrC, etc.), borides (e.g., TiB_2, CrB_2, VB_2, WB, ZrB_2, etc.), oxides (e.g., Al_2O_3, TiO_2, SiO_2, MgO, TiO_2, Y_2O_3, ZrO_2, etc.), or silicides (e.g., $TiSi_2$, $CrSi_2$, $ZrSi_2$, etc.) surrounded by amorphous (a-) matrices (e.g., Si_3N_4, BN, C, etc.). The physical, mechanical, and thermal properties of these hard materials are summarized in **Table 7.1** [22]. The synthesis of such nanocomposite (nc-/a-) coatings critically depends on the ability to co-deposit both the nanocrystalline and amorphous phases, such as Ti–Si–N (nc-TiN/nc-and a-$TiSi_2$/a-Si_3N_4) [2], Ti–Al–Si–N (nc-TiAlN/a-Si_3N_4)[5],W–Si–N(nc-W_2N/a-Si_3N_4)[23],Cr–Si–N(nc-CrN/a-Si_3N_4) [24], Ti–B–C–N (nc-TiB_2 and TiC/a-BN) [25], TiC/DLC (nc-TiC/a-C) [26], WC/DLC (nc-WC/a-C) [27], etc., as schematically presented in **Figure 7.3a**. A variety of hard compounds can be used as the nanocrystalline phases, including nitrides, carbides, borides, oxides, and silicides. Veprek [28] suggested that the

TABLE 7.1

The Physical, Mechanical, and Thermal Properties of Hard Materials

Phase	Crystal structure	Lattice parameters (nm)	Density (g·cm⁻³)	Melting point (°C)	Linear thermal expansion, α (10⁻⁶·K⁻¹)	Thermal conductivity, λ (W·m⁻¹·K⁻¹)	Electrical resistivity (10⁻⁶ Ω·cm)	Enthalpy at 298 K (kJ·mol⁻¹)	Young's modulus (10⁵ N·mm⁻²)	Microhardness (10N·mm⁻²)	Oxidation resistance (×100 °C)
Nitrides											
AlN	hex	0.311/0.498	3.05	2200	6	10	10^{11}	288.9	3.15	1200	13
BN	hex	0.251/0.669	2.25	3000	3.8	284.7	3×10^{14}	252.5	0.9	4400HV	10
CrN	fcc	0.415	6.1	1050	2.3	11.72	640	118–124	4	1800–2100	7–7.5
	cub-B1	0.4149	5.39–7.75	1450	2.3	11.72	640	123.1	3.236	1100	
Cr₂N	hex	0.4760/0.4438	5.9	1500	9.4			30.8	3.138	2250HV	86.3–110.3
HfN	fcc	0.452	13.8	3310	6.9	11.3	26	369.4	3.33–4.8	1700–2000	
Si₃N₄	hex	0.78/0.56	3.44	1900	2.4	20–24'	10^{18}	750.5	2.1	1410HV	12–14
TaN	hcp	0.52/0.29	13.6–13.8	3000	3.6	8.58	128	225.7	5.756	3240	5–8
Ta₂N	hex	0.30/0.493	15.8	3000		10.05	263	270.9		3000	
TiN	cub-B1	0.423	5.21	3220	9.35	30	21.7	336.2	2.512	2400HV	5
VN	fcc	0.41	6.13	2050	8.1	11.3	85–100	147.8	4.6	1520	5–8
ZrN	fcc	0.46	6.93	3000	6	16.75	13.6	365.5	5.1	2000	12
Carbides											
B₄C	rhom	0.5631/1.2144	2.52	2450	6	27.63	10^6	72	4.5	3700	11–14
Cr₃C₂	ortho	1.146/0.552/0.2821	6.68	1900	10.3	18.8	75	88.8	4	1500–2000	12
NbC	fcc	0.45	7.78	3490	6.65	14.24	35–74	139.8	3.4	2400	11
SiC	α:hex	β:0.4360	3.2	2200	5.68	15.49	10^5	71.6	4.8	3500	14–17
	β:fcc	α:0.3–7.3/1–1.5	3.17	2700	5.3	63–155	10^5	73.3	3.9–4.1	1400HV	13–14
TaC	cub-B1	0.4454	14.65	3877	6.04	22.19	25	159.5	2.91	1490	11–14
TiC	cub-B1	0.429–0.433	4.93	3150	7.4	17–23.5	68	179.6	3.22	3200HV	11–14
VC	cub-B1	0.4173	5.36	2770	6.55	4.2	156	105.1	4.34	2760HV	8–11
WC	hex	0.29/0.28	15.7	2600	5.2–7.3	121.42	17	35.2	7.2	2080	8
ZrC	cub-B1	0.4989–0.476	6.51	3400	6.93	20.5	42	181.7	4	2600HV	12

(continued)

TABLE 7.1 (CONTD.)

The Physical, Mechanical, and Thermal Properties of Hard Materials

Phase	Crystal structure	Lattice parameters (nm)	Density (g·cm⁻³)	Melting point (°C)	Linear thermal expansion, α (10⁻⁶·K⁻¹)	Thermal conductivity, λ (W·m⁻¹·K⁻¹)	Electrical resistivity (10⁻⁶ Ω·cm)	Enthalpy at 298 K (kJ·mol⁻¹)	Young's modulus (10⁵ N·mm⁻²)	Microhardness (10 N·mm⁻²)	Oxidation resistance (×100 °C)
Borides											
AlB₂	hex	0.3006/0.3252	3.17	1975				67			
CrB	ortho	0.2969/0.7858/0.2932	6.05	1550				75.4			14–18
CrB₂	hex	0.279/0.307	5.6	2200	11.1		56	94.6	2.15	2250	11–17
HfB₂	hex	0.3141/0.3470	11.01	3200	5.3	430	10	336.6		2800	11–14
MoB₂	hex	0.3/0.31	7.8	2100			45	96.3		1380HV	11–14
NbB₂	hex	0.31/0.33	6.8	3000	7.1–9.6	16.75	32	150.7	2.6	2600	11–14
SiB₆	ortho	1.4470/1.8350/0.9946	2.43	1950	8.3		10⁷		3.3	1910	
TaB₂	hex	0.31/0.33		3150	5.1	21.35	68	209.3	2.62	2200	11–14
TiB₂	hex	0.3/0.32	4.5	2900	6.39	25.96	9	150.7	3.7	3840	11–17
Ti₂B	Tet	0.61/0.46		2200						2500	
VB₂	hex	0.3/0.31	4.8	2400	5.3	16	16	203.9	5.1	2080	13
WB	Tet	0.31/1.7	15.5	2860						3750	
W₂B	Tet	0.56/0.47	16.5	2770	4.7		21.43	163.3		2350	8–14
ZrB	fcc	0.47	6.5	3000						3600	
ZrB₂	hex	0.32/0.35	6.1	3000	6.83	23.03	9.2	326.6	3.5	2200	11
Oxides											
Al₂O₃-α	hex	0.5127	3.99	2043	8	30.1	10²⁰	1580.1	4	2100HV	17
BeO	rhom	0.513	3.9	2030	7.2–8.6	4.2–16.7	10²⁰	1678.5	3.6	2100HV	20
CrO₂	hex	0.2699/0.4401	3	2450	9	264	10²³	569	3	1230–1490HV	17
CrO₃	Tet	0.441/0.291	4.8	170–198				582.8			
Cr₂O₃	ortho	0.573/0.852/0.474	2.81	2440	6.7		10¹³	579.9		1000HV	
HfO₂	rhom	0.536	5.21	2400	5.6			1130.4		2300HV	
	hex	0.495876/1.35942	5.21					1130.4			
	mono	0.512	9.7	2900	10	3	5 × 10¹⁵	1053.4		900	

	Structure	Lattice (nm)							Hardness	Ref.
MgO	cub	0.4208	2850	11.2	36	10^{12}	568.6	3.2	745HV	17
SiO$_2$	quartz	0.4093/0.5393	1703–1729	0.4	1.38	10^{22}	911	0.5–1.0	1130–1260	17
	trigonal	0.421/0.539	1713	0.5–0.75	1.2–1.4	10^{21}	911.5	1.114	1200	
ThO$_2$	*fcc*	0.5859	3250	10	10	10^{16}	1173.1	1.38	950HV	17
TiO	cub-B1	0.417	1750	7.6	11		520		1300	
TiO$_2$	Tet	0.4593/0.2959	1900	4.21–4.25	8	1.2×10^{10}	945.4	2.05–2.80	767–1000HK	
	cub-B1	0.45933/0.29592	1867	9				0.8–2.0	1000HV	
Ti$_2$O$_3$	rhom	0.5454	2130			10^{19}–10^{24}	1433.1		980HV	
Ti$_3$O$_5$	mono	0.9828/0.3776/0.9898	1780				2461			
ZrO$_2$	cub	0.511	2750	7.5–10.5	0.7–2.4	10^{16}	1035	1.63	1200	17
Silicides										
CrSi	cub	0.4629	1550				53.2		1000	
CrSi$_2$	hex	0.442/0.655	1630				100.5		1100	14–18
Cr$_3$Si	cub	0.455	1710				105.5		900–980	
MoSi$_2$	Tet	0.32/0.786	2050	8.4	221.9	21	108.9	3.84	1290	17
NbSi$_2$	hex	0.48/0.66	1950	8.4		6.3	50.2		700	8–11
TaSi$_2$	hex	0.4773/0.6552	2200	8.9/8.8		38	150.7		1410	11
TiSi$_2$	ortho	0.8236/0.4773/0.8523	1520	11.5		18	134.4	2.556	892	11
VSi$_2$	hex	0.46/0.64	1650	11		9.5	95		960	
WSi$_2$	Tet	0.321/0.788	2165	6.5	45	12.5	92.1	5.3	1090	16
ZrSi$_2$	ortho	0.372/1.416/0.367	1700	9.7		161	159.4	2.348	1030HV	8–11

cub: cubic, cub-B1: cubic NaCl-type, *fcc*: face-centered cubic, hex: hexagonal, *hcp*: hexagonal closed-packed, HK: Knoop hardness, HV: Vickers hardness, mono: monoclinic, ortho: orthorhombic, rhom: rhombohedral/trigonal, tet: tetragonal, tri: triclinic

FIG. 7.3   *(a) Schematic diagram of a nanostructured nanocomposite coating proposed by Veprek [28] and (b) HRTEM image and selected area diffraction pattern (SADP) of the nanocomposite Ti–B–N (nc-TiN/a-BN).*

Source: Y. H. Lu, P. Sit, T. F. Hung, H. Chen, Z. F. Zhou, K. Y. Li, and Y. G. Shen, *Journal of Vacuum Science and Technology, B*, 23(2), 449 (2005). With permission.

nanocrystalline grains must be 3~10 nm in size and separated by 1~2 monolayers within an amorphous phase as shown in **Figure 7.2a**. For example, the Ti–B–N nanocomposite, which consists of nanocrystalline TiN (~5 nm in size) in an amorphous BN matrix, has been synthesized and observed by Lu [29], as shown in **Figure 7.3b**.

7.1.3 *Functionally Graded Coatings*

In order to counteract brittle failure and improve fracture toughness, two concepts have been explored. The first involves the use of graded interfaces between the coating and substrate and between layers in the coating. For example, a WC–TiC–TiN (outside layer) graded coating for cutting tools was reported by Fella et al. [30], which showed considerably less wear than single-layer hard coatings used in the cutting of steels. This type of coating is functionally and chemically graded to achieve better adhesion, oxidation resistance, and mechanical properties. One example of how functionally graded architectures improve coating performance is the adhesion of diamond-like carbon (DLC) to steels. DLC, and especially hydrogen-free DLC, has a very high hardness and generally has a large residual compressive stress. The coatings are relatively inert, and adhesion failures of coated steel surfaces were a roadblock to success. This problem was solved through designing and implementing a graded interface between the coating and the substrate. Examples of effective gradient compositions are Ti–TiN–TiCN–TiC–DLC for hydrogenated DLC [31] and Ti–TiC–DLC for hydrogen-free DLC [32]. In the development of the latter composition, the importance of a graded elastic modulus through the substrate coating/interface was highlighted as shown in **Figure 7.4**. The gradual build-up in material stiffness from the substrate with $E = 220$ GPa to the DLC layer with $E = 650$ GPa avoids sharp interfaces

| | FIG. 7.4 | *Schematic diagram of a functionally gradient Ti–TiC$_x$–DLC coating, where chemistry and elastic moduli are transitioned from the metallic substrate to a hard DLC top layer.* |

Material	Hardness	Elastic modulus	Thickness	
DLC at 10^{-5} Pa	70 GPa	650 GPa	400 nm	
DLC at 10^{-1} Pa	43 GPa	450 GPa	100 nm	Ti: 0 at%
Ti$_{0.10}$C$_{0.90}$	25 GPa	290 GPa	25 nm	C: 100 at%
Ti$_{0.25}$C$_{0.75}$	27 GPa	350 GPa	25 nm	
Ti$_{0.30}$C$_{0.70}$	29 GPa	370 GPa	100 nm	
Ti$_{0.50}$C$_{0.50}$	20 GPa	290 GPa	100 nm	Functionally graded
Ti$_{0.70}$C$_{0.30}$	14 GPa	230 GPa	100 nm	
Ti$_{0.90}$C$_{0.10}$	6 GPa	150 GPa	50 nm	
Ti	4 GPa	140 GPa	50 nm	Ti: 100 at%
440 steel	11 GPa	220 GPa	Substrate	C: 0 at%

Source: A. A. Voevodin, M. A. Capano, S. J. P. Laube, M. S. Donley, and J. S. Zabinski, *Thin Solid Films*, 298, 107–115 (1997). With permission.

that can provide places for crack initiation, good chemical continuity, and load support for the hard DLC top-coat. This functionally graded approach can be combined with multilayer and nanocomposite architectures to further enhance tribological properties.

7.2 BACKGROUND OF NANOSTRUCTURED SUPER-HARD COATINGS

Hardness is defined as the resistance to plastic deformation. Plastic deformation of crystalline materials occurs predominantly by dislocation movement under an applied load. Therefore, a higher resistance to dislocation movement of a material will generally enhance its hardness. One approach to obtain high resistance to dislocation movement and plastic deformation is to preclude the formation of stable dislocations. *Super-hard* coatings, with a hardness value in excess of 40 GPa, have attracted significantly increasing interest during the past 10–15 years [33]. A concept for super-hard nanocomposite coatings was suggested by Veprek [34]. The strength and hardness of engineering materials are orders of magnitude smaller than the theoretically predicted values. They are determined mainly by the microstructure, which has to be designed in such a way as to efficiently hinder the multiplication and movement of dislocations and the growth of microcracks. This can be achieved in various ways known from metallurgy, such as solution, work, and grain boundary hardening [35,36]. In this way, the strength and hardness of a material can be increased by a factor of 3–7 times, that is, super-hard materials should form when such an enhancement can be achieved starting from a hard material ($HV > 20$ GPa). Solution and work hardening do not operate in small nanocrystals of about <10 nm because solute atoms segregate to the grain boundary and there are no dislocations. Therefore, we consider the possibilities of extending the grain boundary hardening in poly- and microcrystalline materials, described by the Hall–Petch relationship [37,38], equation (7.1), down to the range of a few nanometers:

$$\sigma_C = \sigma_0 + \frac{k_{gb}}{\sqrt{d}} \tag{7.1}$$

where
    σ_C is the critical fracture stress
    d is the crystallite size
    σ_0 and k_{gb} are constants

Many different mechanisms and theories describe Hall–Petch strengthening [37,38]. Dislocation pileup models and work hardening yield the $d^{-1/2}$ dependence but different formulas for σ_0 and k_{gb}, whereas the grain–grain boundary composite models also give a more complicated dependence of σ_C on d. The strength of brittle materials, such as glasses and ceramics, is determined by their ability to withstand the growth of microcracks. Brittle materials do not undergo any plastic deformation up to their fracture. Their strength or hardness is proportional to the elastic modulus. The critical stress, which causes the growth of a microcrack of size a_0, is given by the general Griffith criterion (equation 7.2).

$$\sigma_C = k_{crack} \sqrt{\frac{2E\gamma_s}{\pi\, a_0}} \propto \frac{1}{\sqrt{d}} \tag{7.2}$$

Here E is the Young's modulus, γ_s is the surface cohesive energy, and k_{crack} is a constant which depends on the nature and shape of the microcrack and on the kind of stress applied [35]. Thus, the $d^{-1/2}$ dependence of the strength and hardness in a material can also originate from the fact that the size a_0, of possible flaws such as voids and microcracks that are formed during the processing of the material, also decreases with decreasing grain size. For these reasons, the Hall–Petch relationship, equation (7.1), should be considered as a semiempirical formula which is valid down to a crystallite size of 20–50 nm (some models predict an even higher limit [39,40]). With the crystallite size decreasing below this limit, the fraction of the material in the grain boundaries strongly increases which leads to a decrease of its strength and hardness due to an increase of *"grain boundary sliding"* [40,41]. A simple phenomenological model (i.e., rule of mixtures) describes the softening in terms of an increasing volume fraction of the grain boundary material f_{gb}, with the crystallite size decreasing below 10–6 nm (equation 7.3) [42].

$$H(f_{gb}) = (1 - f_{gb})H_c + f_{gb}H_{gb} \tag{7.3}$$

Here $f_{gb} \propto (1/d)$. Due to the flaws present, the hardness of the grain boundary material H_{gb} is smaller than that of the crystallites H_c. Thus the average hardness of the material decreases with d decreasing below 10 nm. The first report of a *reverse (or negative) Hall–Petch relationship* was by Chokshi et al. [43]. Later on it was the subject of many studies, with controversial conclusions regarding the critical grain size where a *normal Hall–Petch relationship* changes to a reverse one [39,44]. Various mechanisms of grain boundary creep and sliding were discussed and are described by deformation mechanism maps in terms of temperature and stress [45,46]. Theories of grain boundary sliding are critically reviewed in Ref. [39]. Recent computer simulation studies [41] confirm that the negative Hall–Petch dependence in nanocrystalline metals is due to the grain boundary sliding that occurs due to a large number of small sliding events of atomic planes at the grain boundaries without thermal activation. Although many details are still not understood, there is little doubt that grain boundary sliding is the reason for softening in this crystallite size range. Therefore, a further increase of the strength and hardness with decreasing crystallite size can be achieved only if grain boundary sliding can be blocked by appropriate design of the material. This is the basis of the concept for the design of super-hard nanocomposites, suggested by Veprek [34].

As mentioned in section 7.1.1, another possible way to strengthen a material is based on the formation of nanoscale multilayers consisting of two different materials with large differences in elastic moduli, sharp interface, and small bilayer thickness (lattice period) of about 10 nm [7]. Because this design of nanostructured super-hard materials was suggested and experimentally confirmed before super-hard nanocomposites were developed, we will deal with superlattice coatings in section 7.2.1.

7.2.1 *Nanoscale Multilayer Coatings*

In a theoretical paper published in 1970, Koehler suggested [7] a concept for the design of strong solids, which are nowadays called superlattices. Originally he suggested depositing epitaxial multilayers of two different metals, A and B, having as different elastic constants as possible $E_A < E_B$ but similar thermal expansion and strong bonds. The thickness of the layers should be so small that no dislocation source could operate within the layers. If under applied stress a dislocation, which would form in the softer layer A, would move towards the A/B interface, elastic strain induced in the second layer B with the higher elastic modulus would cause a repulsing force that would hinder the dislocation from crossing that interface. Thus, the strength of such multilayers would be much larger than that expected from the rule of mixture. Koehler's prediction was further developed and experimentally confirmed by Lehoczky who deposited Al/Cu and Al/Ag superlattices and measured their mechanical properties [47]. According to the rule of mixtures, the applied stress which causes elastic strain is distributed between the layers proportional to their volume fractions and elastic moduli. Lehoczky has shown that the tensile stress–strain characteristics measured on multilayers consisting of two different metals displayed a much higher Young's modulus and tensile strength than that predicted by the rule of mixtures, and both of which increased with decreasing thickness of the double layer (lattice period). For layer thicknesses <70 nm, the yield stress of Al/Cu multilayers was 4.2 times larger and the tensile fracture stress was 2.4–3.4 times higher than the values given by the rule of mixture [47]. This work was followed by the work of a number of researchers who confirmed the experimental results on various metal/metal, metal/ceramic, and ceramic/ceramic multilayer systems (see section 7.1.1). In all these cases, an increase in hardness by a factor of 2–4 was achieved when the lattice period decreased to about 5–10 nm. For a large lattice period, where the dislocation multiplication source can still operate, the increase of the hardness and the tensile strength (most researchers measured the hardness because it is simpler than the measurement of tensile strength conducted by Lehoczky) [47] with decreasing layer thickness is due to the increase in the critical stress, which is dependent on multiplicate dislocations such as a Frank–Read dislocation source (equation 7.4):

$$\sigma_C = \frac{Gb}{l_{pp}} \propto \frac{1}{l_{pp}} \tag{7.4}$$

where
   G is the shear modulus
   b is the Burgers vector
   l_{pp} is the distance between the dislocation pinning sites [35]

Usually one finds strengthening dependence similar to the Hall–Petch relationship but with a somewhat different dependence on the layer period (λ^{-n}), instead of the crystallite size d in equation (7.1), with $n = 1/2$ for layers with different slip systems and $n = 1$ for layers with similar slip systems [48]. A more recent theoretical discussion of the Hall–Petch relationship for superlattices was published by Anderson and Li [49]. According to their calculations, strong deviations from continuum Hall–Petch behavior occur when the thickness of the layers is so small that the pileup contains only one to two dislocations.

In a remark added in the proof, Koehler mentioned that the ideas described in his paper would also be valid if one of the layers is amorphous. Recently, several papers have appeared in which one of the layers consists of an amorphous CN_x and a transition metal nitride such as TiN [50] or ZrN [51]. However, with decreasing layer thickness the layered structure vanished and a nanocrystalline composite (i.e., "nanocomposite") structure appeared [50,51]. Such films also exhibit a high hardness of 40 GPa or more.

7.2.2 Single-Layer Nanocomposite Coatings

Using similar ideas for restricting dislocation formation and mobility as used in multilayer approaches to "hardening," nanocomposite coatings can also be super-hard [25,28]. These composites have 3–10 nm crystalline grains embedded in an amorphous matrix and the grains are separated by 1–3 nm. This designs an *architecture* which leads to ultrahard (hardness above 100 GPa) coatings as reported by Veprek and co-authors most recently [2]. The nanocrystalline phase may be selected from the nitrides, carbides, borides, and oxides, as shown in **Table 7.1**, while the amorphous phase may also include metals and DLC as shown in **Figure 7.1a**. The initial model proposed by Veprek to explain hardness in nanocomposites is that dislocation operation is suppressed in small grains

(3–5 nm) and that the narrow space between them (1 or 2 monolayer separation) induces incoherence strains [34,52]. The incoherence strain is likely increased, when grain orientations are close enough to provide interaction between matched but slightly misoriented atomic planes.

In the absence of dislocation activity, Griffith's equation, as shown in equation (7.2), for crack opening was proposed as a simple description of the nanocomposite strength. This equation suggests that strength can be increased by increasing the elastic modulus and surface energy of the combined phases, and by decreasing the crystalline grain sizes. It is noted that elastic modulus is inversely dependent on grain sizes that are in the nanometer range due to lattice incoherence strains and the high volume of grain boundaries [2]. In practice, grain boundary defects always exist, and a 3-nm grain size was found to be close to the minimum limit. Below this limit, a reverse Hall–Petch effect has been observed and the strengthening effect disappears because grain boundaries and grains become indistinguishable and the stability of the nanocrystalline phase is greatly reduced [34,43,53].

Nanocomposites with metal matrixes are in a special category for this discussion. They have been demonstrated to increase hardness, but also have good potential for increasing toughness. Mechanisms for toughening within these systems are discussed in the next section, while mechanisms for hardening are discussed here. The composite strength of ceramic/metal nanocomposites may be described by the following form of the Griffith–Orowan model (equation 7.5) [54] when the dimensions of the metal matrix permit operation of dislocations:

$$\sigma_C = \sqrt{\frac{2E(\gamma_s + \gamma_p)}{\pi\, a}\frac{r_{\text{tip}}}{3d_a}} \tag{7.5}$$

where
  γ_p is the work done during plastic deformation
  r_{tip} is the curvature of the crack tip
  d_a is the interatomic distance

It is noted that the crack tip blunting and the work of plastic deformation considerably improve material strength, while the lower elastic moduli of metals cause a reduction in strength as compared to ceramics. However, in nanocomposites, dislocation operation may be prohibited because the separation of grains is very small. For example, the critical distance l_{pp} for a Frank–Read dislocation source is very small. Matrix dimensions in hard nanocomposite coatings are from 1 to 3 nm, which is well below the critical size for dislocation source operation, even in very soft metal matrixes. Therefore, the mechanical behavior of such nanocomposites can be expected to be similar to that of ceramic matrix composites.

Composite designs that increase elastic modulus and hardness do not necessarily impart high toughness. First, dislocation mechanisms of deformation are prohibited and crack opening is the predominant mechanism for strain relaxation when stresses exceed the strength limit. Second, Griffith's equation does not take into account the energy balance of a moving crack, which consists of the energy required to break bonds and overcome friction losses, potential energy released by crack opening, and kinetic energy gained through crack motion [55]. From crack energy considerations, a high amount of stored stress dictates a high rate of potential energy release in the moving crack. In such conditions, a crack can achieve the self-propagating (energetically self-supporting) stage sooner, transferring into a macrocrack and causing brittle fracture. However, nanocomposites contain a high volume of grain boundaries between crystalline and amorphous phases. This type of structure limits initial crack sizes and helps to deflect, split, and blunt growing cracks.

7.3   NEW DIRECTIONS FOR NANOSTRUCTURED SUPER-TOUGH COATINGS

While super-hard coatings are very important, quite notably for protection of cutting tools, most tribological applications for coatings either require or would receive significant benefit from increased toughness and lower friction. In particular, *high fracture toughness* is necessary for applications where high contact loads and, hence, significant substrate deformations are encountered [27]. A material is generally considered tough if it possesses both high strength and high ductility. High hardness (H) is directly related to high elastic modulus (E) and high yield strength, but it is very challenging to add a measure of ductility to hard coatings. For example, the super-hard coating designs, as stated earlier, prevent dislocation activity, essentially eliminating one common mechanism for ductility. Therefore, designs that increase ductility must also be considered to provide tough tribological coatings. Pharr [56] has suggested that an indication of fracture toughness (i.e., H/E ratio) can be obtained by examining the surface radial cracks created during indentation, described by equation (7.6):

$$K_c = \alpha_1 \left(\frac{E}{H} \right)^{1/2} \left(\frac{P}{c^{3/2}} \right) \tag{7.6}$$

where
   P is the peak indentation load
   c is the radial crack length
   α_1 is an empirical constant related to the indenter geometry

K_c describes the "critical stress intensity" for crack propagation, but it is not an intrinsic parameter that can be used to measure fracture toughness directly. However, it is inversely proportional to fracture toughness. Thus, fracture toughness of coatings would appear to be improved by both a high hardness and a *low elastic modulus*. In this work, the H/E values were calculated and discussed relatively for each coating [57].

7.3.1   *Functionally Graded Multilayer Coatings*

An effective route for improving toughness in multilayers is the introduction of ductile, low elastic modulus alternate layers into the coating structure to relieve

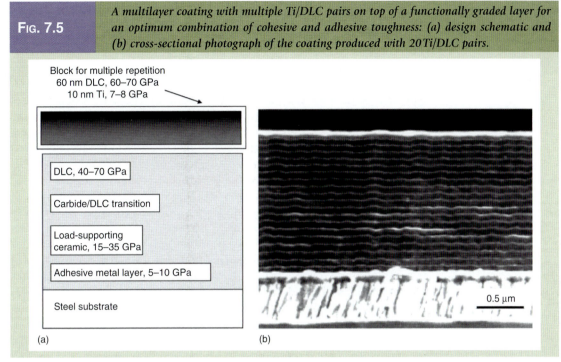

FIG. 7.5 *A multilayer coating with multiple Ti/DLC pairs on top of a functionally graded layer for an optimum combination of cohesive and adhesive toughness: (a) design schematic and (b) cross-sectional photograph of the coating produced with 20 Ti/DLC pairs.*

Source: A. A. Voevodin, S. D. Walck, and J. S. Zabinski, *Wear*, 203–204, 516–527 (1997). With permission.

stress and allow crack energy dissipation by plastic deformation in the crack tip. This approach will result in a decreased coating hardness, but the gain in the fracture toughness improvement may be more important in many tribological applications, excluding coatings for the cutting tool industry. For example, $[Ti/TiN]_n$ multilayer coatings on cast iron piston rings relaxed interface stress and improved combustion engine performance [58]. **Figure 7.5a** shows a schematic of a multilayer $[Ti/DLC]_n$ coating on a graded load support foundation, where the ductile Ti layers in the multilayer stack were graded at every DLC interface to avoid brittle fracture [4]. A cross-sectional photograph of this coating with 20 [Ti/DLC] pairs is shown in **Figure 7.5b**. The ductile Ti layers reduced the composite coating hardness to 20 GPa as compared to a single layer DLC coating, which has a hardness of about 60 GPa. However, due to dramatic improvement in toughness the multilayer coating design permitted operation during sliding friction at contact pressures as high as 2 GPa without fracture failure compared to 0.6–0.8 GPa for a single-layer DLC.

In general, the combination of multilayer and functionally graded approaches in the design of wear protective coatings produces exceptionally tough wear protective coatings for engineering applications. One potential drawback slowing the widespread use of new coatings was the need for reliable process controls to ensure that the correct compositions, structures, and properties are implemented during growth. However, modern process instrumentation and control technologies are able to meet the challenge and permit successful commercialization [59]. Thus, functionally graded and multilayer designs are commonly utilized in the production of modern tribological coatings.

7.3.2 Functionally Graded Nanocomposite Coatings

An alternative to employing multilayers to toughen coatings is embedding grains of a hard, high-yield strength phase into a softer matrix, allowing for increased ductility. This approach has been widely explored in macrocomposites made of ceramics and metals which are known as cermets [60]. This approach was recently scaled down to the nanometer level in thin films made of hard nitrides and softer metal matrixes [28]. A combination of the nanocrystalline/amorphous designs with a functionally graded interface, as shown in **Figure 7.6a**, provides high cohesive toughness and high interface (adhesive) toughness in a single coating. Several examples of tough wear-resistant composite coatings have been reported. Two of them combined nanocrystalline carbides with an amorphous DLC matrix designated as TiC/DLC and WC/DLC composites. In another example, nanocrystalline MoS_2 was encapsulated in an Al_2O_3 amorphous matrix as shown in **Figure 7.6b**. In all cases, the large fraction of the grain boundary phase provided ductility by activating grain boundary slip and crack termination by nanocrack splitting. This provided a unique combination of high hardness and toughness in these coatings. Novel nanocomposite designs suggested by Voevodin and Zabinski [61] for tough tribological coatings are very promising and provide a very attractive alternative to multilayer architectures. One of these novel designs incorporates the *chameleon* approach in which components are added to the coating that provide a low friction coefficient under extreme environmental conditions such as low and high humidity and low and high temperatures [61]. Nanocomposite coatings are more easily implemented, since they do not require precise control in the layer thickness and frequent cycling of the deposition parameters, as is required for fabrication of multilayer

Fɪɢ. 7.6 *(a) Schematic representation of a tough nanocomposite coating design, combining a nanocrystalline/amorphous structure with a functionally gradient interface and (b) TEM image of an Al_2O_3/MoS_2 nanocomposite coating consisting of an amorphous Al_2O_3 ceramic matrix encapsulating 5- to 10-nm inclusions of nanocrystalline MoS_2 grains.*

Source: A. A. Voevodin and J. S. Zabinski, *Composite Science Technology*, 65, 741–748 (2005). With permission.

coatings. They are however relatively recent developments, and suitable scale-up of deposition techniques is currently under intense study.

To prevent tribological failures, there are additional requirements related to the normal (load) and tangential (friction) forces. In general terms, a tough *wear-resistant* coating must support high loads in sliding or rolling contact without failure by wear, cohesive fracture, and loss of adhesion (delamination). A *low friction coefficient* reduces friction losses and may increase load capability. Tribological coatings where a low friction coefficient is also required may be obtained by producing nanocomposite coatings with a mix of hard and lubricating phases, in which a hard primary phase (e.g., nitrides, carbides, or borides, etc.) provides wear resistance and load-bearing capability and a lubricating secondary phase (e.g., a-C, a-Si_3N_4, a-BN, etc.) reduces the friction between two contacting components. Finally, *thermal stability* is required to optimize coating performance and lifetime. The amorphous phases in grain boundaries can act as diffusion barriers (e.g., a-Si_3N_4, a-SiO_2, etc.) for improved thermal stability. For instance, nc-TiN/a-Si_3N_4 coatings with an amorphous Si_3N_4 matrix did not show grain growth at temperatures up to 1050°C as well as super-hardness of about 45 GPa [62]. Moreover, silicon nitride acts as an efficient barrier against oxygen diffusion at the grain boundaries and also by forming an oxidation-resistant SiO_2 surface layer, thus resulting in excellent thermal stability.

In summary, in addition to high hardness, other aspects such as high toughness, low friction coefficient, and high thermal stability are decisive characteristics of nanostructured coatings for their potential as protective tribological coatings.

7.4 PROCESSING TECHNIQUES AND PRINCIPLES

Thin-film deposition is a process in which elemental, alloy, or compound thin films are deposited onto a bulk substrate. The deposition of the thin film may also be coupled with a previous surface treatment modification of the substrate in order to provide the required properties of the total coating or thin-film system. Thin-film deposition of metallic, insulating, conductive, and dielectric materials plays an important role in a large number of manufacturing, production, and research applications. There are a wide range of deposition processes that can be used to produce nanostructured and nanocomposite coatings, based on nitrides, carbides, and oxides, onto different substrate surfaces. The deposition process can be broadly classified into (i) physical vapor deposition (PVD) and (ii) chemical vapor deposition (CVD). These processes are used to deposit a broad range of thin-film materials, including semiconductors, superconductors, insulators, barrier layers, magnetic, optical films, and tribological and wear-resistant coatings, metals, compound, and organics, which play an important role in a large number of manufacturing, production, and research applications, such as protective coatings, optical coatings, microelectronic and optoelectronic devices, and decorative coatings [63]. This section will provide an introduction to the various important deposition techniques as well as some of the current day applications of the films produced.

7.4.1 Plasma Definition

Many vapor deposition techniques take advantage of conducting the process in a plasma medium. Plasma is often referred to as the fourth state of matter. Like the other three states of matter (solid, liquid, and gas), plasma has its own unique properties. A plasma is a gas containing charged and neutral species in varying degrees of excitation, including some or all of the following: electrons, positive ions, negative ions, atoms, and molecules. On average a plasma is electrically neutral, because any charge imbalance would result in electric fields that would tend to move the charges in such a way as to eliminate the imbalance [64].

To generate a plasma, a specific amount of energy needs to be added to separate the gas component molecules (gas breakdown) into a collection of ions, electrons, neutral atoms, etc., such as by applying an electric field or substantially increasing the temperature. Depending on the amount of energy added, the resulting plasma can be characterized as thermal or nonthermal ("cold" plasma). A nonthermal plasma is one in which the mean electron temperature (energy), which usually is of $1-10\,eV$, is higher than that of the bulk gas molecules. This *nonequilibrium* plasma can be generated at low pressures and is used in sputtering, etching, etc. **Figure 7.7a** shows a typical nonthermal plasma used in MS. On the other hand, if the energy is high enough, the ions and electrons are in local thermal (thermodynamic) equilibrium (at the same temperature). This *equilibrium* plasma is usually generated at near and above atmospheric pressures and is used in thermal plasma processing (spraying, heating, melting, etc.). **Figure 7.7b** shows a typical thermal plasma used in thermal plasma spraying.

To sustain a plasma requires that the rate of ionization must balance the loss of ions and electrons from the plasma volume by recombination and diffusion or convection to the plasma boundary. An important parameter of a plasma is the degree of ionization, which is the fraction of the original neutral species which has become ionized. The ionization process will create excited, dissociated, and/or ionized reactive species that are involved in the chemical reactions on the substrate. There are a number of important ionization mechanisms involved in the plasma deposition process. Typical ionization processes and reactions are

FIG. 7.7   *(a) A DC glow discharge used in magnetron sputtering (nonthermal plasma) and (b) a thermal spray plasma used in a thermal plasma spraying process (thermal plasma) [65].*

(a)   (b)

TABLE 7.2	*Summary of the Main Ionization Processes in a Sputter-Discharged Plasma*	
Ionization process	**Reactions**	**Comments**
1. Direct ionization	$e + A \rightarrow A^+ + 2e$	Ions and additional electrons are created.
2. Penning ionization and excitation	$e + A \rightarrow A^m + e$ $A^m + B_2 \rightarrow A^+ + B_2^+ + e$	A metastable species A^m is created by electronic excitation; ions are created when metastable species A^m collides with species B.
3. Dissociative ionization	$e + AB \rightarrow A^+ + B + 2e$	An electron collides with a molecule AB to create ions and radicals.
4. Charge excitation	$A^+ + B \rightarrow A + B^+$	An ion collides with an atom to transfer the charge, where A and B can be the same species.
5. Ion–electron recombination	$A^+ + e \rightarrow A$	Ions and electrons combine to form a neutral species.

summarized in **Table 7.2**. All these processes will increase the ionization rates and excitation rates in the plasma, and this increase is one reason why inert gases like helium and argon are added to process the plasma discharge: they are relatively easy to ionize, while argon is also relatively inexpensive.

7.4.2 *Chemical Vapor Deposition*

The chemical vapor deposition (CVD) process is a popular thin-film deposition technology used to produce a wide range of metal, ceramic, and polymers coatings. In CVD, the material being deposited is generated from chemical vapor precursor species that are decomposed by reduction or thermal decomposition and come into contact with a heated substrate surface, and where they react or decompose forming a solid phase on the hot surface [66]. Reduction is normally accomplished by hydrogen at an elevated temperature, while decomposition is accomplished by thermal activation. A basic CVD process consists of the following steps: (i) diffusion of reactants to the substrate, (ii) adsorption onto the surface, (iii) surface chemical reaction(s) leading to deposition of solid, (iv) gaseous by-products desorption from the surface, and (v) gaseous by-products diffusing into the stream (**Fig. 7.8**).

FIG. 7.8 *Schematic diagram of the CVD process.*

FIG. 7.9 *Schematic drawing of a hot-wall CVD reactor.*

Figure 7.9 shows a typical configuration of a hot-wall CVD deposition system. In CVD, vapor supersaturation affects the nucleation rate of the film whereas substrate temperature influences the rate of film growth. These two factors together influence the extent of epitaxy, grain size, grain shape, and texture. Low gas supersaturation and high substrate temperatures promote the growth of single-crystal films. High gas supersaturation and low substrate temperatures result in the growth of less coherent and possibly amorphous films [67].

CVD encompasses a wide range of reactor and process types. The choice of process/reactor type is determined by the application via the requirements for substrate material, coating material and morphology, film thickness and uniformity, availability of precursors, and cost [66]. Several main CVD reactors that are widely being used are introduced here, and the reader should refer to other references [68–70] for detailed information on other systems.

There are a number of CVD processes that include (i) atmospheric pressure CVD (APCVD), (ii) low-pressure CVD (LPCVD), (iii) plasma-enhanced or -assisted CVD (PECVD or PACVD), and (iv) metal–organic CVD (MOCVD). Atmospheric pressure CVD (APCVD) operates at atmospheric pressure. APCVD has a very high deposition rate and is the simplest in design. The main disadvantages of APCVD are the low purity, poor uniformity of the deposited films and poor step (shape) coverage. In the mid-1970s, it was realized that LPCVD processing operating at medium vacuum (e.g., 1–10 mTorr) could have significant advantage over APCVD systems. By reducing the pressure, it was found that the diffusion coefficient was sufficiently enhanced and that deposition became surface controlled. With the hot-wall system, as shown in **Figure 7.9**, the deposition temperature could be maintained very uniformly, thereby achieving excellent film uniformity [69]. LPCVD processing produces films with excellent purity, uniformity, and good step (shape) coverage, but with lower deposition rates and higher operation temperatures than APCVD reactors. During CVD vapor deposition, many materials have very low vapor pressures and thus are difficult to transport via gases. For example, the availability of suitable precursors for high-Z (atomic number) elements was limited in the growth of ferroelectric thin films due to the fact that the vapor pressure of most high-Z element precursors is too low (e.g., below 1 mTorr at RT) to deliver a sufficient amount of material to the deposition chamber. Consequently, a large-area uniform film would be difficult to achieve due to the depletion of the precursors. One solution is to chemically attach the metal (Ga, Al, Cu, etc.) to an organic compound that has a very high vapor pressure. The CVD process that uses metal–organic source gases is called

MOCVD [71,72]. For instance, MOCVD may use tantalum ethoxide ($Ta(OC_2H_5)_5$), to create tantalum pentoxide (Ta_2O_5), or tetradimethylamino titanium (TDMAT) to create titanium nitride (TiN). The organic–metal bond is very weak and can be broken via thermal means on wafer substrates, thereby depositing the metal while the high vapor pressure organic is pumped away.

CVD is not a "line of sight" process and offers some distinct advantages such as uniform deposition over complex geometries ("conformal" deposition) and large areas, good conformal step coverage, and relatively easy control of stoichiometry of deposited films. However, CVD suffers from limitations in growth of thin films [73,74]. CVD processing is generally accompanied by volatile reaction by-products and unused precursor species. In addition special safety precautions must be taken to insure a minimum of the organic by-products are released into the atmosphere. For example, in MOCVD, as the human body absorbs organic compounds relatively easily, the metal organics are easily absorbed by humans. Once in the body, the weak metal–organic bond is easily broken, thus poisoning the body with heavy metals that often cannot be easily removed by normal bodily functions. Another limitation of APCVD and LPCVD processing is the high deposition temperature. In general, the typical substrate temperature is in the range of 800–1200°C; therefore, the substrates used are limited to a narrow range, such as cemented carbides to minimize the risk of dimensional and microstructural change. The high operation temperature also limits the variety of the materials that can be produced without changing the film properties; for example, a deposition temperature less than 400°C is needed for the deposition of certain silicon thin films used in microelectronic applications.

With respect to overcoming the high process temperature limitations, several modifications of the conventional CVD process have been developed. In recent years, the use of a plasma to dissociate and ionize the gaseous precursors and deposit the coatings can be carried out at much lower temperatures than in conventional CVD. This process is called plasma-enhanced (or -assisted) chemical vapor deposition (PECVD) and provides considerable potential to be utilized in the deposition of semiconductive coatings [75–77]. PECVD reactors also operate under low pressure, but do not depend completely on thermal energy to accelerate the reaction processes. They transfer energy to the reactant gases by using a "glow discharge." The glow discharge used by a PECVD reactor is generally created by applying a radio frequency (RF) field to a low-pressure gas, creating free electrons within the discharge region. The electrons are sufficiently energized by the electric field that gas phase dissociation and ionization of the reactant gases occur when the free electrons collide with the gas species. Energetic species are then adsorbed on the film surface, where they are subjected to ion and electron bombardment, rearrangements, reactions with other species, new bond formation, and film formation and growth. The typical deposition temperature in PECVD can be reduced below 500°C. The relatively low deposition temperature makes PECVD suitable for a wide range of substrates including tool steel and hot work tool steel and Si wafers used in microelectronics so that the deposition temperature is below the tempering temperature of tool steels, therefore maintaining the required microstructure and properties of the substrate and minimizing distortion of the substrate [78].

CVD is capable of producing a wide range of coating materials, including typical nitrides, carbides, and oxides, such as TiN [79,80], TiCN [81], titanium carbide (TiC) [82,83], DLC [84–86] TiB_2 [87], and Al_2O_3 [88]. CVD/PECVD has also been used in a multitude of semiconductor wafer fabrication processes, including the production of amorphous and polycrystalline thin films, for example, amorphous silicon for solar cell [89,90], polycrystalline silicon for gate contact [91], deposition of SiO_2 [92] and silicon nitride [93], and growing single-crystal epitaxial layers [94,95]. It has also been used to produce thick oxides used for isolation between metal interconnects, doped oxides for global planarization, and dielectrics for isolation and capacitors.

7.4.3 *Physical Vapor Deposition*

Compared to the CVD process, PVD can be carried out at lower deposition temperatures and without corrosive products. The substrate deposition temperature in the PVD process is in the range of 150–550°C thereby greatly bordering the substrate selection range. Almost all kinds of tool steel, for example, hot working tool steel and high speed steel, can be used as substrates in the PVD process, and no heat treatment is required after coating deposition due to the low process temperature. A range of desired compositional coatings can be readily produced by PVD that exhibit fine grain size, dense structure, and improved wear properties. In the PVD process, the coating material is vaporized or sputtered in a chamber to produce a flux of atoms or molecules which condenses on the substrate to be coated. Chemical compounds can be deposited using a composite target or by introducing a reactive gas (nitrogen, oxygen, or simple hydrocarbons) containing the desired reactants, which dissociate and ionize in the plasma to form reactive species (e.g., N, O, C) that react with metal atoms sputtered from the PVD source or target [68]. In this section, variations of PVD processes including cathode arc evaporation (CAE), electron beam evaporation, plasma sputtering, ion beam sputtering (IBS), pulsed laser deposition (PLD), and thermal plasma processing will be introduced. Some important developments in plasma sputter deposition including the UBMS, pulsed magnetron sputter deposition, and high-power pulse magnetron sputtering are emphasized later for their important role in enhancing the chemical and/or structural nature of the deposited films. Each of these systems and techniques will be described in this section, as well as some of the current applications of the films produced.

Electron Beam Evaporation.   Electron beam evaporation falls in the catalog of general evaporation, in which a block of the source material to be deposited is transformed into vapor form by means of high-energy electron beam bombardment, and then it is allowed to condense on the substrate [96]. The electron beam evaporation process typically involves the following steps: (1) Generation of an electron beam by an electron gun, which uses thermal or plasma emission of electrons produced by an incandescent filament cathode, (2) emitted electrons are accelerated towards an anode, which usually is the crucible itself holding the evaporation material, by a electric field (kilovolts), (3) beside the electric field, a magnetic field is used to bend the electron trajectory and move (scan) the electron beam across the treated surface allowing the electron gun to be positioned below the evaporation line, (4) vaporization of the material to the volatile state in the form of vapors, and (5) deposition of the vaporized material onto the

FIG. 7.10	*A diagram of the electron beam evaporation equipment.*

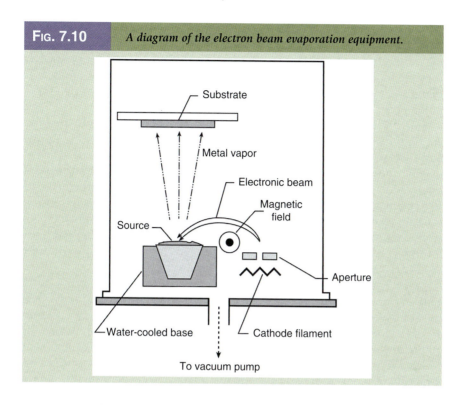

substrate. A diagram of the electron beam evaporation equipment is shown in **Figure 7.10**.

Electron beam evaporation offers several advantages over competing processes. By adjusting the value of the accelerating voltage and the combination of different electric fields, the electron beam can be localized or unfocused, accelerated or retarded, and pulsed [97], thus it is possible to obtain localized heating on the material being evaporated with a range of high densities of evaporation power (several kilowatts). This ability to vary the evaporation power allows precise control of the evaporation rates, from as low as one nanometer per minute to as high as a few micrometers per minute. This technique is an extremely versatile means of depositing uniform high-purity thin films. The materials used for evaporation are available in near limitless shapes and forms, for example, the pellets, slugs, and disks. In addition, elevated temperatures in excess of 3500°C can be used allowing the production of thin-film coatings from pure metals, including materials with high melting point, such as refractory metals W, Ta, C, etc., as well as numerous alloys and compounds. Cooling the crucible avoids contamination problems from heating and degasification. Electron beam evaporation also offers excellent material utilization to other methods, co-deposition and sequential deposition systems, precise film composition, structural and morphological control, uniform low temperature deposition, and freedom from contamination.

However, electron beam evaporation is a "line-of-sight" deposition process, necessitating a two- or three-axis rotation of the components to be coated—and thus special rotational rigs have to be designed within the deposition chamber. The two-axis translational and rotational motion of the shaft helps for coating the outer surface of complex geometries, but this process cannot be used to coat the inner surface of complex geometries. Another potential problem is that filament degradation in the electron gun results in a nonuniform evaporation rate.

Since the introduction of electron beam evaporation in the 1950s, thin-film applications requiring electron beam evaporation are continually increasing. Due to its widespread material availability, very high deposition rate, efficient material utilization, and unmatched film purity and uniformity, electron beam evaporation is employed in the production of wear resistant and thermal barrier coatings (TBC) in aerospace, hard coatings for cutting tools, corrosion resistant coatings, optical films for coating lenses and mirrors, infrared detectors, costume jewelry and filters with anti-reflection, and insulating and resistor films on electronic components for nanotechnology and semiconductor industries [98–102].

Cathodic Arc Evaporation (CAE).   For more than a decade, CAE has been widely used to deposit various kinds of refractory and wear-resistant coatings such as Ti–Al–N [103] and its variants [104] for cutting tools, and CrN [105] coatings for automotive applications. CAE deposition is another evaporation-like process in which a very high current (hundreds of amperes) direct current arc is struck on a metallic cathode surface, where the arc interacts with the cathode surface vaporizing the cathode materials with a high-power density at the contact point [106]. Due to a high-power density on the arc electrodes, the CAE process is characterized by a combination of a high deposition rate and a high degree of ionization ($\geq 90\%$) of evaporated species with high ion energies ($20–150\,eV$), which makes this process a versatile deposition technology for producing well adherent and dense metal and compound films. Due to the nature of the high-current vacuum arc discharge, however, only target materials with good electro-conductivity can be used as evaporation sources. Also materials with a too high or low melting point or poor mechanical strength cannot be used.

The main disadvantage of the cathodic arc deposition is the production of microdroplets (macroparticles) due to the high power density on the cathode. These macroparticles will also become embedded in the films with a typical size from $0.2\,\mu m$ to several micrometers. These macroparticles are undesirable since they will degrade the uniformity of the film surface and consequently functions of the film, especially for thin films used for microelectronics and electro-optics.

The macroparticle emission from the cathode spot can be, to a certain extent, reduced by [107] (i) decreasing the arc current, (ii) decreasing the cathode surface temperature, (iii) using a pure cathode material without gas, and (iv) contamination of the cathode surface by a reaction product with a higher melting point. The macroparticle can also be greatly reduced or removed from the arc plasma by several approaches during plasma transport to the substrate, of which the magnetic filter has been the most successful. There are many designs for macroparticle filters and the most popular design is based on the work by Aksenov et al. in the 1970s [108]. They used a curved magnetic filter with a positive bias of approximately $20\,V$ applied between the filter and anode. The magnetic

| FIG. 7.11 | *A diagram of the cathodic arc system equipped with a magnetic solenoid filter.* |

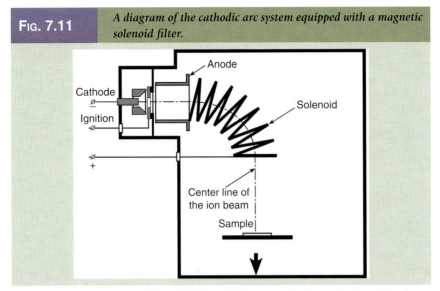

Source: Y. H. Liu, J. L. Zhang, D. P. Liu, T. C. Ma, and G. Benstetter, *Surface and Coatings Technology,* 200(7), 2243–2248 (2005). With permission.

filter can prevent line-of-sight of the macroparticles and guide the vacuum arc plasma by the curved magnetic field to the substrate. A typical cathodic vacuum arc system equipped with a magnetic solenoid filter is shown in **Figure 7.11** [109].

Pulsed Laser Deposition (PLD). Since Dijkkamp and Venkatesan [110] successfully deposited a high-temperature superconductive $YBa_2Cu_3O_7$ film using PLD in 1987, PLD has gained great attention in the past two decades for its ease of use and success in depositing materials of complex stoichiometry that are normally difficult to deposit by other methods, especially multielement oxides.

PLD uses short and high-power laser pulses (typically $\sim 108\,W\cdot cm^{-2}$) to evaporate and ablate material from the surface of a solid target in an ultrahigh vacuum chamber. The primary ablation mechanisms involve many complex physical phenomena such as collisional, thermal, and electronic excitation, and exfoliation and hydrodynamics [111]. The ablated species expand into the surrounding vacuum in the form of a plume containing many energetic species including atoms, molecules, electrons, ions, clusters, particulates, and molten globules, before depositing on the typically hot substrate. Then, the ablated material is deposited through the plasma plume onto the heated substrate surface with nucleation and growth of a thin film. A diagram of the PLD equipment is shown in **Figure 7.12**.

The PLD can occur in the presence of a wide variety of gases, which makes it an extremely versatile technique for preparing a wide range of ceramic oxides, nitrides, metallic multilayers, and various superlattice thin films [112]. Unlike thermal evaporation, which produces a vapor composition dependent on the vapor pressures of elements of the target material, the PLD generates entire/congruent evaporation of the target irrespective of the evaporating point of the constituent elements or compounds of the target that is facilitated by an extremely

FIG. 7.12   *A diagram of the PLD equipment.*

high heating rate of the target surface (10^8 K·s^{-1}). Materials are dissociated from the target surface and ablated out with the same stoichiometry as the target. Therefore, it is generally easier to obtain complex film stoichiometry for multi-element materials using PLD than with other deposition technologies. The conceptually simple design and the versatility of PLD make it highly cost-effective, in that complex multilayer films are relatively straightforward to be produced within a single system by moving various targets into and out of the laser beam focal point. In addition, by using mirrors to change the beam path, several deposition systems can be clustered around a single laser [111]. In the PLD process, due to the short laser pulsed duration (~10 ns) and hence the small temporal spread (<10 ms) of the ablated materials, the deposition rate can be extremely fast (~10 mm·s^{-1}). Consequently a layer-by-layer nucleation is favored and ultrathin and smooth films can be produced.

In spite of the mentioned advantages of PLD, some disadvantages have been identified in use of this deposition technique. One of the major problems is the splashing or the particulate (macroparticles) deposition on the film. The physical mechanisms leading to splashing include surface boiling, expulsion of the liquid layer by shock wave recoil pressure, and exfoliation. The size of particulates may be as large as a few microns. Such particulates will greatly affect the growth of the subsequent layers as well as the electrical properties of the film and therefore should be eliminated. The spot size of the laser and the plasma temperature has significant effects on the deposited film uniformity. The target-to-substrate distance is another parameter that governs the angular spread of the ablated materials.

Another problem is the narrow angular distribution of the ablated species, which is generated by the adiabatic expansion of the laser-produced plasma plume and the pitting of the target surface. These features limit the use of PLD in producing a large-area uniform thin film such that PLD has not been fully deployed in industry. Recently, remedial measures such as inserting a shadow mask to block off the particulates and rotating both target and substrate in order to produce a larger uniform film have been developed to minimize some of the PLD problems.

Applications of the PLD technique range from the production of high-temperature superconducting films, for example, $YBa_2Cu_3O_{7-x}$, $NdBa_2Cu_3O_7$, ferroelectric $BaTiO_3$ and other perovskite thin films [113–116], certain magnetic materials, for example, yttrium iron garnet (YIG) [117,118] and ferromagnetic shape-memory (FSM) alloy Ni–Mn–Ga [119], biocompatible coatings for medical applications, for example, protein [120], pepsin [121], to TBC for turbine blades. In spite of this widespread usage, the fundamental processes occurring during the transfer of material from target to substrate are not fully understood and consequently need further research.

Thermal Plasma Processing of Thin Films. Thermal plasma processing is identified as utilizing high-temperature plasma to deposit metallic and nonmetallic materials in a molten or semimolten state on a substrate [122–124]. Thermal plasma processing may also be referred to as injection plasma processing (IPP), in which different types of processes were developed, such as thermal plasma CVD, plasma flash evaporation, and plasma spray [125]. The plasma spray process is a well-established commercial process that can be operated in normal atmospheric conditions and is referred to as atmospheric plasma spray (APS), or in a vacuum chamber (VPS), or with a protective gas at low pressure (LPPS). However, the basic processing principles are similar in that firstly a high-temperature plasma is initiated from a plasma gas (Ar, N_2, H_2, or He) by applying a high voltage between a cathode and anode (as shown in **Fig. 7.13a**), and material in the form of powder is injected into the high-temperature plasma flame, where it is rapidly heated, melted, and accelerated to a high velocity. The hot material impacts on the substrate surface and rapidly cools forming a coating. A photo showing a thermal plasma spray process at work is presented in **Fig. 7.13b**. Similar to the thermal spray process, plasma flash evaporation also uses the raw

FIG. 7.13 *(a) Schematic diagram of a solution precursor thermal plasma spray process [126] and (b) a photo showing a thermal plasma process at work [127].*

(a)

(b)

powder material as the reactant. However, the powder particle size needs to be fine enough (<10 μm) in an effort to achieve evaporation, instead of only melting the reactant powders.

Plasma spraying has the advantage that it can spray very high-melting point materials such as refractory metals like tungsten, and ceramic, for example, zirconia, which are not suitable for combustion processes. Plasma-sprayed coatings are generally much denser, stronger, and cleaner than the other thermal spray processes. Plasma spray coatings probably account for the widest range of thermal spray coatings and applications and thus making this process the most versatile. Due to the high deposition rate, typical coating thickness ranges between 50 μm to a few millimeters.

However, thermally plasma-sprayed coatings normally contain inhomogeneities such as inter-splat porosity, unmelted particles, and micro-cracks, as shown in **Figure 7.14a** [123]. The presence of interconnected porosity is of prime concern as it limits the effectiveness of the barrier provided by the coating. Another disadvantage of thermal spray is that it is hard to coat complex shaped substrates since it is a line-of-sight process. Postspray polishing is usually needed to obtain the desired surface finish on the surface. LPPS coating can provide coatings with improved microstructure and decreased porosity (**Fig. 7.14b**). The individual Al_2O_3 splats are in the LPPS Al_2O_3 coating shown in transmission electron microscope (TEM) (**Fig. 7.14b**). The splats are seen to have a nanocrystalline Al_2O_3 phase embedded in an ordered amorphous Al_2O_3 phase.

Recently, another new development in thermal plasma processing, plasma chemical vapor deposition (PCVD) process, has been developed. Unlike the previous two-spray process, the reactants fed into the high-energy-density plasma are gaseous and liquid precursors in the process of PCVD [128–131]. With the high temperatures, the precursor material that is injected into the plasma is rapidly vaporized and dissociated and accelerated towards the substrate. The thermal plasma in most thermal plasma processes is initiated by a high voltage DC discharge due to its high energy density, high velocity, and ease of use. However, the steep temperature and velocity gradients occurring in the plasma generally result in nonuniformity in terms of heating, trajectory, and reactions

FIG. 7.14 (a) *Cross-section of a CrC_{75}–$NiCr_{25}$ coating obtained by high-velocity oxy-fuel (HOVF) thermal sprayed coating [123] and* (b) *TEM micrograph of LPPS Al_2O_3. Note the nanocrystalline Al_2O_3 embedded in the amorphous Al_2O_3 in each splat.*

FIG. 7.15	*Scanning electron micrograph (SEM) of carbon nanotubes embedded in carbon black materials, as synthesized by the thermal plasma technique.*

Source: H. Okuno, E. Grivei, F. Fabry, T. M. Gruenberger, J. Gonzalez-Aguilar, A. Palnichenko, L. Fulcheri, N. Probst, and J.-C. Charlier, *Carbon*, 42(12–13), 2543–2549 (2004). With permission.

in addition to the high heat flux to the substrate, which leads to difficulties in controlling the substrate temperature. In recent years, the use of RF and hybrid plasmas has developed rapidly since they offer better control of the plasma to minimize trajectory and reaction variations [125].

Thermal plasma processing is considered to be one of the prime candidates for producing high-temperature thermal barrier coatings for aircraft engines, industrial gas turbine blades, and antiwear and anticorrosion coatings for high-temperature applications due to its relatively low cost, high deposition rate, and high durability [132]. In recent years, thermal plasma processing has been reportedly used to synthesize carbon nanotube and nano-necklaces at relatively low operation temperatures as shown in **Figure 7.15** [124].

Sputter Deposition. The PVD deposition techniques discussed so far are based fundamentally on thermally evaporating materials at elevated (above melting point) temperatures. Another important PVD deposition mechanism is sputtering. Sputter deposition is the deposition of atoms that are vaporized from a solid target or source by bombarding the surface with energetic ions [133]. Since Grove first observed sputtering in a DC gas discharge tube in 1852, sputtering has been widely used for surface cleaning and etching, thin-film deposition, surface and surface layer analysis, sputter ion sources, and for the modification of the properties of thin films (implantation) depending on the ion energy range ($1\,eV$ to $10,000\,keV$) involved in the deposition [134]. The incident particles are usually inert gas ions, but any ion, neutral atom, molecule, or even a photon can be used if the energy is sufficient [135].

As schematically portrayed in **Figure 7.16**, in sputter deposition, the impact of the energetic particles on the target surface kinetically knocks one or more of the surface or near-surface atoms off their equilibrium sites. These atoms, which have received considerable kinetic energy from the initial particles, move deeper into the target material and undergo further collisions. This process continues until eventually causing the ejection of atoms (sputtering) from the target surface [136].

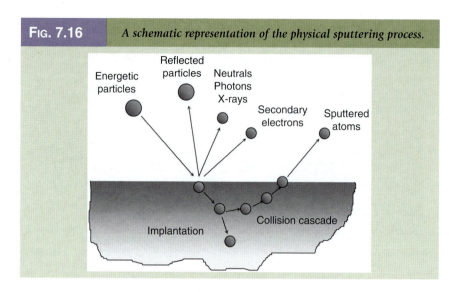

Fig. 7.16   *A schematic representation of the physical sputtering process.*

The interaction of energetic ions with surfaces will also create a variety of interactions besides sputtering, such as generation of secondary electrons, neutrals, photons, x-rays, and implantation of atoms into the substrate (**Fig. 7.16**). These interactions play an important role in the field of surface science, such as the ionization of atomic particles at/near surfaces, the modification of surfaces, and surface analysis. For example, the emission of secondary electrons is essential for maintaining the discharged plasma. Optical emission from a glow discharge during sputtering can be used for process control and chemical analysis in the optical emission spectrometer (OES) [183].

The use of energetic ions in film deposition has numerous advantages in that controlled ion bombardment can lead to significant micro/nanostructural modifications and properties. Ion-assisted deposition of materials can produce coatings with superior properties since the ion assistance provides incident atoms with additional energy. A typical kinetic energy of sputter deposition is $1-10\,eV$, which is orders of magnitude above that of typical evaporated particles. This added energy can modify the nucleation process, improve film adhesion, increase film density, change the film texture, trigger phase changes, influence film stress, and change film microstructure. All these factors can be utilized to produce coatings with improved corrosion resistance, hardness, and optical, physical, and wear properties.

Sputter Deposition I: Ion Beam Sputtering (IBS).   Sputter deposition is typically practiced in either a plasma (**Fig. 7.17a**) or an ion beam (**Fig. 7.17b**) configuration. In plasma systems (**Fig. 7.17a**), a cathode is used which is bombarded by ions from the plasma. If the energy and pressure conditions are appropriate, the cathode (target) material is then sputtered off and can deposit on nearby surfaces. In IBS (**Fig. 7.17b**), a beam of ions is generated using a remote source. This ion beam is then directed onto the target, and atoms are sputtered from the target onto a nearby sample [106].

FIG. 7.17	*Schematic drawing of (a) conventional plasma sputtering and (b) dual ion beam sputtering system.*

During the IBS process, the high-intensity Kaufman ion source is routinely used to generate a high-energy ion beam. The element, alloy, or compound targets are sputtered by the ion beam in a prescribed direction controlled by a magnetic or electrical filed. The sputtered atoms are deposited on a heated or biased substrate to form the thin film. The growing film may also be irradiated with ions from a second ion source, as indicated in **Figure 7.17b**. This technique is termed dual ion beam sputtering and has been used extensively both to modify the film structure and to synthesize compounds. The second ion source may operate with inert or reactive ions with energies ranging from $20\,eV$ to $10\,keV$.

The IBS process can also be operated in the plasma sputter mode. After a pure ion beam has been extracted from an ion source, electrons may be added to the ion beam to form a plasma beam (equal number of positive ions and electrons) which will not diverge and not cause a charge build-up on the target surface. In the Kaufman source these electrons are normally generated from a hot filament ("neutralizer filament"), such as tungsten, that easily generates electrons due to thermal emission. The beam is volumetrically neutral due to the addition of the electrons. Plasma beams have the advantage that the electrons can easily be deflected (steered) by a magnetic or electrostatic field and the ions will generally follow. They can be operated in vacuum and at a high pumping speed. Therefore contamination can easily be controlled. Also the flux and energy of the bombarding particles can easily be monitored and controlled, and insulating surfaces can be sputtered.

The IBS technique offers several advantages. The process of ion beam sputter deposition produces sputtered atoms with high average energy, for example, $3-10\,eV$ (compared to $0.1\,eV$ or so in thermal evaporation). Films made from these atoms show improved properties and adhesion when compared to conventional deposition techniques such as thermal evaporation. Sputter deposition of compound and alloy films is feasible since stoichiometry is normally preserved in the growing film. During the IBS process, the source materials are directly sputtered onto the substrate and the ion beams can be directionally controlled, the films can be deposited with extremely high accuracy and repeatability.

The advantage of the IBS system against a conventional sputtering system lies in the fact that target and substrate can be separated from the plasma compared with the conventional glow discharge sputtering system where substrates are directly exposed to the plasma. This makes the film free from thermal effects and radiation damage by the plasma, resulting in low substrate temperatures [137]. Therefore, IBS deposition can be used in depositing some high-performance transparent conductive oxide (TCO) films such as indium–tin oxide (ITO) and ZnO, which are widely used as electrodes for flat panel displays, solar cells, and electroluminescence (EL) devices [138,139], and narrow-gap semiconductors for thermoelectric power devices, for example, Mg_2Si [137]. The ion bombardment energy, in these latter examples, must be kept low (e.g., $<30\,eV$) to avoid the generation of point (e.g., vacancies, interstitials) and line defects (e.g., dislocations, stacking faults) that will impair the electronic properties of these films.

The IBS technique can also be used as one of the surface modification techniques to reduce surface roughness of materials by selectively detaching atoms and nanoparticles from the surface with the bombarding energetic ions from 1–10 to a few tens of kiloelectron volts onto the film surfaces. This technique can be applied to a surface that needs to have submicrometer surface roughness. Ion beam machining is used for ultraprecision machining of high-melting point and hard, brittle materials where machining depth needs to be precisely controlled. For ultraprecision machining of a large region in IBS processes, a high current density ion beam and a large-area ion source with a uniform beam extraction unit are necessary [140]. However, the small ion beam size (~100 mm diameter) in the IBS limits the substrate size for the film deposition. Therefore, the equipment required is somewhat more expensive when measured in terms of rate and coated area per unit cost.

Sputter Deposition II: Balanced and Unbalanced Magnetron Sputtering. More frequently, the sputter deposition is practiced in a plasma configuration (**Fig. 7.17a**). The simplest system still in use today is the diode sputtering deposition, which consists of the target and anode facing each other. In many cases, the anode is the chamber wall which is grounded, and the cathode is then biased negatively. With the appropriate gas density and an adequate electric field between the anode and cathode, a plasma can be formed by gas breakdown into ions within the chamber. Even though planar diode sputtering deposition is still used today due to its simplicity and the relative ease of fabrication of targets for a wide range of materials, it has several disadvantages. In diode sputtering, not all of the electrons escaping the target contribute to the ionized plasma glow area, thereby resulting in a low deposition rate. The wasted electrons are accelerated away from the target causing radiation and other problems, for example, the heating of the substrate.

Sputtering first achieved widespread use in research and industrial application with the introduction of magnetic assistance [141]. The use of magnetic fields to enhance the sputtering rate led to the term "magnetron sputtering." A MS source addresses the electron problem by placing magnets behind, and sometimes, at the sides of the targets. These magnets capture the escaping electrons and confine them to the immediate vicinity of the target (**Fig. 7.18a**). The ion current (density of ionized atoms hitting the target) is increased by an order of magnitude over conventional diode sputtering systems, resulting in faster deposition rates at lower pressure. The lower pressure in the chamber also helps create a cleaner film and a lower target temperature, enhancing the deposition of high-quality films. Additional modifications to physical sputter deposition have been made to enhance the chemical and/or structural nature of the deposited films.

As mentioned above, the magnetic field behind the MS targets is designed to trap the electrons, and hence, the plasma in the vicinity of the substrate. In traditional planar-balanced magnetron sputtering, the north and south magnets behind the targets are balanced (i.e., equal strength) (**Fig. 7.18a**) [142]. One of the disadvantages of the balanced magnetron source is that the plasma is effectively trapped near the surface of the sputtering target resulting in lower plasma density over large coating volumes (**Fig. 7.18b**). Normally, ion current density (ICD) at the target (cathode) (the current drawn per unit area of a negatively biased substrate) in conventional balanced magnetron sputtering is less than $1\,mA \cdot cm^{-2}$ [142]. The problem was partially solved by adding auxiliary ionization sources or using RF. The other disadvantage is the poor deposition rate compared to thermal evaporation methods.

In recent years many researchers have tried to overcome the disadvantages of MS and increase the plasma density. These developments include the unbalanced magnetron, closed-field configuration of magnetrons, pulsed magnetron sputter deposition, and high-power pulsed DC magnetron sputtering (HPPMS).

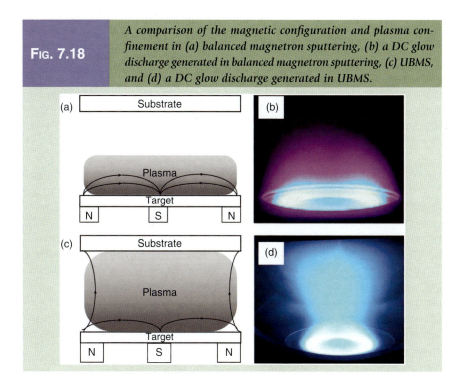

FIG. 7.18   *A comparison of the magnetic configuration and plasma confinement in (a) balanced magnetron sputtering, (b) a DC glow discharge generated in balanced magnetron sputtering, (c) UBMS, and (d) a DC glow discharge generated in UBMS.*

The invention of the unbalanced magnetron by Windows and Savvides in 1986 offered a better solution to enhance the plasma density [143,144]. An unbalanced magnetron uses stronger magnets on the outside than the center resulting in the expansion of the magnetron field lines, electrons, and plasma away from the surface of the target towards the substrate (**Fig. 7.18c**). The effect of the unbalanced magnetic field is to trap fast-moving secondary electrons that escape from the target surface. These electrons undergo ionizing collisions with neutral gas atoms at locations away from the target surface and produce a greater number of ions and further electrons in the region of the substrate, thereby considerably increasing the plasma density and substrate ion bombardment (**Fig. 7.18d**). The ICD in UBMS is increased to $2–10\,mA\cdot cm^{-2}$ [142].

Sputter Deposition III: Closed-Field Unbalanced Magnetron Sputtering (CFUBMS). In the late 1980s an important improvement in MS was developed—termed the closed-field unbalanced magnetron sputtering. The purpose of CFUBMS is to enhance ionization and increase the ICD in MS as proposed by Sproul et al. [145,146] and Tominaga [147]. The ICD can be further increased to $5–20\,mA\cdot cm^{-2}$ compared to the UBMS ($2–10\,mA\cdot cm^{-2}$). [142].

A comparison of a mirrored and a closed-field magnetron configuration [145] is shown in **Figure 7.19**. The major feature in the CFUBMS system was the idea of using unbalanced magnetrons in an arrangement whereby neighboring magnetrons are of opposite magnetic polarity. Using this arrangement, the deposition zone in which the substrates are located is surrounded by linking magnetic field lines (**Fig. 7.19a**). This traps the plasma region, prevents loss of ionizing electrons escaping to the chamber walls resulting in much higher plasma density (ICDs) (**Fig. 7.19b**), and dense, hard, well-adhered coatings by enhanced chemical reaction at the substrate. On the other hand, the magnetic field lines are not closed in the mirrored configuration (**Fig. 7.19c**), thereby resulting in a poor plasma density (**Fig. 7.19d**).

Closed-field systems can be configured using any even number of magnetrons. The use of multiple magnetrons in conjunction with a two-axis or three-axis substrate rotation system allows for the uniform deposition of large and complex shaped components in the closed-field plasma. A diagram showing a four magnetron closed-field unbalanced magnetron sputtering system configuration is presented in **Figure 7.20**.

Sputter Deposition IV: Pulsed Magnetron Sputtering. Ceramic and compound films of various components can be reactively deposited using multiple magnetrons in MS. Reactive sputtering can be used to deposit most thin films with desired composition as a controlled monolithic or compositionally graded structure by control of the power density on multiple targets and partial pressure of the reactive gas [148–150]. Various nitride, carbide, and oxide ceramic coatings such as TiN [151], CrN [152], TiC [153], TiAlN [154], CrAlN [155], Al_2O_3 [156] etc., can be deposited using reactive sputtering from metal targets (titanium, chromium, aluminum, etc.) in a reactive gas atmosphere, for example, N_2/Ar, CH_3/Ar or O_2/Ar gas mixture. When a negative bias is applied to the substrate, ions from the secondary plasma are accelerated towards the substrate and enhance the film growth at the substrate.

A comparison of the magnetic field configurations and plasma photographs in (a) the "mirrored" magnetic field configuration, (b) a DC glow discharge generated in the "mirrored" magnetron sputtering, (c) "closed-field" magnetic field configuration, and (d) a DC glow discharge generated in "closed-field" magnetron sputtering.

However, for deposition of insulating films, the insulating film builds up on the surface of the chamber and/or anode. When the insulating layer on the anode becomes thick, the sputtering discharge becomes unstable. This phenomenon is called "disappearing anode." The target can also charge up quickly due to a nonconductive layer formed on the target at which point the target is said to be "poisoned," which makes sputtering more difficult, decreases the sputtering rate, and increases the power on the target. The charges will generate arcing problems in which microparticles will be ejected from the target and incorporated into the deposited coatings. This condition leads to nonuniform deposition, and inhomogeneities and defects in the coatings, as well as a reduced deposition rate [157].

To overcome this problem, the use of RF sputtering was developed in the 1960s. However, RF sputtering is not used extensively for commercial MS due to its low deposition rate, high cost, and the generation of a high temperature from the self-bias voltage associated with the RF power.

FIG. 7.20   *Diagram showing a four magnetron closed-field unbalanced magnetron sputtering system configuration.*

In recent years, an alternative technique using a pulsed DC plasma has been developed and is called pulsed DC magnetron sputtering. Pulsed DC magnetron sputtering utilizes a pulsed potential to neutralize the positive charge on the target surface and eliminate the arcing by appropriately controlling the pulsing parameters of the target potential. **Figure 7.21** schematically presents some typical target voltage–potential waveforms used in pulsed DC magnetron sputtering. The continuous target voltage in the DC mode (**Fig. 7.21a**) is either turned off periodically in the unipolar mode (**Fig. 7.21b**), or more commonly, switched to a positive voltage (**Fig. 7.21c and d**) in the pulsed DC magnetron sputtering. During the normal pulse-off (target-on) period (τ_{on}), the negative sputtering voltage is applied to the target as in conventional DC sputtering. However, the target negative potential is periodically interrupted by a positive pulse voltage with a period of τ_{rev}. The reversed positive voltage is variable depending on the power supply allowance, which is either reversed to a smaller positive voltage than the nominal negative pulse voltage in the asymmetric bipolar mode (**Fig. 7.21c**) or reversed to the same magnitude of positive voltage as the nominal negative pulse voltage in the symmetric bipolar mode (**Fig. 7.21d**). The duty cycle is defined as the negative pulse time divided by the period of the pulsing cycle (τ_{cycle}) as shown in equation (7.7):

$$\text{Duty cycle} = \frac{\tau_{cycle} - \tau_{rev}}{\tau_{cycle}} \tag{7.7}$$

For a given positive pulse width, the full range of frequencies may not be available due to the power supply limitation, for example, for an Advanced Energy

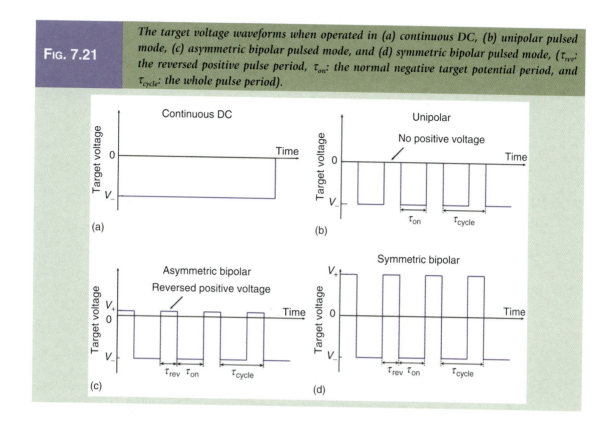

FIG. 7.21 The target voltage waveforms when operated in (a) continuous DC, (b) unipolar pulsed mode, (c) asymmetric bipolar pulsed mode, and (d) symmetric bipolar pulsed mode, (τ_{rev}: the reversed positive pulse period, τ_{on}: the normal negative target potential period, and τ_{cycle}: the whole pulse period).

Pinnacle Plus power supply, the smallest duty cycle is 50% [158]. In practice, the waveforms are virtually never as intended due to nonlinearities of either the plasma or the power supply circuitry. Therefore, the shapes of the resulting power waveforms are complex.

In the closed-field configuration, two or more magnetrons are commonly used for reactive sputtering. All magnetrons can be run in a pulsing condition, thereby resulting in different combinations of pulsing modes. Normally, the magnetron potential can be pulsed in either an asynchronous bipolar mode (**Fig. 7.22a**), in which the two target voltage waveforms are out of phase, or a in synchronous bipolar mode (**Fig. 7.22b**), in which the two target voltage waveforms are in phase. The degree of out of phase in the asynchronous mode is totally dependent on the frequencies and reverse time on each magnetron.

During the reversed positive pulse, the charge built-up on the insulting material during the negative pulse period is discharged to eliminate breakdown and arcing. Therefore, a stable deposition process and smooth coating structure can be obtained. Moreover, with precise control of the partial pressure of reactive gas, high deposition rates can be achieved in depositing insulating films. A comparison of the microstructure of IrO_2 films deposited by P-CFUBMS and normal DC magnetron sputtering is shown in **Figure 7.23**. A dense and smooth surface IrO_2 film was produced by P-CFUBMS. On the other hand, the film deposited by DC magnetron sputtering exhibited considerable microparticles covering the film surface.

FIG. 7.22 *Different target voltage waveforms in a dual magnetron pulsed mode: (a) asynchronous bipolar pulsing and (b) synchronous bipolar pulsing.*

In the sputtering of a metal film, if a bias potential is applied to the substrate, one can control the film properties such as adhesion, texture, morphology, and density of the film. This control is achieved because one can extract ions to bombard the growth front during film growth. However, if the substrate is an insulator, it can get charged up very quickly and one would lose the benefit of ion bombardment, particularly during the initial stages of growth. RF potential can be used to neutralize the substrate. However, the ion energy distribution generated by a RF power is not uniform and is not well controlled. Recently, it has been recognized that using a pulsed bias potential on the substrate can overcome this difficulty and can produce better controlled microstructures and a superior film quality [159].

FIG. 7.23 *SEM photomicrographs of fracture sections of iridium oxide coatings deposited by (a) DC reactive sputtering and (b) pulsed reactive sputtering.*

Sputter Deposition V: High-Power Pulsed DC Magnetron Sputtering (HPPMS).
Recently, a new high-power pulsed DC magnetron sputtering (HPPMS), operating
in a unipolar mode, has attracted wide attention. In this new method, a high-density
plasma is created in front of the sputtering source by using pulsed high-target
power density, ionizing a large fraction of the sputtered atoms [160–162].

In normal DC and pulsed DC magnetron sputtering, the degree of ionization
is low (typically less than 10%), which is due to the low power density (e.g.,
$3\,W \cdot cm^{-2}$) limited by the target overheating from the ion bombardment [163].
However, in HPPMS, the average thermal load on the target is low by operating
the target at high power density (e.g., $3000\,W \cdot cm^{-2}$) in a pulsed condition;
consequently a considerably large fraction of ionized species can be created by the
high probability for ionizing collisions between the sputtered atoms and
energetic electrons. It was reported that the fraction of ionized target species in
high-power pulsed DC magnetron sputtering can be considerably increased up
to 70% or higher compared with less than 10% for conventional DC magnetron
sputtering [164].

The high degree of ionized ions with controlled energy makes it possible to
control the film growth behavior and produce high quality films since energetic
condensation of the film can be easily achieved. The potential of this emerging
new high-power pulsed magnetron sputtering technique includes depositing
fully dense films with equiaxed structure, controlling the orientation of the film,
and depositing thick (maybe up to $20\,\mu m$ or more) films with low residual
stress. As the deposited species are largely ions, it is possible to further control
metal ion energies and their trajectories by biasing the substrate. Alami synthesized Ta films on a Si substrate placed along the wall of a 2-cm deep and 1-cm
wide trench using a mostly neutral Ta flux by conventional DC magnetron sputtering and a mostly ionized Ta flux by HPPMS [162]. Drastic structural changes
were achieved in the films simply by changing the power source, as shown in
Figure 7.24. The Ta thin film grown by HPPMS exhibited a smooth surface and
a dense crystalline structure with grains oriented perpendicular to the substrate
surface (**Fig. 7.24a and c**), whereas the film grown by DC magnetron sputtering
exhibited a rough surface, pores between the grains, and an inclined columnar
structure (**Fig. 7.24b and d**). The improved homogeneity achieved by HPPMS
is a direct consequence of the high ion fraction of sputtered species being
controlled by the bias on the substrate. However, this new technique is in its
developing stage, and the deposition rate and cost of power supply are the most
important challenges that need to be considered.

7.5 GENERAL CONSIDERATIONS AND PRACTICAL ASPECTS OF SPUTTERING DEPOSITION

Sputtering is nowadays considered as a flexible and effective coating method for
surface engineering applications. There are numerous papers and textbooks on
the fundamentals, phenomena, and applications of sputter deposition. The
information contained within this section is not intended to be an all-inclusive

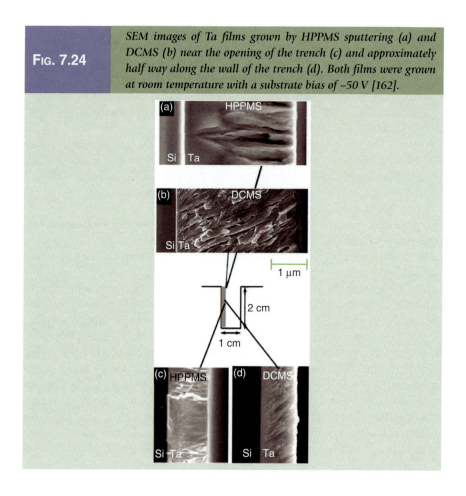

FIG. 7.24 SEM images of Ta films grown by HPPMS sputtering (a) and DCMS (b) near the opening of the trench (c) and approximately half way along the wall of the trench (d). Both films were grown at room temperature with a substrate bias of –50 V [162].

survey of the field. Instead, several aspects which are important practical aspects for the successful deposition of nanostructure and nanocomposite coatings, including reactive gas control with process stability, film structure control, discharged plasma properties and monitoring, and energy-enhanced deposition, will be briefly introduced.

7.5.1 Reactive Sputtering Deposition Process Stability

Compound thin films can be deposited either by direct deposition of a compound source in thermal evaporation, electron beam evaporation, CAE, PLD, sputtering, or by reactive sputtering. However, it should be noted that thermal evaporation of a compound source is often not possible due to noncongruent melting of the compound, while the compound must be conducting for CAE of a compound target. More recently, reactive sputtering has become the deposition of choice of compound thin films. In reactive sputtering deposition, atoms and small molecules are ejected (sputtered) from a target surface under particle bombardment and travel through the plasma discharge to the substrate, where they react with the gas to form a wide variety of compound films on the substrate. Careful control of the reactive gas is important in reactive sputtering. Too low a supply of the reactive gas will cause high rate metallic sputtering, but may give rise to an understoichiometric compound composition of the deposited film. Too high a supply of the reactive gas will allow for stoichiometric composition of the deposited film, but will cause poisoning of the target surface, which may reduce the deposition rate significantly [165].

In reactive sputtering, the reaction takes place both at the cathode/target and substrate surface during the deposition. The reaction at the cathode is usually related to target poisoning and the reaction at the substrate is called a metallic mode [166]. Normally, the sputter yield of the poisoned target (compound material) is substantially lower than the sputter yield of the elemental target material, which causes the deposition rate to exhibit a hysteresis curve with the variation of the flow rates of the reactive gas [165–167].

A typical hysteresis curve is shown in **Figure 7.25** [168]. (a) At low O_2/Ar pressures, the metallic Zr + Y thin films are deposited below the critical oxygen pressure to provide the required oxide stoichiometry. (b) When the partial oxygen pressure increases, the YSZ thin films are deposited through the metallic mode. In the metallic mode, the deposition rates of the yttrium stabilized zirconia (YSZ) are higher than those of Y/Zr metal thin films. (c) Above the critical oxygen pressure, the deposition rate abruptly decreases. (d) The target is poisoned (oxide formation on the metallic target) and the sputtering mode moves to the oxide mode with low deposition rate. (e) When the partial oxygen pressure decreases, the sputtering is still in the oxide mode for a certain oxygen flow rate. (f) The sputtering deposition will return to the original metallic mode at an oxygen pressure below the critical oxygen pressure, when the oxide layer of the target surface is fully removed.

The presence of the hysteresis hampers the process control of the reactive sputtering process. As such several researchers have tried to eliminate the hysteresis

FIG. 7.25	*Hysteresis curve for the reactive sputtering of YSZ thin films with variation of the flow rates of the reactive gas. (The target is Zr alloyed with Y and the reactive gas is a O_2/Ar mixture.)* [168]

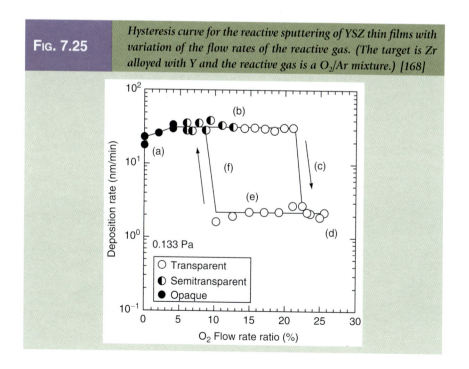

effect. Kadlec et al. [169] suggested that the hysteresis effect can be avoided if the pumping speed of the pumping system is greater than a critical value. The use of very high-speed pumping systems has proved to provide a solution to the problem of reactive sputtering of a metal titanium target to provide a reasonable rate for the preparation of large areas of the titanium oxide. The pumping speed of a system has been shown to affect the process of reactive sputtering of indium–tin oxide [170]. The simple solution requires large pumps, with arrangements of the system to give high gas conductances, which can be difficult to arrange and the cost of the extra pumps is high. For these reasons, the high pumping speed approach is not used very often. Recently, Depla et al. [171] studied the hysteresis behavior during reactive magnetron sputtering of Al_2O_3 using a rotating cylindrical magnetron. They found the hysteresis shifts towards lower oxygen flows when the rotation speed of the target is increased. Sproul et al. [167] demonstrated that using partial pressure control of the reactive gas allows deposition of films in the transition region between the elemental and poisoned states of the target, which leads to higher deposition rates compared to flow control and better film properties.

7.5.2 Film Structure Control (Structure Zone Models)

Thin-film formation is a continuous process in which atoms absorb onto the substrate, nucleate, and grow to form a dense film structure. In a vapor deposition process without plasma assistance, the mode of nucleation and film growth is mainly determined by the thermochemistry of the substrate and film materials and the temperature of the system [172]. There are three steps involved in a thin-film deposition process: (i) production of the appropriate atomic, molecular, or ionic species; (ii) transport of these species to the substrate through a medium; and (iii) condensation on the substrate, either directly or via a chemical reaction to form a solid deposit.

The film nucleation and initial growth can be described by three growth types, as shown in **Figure 7.26** [173]:

a. Island type (Volmer–Weber type): As the binding strength between adatoms is greater than that between adatom and substrate, the clusters grow three dimensionally to form islands.
b. Layered type (Frank–van der Merwe type): As the binding strength between adatoms is less than that between adatom and substrate, nucleated clusters grow layer by layer.

FIG. 7.26   *Three modes of thin-film nucleation and initial growth processes.*

Island growth (Volmer–Weber)   Layer growth (Frank–van der Merwe)   Layer + island (Stranski–Krastanov)

(a)   (b)   (c)

c. Mixed type (Stranski–Krastanov type): A mixed growth mode combines layer growth and island growth.

In this case, after a few monolayers have grown, subsequent layer growth becomes unfavorable and islands are formed. The reasons for the thin-film growth in this mode remain unclear. Factors such as the release of strain energy may trigger the formation of islands. This growth mode is fairly commonly observed in many kinds of thin-film growth systems.

In general, an external input energy is usually necessary to facilitate the formation of the thin films and to improve the quality of the film, which can be divided into three categories: (i) chemically induced mobility, (ii) thermal energy, and (iii) kinetic energy. The chemically induced mobility comes from the nature of the vapor species. The thermal energy originates from heating of the substrate, and the kinetic energy can be obtained from the energetic ion bombardment and the momentum transfer of ion species and neutral adatoms during the deposition of the film. All types of energy assist the migration of the adatoms on the substrate surface and supply the formation energy of the thin films. Generally, most plasma-related deposition systems require a moderate substrate temperature of about 300~400°C to enhance the quality of the thin film. However, thin films with acceptable quality can be prepared at a relatively low substrate temperature, even at room temperature, using high-energy ion bombardment deposition systems. By the addition of a plasma and high-energetic ion species and ion flux (concentration), the film will grow far from its thermodynamic equilibrium state, which makes the nucleation and growth processes much more complicated.

In the past two decades, different structure zone diagrams have been proposed to correlate the microstructure of thin films with the deposition conditions. In the 1960s, efforts were undertaken to generalize the effect of PVD deposition parameters on thin-film structure and properties. The melting temperature T_m of the film was assumed as the basic material parameter and the substrate temperature T was assumed as the main process parameter. Different zone models were developed based on their ratio T/T_m (the homologous temperature consideration), and typical examples include the Movchan–Demchishin model [174] for vacuum vapor deposition, the Grovenor model for thermal evaporation [175], the Thornton model [176] for cathode and magnetron sputtering, and the Messier model [177] for ion beam deposition.

Grovenor et al. [175] examined the microstructure of metal films grown by thermal evaporation. The microstructures were classified into four zones according to their homologous temperatures as depicted in **Figure 7.27a**. At a substrate temperature below $0.15\,T_m$, the film consists of porous columnar grains (Zone I). When the substrate temperature is between 0.15 to $0.3\,T_m$, the film exhibits a transitional structure (Zone T) between Zone I and Zone II. Zone II is a dense columnar structure, resulting from deposited atoms having sufficient surface mobility to diffuse and to increase the grain size. Zone III structure is controlled by bulk diffusion as the substrate temperatures are higher than $0.5\,T_m$.

The most well-known and important zone model is the Thornton model for cathode and magnetron sputtering [176]. In the Thornton model, the sputtering gas pressure is used as a process parameter to determine the film structure

FIG. 7.27   *(a) Grovenor [175], (b) Thornton [176], and (c) Messier [177] structure zone diagrams.*

(**Fig. 7.27b**). Four different zones, including Zone I, Zone T, Zone II, and Zone III were proposed. Zone I ($T/T_m < 0.3$) is a columnar structure with pronounced pores and open columnar boundaries. This formation is due to the weak surface diffusion of atoms in combination with shadowing effects. Zone T is a transition structure that translates the Zone I structure to the Zone II structure, with fine, dense fibrous grains, changing with a rise of temperature to dense columnar grains. Zone T films are denser and have a much smoother surface morphology than the two surrounding zones (Zone I and Zone II). The film structure moves to Zone II when the substrate temperature is higher than about $0.5\,T_m$. Zone II consists of columnar grains separated by distinct boundaries and results from surface diffusion-controlled growth. Lattice diffusion dominates at high substrate temperature, giving rise to the large equiaxed grains of Zone III. For the entire temperature range, a rise in gas pressure causes a shift in the ranges of occurrence of different zones to a higher value of T/T_m ratio. A reduction in sputtering gas pressure during deposition increases the mean free path for elastic collisions between vapor species and the sputtering gas atoms. This behavior leads to higher kinetic energy of the vapor species impinging on the substrate surface, thereby producing a denser microstructure.

For ion beam techniques of coating deposition, Messier et al. [177] modified the Thornton model by replacing gas pressure in the sputtering zone with the energy of ions reaching the substrate surface V_s (eV) (**Fig. 7.27c**). As the bombardment energy increases, the width of Zone T increases at the expense of Zone I. It was also found that inside the columnar structure of Zones I and T, the intrinsic structure can be either polycrystalline or amorphous.

The Thornton and Messier models are widely used today on account of their simplicity and good agreement with industrial practice. A combination of these two models can successfully predict the microstructure of deposited thin films in many cases. However, the effect of ion bombardment rate such as ion/metal flux ratio is not considered or incorporated, which is becoming an increasingly important parameter in plasma-assisted sputtering deposition.

7.5.3 *Sputtering Glow Discharges*

Plasma-assisted material processing includes such techniques as etching, film deposition, and modifications of solid surfaces. Most of the modern PVD processes are operated with additional ion or plasma enhancement to achieve coatings with good adhesion, and mechanical and tribological properties [178]. In this section the plasma refers to a glow discharge that is used for sputter deposition of thin films.

The most common methods for generation of a plasma (breakdown and ionization of a gas) include the DC discharge, pulsed DC discharge, RF and microwave discharge, dielectric barrier discharges, and using electron and laser beams [179]. Here, to describe the plasma architecture we use the simplest DC discharge that is generated in a vacuum chamber between two counter electrodes by applying a DC power across the electrodes to break down the gas (e.g., Ar), as illustrated in **Figure 7.28**. In general, a neutral gas always contains a few electrons and ions due to ionization, for example, produced by cosmic radiation [180]. When a DC potential, V_c, is applied between the cathode and anode, these free charge carriers are accelerated by the electric field and new charged particles may be created when these charge carriers collide with atoms and molecules in the gas or with the surfaces of the electrodes (reactions 1, 2, and 4 in **Table 7.2**). This leads to an avalanche of charged particles that is eventually balanced by charge carrier losses, so that a steady-state plasma develops [179].

The plasma between the cathode and anode consists of different distributions of potential, space charge, and current density, as shown in **Figure 7.28** [181,182]. The edge of the plasma in contact with other surfaces is significantly different from the bulk plasma regions. A dark space or sheath is usually observed adjacent to all surfaces in contact with the plasma. As shown in **Figure 7.28**, the cathode will repel secondary electrons at high velocity away from the cathode, leaving behind the ions with slow mobility. Therefore, the high net positive space charge near the cathode will form a sheath, which is called a "cathode dark space." The cathode dark space is a region of the discharge where the electrical potential drops suddenly between the cathode and the edge of the negative glow and, as a result, there is an extremely low electron density. This lack of electrons results in low levels of excitation of the gas species in the region, and hence the area appears dark. The negative glow region is the brightest part of the discharge where the electric field is close to zero and most ionization collisions

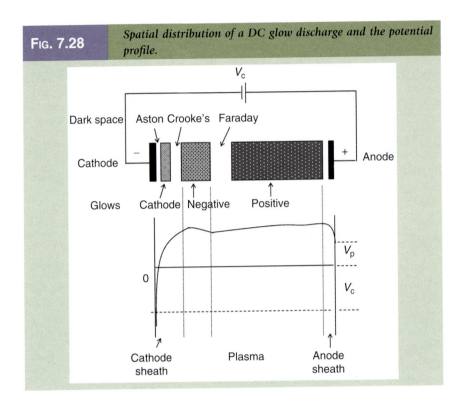

FIG. 7.28   *Spatial distribution of a DC glow discharge and the potential profile.*

take place in this region. Following the negative glow region is the Faraday dark space and the positive glow. In the homogenous positive glow, a constant longitudinal electrical field is maintained. The electrons gain energy in this field and form an electron energy distribution with an appreciable number of energetic electrons for the formation of a sufficiently large number of ions and electrons to balance the charge carrier losses to the wall [179]. However, in a sputter deposition system, the interelectrode separation needs to be small in order to increase the deposition rate. If the anode (substrate) is located in the negative glow region, the positive glow and the Faraday dark space do not exist [180].

In sputter deposition, if an unbiased or grounded substrate is placed in the plasma, it will rapidly charge negatively due to the fact that electrons have greater velocities than ions. The negative surface will attract ions and retard the electrons until the escaping fluxes of ions and electrons are equal. In this steady state, the potential on this surface is known as the floating potential and it is typically negative of the plasma potential and is given by equation (7.8):

$$V_f = \frac{-KT_e}{2e} \ln\left(\frac{m_i}{2\pi m_e}\right) \qquad (7.8)$$

where m_e and m_i are the electron and ion masses and T_e is the electron temperature. Since the ion mass is 3–4 orders of magnitude higher than that of an electron, the floating potential will have a (negative) value several times the electron temperature in volts.

In a vacuum system, the electrons would leave the plasma at a faster rate than ions, ending up at the vacuum chamber walls. The result for the plasma would be a slow increase in the net positive charge. As the plasma charges positively, it becomes less energetically possible for the electrons to leave, because now the walls of the chamber are more negative than the plasma. Eventually, a steady-state condition is reached in which the plasma potential is high enough that the loss rate of electrons is reduced to the same level as the loss rate of ions. In this way the plasma will retain its overall neutrality. The plasma potential V_p, which is now the average potential of the bulk plasma with respect to the chamber, will be on the order of several volts more positive than the chamber potential. Therefore, the plasma will remain the most positive body during a deposition process [64]. As a result of this plasma potential, ions that reach the edge of the plasma are then accelerated with the same voltage to the chamber wall. In the sputter deposition process, a magnetron field is usually applied behind the cathode in an effort to confine the secondary electrons near the cathode region to increase the plasma density, as discussed in more detail in section 7.4.3 under *Sputter Deposition*.

The primary energy loss mechanism for ions in the cathode fall is quantum mechanical through resonant charge exchange collisions. In simple diode discharges, ions move only a relatively small fraction of the cathode fall distance (cathode sheath) before they experience a charge exchange collision and are no longer accelerated by the field. Since this type of reaction does not require physical collision, the neutralized particle continues towards the target while the newly created ion is itself accelerated over a small fraction of cathode sheath. As a result, the average energy of ions incident at the target is much less than $1\,eV$ and the glow discharge sputter yields are less than expected. However, this effect is partially mediated by sputtering due to fast neutrals [183].

7.5.4 *Energetic Enhanced Deposition*

The microstructure and properties of a thin film can be tailored and controlled by incorporating kinetic energy during film deposition. In general, the energy involved in energetic-enhanced deposition includes thermal energy and energetic particle bombardment. Both energies can be used to enhance the adatom mobility and reduce void formation during the film growth stages. For example, for most thin-film applications, a porous Zone I structure is undesirable because it degrades the film's mechanical, tribological, electronic, and optical properties. The film structure can be moved towards the dense, void-free Zone T structure by the use of energy-enhanced deposition [96]. Higher thermal energy can be achieved by increasing the substrate temperature during film deposition. Under raised substrate temperature, the migration rate of adatoms to shadowed regions can be large enough to surpass the rate of void incorporation during film growth [96], thereby increasing the film density. The substrate temperature also has influence on the film grain size, phase change, the formation of crystalline films, and texture alignment. However, many thin-film deposition applications and substrate materials need to be processed under relatively low temperatures. Therefore, the kinetic energy transferred from energetic bombardment in plasma-assisted depositions is widely used.

Various thin-film processing techniques that are based on ion- or plasma-assisted deposition were reviewed in section 7.4. All of these methods have the significant advantage that the structure and properties of thin films may be usefully modified by means of suitably considering the energy of ions/neutral species ranging from a few electron volts up to hundreds of kiloelectron volts impingement on the substrate surface. Some of the particles are not just assisting but they may condense and thereby become part of the growing film. In this section, emphasis will be focused on low energy plasma film deposition techniques, including magnetron sputtering, pulsed magnetron sputtering, and HPPMS, in which the substrate is immersed in a high density plasma to accomplish the deposition with the assistance of energetic bombardment plasma ions with ion energies up to hundreds of electron volts. The means of increasing the ion energy and plasma density, in particular, the generation of high ion energy and ion flux in a pulsed plasma, and the effect of ion bombardment on the growing film structure and properties will be introduced here. As to other high-energy ion/plasma characteristics and processing methods, such as plasma/ion implantation and ion beam deposition, readers are referred to Refs. [68,96].

Effect of Ion-Assisted Deposition.   The ion energy and ion flux within the plasma have been proved to be important factors influencing the growing film structure and properties. A certain level of ion bombardment energies are critical for obtaining high-quality thin films. The impingement of a large amount of energetic ions or atoms upon a substrate surface produces a wide variety of effects during film deposition, such as film densification, decreased grain size, increased hardness and Young's modulus, high compressive stress, reduced roughness, resputtering, and film texture/orientation change depending on the ion energy range.

In general, the ion energies used in magnetron sputtering range up to a few hundred electron volts. The physical structure of a thin film can be changed by the mobility of the adatoms during growth. The enhanced ion energy and ion flux bombardment can transfer kinetic energy to other atoms as the collision sequence develops in time on the surface of the growing film, allowing the surface atoms to move around on the surface (adatom mobility) and find energetically favorable sites. Increasing ion bombardment produces resputtering and forward sputtering of the surface atoms, which fills the voids that would naturally occur along the columnar grains from shadowing effects [148,184,185]. The incident ion transfers kinetic energy to other atoms as the collision sequence develops in time, thereby densifying the film without a need of thermal diffusion.

Figure 7.29 provides a 2D molecular dynamics simulation of structure development versus the kinetic energy carried by atoms perpendicularly incident on a film held at $0\,K$ substrate temperature. The lowest of the three ratios of incident energy E_t to the adatom potential-well depth E_c corresponds to thermal deposition. The lowest normalized impinging energy E_t/E_c is $0.02\,eV$, which corresponds to a thermal deposition process. The highest normalized impinging energy E_t/E_c of $5\,eV$ is characteristic of sputtered energy-enhanced deposition. The effect of added process energy on void filling is obvious [96].

Petrov et al. [186] studied the substrate bias effect on the sputter-deposited TiN film structure. They found a closing of intercolumnar porosity when increasing the substrate bias V_s, from 80 to $120\,V$ at substrate temperature $T_s = 300°C$, as a result of forward sputtering and ion irradiation enhanced adatom mobility (**Fig. 7.30**).

Ion bombardment of the growing film can restrict the grain growth, increase nucleation sites, and permit the formation of nanocrystalline films. The high

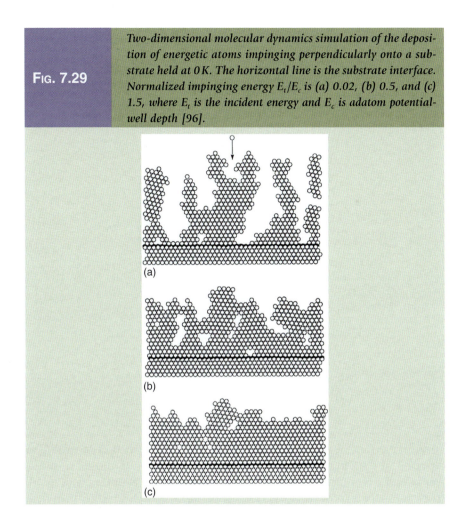

FIG. 7.29 *Two-dimensional molecular dynamics simulation of the deposition of energetic atoms impinging perpendicularly onto a substrate held at 0 K. The horizontal line is the substrate interface. Normalized impinging energy E_t/E_c is (a) 0.02, (b) 0.5, and (c) 1.5, where E_t is the incident energy and E_c is adatom potential-well depth [96].*

surface adatom mobility and diffusivity can increase the surface smoothness of the films by filling in the voids. Forward sputtering places atoms in interstitial sites, which results in compressive residual stresses in the PVD hard coatings [187]. This stress has both a positive effect, increasing the effective hardness of the coating [188], and a negative effect, causing film delamination as the film thickness increases. Energetic ions during ion bombardment can knock off some metal atoms from the substrate or may penetrate the substrate lattice to angstrom levels in addition to cleaning and heating the substrate. This bombardment leads to defects and roughness on the substrate at an atomic level, and this atomic level of roughness is believed to be responsible for the improved adhesion of the coating [189].

Furthermore, the crystallographic orientation of grains can also be influenced by the energy and flux of bombarding ions [190–192]. The ion-assisted reactive sputtering film orientation depends strongly on processing conditions, such as substrate temperature, substrate orientation, substrate bias, film thickness, ion flux, and ion energy. Theoretical modeling of ion energy effects on densification and texture evolution needs to take into account thermodynamic properties and

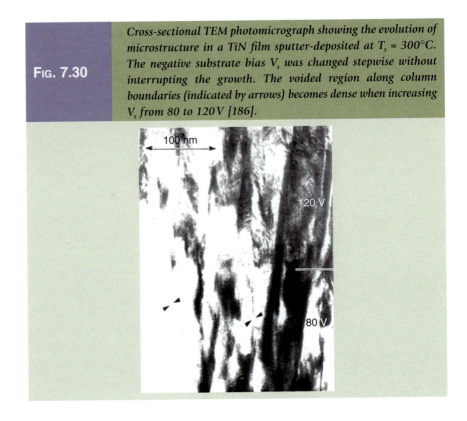

Fig. 7.30 *Cross-sectional TEM photomicrograph showing the evolution of microstructure in a TiN film sputter-deposited at T_s = 300°C. The negative substrate bias V_s was changed stepwise without interrupting the growth. The voided region along column boundaries (indicated by arrows) becomes dense when increasing V_s from 80 to 120 V [186].*

stress and strain energies. In general, it is recognized that the texture evolution observed in PVD thin coatings could be related to the minimization of free energy of the coatings, which is composed of surface and strain energies [192]. In the normal film growth model, the grains will grow on low, rather than high energy surfaces to minimize the surface energy. However, when additional kinetic or thermal energy is incorporated into the deposition process, the adatoms will receive higher momentum transfer and gain higher mobility to move around. At the same time, higher strain energy will be developed in the films, resulting in the change of the film orientations.

However, molecular dynamic computer simulations also predict that there should be optimum ion energy for film densification. Excessive ion energy will result in deeper penetration in the lattice leaving vacancies which cannot be filled by arriving vapor species [193]. Ion bombardment of crystalline films will cause heavy atomic structure damage and, in extreme cases, full amorphization [194] for some materials. **Figure 7.31** shows the defect incorporation as a function of substrate bias V_s in a multilayered TiN film bombarded by argon ions accelerated by V_s = 120, 80, and 40 V at T_s = 700°C [186]. As the substrate bias is increased, the projected range of the penetration of ions becomes larger giving rise to a defect generation at an increasingly larger distance below the growth surface, thereby resulting in dislocation networks and point defect aggregates.

At even higher energies, for example >400 eV, the rate of sputtering (resputtering) of the film surface will exceed the rate of deposition, effectively prohibiting film growth. This range can be characterized as an etch mode, which is important for surface preparation and patterning processes. In the case of polymer

| FIG. 7.31 | *Cross-sectional TEM photomicrograph showing the evolution defect incorporation in a TiN film sputter deposited at $T_s = 700°C$. The negative substrate bias V_s was changed stepwise without interrupting the growth from 120 to 40 V [186].* |

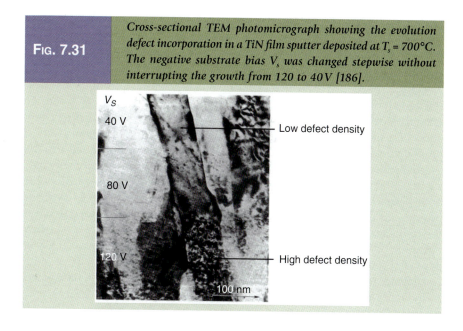

substrates, ions of too high energy and/or too high dose lead to substrate damage by chain-scissoring [195]. Particle energies of a few kiloelectron volts are used to remove material and to produce shallow implant profiles, which is important to some semiconductor processing. Another example of a process in this energy range is plasma nitriding, in which nitrogen ions (N^+) are implanted into the substrate surface forming hard and inert nitride phases in the near-surface region of steel or aluminum alloys [196]. Finally, at the very high end of the ion energy scale, ion implantation at hundreds of kiloelectron volts or some megaelectron volts is conducted routinely, but these processes and applications are beyond the scope of this chapter and will not be considered further.

Energy-Enhanced Deposition Using a Substrate Bias and a Pulsed Plasma. The energetic bombardment in magnetron sputtering can come from the flux of gas ions from the plasma impinging on the biased substrate, from the self energy of the depositing atoms, or from the flux of energetic neutrals that result when using high target power [197]. In general, the threshold ion energy of conventional DC magnetron sputtering is about 10 eV. Since ion energy is the difference between the plasma potential and the potential of the surface that ions bombard, an effective way to enhance the ion bombardment is to apply a negative bias to the substrate to extract ions from the plasma to bombard the growing film during film growth. Extensive studies have demonstrated that substrate bias plays an important role on film growth and microstructural evolution and properties [198–200].

It is important to understand the behavior of plasmas in order to gain a better understanding of the relationship between the deposition process and the film structure and properties [201]. In general, plasma diagnostics refers to the techniques used to gather information about the nature (properties) of the plasma used in a deposition process. Diagnostics help us obtain information about plasma properties in the sputter deposition glow discharge, such as the plasma chemical compositions and species, temperature, plasma density, ion/electron energy distributions, and nonelectronic species. The most commonly used plasma diagnostic techniques can be categorized as optical and electrostatic spectrometry.

Optical spectrometry involves focusing radiation emitted from excited neutral and charged species in the plasma through an optical window on to the entrance slit of an optical spectrometer. A photomultiplier is used to detect radiation of a particular wavelength at the spectrometer exit slit. Alternatively, a photodiode array is mounted at the exit port of the spectrometer and a broad spectral region is detected simultaneously [202,203]. Optical spectrometry uses a spatial and temporal resolution instrument which allows measurement of the bulk plasma properties. It is difficult for optical spectrometry to detect a specific position inside the plasma, for example, to measure the plasma properties near the cathodes or close the substrate. However, the relatively low cost and easy operation of an optical spectrometer make it a useful tool, and is used widely in plasma diagnostics applications.

Langmuir probes are used for monitoring discharge or plasma parameters including spatial distribution of potential, electron density, and electron temperature. The Langmuir probe consists of molybdenum or tungsten electrodes inserted into the plasma. The plasma parameters are estimated by the current–voltage curve of these electrodes. More detailed information on determining the electron density and temperature can be found in Refs. [204,205].

Mass spectrometry is another powerful technique for identifying unknown species, studying ionic species, and probing the fundamental principles of chemical reactions in the plasma. Mass spectrometry is based upon the motion of a charged particle, that is, ion, in an electric or magnetic field. The mass to charge ratio (m/z) of the ion affects this motion. Since the charge of an electron is known, the mass to charge ratio will indirectly measure an ion mass. Mass spectrometry is operated under high vacuum condition. A sample (preferably a gas) is introduced and broken down into charged fragments by electron impact or chemical ionization. The fragments, accelerated by applying a voltage, pass through a mass selector which separates them by their ratio of mass to charge (m/z). The separate fragments are detected and measured as ion current. Under constant conditions, a molecule will break up in the same number of ways, giving a reproducible mass spectrum. Compared to optical spectrometers, mass spectrometers can be placed at any position inside a plasma, thereby providing flexible and specific measurement of the plasma parameters as a function of location in the plasma.

As described in the previous section, pulsed DC magnetron sputtering has been used to eliminate arcing problems during the reactive sputter deposition of insulating films, stabilize the discharge, and reduce the formation of defects in the film. Besides its primary goal of eliminating arcing, it has been found that applying pulsing in DC magnetron sputtering also has benefits on coating structure and properties due to changes in ion energy and ion flux caused by changes in plasma parameters [206–211]. In recent years, in order to understand the nature of the pulsed plasma, time- and space-resolved Langmuir probe and electrostatic quadrupole plasma mass spectrometer measurements have been intensively used to investigate the plasma condition in pulsed DC magnetron sputter deposition [207,211–213]. These investigations have been conducted on various DC magnetron sputter configurations, such as one planar cathode, two planar cathodes facing each other in a mirror configuration, and so on. Observations of some interesting plasma properties and their relationship to the pulsing conditions have been revealed.

One important approach in pulsed plasma examination is using a time-resolved mass spectrometer to elucidate the dynamics of energetic ions in a pulsed DC magnetron plasma and related ion energy distributions (IED) in the plasma. It was found that pulsing provides a wide range of ion energies and ion fluxes in the plasma, and the energy of the energetic species can be up to hundreds of electron volts.

Three different ion energy regions in a pulsed plasma have been well documented [211,212]. **Figure 7.32a** shows a typical time averaged $^{29}N_2^+$ ion energy distribution in a discharged plasma for CrAlN film deposition when the Cr and Al targets were pulsed synchronously at 100 kHz and 5.0 μs. The Al target voltage waveform is shown in **Figure 7.32b**. It can be seen that pulsed ion energies generally consist of three regions, which are also correlated to distinct phases of the discharge voltage.

The low ion energy region "A" in **Figure 7.32a** usually is in the range of 0–10 eV, which corresponds to the negative pulse period (sputtering period) in the target voltage waveform (**Fig. 7.32b**). The plasma potential is several volts above the grounded chamber wall surface potential during this period. The "B" middle ion energy region (20 ~ 50 eV) is the energy gained from the target potential in the reverse positive pulse period (**Fig. 7.32b**), which has an average positive value of a fixed percentage of the nominal sputtering (negative) voltage. The "C" ion energy tail region is the kinetic energy gained from the fast and high positive voltage overshoot at the beginning of the positive pulse period. When the fast and steep positive voltage overshoot is developed on the target, a large electron current is extracted from the bulk plasma towards the target. Due to the slow ion movement, compared with that of electrons, the equilibrium cannot be reached in a short time period, therefore a negative charge density gradient

FIG. 7.32   *(a) A typical time-averaged $^{29}N_2^+$ ion energy distribution in a discharged plasma for CrAlN film deposition when the Cr and Al targets were pulsed synchronously at 100 kHz and 5.0 μs. (b) The Al target voltage waveform.*

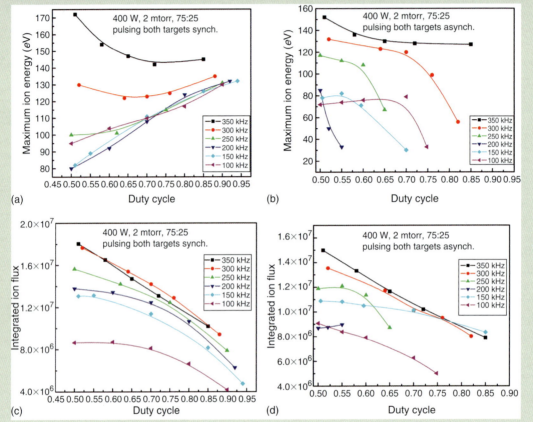

FIG. 7.33 *The maximum ion energy ($^{29}N_2^+$) as a function of the duty cycle at different pulsing frequencies in (a) synchronized mode and (b) asynchronized mode. The integrated ion flux as a function of the duty cycle at different pulsing frequencies in (c) synchronized mode and (d) asynchronized mode [211].*

and a positive charge density gradient will be created in the bulk plasma resulting in an electric field potential being established under this charge gradient. Ions will be accelerated by the electric field potential and will gain energy from it.

The ion energy distribution intensity and the area under the curves in these ion energy regions are relevant to the ion flux. In general, the ion energy distribution and ion flux in a pulsed plasma exhibit strong dependence on a variety of parameters, such as the working pressure, the reactive gas composition, the target power, the pulsing frequencies, and duty cycles [211]. **Figure 7.33a** and **b** exhibit the maximum ion energy evolution ($^{29}N_2^+$) as a function of duty cycle at different pulsing frequencies in the synchronized (waveforms of both targets are in phase) and asynchronized (waveforms of both targets are out of phase) pulsing modes during P-CFUBMS of CrAlN films, respectively. The corresponding integrated ion flux values in the two pulsing modes are plotted in **Figure 7.33c** and **d**. In the asynchronized mode, the maximum ion energy decreases with increasing the duty cycle at most frequencies and drops suddenly to the low energy region at certain duty cycles due to the total separation of the two target positive pulse periods (**Fig. 7.33b**) [211]. The ion flux decreases with an increase in the duty cycle at all frequencies in both the synchronized and asynchronized pulsing modes (**Fig. 7.33c and d**). As the duty cycle decreases (the positive pulse time increases), the cathode is switched proportionally to a positive voltage for a longer period of time, which in turn provides more time for positive ions to stream away from the target. This increased escape time results in a higher number of ions gaining the additional kinetic energy available from the positive potential switch, thereby increasing the flux of higher energy ions at the substrate.

The increased ion energy and ion flux in a pulsed DC plasma will strongly affect the structure and properties of the films. The high ion energy and ion flux will apply additional ion bombardment, increase the mobility of the atoms on the substrate, and reduce the shadowing effect of the columnar structure, thereby changing the film growth microstructure. Muratore and Moore [206] compared the IED of pulsed DC magnetron sputtering and conventional DC magnetron sputtering deposition of TiO films. An increased ion energy up to 140 eV was observed in pulsed DC magnetron sputtering compared with 17 eV in the DC discharge. Increase of crystallographic texture and an 11% increase in hardness were observed in the TiO films processed with the pulsed magnetron sputtering. Backer et al. [214] calculated the ion energy impinging on the pulsed DC magnetron sputtering of Ti and TiO_2 films from the measurements of plasma parameters using a planar Langmuir probe and found that the ion energy values during the positive pulse period increased more than ten-fold the time-averaged value. The high ion energy has an important effect on the film surface roughness. It was found that the titanium films prepared at pulsing frequencies between 100 and 275 kHz exhibited low surface roughness. Higher ion energy at high pulsing frequencies will increase the titanium film surface roughness, which is not desirable. The effect of pulsed ion energy and ion flux bombardment on thin-film microstructure and properties will be illustrated in more detail using the technological example in section 8.3.

References

1. A. A. Voevodin, J. S. Zabinski, and C. Muratore, Recent advances in hard, tough, and low friction nanocomposite coatings, *Tsinghua Science and Technology*, 10(6), 665–679 (2005).
2. S. Veprek, A. Niederhofer, K. Moto, T. Bolom, H.-D. Mannling, P. Nesladek, G. Dollinger, and A. Bergmaier, Composition, nanostructure and origin of the ultrahardness in nc-TiN/a-Si_3N_4/a- and nc-$TiSi_2$ nanocomposites with H_V = 80 to ≥105 GPa, *Surface and Coatings Technology*, 133–134, 152–159 (2000).
3. D. Pilloud, J. F. Pierson, and L. Pichon, Influence of the Silicon concentration on the optical and electrical properties of reactively sputtered Zr–Si–N nanocomposite coatings, *Material Science and Engineering: B*, 131, 36–39 (2006).
4. A. A. Voevodin, S. D. Walck, and J. S. Zabinski, Architecture of multilayer nanocomposite coatings with super-hard diamond-like carbon layers for wear protection at high contact loads, *Wear*, 203–204, 516–527 (1997).
5. I.-W. Park, K. H. Kim, J. H. Suh, C.-G. Park, and M.-H. Lee, Role of amorphous Si_3N_4 in the microhardness of Ti-Al-Si-N nanocomposite films, *Journal of the Korean Physical Society*, 42(6), 783 (2003).
6. L. S. Palatnik, A. I. Il'inskii, and N. P. Sapelkin, *Soviet Physics-Solid State*, 8, 2016 (1967).

7. J. S. Koehler, Attempt to design a strong solid, *Physical Review B*, 2, 547–551 (1970).
8. P. J. Kelly, R. Hall, J. O. Brien, J. W. Bradley, P. Henderson, G. Roche, and R. D. Arnell, Studies of mid-frequency pulsed dc biasing, *Journal of Vacuum Science and Technology, A*, 19, 2856 (2001).
9. L. S. Palatnik and A. I. Il'inskii, Doklady, *Soviet Physics-Technical Physics*, 9, 93 (1964).
10. R. W. Springer and D. S. Catlett, Structure and mechanical properties of Al/Al_xO_y vacuum-deposited laminates, *Thin Solid Films*, 54, 197–205 (1978).
11. B. A. Movchan, A. V. Demchishin, G. F. Badilenko, R. F. Bunshah, C. Sans, C. Deshpandey, and H. J. Doerr, Structure-property relationships in microlaminate TiC/TiB_2 condensates, *Thin Solid Films*, 97, 215–219 (1982).
12. K. K. Shih and D. B. Dove, Ti/Ti-N Hf/Hf-N and W/W-N multilayer films with high mechanical hardness, *Applied Physics Letters*, 61, 654–656 (1992).
13. U. Helmersson, S. Todorova, S. A. Barnett, J.-E. Sundgren, L. C. Markert, and J. E. Greene, Growth of single-crystal TiN/VN strained-layer superlattices with extremely high mechanical hardness, *Journal of Applied Physics*, 62, 481 (1987).
14. M. Larsson, P. Hollman, P. Hedenqvist, S. Hogmark, U. Wahlström, and L. Hultman, Deposition and microstructure of PVD TiN–NbN multilayered coatings by combined reactive electron beam evaporation and DC sputtering, *Surface and Coatings Technology*, 86–87, 351–356 (1996).
15. K. M. Hubbard, T. R. Jervis, P. B. Mirkarimi, and S. C. Barnett, Mechanical properties of epitaxial $TiN/(V_{0.6}Nb_{0.4})N$ superlattices measured by nanoindentation, *Journal of Applied Physics*, 72, 4466–4468 (1992).
16. P. B. Mirkarimi, L. Hultman, and S. A. Barnett, Enhanced hardness in lattice-matched single-crystal $TiN/V_{0.6}Nb_{0.4}N$ superlattices, *Applied Physics Letters*, 57, 2654–2656 (1990).
17. P. B. Mirkarimi, S. A. Barnett, K. M. Hubbard, T. R. Jervis, and L. Hultman, Structure and mechanical-properties of epitaxial $TiNV_{0.3}NB_{0.7}N(100)$ superlattices, *Journal of Materials Research*, 9, 1456 (1994).
18. H. Holleck and H. Schulz, Advanced layer material constitution, *Thin Solid Films*, 153, 11–17 (1987).
19. H. Holleck and V. Schier, Multilayer PVD coatings for wear protection, *Surface and Coatings Technology*, 76–77, 328–336 (1995).
20. K. Yamamotoa, S. Kujimeb, and K. Takahara, Properties of nano-multilayered hard coatings deposited by a new hybrid coating process: Combined cathodic arc and unbalanced magnetron sputtering, *Surface and Coatings Technology*, 200, 435–439 (2005).
21. P. Robinson, A. Matthews, K. G. Swift, and S. Franklin, A computer knowledge-based system for surface coating and material selection, *Surface and Coatings Technology*, 62, 662–668 (1993).
22. R. Riedel, *Handbook of ceramic hard materials*, John Wiley & Sons, Inc., New York (2000).
23. A. Cavaleiro and C. Louro, Nanocrystalline structure and hardness of thin film, *Vacuum*, 64(3), 211–218 (2002).
24. E. Martinez, R. Sanjinés, A. Karimi, J. Esteve, and F. Lévy, Mechanical properties of nanocomposite and multilayered Cr–Si–N sputtered thin films, *Surface and Coatings Technology*, 180–181, 570–574 (2004).
25. I.-W. Park, K. H. Kim, A. O. Kunrath, D. Zhong, J. J. Moore, A. A. Voevodin, and E. A. Levashov, Microstructure and mechanical properties of superhard Ti–B–C–N films deposited by dc unbalanced magnetron sputtering, *Journal of Vacuum Science and Technology, B*, 23(2), 588 (2005).
26. M. Stüber, H. Leiste, S. Ulrich, H. Holleck, and D. Schild, Microstructure and properties of low friction TiC-C nanocomposite coatings deposited by magnetron sputtering, *Surface and Coatings Technology*, 150, 218 (2002).

27. A. A. Voevodin, J. P. O'Neill, S. V. Prasad, and J. S. Zabinski, Nanocrystalline WC and WC/a-C composite coatings produced from intersected plasma fluxes at low deposition temperatures, *Journal of Vacuum Science and Technology, A*, 17(3), 986 (1999).

28. S. Veprek, P. Nesladek, A. Niederhofer, F. Glatz, M. Jilek, and M. Sima, Recent progress in the superhard nanocrystalline composites: Towards their industrialization and understanding of the origin of the superhardness, *Surface and Coatings Technology*, 108–109, 138 (1998).

29. Y. H. Lu, P. Sit, T. F. Hung, H. Chen, Z. F. Zhou, K. Y. Li, and Y. G. Shen, Effects of B content on microstructure and mechanical properties of nanocomposite Ti–B$_x$–N$_y$ thin films, *Journal of Vacuum Science and Technology, B*, 23(2), 449 (2005).

30. R. Fella, H. Holleck, and H. Schulz, Preparation and properties of WC-TiC-TiN gradient coatings, *Surface and Coatings Technology*, 36, 257 (1988).

31. A. A. Voevodin, J. M. Schneider, C. Rebholz, and A. Matthews, Multilayer composite ceramic metal-DLC coatings for sliding wear applications, *Tribology International*, 29, 559 (1996).

32. A. A. Voevodin, M. A. Capano, S. J. P. Laube, M. S. Donley, and J. S. Zabinski, Design of A Ti/TiC/DLC functionally gradient coating based on studies of structural transitions in Ti–C thin films, *Thin Solid Films*, 298, 107–115 (1997).

33. L. Shizhi, S. Yulong, and P. Hongrui, Ti-Si-N films prepared by plasma-enhanced chemical vapor deposition, *Plasma Chemistry and Plasma Processing*, 12(3), 287 (1992).

34. S. Veprek and S. Reiprich, A concept for the design of novel superhard coatings, *Thin Solid Films*, 268, 64 (1995).

35. R. W. Hertzberg, *Deformation and fracture mechanics of engineering materials*, 3rd ed., John Wiley & Sons, Inc., New York (1989).

36. A. Kelly and N. H. MacMillan, *Strong solids*, 3rd ed., Clarendon, Oxford (1986).

37. E. O. Hall, The deformation and ageing of mild steel: III discussion of results, *Proceedings of the Physical Society of London Section B*, 64, 747–753 (1951).

38. N. J. Petch, The cleavage strength of polycrystals, *Journal of the Iron and Steel Institute*, 174, 25 (1953).

39. A. Lasalmonie and J. L. Strudel, Influence of grain size on the mechanical behaviour of some high strength materials, *Journal of Material Science*, 21, 1837 (1986).

40. E. Arzt, Size effects in materials due to microstructural and dimensional constraints: A comparative review, *Acta Materialia*, 46, 5611–5626 (1998)

41. J. Schiotz, F. D. Di Tolla, and K. W. Jacobsen, Softening of nanocrystalline metals at very small grain sizes, *Nature (London)*, 391, 561 (1998).

42. J. E. Carsley, J. Ning, W. W. Milligan, S. A. Hackney, and E. C. Aifantis, A simple, mixtures-based model for the grain size dependence of strength in nanophase metals, *Nanostructured Materials*, 5, 441 (1995).

43. A. H. Chokshi, A. Rosen, J. Karch, and H. Gleiter, On the validity of the Hall-Petch relationship in nanocrystalline materials, *Scripta Metallurgica*, 23, 1679 (1989).

44. R. W. Siegel and G. E. Fougere, Mechanical properties of nanophase metals, *Nanostructured Materials*, 6, 205–216 (1995).

45. M. F. Ashby, A first report on deformation-mechanism maps, *Acta Metallurgica*, 20, 887–897 (1972).

46. S. C. Lim, Recent developments in wear-mechanism maps, *Tribology International*, 31, 87–97 (1998).

47. S. L. Lehoczky, Strength enhancement in thin-layered Al-Cu laminates, *Journal of Applied Physics*, 49, 5479 (1978).

48. S. A. Barnett, Deposition and mechanical properties of superlattice thin films, In *Physics of Thin Films, Vol. 17 Mechanics and Dielectric Properties*, M. H. Francombe and J. L. Vossen, eds., p. 2 Academic Press, Boston, MA (1993).

49. P. M. Anderson and C. Li, Hall-Petch relations for multilayered materials, *Nanostructured Materials*, 5, 349 (1995).

50. W. D. Sproul, Reactive sputter deposition of polycrystalline nitride and oxide superlattice coatings, *Surface and Coatings Technology*, 86–87, 170 (1996).

51. M. L. Wu, X. W. Lin, V. P. Dravid, Y. W. Chung, M. S. Wong, and W. D. Sproul, Preparation and characterization of superhard CN_x/ZrN multilayers, *Journal of Vacuum Science and Technology, A*, 15, 946–950 (1997).

52. S. Veprek, New development in superhard coatings: The superhard nanocrystalline-amorphous composites, *Thin Solid Films*, 317, 449–454 (1998).

53. W. D. Sproul, Multilayer, multicomponent, and multiphase physical vapor deposition coatings for enhanced performance, *Journal of Vacuum Science and Technology, A*, 12(4), 1595–1601 (1994).

54. G. E. Dieter, *Mechanical metallurgy*, McGraw Hill, New York (1976).

55. M. Marder and J. Finberg, How things break, *Physics Today*, 49, 24 (1996).

56. G. M. Pharr, Measurement of mechanical properties by ultra-low load indentation, *Materials Science Engineering, A*, 253, 151(1998).

57. A. Leyland and A. Matthews, On the significance of the H/E ratio in wear control: A nanocomposite coating approach to optimised tribological behaviour, *Wear*, 246, 1 (2000).

58. V. V. Lyubimov, A. A. Voevodin, A. L. Yerokhin, Y. S. Timofeev, and I. K. Arkhipov, Development and testing of multilayer physically vapour deposited coatings for piston rings, *Surface and Coatings Technology*, 52, 145 (1992).

59. A. A. Voevodin, P. Stevenson, J. M. Schneider, and A. Matthews, Active process control of reactive sputter deposition, *Vacuum*, 46, 723 (1995).

60. J. Musil, P. Zeman, H. Hruby, and P. H. Mayrhofer, ZrN/Cu nanocomposite film—A novel superhard material, *Surface and Coatings Technology*, 120–121, 179 (1999).

61. A. A. Voevodin and J. S. Zabinski, Nanocomposite and nanostructured tribological materials for space applications, *Composite Science Technology*, 65, 741–748 (2005).

62. S. Veprek, Conventional and new approaches towards the design of novel superhard materials, *Surface and Coatings Technology*, 97, 15 (1997).

63. J. E. Crowell, Chemical methods of thin film deposition: Chemical vapor deposition, atomic layer deposition, and related technologies, *Journal of Vacuum Science and Technology, A*, 21(5), S88–S95 (2003).

64. S. M. Rossnagel, J. J. Cuomo, and W. D. Westwood, eds., *Handbook of plasma processing technology–Fundamentals, etching, deposition, and surface interactions*, William Andrew Publishing/Noyes, Park Ridge, New Jersey (1990).

65. http://www.gordonengland.co.uk/ps.htm

66. J. R. Creighton and P. Ho, Introduction to chemical vapor deposition (CVD), In *Chemical vapor deposition*, J.-H. Park, ed., ASM International, Materials Park, OH (2001).

67. L. B. Freund and S. Suresh, *Thin film materials: Stress, defect formation and surface evolution*, Cambridge University Press, Cambridge (2003).

68. D. M. Mattox, *Handbook of physical vapor deposition (PVD) processing: Film formation, adhesion, surface*, William Andrew Inc., Westwood New Jersey (1998).

69. A. Sherman, *Chemical vapor deposition for microelectronics: Principles, technology, and applications*, William Andrew Inc., Westwood, New Jersey (1987).

70. J. B. Fortin and T.-M. Lu, *Chemical vapor deposition polymerization: The growth and properties of parylene thin films*, Springer, Norwell, Massachusetts (2003).

71. K. F. Jensen, Chemical vapor deposition, In *Microelectronics processing: Chemical engineering aspects*, D. W. Hess and K. F. Jensen, eds., American Chemical Society, Washington, DC (1989).

72. M. Razeghi, *The MOCVD challenge, volume 2: A survey of GaInAsO-GaAs for photonic and electronic device applications*, CRC Press, Boca Raton, FL (1989).

73. A. Eishabini-Riad and F. D. Barlow, *Thin film technology handbook*, McGraw-Hill, New York (1998).

74. C. Cao, Ferroelectric thin films and applications, In *Ferroelectric thin films and applications in functional thin films and functional materials: New concepts and technologies*, D. Shi, ed., Springer, New York (2003).

75. M. Stoiber, J. Wagner, C. Mitterer, K. Gammer, H. Hutter, C. Lugmair, and R. Kullmer, Plasma-assisted pre-treatment for PACVD TiN coatings on tool steel, *Surface and Coatings Technology*, 174–175, 687–693 (2003).

76. E. Vassallo, A. Cremona, F. Ghezzi, F. Dellera, L. Laguardia, G. Ambrosone, and U. Coscia, Structural and optical properties of amorphous hydrogenated silicon carbonitride films produced by PECVD, *Applied Surface Science*, 252(22), 7993–8000 (2006).

77. N. Martins, P. Canhola, M. Quintela, I. Ferreira, L. Raniero, E. Fortunato, and R. Martins, Performances of an in-line PECVD system used to produce amorphous and nanocrystalline silicon solar cells, *Thin Solid Films*, 511–512, 238–242 (2006).

78. D. Heim, F. Holler, and C. Mitterer, Hard coatings produced by PACVD applied to aluminium die casting, *Surface and Coatings Technology*, 116–119, 530–536 (1999).

79. S. Shimada, Y. Takada, and J. Tsujino, Deposition of TiN films on various substrates from alkoxide solution by plasma-enhanced CVD, *Surface and Coatings Technology*, 199(1), 72–76 (2005).

80. J. Bonitz, S.E. Schulz, and T. Gessner, Ultra thin CVD TiN layers as diffusion barrier films on porous low-k materials, *Microelectronic Engineering*, 76(1–4), 82–88 (2004).

81. K.-T. Rie and J. Wöhle, Plasma-CVD of TiCN and ZrCN films on light metals, *Surface and Coatings Technology*, 112(1–3), 226–229 (1999).

82. Y.-G. Jung, S.-W. Park, and S.-C. Choi, Effect of CH_4 and H_2 on CVD of SiC and TiC for possible fabrication of SiC/TiC/C FGM, *Materials Letters*, 30(5–6), 339–345 (1997).

83. F. Ding and Y. Shi, The study of diamond/TiC composite film by a DC-plasma–hot filament CVD, *Surface and Coatings Technology*, 201(9–11), 5050–5053 (2007).

84. M. Noda and M. Umeno, Coating of DLC film by pulsed discharge plasma CVD, *Diamond and Related Materials*, 14(11–12), 1791–17 (2005).

85. H. Hanyu, S. Kamiya, Y. Murakami, and Y. Kondoh, The improvement of cutting performance in semi-dry condition by the combination of DLC coating and CVD smooth surface diamond coating, *Surface and Coatings Technology*, 200(1–4), 1137–1141 (2005).

86. S. Yoon, H. Yang, R. J. Ahn, and Q. Zhang, Preparation of a-C and DLC-C films using electron cyclotron resonance CVD and RF Bias, *Vacuum*, 49(1), 67–74 (1998).

87. S. H. Lee, K. H. Nam, S. C. Hong, and J. J. Lee, Low temperature deposition of TiB_2 by inductively coupled plasma assisted CVD, *Surface and Coatings Technology*, 201(9–11), 5211–5215 (2007).

88. A. Osada, E. Nakamura, H. Homma, T. Hayahi, and T. Oshika, Wear mechanism of thermally transformed CVD Al_2O_3 layer, *International Journal of Refractory Metals and Hard Materials*, 24(5), 387–391 (2006).

89. H. Sonobe, A. Sato, S. Shimizu, T. Matsui, M. Kondo, and A. Matsuda, Highly stabilized hydrogenated amorphous silicon solar cells fabricated by triode-plasma CVD, *Thin Solid Films*, 502(1–2), 306–310 (2006).

90. S. Faÿ, L. Feitknecht, R. Schlüchter, U. Kroll, E. Vallat-Sauvain, and A. Shah, Rough ZnO layers by LP-CVD process and their effect in improving performances of amorphous and microcrystalline silicon solar cells, *Solar Energy Materials and Solar Cells*, 90(18–19), 2960–2967 (2006).

91. B. Y. Moon, J. H. Youn, S. H. Won, and J. Jang, Polycrystalline silicon film deposited by ICP-CVD, *Solar Energy Materials and Solar Cells*, 69(2), 139–145 (2001).

92. K. Saito, Y. Uchiyama, and K. Abe, Preparation of SiO_2 thin films using the cat-CVD method, *Thin Solid Films*, 430(1–2), 287–291 (2003).

93. J. Perez-Mariano, S. Borros, J. A. Picas, A. Forn, and C. Colominas, Silicon nitride films by chemical vapor deposition in fluidized bed reactors at atmospheric pressure (AP/FBR-CVD), *Surface and Coatings Technology*, 200(5–6), 1719–1723 (2005).

94. M. J. Thwaites and H. S. Reehal, Growth of single-crystal Si, Ge and SiGe layers using plasma-assisted CVD, *Thin Solid Films*, 294(1–2), 76–79 (1997).

95. A. Tallaire, A. T. Collins, D. Charles, J. Achard, R. Sussmann, A. Gicquel, M. E. Newton, A. M. Edmonds, and R. J. Cruddace, Characterisation of high-quality thick single-crystal diamond grown by CVD with a low nitrogen addition, *Diamond and Related Materials*, 15(10), 1700–1707 (2006).

96. D. L. Smith, *Thin-film deposition: Principles and practice*, McGraw-Hill, New York (1995).

97. K. Oczos', The shaping of materials by concentrated fluxes of energy (in Polish). Publications of the Rzeszo'w Technical University, Rzeszo'w (1988).

98. J. Yao, J. Shao, H. He, and Z. Fan, Optical and electrical properties of TiO_x thin films deposited by electron beam evaporation, *Vacuum*, 81(9), 1023–1028 (2007).

99. S.-W. Hsu, T.-S. Yang, T.-K. Chen, and M.-S. Wong, Ion-assisted electron-beam evaporation of carbon-doped titanium oxide films as visible-light photocatalyst, *Thin Solid Films*, 515(7–8), 3521–3526 (2007).

100. C. Rebholz, M. A. Monclus, M. A. Baker, P. H. Mayrhofer, P. N. Gibson, A. Leyland, and A. Matthews, Hard and superhard TiAlBN coatings deposited by twin electron-beam evaporation, *Surface and Coatings Technology*, 201(13), 6078–6083 (2007).

101. A. Lotnyk, S. Senz, and D. Hesse, Epitaxial growth of TiO_2 thin films on $SrTiO_3$, $LaAlO_3$ and Yttria-stabilized Zirconia Substrates by electron beam evaporation, *Thin Solid Films*, 515(7–8), 3439–3447 (2007).

102. R. Al Asmar, J.-P. Atanas, Y. Zaatar, J. Podlecki, and A. Foucaran, Characterization and ellipsometric investigation of high-quality ZnO and $ZnO(Ga_2O_3)$ thin alloys by reactive electron-beam co-evaporation technique, *Microelectronics Journal*, 37(10), 1080–1085 (2006).

103. F.-R. Weber, F. Fontaine, M. Scheib, and W. Bock, Cathodic arc evaporation of (Ti,Al)N coatings and (Ti,Al)N/TiN multilayer-coatings-correlation between lifetime of coated cutting tools, structural and mechanical film properties, *Surface and Coatings Technology*, 177–178, 227–232 (2004).

104. Y. Tanaka, N. Ichimiya, Y. Onishi, and Y. Yamada, Structure and properties of Al–Ti–Si–N coatings prepared by the cathodic arc ion plating method for high speed cutting applications, *Surface and Coatings Technology*, 146–147, 215 (2001).

105. A. M. Merlo, The contribution of surface engineering to the product performance in the automotive industry, *Surface and Coatings Technology*, 174, 21–26 (2003).

106. S. M. Rossnagel, Thin film deposition with physical vapor deposition and related technologies, *Journal of Vacuum Science and Technology, A*, 21(5), S74 (2003).

107. J. Vyskočil and J. Musil, Cathodic arc evaporation in thin film technology, *Journal Vacuum Science and Technology, A*, 10(4), 1740–1748 (1992).

108. I. I. Aksenov, V. A. Belous, V. G. Padalka, and V. M. Khoroshikh, Transport of plasma streams in a curvilinear plasma-optics system, *Soviet Journal of Plasma Physics*, 4, 425–428 (1978).

109. Y. H. Liu, J. L. Zhang, D. P. Liu, T. C. Ma, and G. Benstetter, A triangular section magnetic solenoid filter for removal of macro- and nano-particles from pulsed graphite cathodic vacuum arc plasma, *Surface and Coatings Technology*, 200(7), 2243–2248 (2005).

110. D. Dijkkamp, T. Venkatesan, X. D. Wu, S. A. Shaheen, N. Jisrawi, Y. H. Min-Lee, W. L. McLean, and M. Croft, Preparation of Y-Ba-Cu oxide superconductor thin

films using pulsed laser evaporation from high T_c bulk material, *Applied Physics Letters*, 51(8), 619–621 (1987).

111. D. B. Chrisey and G. K. Hubler, *Pulsed laser deposition of thin films*, John Wiley & Sons, Inc., New York (1994).

112. Z. L. Wang, Y. Liu, and Z. Zhang, *Handbook of nanophase and nanostructured materials*, Springer-Verlag, New York (2002).

113. S.-M. Kim, S. C. Song, and S. Y. Lee, Effect of CeO_2, $BaTiO_3$ and $CeO_2/BaTiO_3$ double buffer layers on the superconducting properties of $Y_1Ba_2Cu_3O_{7-x}$ grown on metallic substrates by pulsed laser deposition, *Physica C: Superconductivity*, 351(4), 379–385 (2001).

114. V. Beaumont, B. Mercey, and B. Raveau, Performant superconducting $NdBa_2Cu_3O_7$ films grown by pulsed laser deposition at "low temperature" in an argon rich atmosphere, *Physica C: Superconductivity*, 340(2–3), 112–118 (2000).

115. N. Scarisoreanu, M. Dinescu, F. Craciun, P. Verardi, A. Moldovan, A. Purice, and C. Galassi, Pulsed laser deposition of perovskite relaxor ferroelectric thin films, *Applied Surface Science*, 252(13), 4553–4557 (2006).

116. W. Biegel, R. Klarmann, B. Stritzker, B. Schey, and M. Kuhn, Pulsed laser deposition and characterization of perovskite thin films on various substrates, *Applied Surface Science*, 168(1–4), 227–233 (2000).

117. N. B. Ibrahim, C. Edwards, and S. B. Palmer, Pulsed laser ablation deposition of yttrium iron garnet and cerium-substituted YIG films, *Journal of Magnetism and Magnetic Materials*, 220(2–3), 183–194 (2000).

118. Y. Nakata, T. Okada, M. Maeda, S. Higuchi, and K. Ueda, Effect of oxidation dynamics on the film characteristics of Ce:YIG thin films deposited by pulsed-laser deposition, *Optics and Lasers in Engineering*, 44(2), 147–154 (2006).

119. A. Hakola, O. Heczko, A. Jaakkola, T. Kajava, and K. Ullakko, Ni–Mn–Ga films on Si, GaAs and Ni–Mn–Ga single crystals by pulsed laser deposition, *Applied Surface Science*, 238(1–4), 155–158 (2004).

120. Y. Tsuboi, M. Goto, and A. Itaya, Pulsed laser deposition of silk protein: Effect of photosensitized-ablation on the secondary structure in thin deposited films, *Journal of Applied Physics*, 89(12), 7917–7923 (2001).

121. G. Kecskeméti, N. Kresz, T. Smausz, B. Hopp, and A. Nógrádi, Pulsed laser deposition of pepsin thin films, *Applied Surface Science*, 247(1–4), 83–88 (2005).

122. L. Pawlowski. *The science and engineering of thermal spray coatings*, John Wiley & Sons, Inc., New York (1995).

123. J. A. Picas, A. Forn, and G. Matthäus, HVOF coatings as an alternative to hard chrome for pistons and valves, *Wear*, 261(5–6), 477–484 (2006).

124. H. Okuno, E. Grivei, F. Fabry, T. M. Gruenberger, J. Gonzalez-Aguilar, A. Palnichenko, L. Fulcheri, N. Probst, and J.-C. Charlier, Synthesis of carbon nanotubes and nano-necklaces by thermal plasma process, *Carbon*, 42(12–13), 2543–2549 (2004).

125. T. Yoshida, The future of thermal plasma processing for coating, *Pure and Applied Chemistry*, 66(6), 1222–1230 (1994).

126. M. Gell, L. Xie, X. Ma, E. H. Jordan, and N. P. Padture, Highly durable thermal barrier coatings made by the solution precursor plasma spray process, *Surface and Coatings Technology*, 177–178, 97–102 (2004).

127. http://www.plasma-group.co.uk/

128. P. R. Strutt, B. H. Kear, and R. Boland, US Patent No. 6025034 (2000).

129. S. D. Parukuttyamma, J. Margolis, H. Liu, C. P. Grey, S. Sampath, H. Herman and J. B. Parise, Yttrium aluminum garnet (YAG) films through a precursor plasma spraying technique, *Journal of American Ceramic Society*, 84, 1906 (2001).

130. J. Karthikeyan, C. C. Berndt, J. Tikkanen, S. Reddy, and H. Herman, Plasma spray synthesis of nanomaterial powders and deposits, *Materials Science and Engineering*, A, 238, 275–286 (1997).

131. N. P. Padture, K. W. Schlichting, T. Bhatia, A. Ozturk, B. Cetegen, E. H. Jordan, Towards durable thermal barrier coatings with novel microstructures deposited by solution-precursor plasma spray, *Acta Materialia*, 49(12), 2251–2257 (2001).

132. X. Ma, F. Wu, J. Roth, M. Gell, and E. H. Jordan, Low thermal conductivity thermal barrier coating deposited by the solution plasma spray process, *Surface and Coatings Technology*, 201, 4447–4452 (2006).

133. J. L. Vossen and W. Kern, eds., *Thin film processes II*, Academic Press, London (1991).

134. E. G. Spencer and P. H. Schmidt, Ion-beam techniques for device fabrication, *Journal of Vacuum Science and Technology*, 8(5), S52 (1971).

135. J. J. Cuomo, S. M. Rossnagel, and H. R. Kaufman, *Handbook of ion beam processing technology: Principles, deposition, film modification, and synthesis*, Noyes Publications, Park Ridge, New Jersey (1989).

136. P. Sigmund, Sputtering by ion bombardment: Theoretical concepts, In *Sputtering by particle bombardment*, I, R. Behrisch, ed., Springer-Verlag, Berlin (1981).

137. T. Serikawa, M. Henmi, T. Yamaguchi, H. Oginuma, and K. Kondoh, Depositions and microstructures of Mg–Si thin film by ion beam sputtering, *Surface and Coatings Technology*, 200, 4233–4239 (2006).

138. S. Iwatsubo, Temperature effect on structure and surface morphology of indium tin oxide films deposited by reactive ion-beam sputtering, *Vacuum*, 80(7), 708–711 (2006).

139. E. Horváth, A. Németh, A. A. Koós, M. C. Bein, A. L. Tóth, Z. E. Horváth, L. P. Biró, and J. Gyulai, Focused ion beam based sputtering yield measurements on ZnO and Mo thin films, *Superlattices and Microstructures*, 42(1–6), 392–397 (2007).

140. B. Y. Kim, J. S. Lee, K.-R. Kim, B. H. Choi, and B. S. Park, Development of ion beam sputtering technology for surface smoothing of materials, *Nuclear Instruments and Methods in Physics Research Section B: Beam Interactions with Materials and Atoms*, 261(1–2), 682–685 (2007).

141. A. Matthews, Plasma-based PVD surface engineering processes, *Journal of Vacuum Science and Technology*, A, 21(5), S224 (2003).

142. R. D. Arnell and P. J. Kelly, Recent advance in magnetron sputtering, *Surface and Coatings Technology*, 112, 170–176 (1999).

143. B. Windows and N. Savvides, Charged particle fluxes from planar magnetron sputtering sources, *Journal of Vacuum Science Technology*, A, 4(2), 196–202 (1986).

144. B. Windows and N. Savvides, Unbalanced dc magnetrons as sources of high ion fluxes, *Journal of Vacuum Science and Technology*, A, 4(3), 453–456 (1986).

145. W. D. Sproul, P. J. Rudnik, M. E. Graham, and S. L. Rohde, High-rate reactive sputtering in an opposed cathode closed-field unbalanced magnetron sputtering system, *Surface and Coatings Technology*, 43, 270 (1990).

146. S. L. Rohde, W. D. Sproul, and J. R. Rohde, Correlations of plasma and magnetic field characteristics to TiN film properties formed using a dual unbalanced magnetron system, *Journal of Vacuum Science and Technology*, A, 9(3), 1178 (1991).

147. K. Tominaga, Preparation of AlN films by planar magnetron sputtering system with facing two targets, *Vacuum*, 41, 1154–1156 (1990).

148. W. D. Sproul, Physical vapor deposition tool coatings, *Surface and Coatings Technology*, 81, 1–7 (1996).

149. W. D. Sproul, M. E. Graham, M.-S. Wong, and P. J. Rudnik, Reactive d.c. magnetron sputtering of the oxides of Ti, Zr, and Hf, *Surface and Coatings Technology*, 89, 10–15 (1997).

150. W. D. Sproul, High-rate reactive DC magnetron sputtering of oxide and nitride superlattice coatings, *Vacuum*, 51, 641 (1998).

151. O. Sánchez, M. Hernández-Vélez, D. Navas, M. A. Auger, J. L. Baldonedo, R. Sanz, K. R. Pirota, and M. Vázquez, Functional nanostructured titanium nitride films obtained by sputtering magnetron, *Thin Solid Films*, 495(1–2), 149–153 (2006).

152. S. Inoue, F. Okada, and K. Koterazawa, CrN films deposited by rf reactive sputtering using a plasma emission monitoring control, *Vacuum*, 66(3–4), 227–231 (2002).

153. E. Kusano, A. Satoh, M. Kitagawa, H. Nanto, and A. Kinbara, Titanium carbide film deposition by DC magnetron reactive sputtering using a solid carbon source, *Thin Solid Films*, 343–344, 254–256 (1999).

154. P. W. Shum, W. C. Tam, K. Y. Li, Z. F. Zhou, and Y. G. Shen, Mechanical and tribological properties of titanium-aluminum-nitride films deposited by reactive closed-field unbalanced magnetron sputtering, *Wear*, 257, 1030–1040 (2004).

155. J. Lin, B. Mishra, J. J. Moore, and W. D. Sproul, Microstructure, mechanical and tribological properties of $Cr_{1-x}Al_xN$ films deposited by pulsed-closed field unbalanced magnetron sputtering (P-CFUBMS), *Surface and Coatings Technology*, 201, 4329–4334 (2006).

156. J. O'Brien and P. J. Kelly, Characterisation studies of the pulsed dual cathode magnetron sputtering process for oxide films, *Surface and Coatings Technology*, 142–144, 621–627 (2001).

157. A. Anders, Fundamentals of pulsed plasmas for materials processing, *Surface and Coatings Technology*, 183, 301–311 (2004).

158. Advanced Energies Inc., *PinnacleTM plus user manual* (2002).

159. E. V. Barnet and T.-M. Lu, *Pulsed and pulsed bias sputtering*, Kluwer Academic Publishers, Norwell, MA (2003).

160. V. Kouznetsov, K. Macak, J. M. Schneider, U. Helmersson, and I. Petrov, A novel pulsed magnetron sputter technique utilizing very high target power densities, *Surface and Coatings Technology*, 122, 290 (1999).

161. K. Macak, V. Kouznetsov, J. M. Schneider, U. Helmersson, and I. Petrov, Ionized sputter deposition using an extremely high plasma density pulsed magnetron discharge, *Journal of Vacuum Science and Technology*, A, 18, 1533 (2000).

162. J. Alami, P. O. Å. Persson, D. Music, J. T. Gudmundsson, J. Bohlmark, and U. Helmersson, Ion-assisted physical vapor deposition for enhanced film properties on nonflat surfaces, *Journal of Vacuum Science and Technology*, A, 23(2), 278 (2005).

163. C. Christou and Z. H. Barber, Ionization of sputtered material in a planar magnetron discharge, *Journal of Vacuum Science and Technology*, A, 18, 2897 (2000).

164. A. P. Ehiasarian, R. New, W.-D. Munz, L. Hultman, U. Helmersson, and V. Kouznetsov, Influence of high power densities on the composition of pulsed magnetron plasmas, *Vacuum*, 65, 147 (2002).

165. S. Berg and T. Nyberg, Fundamental understanding and modeling of reactive sputtering processes, *Thin Solid Films*, 476, 215–230 (2005).

166. S. Berg, T. Nyberg, H.-O. Blom, C. Nender, D. A. Glocker, and S. I. Shah, eds., *Handbook of thin film process technology*, p. A5.3:1 Institute of Physics Publishing, Bristol (1998).

167. W. D. Sproul, D. J. Christie, and D. C. Carter, Control of reactive sputtering process, *Thin Solid Films*, 491, 1–17 (2005).

168. T. Hata, S. Nakano, Y. Masuda, K. Sasaki, Y. Haneda, and K. Wasa, Heteroepitaxial growth of YSZ films on Si (100) substrate by using new metallic mode of reactive sputtering, *Vacuum*, 51, 583 (1998).

169. S. Kadlec, J. Musil, and H. VyskoEil, Hysteresis effect in reactive sputtering: A problem of system stability, *Journal of Physics D: Applied Physics*, 19, 187–190 (1986).

170. R. P. Howson and H. A. Ja'fer, Reactive sputtering with an unbalanced magnetron, *Journal of Vacuum Science and Technology*, A, 10, 1748 (1992).

171. D. Depla, J. Haemers, G. Buyle, and R. De. Gryse, Hysteresis behavior during reactive magnetron sputtering of Al_2O_3 using a rotating cylindrical magnetron, *Journal of Vacuum Science and Technology*, A, 24, 934–938 (2006).

172. A. Anders, Plasma and ion sources in large area coating: A review, *Surface and Coatings Technology*, 200, 1893–1906 (2005).

173. K. Wasa, M. Kitabatake, and H. Adachi, *Thin film materials technology: Sputtering of compound materials*, 1st ed., William Andrew Publishing, Norwich, New York (2004).

174. V. A. Movchan and A. V. Demchishin, Investigation of structure and properties of thick vacuum condensates of nickel, titanium, tungsten, aluminum oxide and zirconium dioxide (in Russian), *Fizika Metallov i Metallovedenye*, 28(4), 83–86 (1969).

175. C. R. M. Grovenor, H. T. G. Hentzell, and D. A. Smith, The development of grain structure during growth of metallic films, *Acta Metalliallia*, 32, 773 (1984).

176. J. A. Thornton, High rate thick film growth, *Annual Review of Materials Science*, 7, 239–260 (1977).

177. R. Messier, A. P. Giri, and R. A. Roy, Revised structure zone model for thin film physical structure, *Journal of Vacuum Science and Technology*, A, 2(2), 500–511 (1984).

178. R. Boxman, P. Martin, and D. Sanders, *Handbook of vacuum arc science and technology*, Noyes Publications, Park Ridge, NJ (1995).

179. H. Conrads and M. Schmidt, Plasma generation and plasma sources, *Plasma Sources Science and Technology*, 9, 441 (2000).

180. M. Venugopalan and R. Avni, Analysis of glow discharges, In *Thin films from free atoms and particles*, K. J. Klabunde, ed., Academic Press, Orlando, FL (1985).

181. A. von Engel, *Ionized gases*, Oxford University Press, London and New York (1965).

182. E. Nasser, *Fundamentals of gaseous ionization and plasma electronics*, John Wiley & Sons, Inc., New York (1971).

183. J. E. Greene, Sputter deposition, AVS short course program, Rocky Mountain Chapter, Golden, CO (2006).

184. I. Petrov, F. Adibi, J. E. Greene, L. Hultman and J.-E. Sundgren, Average energy deposited per atom: A universal parameter for describing ion-assisted film growth? *Applied Physics Letters*, 63, 36–38 (1993).

185. I. Petrov, P. B. Barna, L. Hultman, and J. E. Greene, Microstructure evolution during film growth, *Journal of Vacuum Science and Technology*, A, 21(5), S117–S128 (2003).

186. I. Petrov, L. Hultman, U. Helmersson, J.-E. Sundgren, and J. E. Greene, Microstructure modification of TiN by ion bombardment during reactive sputter deposition, *Thin Solid Films*, 169, 299–314 (1989).

187. H. Ljungcrantz, L. Hultman, and J.-E. Sundgren, Ion induced stress generation in arc-evaporated TiN films, *Journal of Applied Physics*, 78(2), 832 (1995).

188. D. T. Quinto, Mechanical property and structure relationships in hard coatings for cutting tools, *Journal of Vacuum Science and Technology*, A, 6, 2149 (1988).

189. S. PalDey and S. C. Deevi, Single layer and multilayer wear resistant coatings of (Ti,Al)N: A review, *Materials Science and Engineering*, A, 342, 58–79 (2003).

190. J. Pelleg, L. Z. Zevin, and S. Lungo, Reactive-sputter-deposited TiN films on glass substrates, *Thin Solid Films*, 197, 117–128 (1991).

191. J. E. Greene, J.-E. Sundgren, L. Hultman, I. Petrov, and D. B. Dergstrom, Development of preferred orientation in polycrystalline TiN layers grown by ultrahigh vacuum reactive magnetron sputtering, *Applied Physics Letters*, 67, 2928 (1995).

192. J. P. Zhao, X. Wang, Z. Y. Chen, S. Q. Yang, T. S. Shi, and X. H. Liu, Overall energy model for preferred growth of TiN films during filtered arc deposition, *Journal Physics D: Applied Physics*, 30, 5–12 (1997).

193. T. Mueller, A. Gebeshuber, R. Kullmer, C. Lugmair, S. Perlot, and M. Stoiber, Minimizing wear through combined thermochemical and plasma activated diffusion and coating process, *Materiali in Tehnologije*, 38, 6 (2004).

194. A. R. Gonzalez-Elipe, F. Yubero, J. P. Espinos, A. Caballero, M. Ocana, J. P. Holgado, and J. Morales, Amorphisation and related structural effects in thin films prepared by ion beam assisted methods, *Surface and Coatings Technology*, 125, 116–123 (2000).

195. R. Rank, T. Wuensche, M. Fahland, C. Charton, and N. Schiller, Adhesion promotion techniques for coating of polymer films, 47th Annual Technical Conference, Society of Vacuum Coaters, Dallas, TX, p. 632 (2004).

196. W. Möller, S. Parascandola, T. Telbizova, R. Günzel, and E. Richter, Surface processes and diffusion mechanisms of ion nitriding of stainless steel and aluminium, *Surface and Coatings Technology*, 136, 73 (2001).

197. B. Windows, Issues in magnetron sputtering of hard coatings, *Surface and Coatings Technology*, 81, 92–98 (1996).

198. S. K. Karkari, A. Vetushka, and J. W. Bradley, Measurement of the plasma potential adjacent to the substrate in a midfrequency bipolar pulsed magnetron, *Journal of Vacuum Science and Technology*, 21, L28 (2003).

199. M. Audronis, A. Leyland, P. J. Kelly, and A. Matthews, The effect of pulsed magnetron sputtering on the structure and mechanical properties of CrB_2 coatings, *Surface and Coatings Technology*, 201, 3970 (2006).

200. J.-W. Lee, S.-K. Tien, and Y.-C. Kuo, The effects of pulse frequency and substrate bias to the mechanical properties of CrN coatings deposited by pulsed DC magnetron sputtering, *Thin Solid Films*, 494, 161 (2006).

201. J. E. Greene, Optical spectroscopy for diagnostics and process control during glow discharge etching and sputter deposition, *Journal of Vacuum Science and Technology*, 15, 1718 (1978).

202. A. Belkind, A. Freilich, J. Lopez, Z. Zhao, W. Zhu, and K. Becker, Characterization of pulsed DC magnetron sputtering plasmas, *New Journal of Physics*, 7, 90 (2005).

203. N. B. Dahotre and T. S. Sudarshan, eds., *Intermetallic and ceramic coatings*, Marcel Dekker, Inc., New York (1999).

204. I. Langmuir and H. M. Mostt-Smith, *General Electric Review*, 27(449), 583 (1924).

205. M. A. Lieberman and A. J. Lichtenberg, *Principles of plasma discharges and materials processing*, John Wiley & Sons, Inc., New York (1994).

206. C. Muratore, J. J. Moore, and J. A. Rees, Electrostatic quadrupole plasma mass spectrometer and langmuir probe measurements of mid-frequency pulsed DC magnetron discharges, *Surface and Coatings Technology*, 164, 12 (2003).

207. J. W. Bradley, H. Bäcker, P. J. Kelly, and R. D. Arnell, Space and time resolved langmuir probe measurements in a 100 kHz pulsed rectangular magnetron system, *Surface and Coatings Technology*, 142–144, 337–341 (2001).

208. J. W. Bradley, H. Bäcker, P. J. Kelly, and R. D. Arnell, Time-resolved langmuir probe measurements at the substrate position in a pulsed mid-frequency DC magnetron plasma, *Surface and Coatings Technology*, 135, 221 (2001).

209. P. S. Henderson, P. J. Kelly, R. D. Arnell, H. Bäcker, and J. W. Bradley, Investigation into the properties of titanium based films deposited using pulsed magnetron sputtering, *Surface and Coatings Technology*, 174–175, 779 (2003).

210. M. Mišina, J. W. Bradley, H. Bäcker, Y. A. Gonzalov, S. K. Karkari, and D. Forder, Investigation of the pulsed magnetron discharge by time- and energy-resolved mass spectrometry, *Vacuum*, 68, 171 (2003).

211. J. Lin, J. J. Moore, B. Mishra, W. D. Sproul, and J. A. Rees, Examination of the pulsing phenomena in pulsed-closed field unbalanced magnetron sputtering (P-CFUBMS) of Cr–Al–N thin films, *Surface and Coatings Technology*, 201, 4640 (2007).

212. J. W. Bradley, H. Bäcker, Y. Aranda-Gonzalvo, P. J. Kelly, and R. D. Arnell, The distribution of ion energies at the substrate in an asymmetric bi-polar pulsed DC magnetron discharge, *Plasma Sources Science and Technology*, 11, 165–174 (2002).

213. J. Lin, J. J. Moore, B. Mishra, M. Pinkas, W. D. Sproul, and J. A. Rees, Effect of asynchronous pulsing parameters on the structure and properties of CrAlN films deposited by pulsed closed field unbalanced magnetron sputtering (P-CFUBMS), *Surface and Coatings Technology*, 202, 1418–1436 (2007).
214. H. Backer, P. S. Henderson, J. W. Bradley, and P. J. Kelly, Time-resolved investigation of plasma parameters during deposition of Ti and TiO2 thin films, *Surface and Coatings Technology*, 174–175, 909–913 (2003).

Problems

7.1 List at least five reasons and corresponding examples to employ surface engineering and coatings in industrial applications.

7.2 List typical microstructure designs involved in nanostructured and nanocomposite coatings. Explain the key features and critical requirements for each structural design.

7.3 Define the Hall–Petch and reversed Hall–Petch relationships. Explain briefly (i) why the hardness of nanostructured films will increase when the grain size is reduced, (ii) why the hardness may start to decrease when the grain size is below 10 nm in nanocrystalline films, and (iii) why the hardness can be further enhanced in the nanocomposite coating design even though the nanocrystalline size is below 10 nm?

7.4 Explain the relationship between the bilayer period and the properties (e.g., hardness) of the nanoscale multilayer coatings. List at least three possible mechanisms of hardness enhancement in the superlattice coatings.

7.5 Explain the nanocomposite coating design. List at least five nanocomposite coating systems based on the information in the text and other research papers, and identify the nanocrystalline and amorphous phases in each coating system.

7.6 Explain why high toughness is as important as hardness for the industrial application of nanostructured coatings. Explain the relationship between the hardness and fracture toughness of nanostructured coatings.

7.7 Distinguish between physical and chemical vapor deposition using technical examples. Prepare a table to summarize the advantages and disadvantages of different PVD and CVD techniques, for example, electron beam evaporation, cathodic arc evaporation, PLD, thermal plasma processing and sputter depositions.

7.8 Define thermal and cold plasmas. Use cartoon drawings to explain the main ionization mechanisms in a sputter-discharged plasma.

7.9 Compare the characteristics of balanced, unbalanced, and closed-field unbalanced magnetron sputtering techniques. Explain why a higher plasma density and ICD can be achieved if the magnetrons are designed as unbalanced and in closed-field configuration.

7.10 Describe the mechanism of arc formation during DC reactive sputtering of Al in an Ar and O_2 atmosphere. Explain the principles of pulsed magnetron sputtering for suppressing the arc problem. Consider an Al_2O_3 layer of area A and thickness h. Calculate the frequency of target pulsing required for arc suppression given the target current density ($J = 10 \, mA \cdot cm^{-2}$), dielectric constant of Al_2O_3 film ($\varepsilon_r = 10$), and dielectric breakdown electric field ($E = 10^5 \, V \cdot cm^{-1}$). (Hint: The capacitance C and the electric field E that build up across the layer are $C = \varepsilon_r \varepsilon_0 A/h$ and $E = q/(Ch)$, respectively.)

7.11 Calculate the duty cycles and draw the voltage waveforms for the following symmetrically pulsed plasma using a sputtering voltage of 400 V: (1) 100 kHz and 1.0 μs, (2) 100 kHz and 5.0 μs, (3) 300 kHz and 1.0 μs, and (4) 300 kHz and 1.4 μs.

7.12 Explain the principles of the high-power pulsed DC magnetron sputtering (HPPMS) technique. Prepare a table showing the comparisons between HPPMS, conventional DC, and pulsed DC magnetron sputtering under various aspects, for example, the degree of ionization, plasma density, target power density, etc.

7.13 Assuming you are producing Al_2O_3 coatings using a magnetron sputtering system in Ar and O_2 gas mixture, list potential procedures to minimize the target poisoning effect in the hysteresis curve.

7.14 The Zone structure model proposed by Thornton established relationships between the sputtered coating structure and surface morphology to the pressure and substrate temperature, respectively. Describe the key features of Zone I, Zone T, Zone II, and Zone III structures of the sputtered coatings, and explain why and how a change of the pressure and substrate temperature will change the coating structure and morphology.

7.15 In a DC planar magnetron sputtering system, the cathode fall distance (L) can be estimated from Child's law:

$$L^2 = \left(\frac{4\varepsilon_0}{9}\right)\left(\frac{2e}{m}\right)^{1/2}\left(\frac{V^{3/2}}{J}\right)$$

where

ε_0 (8.85×10^{14} F·cm^{-1}) is the permittivity of vacuum

J is the target current density in mA·cm^{-2}

e/m is the charge/mass ratio of the extracted ions

V/J a target operation voltage of 1000 V and a target current density of 1 mA·cm^{-2}

i. Use Child's law to estimate the cathode fall distance (L) for Ar DC sputtering.

ii. If the cathode to anode spacing is 10 cm, determine the magnetic field that needs to be applied to trap electrons within 0.5 cm of the target?

7.16 What is a floating substrate potential? Explain how the floating potential is calculated in a DC discharged plasma.

7.17 What is the ion energy range in a conventional DC discharged plasma? Explain how the ion bombardment on the growing film in DC magnetron sputtering may be increased. Explain why a wide range of ion energies will be produced in a pulsed DC discharged plasma in reference to the features in a pulsed target voltage waveform.

7.18 Explain why ion bombardment is important in a magnetron sputtering process. Explain the effect of substrate bias and/or pulsed ion energetic bombardment on the following structure and properties of the films: (a) composition, (b) texture, (c) density, (d) grain size, (e) surface roughness, (f) residual stress, and (g) defect.

Applications of Thin Films

John J. Moore, Jianliang Lin, and In-Wook Park

In thinking about nanotechnology today, what's most important is understanding where it leads, what nanotechnology will look like after we reach the asssembler breakthrough.

K. Eric Drexler

THREADS

Chapter 8 is a continuation of *chapter 7* and provides examples of applications of the thin film technology gleaned from it. With this chapter, we finish the *Mechanical Nanoengineering* division of the text and move on to the *Chemical Nanoengineering* division.

8.0 TECHNOLOGICAL APPLICATIONS OF THIN FILMS

In the previous chapter, the concept, classification, and technical properties of nanocomposite and nanostructured thin films and coatings have been introduced. In addition, important and popular thin-film deposition techniques and principles have been reviewed. For a variety of applications, different structures and properties of a nanostructured and nanocomposite coating system are required, leading to specific coating design and correlated deposition process parameter control. In general, nanocomposite coatings can demonstrate different mechanical, electrical, optical, electrochemical, catalytic, and structural properties than those of each individual component [1]. This multifunctional behavior is closely connected to the film structure. The structure, however, depends on the phases, chemical composition, and the arrangement of the phases in the material and this in turn is strongly governed by the deposition process.

In this section, some technological examples of nanostructured and nanocomposite coatings will be illustrated, and which been used widely in different areas, such as high-temperature self-lubricating coatings for aerospace applications, high hard and tribological coatings for pressure die casting die protection, and high hard and toughness tribological coatings for bearing protection.

8.1 UNBALANCED MAGNETRON SPUTTERING OF TI–AL–SI–N COATINGS

Nanocomposite coatings based on nanocrystalline hard transition metal carbide, for example, nc-TiN imbedded in a solid amorphous carbon matrix, a:Si_3N_4, have been shown to enhance the film hardness and toughness while maintaining low sliding friction coefficients. These coatings have significant applications as a protective layer for roller or sliding bearings and gears in the automotive industry.

Quaternary Ti–Al–Si–N coatings have been prepared by a hybrid coating system, where cathodic arc evaporation (CAE) was combined with a magnetron sputtering technique. Various analyses (e.g., high resolution transmission electron microscopy [HRTEM], x-ray photoelectron spectroscopy [XPS], x-ray diffraction [XRD]) revealed that the synthesized Ti–Al–Si–N coatings exhibited nanostructured composite microstructures consisting of solid-solution (Ti, Al, Si)N crystallites and amorphous Si_3N_4. The Si addition caused the grain refinement of (Ti, Al, Si)N

FIG. 8.1 *Dark-field TEM and SEM images of Ti–Al–Si–N coatings containing (a) 0, (b) 4, (c) 9, and (d) 19 at% Si, respectively.*

Source: J. Takadoum, H. Houmid-Bennani, and D. Mairey, *Journal of European Ceramic Society*, 18, 553 (1998). With permission.

crystallites and their uniform distribution with percolation phenomenon of amorphous silicon nitride similar to that of the Si effect in TiN films [2]. **Figure 8.1** shows dark-field transmission electron microscope (TEM) images of Ti–Al–Si–N coatings containing different amounts of Si. The (Ti, Al)N crystallites (**Fig. 8.1a**) appear to be large grains with a columnar structure. The (Ti, Al)N crystallites became finer with a uniform distribution as the Si content was increased.

In **Figure 8.1b**, the crystallites were embedded in an amorphous matrix. These crystallites were revealed to be solid-solution (Ti, Al, Si)N phases of typical face-centered cubic (*fcc*) crystal structure from the electron diffraction patterns. The (Ti, Al, Si)N crystallites showed a partly aligned microstructure penetrated (*percolated*) by an amorphous phase, but were not distributed homogeneously in the amorphous matrix. The microstructure, however, changed to that of a nano-composite, having much finer (Ti, Al, Si)N crystallites (approximately 10 nm) and uniformly embedded in an amorphous matrix as the Si content in films increased to 9 at% (**Fig. 8.1c**). Such a microstructure as shown in **Figure 8.1c** was in agreement with the concept of a nanocomposite architecture suggested by Veprek et al. [3]. Therefore, the Ti–Al–Si–N coatings with Si content of 9 at% exhibited maximum hardness among the experimental conditions. On the other hand, at a higher Si content of 19 at% (**Fig. 8.1d**), (Ti, Al, Si)N crystallites decreased (~3 nm) and the film consisted mainly of the amorphous phase.

Figure 8.2a shows the nanohardness and Young's modulus of Ti–Al–Si–N coatings with various Si contents and average grain sizes. As the Si content increased, the nanohardness and Young's modulus of the Ti–Al–Si–N coatings steeply increased, and reached maximum values of ~55 and 650 GPa at Si content of 9 at%, respectively, and then dropped again with further increase of Si content. The reason for large increases in hardness and Young's modulus of Ti–Al–N with Si addition is due to the grain boundary hardening both by strong cohesive energy of interphase boundaries and by Hall–Petch strengthening derived from crystal size refinement, as mentioned in section 7.3, which were

FIG. 8.2 *(a) Nanohardness and Young's modulus and (b) friction coefficients of nanocomposite Ti–Al–Si–N [nc-(Ti, Al)N/a-Si$_3$N$_4$] coatings as a function of Si content.*

simultaneously caused by the percolation of amorphous Si$_3$N$_4$ (H = ~22 GPa, E = ~250 GPa) [2] into the Ti–Al–N film. Another possible reason would be due to solid-solution hardening of crystallites by Si dissolution into Ti–Al–N. The maximum hardness value at the silicon content of about 9 at% results from the nanosized crystallites and their uniform distribution embedded in the amorphous Si$_3$N$_4$ matrix as shown in **Figure 8.2**. On the other hand, the hardness reduction with further increase of Si content after maximum hardness as shown in **Figure 8.2a** has been explained with the thickening of amorphous Si$_3$N$_4$ phase with increase of Si content [4]. When the amorphous Si$_3$N$_4$ is increased, the ideal interaction between nanocrystallites and the amorphous phase is lost, and the hardness of the nanocomposite becomes dependent on the property of the amorphous phase. On the other hand, Young's modulus, which must be related with density and atomic structure of the film, also largely increased from 470–670 GPa with Si addition. This latter result was attributed to the densification of Ti–Al–Si–N films by filling the open structure of the Ti–Al–N grain boundaries with amorphous silicon nitride. Young's modulus reduction with further increase of Si content above 9 at% was explained by the increase of volume fraction of the amorphous Si$_3$N$_4$ phase, which has a lower atomic density than the crystalline (Ti, Al, Si)N phase.

Figure 8.2b shows the friction coefficients of the Ti–Al–N, Ti–Al–Si(9 at%)–N, and Ti–Al–Si(31 at%)–N films against a steel ball counterpart. The average friction coefficient of the film decreased from 0.9 to 0.6 with increasing Si content. This result is caused by tribo-chemical reactions, which often take place in many ceramics [5], for example, Si$_3$N$_4$ reacts with H$_2$O to produce a SiO$_2$ or Si(OH)$_2$ tribo-layer [6]. The products of SiO$_2$ and Si(OH)$_2$ are known to play the role of a self-lubricating layer.

Figure 8.3 shows the surface morphologies of the wear track and composition analyses for the wear debris after a sliding test. The surface morphology of the wear track for the Ti–Al–N film was rough, and the width of the wear track was

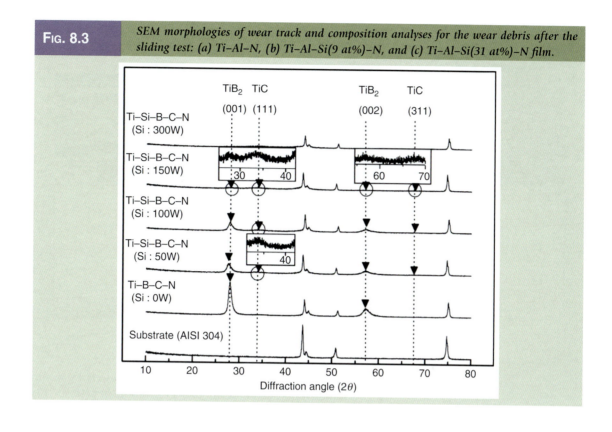

SEM morphologies of wear track and composition analyses for the wear debris after the sliding test: (a) Ti–Al–N, (b) Ti–Al–Si(9 at%)–N, and (c) Ti–Al–Si(31 at%)–N film.

relatively narrow as shown in **Figure 8.3a**, whereas, the surface morphology for the Ti–Al–Si(9 at%)–N film was relatively smooth and the width of wear track was wide (**Fig. 8.3b**). This result is due to the adhesive wear behavior between the hard film (~55 GPa) and the relatively soft steel (~700 $H_{v0.2}$). Thus, the steel ball is worn and smeared onto the Ti–Al–Si(9 at%)–N film having higher hardness (~55 GPa). On the other hand, the surface morphology of wear track for the Ti–Al–Si(31 at%)–N film was very smooth, and the width of the wear track narrowed again as shown in (**Fig. 8.3c**). This reflects that the formation of a self-lubricating tribo-layer such as SiO_2 or $Si(OH)_2$ was activated on increasing the Si content. From energy dispersive x-ray spectroscopy (EDS) analyses of the wear debris (**Fig. 8.3a and b**), the peak intensity of Fe for the Ti–Al–Si(9 at%)–N film was higher than those for the Ti–Al–N film, and the peak intensities of Ti and Al were reversed for the two films. Similar EDS results were found between Ti–Al–Si(9 at%)–N and Ti–Al–Si(31 at%)–N films. This indicates that the harder film was more wear resistant against the steel counterpart.

8.2   UNBALANCED MAGNETRON SPUTTERING OF TI–SI–B–C–N COATINGS

Figure 8.4 shows the x-ray diffraction patterns of Ti–B–C–N and Ti–Si–B–C–N films on AISI 304 stainless steel substrates with various Si target powers at a

FIG. 8.4 *XRD patterns of Ti–B–C–N and Ti–Si–B–C–N films on AISI 304 stainless steel substrates with various Si target powers at a fixed TiB₂–TiC composite target power of 700 W.*

fixed TiB_2–TiC composite target power of 700 W. At a Si target power of 0 W, the diffraction pattern of the Ti–B–C–N film exhibited a crystalline hexagonal TiB_2 phase with preferred orientations of (001) or (002) crystallographic planes. Any XRD peaks corresponding to the crystalline TiC or TiN phase were not observed from the Ti–B–C–N diffraction pattern. As the Si target power was increased, the diffraction peak intensities of TiB_2 (001) and (002) gradually reduced and completely disappeared at the Si target power of 300 W. At the Si target power of 50 and 100 W, the TiB_2 peaks corresponding to the same (001) and (002) planes as well as small TiC peaks for (111) and (311) crystallographic planes were present. And, at a Si target power of 150 W, very small diffraction TiB_2 peaks for (001) and (002) as well as TiC peaks for (111) and (311) were observed. Furthermore, at the highest Si target power of 300 W, the XRD pattern presented no diffraction peaks for the film, indicating that the film is comprised mainly of an amorphous phase. The gradual changes in the XRD patterns of Ti–Si–B–C–N films with Si additions into Ti–B–C–N are similar to the case of N addition into Ti–B–C, as previously reported by the authors for the Ti–B–C–N nanocomposite system [7].

In the report, it was revealed that the crystallites in Ti–B–C–N films were composed with solid-solution (Ti, C, N)B_2 and Ti(C, N) crystallite (~10 nm in size). Addition of nitrogen into the Ti–B–C film led to grain refinement of (Ti, C, N)B_2 and Ti(C, N) crystallites, and their distribution is coupled with a percolation phenomenon of amorphous BN and carbon phase. In order to determine the chemical composition and to investigate the bonding status of the Ti–Si–B–C–N coating, x-ray photoelectron spectroscopy (XPS) was performed on Ti–B–C–N and Ti–Si–B–C–N coatings deposited by unbalanced magnetron sputtering from a TiB_2–TiC composite target and a Si target at different Si target powers. **Figure 8.5a** provides the content of each element in the Ti–Si–B–C–N coating

Fig. 8.5 *XPS data for (a) content of Ti, Si, B, C, N, and O, and XPS spectra of (b) Si 2p, (c) C 1s, and (d) O 1s for Ti–B–C–N with Si target power of 0 W and Ti–Si–B–C–N coating with Si target power of 300 W (14.2 at% Si in film).*

with two different Si target powers and a fixed TiB_2–TiC target power of 700 W. As the Si target power was increased, the Si content was increased in the Ti–Si–B–C–N film from 0 to 14.2 at%. The boron content steeply decreased. **Figure 8.5b–d** present the XPS spectra binding energies of Si, C, and O for the Ti–B–C–N and Ti–Si(14.2 at%)–B–C–N coatings. For the Si 2p region (**Fig. 8.5b**), $TiSi_2$ is clearly present with smaller amounts of SiC and SiB_4. The C 1s region (**Fig. 8.5c**) confirms the presence of $Ti_xB_yC_z$ components in amorphous free carbon and CN_x. With the addition of Si into the Ti–B–C–N coating, the free-carbon peak intensity significantly decreased. The large decrease in the free-carbon peak intensity of Ti–B–C–N with 14.2 at% Si addition is most likely the result of formation of SiC as shown in **Figure 8.5b**.

Based on the results from the XRD and XPS analyses, it is concluded that the Ti–Si–B–C–N coatings are nanocomposites consisting of nanosized $(Ti,C,N)B_2$ and Ti(C,N) crystallites embedded in an amorphous $TiSi_2$ and SiC matrix including some carbon, SiB_4, BN, CN_x, TiO_2, and B_2O_3 components.

Figure 8.6 presents the nanohardness and H/E values of the Ti–Si–B–C–N coatings as a function of Si target power. The hardness of the Ti–Si–B–C–N

FIG. 8.6

Nanohardness and H/E values of Ti–Si–B–C–N coating as a function of Si target power.

coating decreased from ~42 GPa at 0 W Si target power to ~36 GPa at 50 W Si target power. The hardness was constant at about 35 GPa from 50 to 150 W Si target power and decreased again with a further increase in Si target power to about 25 GPa at a Si target power of 300 W (14.2 at% Si in coating). The decrease in hardness of Ti–B–C–N with 50 W Si target power is most likely due to a reduction in hard TiB_2-based crystallites. The supporting evidence is shown in **Figure 8.5a**, where boron content decreases with increasing silicon content. However, the Ti–Si–B–C–N coating with a Si target power from 50 to 150 W exhibited a high hardness of about 35 GPa. The reason for maintaining the high hardness (~35 GPa) in Ti–Si–B–C–N coatings with a small amount of Si is most likely due to percolation of amorphous $TiSi_2$ and SiC in the grain boundaries. Veprek et al. [8] have found that ultra-hardness ($80 GPa \leq H_v \leq 105 GPa$) is achieved when the nanosized and/or amorphous $TiSi_2$ precipitate in the grain boundaries in their Ti–Si–N (nc-TiN/a-Si_3N_4/a- and nc-$TiSi_2$) nanocomposites. On the other hand, the hardness reduction with further increase in Si target power above 200 W can be explained by either an increase in the soft amorphous $TiSi_2$, SiC, and SiB_4 phases or reduction of hard TiB_2-based crystallites.

When the amount of amorphous phase is increased the ideal interaction between nanocrystallites and the amorphous phase can be lost, and the hardness of the nanocomposite becomes dependent on the property of the amorphous phase [9]. In addition, H/E values, the so-called *elastic strain to failure criterion*, were calculated from the obtained hardness (H) and Young's modulus (E). Recently, Leyland and Matthews [10] have suggested that a high H/E value is often a reliable indicator of good wear-resistance. In **Figure 8.6**, the H/E values exhibited a similar tendency with hardness. As the Si target power increased, the H/E value of Ti–Si–B–C–N coatings decreased from ~0.105 to 0.087. From the results of hardness and H/E, it can be suggested that the Ti–Si–B–C–N coating

FIG. 8.7	*(a) Friction coefficients and (b) wear rates of Ti–Si–B–C–N coating against WC-Co ball as a function of Si target power.*

(a)

(b)

with Si target power up to 150 W could provide a better wear-resistance with higher fracture toughness than that of Ti–Si–B–C–N coatings with Si target power above 200 W.

Figure 8.7a shows the friction coefficients of the Ti–Si–B–C–N coating against a WC-Co ball as a function of Si target power. The average friction coefficient of the Ti–Si–B–C–N coating rapidly decreased by increasing the Si target power and showed a minimum value of ~0.15 at a Si target power of 50 W, and then rebounded with further increase in Si target power above 100 W. This large decrease in the friction coefficient of the Ti–Si–B–C–N coating with 50 W Si target power is most likely due to the formation of smooth solid-lubricant tribo-layers formed by tribo-chemical reactions during the sliding test. For example, silicon compounds such as $TiSi_2$, SiC, or SiB_4 in the coating react with ambient H_2O and oxygen to produce SiO_2 or $Si(OH)_2$ tribo-layers. These by-products of SiO_2 and $Si(OH)_2$ have been known [5] to play the role of a self-lubricating layer. This Si effect on tribological behavior in nanocomposites has also been found [11] by other authors. In addition, the carbon and hydroxide phases in the coating, as shown in **Figure 8.5c** and **d**, with small Si content may also contribute to the minimum friction coefficient value.

However, the friction coefficient slightly increased with further increase of Si content in the coating. The increase in friction coefficient with increased target power is most likely due to a reduction in the free-carbon and hydroxide phases, which are self-lubricating phases, in the Ti–Si–B–C–N coating as shown in **Figure 8.5c** and **d**. **Figure 8.7b** represents the wear rates of the Ti–Si–B–C–N coating on AISI 304 substrates as a function of Si target power. The wear rate of the Ti–Si–B–C–N coating slightly increased from ~3×10^{-6} $mm^3 \cdot N^{-1} \cdot m^{-1}$ at 0 W to ~10×10^{-6} $mm^3 \cdot N^{-1} \cdot m^{-1}$ at a Si target power around 200 W. These very low wear rates would be due to the adhesive wear behavior between the hard coating (~35 GPa) and relatively soft WC-Co (~22 GPa) ball. On the other hand, at the Si target power above 250 W, the wear rate of the Ti–Si–B–C–N coating steeply increased to about 500×10^{-6} $mm^3 \cdot N^{-1} \cdot m^{-1}$ at a Si target power of 300 W. This large increase in wear rate would be due to the abrasive wear behavior between the relatively soft coating (~25 GPa) and the WC-Co ball. Combining the results of the H/E (**Fig. 8.6**) values and wear rates (**Fig. 8.7b**), the Ti–Si–B–C–N coating with a higher H/E of above ~0.090 had a better wear-resistance against the WC-Co ball in agreement with Leyland and Matthews [10].

8.3 PULSED CLOSED FIELD UNBALANCED MAGNETRON SPUTTERING OF CR–AL–N COATINGS

One of the most important applications for tribological nanostructured coatings is as a protective layer to improve multiple properties of the working surface of a bulk material, or tool, used in an aggressive environment. Development of transition metal nitride thin films (e.g., TiN, TiAlN, CrN, CrAlN, etc.) has been widely documented as a means to increase productivity and tool life in material-forming processes, such as die casting, metal forming, plastic molding, glass forming tool, and machining/cutting applications. In these applications, however, the coating material often experiences extreme mechanical, thermal, and chemical loading conditions. For example, in the modern high-pressure die casting process, a molten aluminum alloy at temperatures ranging from 670 to 710°C is injected into the die cavity at high velocities of the order of 30–100 m · s⁻¹. The injection pressures are of the order of 50–80 MPa, with a temperature gradient of around 1000°C · cm⁻² [12]. In the machining of steels, stainless steels, and cast irons with coated cemented carbide tools, the cutting edges are worn according to different wear mechanisms at high cutting speeds; the amount of heat generated in the cutting zone is considerable and the temperature at the cutting edge of coated cutting tools may exceed 1000°C in an ambient oxidation environment [13].

The demand for continual improvement of hard coatings leads to the need to develop multifunctional hard tribological coatings, which can provide a wide range of properties. The successful application and improvement of these tribological hard coatings strongly depends on the microstructure (nanostructure) design and deposition process control.

The CrAlN ternary compound film is a very promising die coating candidate that shows high toughness and hardness, good wear resistance, and excellent oxidation resistance combined with corrosion resistance [14–17]. It was found that the aluminum content in the film plays a significant role in determining the structure and properties of the Cr–Al–N coatings. The incorporation of aluminum into the B1 cubic CrN lattice will lead to the precipitation of a B4 hexagonal AlN phase if the solubility limit of AlN in the coating is exceeded. The formation of the wurtzite hexagonal structure is not desired due to its low hardness and poor ductility [18]. It has been predicted that CrN shows the highest solubility of 77.2% for aluminum among the transition nitrides with a B1 cubic structure [19,20]. Therefore, it is possible to add a large amount of aluminum into CrAlN films without changing the cubic phase, thereby extending the CrAlN film oxidation resistance temperature while maintaining good mechanical and

FIG. 8.8 — *The GIXRD patterns for $Cr_{1-x}Al_xN$ films with different aluminum concentrations.*

tribological properties. **Figure 8.8** shows the change in the crystal structure of $Cr_{1-x}Al_xN$ deposited by CFUBMS as a function of aluminum content. The B4 hexagonal AlN structure was observed when the aluminum content is at 64 at% [14]. The coexistence of cubic and hexagonal phases is observed when the aluminum concentration in the film is at or beyond 64.0 at%.

Figure 8.9 provides the cross-sectional scanning electron microscope (SEM) photomicrographs of CrN and $Cr_{0.415}Al_{0.585}N$ films. The columnar grain size of CrN is about 100 nm (**Fig. 8.9a**). A significant decrease in the grain size was observed with an increase in the aluminum concentration in the films (**Fig. 8.9b**). The $Cr_{1-x}Al_xN$ films' hardness and Young's modulus values are plotted as a function of aluminum concentration in **Figure 8.10**. The results show that the CrN film has a hardness of 25.0 GPa. As aluminum is incorporated into the film, the

FIG. 8.9 — *Scanning electron micrographs of cross sections of (a) CrN and (b) $Cr_{0.415}Al_{0.585}N$ films on a silicon substrate [14].*

| FIG. 8.10 | *Nanoindentation hardness and Young's modulus of $Cr_{1-x}Al_xN$ films with different aluminum concentrations.* |

hardness and Young's modulus values of the $Cr_{1-x}Al_xN$ films increase, and both reach the maximum values of 36.3 and 370 GPa, respectively, at 58.5 at% aluminum concentration. A further increase in aluminum concentration results in a decrease in both hardness and Young's modulus. The increase of the film hardness is possibly related to the decrease in the grain size and a denser film structure (**Fig. 8.9b**).

According to the "Hall–Petch" relationship (equation 7.1), the hardness of the material increases with decreasing crystallite size, especially prominent for grain sizes down to tens of nanometers. In addition, the incorporation of aluminum into the CrN lattice will decrease the lattice parameter because of the smaller atomic size of aluminum atoms compared with the Cr atom. This behavior will increase the covalent energy in the films, because the interatomic distance d, is related to the covalent bandgap E_h according to the expression $E_h = Kd^{-2.5}$. Furthermore, addition of aluminum in CrN increases the covalent bonding, as CrN is a largely metallically bonded material, while AlN is predominantly covalently bonded [21]. Therefore, the increase in hardness of $Cr_{1-x}Al_xN$ films with increasing aluminum content is probably related to the above grain size and bonding energy effect.

The steady-state coefficient of friction (COF) values and calculated wear rates of the $Cr_{1-x}Al_xN$ nanostructured films are plotted as a function of aluminum content in **Figure 8.11**. It was found that the CrN film exhibits a lower COF value (0.28) than all of $Cr_{1-x}Al_xN$ films. Adding a small amount of aluminum into $Cr_{1-x}Al_xN$ films ($x = 0.22$) resulted in a sudden increase in COF to 0.55. Further increasing the aluminum content in the film, the COF of $Cr_{1-x}Al_xN$ films decreased to a low value of 0.37 at $x = 0.585$ and then started to increase up to 0.45 at $x = 0.68$. The wear rate exhibits a similar trend compared with the COF change in $Cr_{1-x}Al_xN$ films.

Controlled ion bombardment of growing thin films can be used to modify and improve the film structure and properties. Higher energetic species (up to

FIG. 8.11   *COF and wear rate as a function of the aluminum concentration in $Cr_{1-x}Al_xN$ films.*

hundreds eV) found in the plasma by pulsing the target(s) in magnetron sputtering has been discussed in section 7.5.4. In this section, the effect of the pulsed ion energy and ion flux on the P-CFUBMS-deposited CrAlN film structure and properties will be illustrated.

Figure 8.12a shows the $^{29}N_2^+$ IEDs for pulsing both Cr and Al targets asynchronously at 100 and 350 kHz with different reverse times, respectively. Different

FIG. 8.12   *(a) IEDs of $^{29}N_2^+$ for pulsing Cr and Al targets asynchronously at 100 and 350 kHz with different reverse times in P-CFUBMS-deposited CrAlN films. (b) The integrated ion fluxes in three ion energy regions at different pulsing conditions in asynchronous modes [22].*

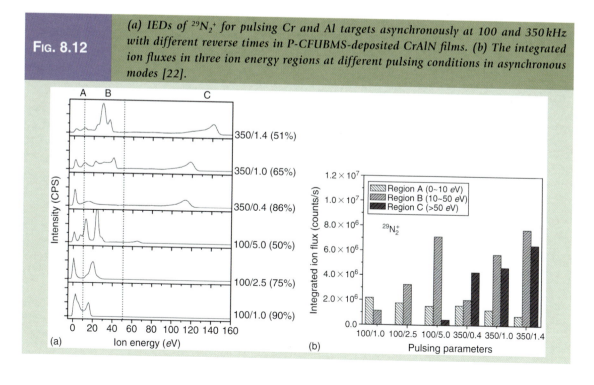

maximum $^{29}N_2^+$ energies with a range from 22 to 150 eV were observed in the discharged plasma in accordance with different asynchronous pulsing parameters (**Fig. 8.12a**) [22]. The ion fluxes of $^{29}N_2^+$ corresponding to the three pulsed ion energy regions ("A", "B", "C") under different pulsing conditions are shown in **Figure 8.12b** [22]. As can be seen, the ion fluxes in the ion energy region "A" decrease with an increase in the reverse time under the same pulsing frequency. On the other hand, the ion fluxes in ion energy regions "B" and "C" exhibit the reverse trend, in which they increase when the reverse time increases under the same pulsing frequency.

The wide range of pulsed ion energies and ion fluxes in a pulsed plasma have a significant influence on the film surface morphology and microstructure. Three-dimensional atomic force microscopy (AFM) images and cross-sectional SEM micrographs of Cr–Al–N films deposited at different asynchronous pulsing conditions are shown in **Figures 8.13** and **8.14**, respectively.

The change in film surface roughness and morphology can be explained by the ion energy/ion flux change in the plasma. The low ion energy and ion flux in the 100 kHz and 1.0 μs pulsed plasma lead to a low nucleation density due to the low mobility of the adatoms on the substrate, developing coarser grains with a high density of sub-grains and large cell boundaries (**Fig. 8.13a**). This is consistent with an open columnar structure, corresponding to zone 1 structure in the Thornton Zone model [24] as seen in the SEM image (**Fig. 8.14a**). This film exhibited a mean surface roughness of 5.45 nm.

The film that was deposited in the 100 kHz and 5.0 μs condition exhibits a very dense grain structure and the smoothest surface with a surface roughness of 1.04 nm (**Figs. 8.13b** and **8.14b**). This may be attributed to the significant increase of ion flux in the "B" (middle) ion energy region, and also the total ion energy reached 122 eV. In this case, a large flux of ions with middle energy level bombard the substrate surface, significantly enhancing the adatom mobility and diffusivity by the momentum transfer from the impingement. The highly

FIG. 8.13 *2- and 3-D AFM surface morphologies of Cr–Al–N films deposited at different asynchronous pulsing conditions with different maximum $^{29}N_2^+$ ion energies: (a) 100 kHz and 1.0 μs (E_{total} = 72 eV), (b) 100 kHz and 5.0 μs (E_{total} = 122 eV), (c) 350 kHz and 1.0 μs (E_{total} = 180 eV). (E_{total} = pulsed ion energy + substrate bias (−50 V), R_a = surface roughness).*

R_a = 5.45 nm R_a = 1.04 nm R_a = 2.01 nm

5 μm   5 μm   5 μm

(a)   (b)   (c)

Source: J. Lin, J. J. Moore, B. Mishra, M. Pinkas, W. D. Sproul, and J. A. Rees, *Surface and Coatings Technology*, (2007). With permission.

FIG. 8.14

Cross-sectional SEM micrographs of Cr–Al–N films deposited at different asynchronous pulsing conditions with different maximum $^{29}N_2^+$ ion energies: (a) 100 kHz and 1.0 μs (E_{total} = 72 eV), (b) 100 kHz and 5.0 μs (E_{total} = 122 eV), and (c) 350 kHz and 1.0 μs (E_{total} = 180 eV).

Source: J. Lin, J. J. Moore, B. Mishra, M. Pinkas, W. D. Sproul, and J. A. Rees, *Surface and Coatings Technology*, (2007). With permission.

mobile adatoms can move or diffuse into the inter-grain voids under high energy ion bombardment, and a denser structure is attained.

The pulsed ion energy and ion flux is further enhanced in the 350-kHz pulsed plasma (**Fig. 8.12**). However, it is noted that the ion energy increase is largely from the "C" high ion energy region while the contribution from the "A" low ion energy region is decreased compared to the 100-kHz conditions. In the cross-sectional field emission scanning electron microscopy (FESEM) micrographs of the 350-kHz and 1.0-μs films (**Fig. 8.14c**), a different microstructure compared to those in 100-kHz pulsing conditions is revealed. Columnar grains are still observed but "renucleation" (interruption of film growth due to high ion energy bombardment followed by localized growth) on the individual column is seen throughout the film. The incoming high energy (>180 eV) ion bombardment can disrupt the continued growth of the columnar grains. Thus, only few grains grew throughout the film thickness and short columnar grains were formed (**Fig. 8.14c**), while the film surface roughness increased to 2.25 nm by kinetic roughening (**Fig. 8.13c**).

A comparison of the cross-sectional TEM micrographs and corresponding selected area electron diffraction (SAED) patterns of Cr–Al–N films deposited at 100 kHz and 5.0 μs and 350 kHz and 1.0 μs are presented in **Figure 8.15a** and **b**, respectively. In both films the energetic-enhanced deposition conditions resulted in the formation of a dense nanocrystalline structure. The average grain size in the 100-kHz and 5.0-μs film is 20–50 nm (**Fig. 8.15a**). The SAED pattern displays a typical form of a nanocrystalline material consisting of the *fcc* (Cr, Al) N phase. No amorphous rings are observed in the SAED pattern. In the film deposited at 350 kHz and 1.0 μs, the arc- and spot-shaped SAED pattern indicate that a bi-model grain size is formed in this high-energy bombarded film. The small grains are formed on renucleation sites along the larger columnar grains, as shown in **Figure 8.15b**. However, this high ion energy may be excessive for the film growth, in that it can cause increased point (vacancy) and line (dislocations) defect density and intensive resputtering and a consequent decrease in nucleation sites. The intergranular residual damage is the most prevalent defect

FIG. 8.15 *Cross-sectional TEM photomicrographs of Cr–Al–N films deposited at different asynchronized pulsing conditions. (a) 100 kHz, 5.0 μs and (b) 350 kHz, 1.0 μs.*

in the 350-kHz and 1.0-μs Cr–Al–N film, which was subjected to excessive ion bombardment as shown in **Figure 8.16a**. The grains contain an appreciable amount of lattice defects visible by the speckled contrast within the columnar grains. Edge dislocations that compensate for the high strain within the columnar grains are revealed in **Figure 8.16b**, which is a high-resolution TEM micrograph of the distorted lattice of the Cr–Al–N film.

High-resolution lattice images of Cr–Al–N films deposited at 100 kHz and 5.0 μs and 350 kHz and 1.0 μs are shown in **Figure 8.17a** and **c**, respectively. Fourier transform filtered images of the same areas are shown in **Figure 8.17b** and **d**, respectively. From the filtered lattice images, the film deposited at 100 kHz and

FIG. 8.16 *Cross-sectional TEM photomicrograph showing (a) high dislocation density within the columnar grains and (b) high-resolution TEM photomicrograph showing the distorted lattice and low-angle grain boundaries, in the Cr–Al–N film deposited in asynchronized mode at 350 kHz and 1.0 μs.*

Fig. 8.17	*(a) High-resolution TEM lattice images of Cr–Al–N film deposited at 100 kHz and 5.0 µs, (b) the fast Fourier transform (FFT) filtered image of (a), (c) high-resolution TEM lattice images of Cr–Al–N film deposited at 350 kHz and 1.0 µs, and (d) the FFT filtered image of (c).*

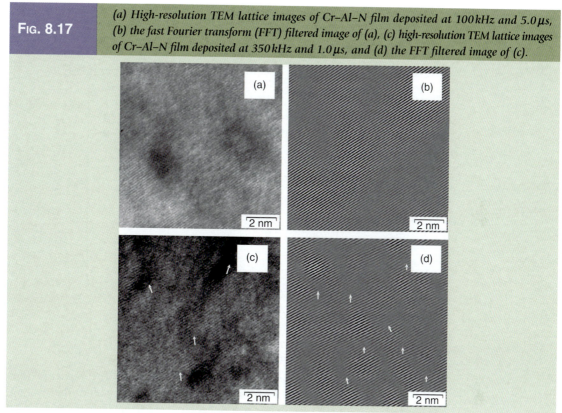

Source: J. Lin, J. J. Moore, B. Mishra, M. Pinkas, W. D. Sproul, and J. A. Rees, *Surface and Coatings Technology*, (2007). With permission.

5.0 µs exhibits a uniform lattice with few defects. On the other hand, many defects in the form of edge dislocations (indicated by arrows) are observed in the 350-kHz and 1.0-µs film.

The evolution of the film microstructure at controlled ion bombardment by pulsing the targets has significant influence on the film mechanical and tribological properties. The hardness and Young's modulus of Cr–Al–N films deposited at different asynchronous pulsing conditions are plotted in **Figure 8.18**. The film hardness increased from 34 to 48 GPa when the total ion energy in the discharged plasma increased from 72 to 200 eV accordingly. The Young's modulus of the films exhibits a similar trend. The increased hardness of the films deposited with pulsed ion energy and ion flux bombardment can be explained by two features: (i) improved density and decreased (nanocrystalline) grain size and (ii) development of internal residual stress and large defect densities.

The maximum ion energies were increased from 72 to 122 eV when the reverse time was increased from 1.0 to 5.0 µs in asynchronized mode at 100 kHz. At the same time, the main ion flux contributed from the "B" ion energy region

FIG. 8.18

The hardness and Young's modulus values of Cr–Al–N films deposited in asynchronized mode as a function of maximum ion energy in the plasma.

(10~30 eV) increases correspondingly. This controlled middle ion energy bombardment can effectively increase adatom mobility, enhance film density, and decrease the columnar grain size without greatly increasing the defect density in the film (**Fig. 8.15a**). The increased hardness values from 34 to 41 GPa in these films mainly result from the film structural improvements. Nevertheless, when the Cr–Al–N films were deposited in an asynchronized mode at 350 kHz, the excessive ion energy bombardment from the "C" high ion energy region will result in high defect incorporation as well as high residual stress in the films. In general, a high defect concentration in a compressively stressed material will restrict the plastic flow, and thus be a contributing factor in enhancing the hardness [23]. Therefore, further increase in the hardness (41~48 GPa) in these excessively ion bombarded Cr–Al–N films is possibly related to the strain hardening and a high defect density.

The steady-state COF values and calculated wear rates of the Cr–Al–N films deposited in the asynchronized mode with different maximum ion energies are plotted in **Figure 8.19**. As can be seen, the COF values and wear rate of Cr–Al–N films increased with an increase in the maximum ion energy. The films deposited at 100-kHz pulse frequency exhibit low COF values in the range of 0.38 to 0.46. However, the high point and line defect densities and high residual stress incorporated into the films deposited at 350-kHz conditions can decrease the toughness and increase the brittleness of the film. Cr–Al–N films deposited at 350-kHz pulse frequency exhibit decreased wear resistance, as shown in **Figure 8.19**.

The technological example of P-CFUBMS of CrAlN films demonstrates the importance of film composition design and the deposition process control for

| FIG. 8.19 | *COF and wear rate of Cr–Al–N films deposited in asynchronized mode with different maximum ion energy.* |

obtaining high quality films. For the pulsed magnetron sputtering, there is an advantage to maintain the maximum pulsed ion energy less than 120 eV and increase the ion flux in the "A" and "B" middle ion energy regions (10~70 eV) for obtaining improved film nanostructure and tribological properties. If the pulsed ion energies are excessive (>120 eV), there will be an increase in point and line defects, introduced in the form of high residual stress in the crystalline structure.

8.4   CONCLUDING REMARKS

The main purpose of chapters 7 and 8 is to provide a basic level of understanding of the concept, background, and processing of nanostructure and nanocomposite thin films. The chapter discusses how nanostructure and nanocomposite thin films can result in improved performance to meet the required applications, such as high wear resistance, low friction coefficient, self-lubrication, high oxidation, and/or corrosion resistance. These chapters also provide a review of some of the thin-film deposition techniques which are widely used to process nanostructure and nanocomposite thin films. Approaches to control the film chemistry and ion energy (ion flux) in tailoring the structure and properties of the films to meet specific tribological applications are emphasized. A number of multicomponent, nanostructured coating examples processed using various deposition processes were given to demonstrate the relationship among processing, structure, properties, and performance: Ti–Al–Si–N, Ti–Si–B–C–N, and Cr–Al–N.

References

1. P. M. Ajayan, L. S. Schadler, and P. V. Braun, *Nanocomposite science and technology*, WILEY-VCH Verlay, GmbH Co. KGaA, Wienheim (2003).

2. M. Diserens, J. Patscheider, and F. Lévy, Mechanical properties and oxidation resistance of nanocomposite TiN–SiN$_x$ physical-vapor-deposited thin films, *Surface and Coatings Technology*, 120–121, 158 (1999).

3. S. Veprek and S. Reiprich, A concept for the design of novel superhard coatings, *Thin Solid Films*, 268, 64 (1995).

4. S. H. Kim, J. K. Kim, and K. H. Kim, Influence of deposition conditions on the microstructure and mechanical properties of Ti–Si–N films by DC reactive magnetron sputtering, *Thin Solid Films*, 420–421, 360 (2002).

5. S. Wilson and A. T. Alpas, Tribo-layer formation during sliding wear of TiN coatings, *Wear*, 245, 223 (2000).

6. J. Takadoum, H. Houmid-Bennani, and D. Mairey, The wear characteristics of silicon nitride, *Journal of European Ceramic Society*, 18, 553 (1998).

7. I.-W. Park, K. H. Kim, A. O. Kunrath, D. Zhong, J. J. Moore, A. A. Voevodin, and E. A. Levashov, Microstructure and mechanical properties of superhard Ti–B–C–N films deposited by dc unbalanced magnetron sputtering, *Journal of Vacuum Science and Technology*, B, 23(2), 588 (2005).

8. S. Veprek, P. Nesladek, A. Niederhofer, F. Glatz, M. Jilek, and M. Sima, Recent progress in the superhard nanocrystalline composites: Towards their industrialization and understanding of the origin of the superhardness, *Surface and Coatings Technology*, 108–109, 138 (1998).

9. J. Patscheider, T. Zehnder, and M. Diserens, Structure–performance relations in nanocomposite coatings, *Surface and Coatings Technology*, 146, 201 (2001).

10. A. Leyland and A. Matthews, On the significance of the H/E ratio in wear control: A nanocomposite coating approach to optimised tribological behaviour, *Wear*, 246, 1 (2000).

11. J. Xu and K. Kato, Formation of tribochemical layer of ceramics sliding in water and its role for low friction, *Wear*, 245, 61 (2000).

12. S. Gopal, A. Lakare, and R. Shivpuri, Evaluation of thin film coatings for erosive-corrosive wear prevention in die casting dies, In *Surface modification technologies XII*, T. S. Sudarshan, K. A. Khor, and M. Jeandin, eds., ASM International, Materials Park, Ohio (1998).

13. H. O. Gekonde and S. V. Subramanian, Tribology of tool–chip interface and tool wear mechanisms, *Surface and Coatings Technology*, 149, 151 (2002).

14. J. Lin, B. Mishra, J. J. Moore, and W. D. Sproul, Microstructure, mechanical and tribological properties of Cr$_{1-x}$Al$_x$N films deposited by pulsed-closed field unbalanced magnetron sputtering (P-CFUBMS), *Surface and Coatings Technology*, 201, 4329–4334 (2006).

15. J. C. Sánchez-López, D. Martínez-Martínez, C. López-Cartes, A. Fernández, M. Brizuela, A. García-Luis, and J. I. Oñate, Mechanical behaviour and oxidation resistance of Cr–Al–N coatings, *Journal of Vacuum Science and Technology*, A, 23, 4 (2005).

16. J. Lin, B. Mishra, J. J. Moore, W. D. Sproul, and J. A. Rees, Effects of the substrate to chamber wall distance on the structure and properties of CrAlN films deposited by pulsed-closed field unbalanced magnetron sputtering (P-CFUBMS), *Surface and Coatings Technology*, 201, 6960 (2007).

17. C. Brecher, G. Spachtholz, K. Bobzin, E. Lugscheider, O. Knotek, and M. Maes, Superelastic (Cr,Al)N coatings for high end spindle bearings, *Surface and Coatings Technology*, 200 1738 (2005).

18. Y. Sun, Y. H. Wang, and H. P. Seow, Effect of substrate material on phase evolution in reactively sputtered Cr–Al–N films, *Journal of Materials Science*, 39, 7369–7371 (2004).

19. A. Sugishima, H. Kajioka, and Y. Makino, Phase transition of pseudobinary Cr–Al–N films deposited by magnetron sputtering method, *Surface and Coatings Technology*, 97, 590 (1997).

20. Y. Makino and K. Nogi, Synthesis of pseudobinary Cr–Al–N films with B1 structure by RF-assisted magnetron sputtering method, *Surface and Coatings Technology*, 98, 1008 (1998).

21. H. Holleck, Material selection for hard coatings, *Journal of Vacuum Science Technology*, A, 4, 2661 (1986).

22. J. Lin, J. J. Moore, B. Mishra, M. Pinkas, W. D. Sproul, and J. A. Rees, Effect of asynchronous pulsing parameters on the structure and properties of CrAlN films deposited by pulsed closed field unbalanced magnetron sputtering (P-CFUBMS), *Surface and Coatings Technology*, 202, 1418 (2008).

23. L. Karlsson, A. Horloing, M. P. Johansson, L. Hultman, and G. Ramanath, The influence of thermal annealing on residual stresses and mechanical properties of arc-evaporated TiC_xN_{1-x} ($x = 0$, 0.15 and 0.45) thin films, *Acta Materialia*, 50, 5103 (2002).

Problems

8.1 Give three examples of nanostructured tribological coatings used in industry. Select one of your interest to prepare a case study under the following items from the literature to study the process–structure–property relationship: (a) deposition technique and parameters, (b) chemical composition, (c) phase structure, (d) grain size and microstructure, (e) properties (e.g., hardness, adhesion, tribological, stability with temperature, corrosion resistance, surface roughness, stress and defect, etc.), and (f) on-going research.

8.2 The ball-on-disk wear test is a widely used technique to evaluate the wear properties of tribological films. Calculate the wear rate of the film expressed in $mm^3 \cdot N^{-1} \cdot m^{-1}$, given the following test parameters: tests were carried out along a circular track of 12-mm diameter under a load of 5 N and at a constant sliding speed of $40 \, mm \cdot s^{-1}$, for sliding distances up to 1000 m. After the wear tests, the average cross-sectional area of the wear track in the film was determined to be $150 \, \mu m^2$ using a surface profilometer.

8.3 Briefly describe the possible effects of the following deposition parameters on the deposition process and the structure and properties of films: (a) working pressure, (b) deposition temperature, (c) substrate bias, (d) target power, and (e) the gas flow rate.

8.4 List at least two characterization techniques which can be utilized to determine each of the following properties of nanostructured coatings, and explain the advantages and limitations of each technique: (a) grain size, (b) chemical compositions, (c) crystalline phase, (d) stress, (c) surface morphology, (d) cross-sectional morphology, (e) texture, and (f) hardness.

Section 4

Chemical Nanoengineering

NANOCATALYSIS

Scott W. Cowley

I hence will name it the catalytic force of the substance, and I will name decomposition by this force catalysis. The catalytic force is reflected in the capacity that some substances have, by their mere presence and not by their own reactivity, to awaken activities that are slumbering in molecules at a given temperature.

JÖNS JACOB BERZELIUS, 1836

FIG. 9.0

Jöns Jacob Berzelius was one of the pioneers of modern chemistry. He is responsible for elucidating the "law of definite proportions" and for developing a table based on atomic weights. He was the first to use the phrase "organic chemistry." He is known for stating the following about catalysis: "Thus it is certain that substances have the property of exerting an effect quite different from ordinary chemical affinity, in that they promote the conversion without necessarily participating in the process with their own component parts."

THREADS

Chapter 9 presents a generalized discussion about catalysis by nanoparticles. This chapter is the first chapter in the *Chemical Nanoengineering* division of the book. Catalysts have been with us for a long time. Over the past 50 years, catalysts have made many industrial processes successful. Catalysts are expected to be enabled and enhanced by applications of nanotechnology—whether in the form of new syntheses, better characterization, or new applications.

The next chapter, *chapter 10*, in this division explores the domain of nanocomposites. We have already had a discourse about nanocomposites in *chapters 7* and *8*, especially as they apply to thin films. Polymer chemistry and the chemical modification of carbon nanotubes in particular team up to present an entirely new and innovative chapter in our already rich experience with composite materials.

Following *chapter 10*, *chapters 11–13* delve into biological aspects of nanotechnology and applications. *Chapter 14* is a member of the *Biological Nanoengineering* division and presents discussions about the environment.

9.0 Introduction to Catalytic and Nanocatalytic Materials

9.0.1 The Importance of Catalysis in a Modern Society

Catalysts have an enormous impact on the social and economic structure of our world today. It is projected that by the year 2010, approximately $12 billion worth of catalyst materials will be used each year to produce hundreds of billions of dollars worth of goods on a worldwide basis [1]. Catalysts are used to make fertilizers, fuels, chemicals, medicines, textiles, and many other important products. For example, the Haber process, developed by Fritz Haber in 1913, uses an iron-based catalyst to convert nitrogen and hydrogen gases into ammonia. When ammonia is combined with nitric or sulfuric acid, then ammonium nitrate or ammonium sulfate fertilizers are produced. The utilization of fertilizers results in an increased production of food, which supports a much larger population. A similar example can be given for the catalytic conversion of petroleum products into the various hydrocarbon fuels needed for our farming, manufacturing, and transportation industries, thus bringing people, food, and goods to the market place. Catalysts also play a major role in improving our quality of life from an environmental perspective. For example, a catalyst containing platinum, palladium, rhodium, cerium, and other compounds is used to control the emission levels of toxic gases emitted from our automobiles.

9.0.2 What Is a Catalyst?

The term "catalyst" was first used by the Swedish chemist Jöns Jacob Berzelius in his published work on the catalytic decomposition of hydrogen peroxide in the *Edinburg Philosophical Journal* in 1836. He observed that a catalytic substance increased the rate of reaction without being changed or being consumed itself. However, the exact function of the catalyst remained a mystery to Berzelius for he described it as "an inherent force whose nature is still unknown."

Advances in scientific methods and analytical techniques have greatly improved our understanding of how a catalyst works. Today we know that a catalyst increases the rate of reaction by lowering the activation energy (E_a) required to convert reactants into products. The catalyst may actually be consumed during the reaction, but it is always regenerated by the end of the reaction cycle. Once regenerated, the catalyst is available to participate in another reaction cycle. Consequently, only a small amount of catalyst is needed to convert a large amount of reactant into product. Thus catalytic materials become very important when considering if a chemical process is "green" or "sustainable." Consider the thermal decomposition of formic acid (HCO_2H) into water (H_2O) and carbon monoxide (CO), or into hydrogen (H_2) and carbon dioxide (CO_2) as shown in equation (9.1).

$$CO + H_2O \leftarrow HCO_2H \rightarrow H_2 + CO_2 \tag{9.1}$$

At room temperature, ~21°C, the reaction rate is nearly immeasurable, because only a very small population of molecules has sufficient energy to react, that is, they have insufficient energy to overcome the activation energy (E_a) barrier that is required for a chemical change, as shown in **Figure 9.1**. The formic acid molecules must collide with sufficient energy to go through the transition states that

FIG. 9.1   *Schematic of the thermal and catalytic decomposition of formic acid into carbon dioxide and hydrogen or into carbon monoxide and water.*

are shown at the top of the energy curve in **Figure 9.1**, and form the products given in equation (9.1). If the temperature is increased, a larger population of molecules will have enough energy to overcome the activation energy barrier and the reaction will proceed at a measurable rate. However, now there is enough energy to cross the reaction barriers in both directions resulting in a nonselective process. The use of a catalyst can dramatically alter this situation by selectively lowering the activation energy for one pathway but not for the other. Platinum selectively decomposes formic acid into hydrogen and carbon dioxide, while aluminum oxide gives carbon monoxide and water as the major products. In summary, a catalyst accomplishes the following:

1. It lowers the activation energy of the reaction, thus it speeds up the rate of the reaction. A catalyst does not change the thermodynamic equilibrium of a reaction; it only increases the speed for which equilibrium is reached. In addition, a lower activation energy means that a lower reaction temperature can be used.
2. It participates in the reaction chemistry, but it is always regenerated and is available for other reaction cycles. Therefore, only a small amount of catalyst is required for the overall reaction process.
3. It can increase the selectivity for a given reaction by lowering the activation energy of one pathway over that of another.

9.0.3 *The Nano Perspective*

Only a portion of the catalysts used today can be classified as nanocatalysts. The term nanocatalyst is defined as a material that has catalytic properties on at least one nanoscale dimension. The concept of a "nanocatalyst" is not new to scientists and engineers working in this field. Although the term "nano" was not commonly used until more recent times, researchers traditionally focused on producing very small particles of active catalytic agents in order to maximize the reaction efficiency and to reduce the overall cost of the chemical process. Recent advances in synthetic methods for the production of nanomaterials has produced new nanocatalysts with novel properties and reactivity. Traditional commercial nanocatalysts such as enzymes, zeolites, and transition metal nanocatalysts represent about 98% of the global nanocatalyst market. Newer nanocatalyst materials account for the other 2%. The global market for nanocatalysts is forecast to be $5 billion by the year 2009, with the newer novel nanocatalyst materials expanding to about 7% of the market.

A simple schematic of a commercial platinum catalyst is shown in **Figure 9.2**. The functioning catalyst consists of small nanosized (1–5 nm) metal crystallites supported on a porous metal oxide support. In this case, the active catalyst is platinum and the porous support is aluminum oxide (alumina). The support plays several important roles. Since platinum is a very expensive metal, it is important that every platinum atom be involved in the reaction in order to recover the cost of the platinum in reasonable time. Since the chemical reaction occurs only at the surface of the metal crystallite, any atom in the interior of the crystallite is inaccessible and provides no value to the reaction process. In other words, they cost the user money but don't pay for themselves. Therefore, it is beneficial to have as many atoms on the surface of the crystallite as possible to increase the benefit-to-cost ratio of the catalyst. The porous support provides an inexpensive, but high surface area, base on which to disperse the platinum crystallites. In some cases the support is inert and in other cases it plays an important role in the chemistry as well. Support materials commonly consist of silica (SiO_2), alumina (Al_2O_3), activated carbon, amorphous silica–alumina (SiO_2–Al_2O_3), and crystalline zeolites (SiO_2–Al_2O_3), and have surface areas of 100–$1000\,m^2 \cdot g^{-1}$. In addition the high surface area of the support permits the catalyst to be quite compact, but able to

FIG. 9.2	Schematic of a supported platinum metal catalyst.

Source: R. Farrauto and C. Bartholomew, *Fundamentals of industrial catalytic processes,* John Wiley & Sons, (2006). With permission.

| FIG. 9.3 | *A transmission electron microscope (TEM) image of rhodium metal nanocrystallites on a nonporous silica microsphere.* |

30 nm

Source: S. Chakraborti, A. K. Datye, and N. J. Long, *Journal of Catalysis*, 108, 444–451 (1987). With permission.

handle a large volume of reactant. A TEM image of rhodium metal nanocrystallites on a nonporous silica microsphere is shown in **Figure 9.3** [3].

There are other benefits to making the metal crystallite smaller. The material properties, such as optical, magnetic, and surface chemistry of the crystallite, change dramatically once they are in the nanocrystallite range. For example, gold is a very inert metal and consequently makes a poor catalyst. Gold particles, which are smaller than 20 nm, are purple in color rather than the traditional yellow color, and are capable of oxidizing carbon monoxide into carbon dioxide at room temperature.

9.1   FUNDAMENTALS OF CATALYSIS

9.1.1   *Adsorption of a Molecule on a Catalyst Surface*

An atom in the interior of a crystallite may have as many as 12 nearest neighbors and is fully satisfied in terms of its bonding needs. However, as shown in **Figure 9.4**, surface atoms have fewer nearest neighbors. A terrace surface atom will have nine, an edge atom six, and a corner surface atom four nearest neighbors. Thus the bonding needs of these surface atoms have not been met, resulting in a higher energy at the surface than in the interior. Consequently, surfaces will form bonds with molecules that they come in contact with in order to lower the overall energy of the crystallite. If a bulk metal is cleaved to expose a new active surface, it will immediately bond to the molecules in air to satisfy its bonding needs. **Figure 9.5** [4] shows a high-resolution transmission electron microscope image of Pd nanocrystallites on a silica surface.

FIG. 9.4

Illustration of metal crystallites on a support surface. Top views of crystalline structures on a substrate support material. The lack of nearest neighbors results in higher surface energy that translates into catalysis. Terrace surface atoms have nine nearest neighbors. Edge atoms have six and corner atoms are extremely undersaturated with only four nearest neighbors. Other forms of surface species also exist.

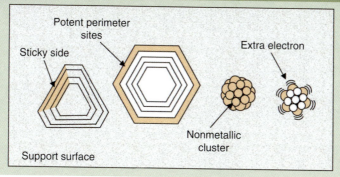

Since oxygen forms a strong bond with most metals, it reacts faster with the metal surface than the other molecules present in air, and soon a monolayer of oxygen atoms are bonded to the surface. This adsorption process is called chemical adsorption or chemisorption. The speed at which chemisorption occurs depends on the metal–oxygen bond strength for the various metals. As one might imagine, the fewer neighbors that a surface atom has, the more reactive it will be. Thus, the strength of bonding to the metal surface occurs in the order of corner atoms > edge atoms > terrace atoms. We can take advantage of this phenomenon

FIG. 9.5

A HRTEM image of palladium nanocrystallites on a silica support.

Source: H.-Y. Lee, S. Ryu, H. Kang, Y.-W. Jun, and J. Cheon, *Chemical Communication*, 1325–1327 (2006). With permission.

FIG. 9.6 *Illustration of the dissociative adsorption of hydrogen molecules into hydrogen atoms on a nickel metal surface.*

Physically adsorbed hydrogen molecule   Transition state   Chemisorbed hydrogen atoms

Source: G. C. Bond, *Heterogeneous catalysis:* Principles and applications, Oxford Press, Oxford, (1974). With permission.

by generating an active metal surface in the presence of reactant molecules in a closed system, rather than air. **Figure 9.6** [2] illustrates the adsorption of hydrogen onto a nickel surface. As hydrogen molecules approach the surface the atoms in the molecule begin to form hydrogen–metal bonds and simultaneously break the hydrogen–hydrogen bond, resulting in the dissociation of the hydrogen molecule into hydrogen atoms on the nickel surface, as shown in equation (9.2)

$$H_2 + 2Ni_{(S)} \rightarrow 2HNi_{(S)} \tag{9.2}$$

where

$Ni_{(S)}$ represents a surface nickel atom

$HNi_{(S)}$ represents a hydrogen atom bonded to a nickel surface atom

9.1.2 *Adsorption Theory*

In order for a chemical reaction to occur on the surface of a catalyst, all the reactants must adsorb onto the surface and all the products must desorb. In 1916, Irving Langmuir published his theoretical work on the adsorption of gases on solid surfaces, and in 1936 he received the Nobel Prize for his "discoveries and investigations in surface chemistry." His theory was based on the following four precepts:

1. The surface of the solid catalyst is uniform, thus all surface bonding sites are equivalent and interact with each adsorbing gas molecule in the same way.
2. The mechanism for bonding the gas to the surface is the same for every surface site.
3. Each adsorbed molecule behaves as an independent species and does not interact with other adsorbed molecules on the surface.
4. As the partial pressure of the adsorbate gas (P_A) is increased, the population of gas molecules adsorbed on the surface is also increased

until a maximum coverage is obtained. This maximum coverage is referred to as a monolayer. Thus, the moles of gas, which are adsorbed on the surface, are directly proportional to the number of adsorption sites on the surface.

Langmuir's precepts are idealistic and exceptions to the precepts are well known. Although more complex theories have evolved to address these discrepancies, his theory serves as a good model for introducing the student to the concepts of adsorption. Let's assume that a clean metal surface is placed into a closed vessel containing only the adsorbate gas and the system is allowed to reach equilibrium, and according to equation (9.3)

$$A_{(g)} + S \underset{k_2}{\overset{k_1}{\longleftrightarrow}} AS \tag{9.3}$$

where

$A_{(g)}$ represents the adsorbate gas
S represents an unoccupied surface bonding site
AS represents a surface site occupied by or chemically bound to an adsorbate molecule

When the partial pressure of A is increased, the equilibrium is shifted to the right, resulting in a higher population of adsorbed species. The rate of adsorption (r_a) and rate of desorption (r_d) can be expressed in terms of the partial pressure of adsorbate (P_A), the adsorption rate constant k_1, the number of unoccupied sites (S), the desorption rate constant k_{-1}, and the number of occupied sites (AS), as given in equations (9.4) and (9.5).

$$r_a = k_1 P_A S \tag{9.4}$$

$$r_d = k_{-1} AS \tag{9.5}$$

At equilibrium, the rate of adsorption equals the rate of desorption and

$$k_1 P_A S = k_{-1} AS \tag{9.6}$$

or

$$AS = K_A P_A S, \qquad \text{where } K_A = k_1 / k_{-1} \tag{9.7}$$

The total number of surface sites (S_T) is equal to the number of unoccupied sites (S) plus the number of occupied sites (AS).

$$S_T = S + AS = S + K_A P_A S \tag{9.8}$$

Thus

$$\theta_A = AS / S_T = K_A P_A S / (S + K_A P_A S) \tag{9.9}$$

$$\theta_A = K_A P_A / (1 + K_A P_A) \tag{9.10}$$

Equation (9.10) is known as the Langmuir isotherm equation, and θ_A represents the fraction of surface sites occupied by adsorbent A. When θ_A is plotted versus

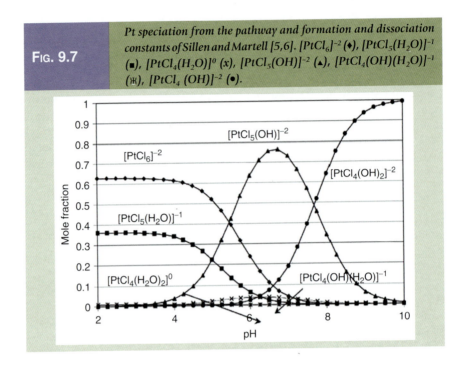

FIG. 9.7 *Pt speciation from the pathway and formation and dissociation constants of Sillen and Martell [5,6]. [PtCl₆]⁻² (♦), [PtCl₅(H₂O)]⁻¹ (■), [PtCl₄(H₂O)]⁰ (x), [PtCl₅(OH)]⁻² (▲), [PtCl₄(OH)(H₂O)]⁻¹ (✳), [PtCl₄ (OH)]⁻² (●).*

the P_A, a Langmuir isotherm is obtained, as shown in **Figure 9.7**. If Langmuir's precepts are valid, then a monolayer of A is formed on the surface when the P_A is large enough to make θ_A equal to 1.0. At this point no further adsorption is possible. Information about the bond strength and the effect of temperature on the adsorption equilibrium are contained in K_A. If the temperature is kept constant, then the magnitude of the K_A value reflects the strength of the adsorbate–surface bond.

In **Figure 9.7**, the shape of the adsorption curves correlates to progressively weaker adsorbate–surface bonding as $K_{A1} > K_{A2} > K_{A3}$. For example, the relative K_A values for the adsorption of hydrogen onto nickel, iron, or copper are $K_{A(Ni)} > K_{A(Fe)} > K_{A(Cu)}$. If one is interested in the hydrogenation of ethene ($H_2C=CH_2$) to ethane ($H_3C=CH_3$), as shown in equation (9.11), then nickel metal would be a good candidate and copper a poor one. Because Ni is the catalyst, it is not included in the overall reaction equation, but written over the arrow, indicating it is not consumed during the reaction.

$$H_2C = CH_2 + H_2 \xrightarrow{\text{Ni}} H_3C - CH \qquad (9.11)$$

If the surface is kept constant, then the shape of the curve represents the effect of temperature on the adsorption equilibrium, that is, as the temperature is increased the adsorption equilibrium shifts further to the left, thus K_{A1} would represent the lowest temperature and K_{A3} the highest. Since most adsorptions are nearly always exothermic, the adsorbate tends to desorb from the surface as the temperature is increased.

In reality the four precepts suggested by Langmuir are not often valid. Crystallite surfaces tend to be nonuniform as shown in **Figure 9.4**. The mechanism for adsorption can vary depending on the type of sites available on the surface. For example, carbon monoxide can adsorb in a linear fashion to a single-surface metal atom or it can form a bridge across two metal atoms. In many cases, adsorbate molecules can interact with each other. Depending on the temperature and the strength of the adsorbate–surface bond versus adsorbate–adsorbate interactions, multiple layers of adsorbate can form. This kind of multiple layer adsorption is best described by the BET equation published by Stephen Brunauer, Paul Emmett, and Edward Taylor in 1938.

9.1.3 Surface Reactions

Once reactant molecules adsorb onto a surface, they must undergo a reaction to form products. Eventually the products desorb and the catalyst surface is restored to its initial chemical state (regenerated) and is ready for the next reaction cycle. The reaction conditions that produce the optimum product yield are determined by varying the reaction temperature, pressure, and catalyst contact time with the reactant. The catalyst contact time can be changed by changing the flow or agitation rate in the vessel. One of the objectives of a researcher in this field is to understand the reaction mechanism. If some knowledge of the reaction mechanism is available, then the catalyst formulation can be improved using sound scientific principles, rather than relying on a trial and error approach. Details of the surface reaction chemistry can be obtained by doing a kinetic study combined with a variety of surface and bulk analytical methods. This section focuses on the kinetic approach and section 9.3 focuses on the most common analytical methods that are used.

In his early career, Cyril Hinshelwood published his work (1921) on molecular kinetics at surfaces, using the precepts of Langmuir. Today it is referred to as the Langmuir–Hinshelwood mechanism. The overall reaction given in equation (9.12) is used to illustrate this mechanism, where $A_{(g)}$ and $B_{(g)}$ represent the gaseous reactants and $C_{(g)}$ represents the gaseous product. Although this equation represents the overall stoichiometric reaction, it does not include all the intermediate steps of the reaction.

$$A_{(g)} + B_{(g)} \rightarrow C_{(g)} \tag{9.12}$$

Hinshelwood assumed that following reaction sequence took place on the surface of the catalyst:

$$A_{(g)} + S \rightarrow AS \text{ (associative adsorption of } A \text{ onto vacant site } S) \tag{9.13}$$

$$B_{(g)} + S \rightarrow BS \text{ (associative adsorption of } B \text{ onto vacant site } S) \tag{9.14}$$

$$AS + BS \rightarrow CS + S \text{ (surface reaction)} \tag{9.15a}$$

$$CS \rightarrow C_{(g)} + S \text{ (desorption of } C \text{ from } S \text{ to regenerate vacant site)} \tag{9.15b}$$

and that the reactants A and B were associatively adsorbed, that is, they were adsorbed without the molecules breaking apart on the surface. If we assume that the adsorption steps are fast and in equilibrium and that the surface reaction is the slow or rate-limiting step, a rate expression can be written as follows:

$$\text{Rate} = k_3(AS)(BS) \tag{9.16}$$

In order for this rate expression to be meaningful, it must be written in terms that can be measured experimentally. The values AS and BS are very difficult, if not impossible, to measure under experimental conditions; therefore, the rate expression must be rewritten in terms of quantities that can be measured, such as the concentration or partial pressures of reactants and products, that is, C_A, C_B, and C_C, or P_A, P_B, or P_C. If adsorption is fast and reversible, then it reaches equilibrium quickly and can be written as follows:

$$K_A = \frac{(AS)}{(P_A S)} \text{ or } (AS) = K_A P_A S \tag{9.17}$$

$$K_B = \frac{(AS)}{(P_B S)} \text{ or } (AS) = K_B P_B S \tag{9.18}$$

$$K_C = \frac{(AS)}{(P_C S)} \text{ or } (AS) = K_C P_C S \tag{9.19}$$

Substituting for AS and BS, we can express the rate expression in terms of partial pressures of A (P_A) and B (P_B):

$$\text{Rate} = k_3(K_A P_A S)(K_B P_B S) = k_3(K_A P_A)(K_B P_B)S^2 \tag{9.20}$$

The number of vacant sites (S) is another experimental value that is very difficult to measure, but the total number of surface sites (S_T) can be measured using adsorption techniques. At any point during the reaction, S_T is equal to the sum of the vacant and occupied sites as given in equation (9.21):

$$S_T = S + AS + BS + CS = S(1 + K_A P_A + K_B P_B + K_C P_C) \tag{9.21}$$

$$S = \frac{S_T}{(1 + K_A P_A + K_B P_B + K_C P_C)} \tag{9.22}$$

$$\text{Rate} = \frac{k_3(K_A P_A)(K_B P_B)S_T^2}{(1 + K_A P_A + K_B P_B + K_C P_C)^2} \tag{9.23}$$

$$\text{Rate} = \frac{k'(P_A P_B)S_T^2}{(1 + K_A P_A + K_B P_B + K_C P_C)^2} \tag{9.24}$$

It is important to note that under the reaction conditions normally used, the adsorbed species AS and BS are mobile and move from site to site. When AS and BS are on adjacent sites and collide with sufficient energy to overcome the E_a of the reaction, then a surface reaction occurs and the product CS is formed. The example above is of course very idealized. For real reactions the molecules may actually dissociate into smaller fragments on the surface. These fragments can recombine to form new products. The structure and relative ratio of these fragments on the surface is controlled by the strength of bonding on the catalyst, the partial pressure of the reactants, and the reaction temperature. The hydrogenation of carbon monoxide serves as a good example of this. When copper is used as a catalyst material, it is unable to dissociate carbon monoxide into carbon and oxygen atoms, thus carbon monoxide adsorbs as an intact molecule on the surface. Hydrogen on the other hand dissociates into hydrogen atoms. When adsorbed hydrogen atoms collide with an adsorbed carbon monoxide molecule, a reaction occurs and methanol is produced. The net methanol synthesis reaction is shown in equation (9.25):

$$CO + 2H_2 \xrightarrow{\text{Cu}} CH_3OH \text{ (net reaction with Cu catalyst)} \qquad (9.25)$$

A simplified mechanism is given below:

$$CO + S \rightarrow COS \text{ (associative adsorption of CO)} \qquad (9.26)$$

$$H_2 + 2S \rightarrow 2HS \text{ (dissociative adsorption of } H_2) \qquad (9.27)$$

$$COS + 2HS \rightarrow H_2COS + S \text{ (surface reaction)} \qquad (9.28)$$

$$H_2COS + 2HS \rightarrow CH_3OHS + S \text{ (surface reaction)} \qquad (9.29)$$

$$CH_3OHS \rightarrow CH_3OH + S \text{ (desorption of methanol product)} \qquad (9.30)$$

On the other hand, if nickel is used as the catalyst then a different outcome is obtained and carbon monoxide and hydrogen are converted into methane. Nickel can dissociate (break apart) the CO molecule on the surface and copper cannot. Both metals can dissociate hydrogen into hydrogen atoms. Hydrogen is so strongly adsorbed onto nickel that each time a carbon monoxide molecule adsorbs and breaks apart it is surrounded by a very large population of hydrogen atoms and the most probable reactions are multiple collisions with hydrogen atoms to eventually form methane. The adsorbed oxygen atom shares the same fate and is quickly converted into water. The net methane synthesis reaction is shown in equation (9.31) and a simplified mechanism is presented in equations (9.32)–(9.37).

$$CO + 3H_2 \xrightarrow{\text{Ni}} CH_4 + H_2O \text{ (net reaction with Ni catalyst)} \qquad (9.31)$$

$$CO + 2S \rightarrow CS + OS \text{ (dissociative adsorption of CO)} \qquad (9.32)$$

$$H_2 + 2S \rightarrow 2HS \text{ (dissociative adsorption of } H_2) \qquad (9.33)$$

$$CS + 2HS \rightarrow H_2CS + S \text{ (surface reaction)} \qquad (9.34)$$

$$H_2CS + 2HS \rightarrow CH_4 + 2S \text{ (formation and desorption of } CH_4) \qquad (9.35)$$

$$2HS + OS \rightarrow H_2OS + S \text{ (surface reaction)} \qquad (9.36)$$

$$H_2OS \rightarrow H_2O + S \text{ (desorption of water)} \qquad (9.37)$$

Iron can also dissociate carbon monoxide into carbon and hydrogen atoms, but it does not adsorb hydrogen nearly as well as nickel. Therefore, carbon species have an opportunity to collide with each other as well as with hydrogen atoms. This process is known as the Fisher–Tropsch synthesis of hydrocarbons and is currently being used to synthesize gasoline and other hydrocarbon products at the Sasol plant in South Africa.

$$n\text{CO} + (n+2)\text{H}_2 \xrightarrow{\text{Fe}} (\text{CH}_2)_n + \text{H}_2\text{O} \text{ (net reaction with Fe catalyst)} \tag{9.38}$$

$$\text{CO} + 2\text{S} \rightarrow \text{CS} + \text{OS} \text{ (dissociative adsorption of CO)} \tag{9.39}$$

$$\text{H}_2 + 2\text{S} \rightarrow 2\text{HS} \text{ (dissociative adsorption of H}_2) \tag{9.40}$$

$$\text{CS} + 2\text{HS} \rightarrow \text{H}_2\text{CS} + \text{S} \text{ (surface reaction)} \tag{9.41}$$

$$\text{H}_2\text{CS} \rightarrow \text{SCH}_2\text{CH}_2\text{S} \text{ (surface reaction)} \tag{9.42}$$

$$n\text{H}_2\text{CS} \rightarrow \text{S}(\text{CH}_2)_n\text{S} + (n-2)\text{S} \text{ (surface reaction)} \tag{9.43}$$

$$\text{S}(\text{CH}_2)_n\text{S} + 2\text{HS} \rightarrow \text{CH}_3(\text{CH}_2)_{(n-2)}\text{CH}_3 \text{ (product formation and desorption)} \tag{9.44}$$

As one can see, the proper selection of catalyst materials and reaction conditions is critical in determining the reaction pathway and thus the end product.

9.2 SYNTHESIS

9.2.1 *Synthesis Requirements*

The primary synthetic objectives, for a commercially viable catalyst, are to make a catalyst that is highly active, highly selective for the desired product, mechanically durable, thermally stable, long lived, and cost-effective. It can be quite a challenge to meet all of these requirements. To illustrate this philosophy, let's examine the components of a catalytic converter on an automobile to reduce exhaust emissions. The catalytic converter consists of a ceramic monolith or honeycomb, as shown in **Figure 9.6**, that is coated with a fine powder of highly porous ($\sim150\,\text{m}^2\cdot\text{g}^{-1}$) gamma aluminum oxide, which is also known as gamma-alumina (γ-Al_2O_3). The ceramic honeycomb gives mechanical strength to withstand the vibrations and thermal variations experienced inside the catalytic converter. Its open structure provides uniform flow and good contact with the combustion gases exiting from the engine. However, it has a very low surface area and surface features that don't permit good dispersion or adhesion of the active catalyst components, which are platinum, palladium, and rhodium. The alumina is added because it bonds well to the ceramic and provides a surface that gives good binding and dispersion of the active catalyst metals. Dispersion means the metal or metal oxide crystallites of the active catalytic material are small and is defined by the following equation:

$$\%D = \frac{M_S}{M_T} \times 100 \tag{9.45}$$

where

%D represents the percent dispersion

M_S represents moles of metal atoms on the particle surface

M_T is the total moles of metal atoms in the particle

For nanocatalyst particles the value of %D approaches 1, that is, nearly every atom in the particle is exposed to the surface. Since only atoms on the surface can participate in a catalytic process, a high metal dispersion corresponds to a high catalytic activity.

In this example, the active catalytic components are platinum, palladium, and rhodium. These are very expensive elements. If a company wishes to recover the cost of the initial expense of the catalyst, it is important that as many platinum atoms as possible contribute to making the desired product. For a 200-μm-sized platinum metal particle, only the atoms on the surface contribute to the reaction chemistry. However, in this case there are more platinum atoms in the interior of the particle than there are on the surface. The interior atoms cost money, but provide no catalyst benefit to the user. The smaller the catalyst particle the higher its effectiveness, that is, the more atoms that are on the surface participating in the reaction, and the catalyst pays for itself in a shorter time.

One of the objectives of catalyst synthesis is to generate active catalyst particles that have diameters in the nanometer range. Such small particles will have nearly every atom exposed to the surface. Although this is the desired outcome, it is difficult to achieve because atoms on the surface are higher in energy than those on the interior. The more atoms that are exposed to the surface, the higher the overall energy of the particle. Thermodynamics dictates that larger particles are more stable than smaller ones. Thus thermodynamics drives the synthesis in the direction of larger particles, which is opposite to the desired objective. The challenge is to prepare a catalyst with nanosized particles that are stable under the conditions used during the reaction. The next section will discuss how that is done.

9.2.2 Example of a Conventional Synthetic Technique

There are many ways to prepare a catalyst with nanosized particles. This section will focus on the more common methods for making a commercial catalyst material. There are four traditional ways for preparing a catalyst. The first and most common method is called the impregnation or incipient wetness technique. In this method, an aqueous solution containing the soluble metal salt of the active catalyst, such as $(NH_3)_4Pd(NO_3)_2$, $PtCl_4$, $PlCl_2$, $Ni(NO_3)_2$, $HAuCl_4$, etc., is coated (impregnated) onto the surface of a highly porous support material, such as alumina (Al_2O_3), silica (SiO_2), or silica–alumina (SiO_2–Al_2O_3). The catalyst is dried at ~120°C to remove excess water and then heated to ~500°C in air, also known as calcining, to decompose the metal salt into its active metal or metal oxide form. For example, platinum and gold decompose directly into the metal form under calcining conditions, while palladium and nickel form a metal oxide. Different forms of the catalyst are used for different types of reactions. For example, a hydrogenation reaction requires the metal form and an oxidation reaction usually requires the oxide form. If the metal form of the catalyst is required, then it can be reduced to the metal form in a subsequent step using a reducing agent, such as hydrogen.

The key to forming metal or metal oxide particles in the nanometer range lies in the details of the preparation. The chemistry of the support surface plays a very important role in generating such small catalyst particles. When exposed to an aqueous solution, a porous metal oxide support has both positive and negative charges on its surface. The relative population of positive or negative charges depends upon the pH of the solution and the isoelectric point (IEP) of the support. The IEP is defined as the pH of the solution where the population of positive and negative charges is equal and the net charge on the surface is zero. The IEP for silica is zero at a pH of approximately 2, while that of γ-Al_2O_3 is zero at a pH of approximately 7. If one is making a catalyst that requires platinum, then a soluble platinum salt, such as $(NH_3)_4Pt(NO_3)_2$ or H_2PtCl_6, is used. When the salt is dissolved into water, ions as shown in equations (9.46) and (9.47) are formed.

$$(NH_3)_4Pt(NO_3)_2 \rightarrow (NH_3)_4Pt^{2+} + 2(NO_3)^- \tag{9.46}$$

$$H_2PtCl_6 \rightarrow [PtCl_6]^{2-} + 2H^+ \tag{9.47}$$

In the first example, the platinum ions have a positive (+2) charge and in the second they have a negative (–2) charge. The chloroplatinic acid (H_2PtCl_6) solution has a pH around 2.5, which means the silica surface is slightly negatively charged. Consequently, the negatively charged platinum compound $[PtCl_6]^{-2}$ tends to aggregate with itself rather than be dispersed on the surface, resulting in metal particles that are larger than the desired nanoparticle range. When alumina is used as the support, the surface is predominantly positively charged at a pH of 2, and thus the platinum ions are highly dispersed on the alumina surface, rather than aggregating into metal salt crystallites on the surface. As the catalyst is dried and calcined very small metal particles result. A similar result can be obtained for a silica support by modifying it with a small amount of lanthanum oxide (lanthana). The resulting surface has an IEP value of zero at a pH of approximately 6. In the case of $PtCl_4$, the nature of the dissolved species and its charge are pH dependent (see **Fig. 9.7**).

It is easily seen that the IEP of the support surface and the charge of the metal ion in solution are factors that can be manipulated to optimize the dispersion of the metal on the surface during preparation of the catalyst and can lead to the generation of nanosized particles. Other preparation methods include coprecipitation, ion exchange, and chemical vapor deposition. Details of these synthetic methods are presented in other references.

9.2.3 Nontraditional Methods for Preparing Nanocatalysts

The problem with conventional catalyst preparations is the lack of uniformity of the catalyst particles produced. One obtains a range of particle sizes that is solely dependent on the random aggregation of metal atoms on the support surface. Newer methods of catalyst preparation permit the formation of nanometal clusters of a well-defined size and composition, which results in a material which demonstrates a highly size-dependent catalyst activity and selectivity. The control of particle size introduces a new dimension of control in the catalyst performance that is not achievable using traditional synthetic methods. These techniques involve atomic layer deposition, laser ablation, and high vacuum techniques. For example, nanosized clusters of metals or metal alloys can be generated using vacuum techniques and mass spectrometry to selectively deposit metal clusters of a uniform size. These clusters demonstrate unique adsorption and catalytic properties and provide the basis for obtaining a better understanding of how nanosized particles affect the fundamental catalytic performance.

Recent work at the University of Georgia and the Technical University of Munich have discovered that gold nanoparticles are negatively charged. This phenomenon is thought to be responsible for aiding the low temperature adsorption and oxidation of carbon monoxide into carbon dioxide. A conventional catalytic converter must be heated by the engine exhaust to a temperature, called the light-off temperature, where the temperature is sufficient to sustain the oxidation reactions needed for pollution control, especially hydrocarbons and carbon monoxide being oxidized to less toxic substances, such as carbon dioxide and water. A significant amount of pollution occurs between the cold start-up of the engine and the catalyst light-off time. A low temperature catalyst would make a tremendous impact on this problem.

Although new nanocatalyst materials have tremendous economic potential, there are still many problems to be resolved. Uniform nanocatalyst particles tend to be expensive to produce and are often unstable at the reaction conditions normally employed in many catalytic processes. New nanomaterials have provided the scientific community with a much improved understanding of the fundamental workings of a catalyst. Incorporating this new knowledge into an economically viable catalytic material is the challenge to be met by current and future scientists.

9.3 Catalyst Characterization

9.3.1 Overview

It is important to know the bulk and surface composition of a catalyst once it has been prepared. For a new catalyst formulation, this important information allows the researcher to accurately reproduce the synthesis and to better understand how the catalyst actually performs its catalytic function. With such knowledge, the researcher can use scientific principles to design a better catalyst in the future, rather than relying on trial and error methods of discovery. Some industrial catalysts have gradually evolved over decades. However, the introduction of modern analytical methods has greatly shortened the development time for new catalyst formulations that are more active, selective, and cost-effective.

Catalysts usually have a limited lifetime and begin to show a gradual or even rapid deactivation under certain reaction conditions. Analytical methods can provide valuable information about the cause of this deactivation. If the problem is identified, then new formulations can be designed to prolong the catalyst lifetime or reaction conditions can be selected to avoid or slow down the deactivation process. For example, platinum or palladium nanoparticles can react with trace impurities in the reaction feedstock, such as hydrogen sulfide, to form a stable platinum or palladium sulfide surface that is catalytically inactive. Surface analysis can identify the chemical culprit and then the researcher can take appropriate action to minimize or eliminate this problem. Today there are numerous analytical techniques available to scientists and engineers in this field. In the interest of brevity, only a few of the more commonly used methods will be discussed here.

9.3.2 *Bulk Characterization Techniques*

Although all of the chemistry takes place at the solid–gas or solid–liquid interface of the catalyst surface and the reaction medium, it is useful to know the bulk composition of the catalyst as well. There are a number of important questions that need to be answered. What is the bulk formula of the catalyst? Is the solid crystalline or amorphous? If it is crystalline, what is the crystalline phase? Is a particular crystalline phase necessary for the catalyst to express its unique activity? What effect does the size of the nanoparticle has on the catalyst performance? As mentioned previously, gold has always been considered a poor catalyst. It is used as jewelry and in many scientific applications where an inert surface is required, because its surface is so stable. However, nanoparticles of gold are able to participate in chemical reactions that larger particles cannot.

Three important bulk characterization techniques will be discussed here. They are inductively coupled plasma (ICP), x-ray diffraction (XRD), and high-resolution transmission electron microscopy (HRTEM).

Inductively Coupled Plasma.   ICP is a method where the sample is dissolved in acid or base to make a solution of known concentration. The solution is fed into a plasma flame and atomized. The excited atoms or ions emit radiation that is characteristic of that element. A detector identifies the emission frequencies, a computer data collection system identifies the elements that produced those frequencies, and the signal intensity is used to determine the amount of each element present in the original sample. This information is used to determine the amount of each element present in the catalyst sample and to identify any trace elements that may be present. The technique can detect most elements in parts per million or parts per billion amounts.

X-ray Diffraction.   XRD is used to determine the crystal structure of the bulk material and in some cases can be used to show changes in catalyst particle size. It has a limit of about 5 nm, so it is not useful for determining the size of nanoparticles. However, it is still a useful tool in observing the success of a preparation. **Figure 9.8** [11] shows the XRD pattern for a catalyst prepared using $(NH_3)_4Pt(NO_3)_2$ and $PlCl_4$. The SiO_2 support has been treated with La_2O_3 to change the IEP of the surface from approximately 2.0 to 5.3. This means the SiO_2 support surface has a net positive charge for the pH of the solution used during the catalyst preparation. In the case of the $(NH_3)_4Pt(NO_3)_2$ compound, the platinum species has a positive charge and is poorly adsorbed onto the support surface, resulting in aggregation and the production of large Pt particles in the finished catalyst. The XRD pattern shows two peaks that are characteristic of Pt metal on the support surface and with an average particle diameter of

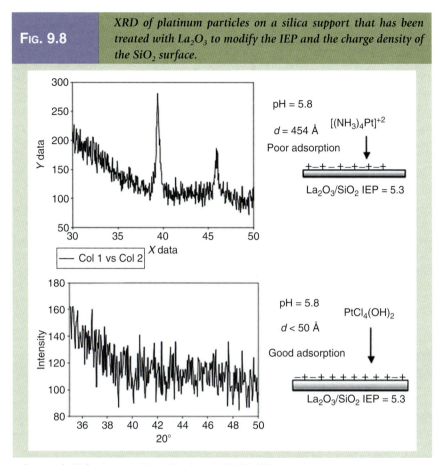

Fig. 9.8 *XRD of platinum particles on a silica support that has been treated with La₂O₃ to modify the IEP and the charge density of the SiO₂ surface.*

Source: S. W. Cowley et al., Unpublished results (2007). With permission.

about 450 Å. The use of PtCl₄ results in a negatively charged platinum species in solution, resulting in good adsorption on the surface, and generates Pt metal particles smaller than 5 nm, which cannot be observed by XRD. Notice the absence of Pt peaks in the XRD data.

High-Resolution Transmission Electron Microscopy. HRTEM can be used to observe metal particles on a variety of metal oxide supports. The dark spots in **Figure 9.3** represent small particles of Rh metal on a silicon oxide support. By measuring the diameter of the metal particles, as compared to the 30 nm scale in the micrograph, one can obtain a good idea of the average particle diameter in the sample. This technique is employed to obtain size information about nanosized particles. By using a higher-energy electron beam, much higher resolution is obtained and details of the crystal structure or arrangement of the atoms in the nanoparticles can be seen. **Figure 9.5** is an example of a HRTEM image. The nanoparticles appear as rafts of atoms on the support surface rather than a sphere-like particle. This shape presents a highly active surface and explains the higher activity often observed for nanosized metal particles.

9.3.3　Surface Characterization Techniques

There are a large number of surface techniques that can be used to characterize the surface of the catalyst. A few examples are x-ray photoelectron spectroscopy (XPS), atomic force microscopy (AFM), high-resolution electron energy loss spectroscopy (HREELS), chemisorption, and temperature-programmed desorption (TPD). Only one method will be discussed here to give the reader a feel of the kind of information that can be obtained by spectroscopic methods. XPS is a true surface analytical method, where the elemental composition of the surface can be obtained. Remember the reaction takes place at the surface of the catalyst particle, so detailed information about the surface is essential if we are to determine the nature of the active catalyst species on the surface of the catalyst particle, how reactant molecules adsorb onto that surface, and how adsorbed molecules undergo transformation into products. XPS focuses a small beam of x-rays onto the sample, which results in the ejection of core electrons from surface atoms with very discrete energies. An energy analyzer measures the kinetic energy of the escaping electron and identifies the corresponding element from which it came. This is achieved by using the following equation:

$$h\nu(\text{photon}) = E_\text{B} + E_\text{K} \tag{9.48}$$

FIG. 9.9　*(a) XPS spectra of $(NH_3)_4Pt(NO_3)_2$ adsorbed onto SiO_2 and (b) La_2O_3-treated SiO_2.*

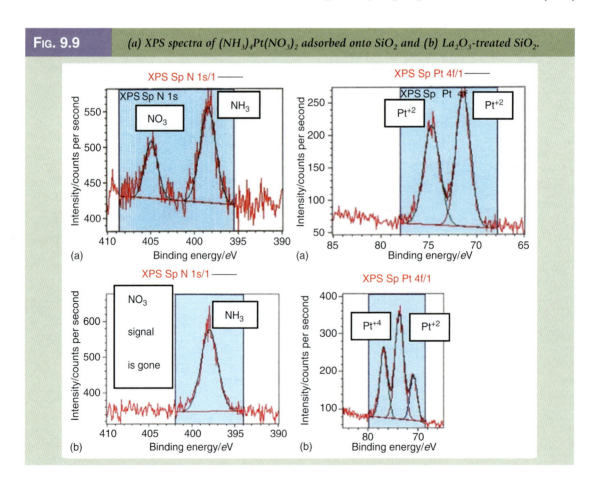

where

hv is the x-ray photon energy that is transferred to a core electron in a surface atom

E_B is the binding energy of the atom

E_K is the kinetic energy of the electron after it leaves the atom

E_B is the energy required to remove a core electron from an atom to a distance where the electron can no longer feel the attractive force of the nucleus. E_K is the residual kinetic energy of the electron after it has broken free of the binding force of the atom. Only electrons located in the top few layers of the sample can escape their host atoms and reach the surface without undergoing collisions with other atoms and loosing some of the original photon energy transferred to them by the x-ray beam. The discrete kinetic energy (E_K) of the escaping electrons is measured and quantified by the XPS instrument. Since hv and E_K are now known, the E_B value can be calculated. Each atom has a unique E_B value and in this fashion the elements present on the surface and their amounts are determined. In addition, small changes in the E_B give clues as to the oxidation state of the metal. **Figure 9.9** shows important chemical changes that occur during the synthesis of platinum nanoparticles on a SiO_2 surface and a La_2O_3-treated SiO_2 surface.

This information is used to further our understanding of the chemical transformation responsible for the formation of nanosized catalyst particles and how those particles function as catalysts. For more detail concerning catalysts, please refer to the excellent sources listed in the references section of this chapter [7–10].

Acknowledgments

Scott W. Cowley is an associate professor in the Department of Chemistry & Geochemistry at the Colorado School of Mines in Golden, Colorado. His research interestes include heterogeneous catalysis, surface science and analysis, surface analysis, solid sorbents for environmental applications, oxide film growth and corrosion, xerography and laser printer materials, and photoconducting films. Dr. Cowley was program co-chair of the North American Catalysis Society in 1995, chairman of the Rocky Mountain Fuel Society (1991–1995), and chairman of the Western States Catalysis Club (1996–2000).

Scott Cowley received a BS in chemistry from Utah State University in 1967, and a MS in physical organic chemistry from Utah State in 1972, under the direction of Grant Gill Smith, studying the "High Temperature Thermal Decomposition of Organic Compounds." He received a PhD in 1975 from Southern Illinois University in physical organic chemistry, under the direction of Gerard V. Smith studying the "Mechanism for the Catalytic Hydrodesulfurization of Thiophene." Dr. Cowley then served in a postdoctoral position at the University of Utah working on catalysis with Frank Massoth studying the "Mechanism of Catalytic Hydrodesulfurization and Catalytic Fischer-Tropsch Synthesis of Olefins." During 1976–1979 he was an assistant professor of fuels engineering at the University of Utah. He has been at CSM in the Department of Chemistry since 1979.

References

1. R. J. Farrauto and C. H. Bartholomew, *Fundamentals of industrial catalytic processes*, John Wiley and Sons, New York (2005).
2. G. C. Bond, *Heterogeneous catalysis: Principles and applications*, Oxford Press, Oxford (1974).
3. S. Chakraborti, A. K. Datye, and N. J. Long, Oxidation reduction treatment of rhodium supported on non-porous silica spheres, *Journal of Catalysis*, 108, 444–451 (1987).
4. H.-Y. Lee, S. Ryu, H. Kang, Y.-W. Jun, and J. Cheon, Selective catalytic activity of ball-shaped Pd@MCM-48 nanocatalysts, *Chemical Communications*, 1325–1327 (2006).
5. L. G. Sillen and A. E. Martell, The stability constant of metal ion complex, Special Publication No.25 (Suppl.1), The Chemical Society, Burlington House, London (1971).
6. W. A. Spieker, J. Liu, J. T. Miller, A. J. Kropf, and J. R. Regalbuto, An EXAFS study of the coordination chemistry of hydrogen hexachloroplatinate(IV): 1. Speciation in aqueous solution, *Applied Catalysis A: General*, 232, 219–235 (2002).
7. W. A. Spieker, J. Liu, X. Hao, J. T. Miller, A. J. Kropf, and J. R. Regalbuto, An EXAFS study of the coordination chemistry of hydrogen hexachloroplatinate(IV): 2. Speciation of complexes absorbed onto alumina, *Applied Catalysis A: General*, 243, 53–66 (2003).
8. J. C. Vickerman, ed., *Surface analysis—The principle techniques*, John Wiley & Sons, Chi Chester, New York, Weinheim, Brisbane, Singapore, Toronto (1997).
9. D. T. Wickman, B. W. Logsdon, S. W. Cowley, and C. D. Butler, A TPD and XPS investigation of palladium on modified alumina supports used for the catalytic decomposition of methanol, *Journal of Catalysis*, 128, 198–209 (1991).
10. S. Youngwilai, Ph.D. Thesis, The chemical modification of silica surfaces to enhance the dispersion and activity of platinum metal crystallites, Department of Chemistry, Colorado School of Mines, Golden (2003).
11. S. W. Cowley et al., Unpublished results (2007).

NANOCOMPOSITES AND FIBERS

Make the workmanship surpass the materials.

OVID
Metamorphoses, 1st Century B.C.

Human wisdom is the aggregate of all human experience, constantly accumulating, selecting and reorganizing its own materials.

JOHN STORY

*C*hapter 10

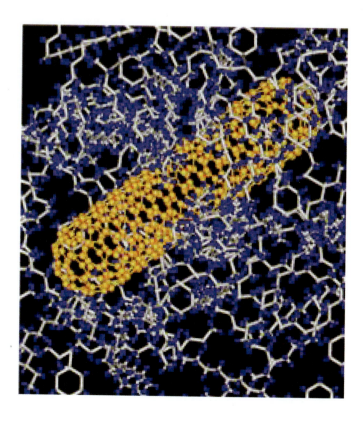

Chapter 10 is the second chapter in the *Chemical Nanoengineering* division of the text, following our discussion of catalysis in *chapter 9*. *Nanocomposites* are a great part of the new wave of nanomaterials—inorganic, organic, biological, and any combination and permutation thereof. Most materials in nature are composites. Composite materials permeate our synthetic products as well. The reason for composites is straightforward: combinations of pure materials and phases result in a material with superior performance. The reason for nanomaterials is also straightforward: inclusion of nanoscale materials in bulk material matrices result in composites with superior properties. We provide a brief review of engineering materials, engineering properties,

and mechanical, electrical, and thermal test methods. All of these categories apply to nanomaterials and to nanomaterials embedded within a matrix.

The importance of composite materials transcends our technology. There is not much that isn't a composite material. Extending this theme further, there is not much in biology that is not made of composite materials, more specifically, nanocomposite materials. Following this chapter, the biologically inclined should feel at home in the next four chapters. Bio-nano (or nano-bio) technology and chemistry, biomimetics, and nanomedical topics are presented. The final chapter, *chapter 14*, addresses environmental applications of nanotechnology to round off the text.

10.0 NANOCOMPOSITES AND FIBERS

It will not be through our development of new materials that lift our technology to the next level but rather through combinations of new materials. Composite materials are able to dodge the tradeoffs that restrict the performance of our traditional engineering materials. The addition of nanoscale materials to composites furthers this trend by enabling new materials that are able to dodge the tradeoffs that restrict the performance of our traditional composite engineering materials. Polymeric composites influenced our technology (and way of life) decades ago just as nanocomposites are expected to influence our technology and way of life for decades to come.

Nanocomposites consist of bulk material matrix with a reinforcing phase of one or more nanomaterials. In general, mechanical testing of nanocomposites has shown mixed results [1]. Because nanomaterials are small, the high surface-to-volume ratio and concomitant high surface energy (e.g., high interfacial energy) should promote facilitative bonding with the matrix material. Such enhanced bonding should afford the composites filled with nanomaterials superior performance over those filled with micrometer-sized materials [2]. The interphase layer plays a critical role in the mechanical property performance of the nanocomposite. If, for example, in the interphase region, the polymer chain has limited capacity to bond with the nanoparticles, then the density of the interphase region is expected to be lower than that of the bulk matrix. The condition is exacerbated if a high volume fraction of filler is required. One of the prime driving forces behind the use of nanomaterial fillers is the overall reduction in the amount of filler required. Micrometer-sized fillers, for example, require V_f above 20% in order to be effective. Therefore, the objective of composite researchers is two-fold: (1) reduction of the amount of filler, and (2) enhancement of bonding between the matrix material and the nanomaterial inclusion. Both numbers one and two are obtainable with nanotechnology.

Nanotechnology impacts both the inclusion and the host. What is it about nanoparticles (zero-, one-, and two-dimensional) that make them excellent inclusion materials in composites? Then, why do polymers, metals, or ceramics made of nanoscale building blocks and then transformed into a macroscopic material demonstrate superior properties? How do nanomaterials affect the structure–property relationship between the filler and matrix? Traditional polymer chemistry has taught us about the relationship between molecular weight, chain length, architecture, fillers, ordering, functional groups, and scaffolds with solubility, rheology, mechanical properties, and other chemical behavior. Also the effects of solvent, size, chemistry, and fabrication conditions are relatively well understood for polymers. What then happens to such baseline polymers when nanomaterials are inserted? How do mechanical, physical, and chemical properties respond to nanometer-sized inclusions of varying size, shapes, compositions, and orientations—whether in the form of particles, tubes, porous materials, discs, wires, or blocks?

Polymer nanocomposites by definition are multicomponent systems in which the major component is a polymer and the minor component is a nanomaterial with one dimension below 100 nm [3]. Properties of nanocomposites filled with small levels of nanoparticle fillers (1%–5%) are comparable or better than materials that contain a larger volume fraction of similar but conventional-sized materials (15%–40% or more) [3]. Smaller amounts of fillers in nanocomposites enhance processing and reduce overall weight of the composite. Other properties of nanomaterial-filled composites include enhanced optical clarity, self-passivation, and increased flame retardation, oxidation, and ablation resistance [3]. Automotive parts, coatings, and flame retardants are some commercial avenues already utilizing the unique properties of these nanocomposites [3]. Nanoscale building blocks are individually remarkable because of several reasons: they are free of defects, reactive, and have unusual physical properties. It is however not a simple task to keep these remarkable properties intact in a macroscale composite [4].

Please note that we will not focus on nanocomposite materials in general. In this chapter, for example, we do not intend to discuss SAMMS (self-assembled monolayers on mesoporous substrates) or OPVs (organic photovoltaic solar cells) although these are most certainly considered to be nanocomposite materials. We do focus on traditional polymer matrices that incorporate nanoscale materials. As a second note, we limit our discussion to composite mechanical properties although brief excursions are made into thermal, electronic, and optical domains, especially when presenting literature and data concerning carbon nanotubes.

Most polymers are either thermoplastics (e.g., heavily cross-linked long-chain molecules like moldable polyethylene that softens on heating), thermosets (e.g., epoxy resins that are 3-D covalent bonded network, light-weight, rigid, thermally stable composites that harden upon application of heat), elastomers and rubbers (e.g., linear memory polymers, occasionally cross-linked), and natural polymers (e.g., cellulose, lignin, and proteins).

10.0.1 *Background*

A composite material (from the Latin *compositus* "to put together," from *com* "together" + *ponere* "to place") is an intimate combination of two or more materials that differ in composition, structure, and form. A composite material is a heterogeneous engineering material with superior properties in which one phase is dispersed within another—the phases bonded mechanically or chemically. For example, the tempering of steel with highly dispersed finely divided carbides imparts an unusual combination of strength and toughness to the steel—resulting in an in situ composite with fine structure [5]. Although the constituents of a composite act together, they act together to perform unique mechanical, chemical, thermal, optical, or electrical tasks that individually would not be possible. Structural concrete, for example, consists of steel reinforcement embedded within the concrete matrix—the steel bars are responsible for mitigating tension loads while the concrete is responsible for mitigation of compression loads.

A composite material exhibits a combination of functions and materials working in concert. Most composites have a reinforcing element, (or filler) dispersed within a matrix (or binder). In traditional composites, distinct identifiable interfaces exist that retain the chemical identity of each component and do not merge into one another. In other words, the phases of a composite do not react with or dissolve into each other. The reinforcing element is usually a fiber (e.g., made of glass or carbon). The direction of the reinforcement or property vector is also a factor in the design of the material. Reinforcement, for example, can also be made to be isotropic or anisotropic. Thermal or electrical conductivity can be made to be isotropic or anisotropic. The same is true for optical properties.

Constituents of a composite technically should not dissolve or blend within each other. In other words, each constituent within a composite retains its original identity—steel remains steel, concrete remains concrete, carbon fibers remain as carbon fibers. Composites, therefore, are distinctly different from the pure classes of engineering materials like metals and semiconductors, ceramics and glasses, and polymers and plastics. Composite materials consist of one or more material that is integrated to form a unique material—one that demonstrates enhanced properties above and beyond those of its individual components. In addition to the preexisting properties of the dispersed component and the matrix, the physical properties of the composite depend on the geometry, size, shape, roughness, and orientation of the dispersed component and the nature of the interface between the minority inclusion and the host matrix material.

Brief History of Synthetic Composites.   Perhaps one of the first synthetic composite materials was the straw-reinforced mud brick developed by early humans. The straw added tensile properties while the bricks provided mass and thermal and compression resistance—an excellent combination for a dwelling. Modern reinforced concrete is based on the same concept although component material physical properties have certainly been improved over several millennia. The Mongols of the twelfth century developed small and powerful bows by combining several natural products: tendons (for tension) and horn sheaths from cattle (to prevent compression), bamboo (core structure), and silk (soaked in resin to wrap the structure). The strength of the Mongol bows compared favorably to modern composite bows.

In the mid to late 1800s, chemists developed the first polymers and invented cross-linked synthetics such as celluloid, melamine, and Bakelite. Polyester arrived on the scene in the 1930s and glass-reinforced polymer composites soon followed. Carbon black and fumed silica have been added to components in plastics since the invention of tires. Embedding nanoparticles in tires has improved their performance. Starting with larger embedded particles and proceeding to smaller inclusions, tensile strength was improved. The following particle inclusion size with tensile strength (in parentheses) demonstrates this trend. The tensile strength is given in parentheses: 250–350 nm (10 MPa), 70–100 nm (15 MPa), 30–35 nm (22 MPa), and 20–25 nm (25 MPa).

Roger Bacon in 1958 at the Parma Technical Center of Union Carbide near Cleveland, Ohio was the first to fabricate a high-performance carbon fiber [6]. He experimented with arc-discharge fabrication techniques to produce carbon whiskers a few to greater than 5 μm in diameter and ca. 3 cm in length. The structure of the whiskers revealed a hierarchy of concentric tubes, each in the form of a rolled up scroll (e.g., a multiwalled carbon nanotube). Fibers made of the whiskers demonstrated tensile strength of 20 GPa and Young's modulus on the order of 700 GPa. The fibers consisted of continuous graphite sheets rolled into scrolls to make a filament. The whiskers were grown under conditions of 92 atm of Ar at 3900 K in a DC electric arc. The electrical room temperature resistivity of the tubes was 65 μΩ·cm—a value typical for crystalline graphite [7]. Although T.A. Edison baked cotton and other organic filaments at high temperature to produce carbon filaments back in the late 1870s and that the U.S. Navy produced carburized fabric from rayon and cotton for missile heat shields in 1957, Bacon was the first person to unwittingly produce carbon nanotubes.

Heat treatment of rayon at temperatures as high as 3000 K yielded high performance fibers and cloths. Stretching the polymer fabric at high temperatures (e.g., hot stretching) resulted in orienting the graphitic layers along the long axes of the fibers. By 1965, fibers (like Thornel 25) with Young's modulus of ca. 170 GPa were attained. Carbon fiber reinforced plastic (CFRP) composites consist of layers of carbon fiber cloth in a mold that is in the shape of a structure (e.g., an aircraft wing component). The orientation and weave of the carbon fiber cloth is selected to optimize mechanical properties of the final product. The mold is then filled with an epoxy polymeric material and cured in an autoclave (to outgas) at a predetermined temperature. The resulting product is a lightweight material with good stiffness and tensile properties.

The fuselage of Boeing's new *787 Dreamliner* is composed of CFRP. The new fuselage design is lighter in weight and is expected to show better mechanical performance (fatigue resistance). The 787 is the first airliner to use composite materials for most of its construction—over $6 billion in carbon fiber is expected to go into the new planes. The plane consists by weight of 50% composite, 20% aluminum, 15% titanium, 10% steel, and the remainder other materials—over 35 tons of CFRP of which 23 tons are carbon fibers. By volume, the plane will consist of 80% composite materials [8].

Natural Composites. Natural composites surround us every day—mainly in the form of wood, clays, and paper. Biological nanocomposites dwell within us and within all of life. Bone, spider silks, sea urchin skin, cellulose, nacre (mother of pearl), teeth, and a plethora of other natural products made of combinations of proteins, lipids, carbohydrates, nucleotides, and all other biological subunits form, assemble, and merge to create some of the most remarkable materials to be found. Yes, all biological materials are manufactured from the bottom up and we challenge you to name structural materials in living things that are not nanocomposites. We shall discuss a few types later in the chapter.

10.0.2 *Overview of Engineering Materials*

Keep in mind that nanomaterials, and nanocomponents within composites, are nanosized versions of the basic classes of engineering materials: metals, alloys, and semiconductors; ceramics and glasses; and polymers and inorganic carbon-based materials. The world of nano also includes a vast array of biological materials. Composites usually consist of a majority (matrix) material and a minority (inclusion or dispersed phase) material. Basic engineering materials are reviewed below.

Ceramics.   Early humans made use of ceramics in the form of crystalline stone tools and eventually pottery well before the use of metals [9]. Hardness, temperature resistance, compression resistance, and strength are parameters used to characterize ceramics. However, brittleness and the tendency to fracture in a brittle way under tensile, shear, or impact stresses are undesired qualities in modern materials. It is these exact traits however that helped us get the edge over animals and kickoff the Stone Age. The first tools with sharp edges were formed by cleavage of appropriate "hard" materials, for example, flint, a sedimentary cryptocrystalline form of quartz [10]. Quartz is crystalline SiO_2 that is harder than flint and therefore served better as early hammers and grinders.

Ceramics are composed of combinations of metallic and nonmetallic elements. Examples include aluminum oxides (Al_2O_3), titania (TiO_2), zirconia (ZrO_2), silica (SiO_2), and sodium oxide (Na_2O) [9,11]. Other kinds of ceramics include various metal carbides like silicon carbide (SiC), titanium carbide (TIC), boron carbide (BC), and tungsten carbide (WC); nitrides like silicon nitride (Si_3N_4), titanium nitride (TiN), and boron nitride (BN); and borides like titanium boride (TiB_2) [9]. Ceramic materials are able to serve either as the matrix host material in a composite or as the minority inclusions in nanocomposites—whether in a metal or polymer matrix. Traditional uses of ceramics include their applications as refractory materials.

Recently, ceramics have made inroads into electrical, magnetic, piezo, optical, thermal, and high-temperature superconducting (YBa_2CuO_7) devices in the form of magnetic insulators, capacitors, high-frequency circuits, sensors, and transducers that convert optical, thermal, mechanical, and electrical stimuli into analytical signals [9]. Nanoscale ceramic materials are used as catalysts, photoconductors (charge carrier and separation), and template materials that are revolutionizing device fabrication.

Metals and Alloys.   Metals are highly reflective, thermally conducting, and malleable due to the presence of highly mobile surface electrons. Advantageous characteristics of metals include high electrical conductivity, thermal conductivity, tensile strength (strength), high modulus of elasticity (stiffness), and toughness (fracture resistance) [9,11]. Fabrication is facilitated by metals' inherent ductility and their tendency towards plastic deformation before failure. The average density of metals is ca. $7.5 \, g \cdot cm^{-3}$ [9], considered to be a disadvantage in the aerospace industry.

Nanometals are traditionally exploited in a variety of catalytic applications, for example, catalytic converters in automobiles and the various steam reforming and polymerization processes that abound in today's polymer technology. Nanometals are inserted into a composite if layer-specific optical, electrical, or thermal properties are desired. Metals are often mixed with ceramics to produce high-performance composite materials. Metals made of nano rather than micrograins have shown superior performance. Descriptions of metal structure and properties, along with those of other engineering materials, are found in Ref. [12].

Semiconductors. We are all quite familiar with the impact made on today's technology by semiconductors. Ranging from pure materials like silicon to oxides like titania and complex semiconductors like CIGS (copper–indium–gallium–selenide), highly integrated and complex electronic applications all contain semiconductor materials. Nanosized semiconductors already have had major impacts on the computer industry. The ultimate nano-version of semiconductors is the quantum dot. Semiconductors have found their way into nanocomposites with optical, electronic, or sensing functions. Gold–semiconductor nanocomposite hybrids can be formed by citrate stabilization of Au from $HAuCl_4$ with borohydride reduction and coupling to CdS, CuS, NiS, or PbS semiconductors [13]. Other examples include Au nanometal clusters embedded in an amorphous Si matrix (1-nm layers) to generate a unique optical response based on the surface plasmon [14], metal–semiconductor nanocomposite layered optical fibers [15], and next generation solar cells based on semiconductor nanocomposites [16]. In this chapter, we do not focus on optical properties are based on these technologies solar cells are discussed in more detail in chapter 14.

Polymers. Polymers are lightweight materials (long-chain hydrocarbons) that have found applications in every major industry. Polymers, consisting mainly of carbon and light elements like N, O, P, S, F, and Cl, have a low average density of ~1.5 $g \cdot cm^{-3}$ [9]. There are three general classes of polymers: thermosets (polyurethanes, phenolic resins, epoxy resins, polyesters, and vinyl esters), thermoplastics (polyethylene, nylon, ABS, PI, PP, PC, and PEEK), and elastomers (silicones and EPDM). Thermoset polymers are stronger than thermoplastic polymers (polyethylene or polyvinyl chloride) [2].

Organic polymers are relatively easy to fabricate and have a high strength-to-mass ratio. One disadvantage of polymers is long-term instability under loads. Polymers, however, form the perfect matrix for a composite. Polymers can be easily mixed to form new materials. The incorporation of carbon or glass has created materials that overcome disadvantages like lack of stiffness, lack of hardness, and lack of tensile strength inherent to most organic polymers. At the nanoscale, the block-co-polymer has become an indispensable nanomaterial template and matrix element. The carbon nanotube-reinforced polymer is just one reason why nanocomposites are such a hot area today.

Inorganic Carbons.   Inorganic carbon materials like graphite, carbon blacks, fullerenes, carbon nanotubes, and diamondoids have made important contributions to the composite materials industry—especially from the contribution of the carbon reinforced epoxy resin composite. The spectrum of inorganic carbons ranges from the humble pencil lead, to the strongest material known to science, the carbon nanotube. It is no wonder that researchers are fervently trying to incorporate carbon nanotubes into composites with all the major classes of engineering materials as host. A special section in the chapter is devoted to carbon fibers, whiskers, multiwalled, and single-walled carbon nanotubes.

Materials from Biology.   Diatoms have been used as filler materials for centuries. Biologically based materials such as nucleotides, carbohydrates, proteins, and lipids have shown value as components in composites. There is great effert underway to copy nature's materials. For example, conductive self-assembled DNA composite membranes supported on nanoporous polycarbonate track-etched substrates showed enhanced mechanical properties, proton conductivity (8.0×10^{-2} S$\cdot$cm^{-1}), and decreased methanol permeability (10 wt% @ 5.0×10^{-7} cm$^2 \cdot$s^{-1}) for 1.6 mg$\cdot$cm^2 DNA loading [17]. Water soluble composites were prepared with C_{60} and saccharose, fructose, or dextrans by mechanochemical procedures [18]. Lipid-based nanocomposites are formed by a simple beaker immersion method [19]. Formation of nanoparticle–lipid hybrid films is accomplished with electrostatic control, starting with the deposition of a lipid on a solid substrate [19]. Ordered nanocomposites can be formed from proteins. Proteins are nontoxic, don't require organic solvents to disperse, can be poured into molds, and are easily functionalized [20].

10.0.3   *Types of Composite Materials and Generic Structures*

We all are exposed to many kinds of composites on a daily basis. We can also classify them in numerous ways—by composition, by inclusion size, by inclusion morphology, and so on and so forth. Particulate zero-dimensional (nanosized) and three-dimensional (macroscopic) composite products are represented by caulking compounds (latex, acrylic, silicone, calcium carbonate, talc, and others). Fishing lines consist of hollow glass spheres or tungsten embedded in poly(vinyl chloride). Composites with fibrillar or filamentous (one-dimensional) composite products are found in surf boards, automobile skin (polyesters, epoxides impregnated with glass fibers), and sporting equipment such as tennis racquets, bicycles, and golf clubs. Laminate composites (two-dimensional) include chemically retardant safety gear and armored windows made of glass–ceramics or ceramic–organic polymers. Just take a look around and you will find that many of our products contain some kind of composite material.

They can be classified according to compositional variables: by the type of majority component (major engineering materials), by the type of minority components (once again, members of the basic types of engineering materials), by the number of minority components, by the amount of majority component, by the amount of minority component, by the form of the minority component (fibrillar, particulate, laminate), and finally, by the dispersion of minority components (bulk, agglomerate, surface enriched, etc.) [5].

They are also classified according to type and inclusion size as did Brooks et al: (1) natural composites (wood, bone, bamboo, and muscle tissue), (2) microcomposite materials (metal alloys, toughened and reinforced thermoplastics, sheet and molding compounds, aligned or random continuous fibers, particulates), and (3) macrocomposite materials (reinforced concrete, tennis racquets, skis etc.) [5]. To this we add a fourth category, nanocomposite materials (carbon nanotube reinforced polymers).

Others categorize composites by the nature of the matrix material (e.g., metal, ceramic, or polymer). **Table 10.1** lists selected synthetic composites. Nature gives us abundant examples of nanocomposite materials. Some have been discussed in *Introduction to Nanoscience* [21] and more will be discussed in chapter 12.

Types of Nanocomposites. Nanocomposites consist of traditional matrix materials that happen to have nanomaterial fillers. The nanomaterial inclusions

TABLE 10.1	*Synthetic Composite Materials*
Synthetic composite	**Description**
Concrete	A ceramic + ceramic composite, ceramic (quicklime, CaO) cement (usually Portland cement) + ceramic (fly ash, aggregate), hardens after a third component (water) is added; reinforced concrete is a higher level composite in which steel bars (rebar) are added to provide tensile strength. The composite has the best of both worlds: compression strength is provided by the cementious material and tensile properties by the steel reinforcement.
Carbon fiber reinforced plastic (CFRP)	Graphitic carbon threads consisting of thousands of carbon filaments a few micrometer in diameter; plastic material is usually thermosetting materials like epoxy resin but others like polyester, vinyl esters, and nylon are used; carbon fibers can be made from other polymers like polyacrylonitrile that is heated to form graphitic fibers; tensile strength greater than 5 GPa and modulus of elasticity greater than 500 GPa are common.
Fiberglass	Fiber-reinforced polymer (FRP); glass fiber + polymeric support; silica-based fibers mostly $[(SiO_2)_n]$ but also contains Al_2O_3, Fe_2O_3, CaO, MgO, Na_2O, TiO_2, ZrO_2; formed by extrusion through nozzles to form an amorphous solid (25 μm diameter fibers); polymer matrix materials include epoxies (high strength), polyesters (for generalized structures), phenolics (high temperature), and silicones (electrical applications). Composites of continuous fibers, discrete fibers, or woven fabric offer different mechanical properties.
Other fiber-matrix composites	Para-aramid (Kevlar) + epoxy or polyester Metal-matrix: B + Al; Al_2O_3 + Al; SiC + Al; SiC + Ti Ceramic matrix: SiC + Al_2O_3; SiC + Si_3N_4

can be zero-dimensional (metal oxides, carbon black, polymeric colloids, carbides like SiC and WC, boron and silicon nitrides, and block copolymers), one-dimensional (carbon nanotubes, metal oxide whiskers and nanotubes, boron nitride nanotubes, and linear block copolymers), two-dimensional nanolayered materials (phyllosilicates, nanoclays, hydrotalcites, and 2-D block copolymers) and three-dimensional networks (block copolymers).

Basic Composite Structure.   In order to understand the mechanical and physical properties of composite materials, knowledge of macro-, micro-, and nanostructure is required beforehand. Mechanical properties include tensile strength, ductility, toughness, hardness, impact resistance, resilience, fatigue resistance, and creep and many more—not all are considered to be independent of one another. Physical properties include the glass transition temperature (T_g), and permeability and dielectric, optical, thermal, and electrical properties. Chemical properties include reactivity, corrosion resistance, cross-linked structure, intermolecular bonding, polymerization, and biochemical compatibility.

Average properties of traditional composites can be estimated similar to the way electrical resistance is evaluated, for example, in series. With $x \leq 1$ representing the inclusion fraction

$$\frac{1}{\text{Composite property}} = \frac{(x)(\text{Inclusion})}{\text{Inclusion property}} + \frac{(1-x)(\text{Host})}{\text{Host property}} \quad (10.1)$$

This relation, however, may not apply appropriately to nanocomposites for several reasons: filler and polymer dimensions are on comparable size scales, interfacial areas are immense, and distances between filler and polymer phases are separated by mere nanometers. All these factors imply that the properties of the interfacial-bound polymer regions would most likely dominate the properties of the composite rather than any weighted average of bulk properties of the two phases. In addition, nanometer properties most likely differ drastically from those of bulk material counterparts. Nanometer-scale inclusions would also, due to their intimate relationship with polymers at the atomic scale, impart effects on chain morphology and conformation. The conclusion? It would be dangerous to evaluate nanocomposites based on traditional composite theory alone.

Inclusion Configurations in Polymer Hosts.   Simplistically speaking, physical properties of composite materials are either isotropic (independent of direction) or anisotropic (dependent on the direction of the applied force, thermal stress, or optical probe—a.k.a. orthotropic). Most metals and ceramic composites have isotropic mechanical properties. Filler materials can occupy several predetermined orientations within a polymeric host material (**Fig. 10.1**). Physical properties depend on not only the type of material of the inclusion but also its aspect ratio, size, shape, concentration, the presence of interpenetrating networks, porosity (if applicable), and orientation.

10.0.4   *The Nano Perspective*

We have repeated over and over how nanomaterials possess unique and even remarkable properties. What happens to these unique properties and phenomena

FIG. 10.1

Various inclusion configurations in polymer matrices are depicted. (a) Randomly orientated group of small fibers (e.g., multiwalled carbon nanotubes) is dispersed throughout the polymer matrix. (b) Long strands of fibers are dispersed randomly. (c) Composite material with parallel fibers oriented perpendicular to the plane of the composite afford good compression resistance. (d) Fibers oriented along the load-bearing (tensile) axis of the material enhance tensile properties. Long, uniform, unbundled, evenly dispersed single-walled carbon nanotubes embedded in such a polymer would consist of a holy grail for composite researchers. (e) Micrometer or nanofibers crosshatched in the form of a textile offer strengthening properties along more directions in the composite. (f) Composite with nanoparticle inclusion. Depending on the concentration and type of inclusion, the composite may exhibit unique electrical conductivity or optical properties.

when a nanomaterial is embedded within a bulk material? Or when nanomaterials interact with other nanomaterials to form a bulk material, what are the properties of the resulting homogeneous material or composite? What happens when nanomaterials are woven together to make a fiber and then serve as a structural element in a composite? As is a usual theme in these texts, we refer to a lesson from nature.

Why Nanomaterials? Nanomaterials have remarkable, and oftentimes unusual, physical and chemical properties as well as increased surface area and interfaces. Composites, therefore, that contain such nanomaterials are in turn expected to outperform composites made of bulk materials. Composites that contain nanomaterials are called nanocomposites. The possibility that nanomaterials combine in ways that are unknown to either parent material in the bulk phase is also likely. We have learned that in composites, the merged materials retain their identity and properties. In nanocomposites, materials remaining separate and distinct in the nanocomposite may not adhere to the fundamental definition given earlier. By binding a nanoparticle or nanofiber, do you change it? Are we confronted with a Heisenberg-like dilemma consisting of a complementary pair of binding and properties? The problem faced by engineers is how to incorporate such materials without losing those amazing characteristics. How does one bind carbon nanotubes into a polymer matrix without compromising tensile strength, stiffness, and electrical and thermal conductivity?

We restrict the scope of our discussion in this chapter to nanocomposite materials defined by one or more nanoscale materials embedded within a matrix material. To further squeeze the definition, let's continue to define nanocomposites as materials that are intended to serve as a structural material first but may have enhanced electrical and thermal properties. In the broadest sense, nearly every nanomaterial, beyond that of the standalone material, is in one way, shape, or form a nanocomposite. A monolayer of silica particles on top of an ITO surface is then, by application of this broad definition, a nanocomposite. A quantum dot tethered to a carbon nanotube is then also a nanocomposite, but let's not split filaments over this arbitrary demarcation.

10.1 Physical and Chemical Properties of Materials

A brief review of mechanical, electrical, and thermal properties and testing is provided. Engineering materials undergo engineering testing; nanocomposites are no different in that regard. We just need to learn more about them to understand why exactly physical properties are different and enhanced. Nanocomposites are expected to improve stiffness, strength, toughness, and other mechanical properties—density, permeability, thermal expansion, conductivity, and other physical properties—all at lower costs. We need to understand why this prediction is true or why it will prove to be true.

10.1.1 *Mechanical Properties*

Mechanical properties of a bulk material are the sum total of physical and chemical properties measured at the macroscale. A physical device pulls, compresses, pushes, indents, or impacts to determine the mechanical performance of a material. Failure analysis, however, is very much a microscopic process in which the root cause of poor performance or failure is rooted out at the atomic level. As it turns out, all mechanical properties valid at the macro level transpose directly to the nano level—via some scaling effect of course. Let's review a few mechanical properties and associated terminology.

Stress. Stress is the measure of average force exerted per area: $F \cdot m^{-2}$. The units of stress are commonly given in terms of pascal (usually associated with mega or giga although now nano and tera are making headway). Stress, σ, is defined as

$$\sigma = \frac{F}{A_o} \tag{10.2}$$

where
    F is force in newtons
    A_o is the initial area of the material cross section in m^2 [22]

Linear stress is a rank-two tensor and is represented by a $[3 \times 3]$ matrix. There are several kinds of stress: tensile, compressive, shear, bending, and torsional, to name a few. A tensile stress situation is shown in **Figure 10.2**.

Strain. Tensile strain is caused by tensile stress. Tensile strain is a dimensionless geometrical expression of deformation—the extent to which a body is distorted after application of a deforming stress. It is the ratio of the extension or compression (in terms of length, area or volume) of a material to its original dimensions. Linear strain is measured as a change in length along a line or the change in angle between two lines. Other forms of strain are plane, shear, and volumetric strain. The formula for linear strain is given below:

$$\varepsilon = \frac{\Delta L}{L_o} \tag{10.3}$$

where
    L_o is the initial length
    ΔL is the change in length under stress

FIG. 10.2 *Typical tensile stress configuration. Equivalent forces are applied symmetrically to an original material (tan) as the material is stretched. In compression tests, the forces are applied inward and the deformation occurs in the z-direction with regard to the graph.*

Tensile Strength.   Tensile strength is an intensive property and is the measure of the level of tensile stress required to pull a material apart [23]. Yield strength is the point at which elastic deformation gives way to permanent, or plastic, deformation. The ultimate strength is the maximum stress that the material is subjected, and breaking strength is the point where rupture occurs in the material (**Fig. 10.3**).

Young's Modulus, E.   Young's modulus (a.k.a. modulus of elasticity or tensile modulus) is a measure of stiffness. It is the measure of the rate of change of stress

FIG. 10.3

A simple tensile test is able to reveal mechanical properties of a generic metal. The modulus of elasticity (Young's Modulus) is determined from the initial slope of the stress–strain curve. Elastic deformation is temporary deformation and the linearity of this region reflects Hooke's law. Young's modulus is also referred to as the stiffness of the material. The region past the elastic region is the point of plastic deformation or permanent deformation. The original shape of the material cannot be recovered after it is mechanically stressed beyond this region. The yield strength represents the minimum amount of stress necessary to generate a small amount of the permanent deformation [11]. The ultimate tensile strength is reflected by the highest point of the curve, for example, the point of maximum stress. Toughness is defined as the total area under the curve and is a reflection of a combination of mechanical properties—overall reflecting the ability of a material to absorb energy until fracture [24]. Ductility is the percent elongation at fracture. Resilience is the ability of a material to absorb energy and is calculated from the area under the curve to the elastic limit point [24]. A curve such as this is also dependent upon the rate at which the test is conducted. Hardness, not shown in the graph, is the ability of a material to withstand local plastic deformation.

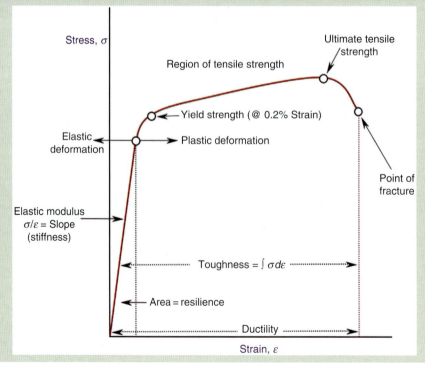

with strain and is determined from the slope of stress–strain curves extracted during tensile tests. Young's modulus is a numerical evaluation of Hooke's law, the ratio of stress to strain or the measure of resistance to elastic deformation. Stress and strain are related through an expression related to Hooke's law

$$\sigma = E\varepsilon \tag{10.4}$$

where E is a proportionality constant equal to $FL/A\Delta L$

$$E = \frac{\sigma}{\varepsilon} = \frac{F \cdot A_o^{-1}}{\Delta L \cdot L_o^{-1}} = \frac{F \cdot L_o}{A_o \cdot \Delta L} \tag{10.5}$$

where E is the Young's modulus in terms of force per unit area or pressure ($F \cdot m^{-2}$ or Pa).

Young's modulus is based on Hooke's simple spring law where

$$F = \left(\frac{E \cdot A_o}{L_o}\right)\Delta L = kx \tag{10.6}$$

Young's modulus is used to determine the length that a material will stretch under tension or the load under which a material will buckle under compression. The Young's modulus (in GPa) of selected common materials is given in **Table 10.2**.

A typical stress–strain curve for a typical metal was shown in **Figure 10.3**. A standardized test coupon (usually in the shape of a dog bone) is clamped at both ends while a tensile (pulling) stress is applied. The cylindrical cross-sectional area of the tensile specimen is recorded. Significant kinds of data can be extracted from such a simple test: Young's modulus (elastic deformation, stiffness), plastic deformation, tensile strength, resilience, ductility, toughness, yield strength, and point of fracture.

Loss Modulus. Loss modulus is an engineering term that describes the dissipation of energy into heat as a material deforms under tensile or shear mechanical testing in viscoelastic solids. It is often referred to as the damping factor (in Pa). Although beyond the scope of this chapter, the loss modulus is the imaginary component of the storage modulus. The storage modulus measures the level of stored energy (the elastic portion) and the dissipation of heat (of the viscous portion). Loss modulus is often correlated with shear modulus (modulus of rigidity or the shear stress over shear strain) from which the viscosity of material can be determined. Opposing forces are often applied to a material upon testing. Elastic deformation under shear stress is shown in **Figure 10.4**.

Poisson's Ratio. Materials stretched along the z-axis tend to contract along the x- and y-axis (**Fig. 10.5**). Poisson's ratio is the ratio of the relative contraction strain over that of the relative extension (axial) strain. For example

$$v = -\frac{\varepsilon_x}{\varepsilon_z} \tag{10.7}$$

TABLE 10.2	Young's Modulus of Common Materials	
General engineering material	**Example**	**Young' modulus (GPa)**
Metal	Steel	200
	Cu	115
	Brass	110
	Al	69
Ceramics	Glass	60–90
	Tungsten carbide	550
	Silicon carbide	450
Carbon materials	Diamond	1100
	Graphite	
	Single-walled carbon nanotubes	1000+
Fibers	C-glass	69
	E-glass	72.4
	S-glass	85.5
	Graphite fiber	340–380
	Al_2O_3 whisker	430
	SiC fiber	430
	Boron filament	410
Polymers	Epoxy	6.9
	Polyester	6.9
	Low-density polyethylene	<0.2
	High-density polyethylene	1.4
	Nylon	3–7
Composites	Tendon	1.5
	Bone (compression)	9.4
	Bone (tensile)	16
	Pine wood (along grain)	9.0
	Oak wood (along grain)	11
	Plywood	4.0
	Rubber	<0.1
	CFRP	125–150
	Concrete (compression)	41
	W-filament ($V_f = 0.50$) strengthened Cu	260
	W-particle ($V_f = 0.50$) strengthened Cu	190
	E-glass fibers ($V_f = 0.73$) in epoxy	56
	Al_2O_3 whiskers ($V_f = 0.14$) in epoxy	41
	C fiber ($V_f = 0.67$) in epoxy	221
	Kevlar fibers ($V_f = 0.82$) in epoxy	86
	Boron filaments ($V_f = 0.70$) in epoxy	210–280

FIG. 10.4 *Elastic deformation along the x-axial direction due to shear stress by opposing forces is depicted.*

FIG. 10.5 *From Poisson's ratio, the characteristic contraction perpendicular to the tensile stress is defined.*

where ε_x and ε_z are the strains in the x and z directions. Poisson's ratio is a fundamental engineering relationship that describes the elastic behavior of materials.

Average Properties.   Composites consist of more than one material. How then does one predict the mechanical properties of the composite from the known mechanical properties of the individual components? The mechanical properties of the composite are also dependent upon the geometric configuration of the structural materials—from the micro to the nanostructure—and how well the inclusions are bound to the matrix. Assuming tight binding, simple averaging of component mechanical properties yields a good first approximation of the mechanical properties of the composite.

The simplest case is that of a circular composite consisting of intimately bonded fibers running parallel through the length of the material [11]. Shackelford provides a good definition of average mechanical properties [11]. If a load is applied to a cylindrical composite parallel to the fibers (uniaxial stressing), an isostrain condition is appropriate for consideration. First of all, the strain for the composite material and its components is given by

$$\varepsilon_c = \frac{\sigma_c}{E_c} \quad \text{and} \quad \varepsilon_m = \frac{\sigma_m}{E_m} \quad \text{and} \quad \varepsilon_f = \frac{\sigma_f}{E_f} \tag{10.8}$$

where the subscripts c, m, and f correlate to the composite, matrix, and fiber, respectively, and E_c, E_m, and E_f are the bulk modulii for each component. For the isostrain condition

$$\varepsilon_c = \varepsilon_m = \varepsilon_f \tag{10.9}$$

because the composite and its components are integrally linked. The sum total of the load F_c is equal to the sum of the loads on the individual constituents

$$F_c = F_m + F_f \tag{10.10}$$

Since $\sigma = F/A$ (from equation 10.2, page 419)

$$\sigma_c A_c = \sigma_m A_m + \sigma_f A_f \tag{10.11}$$

This formulation is commonly referred to as the "rule of mixtures." The ROM works well for calculating the density of a composite. Combining equations (10.7), (10.8), and (10.9) yields

$$E_c \varepsilon_c A_c = E_m \varepsilon_m A_m + E_f \varepsilon_f A_f \tag{10.12}$$

Dividing by A_C and ε_c (obtained from equation 10.12) and noting that A_m/A_C and A_f/A_C are the same as volume fraction V_m and V_f, respectively, the Young's modulus of the composite is equal to the sum of the Young's moduli of the components (where $V_m + V_f = 1$)

$$E_c = V_m E_m + V_f E_f \tag{10.13}$$

For the *isostress* condition, in which the fibers are loaded perpendicular to the axis of the cylindrical composite, the following relationship applies

$$\frac{1}{E_c} = \frac{V_m}{E_m} + \frac{V_f}{E_f} \tag{10.14}$$

A good question to ask at this time is the following: "Does elastic modulus scale with size as implied by the above formulae to the nanomaterials?"

Composite mechanical properties are also analyzed via micromechanics in which analysis of individual phases on a multi-axial level takes place. Two kinds, the "method of cell" and the Mori–Tanaka averaging scheme, are used to predict mechanical behavior of composites. Fiber interaction effects relating inclusion strain to average matrix strain can be predicted by the Mori–Tanaka method.

Hardness Testing. Hardness is a test applied to materials in the solid phase. Engineering parameters such as elasticity, plasticity, viscosity, viscoelasticity, strength, strain, brittleness, ductility, and toughness can be gleaned from hardness testing [25]. Hardness values are dependent on the type of test applied and the geometry of the probe. Hardness is the ability of a material to resist permanent deformation induced by an applied force. There are several types of hardness testing: Rockwell, Knoop, Vickers, and Brinell are just a few of the most common ones. Scratch, indentation, and rebound hardness are some types of engineering definitions associated with applied hardness. In scratch hardness, resistance to fracture and plastic deformation due to friction applied via a sharp probe are measured. Indentation hardness measures the resistance to plastic deformation due to a persistent load delivered via a sharp probe. Rebound hardness measures elasticity from the height of recoil of an object dropped from a predetermined distance.

Hardness can be defined at the macro-, micro-, or nanoscales according to the level of applied force and the extent of the displacement. Macrohardness measurements are facilitative means of obtaining engineering values, especially in the case of metals. Microindenters function by continuous measurement of penetration depth induced by an applied load. Nanoindenters apply forces on the order of 1 nN. The indenter is driven into the sample by the action of a magnetic coil fixed to an indenter assembly. A capacitance displacement gauge monitors penetration depth.

The Vickers hardness test utilizes a diamond probe with a 136° angle between opposite faces of the indenter (or 22° between the indenter face and the surface) (**Fig. 10.6**) [26].

The area of the indentation is determined from

$$A = \frac{d^2}{2\left(\dfrac{\sin 136°}{2}\right)} \approx \frac{d^2}{1.854} \tag{10.15}$$

$$HV = \frac{F}{A} \approx \frac{1.854F}{d^2} \tag{10.16}$$

where HV is in terms of $kg \cdot F \cdot mm^{-2}$ ($1\,kg \cdot F \cdot mm^{-2} = 9.807\,MPa$). Vickers hardness and other physical and mechanical properties are listed in **Table 10.3**.

Good correlation was achieved between nanoindentation tests and the standard Vickers hardness tests conducted on model ferritic alloys [27]. Nanoindenter loads of 0.05 g (Nanoindenter-II, Nano Instruments, Inc., Oak Ridge) were compared to conventional Vickers microhardness derived from 200 to 500 g applied loads [27]. Two methods of testing were used: one in which constant displacement depth was measured and the other in which a constant load was applied to all specimens. Nanohardness data was corrected to mitigate the differences between projected and actual indenter area. Overall, good correlation was achieved between the two techniques for hardness values in the range of 0.7 and 3 GPa. Tensile property measurements were also conducted on all specimens. The linear correlation between Vickers hardness and yield strength is apparently valid for data derived by the nanoindentation technique [27].

Rheology and T_g. Differential scanning calorimetry (DSC, a good way to determine the glass transition temperature, T_g) and rheological methods are

FIG. 10.6	*A Vickers hardness tip is depicted. The tip is made of diamond in the shape of a square-based pyramid. Consistency is obtained in that impressions are similar regardless of the load applied. The Vickers Pyramid Number (HV) is determined from the ratio F/A where F is the applied force and A is the surface area of the indentation.*

Source: Image from Wikipedia commons, en.wikipedia.org/wiki/Vickers_hardness_test (2008). With permission.

TABLE 10.3	*Engineering Properties of Elemental Materials*							
	Vickers hardness (MPa)	Mohs hardness (MPa)	Young's modulus (GPa)	Poisson ratio	Thermal conductivity $[W \cdot (m \cdot K)^{-1}]$	Thermal expansion $(10^{-6} K^{-1})$	Electrical conductivity $(10^7 S \cdot m^{-1})$	Density $(g \cdot cm^{-3})$
Fe	608	4	211	0.29	80	11.8	1.0	7.874
Al	167	2.75	70	0.35	235	23.1	3.8	2.7
Au	216	2.5	78	0.44	320	14.2	4.5	19.3
Ag	251	2.5	83	0.37	430	18.9	6.2	10.49
Cu	369	3	130	0.34	400	16.5	5.9	8.92
Bi	N/A	2.25	32	0.33	8	13.4	7.7×10^5	9.78
Ca	N/A	1.75	20	0.31	200	22.3	2.9	1.55
Sn	N/A	1.5	50	0.36	67	22	9.1×10^6	7.31
Be	1670	5.5	287	0.032	190	11.3	2.5	1.848
Mo	1530	5.5	329	0.31	139	4.8	2.0	10.28
W	3430	7.5	411	0.28	170	4.5	2.0	19.25

Note: Vickers hardness of the elements: periodictable.com/Elements/004/data.html.

good means of probing the interfacial area. The T_g is sensitive to the level of nanoparticles that have either an attractive (decreased mobility) or repulsive (increased mobility) reaction with the polymer matrix [28]. Rheology is the study of materials that deform or flow. With more filler, the polymer mobility and T_g (relaxation) are influenced.

Tribology. Tribological behavior is also affected by nanofillers. Dry lubrication of bearings is accomplished with homogeneous polymers like poly(tetrafluoroethylene), a.k.a. PTFE (e.g., they are nonabrasive with low friction), but they present issues with wear resistance. When micrometer-sized hard fillers are added to increase wear resistance, hardness, modulus, and strength, abrasion factors begin to emerge [28]. What is required is an engineered material that provides the best of both worlds—enter the nano-solution. PTFE does not wear well. If 15-wt% micrometer-sized fillers are added, two orders of magnitude improvement are observed but at a cost—an increase in the dry sliding coefficient [29]. Tests were conducted on four kinds of 304-stainless steel surfaces: electropolished, wet-sanded, lapped, and dry-sanded [30]. PTFE (2–20 μm grain size) + alumina (44, 80, or 500 nm) composites were prepared. Filling rates were varied from 0, 1, 5, and 10 wt%. The lapped surface showed the best results of the four. If up to 10 wt% of γ-alumina nanoparticles are added to the polymer, three orders of magnitude (OOM) reduction in wear rate is observed and 4-OOM reduction is observed if 1-wt% γ-alumina nanoparticles are mixed with the PTFE [30]. Other tests using α-alumina showed even better performance. Apparently α-alumina leads to stronger interfacial interactions with the PTFE polymer than does γ-alumina [31]. Apparently nanofillers (especially the 4-OOM filler at 1 wt%) stabilize the phase of PTFE that is tougher and fibrillates more easily and these conditions promote the formation of "well-adhered transfer films," a phase that is not normally available at room temperature [28]. Fibrillation in this case fills cracks better thereby reducing the formation of wear debris. DSC analysis revealed that nanofilled PTFE had a higher melting temperature than unfilled PTFE.

Mechanical Testing at the Nanoscale. Failure in materials usually involves some kind of fracture, and as a result, most failure analysis involves examination of the results of mechanical loading tests—tensile overload, fatigue, or creep [5]. However, we do not focus on failure analysis in this chapter. Regardless, most failures can be traced to atomic or molecular phenomena—phenomena now that are closer to the nanoscale in size. The question to be asked at this stage is how exactly does one test for nanomaterial mechanical properties? Apparently an entire new arsenal of nanomechanical testers has been invented that involve the use of AFM probes and nanoindenters, for example. Although containing nanomaterials, a nanocomposite is realized as a macroscopic, graspable material and is therefore amenable to the same kinds of mechanical testing that is perpetuated on traditional composites.

Transposition of bulk properties to nanoscale materials may be appropriate in some cases, depending on size, but the practice of doing so may be fraught with pitfalls as well. Bulk optical properties, for example, are used to describe the optical response of colloidal particles (ca. $10 < d < 100$ nm). However, for particles below 10 nm in size where surface effects and quantum effects start to exert influence, caution must be inserted into the equation. In addition, nano-materials are purported to have fewer defects. This too is expected to affect their material properties.

10.1.2   *Thermal Properties*

Material thermal properties of concern to engineers come in the form of thermal conductivity, the ability of a material to conduct heat, and a parameter known as the coefficient of thermal expansion (CTE), the microscopic expansion of a material (strain) under thermal stress.

Coefficient of Thermal Expansion.   CTE is the change in dimension (linear, area, or volume) of a material in response to a change in temperature. The degree of expansion (or the fractional increase in strain) divided by the change in temperature is known as the coefficient of thermal expansion. Thermal expansion for fibers, rods, or cables is expressed by the ratio of strain along their length

$$\varepsilon_{\text{Thermal}} = \frac{L - L_o}{L_o} = \frac{\Delta L}{L_o} \tag{10.17}$$

Thermal expansion is proportional to the change in temperature

$$\varepsilon_{\text{Thermal}} \propto T - T_o = \Delta T \tag{10.18}$$

Heat is stored in materials during heat transfer in several ways—in the form of translational, rotational, vibrational, or electronic energies. Carbon materials, in general, are characterized by low coefficients of thermal expansion. This is one of the reasons such materials perform well in spacecraft that is exposed to temperature extremes. Polymers, on the other hand, have higher coefficients of thermal expansion than do metals. Ceramic materials possess some of the lowest thermal coefficients of expansion (**Table 10.4**).

Thermal expansion can be isotropic or anisotropic. Crystals, for example, may expand along one crystalline direction more so than another. Graphite expands more along one direction (x, y) of its plane than along the other, the z-direction. The thermal coefficient of expansion is expressed in linear, areal, or volumetric forms as

$$\alpha_L = \frac{\Delta L}{L_o \Delta T}, \quad \alpha_A = \frac{\Delta A}{A_o \Delta T}, \quad \text{and} \quad \beta_V = \frac{\Delta V}{V_o \Delta T} \tag{10.19}$$

CTE is measured via interferometry, a procedure that analyzes the changes in the interference pattern of monochromatic light. Each shift in the fringe pattern corresponds to a change in L of 0.5λ the laser wavelength. Plots of strain versus temperature, and the slope of the strain–temperature curve yield the instantaneous value of CTE. Quartz dilatometry is used to determine CTE of materials that are expected to have large changes in dimension under thermal stress. CTE mismatches lead to mechanical failure of materials even if CTE values have apparent small values. Therefore, careful selection of materials with regard to CTE must be done during the design phase of an engineering material.

TABLE 10.4	Linear Coefficients of Thermal Expansion (CTE) of Common Materials	
General engineering material	**Example**	**CTE ($10^{-6} \cdot K^{-1}$ or $10^{-6} \cdot C^{-1}$)**
Metals	Carbon steel A36	11.7
	Stainless steel alloy 304	17.2
	Cast iron G1800	11.4
	Stainless steel alloy 440A	10.2
	W (refractory metal)	4.5
	Mo (refractory metal)	4.9
	Ta (refractory metal)	6.5
	Au (noble metal)	14.2
	Ag (noble metal)	19.7
	Pt (noble metal)	9.1
	Ni (nonferrous metal)	13.3
	Zn (nonferrous metal)	23.0
	Zr (nonferrous metal)	5.9
	Al 7075 (nonferrous metal)	23.4
	Bronze (nonferrous metal)	18.0
Semiconductors	GaAs	5.9
	Si	2.5
Ceramics	Aluminum oxide	7.0
	SiC	4.6
	Zirconia	9.6
	Soda lime glass	9.0
	Si_3N_4	2.7–3.1
Carbon materials	Graphite	2.0–2.7
	Diamond	0.11–1.23
Polymers	Butadiene–acrylonitrile, styrene–butadiene, and silicone (elastomers)	220–270
	Phenolics (thermosets)	122
	Polyesters (thermosets)	100–180
	Epoxies (thermosets)	80–117
	Low-density polyethylene (LDPE)	180–400
	Polystyrene (PS)	90–150
	Polymethyl methacrylate (PMMA)	90–162
	Polyvinyl chloride (PVC)	90–180
	Teflon (PTFE)	126–216
Fibers	Aramid (Kevlar 49)	-2.0 (longitudinal)
		60 (transverse)
	Carbon fiber	-0.6 (longitudinal)
	Polyacrylonitrile (PAN)	10.0 (transverse)
	E-glass	5.0

(continued)

TABLE 10.4 (CONTD.)	Linear Coefficients of Thermal Expansion (CTE) of Common Materials	
General engineering material	**Example**	**CTE ($10^{-6} \cdot K^{-1}$ or $10^{-6} \cdot C^{-1}$)**
Composites	Concrete	10.0–13.6
	Aramid fiber/epoxy matrix $V_f = 0.6$	−4.0 (longitudinal)
		70 (transverse)
	High modulus carbon fiber/epoxy	−0.5 (longitudinal)
		32 (transverse)
	E-glass fibers/epoxy resin	6.6
	Douglas fir	3.8–5.1 (grain parallel)
		25.4–33.8 (grain perpendicular)
	Oak	4.6–5.9 (grain parallel)
		30.6–39.1 (grain perpendicular)
	SiC–Metal	0.4 [32] (significantly less than pure metal)
	AlN fiber-reinforced polystyrene	0.3 times less than pure PS

Source: www.stormcable.com/uploads/Thermal_expansion_data_table_tb06.pdf (viewed 2008).

Thermal conductivity. Thermal conductivity is the ability of a material to conduct heat—the specific heat flux that flows through a material if a certain temperature gradient is in place. Heat conduction is defined by Fourier's law:

$$k = -\frac{dQ/dt}{A(dT/dx)} = \frac{\Delta Q/\Delta t}{A(\Delta T/\Delta x)} \tag{10.20}$$

where
  $dQ/\Delta dt$ is the rate of heat flux (transfer) across an area A in $J \cdot s^{-1}$
  κ is the thermal conductivity coefficient of a specific material
  A is the area
  x is the thickness of the conducting surface

For a flat slab under steady-state heat conduction, the differentials become average values (RHS of equation 10.20). The units of k are $J \cdot (s \cdot m \cdot K)^{-1}$ or more commonly, $W \cdot m^{-1} \cdot K^{-1}$ ($1\,W = 1\,J \cdot s^{-1}$). In general, metals are good conductors of heat and polymers are good insulators (**Table 10.5**).

10.1.3 Electronic Properties

Nanomaterials cover the entire gamut of electronic phenomena. Nanomaterials can act as conductors, semiconductors, and insulators. Carbon nanotubes in particular offer engineers a wide range of properties that are amenable to use in advanced materials—composites in particular, whether in the form of shielding materials or those designed to conduct electrically in an anisotropic fashion. We shall review a few of the basic engineering parameters that we need to be aware when discussing electronic and magnetic properties.

TABLE 10.5	*Thermal Conductivities of Common Materials*	
General engineering material	**Example**	**Thermal conductivity ($W \cdot m^{-1} \cdot K^{-1}$)**
Metals	Carbon steel	54
	Stainless steel	16
	Al	250
	Co	69
	Au	310
	Ag	429
	Cu	401
Ceramics	Quartz	3
	Aluminum oxide	30
	Granite	1.7–4.0
	Mica	0.71
	Glass	1.05
	Plaster board gypsum	0.17
	TiC (100°C)	25
	Silica glass (100°C)	2.0
	Soda lime glass (100°C)	1.7
	ZrO_2 (100°C)	2.0
	Porcelain (100°C)	1.7
Carbon materials	Diamond	900–2320
	Graphite (100°C)	180
	Carbon nanotubes	1800–6000
Polymers	Nylon	0.25
	High-density polyethylene (HDPE)	0.42–0.51
	Polystyrene (PS)	0.033
	Polypropylene (PP)	0.1–0.22
	Polyvinyl chloride (PVC)	0.19
Composites	Leather	0.14
	Balsa wood	0.048
	Cement	0.29
	Cork	0.043
	Cotton	0.03
	Fire-clay brick	1.4
	Fiberglass insulation	0.04
	Oak	0.17
	Vermiculite	0.058

Resistivity and Conductivity.   Electrical resistance depends on the composition and geometry of a material. For example, resistance Ω increases with sample length L and decreases with sample area A [11]. Resistivity ρ, on the other hand, is a parameter that is independent of sample geometry with units of $\Omega \cdot m$

$$\rho = \frac{\Omega A}{L} \qquad\qquad (10.21)$$

Conductivity σ is the reciprocal of resistivity

$$\sigma = \rho^{-1} \tag{10.22}$$

Another term that is encountered frequently when discussing nanocomposite electrical properties is the *percolation limit*, or the correlation between geometrical and electrical connectivity. The percolation limit is defined as the minimum concentration of filler required to support electrical conduction in a composite. Particle contact and percolation require a large volume fraction when the metal particles and the polymer grains are of comparable size. When the conducting particles are small relative to the grain size of the polymer, they are forced into interstitial (interfacial) sites and in contact with one another within a lower volume. This configuration constitutes a "low percolation limit," another advantage of small size. The geometrical and electrical onset of connectivity may not occur simultaneously in some cases but their lack of correlation may be explained by the tunneling phenomena. **Table 10.6** lists the conductivity of selected metals, semiconductors, and insulators in terms of $\Omega^{-1} \cdot m^{-1}$ (siemens per meter, $S \cdot m^{-1}$, also indicated as $S \cdot cm^{-1}$) [11,33,34].

10.1.4 *Chemical Properties*

The chemistry of nanocomposites takes place within and during its manufacture, not after its introduction to its application environment. Structural nanocomposites are designed to be inert. We do not want composite materials that react with radiation, humidity, erosion, acids, corrosive agents, toxic gases, sequestered liquids, or any other chemical or physical process. We want our structural materials to retain their integrity under a variety of conditions. If there is any surface chemistry applied to nanocomposites, it is in the form of protective materials like sealants, paints, and other coatings. These treatments are not discussed in this chapter.

This chapter is placed in the *Chemical Nanoengineering* division of the text. Although some mechanical, electrical, thermal, and optical properties are discussed, we are specifically interested in the interfacial chemistry between the nanophase and the bulk matrix phase of the composite. What chemistry is required, for example, to render a single-walled carbon nanotube reactive to potential matrix bonding?

Chemical Modification. Nanomaterials have high surface energy. In general, due to this excess surface energy, nanomaterials seek stabilization by many methods. Chemical stabilization is one means by which nanomaterials achieve this goal. Gold-55 clusters, for example, require a ligand shell to help stabilize its geometry or otherwise agglomeration results. Many kinds of nanoparticulate fillers in composites have reactive surfaces, like silicates, that react readily with the polymer matrix. Others, such as nanogold, require a ligand shell in order to interface strongly with the matrix material of the composite. Ligands may provide specially designed functional groups that bind with side-chain counterparts of a polymer. These processes are highly chemical in nature and will be discussed throughout the remainder of the chapter.

TABLE 10.6	*Electrical Conductivity of Selected Engineering Materials*	
General engineering material	**Example**	**Conductivity ($S \cdot m^{-1}$ or $\Omega^{-1} \cdot m^{-1}$)**
Metals	Al	3.54×10^7
	Cu	5.8×10^7
	Fe	1.03×10^7
	Au	4.26×10^7
	Steel	5.7–9.4×10^6
Semiconductors	Ge	2.0
	Si	0.40×10^{-3}
	PbS	38.4
	Indium–tin oxide (ITO)	$\sim 1 \times 10^4$ $S \cdot cm^{-1}$
Carbon materials	Graphite	1.28×10^5
Ceramics	Aluminum oxide	10^{-10} to 10^{-12}
	Borosilicate glass	10^{-13}
Polymers	Polyethylene	10^{-13} to 10^{-15}
	Nylon 66	10^{-12} to 10^{-13}
	Doped conductive polymers: polyaniline (PANI), polypyrrole (PPy)	10^{-5} to 10^2
	Polypyrrole doped with AsF_3	$\sim 10^3$ $S \cdot cm^{-1}$
	Polypyrrole doped with I_2	$\sim 10^2$ $S \cdot cm^{-1}$
	Polyacetylene doped with AsF_3	$\sim 10^5$ $S \cdot cm^{-1}$
	Polyacetylene doped with I_2	$\sim 10^4$ $S \cdot cm^{-1}$
Composites	K-intercalated graphite (10%)–polystyrene [35]	1.3×10^{-1} $S \cdot cm^{-1}$
	Graphite (10%)–polystyrene [35]	5.0×10^{-3} $S \cdot cm^{-1}$
	Br-intercalated graphite fiber–epoxy [36]	1.1×10^{-1} $S \cdot cm^{-1}$
	Graphite fiber–epoxy [36]	2.2×10^{-3} $S \cdot cm^{-1}$
	Graphene–silica ($V_f = 0.11$) spun composite	0.45 $S \cdot cm^{-1}$ [37]
	MWNT–silica ($V_f = 0.093$) spun composite	0.57 $S \cdot cm^{-1}$ [37]

Note: Data from *Materials Properties Tables: Electrical Conductivity and Resistivity*, NDT Resource Center, Iowa State University, www.ndt-ed.org/GeneralResources/MaterialProperties/ET/et_matlprop_index.htm.

Intermolecular Interactions. Intermolecular interactions play a major role in nanoinclusion–matrix interfacial chemistry. The entire gamut of these "noncovalent, nonmetallic, and nonionic bonding" participates in interfacial bonding with matrix polymeric materials. Hydrogen bonds, van der Waals, dative bonds, π-stacking, and aromatic dipole–dipole, dipole-induced dipole, and induced dipole–induced dipole; capillary, and hydrophobic interactions all are capable of binding to polymers. Some nanoparticles actually are able to catalyze or direct the formation of the polymer during synthesis of the nanocomposite.

Chemical Modification of Carbon Nanotubes. Some kinds of nanomaterials do not require stabilization by chemical or other means. Single-walled carbon nanotubes, for example, are kinetically stable under room ambient conditions.

However, they do not make for good fillers in nanocomposites unless some chemical modification takes place on their surface. The modification is required so that integral contact with a host matrix material, usually a polymer, is possible. The chemical bonds thus formed are able to conduct load transfer from the host bulk majority component to the carbon nanotubes—the whole idea behind fiber reinforcement in a composite.

10.2 NATURAL NANOCOMPOSITES

Biomimetic technology is discussed in detail in chapter 12 and we do not wish to steal any of its thunder; however, we cover a few notable examples of natural nanocomposites found in nature that serve as excellent models for our technology. A few selected examples of natural nanocomposites are listed and described in **Table 10.7**. Every hierarchical structure found in living things is a nanocomposite. Perhaps we can extend this expression to include "every structure in nature is a nanocomposite." What do you think?

10.2.1 *Skin of the Sea Cucumber*

Lessons from the Humble Sea Cucumber.   Recently, S.J. Rowan and C. Weder of Case Western Reserve University have developed a material (with a low-modulus matrix) that can harden or soften depending on its immediate environment—a process known as chemoresponsive mechanic adaptability [39]. When threatened, the sea cucumber, an echinoderm that is able to rapidly and reversibly control the stiffness of its inner skin membrane (dermis), is able to transform its outer skin into a hard shell-like protective material in seconds—ten-times harder than the relaxed version. The reaction is due to enzymes that are able to bind protein fibers together. Another set of enzymes allows the skin of the sea cucumber to return to its relaxed state [39]. The humble sea cucumber serves as the model for the new nanopolymeric material.

   The material developed by Rowan et al. is able to harden to a level that is greater than 2500 times its softer state—reversal of which occurs after soaking in water. The hardening of the material is due to the collective action of cellulose fiber matrix held together with hydrogen bonds. When wetted, water is able to hydrolyze bonds between fibers and the material undergoes relaxation. When dried, the material reforms hydrogen bonds and that process induces stiffening of the membrane. The elastic host polymer and stiff cellulose nanofibers demonstrates reversible reduction in tensile modulus by a factor of 40—from 800 to 20 MPa—during cycling of the material in tests. Polymers in which thermal transitions corresponded to simulated physiological conditions exhibited more dramatic tensile modulus swings—4.2 to 1.6 MPa, a factor of 2625 [39]. The material, once commercialized, could be used in brain implants that reduce inflammation or into clothing that can transform into armor.

10.2.2 *Hard Natural Nanocomposites*

The More Humble Mollusk.   The shells of mollusks protect their soft bodies from predators and other traumas inflicted by nature. The composite, generally

TABLE 10.7	*Natural Composite Materials*
Natural composite	**Description**
Cellulose (wood)	Polysaccharides of linear polymers of D-glucose subunits linked by $\beta(1\rightarrow 4)$ glycosidic bonds; found in cell walls of plants; composite formed from flexible cellulose + stiff lignin.
Chitin	Nitrogenous polysaccharides $[(C_8H_{13}O_5N)_n]$ in linear configuration (*N*-acetyl-D-glucose-2-amine subunits) that form $\beta(1\rightarrow 4)$ like cellulose; chitin embedded in a proteinaceous matrix forms the exoskeleton composite material of insects; hydrogen bonding between adjacent polymers gives the material increased strength.
Bone	Parallel collagen triple helices in staggered array; 40-nm gaps at ends of tropocollagen serve as nucleation centers for hydroxyapatite, $[Ca_{10}(PO_4)_6(OH)_2]$, forming a composite material that is very hard (due to the mineral part) but elastic and fracture resistant (due to the collagen part).
Silk	β-Pleated sheet proteinaceous materials made up of gly–ser–gly–ala–gly (gly at every other position) monomer subunits; fibers have high tensile strength and are resistant to stretching. Spider silk is a composite of a gel core + surrounding solid casing of aligned molecules, usually consisting of alternating gly-ala or just ala amino acids. Spider silk (2–4 μm diameter) tensile strength is on the order of high-grade steel but a better strength-to-weight ratio; capable of 40% stretch without rupture which makes the material ductile with high toughness; β-sheets stack to form crystalline and amorphous domains, and it is the relationship between these two domains in the composite that give spider silk its remarkable mechanical properties.
Collagen	Modified α-helical protein materials consisting of repeating monomers of glycine–proline–hydroxyproline found in triple-helical structure; glycine at every third position; chains are stabilized by steric repulsion due to pyrrolidone rings of proline and hydroxyproline; three chains are hydrogen bonded to each other. Collagen + keratin gives skin its integrity and elasticity.
Nacre	Aragonite (ortho-calcium carbonate, $CaCO_3$) + biological macromolecules; a.k.a. mother of pearl; organic–inorganic composite material that is strong and resilient. Composition is of hexagonal platelets of aragonite 10–20 μm in width and 0.5-μm thick in parallel lamellar arrangement; the sheets are separated by elastic biopolymers made of complex polysaccharides like chitin and lustrin and β-sheet proteins found in silks; due to this configuration, transverse crack propagation is inhibited.
Diatoms	Cell walls consist of nanostructured polymerized silicates and biomolecules (frustulins on surface and silaffin polypeptides embedded within the silica) for extraordinary mechanical stability; the diatom cell wall consists of crystalline and amorphous regions that gives the composite excellent mechanical properties [38].

known as nacre, consists of calcium carbonate (the ceramic phase, polymorphs of $CaCO_3$, aragonite, or calcite) and an organic binder. One kind of organic constituent is the highly ordered layered β-chitin proteins arranged in parallel arrays within interlamellar sheets. Another kind is similar in structure to silk in the form of an amorphous gel phase. The organic phase is responsible for stabilizing the metastable aragonite and for directing the morphology and orientation of the crystals [40].

Comparison of Hard Natural Materials.   A comprehensive study was conducted to verify the mechanical properties of some natural materials. J. D. Currey et al.

in 2000 found that nacre outperformed highly mineralized bone because of its extremely well-ordered microstructure. The organic material in nacre forms a nearly continuous enclosure around all the aragonite platelets (~500 nm thick), and is designed for toughness. This is surprising because normal bone contains more organic material (mostly collagenacious) than does nacre (~1%). The rostral bone of the whale, however, does have similar organic phase content within its bony structure. The compositions were as follows: bovine bone (65% mineral, 25% organic, and 10% water), rostrum bone (96% mineral, 1% organic, and 3% water), and nacre (98% mineral, 1% organic, and 1% water) [41]. Young's modulus for the three materials was found to be 20, 40, and 34 GPa respectively; bending strength of 220, 55, and 210 MPa, respectively, and hardness of 55–70, 227, and 200 kg·m^{-2}, respectively [41].

The highly mineralized content of the rostrum bone resulted in greater stiffness (increases Young's modulus) and stronger in bending but loss of strength and toughness compared to bovine bone. Although nacre has a highly mineralized content, the loss of strength and toughness was not observed. The organic layer acts as a toughening device. According to Currey et al., there is a mystery associated with nacre and it is the following: why was the material never modified by the pressure of natural selection to have a larger component of organic materials that would impart more strength and toughness while still remaining quite stiff?

Nanostructured Inorganic Framework/Polymer Nanocomposites. Inspired by the structure of natural nacre and bone—natural composite materials that are composed of alternating layers of soft and hard materials in a periodic array—mechanically robust, multifunctional silica/polymer nanocomposites were prepared by a straightforward self-assembly/evaporation process [42]. Depending on process parameters, Y. Yang et al. formed nanostructured, conjugated polymers of hexagonal, cubic, or lamellar structures consisting of an inorganic framework that protects, stabilizes, and orients the polymer while providing mechanical and chemical stability [42]. Polymerizable amphiphilic diacetylene (PDA) surfactant molecules served as structure-directing agents and as the monomeric precursor material. By varying the size of the headgroup of the diacetylene surfactant (oligoethylene glycol, EO), it was possible to control the mesostructure of the resultant polymer. Addition of a silica inorganic host influenced the polymerization of the PDA. The process is generally known as a hybrid organic/inorganic self-assembly process.

The synthesis began with coupling DA with diethylene glycols to form DA-OE$_n$ where $n = 3, 4, 5,$ or 10. Tetraethyl orthosilicate (TEOS, [Si(OC$_2$H$_5$)$_4$]) provided the source of the silica. HCl catalyst in tetrahydrofuran (THF) and water was all that was required to generate the nanocomposite. The size of the head group played a major role in directing the final morphology of the composite: $n = 3$ produced lamellar composites, $n = 5$ produced hexagonal structured composites, and $n = 10$ formed cubic composites [42].

Mechanical properties were determined by the nanoindentation procedure. The Young's modulus ranged from approximately 9–5 GPa and the hardness from 0.6 to 0.4 GPa—both values comparable to calcined mesoporous silica films that are used routinely in microelectronic devices. The membranes were also impermeable to gases indicating that there is no significant porosity within its structure.

10.3   CARBON FIBERS AND NANOTUBES

The importance of carbon fibers was mentioned briefly in section 10.0.1. Although, carbon fibers, whiskers, rods, filaments, and nanotubes have been fouling catalytic processes for decades, the discovery of nanosized carbon tubes has opened the doors to a whole new generation of advanced materials.

Carbon Fibers.   A fiber (from the Latin *fibra* "a fiber, filament," related to *filum* "thread") is a high aspect ratio material ($L \gg D$). There are many classes of carbon filamentous materials and others based on inorganic materials like alumina, boron nitride, and silicon carbide. Conventional carbon fibers range in diameter from a few to 10 μm (and larger depending on the application). Relatively solid carbon nanofibers (CNF) have diameters on the order of 50–200 nm. MWNTs, as we know well by now, have diameters between 10 to 20 nm and are as large as 50 nm. The diameter of SWNTs is limited to a few nanometers. The structure of CNFs, MWNTs, and SWNTs varies depending on fabrication conditions. Fiber nanostructure, for example, is demonstrated in the form of cones, cups, or plates.

Although we have implicitly touched upon fiber technology in previous sections in this chapter, we go ahead to provide a more complete treatment of the subject matter, of nanofibers and nanotubes based on carbon in particular. A nanofiber or nanotube is defined as a filament (hollow or solid) with diameter less than 100 nm—although we can safely extend this boundary to elongated materials with diameter less than 1 μm. Large diameter filaments are characterized by solid morphology (e.g., no central canal).

Nanofibers are produced in several ways—interfacial polymerization, electrospinning, and weaving to name a few. Nanotubes are most efficiently produced by CVD techniques. Nanofibers have multiple functions and applications beyond enhancing mechanical and physical properties of polymer composites—as filters, in medical devices, composites, garments, insulation, and energy storage to name a few. We define fibers as those materials that are not necessarily carbon nanotubes but can contain carbon nanotubes. Graphitic nanofibers range in microstructure from solid herring-bone to stacked graphitic platelets—parallel or perpendicular to the fiber axis.

Carbon Nanotubes.   Carbon nanotubes and their ilk once again are the headliners in a chapter of a nanotechnology text. That should be of no surprise by now. They are amazing materials with amazing properties and why not place them within a polymeric matrix with the intent of fabricating a super-composite. Carbon fibers have been used to reinforce polymers for several decades—with the earliest efforts linked to the aerospace industry in the mid- to late-twentieth century. One of the holy grails of aerospace design is to incorporate materials that are stronger (higher tensile strength) but lighter, for example, higher strength to weight ratio. Carbon-reinforced resins and carbon nanotube reinforced resins offer an optimistic route to achieve that goal.

Carbon nanotubes have certainly changed the landscape with the advent of nanotube-reinforced composites, creating a new generation of excitement. According to E.V. Barrera of Rice University in Texas, advanced applications of composites are expected in many areas: radiation protection, heat dissipation, static discharge capacity, high strength-lightweight parts and housings, heat engine components, deicing coatings, lightning protection, stress sensors, organic LEDs, electrically conductive ceramics, paintable polymers, antifouling paints, UV protection coatings, and overall corrosion protection [43].

Most research today is limited by the availability of carbon nanotubes. Berrara et al. go on to say that, and we quote indirectly, that a dearth of research materials limits the concentration of NTs in composites to low volume fraction—ca. <10% and that research has focused on nanotube dispersion, untangling, alignment, bonding, molecular distribution, and retention of nanotube properties [43]. However, it is known from theoretical studies that NT fractions of 40%–50% show "broader promise" [43].

Superior NT properties take form in many configurations. For example, the percolation limit (i.e., the minimal concentration of inclusion materials to maintain electrical conduction) can be as low as 1wt% because of the small dimension of the nanotube [43,44,45]. In addition to concentration, alignment of carbon nanotubes is also problematic at this time.

If you were to sell carbon nanotubes, what would be the content of your pitch? Perhaps tout that CNTs have tensile strength 10–100 times and an elastic modulus 5–10 times that of the best steel at one-fifth the weight; or that the electrical current capacity of CNTs can be 1000 times better than copper wire; or its thermal conductivity twice that of diamond; or how about its thermal stability up to 2800°C in vacuum; or that it can be a conductor, semiconductor, or insulator depending solely on its structure without doping or chemical modification. Is this too good to be true? Yes indeed, carbon nanotubes are the strongest, stiffest, and lightest materials known to science.

Performance of Advanced Carbon Composite Materials. More and more demand is placed on developing advanced materials that conform to the extreme environments encountered by tomorrow's aircraft. The development of nanocomposites that can greatly improve the strength, ablation resistance, stealth, thermal performance, and radiation hardening characteristics of materials is imperative for the next generation of aircraft. Radiation-hardened materials (and devices) have the ability to withstand damage or malfunctions caused by high-energy particles (cosmic rays, high energy protons and electrons, solar particle events, high-energy electrons from the Van Allen radiation belt, etc.) and electromagnetic radiation (ultraviolet, infrared, and x-rays) for applications in space, high altitude flight, or in or near nuclear reactors. Radiation and high-energy particles cause electron ejection, lattice displacements, depletion of minority carriers, recombination, and ionization. Neutrons, for example, can cause atomic displacement that disrupts the structure, and thereby reduces performance.

10.3.1 Types of Fibers, Whiskers, and Nanotubes

Fibers. The fiber filament offers one-dimensional reinforcement to a composite- or, it can be manufactured as a stand-alone material. It is with the carbon nanotube that we are most interested.

Non-Wovens. Another class of engineered fibrous materials is known as *non-wovens*. Non-wovens are materials that are not specifically woven or knit. The most prominent example of a non-woven material is paper; another is felt. Non-wovens, therefore, are engineered fabrics that take form as sheets or webs. Entanglement in non-wovens is generated mechanically, thermally, or chemically. Non-wovens act as absorbents, liquid repellants, flame retardants, and filters. Although non-wovens are capable of being bound internally by the action of adhesives, serrated mechanical needles, binders, or polymer melts, the strength of non-wovens does not compare to other fiber-reinforced materials [46].

Carbon Nanotubes. We are all familiar with the exceptional properties exhibited by the legendary eighth century Damascus swords—swords that possessed characteristic wavy banding patterns (called *damask*), incredibly sharp blades, and extraordinary mechanical properties [47,48]. Recent electron micrographic studies have revealed that the sword contained an ultrahigh content of carbon, nanowires (iron-based cementite), and carbon nanotubes [47,48]. The bundles of nanotubes and nanotube-encased cementite run parallel to the blade's surface. Softer steel is found between the nano-inclusions. The result is a blade that is strong and flexible. Following an etching process, wavy lines, consisting of nano-structured components sticking out from the blade's edge (e.g., sawtooth-like) became apparent.

Although Sumio Iijima of NEC Corporation is formally credited with the discovery of carbon nanotubes in 1993, two relatively unknown Soviet scientists, L. V. Radushkevich and V. M. Lukyanovich, were the first to formally describe hollow, nanometer dimension carbon tubes (**Fig. 10.7**) [49]. Since then, carbon nanotubes have been discovered and rediscovered until their "formal discovery" was finally attributed to S. Iijima et al. in 1991 [50,51]. Interestingly, the objective of carbon nanotube work over the past 50 years or so just before the 1990s and the advent of the *Nano Age* was specifically to reduce nanotube formation in order to prevent coking and fouling of catalysts used in other processes.

S. Iijima, and simultaneously D. Bethune et al. of IBM, discovered the first SWNTs in 1993 [52,53].

Elastic Modulus of Carbon Nanotubes. The theoretical Young's modulus is predicted to be from 1 to 5 TPa (that's terapascal!) whereas the best stainless steel and Kevlar values are in the range of 200 and 250 GPa, respectively. SWNTs show ca. 24% elongation of the materials at breaking as compared to 50% maximum for steel and around 2% for Kevlar. Because carbon nanotubes have such a low density, $1.3-1.4 \, \mathrm{g \cdot cm^{-3}}$, the specific strength of CNTs can be as high as 4.8×10^7 $\mathrm{N \cdot m \cdot kg^{-1}}$, much better than high carbon steel ($1.54 \times 10^5 \, \mathrm{N \cdot m \cdot kg^{-1}}$).

Fig. 10.7

The abstract from a seminal paper in the 1952 edition of the Journal of Physical Chemistry of Russia *describing carbon nanotubes with 50-nm diameter from CO at 600°C shown. The discovery by L.V. Radushkevich and V.M. Lukyanovich remained obscure due to the Cold War climate of the time. Sumio Iijima of NEC Corporation of Japan went on to formally describe multiwalled carbon nanotubes in 1991 [49,50,51].*

1952   ЖУРНАЛ ФИЗИЧЕСКОЙ ХИМИИ   т. XXVI, вып. 1

О СТРУКТУРЕ УГЛЕРОДА, ОБРАЗУЮЩЕГОСЯ ПРИ ТЕРМИЧЕСКОМ РАЗЛОЖЕНИИ ОКИСИ УГЛЕРОДА НА ЖЕЛЕЗНОМ КОНТАКТЕ

Л. В. Радушкевич и В. М. Лукьянович

Данная работа возникла в связи с электронно-микроскопическим изучением структуры различных адсорбентов, главным образом активных углей, графита и т. п. При исследовании препаратов углерода мы обратили внимание на сажу, получающуюся при разложении окиси углерода на железном контакте при температуре около 600° С. Так как сажа из окиси углерода изучалась адсорбционными методами и для нее была определена удельная поверхность по изотерме адсорбции, то представлялось интересным проверить эти результаты путем непосредственного измерения размеров частиц. Но уже первые наблюдения, сделанные нами [1], показали, что образующийся из СО углерод имеет весьма своеобразную структуру, до настоящего времени никем не описанную, и поэтому, естественно, наше внимание было перенесено на систематическое изучение этой структуры, а также на условия ее образования.

Young's modulus was determined from thermal vibration amplitude under TEM observation. T. Ebbesen of NEC Corporation measured the deflection of the tips of SWNTs as temperature was increased from 300 to 600°C. The amplitude of the vibration provided a means to assess the elasticity of the nanotubes—demonstrating high bending stiffness. Young's modulus is also calculated by an AFM method (**Fig. 10.8**) [54]. In this case, the AFM probe tip applies pressure to the distal end of a secured carbon nanotube and a correlation between applied force and material deflection is ascertained. Application of Euler's small deflection equation yields the Young's modulus

$$E = \frac{FL^3}{3\delta I} \tag{10.23}$$

where δ is the deflection of the cantilever (the carbon nanotube) at the point of contact, L is the length of the nanotube (between the fixed position and the point of force), F is the force applied by the AFM tip, and E is the Young's modulus as before. I is the areal moment of inertia of the cross section of the nanotube about its axis

$$I = \frac{\pi\left(r_o^4 - r_i^4\right)}{4} \tag{10.24}$$

where r_o and r_i are the outer and inner radii of the elastic cylinder, respectively [55].

FIG. 10.8 *Static mechanical measurement of Young's modulus for a single-walled carbon nanotube. A force supplied by an AFM probe tip is applied at the distal end of a fixed carbon nanotube. The deflection δ is measured and the value of E is calculated from Equation 10.24 above. The thickness of the wall of the SWNT is on the order of 0.34 nm (e.g., $r_o - r_i = 0.34$ nm for a SWNT).*

Why do carbon nanotubes have such extraordinary stiffness and strength? The strength is due to the strength and adaptability of the carbon–carbon bond, one of the strongest in nature. Stiffness is due to the phenomenon of hybridization. The carbon–carbon bonds in the hexagonal carbon rings are a blend of sp^2 and sp^3 character. When stressed, the bonds can rehybridize to accommodate the stress (increased bond angle by assuming more sp^3 character). Once the stress is removed, recovery to the original configuration occurs. The degree of rehybridization is a function of the degree of applied stress. How do these remarkable materials perform inside a polymer matrix? The answer depends on how well the CNTs are linked to the matrix elements.

Thermal Properties. *Carbon allotropes* show a variety of thermal properties. The thermal conductivity for diamond ranges from 900 to 2320 $W \cdot m^{-1} \cdot K^{-1}$, a value five times better than silver. Thermal conductivity in carbon nanotubes depends on the temperature and the large phonon mean free path. In 1999, Hone et al. determined that the thermal conductivity of a single SWNT rope at room temperature could vary from 1800 to 6000 $W \cdot m^{-1} \cdot K^{-1}$ [56]. Goddard et al. determined the thermal conductivity of a (10,10) tube approached 2980 $W \cdot m^{-1} \cdot K^{-1}$ [57]. How would a carbon nanotube fiber behave in a polymer composite?

10.3.2 Synthesis of Fibers and Nanotubes

Carbon nanotubes can be fabricated by laser vaporization (quartz tube at 1200°C with 1.2% CO–Ni alloy graphite target), carbon arc method (500 torr He, 20–25 V dc @ 50–120 A; no catalysts for MWNTs but Co, Ni, and Fe nanocatalysts are required for SWNT graphite targets), and chemical vapor deposition

(600–1000°C, argon atmosphere, carbon gas sources like methane, propylene, or acetylene and supported catalysts Co, Ni, or Fe on alumina or floating ferrocene and thiophene).

Polymer-Based Fibers. Polymer-based nanofibers are fabricated by several techniques: drawing, template synthesis, phase separation, self-assembly, weaving, melt spinning, and electrospinning [58]. Each technique, just like with anything else, has a complete set of advantages and disadvantages.

Melt Spinning. Melt spinning is the preferred method to manufacture polymeric fibers. The polymer is first melted and then pumped through a spinneret that has thousands of holes. Cooling takes place before the fibers are collected on a take-up wheel. Fibers can be stretched later to influence the structure and orientation. Typically, fibers on the order of 5–70 μm diameter are formed. Electrospinning is another fiber-generating technique that has gained popularity.

Drawing. Drawing involves the use of a pipette (a few micrometers in diameter) dipped into a viscoelastic solution (e.g., a droplet containing citrate molecules) and withdrawn rapidly (@ $100\,\mu m \cdot s^{-1}$). A nanofiber is pulled during withdrawal and deposited on a collector surface. Minimum equipment is required but drawing is a discontinuous process and scale-up would therefore become somewhat problematic [58].

Template Synthesis. Template synthesis is a generic means to produce nanowires from nearly any engineering material or combination of materials. Advantages include the manufacture of monodisperse fibers with specific diameter. Fiber length, however, is limited by the aspect ratio of the template pocket or template tube.

Phase Separation. In phase separation, nanofibrous poly(L-lactic)acid fibers (PLLA) are formed in a five-step process: (1) polymer dissolution of PLLA in tetrahydrofuran solvent to form a 1% w/v mixture, (2) gelation at –18°C, (3) solvent extraction with distilled water, (4) water removal and freezing, and lastly (5) freeze drying to produce a porous nanofibrous structure [59]. Advantages include tailored design of mechanical properties by manipulation of the polymer concentration, but a major disadvantage is that the process is limited to specific polymers.

Self-Assembly. *Self-assembly* is becoming exceedingly popular and is an excellent way to fabricate really small fibers but it often involves rather complex chemistry [58].

Electrospinning. Electrospinning provides an effective alternative to nanofiber manufacture and is exceedingly popular. Electrospun fibers with diameters of 250 nm have been a part of industrial materials for more than 20 years. Electrospinning employs electrostatic and mechanical force to spin fibers from the tip of a finely tuned orifice called a spinneret. The spinneret is held at a positive or negative charge by means of a DC power supply. The key principle involves the balance between the electrostatic charge and the surface tension of the polymer solution. Heat may be applied to keep the polymer in liquid form. When the electrostatic repelling force overcomes the surface tension, the viscoelastic liquid essentially spills forth from the spinneret to form finely divided but continuous filament. The filament is subsequently collected onto a rotating or stationary collector of opposite charge. The distance between the spinneret tip and the rotating collector is between 15 and 30 cm. Fibers ranging from 10 nm to 1 μm can be produced by this process [46].

The following polymer–solvent pairs (solvent in parentheses) are employed in electrospinning processes: nylon-6 and nylon-66 (formic acid), polyacrylonitrile (dimethylformamide, DMF), polyester (PET), (trifluoroacetic acid/dimethyl chloride), polyvinyl alcohol (PVA) (water), polystyrene (PS) (DMF/toluene), polyaramide (sulfuric acid), and polyimides (phenol) [46].

Commercial Synthesis of Carbon Fibers.   Carbon fibers (a.k.a. graphite fibers or carbon whiskers) can be formed under pyrolytic conditions that utilize relatively large catalytic particles or by chemical and physical modification of polymers such as polyacrylonitrile (PAN). For example, long chains of PAN are drawn and subsequently aligned during a drawing process to form continuous filaments. The filaments are then heated to temperatures above 300°C in air and oxidized to disrupt the hydrogen bonding [6]. The oxidized form of PAN is then placed in an inert atmosphere (e.g., argon) and heat treated at temperatures 1500–3000°C to stimulate graphitization. Carbon precursor polymeric materials heated between 1500 and 2000°C possess high tensile strength because of the turbostratic carbon structure (due to carbonization). Carbon precursors heated to temperatures between 2500 and 3000°C have a highly graphitic structure and exhibit Young's modulus near 500 GPa [6].

Synthesis of Multiwalled Carbon Nanotubes.   Carbon nanotubes (CNT) form under a variety of conditions—specifically, any environment in which there is a carbon source, anaerobic (pyrolytic) conditions, high temperature, and/or the presence of catalytic materials. Methods to fabricate carbon nanotubes include electric arc-discharge, laser ablation, solar furnace, and various forms of chemical vapor deposition (CVD). Arc-discharge and laser ablation methods are limited with regard to practical upscale as both are relatively energy intensive. The most practical means to generate nanotubes is afforded by chemical vapor deposition. Flame synthesis has become a most profitable commercial synthetic method.

The use of MWNT fillers to strengthen polymers and ceramics has become a $200 million dollar or more industry worldwide. Extremely effective means of producing pure multiwalled carbon nanotubes by a flame process are currently in use. For example, catalyst particles are placed on a support or seeded into a flame. Catalyst particles of Ni are generated by a thermal evaporation/condensation process with subsequent entrainment into a flame. The carbon source gas in many procedures consists of an ethylene/H_2 mixture to produce MWNTs [60]. Nanofibers are produced with a ternary gas mixture of $CO/C_2/H_2$ with Ni at 700°C [60]. Flame synthesis is capable of producing MWNTs in large quantities at a reduced cost [61]. A thermal evaporation technique is used to create the catalyst nanoparticles of Fe or Ni through gas condensation followed by entrainment into the flame. Each system yields consistent results, with CO/H_2 mixtures generally yielding single-walled nanotubes (SWNTs) with Fe while C_2H_2/H_2 mixtures usually produce multiwalled nanotubes (MWNTs) with Ni. A ternary gas mixture of $CO/C_2/H_2$ produces a better yield of nanofibers than either a CO/H_2 or C_2H_2/H_2 mixture at 700°C with a Ni catalyst [61]. A combination or perhaps a synergy between thermal—plus adsorbate-induced restructuring and adsorbate–particle steric factors affect particle structure and reactivity [61].

Commercial Synthesis of SWNTs.   Several processes are available to manufacture SWNTs at levels viable for commercialization. Two of the most effective include the flame process and the HiPco process.

Flame Synthesis of SWNTs.   A current commercially viable process to synthesize SWNTs is similar to the flame method discussed earlier for MWNT production. However, in this case, in place of a Ni catalyst with ethylene/H_2, Fe is used in a CO/H_2 gas mixture to generate SWNTs [60]. SWNTs have been detected in the postflame region (height-above-burner in millimeters, HAB) of a $C_2H_2/O_2/Ar$ flame at 50 torr with iron pentacarbonyl vapor. The iron pentacarbonyl decomposes in the flame to form the metal catalyst particles 5–10 nm in diameter in the 10- to –40-mm region of the flame [61]. After ca. 30 ms, incipient tubes were observed at 30-mm HAB, and nanotubes were detected after 40-mm HAB. Cluster (bundling) occurred between 40- and 70-mm HAB, and the growth rate was determined to be on the order of $10 \, \mu m \cdot s^{-1}$ [61].

SWNTs are also produced in oxy-fuel inverse diffusion flames of high fuel-rich stoichiometric mixture fraction [62]. Fuel rich regions are devoid of soot and polyaromatic hydrocarbons (PAH). In an inverse diffusion flame, oxygen is introduced into the center of the flame and fuel around the periphery of the oxidizer (**Fig. 10.9**). As a result, carbon nanotubes are formed in the periphery of the flame and are not exposed to oxygen. The SWNTs on the order of $1 \, \mu m$ in length have been produced under such conditions.

Companies such as Nano-C, Inc., of Westwood, Massachusetts have developed processes to produce SWNTs by flame technology without the need for expensive gases and external energy supplies with facilitative scalability [63]. The flame, generated by premixed benzene–oxygen, can be adjusted with high specificity to produce SWNTs with low levels of contamination (no amorphous carbons, spherical fullerenes, and MWNTs) [63]. A catalyst precursor, iron pentacarbonyl $Fe(CO)_5$, is added to the nonsooting flame to form SWNTs [63].

If no catalyst is added to the mix, fullerenes are produced. Equivalence ratios φ are defined as the fuel-oxidizer ratio divided by the stoichiometric fuel-oxidizer ratio corresponding to the conversion of all hydrocarbons into CO_2 and H_2O. Acetylene and ethylene, for example, form SWNTs in the equivalence ratio range of 1.7–3.8. Good quality nanotubes are formed under conditions of acetylene fuel and O_2 oxidizer ($\varphi = 1.6$), Ar diluent at 18 mol%, $Fe(CO)_5$ catalyst precursor forming metal concentration of 6000 ppm, 50 torr, $30 \, cm \cdot s^{-1}$ gas flow, HAB at 50 mm, and temperature profile at 1800 K @ 10 mm and 1500 K @ 80 mm [61].

HiPco Process.   Large-scale production of single-walled carbon nanotubes is demonstrated by the HiPco process—short for "high pressure carbon

Fig. 10.9	Left: *Inverse diffusion flame is depicted. The flow of fuel is introduced around an oxidizer (usually air) core. In this configuration, SWNTs do not pass through the oxidation zone of the flame. SWNTs are collected downstream on a device such as a cold finger.* Right: *HAB versus φ shows regions of metal formation, nanotube conception, nanotube aggregation, and soot formation.*

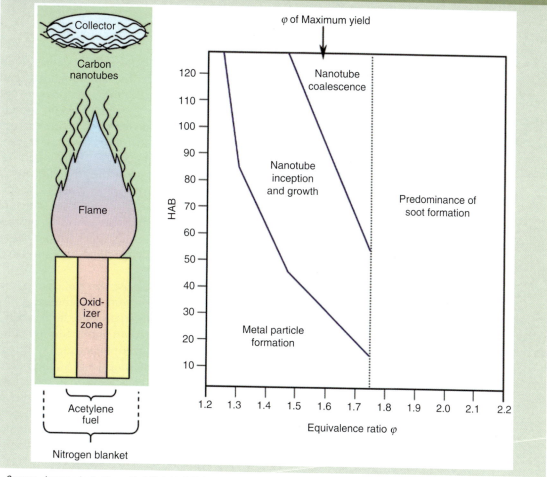

Source: Image adapted from M. J. Height, J. B. Howard, and J. B. Vandersande, Method and apparatus for synthesizing filamentary structures, US Patent 7335344, Publication Date: February 26 (2008).

monoxide." HiPco is a gas-phase CVD process operating under 30–50 atm and 900–1100°C. The carbon source material is flowing carbon monoxide (CO) and iron pentacarbonyl is the catalyst precursor material. Fe clusters are formed in situ from the decomposition (above 300°C) of the $Fe(CO)_5$ and subsequent condensation of reduced iron atoms in the gas stream. CO disproportionation occurs according to the following reaction (similar to the industrial Boudouard reaction)

$$2CO \rightarrow CO_2 + C_{SWNT} \qquad (10.25)$$

Purity on the order of 97-mol% SWNTs and 3-mol% Fe is achieved by this process [64]. Standard production conditions are as follows: 30 atm, 1050°C, CO volume flow rate @ 8.4 L·min^{-1} (or 250 standard liters per min mass flow; slm), and 0.25 torr Fe(CO)$_5$ @ 1.4 8.4 L·min^{-1} (42 slm) for 24–72 h. The rate of production of SWNTs under these conditions was ca. 450 mg·h^{-1} or 10.8 g·day^{-1} [64].

10.3.3   *Chemical Modification of Carbon Nanotubes*

Although nanotube fillers in metals seem to work quite well, the same cannot be said for nanotubes in polymer matrices. In order for carbon nanotubes to become effective components in a composite (as we know, pure tubes are not easy to integrate within the polymer matrix due to their apparent "atomic smoothness"), the tubes must be chemically modified to provide sites that are capable of cross-linking with the polymer. There exist two generalized approaches to chemical modification of carbon nanotubes: (1) covalent bonding to the structure of the nanotube in which the alteration of nanotube properties occurs due to disruption of the resonance-stabilized conjugate structure, and (2) modification of nanotubes by weak molecular forces—in which the basic structure and properties of the nanotube are not significantly perturbed. Numerous chemical procedures of both kinds are available. Tubes are either synthesized in pure form or purified post de facto, then chemically modified, and integrated within the polymer matrix.

Depending on the application, tubes of monodisperse size and chirality are ideally desired (much progress has been made in recent years to synthesize, purify, and/or separate tubes to achieve this objective). Applications range from anticorrosion paints (high purity and uniformity not required) to thin conductive polymer films (higher level of purity and orientation required) to potential structural materials exposed to high stress situations (e.g., aircraft wings) that require longer, uniform, and highly pure tubes. One technological challenge faced by composite engineers is addressing the solubility issues associated with nanotube bundles—in general, the larger the bundle, the worse the solubility. Bundled nanotubes, in any event, are not desirable due to the "sword-in-the-sheath" effect, for example, that tubes found in the middle of bundles would easily slide out of the engineering position because the only force acting to hold them in place are van der Waals forces, not strong enough to overcome high mechanical external shear stresses. The same issues confront the incorporation of MWNTs because internal tubes are not held as strongly (only by van der Waals forces).

Chemical modification.   Chemical modification techniques are summarized and discussed more intensely in *Introduction to Nanoscience*, chapter 9. Covalent derivatization occurs by well-known organic chemical methods: addition that includes cycloaddition, nucleophilic substitution, radical addition, electrophilic addition, hydrogenation, etc. Derivatization via intermolecular interactions occurs by well-known intermolecular reactions such as van der Waals, π-interactions, hydrophobic interactions, hydrogen bonding (once derivatized), and electrostatic methods, etc.

Covalent procedures include *oxidation* of terminal ends or sidewalls by treatment with strong acid at elevated temperature under reflux conditions. Addition of –CO_2H groups activate CNTs for further substitution. *Halogenation* occurs with F_2 by 1,2-addition or 1,4-addition at 150–400°C. Further derivatization is achieved with other well-known chemical processes such as alkylation with Grignard and organolithium agents as well as diols and diamines. Substitution with terminal amino groups render aminoalkylated CNTs soluble in water. Cycloaddition with Cl_2 occurs with application of CH_2Cl_2. *Hydrogenation* is achieved with Li metal and methanol in liquid NH_3, glow discharge, or atomic bombardment. Nanotube walls appear corrugated following hydrogenation. Stoichiometry of hydrogenated nanotubes is $C_{11}H$. Hydrogenated tubes are stable to 400°C. *Cycloaddition* occurs readily with carbenes, nitrenes, azomethine ylids (that form pyrrolidine fused rings), and nitrile imines. Pyrrolidine rings are useful centers for further reactions. Examples include Diels–Alder and dipolar cycloaddition. Radical addition methods by thermochemical, photochemical methods are also utilized for functionalization. Ozonolysis, mechanochemical functionalization, and plasma activation are other means to modify the chemistry of nanotubes.

Noncovalent nanotube functionalization methods include polymer wrapping. MWNT and SWNT polystyrene (PS), poly(vinyl alcohol) (PVA), polyhydroxyaminoether (PHAE), and epoxy thermoset composites are produced following solvent evaporation. Some problems include aggregation of tubes limits solubility. Low loading due to saturation at 1%–2% results. Other noncovalent methods include polymer grafting "To" and polymer grafting "From." Grafting to oxidized tubes is also practiced.

Chemical Modification of SWNTs. Although SWNTs have superior mechanical, thermal, and electrical properties, realization of their advantages in composites requires special preparation of the nanotubes. By chemical functionalization, the interfacial binding between the polymer host and the SWNT becomes enabled—hopefully without compromising the properties of the nanotube. Without such chemical preparation, the SWNT is not able to transfer exterior loads encountered by the composite.

Functionalization has other advantages. Neat SWNTs are not very soluble. Depending on the chemical nature of the functional group, SWNTs can be dispersed in aqueous as well as organic solvents. Homogenious dispersal of SWNTs within a polymer results in better overall mechanical performance of the composite.

Why is this important? SWNTs usually exist in the form of tightly bound ropes—aggregations of van der Waals bonded SWNTs producing bundles that can be on the order of micrometers in thickness. Such ropes are virtually insoluble in any solvent. The mechanical properties of polymers that contain SWNT ropes are not enhanced to the degree theoretically possible. For example, SWNTs located within the rope will have a tendency to slide over one another, similar to graphite planes. Chemical modification occurs by linking to SWNTs with strong covalent bonds or by linking with intermolecular interactions.

FIG. 10.10 Left: *Chemical modification of SWNT: (1) oxidation with strong acids, reflux, temperature > 100°C, to produce carboxylic acid reaction sites –COOH, (2) chemical reaction with acetyl amide to form reactive terminal group, (3) and (4) addition of free radical initiator and styrene monomer to initiate polymerization process, (5) propagation and subsequent termination of the polymerization process to form a polystyrene. Right: Chemically modified SWNTs aligned in an electric field are cross-linked by polystyrene chains. Anisotropic mechanical properties result from such alignment [43].*

Source: Image adapted from E. V. Barrera and K. Lozano, *Journal of Materials*, 32, 38–42 (2000).

Oxidation of SWNTs by treatment in strong acids at elevated temperatures under reflux conditions yields open-ended SWNTs stabilized by carboxyl groups (**Fig. 10.10**).

10.3.4   Carbon Nanotube Applications

From yarns to "beds of nails," there seems to be no limit in the applications of carbon nanotubes.

SWNT Bed of Nails.   SWNTs have been arranged in a configuration known as a "bed of nails" [65]. The SWNTs were packed in a hexagonal close-packed 2-D triangular lattice structure. The length of the tubes (membrane thickness) was on the order of 75 nm and the density of tubes in the membrane was ca. $10^{14} \cdot cm^{-2}$. The diameter of the tubes is tunable between 0.4 to 3 nm, with applications in batteries (intercalation of Li$^+$), Pt nanoparticle dispersion, and molecular transport with high sensitivity and flux [65]. The membranes were formed by application of the FIB (focused ion beam) mill perpendicular to a previously prepared neat fiber of SWNTs.

Carbon Nanotube Woven Fibers.   Perhaps one of the most amazing break-throughs in nanotechnology (and there have been many) occurred in the laboratory of R.H. Baughman et al. at the University of Texas at Dallas' NanoTech Center (**Fig. 10.11**). Baughman and his team have managed to weave 100 m (that's meters!) of nanotube composite fibers that exhibit better toughness than

FIG. 10.11   *Woven nanotube yarns from the Baughman group at the NanoTech Center at the University of Texas at Dallas. (a) Fibers/yarns are woven in a way similar to the way early humans did it—by rolling random strands until a fiber is formed. (b) Several weaving and braiding steps are required to make the yarn seen at the left. 10-nm diameter agglomerated MWNTs are woven into 30-nm diameter bundles. The 30-nm bundles form a yarn that is 1 µm in diameter. The next step is to form a plied yarn that is 6 µm in diameter. And finally, the yarn is braided to form the image in the figure to the left—40 µm or 0.04 mm in diameter. (c) A fabric is woven from the yarns, representing a higher level in the nanotube hierarchy. These accomplishments are absolutely amazing.*

Sources: (a, b) M. Zhang, K. R. Atkinson, and R. H. Baughman, *Science*, 306, 1358–1361 (2004); (c) A. B. Dalton, S. Collins, E. Munoz, J. M. Razal, V. H. Ebron, J. P. Ferraris, J. P. Coleman, B. G. Kim, and R. H. Baughman, *Nature*, 423, 703 (2003). With permission.

any natural or synthetic organic fiber [66]. The composites consist of carbon nanotubes in a PVA (polyvinylalcohol) matrix. Toughness is the energy required, in this case, to rupture a fiber. Spider silk, for example, is five times tougher than steel of the same mass. These carbon nanotube composite fibers are 20x tougher than steel, 17x as tough as Kevlar, and 4x as tough as spider silk.

Kaili Jiang et al. at the Department of Physics at Tsinghua University in Beijing developed a method to "self-assemble" yarns made of carbon nanotubes by drawing them out from super-aligned planar arrays [67]. The process is actually similar to the way wool was drawn and spun since antiquity. String was made in Paleolithic times (ca. 20,000 years ago) by rolling tufts of animal hair up and down one's thigh into a fiber. The length of the fiber was determined by the amount of hair added. The process also works with grass. Just take some grass and roll it back and forth in your hands. You'll end up getting a strand in which the individual blades of grass are intertwined.

Ray Baughman embellished the method by introducing a twist during the spinning to make them "multi-plied" and torque stabilized [68]. MWNT sheets can be spun at a rate of $7\,\text{m}\cdot\text{min}^{-1}$ from "nanotube forests." The rate can be increased to $30\,\text{m}\cdot\text{min}^{-1}$ if lower quality (wool) is required. The MWNT sheets initially form as an aerogel that eventually condense into sheets as thin as 50 nm. A process called densification is required to impart strength to the sheet. Densification is accomplished by twisting the yarn. In the aerogel phase, the material has no strength. Densification, in addition to strengthening the material, aligns the nanotubes along the axial direction of the spun yarn. The process can be applied to MWNTs, DWNTs, and SWNTs.

According to Baughman, the yarns possessed some quite remarkable properties: strength greater than 460 MPa, hysteretical deformation with 48% energy damping, tough as fibers in bullet-proof vests, and no degradation at 450°C following immersion into liquid nitrogen. Yarn strength increased with polymer infiltration as demonstrated by high creep resistance and high electrical conductivity test results [68]. The physical properties as one might imagine, would be enhanced if longer nanotubes can be utilized in the yarn. Baughman et al. have grown nanotube forests on the order of several millimeters in height (Zhang). For each kilogram of yarn, it is estimated that over 3 billion kilometers of nanotubes must be incorporated into the yarn. Apparently the rate-limiting step in the process is not the weave step. The rate-limiting aspect of the process is the rate of CVD growth of carbon nanotubes. Scale-up is not expected to be a problem for this nanomaterial. The strength of such sheets is better than the best strength that steel can afford. The power of nano! Truly incredible.

The nanotube yarns can be configured into knots. Unlike conventional yarns, knots in nanotube yarns do not lose strength [69]. The nanotube yarns (and yarns in general) reversibly dissipate mechanical energy. Twist-induced reinforcement of the strength of the yarn is approximated by Hearle's equation:

$$\frac{\sigma_Y}{\sigma_F} \approx \cos^2 \alpha (1 - k \operatorname{cosec} \alpha) \tag{10.26}$$

$$k = \sqrt{\frac{dQ}{\mu}} \Big/ 3L \tag{10.27}$$

σ_Y / σ_F is the ratio of yarn strength to the strength of the fiber; α is the angle of the helix with respect to the axis of the yarn; d is the fiber diameter; μ is the coefficient of friction, L is the fiber length and Q is the fiber migration length. Strength is less for a yarn than a fiber making up the yarn. According to this equation, the $(1 - k \cosec \alpha)$ term describes the transfer of tensile stress to transverse stress, for example, the locking of fibers [69]. To get the best performance, decreasing the nanofiber diameter d, increasing the fiber length L, and/or increasing the coefficient of friction μ need to be done.

Applications of the yarn are straightforward as structural reinforcement elements, but applications of the textile (sheet) are potentially quite unusual. These include polarized incandescent light, microwave plastic welding, electrodes (transparent, elastic), conducting appliques (decorative fabrics that are optically transparent, electrically conducting, and microwave absorbing), and flexible OLEDs [69].

Cellular Foams.   Condensed carbon nanotube free-standing foams with shock absorbing structural reinforcement behavior were synthesized by capillary-driven assembly [70]. Vertically aligned MWNT arrays were grown on patterned silica Si(100) substrates by CVD at 800°C. The nanotube arrays were then exposed to oxygen plasma in a glow-discharge chamber for 10 min. The MWNTs did not degrade in the chamber. The product was then immersed in solvents such as acetone, toluene, DMF, THF, or methanol. Evaporation of various liquids occurred at room temperature from the interstices to form the foams. Annealing at 800°C did not change the conformation. Foams with intricate open cellular structure were formed by the collapse and condensation of the nanotubes. These advanced materials demonstrated flexibility and good mechanical strength [70].

10.4   Organic Polymer Nanocomposites

What is it about nanomaterials (zero-, one-, and two-dimensional) that make them excellent inclusion materials in composites? How do nanomaterials, as opposed to fillers with larger size, affect the structure–property relationship between filler and matrix, for example, the interfacial area? Traditional polymer chemistry has taught us about the relationship between and among molecular weight, chain length, architecture, fillers, ordering, and functional groups and scaffolds with solubility, rheology, mechanical properties, and other chemical behavior. Also the effects of solvent, size, chemistry, and fabrication conditions are well understood. What then happens to such baseline polymers when nanomaterials are inserted? How are mechanical, physical, and chemical properties impacted by nanometer-sized inclusions of varying size, shapes, compositions, and orientations—whether in the form of particles, tubes, porous materials, discs, wires, or blocks?

Nanoscale building blocks are individually remarkable because of several reasons. They are free of defects, reactive, and have unusual physical properties. It is however, not a simple task to keep these remarkable properties intact in a macroscale composite.

10.4.1 Introduction to Polymers

We start this section on nanocomposites with polymers based on organic carbon. Polymer chemistry was introduced in *Introduction to Nanoscience* [21]. There are many kinds of polymers, but our discussion early on is limited to those based on organic carbon. Many inorganic materials are capable of polymeric reactions that connect building blocks into chain and network configurations. Examples of inorganic polymers include silicates, silicones, and glasses [11].

Brief History of Organic Polymers. The ancient Mayans used extract of rubber trees to make balls. Goodyear in 1839 improved the performance of rubber products by treating the polymer at 132°C in the presence of sulfur, a process called vulcanization that results in cross-linking of polymer molecules. In 1907, the first synthetic plastic, called Bakelite, was used as an insulating material due to its hardness and high heat resistance properties. In 1917, the molecular structure of cellulose was determined by an x-ray diffraction technique. In 1920, Nobel Laureate Hermann Staudinger revealed that polymers consisted of long chains made of molecular repeating units called monomers. Poly(vinyl chloride), or PVC, made the scene in 1927, polystyrene in 1930, nylon in 1938, and poly(ethylene) in 1941. Moldable high-temperature polymers were developed by 1970 and Kevlar was invented in 1971. Kevlar is an example of a fiber-reinforced polymer with extreme hardness and excellent temperature resistance (<300°C). By 1976, polymers and plastics overtook steel as the most widely used material in the United States.

Polymers offer an easily attainable, commercially viable route to stronger, lighter, less expensive, and versatile materials. Polymers are relatively easy to work with and do not require extreme conditions during their synthesis. Raw materials are readily available (at least until oil runs out). It is also relatively easy to embed inclusion materials within the matrix of a polymeric host. The rich chemistry of polymers allows us nearly an unlimited store of reactions to draw from to fabricate a composite.

Selection of the monomer, modification of side chains, various cross-linking strategies, materials concentration, and external control parameters such as temperature and pressure yield nearly an unlimited store of polymer materials. It is no wonder that polymers (composites made from them), and nanocomposites in particular, will establish a new class of materials with future promise in the field of high-performance materials.

Devising the order of presentation in this section was somewhat challenging. We decided that presenting polymer nanocomposites (PNCs) first, in a general sense, was a good idea because so many classes of nanocomposites are based on organic polymeric host matrices. There are, for example, clay–polymer nanocomposites, metal-polymer nanocomposites, and carbon fiber-reinforced polymer nanocomposites, just to name a few. In addition, several characteristics of polymers need to be understood before we proceed into sections describing other types of nanocomposites, and that terms associated continually with polymers need to be defined.

Polymerization. Although we will not spend too much time in this area, the basics are reviewed briefly. Polymerization is a process by which small chemical units (called monomers) are linked together by means of a chemical reaction. There are two basic fabrication schemes: addition (or chain) growth and step (or condensation) growth. Chain growth relies on the carbon–carbon double bond across which addition occurs readily, and condensation growth relies on molecules with bifunctional active groups that condense (react) to form water. Proteins are assembled via a condensation process to form peptide bonds. Chain growth proceeds in the presence of free radical, anionic, or cationic catalysts. Initiation, propagation, and termination (terms that are relatively self-explanatory) are the three general phases of this kind of polymerization.

If different kinds of monomers exist within the polymeric soup, a *copolymer* is produced that is similar in concept to a metal alloy [11]. *Block copolymers* (materials extremely important in nanotechnology) consist of ordered domains of repeating units that arise from different monomers. Polymeric templates and other nanostructures are based on block copolymer materials. *Polymer blends* consist of mixtures of preexisting polymers. *Linear polymers*, as the term implies, are made of long-chain hydrocarbons that can be linked. The degree of polymerization *n* is derived from the ratio between the molecular weight of the polymer and monomer, respectively. Chain length is approximated by the following relation:

$$L_{chain} = N_{CC} L_{bond} \sin\left(\frac{109.5°C}{2}\right) \tag{10.28}$$

where

N_{CC} is the number of carbon bonds
L_{bond} is the bond length (0.154 nm)

Placement of side groups can be *isotactic* (all on one side), *syndiotactic* (alternating), or *atactic* (placed irregularly) [11]. The kind of placement influences the overall structure of the polymer as well as the potential for cross-linking. *Branching* is a consequence of reaction conditions, monomer type, and side group steric factors.

Thermoplastic polymers consist of high molecular weight linear chains that are branched (no cross-linking) or unbranched that soften when heated. Conversely, thermoplastics freeze into a brittle state when cooled sufficiently. The chains of thermoplastic polymers interact primarily by van der Waals forces (polyethylene), dipole–dipole interactions, and hydrogen bonds (nylon) or aromatic stacking (polystyrene). Other examples include addition polymers such as polypropylene. Thermoplastic polymers are characterized by a parameter known as the glass transition temperature T_g, the temperature at which flexibility is attained. Thermoplastic polymers, in general, can be remelted.

Thermoset polymers, on the other hand, exhibit better mechanical properties due to higher cure temperatures and extended cross-linking. Most thermoset plastics contain a 3-D network of covalently bonded molecules. Curing of thermosets, therefore, is an irreversible process. Thermosets are used in structural applications and as adhesives. Bakelite (insulators), vulcanized rubber (tires), epoxy resins (graphite reinforced plastics), melamine resin (hard coatings), urea-formaldehyde (plywood), and polyimides (printed circuit boards) are some examples of thermoset plastic polymers. Thermosets exhibit

| Fig. 10.12 | *Organic polymers can be:* Top: *layered (laminate configuration);* Middle: *branched; or* Bottom: *highly cross-linked.* |

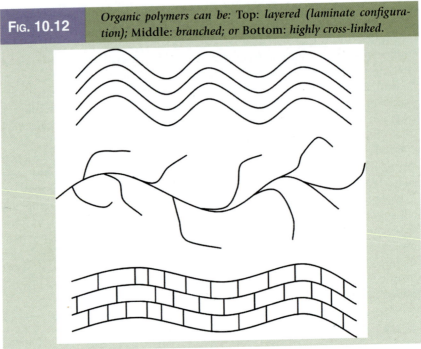

Source: C. R. Brooks and A. Choudry, *Failure analysis of engineering materials,* McGraw-Hill, New York (2002). With permission.

high thermal stability, rigidity, and dimensional stability, are resistant to creep and deformation under load, have high strength-to-mass ratios when compared to metals, and demonstrate high electrical and good thermal resistance properties.

Basic Types of Polymers.   The basic types of polymer structure are shown in Figure 10.12.

10.4.2   *Interfacial Area*

A significant resource that helped place this section into focus was provided by an excellent article by Linda S. Schadler et al. of the Rensselaer Polytechnic Institute (RPI) [28]. By definition, polymer nanocomposites contain fillers with sizes between 1 and 100 nm—along one or more dimensions. Such a small size is a major advantage to nanocomposites due to a dramatic increase in interfacial contact area over composites with larger fillers or fibers (please refer to example 10.1). The result is a composite matrix that has a significant percent of its volume directly associated with the interfacial area. Then, as stated earlier, the mechanical properties of the composite are dominated by the tightly bound interfacial phase. There are many other compelling reasons that structural perfection is achieved as the reinforcing elements approach smaller dimensions. Interestingly, the creation of such a large interfacial volume requires less filler material and not more (e.g., lower loading). For example, composites with micrometer-sized

EXAMPLE 10.1   *Surface Area and Fill Volume*

Conduct a first approximation of specific surface area of a spherical gold nanoparticle filling material of (a) $1\,\mu m$, (b) $100\,nm$, and (c) $1\,nm$ diameter in a generic polymer matrix of $V_{polymer} = 1\,cm^3$ if the fill factor ε is 4.5 vol%. The density ρ of gold is $19.3\,g\cdot cm^{-3}$.

Solution:

Surface area in m^2, $A_p = 4\pi r_p^2$

Specific surface area in $m^2 \cdot g^{-1}$, $S = \dfrac{A_p}{\rho V_p}$

Total surface area in m^2, $A_T = \sum N_p A_p$

Particle volume in m^3, $V_p = \frac{4}{3}\pi r_p^3$

Volume of fill factor, $V_{ff} = V_{polymer}\,\varepsilon = 1\,cm^3\;0.045 = 4.5 \times 10^{-8}\,m^3$

(a) Number of 1-μm particles $\rightarrow N_{1\mu m} = \dfrac{\varepsilon}{V_{1\mu m}} = \dfrac{4.5 \times 10^{-8}\,m^3}{5.2 \times 10^{-19}\,m^3} = 8.6 \times 10^{10}$ particles

Specific surface area: $S_{1\mu m} = \dfrac{3.1 \times 10^{-12}\,m^2}{19.3\,g\cdot cm^{-3}\left(\dfrac{100\,cm}{m}\right)^3 \left(5.2 \times 10^{-19}\,m^3\right)} = 0.31\,m^2 \cdot g^{-1}$

Total surface area: $A_{T-1\mu m} = \left(8.6 \times 10^{10}\right)\left(3.1 \times 10^{-12}\,m^2\right) = 0.27\,m^2$

(b) Number of 100-nm particles $\rightarrow N_{100\,nm} = 8.6 \times 10^{13}$ particles
Specific surface area: $S_{100\,nm} = 3.1\,m^2 \cdot g^{-1}$
Total surface area: $A_{T-100\,nm} = 2.7\,m^2$

(c) Number of 1-nm particles $\rightarrow N_{1\,nm} = 8.6 \times 10^{19}$ particles
Specific surface area: $S_{1\,nm} = 313\,m^2 \cdot g^{-1}$
Total surface area: $A_{T-1\,nm} = 277\,m^2$

What would the specific surface area be if the nanoparticle was not gold but rather titania $(\rho = 4.23\,g\cdot cm^{-3})$?

fillers require loadings on the order of 60 vol% as opposed to < 5 vol% with nanoparticles [28]. This allows for great flexibility in material design but simultaneously makes predictions about mechanical properties more difficult than for traditional composites.

Nanomaterials, nanoparticles in particular, have several inherent advantages due to their small size over their larger counterpart materials. First, collective surface area can be enormous leading to enormous interfacial linking; second, nanoparticles, due to their small size, have access to sites not available to larger particles; third, nanoparticles have higher surface-to-volume ratio leading to increased reactivity; lastly, nanoparticles can be more soluble. Therefore, the interfacial polymer chain structure, mobility, kinetics, and local chemistry all differ from the bulk polymer material. The structure of the interphasic region is different from the both polymer and the filler material. Obviously, if the interfacial region is homogeneous and covers a large surface area, the properties of the interphase region control the behavior of load transfer in the composite. Because nanomaterials have large surface area, the interphase, depending on the amount of filler material, is capable of dominating the mechanical behavior of the nanocomposite. There is still much to be learned about the interphase.

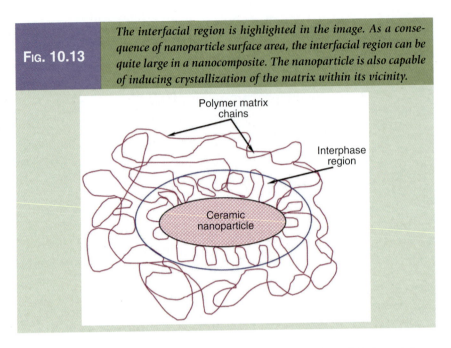

FIG. 10.13 *The interfacial region is highlighted in the image. As a consequence of nanoparticle surface area, the interfacial region can be quite large in a nanocomposite. The nanoparticle is also capable of inducing crystallization of the matrix within its vicinity.*

As stated previously, nanoscale fillers are relatively free of defects. Micrometer-sized fillers, on the other hand, scale with critical crack sizes that are known to cause mechanical failure [28]. Nanosized fillers would not be subjected to this form of mechanical failure. Therefore, a large volume of interfacial area that is radically different from that of the matrix host polymer material leads to better properties in this sense [28,71,72]. Crystallinity, mobility, chain conformation, molecular weight, chain entanglement density, charge distribution, the phases present, and cross-link density due to migration of small molecules where appropriate are some characteristics of the polymer at the interface that are impacted by the presence of the nanofiller, according to Schadler of RPI [28] (**Fig. 10.13**).

10.4.3 *Nanofilled Composite Design, Synthesis, and Properties*

Composite design requires consideration of the filler and the matrix. What feature do you wish to improve? Elongation? Toughness? Storage modulus? Tensile strength? Impact resistance? How does one go about fabricating a composite with enhanced properties? What is your choice of matrix? Choice of filler? Choice of chemical modification? Fiber length? Fiber diameter? Percent inclusion? All in all, chemical engineers must be ecstatic when confronted with all these options, permutations, and possibilities. It would indeed seem that anything one tries would result in a new material. We provide a few examples of nanocomposite design, synthesis, and performance. Although SWNTs, MWNTs, and carbon nanofiber fillers possess superior mechanical, thermal, and electronic properties than those of polymer hosts, their addition to polymers have not always improved the mechanical properties to the extent possible [66]. Many factors have to be considered.

Nanoparticles. There are many examples of nanofillers with enhanced interaction zone (large interfacial, larger interfacial volume, and better particle–polymer bonding) that show improved material properties. Addition of 1.5-vol% nanosilica to cross-linked polyethylene nanocomposites improved electrical endurance strength (breakdown strength and voltage endurance) by an order of magnitude compared to that of micrometer-sized fillers due to better local conduction and electron scattering [28,73]. The dielectric permittivity, space charge distribution, and dynamics also were significantly altered in the nanopolymer [28,73].

Nanofibers. A strong correlation between polymer ordering at the interface and reinforcement (Young's modulus) was determined for nanocomposites with multi-walled nanotubes in PVA [28,74]. The reinforced polymers showed significant increase in Young's modulus over controls. PVA is a polymer that is capable of "nanotube-induced" ordering [28,74]. The experimental data exceeded FOM and Mori–Tanaka expectations (ideal input parameters of straight aligned nanotubes, nanotube-matrix bonding, and nanotube properties) [28,74]. Interfacial effects are not included in these models. According to Jonathan N. Coleman et al. [74], the results are due to enhanced interfacial phenomena:

> Rather than acting as intrinsically stiffer reinforcing agents, our results suggest that the major role played by the nanotubes in improving the mechanical properties of composites is to nucleate an ordered polymer coating. It is the presence of this stiff ordered phase that dominates the reinforcement mechanism

The extent of the interface and other characteristics can be tuned by specific chemical modification of the carbon nanotubes. As a matter of fact, it was shown that the addition of MWNTs in semicrystalline polymers resulted in fibers with a higher specific energy to fracture than even Kevlar [75,76].

γ-Alumina Fillers. In poly(ethylene terephthalate) (PET) thermoplastic composites, γ-alumina nanofiller was able to alter the crystalline content of the PET matrix. At 1-wt% (<1-vol%) loading, the spherical γ-alumina was dispersed efficiently and nucleation and growth were inhibited relative to PET controls. With increased levels of filler and subsequent agglomeration, the γ-alumina stimulated nucleation and growth of PET, controlled the lamellar structure, and decreased the tendency of PET fibers to fibrillate (the tendency via abrasive action to develop smaller fibers) [28]. Although surface fibrillation is a good thing in adhesives, longitudinal fibrillation in composites may result in low composite strength under applied transverse stress [29].

What's the bottom line you ask? The structure and properties of the interface are different from the bulk but also are able to dominate the behavior of the nanocomposite leading to improved mechanical behavior [28].

10.4.4 *Enhanced Polymer Nanocomposites*

Interfacial Friction Damping. Interfacial shear, a mechanical phenomenon that is detrimental to high stiffness and strength in a composite, is beneficial to applications that require mechanical damping. Interfacial slippage may result between nanotubes in bundles, concentric nanotubes in MWNTs, and between nanotubes and the coordinating polymer chain [77,78]. Damping materials are required for acoustic (noise) and vibration suppression in dynamic systems [77]. Viscoelastic polymer-based damping systems suffer from compactness issues, unreliability, a high weight hit, low thermal conductivity and poor performance at elevated temperatures [77]. Nanotubes in polymers offer increased interfacial area contact (better load-mass transfer), high aspect ratio, and low mass density. These features allow for interfacial sliding of nanotubes for enhanced energy dissipation, seamless integration, and low-weight penalty without losing the mechanical properties of the nanotubes or structural integrity of the composite [77].

Tube-within-a-tube sliding was demonstrated by MWNT nanocomposites, but the required loading fraction was too high, ~50% to induce sliding. Apparently weak van der Waals forces between concentric tube layers are an ineffective means of dissipating energy [77,78]. Direct shear testing of epoxy thin films with dense packing of MWNTs revealed strong viscoelastic behavior with up to a 1400% increase in damping (ratio) compared to the baseline epoxy material. The mechanical properties (strength and stiffness) of the polymer remained intact. The interfacial sliding in MWNT composites was due to nanotube–nanotube interfaces [77].

J. Suhr et al. of Rensselaer Polytechnic Institute demonstrated the effect of temperature on SWNT inclusion/polycarbonate composites. They found that the NT-polymer interface activated at low dynamic strain levels of ~0.35% if the temperature was raised to 90°C. The increased mobility of the polymer chain backbone and thermal relaxation of "radial compressive stresses at the NT-polymer interface" were responsible for the enhanced damping, according to researchers [77]. Composite beams showed under dynamic cycling load tests greater than 1000% increase in the loss modulus of the nanocomposite with 2-wt% fraction SWNT filler. The damping increase was due to frictional sliding at the nanotube–composite interfacial contact areas without sacrificing the mechanical strength and stiffness of the composite [77].

Synthesis of SWNT-Polycarbonate Nanocomposites.   HiPco SWNT bundles were first oxidized by sonication in nitric acid to generate carboxylic acid groups on the SWNT walls. Carboxylic acid groups enhance exfoliation of SWNT bundles. The exfoliation, however, was not complete as 35-nm diameter rope bundles were detected after treatment with nitric acid. Modified SWNTs were transferred to tetrahydrofuran (THF), sonicated, and then mixed with polycarbonate (also in THF) in a predetermined proportion (to a maximum of 1.5 wt%) required for the nanocomposite (**Fig. 10.14**). Dipole–dipole intermolecular interactions between the carboxylic acid groups on SWNTs and the carbonate groups of the polymer initiated good dispersion within the matrix [77].

The composite mixture precipitated immediately after resonication and drop-wise introduction into anhydrous methanol (the antisolvent for polycarbonate) in a 1:5 THF:methanol ratio. Following filtration and drying under vacuum, the material was placed in a standard tensile test mold and hot-pressed (205°C). Baseline polycarbonate control specimens (without the nanotube filler materials) were made by the same procedure [77]. Uniaxial dynamic cyclic loading tests were conducted from –60 to 90°C, maintained for 10 min at each ten degree increment of temperature within that range. The storage and loss modulus of the nanocomposite test coupons and controls were calculated.

FIG. 10.14   *SWNT–polycarbonate component is depicted. The sliding occurs at the interface between the two materials of the nanocomposite.*

Source: Image adapted from J. Suhr, W. Zhang, P. Ajayan, and N. Koratkar, *Nano Letters*, 6, 219–223 (2006). With permission.

Interfacial Slip Measurements.   A novel "nano means" to measure interfacial slipping by AFM was accomplished by J.D. Whittaker et al. in 2006 [79]. A suspended single nanotube across a lithographically formed trough was used to study the nanotube–silicon dioxide adhesion forces (**Fig. 10.15**). Trenches, 300–400 nm in width and 40–50 nm in depth, were made of silica fabricated by e-beam lithography and dry etching techniques. Tubes were embedded in oxide

FIG. 10.15   *The Veeco probe used in this study possessed a nominal tip curvature of 20 nm. The spring constant of the probe was 0.24 N·m⁻¹. The tip curvature and spring constant of a MikoMasch probe were 10 nm and 0.14 N·m⁻¹, respectively. The probe was pushed onto the nanotube surface. Deflection of the tube and slippage from the anchor points on the trough were monitored.*

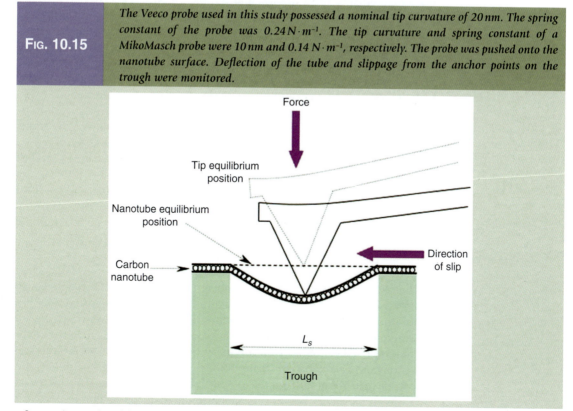

Source: Image adapted from J. D. Whittaker, E. D. Minot, D. M. Tannenbaum, P. L. McEuen, and R. C. Davis, *Nano Letters*, 6, 953–957 (2006). With permission.

by a CVD technique. Measures were taken to ensure that no oxide was applied to the suspended portion of the nanotube. Pressure was applied via an AFM tip to the nanotube (**Fig. 10.15**). The slipping force depends on the geometry of the nanotube–oxide interface by measuring the degree of slipping of different nanotube–oxide contact lengths. Measurements were conducted on nanotubes on bare silica and those embedded in silica.

Apparently, 7–8 nN applied tension are required to cause slippage along the SiO_2 surface, but an embedded nanotube requires 10 nN of applied tension before slippage occurs [79]. Also, the researchers found that nanotubes stretch up to 10%, confirming results obtained by other laboratories obtained by other means. A truly nanometer–scale method to measure interfacial structures!

Composite Toughening Mechanisms with CNFs. Modified carbon nanofibers (MCNFs) were chosen as the filler material due to their facilitated mass production and anisotropic shape. CNF diameter (50–200 nm) lies between SWNTs (a few nm), MWNTs (20–40 nm), and conventional carbon graphitic fibers (1–10 μm). It was found that although CNFs did not influence the modulus and strength of the matrix, they most certainly were able to improve the toughness of the composite (the elongation-to-break ratio) [80].

Ultrahigh molecular weight polyethylene (UHMWPE) was chosen as the polymer matrix because it is made of very long chains with MW in the millions, has a high entanglement density, and has high modulus and tensile strength. Long chains improve the load transfer to the polymer backbone by strengthening intermolecular interactions. As a result, UHMWPE has the highest impact strength of any thermoplastic. UHMWPE is also known as high-modulus or high-performance polyethylene.

Modification of CNFs was accomplished by oxidation of the surface with potassium chlorate and sulfuric acid to produce carboxylic acid and hydroxyl groups. The oxidized CNFs were then dispersed in octadecylamine $[CH_3(CH_2)_{18}NH_2]$ at 200°C for 20 h [80]. Bonds are formed between the amine group of octadecylamine and the carboxylic acid via a condensation reaction. These long hydrocarbon chains are then able to interact with the long chains of the polymer matrix.

The MCNF acted as solvent carriers in the stiff polymer matrix, and the short hydrocarbon chains grafted to the fibers were able to plasticize the UHMWPE chains in the interfacial region. This resulted in "interfacial flow under stretching and enhanced the elongation-to-break ratio" [80]. Untreated CNFs in UHMWPE did not demonstrate the enhanced toughness qualities [80].

UHMWPE nanocomposites exhibited increased toughness due to inclusion of chemically modified carbon nanofibers (MCNFs) [80]. The toughness of melt-pressed nanocomposite films containing 0.2 and 5-wt% MCNF was 10× that of the pure polymer. The nanocomposite at 118°C also showed a factor of 2× better toughness than the pure polymer composite. According to the researchers, the effect of the mobile hydrocarbon layers at the UHMWPE/MCNF interface was enough to overcome the barrier of high chain entanglements of the solid UHMWPE matrix that induced the toughening of the composite. The interfacial layer was on the order of 10–20 nm, and the induced interfacial flow resulted in a large elongation-to-break ratio (>500%) [80].

Enhanced Storage Moduli of SWNT-PS Nanocomposites. Storage modulus is the measure of energy stored in a sample during frequency cycle testing and the loss modulus is the energy lost. We stay with carboxylic acid–octadecylamine modification but this time choose SWNTs as our reinforcing material and change the polymer matrix to polystyrene—all part of the process of chemical engineering design and exploitation of numerous options. In this case, octadecylamine (ODA) or amino-terminated polystyrene were grafted onto oxidized SWNTs [81]. The SWNTs were purified in HNO_3 and their lengths shortened (why?). Following modification, the SWNTs were rendered soluble in dichloromethane and aromatic organic solvents. The chemically modified tubes were then mixed with polystyrene.

SWNTs were modified in $SOCl_2$ in DMF at 70°C for 24 h. Amino-terminated PS or ODA was mixed for 5 h at 90°C. The acid chloride groups on the modified SWNTs acted as linkers to the PS or ODA. The grafted SWNTs (1–3 wt%) were then mixed with PS in benzene, the mixture cast as a test coupon and dried at 90°C for 7 days [81].

The storage modulus of the polymer composite containing unmodified but cut SWNTs was smaller than that of the pure polymer. This is due to the lack of adequate dispersion of SWNTs in the PS-polymer and that interaction between the PS matrix and SWNT was poor. This resulted in slippage. The PS-SWNT showed better storage modulus than did the ODA-SWNT due to better compatibility between PS moieties. Both modified filler composites showed nonterminal behavior characteristic of good dispersion [81].

Flexible Electronic Applications. Pulickel Ajayan of the Department of Physics at Rensselaer Polytechnic Institute is one of the major contributors to nanotube composite research. One example, of many, of his work involves the development of a flexible hybrid composite structure using MWNT arrays in a poly(dimethylsiloxane) (PDMS) matrix [82]. The PDMS matrix (excellent conformal filling properties) with the dense aligned and patterned nanotube network retains robustness under high stress conditions (tensile and compressive strain) while maintaining electrical conductivity. SWNTs are first grown on a patterned SiO_2–Si substrate by CVD of ferrocene and xylene at 800°C. The array consists of circular domains of vertically aligned MWNTs 500 µm in diameter ~1 mm apart spread in a triangular 2-D array. PDMS is poured over the array and cured at 100°C for 10 h. The PDMS-composite is then peeled off the substrate to form a free-standing film ca. 100-µm thick [82]. Use of these materials is expected in strain gauges, field emission devices, and tactile and gas sensors. This is a different kind of composite—one that has minimal interfacial component compared to others we have discussed. It also has a different mission, different design parameters, different materials, different synthesis, etc.

Alumina/Magnetite and PMMA/PS Nanocomposites. Mechanical properties of alumina (Al_2O_3) and magnetite (Fe_3O_4) nanoparticles embedded in poly(methylmethacrylate) (PMMA) and polystyrene (PS) polymer matrices were analyzed by tensile testing, dynamic mechanical analysis (DMA), and nanoindentation. Overall, the mechanical tests showed that the nanocomposite systems possessed worse mechanical properties than the respective pure polymer systems. The cause of the poor mechanical performance of the nanocomposites was credited to poor interaction of the nanoceramic materials with the polymer matrix. In this case, it is not enough just to dump nanoparticles into a polymer matrix. Some consideration must be invested into chemical preparation of the inclusion.

Ceramic-Polymer Nanocomposites. The mechanical properties of poly(vinyl alcohol) (PVA) matrix nanocomposites with SiC or Al_2O_3 nanowires have demonstrated remarkable increase (to a level of 90%) in elastic modulus and increased strength above the native polymer at levels as low as 0.8 vol% [83]. S. R. C. Vivekchand et al. claim that the enhanced stiffness was due to the nanowire-induced crystallization of the polymer due to high aspect and surface-to-volume ratios of the nanowires and possible in-plane alignment of the nanowires. Increased strength was due to "significant pull-out of the nanowires" and the corresponding stretching of the matrix as a function of complete wetting by the polymer of the nanowires [83].

10.5   METAL AND CERAMIC NANOCOMPOSITES

10.5.1   *Metal Nanocomposites*

Metal Alloys. The formation of a bulk metal from nano-sized constituents improves the mechanical properties of the metal [84]. These materials are not necessarily nanocomposites. They are simply pure materials that are comprised of the nanoparticulate metal. Nanostructured copper films (27-nm grain size) formed by electrodeposition showed enhanced yield strength at 119 MPa [84]. Steel, on the other hand is hardened by the addition of carbon. In this case, nanoparticulate iron carbide filling the interstitial spaces between grains is able to prevent the mobility of dislocations and thereby prevent mechanical failure. In general, materials that contain smaller grains, that is, nanograins, have more grain boundaries that are able to block the propagation of mobile dislocations. The result of nanograin-embedded materials is increased strength with loss of ductility (e.g., materials are more brittle under tension) [84].

Alloys are composite materials—although the boundary is stretched somewhat because both components are metals—but they are composite materials nonetheless. In one example, nanoparticulate Fe and Cu fabricated by the process of ball milling were compacted and consolidated by the action of a tungsten-carbide die at 1 GPa for 24 h and then heat-treated at 400°C at a pressure of 870 MPa (sinter-forging) [84,85]. The grains of the alloy ($Fe_{85}Cu_{15}$ or $Fe_{60}Cu_{40}$) ranged from 20 to 70 nm. Overall, alloying at the atomic level obtained a homogeneous microstructure of nanocrystalline grains in the consolidated product [85]. Although its elastic modulus was similar to that of pure iron of grain size 50–150 μm, fracture occurred at a level of stress five times higher in the nanostructured alloy. The materials also exhibited very high strength under compression indicating low flaw populations with the attainment of nanoscale grain structure [85].

Metal Matrix Composites. These kinds of composite materials are complementary to cermets. In cermets, the matrix material is a ceramic and the filler is a metal. In metal matrix composites, the opposite case is in effect: the matrix is a metal and the filler is a ceramic. Aluminum–carbon fiber hybrids are examples of a metal matrix composites. Reinforcement can be continuous or discontinuous. Discontinuous matrix elements result in a nanocomposite with isotropic mechanical, thermal, or electronic properties. Monofilamental wires of SiC or carbon fiber are embedded within the metal oriented along one or another axis to provide anisotropic mechanical properties.

10.5.2 *Inorganic Nanofibers*

Silicon Carbide. SiC nanostructures have many potential applications ranging from high-strength, temperature resistance, and extremely hard materials. The Si–C bond ($318 \, kJ \cdot mol^{-1}$) is nearly as strong as the C–C bond ($346 \, kJ \cdot mol^{-1}$). The most stable form of SiC is the β-form that adopts a diamond-like structure.

SiC nanotubes and SiC-coaxial nanotubes can be formed from carbon nanotube (CNT) templates by a process called "shape memory synthesis," a template method that exploits a gas–solid phase reaction. The thermal decomposition of gaseous SiO results in the decomposition of Si on the carbon nanotube. Following heating, the Si-CNT is converted into a tube of SiC while releasing CO gas [86]. Similar structures were obtained by reacting carbon nanotubes with Si powder at 1200°C [87]. Sun et al. prepared β-SiC, biaxial SiC–SiO_2, and multiwalled SiC with intertube spacing between 3.5 and 4.5 Å [87]. Concentric SiC tubes were formed layer-by-layer via Si diffusion [87]. SWNTs were also used as the carbon source and template material to form SiC nanotubes. The thermally induced template reaction consisted of vaporized Si in N_2 or NH_3 carrier gas to form SiC nanofibers and nanotubes [88]. Depending on the duration, altered forms of SiC were generated [88].

SiC nanofibers can be synthesized from polymer blend precursors and then subjected to a melt spinning technique to form them into filaments. Polycarbosilane (PCS) with chemical structure [$-SiH(CH_3)CH_2-$], the SiC precursor, is first finely dispersed in phenol–formaldehyde (PF) resin, the carbon precursor, in the proportion of 3:7 (PCS:PF). Fibers were formed by continuous melt-spinning process. After soaking in acid for stabilization (e.g., to render them infusible) and heating at 1000°C, fibers were embedded in a carbon matrix (from the phenolic resin) and nanofibers were derived from the PCF precursor. Treatment with nitric acid removed the carbonized matrix and released nanofibers that were collected after filtration. The fibers were of amorphous structure with a large concentration of oxygen compounds—ca. 100 nm in diameter and over 100 μm in length. Further heat treatment at 1500°C transformed the amorphous oxygenated material into pure β-SiC [89].

Boron. Boron nanofibers 20- to 100-nm diameter form another class of important engineering fibers. Catalyst-free growth of α-tetragonal B tubes was achieved by pyrolysis of diborane at 630–750°C in 200 mTorr [90]. Boron is attractive because it has a high melting point (2200°C), low density ($2.34 \, g \cdot cm^{-3}$), high hardness (2900 Knoop hardness), and high Young's modulus (380–400 GPa). Boron imparts stiffness, toughness, and strength to nanocomposites [90].

Boron Nitride Nanotubes and Fibers. BN is isoelectronic with carbon, and therefore one would expect BN compounds to be flexible in terms of configurations.

For example, there are diamond and graphite analogs with BN. BN fibers can be prepared by decomposition of borazine fibers and boron oxide in nitrogen at 1800°C or from the thermal decomposition of cellulose fibers in the presence of boric acid at 1000°C. BN fibers find great utility in metal matrix composites.

10.5.3 Cermets

A cermet is a composite material composed of a ceramic and a metal. Ultimately, the objective of cermets is to generate a material with high temperature resistance, chemical and oxidation resistance and hardness (typical of ceramics) that also demonstrates some level of plastic deformation and high thermal conductivity typical of metals. In cermets, the metal (e.g., Ni, Co, or Mo, usually $V_f < 20\%$) serves as the filler material. Ceramic oxides often used for cermet matrices include alumina, tungsten carbide, borides, and various oxides like MgO and BeO [91]. Cermets also include a class of materials in which the outermost coating of a metal is modified to resemble a ceramic like layer. In the 1950s in Siberia, such layers were accidentally formed on drill bits during deep drilling operations. The cermet layer afforded the bit reduced friction, renewability and enhanced hardness, and smoothness—all due to nanoparticulate ceramic materials embedded at the metal surface. Common applications of cermet materials are found in turbine blades and other components exposed to high temperature in jet engines and ceramic to metal joints and seals.

A good example of a cermet is the tungsten carbide–cobalt cutting tool—a high hardness carbide ($0.6 < V_f < 0.9$) embedded in a ductile metal matrix like cobalt. The function of the carbide moiety is to cut hardened steel while the metal provides toughness and prevents crack propagation caused by particle-to-particle contact by the brittle carbide phase [11]. Both materials possess refractory properties and can therefore withstand high temperatures [11]. In WC-Co nanocomposites in general, the hardness of the cermet is inversely proportional to the size of the grain and fracture toughness is inversely proportional to the hardness—finer grain size leads to lower toughness. However, in nanostructured cermets, the mechanisms of strengthening are different due to the large volume fractions of grain boundaries. Therefore, it is expected that the fracture toughness of WC-Co composites will actually improve as grain size approaches nanodimensions [92]. Dramatic reduction in flaw size is expected. For example, with grain size less than 30 nm, flaws are expected to be a few nanometers in size, a condition beneficial for fracture toughness that is independent of the hardness of the material. Secondly, the increased amount of interfaces between WC nanograins and the Co metal enhances overall toughness by increasing the direction of crack path through the metal–ceramic interfaces rather than through conventional pathways. Lastly, an enhanced number of interfaces reduce the probability of conventional crack propagation (via dislocations) to that of sliding and short-range diffusion and other interface-dependent mechanisms [92]. These factors become more pronounced as both the ceramic and metal phase approach nanoscale [92].

Strengthened cermets ($V_f < 0.15$) consist of nano- to micrometer–sized particulates of oxide particles embedded within a metal. The oxides serve to strengthen the metal by preventing dislocation mobility—as discussed earlier.

The tensile strength, for example, can be increased by a factor of four times in aluminum if aluminum oxide ($V_f = 0.10$) is dispersed within the matrix of the metal [11].

10.5.4   Concrete

Concrete is an isotropic composite material that is comprised of cement. Cement is a siliceous binding material that is moldable when wet and hard when dried and a material that is able to bind other materials together. Cement (from the Latin *caementa* "stone chips for mortar"; from *caedra* "to cut down, chop, hew, fell") was a term used by the Romans to characterize materials that looked like small stones—*opus caementium*. Early concrete consisted of crushed stones held together with lime as a binder. Volcanic ash, pulverized rocks, and burnt lime were mixed together to form cement [93]. Hydraulic cements harden after mixing with water due to chemical reactions. Portland cement, the most common form of the material, consists of limestone, clay minerals, and gypsum treated at high temperatures to alter its chemistry.

Carbon Fiber and Nanotube-Reinforced Concrete.   Because carbon fibers, and now carbon nanotubes, have such unique mechanical properties, why not use them in concrete? Carbon fiber reinforced concrete (CFRC) has been in existence for several decades (**Figure 10.16**). CFRCs contain 3 to 15 vol% short carbon fibers embedded in concrete. The carbon fibers originate from either pitch-based or PAN-based carbon fiber. The application of carbon nanotubes in concrete is a natural evolutionary progression of CFRC technology. The modulus of elasticity (theoretically > 1 TPa), yield stress (between 20 and 60 GPa), and elastic strain properties (ca. 10%) of CNTs make them perfect candidates to bolster the already good mechanical properties of concretes in existence today [94].

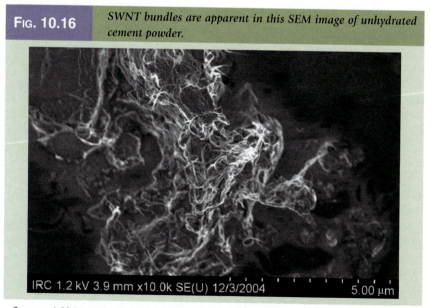

| FIG. 10.16 | *SWNT bundles are apparent in this SEM image of unhydrated cement powder.* |

IRC 1.2 kV 3.9 mm x10.0k SE(U) 12/3/2004   5.00 μm

Source: J. Makar et al., Canadian Research Council. With permission.

Commercial grade SWNTs were dispersed by application of sonication in isopropanol. Portland Type 10 was added under constant sonication to produce a 0.02-wt% SWNT-cement ratio. Following 4 h of mixing, the isopropanol was evaporated and the resultant mixture was ground into fine particles. No hydration was allowed during these stages of preparation. Mixtures of water, CNT-concrete, and superplasticizer were prepared and placed in elastic moulds and allowed to completely hydrate to form concrete nanocomposites. The unhydrated cement powder is shown in **Figure 10.16** [94].

SEM images show how SWNT fibers (actually bundles) were able to bridge cracks ca. 500 nm in width in the matrix. Do SWNTs actually reinforce the concrete? It appeared that the SWNTs were pulled into the cracks (pull-out). Such pulling may be an indication of poor adhesion of SWNTs to the cement matrix. However a fivefold increase in fracture toughness and corresponding flexural toughness indicates otherwise. The cured concrete nanocomposites were subjected to Vickers hardness testing and showed higher microhardness values. Vickers hardness, as indicated earlier, can be directly correlated to elastic modulus and the compressive strength of the cement [94].

Nanocomposite Cements.   Material performance can be enhanced with nanotechnology. I think we all agree that this is fast becoming a true statement. Concrete, as we now begin to understand, is very much a ceramic–ceramic nanocomposite. The addition of TiO_2 to cement, another ceramic, although not affording mechanical strength per se, is able to reduce the level of pollutants like NO_x, aromatics, ammonia, and aldehydes in the atmosphere near and around concrete structures like bridges and buildings [95]. The reduction in NO_x percentage over a period is absolutely dramatic—from 100% (normalized) NO_x to 0% in about 6 h time [95].

There are many ways to improve the performance of cements and concrete with nanotechnology. Some include enhancements during the curing cycle (e.g., hydration efficiency), and other ways utilize the incorporation of reinforcing elements (as given in the example given above with carbon nanotubes) and enhancement of bonding between preexisting components of the cement. Some of the most significant developments are expected to be in the areas of superplasticity and high-strength fibers for energy-absorbing capability [96]. The addition of nanosilica is part of a trend to incorporate smaller and smaller particles (from fly ash to fumed silica and now nanosilicas) into ocncretes [96].

Nanoparticles lead to increased flexural and compressive strength in various formulations of cement and concrete because of the following: (1) Excellent dispersion of nanoparticles increases the viscosity of the mixture that helps suspend cement grains and aggregates allowing for better "workability;" (2) nanoparticles serve to fill voids between grains that immobilize "free-water" (known as the filler effect); (3) well-dispersed nanoparticles act as centers of crystallization for cement hydrates and this process accelerates the hydration mechanism; (4) nanoparticles catalyze the formation of smaller crystals of $Ca(OH)_2$, calcium silicate hydrates (a.k.a. CSH); (5) nanosilicates promote the *pozzolanic reaction* that consume $Ca(OH)_2$ to form more CSH; (6) nanoparticles enhance the "contact zone" resulting in better bonds between aggregate and cement phases (e.g., enhanced surface area effects); and (7) nanoparticles add crack arrest and interlocking effects between slip planes. This results in better toughness and shear, tensile, and flexural strength [96].

Nanotechnology is expected to stimulate research in many areas: accelerated hydration, mechanico-chemical activation of cement ingredients, nanoparticle-reinforced binders, nanoengineered internal bonding, nacre-like structures, superplasticizers for better workability, binders with humidity-sensitive moisture delivery to avoid cracking, reduced level of binder materials, self-healing materials, and self-cleaning, purifying materials [96].

10.6 Clay Nanocomposite Materials

Nanocomposites composed of clays and polymers demonstrated improved mechanical and thermal properties over control samples [97,98]. The enhanced properties are related to the degree of dispersion and the degree of exfoliation (dispersion of platelets) of the tactoids (clay platelet stacks) in the polymer matrix [99]. The degree of exfoliation is dependent on the coating of the platelet (usually an organic monolayer) and the solvent. Sodium montmorillonite is a crystalline aluminosilicate with platelet morphology. In this section, we apply clays as the minority inclusion material. We could have easily placed clay-polymer nanocomposite text in either the ceramic section or the polymer section. Because they form such unique nanocomposites, we decided to place them in a special section dedicated to them.

Clays are naturally occurring silicate nanomaterials composed of fine particles that interact dramatically with water—giving it a variable range of plasticity. When hardened, clay becomes a ceramic. There are a few classes of natural clays of importance to scientists: kaolinite, montmorillonites, illites, bentonites, and chlorites. Laponite is a synthetic smectite clay that resembles the natural clay hectorite. Synthesis is accomplished by combining salts of Na, Mg, and Li with sodium silicate. The resultant powder consists of nanoparticulates 0.92-nm thick and 25-nm across. Laponite forms good composites with carboxymethylcellulose and is used in making a variety of glazes.

In clays, the bonds between atoms in a layer are very strong but the bonds between layers are much weaker. The clay layer consists of three subunits consisting of an octahedral central layer (Al^{3+}) sandwiched between two tetrahedral layers consisting of silica (silicon and oxygen). The layer may be charged if other cations are substituted in its structure (e.g., Mg^{2+} for Al^{3+}) to generate a negative charge or substitution of Si^{4+} with Al^{3+} to yield an overall positive charge. Negative charges can be neutralized by the introduction of hydrated Na^+, K^+, or Ca^{2+} in the interlayer region of the clay [100] (**Fig. 10.17**).

Clay tactoids, stacks of platelets, in contact with monomers are able to form three distinct nanocomposites (**Fig. 10.18**). In the first configuration, the tactoid remains intact because it is coordinated all around by the polymer. In the second one, intercalation occurs in which the polymer is able to squeeze in between the nanoscale thin layers of clay. In the third configuration, complete exfoliation of the tactoid occurs as individual platelets are coordinated within the polymer matrix.

10.6.1 *Polypropylene–Clay Nanocomposites*

In order for clay platelets to disperse in a hydrophobic polymer, the surface of the clay must be modified. Clays naturally are water soluble polar polymeric

A schematic rendition of the structure of a generic clay with polymer intercalation is depicted. Clays are known to intercalate water and swell to enormous dimensions. For a sodium montmorillonite clay, the primary structure consists of a layer of aluminum hydroxide between two layers of silicate. The overall chemical structure is $Na_{0.33}[(Al_{1.67}Mg_{0.33}) Si_4O_{10}(OH)_2] \cdot nH_2O$. The platelets have a net negative charge and are weakly bound via electrostatic forces with an interlayer of hydrated Na^+, Li^+, Ca^{2+}, Fe^{2+}, or Mg^{2+} cations. The thickness of each plate is 1 nm but laterally can achieve dimensions of 1 µm [99]. Stacks of platelets are called tactoids. The space between layers is called the gallery. The particles are hydrophilic but the surface can be altered through chemical substitution to render the particles hydrophobic (via an organic cation substitution reaction) [99]. Following surface modification, the tactoids are completely exfoliated in chloroform or trichloroethylene. The surface modified layer is not shown in the image.

FIG. 10.17

Intercalated nanocomposite structure

Tetrahedral silicate layer
Octahedral aluminum layer
Tetrahedral silicate layer

Polymer precursor

Charge stabilizing cations

materials. Modification of the surface can occur with application of alkyl ammonium chlorides in which there are anywhere from a few to 18 carbons in the chain. The intercalated chains are able to expand the gallery spacing to 2.4 nm [101]. C_{18} hydrocarbon derivatized montmorillonite is called C_{18}-mmt. Addition of maelic anhydride modified polypropylene (PP-MA) compatibilizer to C_{18}-mmt results in intercalated clays with gallery spacing greater than 3 nm. Exfoliation of the clay was accomplished with addition of the PP-homopolymer. The best-case scenario requires that the tactoids become completely exfoliated (e.g., no correlation between platelets). If this is the case, interfacial area interaction with the polymer is maximized. The chemistry involved in this process is obviously complex. Physical parameters such as temperature also influenced the microstructure of the nanocomposite.

Inorganic fillers such as talc, calcium carbonate, glass fibers, glass beads, and mica are traditional means of improving mechanical properties of polymers. Larger particles act as stress concentrators that initiate cracks and degrade strength and toughness. The idea behind the use of nanoparticles is to enhance stiffness and strength while simultaneously improving toughness [101]. Clay platelets appeared to be a perfect solution because of their small size, large aspect ratio, and ability to be oriented within the polymer. In addition, enhanced

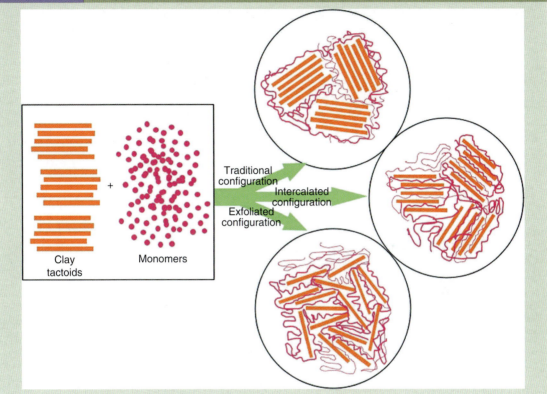

FIG. 10.18

A basic clay–polymer nanocomposite is displayed. Three possible configurations are possible: Top: a conventional nanocomposite in which the polymer envelopes the entire clay tactoid; *middle:* intercalation of the polymer between the lamellae of the clay offers a new set of enhanced properties; *and bottom:* exfoliation of the tactoid occurs in which randomly oriented clay platelets are individually enveloped in the polymer matrix. Each configuration is expected to exhibit different mechanical properties.

properties could be achieved with a lower volume fraction of filler material. This was demonstrated by Young's modulus of polypropylene-composite materials (**Fig. 10.19**).

PP-MA-mmt nanocomposites have also demonstrated a flame-retardant character. The reason for this remarkable attribute is the presence of a char layer on the surface. The char layer was found to have a high concentration of mmt [101].

10.6.2 Montmorillonite Clay Nanocomposites

Organically modified nanometer scale layered magnesium aluminum silicate platelets are added to polymeric materials to improve mechanical and physical properties. The silicate platelets are 1 nm in thickness and 70–150 nm across. The platelet surface is hydrophobically modified with a monolayer that allows for complete dispersion and miscibility within thermoplastic matrix polymeric materials such as acrylics, styrene/acrylics, and vinyl acetate copolymers. Addition

FIG. 10.19 *Young's modulus of polypropylene composite materials as a function of filler wt%. The nanocomposite showed higher modulus of elasticity with less filler material—the power of nano [101].*

of the clay enhances flexural and tensile modulus, lowering the coefficient of linear thermal expansion of the polymer while increasing the flame-retardant capacity of the composite.

Interfacial Area in Clay–Polymer Nanocomposites. Advantage of interfacial area is demonstrated clearly in clay–polymer nanocomposites. Layered silicates in polymers have interfacial surface area on the order of $700 \, m^2 \cdot g^{-1}$. Add to this that the distance between 1-nm thick plates is a mere 7 nm—and this achieved with a volume loading of 7 vol%. The morphology and resultant physical properties of the composite are significantly influenced, if not dominated, by the platelets.

10.6.3 *Halloysite Nanotube Clay Composites*

Halloysites nanotubes are another form of clay with the chemical composition of $Al_2Si_2O_5(OH)_4$. The tubes have diameter less than 100 nm (range from 40 to 200 nm) and length ranging from 500 nm to 1.2 µm. Halloysites are formed by the surface weathering of aluminosilicate minerals [102]. According to Paul Schroeder, in the figure, images of halloysites are depicted. Well-formed halloysite tubes occur naturally in surface hot springs (hydrothermal) environments. The mineral is made from a 0.7 nm structure comprised of silica tetrahedra and water sheets. The sheet misfits cause the layers to curl into tube-shaped morphology. The inside diameter averages 150 nm. The samples depicted in the figure were acquired from northwestern Turkey (**Fig. 10.20**). In 2007, halloysite nanotubes (HNT) were successfully functionalized in polypropylene at levels of 3%–5%. The modulus of the nanocomposite was measured to be twice that of the polymer alone [102]. In addition to an ingredient in polymers, halloysites can also be coated with metal for electronic applications.

In another study, halloysite nanotubes were made to react with 2,5-bis (2-benzoxazoyl)thiophene (or BBT) for a role as reinforcing elements in

FIG. 10.20	*Natural halloysite clays are depicted. Notice the hollow features at the end of the nanotubes.*

Source: Paul A. Schroeder, Department of Geology, University of Georgia. With permission.

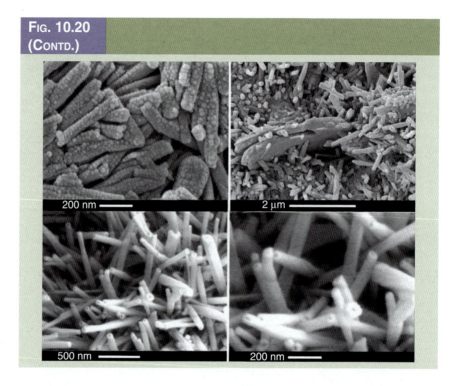

Fig. 10.20 (Contd.)

polypropylene (PP) nanocomposites [103]. Clays tend to interact with organic materials via electron transfer mechanisms. This was confirmed in the BBT–HNT couple. Because of this, it is possible to enhance the interfacial properties of clay–polymer composites. BBT is capable of donating electrons to the positively charged HNTs and is therefore the interfacial modifier of the halloysite. Fibril formation of BBT is induced in the presence of HNTs in the nanocomposite. BBT fibrils of high crystallinity are formed under melt shearing. BBT-HNT-PP nanocomposites showed substantially increased tensile and flexural properties due to the presence of the crystallites [103]. HNTs were also complexed with polyvinyl alcohol (PVA [104]. Solutions of PVA and HNTs were prepared with ultrasonication. Particle size and distribution of HNTs was independent of the ratio between HNT and PVA; however, too much HNT depressed the crystallinity temperature (at T_g).

References

1. P. H. T. Vollenberg and D. Heikens, Particle size dependence of the Young's modulus of filled polymers: I. Preliminary experiments, *Polymer*, 30, 1656–1662 (1989).
2. M. Z. Rong, M. Q. Zhang, S. L. Pan, B. Lehmann, and K. Friedrich, Analysis of the interfacial interactions in polypropylene/silica nanocomposites, *Polymer International*, 53, 176–183 (2003).
3. K. I. Winey and R. A. Vaia, eds., Polymer Nanocomposites, *MRS Bulletin*, 32, Single Issue, April (2007).

4. P. Podsiadlo, A. K. Kaushik, E. M. Arruda, A. M. Waas, B. S. Shim, J. Xu, H. Nandivada, B. G. Pumplin, J. Lahann, A. Ramamoorthy, and N. A. Kotov, Ultrastrong and stiff layered polymer nanocomposites, *Science*, 318, 80–83 (2007).

5. C. R. Brooks and A. Choudry, *Failure analysis of engineering materials*, McGraw-Hill, New York (2002).

6. Carbon Fiber, en.wikipedia.org/wiki/Carbon_fiber (2008).

7. R. Bacon, Growth, structure and properties of graphite whiskers, *Journal of Applied Physics*, 31, 283–290 (1960).

8. Boeing 787, en.wikipedia.org/wiki/787_Dreamliner (2008).

9. M. Ohring, *Materials science of thin Films: Deposition and structure*, Academic Press, San Diego, CA (2002).

10. Flint, en.wikipedia.org/wiki/Flint#cite_note-0 (2008).

11. J. F. Shackelford, *Introduction to materials science for engineers*, 4th ed., Prentice-Hall, Upper Saddle River, NJ (1996).

12. J. F. Shackelford, *Introduction to materials science for engineers*, 6th ed., Prentice Hall, Upper Saddle River, NJ (2004).

13. M. S. Bakshi, P. Thakur, S. Sachar, and T. S. Banipal, Synthesis of nanocomposite gold-semiconductor materials by seed-growth method, *Materials Letters*, 61, 3762–3767 (2007).

14. K. S. Lee, I. H. Kim, T. S. Lee, B. Cheong, J. G. Park, J. G. Ha, and W. M. Kim, Dielectric confinement and the surface plasmon damping in Au: semiconductor nanocomposite thin films, *Surface and Coatings Technology*, 198, 51–54 (2005).

15. A. K. Sharma and B. D. Gupta, Metal-semiconductor nanocomposite layer based optical fibre surface plasmon resonance sensor, *Journal of Optics A: Pure and Applied Optics*, 9, 180–185 (2007).

16. M. Valdésa, M. A. Frontini, M. Vázquez, and A. Goossens, Low-cost 3D nanocomposite solar cells obtained by electrodeposition of $CuInSe_2$, *Applied Surface Science*, 254, 303–307 (2007).

17. J. Won, S. K. Chae, J. H. Kim, H. H. Park, Y. S. Kang, and H. S. Kim, Self-assembled DNA composite membranes, *Journal of Membrane Science*, 249, 113–117 (2005).

18. L. S. Litvinova, A. V. Gribanov, M. V. Mokeev, and V. N. Zgonnik, Physicochemical properties of water-soluble fullerene C60-carbohydrate composites, *Russian Journal of Applied Chemistry*, 77, 438–440 (2004).

19. V. M. Rotello, ed., Nanoparticles: Building blocks for nanotechnology, Plenum Press, New York (2003).

20. N. K. Cygan, A. Rice-Ficht, and K. Carson, Fabrication of protein nanocomposites, Presentation, Texas Institute for Intelligent Bio-Nano Materials and Structured for Aerospace Vehicles, TiiMS 3rd Annual Meeting, August (2005).

21. G. L. Hornyak, J. Dutta, H. F. Tibbals, and A. K. Rao, *Introduction to nanoscience*, CRC Press, Boca Raton, FL (2008).

22. Stress, en.wikipedia.org/wiki/Stress_(physics) (2008).

23. Tensile strength, en.wikipedia.org/wiki/Tensile_strength (2008).

24. R. Kelsall, I. Hamley, and M. Geoghegan, *Nanoscale science and technology*, John Wiley & Sons, Ltd., Chichester (2005).

25. Hardness, en.wikipedia.org/wiki/Hardness_%28materials_science%29 (2008).

26. Vickers hardness, en.wikipedia.org/wiki/Vickers_hardness_test (2008).

27. R. E. Stoller and P. M. Rice, Correlation of nanoindentation and conventional mechanical properties, *Materials research Society Symposium–Proceedings*, MRS (2000).

28. L. S. Schadler, L. C. Brinson, and W. G. Sawyer, Polymer nanocomposites: A small part of the story, *Journal of Materials*, 59, 53–70 (2007).

29. D. L. Burris and W. G. Sawyer, Improved wear resistance in alumina-PTFE nanocomposites with irregular shaped nanoparticles, *Wear*, 260, 915–918 (2006).

30. D. L. Burris and W. G. Sawyer, Tribological sensitivity of PTFE/alumina nano-composites to a range of traditional surface finishes, *Tribology Transactions*, 48, 147–153 (2005).

31. W. G. Sawyer, K. D. Freudenberg, P. Bhimaraj, and L. S. Schadler, A study on the friction and wear behavior of PTFE filled with alumina nanoparticles, *Wear*, 254, 573–580 (2003).

32. S. Yu, P. Hing, and X. Hu, Thermal expansion behaviour of polystyrene-aluminium nitride composites, *Journal of Physics D: Applied Physics*, 33, 1606–1610 (2000).

33. C. A. Harper, ed., *Handbook of materials and processes for electronics*, McGraw-Hill, New York (1970).

34. J. K. Stanley, *Electrical and magnetic properties of metals*, American Society for Metals, Park, Ohio (1963).

35. H. Kim, H. T. Hahn, L. M. Viculis, S. Gilje, and R. B. Kaner, Electrical conductivity of graphite/polystyrene composites made from potassium intercalated graphite, *Carbon*, 45, 1578–1582 (2007).

36. D. A. Jaworske, R. D. Vannucci, and R. Zinolabedini, Mechanical and electrical properties of graphite fiber-epoxy composites made from pristine and bromine intercalated fibers, *Journal of Composite Materials*, 21, 580–592 (1987).

37. S. Watcharotone, D. A. Dikin, S. Stankovich, R. Piner, I. Jung, G. H. B. Dommett, G. Guennadi, S.-E. Wu, S.-F. Chen, C.-P. Liu, S. T. Nguyen, and R. S. Ruoff, Graphene-silica composite thin films as transparent conductors, *Nano Letters*, 7, 1888–1892 (2007).

38. E. Baeuerlein, P. Behrens, and M. Epple, eds., *Handbook of biomineralization: Biomimetic and bioinspired chemistry*, Wiley-VCH, Weinheim (2007).

39. J. R. Capadona, K. Shanmuganathan, D. J. Tyler, S. J. Rowan, and C. Weder, Stimuli-responsive polymer nanocomposites inspired by the Sea Cucumber Dermis, *Science*, 319, 1370–1374 (2008).

40. B. Pokroy, E. Zolotoyabko, and N. Adir, Purification and functional analysis of a 40kD protein extracted from the *Strombus decorus persicus* mollusk shell, *Biomacromolecules*, 7, 550–556 (2006).

41. J. D. Currey, P. Zioupos, P. Davies, and A. Casinos, Mechanical properties of nacre and highly mineralized bone, *Proceedings of the Royal Society of London Series B-Biological Science*, 268, 107–111 (2000).

42. Y. Yang, Y. Lu, M. Lu, J. Huang, R. Haddad, G. Xomeritakis, N. Liu, A. P. Malanoski, D. Sturmayr, H. Fan, D. Y. Sasaki, R. A. Assink, J. A. Shelnutt, F. van Swol, G. P. Lopez, A. R. Burns, and C. Jeffrey, Brinker functional nanocomposites prepared by self-assembly and polymerization of diacetylene surfactants and silicic acid, *Journal of American Chemical Society*, 125, 1269–1277 (2003).

43. E. V. Barrera, M. L. Shofner, and E. L. Corral, Applications: Composites, In *Carbon nanotubes: Science and applications*, M. Meyyappan, ed., pp. 253–275, CRC Press, Boca Raton, FL (2005).

44. V. A. Davis et al., Phase behavior and rheology of SWNTs in superacids, *Macromolecules*, 37, 154–160 (2004).

45. J. Fournier et al., Percolation network of polypyrole in conducting polymer composites, *Synthetic Metals*, 84, 839–840 (1997).

46. L. C. Wadsworth and K. Duckett, *Nonwovens science and technology II*, www.engr.utk.edu/mse/Textiles/Nanofiber%20Nonwovens.htm (2004).

47. M. Reibold, P. Paufler, A. A. Levin, W. Kochmann, N. Pätzkel, and D. C. Meyer, Materials: Carbon nanotubes in an ancient Damascus sabre, *Nature*, 444, 286 (2006).

48. M. Inman, Legendary swords' sharpness, strength from nanotubes, study says, *National Geographic News*, news.nationalgeographic.com/news/2006/11/061116-nanotech-swords.html (2006).

49. L. V. Radushkevich and V. M. Lukyanovich, O strukture ugleroda obrazujucegosa pri termiceskom razlozeni okisi ugleroda kontakte, *Zhurnal Fizicheskoi Khimii*, 26, 88–95 (1952).

50. S. Iijima, Helical microtubules of graphite carbon, *Nature*, 354, 56–58 (1991).

51. M. Monthioux, Who should really get the credit for the discovery of carbon nanotubes? *Carbon*, 44, 1621–1624 (2006).

52. S. Iijima and T. Ichihashi, Single-shell carbon nanotubes of 1 nm diameter, *Nature*, 363, 603–605 (1993).

53. D. S. Bethune, C. H. Kiang, M. S. De Vries, G. Gorman, R. Savoy, J. Vazquez, and R. Beyers, Cobalt catalysed growth of carbon nanotubes with single-atomic-layer walls, *Nature*, 363, 605–607 (1993).

54. E. W. Wong, P. E. Sheehan, and C. M. Lieber, Nanobeam mechanics: elasticity, strength, and toughness of nanorods and nanotubes, *Science*, 277, 1971–1975 (1997).

55. R. Saito, G. Dresselhaus, and M. S. Dresselhaus, *Physical properties of carbon nanotubes*, Imperial College Press, London (1998).

56. J. Hone, M. Whitney, C. Piskoti, and A. Zettl, Thermal conductivity of single-walled carbon nanotubes, *Physical Review B*, 59, R2514–R2516 (1999).

57. J. Che, T. Cagin, and W. A. Goddard III, Thermal conductivity of carbon nanotubes, *Nanotechnology*, 11, 65–69 (2000).

58. S. Ramakrishna, K. Fujihara, W.-E. Teo, T.-C. Lim, and Z. Ma, *An introduction to electrospinning and nanofibers*, World Scientific Publishing Co., Pte., Ltd., London (2005).

59. P. X. Ma and R. Zhang, Synthetic nano-scale fibrous extracellular matrix, *Journal of Biomedical Materials Research*, 46, 60–72 (1999).

60. R. L. Vander Wal and T. M. Ticich, Flame and furnace synthesis of single-walled and multi-walled carbon nanotubes and nanofibers, *Journal Physical Chemistry B*, 105, 10249–10256 (2001).

61. M. J. Height, J. B. Howard, and J. B. Vandersande, Method and apparatus for synthesizing filamentary structures, US Patent 7335344, Publication Date: February 26 (2008).

62. C. J. Unrau, R. L. Axelbaum, P. Biswas, and P. Fraundorf, Synthesis of single-walled carbon nanotubes in oxy-fuel inverse diffusion flames with online diagnostics, *Proceedings of the Combustion Institute*, 31, 1865–1872 (2007).

63. H. Richter, P. M. Jardim, J. B. Vander Sande, and J. B. Howard, Synthesis of fullerenic materials by controlled premixed combustion, *Nano-C Poster Symposium*, Boston, MA (2006).

64. M. J. Bronikowski, P. A. Willis, D. T. Colbert, K. A. Smith, and R. E. Smalley, Gas-phase production of carbon single-walled nanotubes from carbon monoxide via the HiPco process: A parametric study, *Journal of Vacuum Science and Technology, A*, 19, 1800–1806 (2001).

65. Y. Wang, S. Da, M. J. Kim, K. F. Kelly, W. Guo, C. Kittrell, R. H. Hauge, and R. E. Smalley, Ultrathin "bed of nails" membranes of single-wall carbon nanotubes, *Journal of American Chemical Society*, 126, 9502–9503 (2004).

66. A. B. Dalton, S. Collins, E. Munoz, J. M. Razal, V. H. Ebron, J. P. Ferraris, J. N. Coleman, B. G. Kim, and R. H. Baughman, Super-tough carbon-nanotube fibres, *Nature*, 423, 703 (2003).

67. K. Jiang, Q. Li, and S. Fan, Nanotechnology: Spinning continuous carbon nanotube yarns, *Nature*, 419, 801 (2002).

68. M. Zhang, K. R. Atkinson, and R. H. Baughman, Multifunctional carbon nanotube yarns by downsizing an ancient technology, *Science*, 306, 1358–1361 (2004).

69. R. H. Baughman, M. Zhang, S. Fang, A. A. Zakhidov, M. Kozlov, S. B. Lee, A. E. Aliev, C. D. Williams, and K. R. Atkinson, Solid state processing, structure and

multifunctional applications of carbon nanotube yarns and transparent sheets, Presentation, Science and technology without boundaries, Baughman.pdf (viewed 2008).

70. N. Chakrapani, B. Wei, A. Carrillo, P. M. Ajayan, and R. S. Kane, Capillarity-driven assembly of two-dimensional cellular carbon nanotube foams, *Proceedings of National Academy of Science*, 101, 4009–4012 (2004).

71. P. M. Ajayan, L. S. Schadler, and P. V. Braun, *Nanocomposite science and technology*, Wiley-VCH Verlag, Weinheim, Germany (2004).

72. R. A. Vaia, Polymer nanocomposites open a new dimension for plastics and composites, *AMPTIAC Newsletter*, 6, 17–24 (2002).

73. M. Roy, J. K. Nelson, R. K. MacCrone, L. S. Schadler, C. W. Reed, and R. Keefe, Polymer nanocomposite dielectrics-the role of the interface, *IEEE Transactions on Dielectrics Electrical Insulation*, 12, 629–643 (2005).

74. J. N. Coleman, M. Cadek, K. P. Ryan, A. Fonseca, J. B. Nagy, W. J. Blau, and M. S. Ferreira, Reinforcement of polymers with carbon nanotubes: The role of an ordered polymer interfacial region. Experiment and modeling, *Polymer*, 47, 8556–8561 (2006).

75. S. Ruan, P. Gao, and T. X. Yu, Ultra-strong gel-spun UHMWPE fibers reinforced using multiwalled carbon nanotubes, *Polymer*, 47, 1604–1611 (2006).

76. F. T. Wallenberger and N. E. Weston, *Natural fibers, plastics and composites*, Kluwer Academic Publishers, Dordrecht, the Netherlands (2004).

77. J. Suhr, W. Zhang, P. Ajayan, and N. Koratkar, Temperature activated Interfacial friction damping, in carbon nanotube polymer composites, *Nano Letters*, 6, 219–223 (2006).

78. N. A. Koratkar, J. Suhr, A. Johsi, R. S. Kane, L. S. Schadler, P. M. Ajayan, and S. Bertolucci, Characterizing energy dissipation in single-walled carbon nanotube polycarbonate composites, *Applied Physics Letters*, 87, 063102 (2005).

79. J. D. Whittaker, E. D. Minot, D. M. Tannenbaum, P. L. McEuen, and R. C. Davis, Measurement of the adhesion force between carbon nanotubes and a silicon dioxide substrate, *Nano Letters*, 6, 953–957 (2006).

80. X. Chen, K. Yoon, C. Berger, I. Sics, D. Fang, B. S. Hsiao, and B. Chu, In-situ x-ray scattering studies of a unique toughening mechanism in surface-modified carbon nanofiber/UHMWPE nanocomposite films, *Macromolecules*, 38, 3883–3893 (2005).

81. H. T. Ham, C. M. Koo, S. O. Kim, Y. S. Choi, and I. J. Chung, Chemical modification of carbon nanotubes and preparation of polystyrene/carbon nanotube composites, *Macromolecular Research*, 12, 383–390 (2004).

82. Y. J. Jung, S. Kar, S. Talapatra, C. Soldano, G. Viswanathan, X. Li, Z. Yao, F. S. Ou, A. Avadhanula, R. Vajtai, S. Curran, O. Nalamasu, and P. M. Ajayan, Aligned carbon nanotube-polymer hybrid architectures for diverse flexible electronic applications, *Nano Letters*, 6, 413–418 (2006).

83. S. R. C. Vivekchand, U. Ramamurty, and C. N. R. Rao, Mechanical properties of inorganic nanowire reinforced polymer–matrix composites, *Nanotechnology*, 17, S344–S350 (2006).

84. C. P. Poole and F. J. Owens, *Introduction to nanotechnology*, Wiley-Interscience, John Wiley & Sons, Inc., Hoboken, NJ (2003).

85. L. He, L. F. Allard, K. Breder, and E. Ma, Nanophase Fe alloys consolidated to full density from mechanically milled powders, *Journal of Materials Research*, 15, 904–912 (2000).

86. N. Keller, C. Pham-Huu, G. Ehret, V. Keller, and M. Ledoux, Synthesis and characterisation of medium surface area silicon carbide nanotubes, *Carbon*, 41, 2131–2139 (2003).

87. X.-H. Sun, C.-P. Li, W.-K. Wong, N.-B. Wong, C.-S. Lee, S.-T. Lee, and B.-K. Teo, Formation of silicon carbide nanotubes and nanowires via reaction of silicon (from disproportionation of silicon monoxide) with carbon nanotubes, *Journal of American Chemical Society*, 124, 14464–14471 (2002).

88. M. H. Rümmeli, E. Borowiak-Palen, T. Gemming, M. Knupfer, K. Biedermann, R. J. Kalenczuk, and T. Pichler, On the formation process of silicon carbide nano-phases via hydrogenated thermally induced templated synthesis, *Applied Physics A*, 80, 1653–1656 (2005).

89. Z. Correa, H. Murata, T. Tomizawa, and A. Oya, Preparation of SiC nanofibers by using the polymer blend technique, *Advanced Science and Technology*, 51, 60–63 (2006).

90. T. T. Xu, J.-G. Zheng, N. Wu, A. W. Nicholls, J. R. Roth, D. A. Dilkin, and R. S. Ruoff, Crystalline boron nanoribbons: Synthesis and characterization, *Nano Letters*, 4, 963–968 (2004).

91. Cermet, en.wikipedia.org/wiki/Cermet (2008).

92. P. Seegopaul and Z. Fang, Tungsten-carbide-cobalt nanocomposites: Production and mechanical properties, In *Dekker Encyclopedia of nanoscience and nanotechnology*, Vol, 5, J. A. Schwartz, C. I. Contescu, and K. Putyera, eds., pp. 3943–3952, CRC Press, Boca Raton, FL (2004).

93. Cement, en.wikipedia.org/wiki/Cement (2008).

94. J. Makar, J. Margeson, and J. Luh, Carbon nanotube/cement composites—early results and potential applications, National Research Council of Canada, 3rd International Conference on Construction Materials: Performance, Innovations and Structural Applications, Vancouver, BC, August (2005).

95. L. Cassar, Presentation, CTG Italcementi Group, Italy, 2nd International Symposium on Nanotechnology in Construction, Bilboa, Spain, November (2005).

96. K. Sobolev and M. Ferrad-Gutiérrez, How nanotechnology can change the concrete world: Part 2, *American Ceramic Society Bulletin*, 11, 16–19 (2005).

97. A. Mourchid, E. Lecolier, H. Van Damme, and P. Levitz, On viscoelastic, birefringent, and swelling properties of laponite clay suspensions: Revisited phase diagram, *Langmuir*, 14, 4718–4723 (1998).

98. J. W. Gilman, C. L. Jackson, A. B. Morgan, R. Harris, Jr., E. Manias, E. P. Giannelis, M. Wuthenow, D. Hilton, and S. H. Phillips, Flammability properties of polymer-layered-silicate nanocomposites: Polypropylene and polystyrene nanocomposites, *Chemistry of Materials*, 12, 1866–1873 (2000).

99. D. L. Ho and C. J. Glinka, Effects of solvent solubility parameters on organoclay dispersions, *Chemistry of Materials*, 15, 1309–1312 (2003).

100. C. Greenwell, *The clay-polymer project*, www.exclaim.org.uk/CP_Intro.html (2005).

101. F. M. Mirabella, Jr., Polypropylene and thermoplastic olefin nanocomposites, In *Dekker Encyclopedia of nanoscience and nanotechnology*, Vol. 4, J. A. Schwartz, C. I. Contescu, and K. Putyera, eds., pp. 3015–3030, CRC Press, Boca Raton, FL (2004).

102. Halloysite nanotubes, *Natural Nano*, www.naturalnano.com (2008).

103. M. Liu, B. Guo, Q. Zou, M. Du, and D. Jia, Interactions between halloysite nanotubes and 2,5-bis(2-benzoxazolyl) thiophene and their effects on reinforcement of polypropylene/halloysite nanocomposites, *Nanotechnology*, 19, 205709–205718 (2008).

104. M. Liu, B. Guo, M. Du, and D. Jia, Drying induced aggregation of halloysite nanotubes in polyvinyl alcohol/halloysite nanotubes solution and its effect on properties of composite film, *Applied Physics A: Materials Science and Processing*, 88, 391–395 (2007).

Problems

10.1 What is the maximum strain energy that can be stored in a tendon with $E_o = 1.5$ GPa and density equal to 1120 kg·m⁻³?

10.2 Everything else being equal and occupying equal volumes (volume fractions), how much more surface area do nanotubes with diameter 2 nm have compared to 2-μm diameter tubes? or 20-μm diameter fibers?

10.3 Derive Equation 10.14, $\dfrac{1}{E_c} = \dfrac{V_m}{E_m} + \dfrac{V_f}{E_f}$, assuming cylindrical geometry and that the load is perpendicular to the axis of the fiber composite. What other formulas do the isostrain and isostress relationships resemble? (Hint: think electricity).

10.4 The coefficient of thermal expansion for a thermoplastic sheet is 7.56×10^{-5}·C⁻¹. How much will a 4×8 foot sheet of the material expand in the long dimension when taken out of a freezer at 30°F to room temperature (70°F)?

10.5 Determine the specific surface area of a clay that is 1 nm in thickness and 1 μm in lateral dimensions. The density of the clay is 2.65 g·cm⁻³.

10.6 Averaging of properties for composites seems to work well for macroscopic to micrometer-sized materials. Do the formulae apply to nanomaterials? Why or why not?

10.7 Boeing's new *Stratoliner 787* is made primarily of composites based on carbon fibers. What advantages does this material provide over conventional aluminum materials used in planes for many decades?

Section 5

Biological and Environmental Nanoengineering

NANOBIOTECHNOLOGY

Nanotechnology allows us to do things we are unable to do otherwise. Nano-technology gives us a 'set of tools.'

DR. OMID FAROKHZAD
Harvard Medical School

FIG. 11.0 *Nanobiotechnology combines the functionality of biological nanomachinery with the power and speed of electronics. This conceptual image symbolizes the integration of macromolecular nanomachinery with microelectronics.*

THREADS

Chapter 11 presents a short introduction to nano-biotechnology, which is the application of nano-technology to biological systems to achieve practical engineering results. These applications include separation and identification of biological products, modification of biological materials to improve their useful properties, and the use of biological molecules to create sensors. Nanoscale materials, sensors, and devices can be and are being designed based on almost any biological system, and inorganic nanodevices can be applied to many biological applications. In this selective introduction to a very broad and rapidly expanding field, we focus on a few examples which strongly illustrate the role of nanoscale phenomena in creating new types of devices and materials. To distinguish this focus area, we use the term nano-biotechnology.

As in instructive example, in this chapter we take one biological system and one nanoscale physical system and show how they can be integrated to produce a nano-biosensor. From the many possibilities, we chose the biological immune system and the physical nanoscale cantilever. In the course of this discussion, we cover considerably more detail on the immune system than is immediately needed for application to the sensors, in order to give an understanding of the depth and complexity of biological systems, and to give the student some resources for exploring other possibilities. In addition, we briefly survey other types of sensors and some other nano-scale biodevices and materials.

11.0 INTRODUCTION TO NANOBIOTECHNOLOGY

Nanobiotechnology is an emerging technology on the interface between bio-technology and nanotechnology. Biotechnology manipulates the molecules and processes of biological systems to gain practical improvements, for example, in producing harvestable yields of useful drugs and antibodies, or in developing indicators and sensors based on the interaction of antibodies with substrates. Nanobiotechnology uses nanoscale mechanical, electrical, optical, and magnetic effects in biological systems to create new types of devices and applications which are beyond the scope of previous molecular biology- and biochemistry-based biotechnologies. As a result, the ability to sense and detect the state of biological systems and living organisms is being radically transformed.

11.0.1 Definitions

Because of the rapid rise of new developments in the interface between nanotechnology and biology, it is useful to go over some definitions of terms. Besides the well established but not always clearly defined term "biotechnology," new terms are coming into use, including "bio-nanotechnology," "biomolecular nanotechnology," "biomedical nanotechnology," and others.

11.0.2 Biotechnology

Biotechnology generally refers to the manipulation of key biological systems such as the DNA–RNA encoding and synthesis of proteins and the targeting of antigens by antibodies [1–3]. The understanding of the molecular basis of these fundamental biological processes led to the ability to modify and exploit their operation. Biotechnology utilizes enzymes, viruses, biochemistry, and biophysics to achieve its results. DNA can be modified to produce useful proteins; antibodies can be modified to include indicators or "reporters" such as fluorescent chemical groups, which change to indicate the presence of a selected antigen. DNA can be interfaced to photochemical or electrochemical sensors to detect the presence of complementary strands of DNA or RNA, thus providing a means of detecting and analyzing genetic material in unknown samples, or identifying a disease agent by its DNA. Biotechnology includes the modification of DNA to produce unique genotypes for animals and plants, and cloning of organisms to reproduce their genotypes. Biotechnology is being impacted by nanotechnology, as seen by many examples elsewhere in this book.

11.0.3 Bio-Nanotechnology

Bio-nanotechnology has emerged as the manipulation and exploitation of proteins, the natural nano-engines of biological systems. This is a vast field which is growing naturally out of the discipline of protein chemistry, and has many exciting possibilities [4].

11.0.4 Biomolecular Nanotechnology

Biomolecular nanotechnology is an emerging field which is being defined more broadly than protein bio-nanotechnology to include the most sophisticated application of biotechnology, nanofabrication, materials science, nanoelectronics, biochemistry, and biophysics to design useful and interesting structures on the nanoscale. Biomolecular nanotechnology includes using DNA as a template for molecular-scale computational engines, and adapting muscle and flagella mechanisms to make artificial organic nano-engines, for example [5].

11.0.5 *Biomedical Nanotechnology*

Biomedical nanotechnology is the application of nanotechnology to biomedical applications such as diagnostics and high-throughput screening, drug delivery, artificial tissues for implants and prostheses, and medical imaging. We will survey some of these areas in the next chapter of this book. Other areas of interface between nanotechnology and biology, such as modification of natural nanomaterials and biomimetic design of nanodevices, are covered elsewhere in this and the companion books [6–17].

11.0.6 *Nanobiotechnology*

This brings us to *nanobiotechnology*, the subject of this chapter, which we define as the application of nanotechnology to the above fields in ways which exploit phenomena that are unique to the nanoscale. As an example of this class of applications, we will discuss cantilever sensors, whose principles of operation depend upon the ratio of surface to volume that is characteristic of the nanoscale.

In order to make a nano-biosensor with a nanoscale cantilever, we utilize the biological immune system and integrate a selected portion of it with a nanoscale physical mechanism. The main example in this chapter will be to show how we integrate the molecular action of the biological immune system with the nanoscale physics and chemistry of absorption on the surfaces of nanoscale cantilevers, while using a variety of electronic and optical means to translate the results into a measurable form. This example is just one of a number of ways that nanoscale phenomena can be exploited to create sensors capable of responding to very small inputs. In the next section, we briefly describe the biological immune system before proceeding to the physical description of nanoscale cantilevers.

11.1   THE BIOLOGICAL IMMUNE SYSTEM

The biological immune system is a primary example of how molecular recognition works at the nanolevel. The macromolecules of the immune system are directly harnessed in nanotechnology devices and as models for making similar artificial molecular recognition nanomachinery.

Living organisms fight off invasions of disease agents with an array of complex cellular and molecular defenses called the immune system [18–20]. Biological immunity works through molecular pathways which control recognition and signaling in cells. We have introduced some basic aspects of gene expression and molecular biology in chapter 14 of *Introduction to Nanoscience*. We will now go further into the details of how these nanoscale systems work in the immune system.

11.1.1 *Natural Molecular Recognition*

In the chapter in *Introduction to Nanoscience* on molecular nanobiology, we discussed how DNA and RNA act as molecular templates to reproduce and transmit information. In this section we examine another example of molecular recognition, the immune system. The immune system, especially of vertebrates, is a set of highly developed nanoscale mechanisms of molecular recognition, which can be utilized to create drugs, probes, and diagnostic devices on the molecular and nanoscale levels. Humans and higher vertebrates have two levels of immunity which work together to defend against infectious pathogens: the *innate* and *adaptive* immune systems.

Vaccines and the Discovery of the Immune System.   Long before the birth of scientific medicine, there was some cultural awareness of protection by inoculation. In Thucydides's historical account of the Athenian plague in 430 B.C., he noted that those who survived became immune. (His account is also noteworthy as one of the early instances of careful and systematic descriptions of symptoms.) It was also known that exposure to some types of illness did not confer future protection, for example the bubonic plague. Classical medicine recognized the effects on the body of illness, and the main symptoms are still a standard part of medical diagnosis taught to every medical student today: *rubor* (redness), *calor* (warmth), *tumor* (swelling), and *dolor* (pain). These are now recognized as inflammation due to the body's immune responses.

Various forms of inoculation (defined as the introduction of an infectious agent into the body, or into a culture medium) through the skin or nose were part of the ancient medical practices of India, China, Persia, and primitive folk medicines. The disastrous effects of the smallpox epidemics of the seventeenth and eighteenth centuries led to trials of inoculation (also called variolization in the case of smallpox) by infecting a scratch on the skin with material from another person with the disease. This risky procedure became obsolete after the late eighteenth century when the English physician Edward Jenner demonstrated that serum from the weaker, related disease cowpox could be used to inoculate successfully against the much more dangerous smallpox. Widespread use of vaccination after a long period of gaining acceptance eventually led to the eradication of smallpox late in the twentieth century.

By the late nineteenth century Robert Koch showed that infectious diseases are caused by microorganisms, and Louis Pasteur developed vaccines against cholera and rabies. In 1890 Emil von Behring and Shibasaburo Kitasato discovered *antibodies* that immobilized specific pathogens in the serum of vaccinated individuals. In the meantime Elie Metchnikoff discovered that many microorganisms could be engulfed and digested by phagocytic cells, which he called macrophages. Much to the astonishment of the scientific world of the day, these amoeba-like cells were shown to circulate in the bloodstream, seeking out and attacking pathogenic microbes, while not harming the sister cells of their own organism. Like amoebas, the macrophages demonstrate chemotaxis, motion directed by a chemical trail or gradient.

During the next century these and other white blood cells were studied and characterized microscopically, and antibodies were studied chemically and in their effects on cell cultures, as answers were sought to how the body's immune cells could recognize invaders and distinguish them from itself. Advances were made in vaccines and in using white blood cells to diagnose disease. Studies of rejection of blood transfusions in humans, and tissue grafts among inbred, genetically homogeneous mice helped understand the mechanisms for generation of antibodies and their relation to the genetic machinery of the cells. With the advent of molecular biology on the nanoscale, the beauty and complexity of the immune system has become apparent as the workings of intricate biomolecular nanomachinery [21].

11.1.2   The Innate Immune System

The *innate immune system* reacts to a wide range of innately recognized foreign bodies with a rapid cascade of reactions that requires no learning period or prior exposure to the pathogens. The first response to injury or invasion is inflammation, turned on by chemical signals released by damaged cells.

The most striking instruments in the armory of innate immunity are specialized amoeba-like white blood cells called *phagocytes* that engulf and digest microorganisms and foreign materials in the body. In the human immune system these phagocytic cells are called *macrophages*. In general, the phagocytes attack anything that is recognized as not being part of the organism ("not self").

The innate immune system includes a number of other types of specialized cells that attack pathogens using different strategies, and are recruited to the site of attack or inflammation by chemical signaling. Some of these cells also coordinate responses with the second, adaptive, level of defense.

The innate immune system works through many mechanisms. The most interesting aspect of innate immunity in terms of nanoscale molecular recognition is the way in which white blood cells sense and recognize pathogens. Receptors on the surface of white blood cells recognize repetitive molecular patterns characteristic of bacterial cell walls. These receptors are called Toll-like receptors, named after the type of genes that express them. The first example of this gene was found in studies of the fruit fly, *Drosophila*, and related genes are found in a wide range of organisms. The Toll-like proteins recognize molecular patterns that are common to many pathogens, such as the spacing of sugar units in the polymeric polysaccharide cell walls of bacteria. The receptor proteins attach firmly only when the pattern spacing is a match, leaving normal cell sugar metabolism undisturbed. These antigen proteins occur both as freely secreted molecules, which mark bacteria for destruction by macrophages, and as trans-membrane proteins, which bind defensive cells to the invading bacteria to surround it and in some cases to signal for the initiation of adaptive responses.

11.1.3   The Adaptive Immune System

The second part of the immune system is *adaptive immunity*, which is launched as an *immune response* to a particular pathogen. Adaptive immune response is mediated by the generation of *antibodies* which bind to a specific pathogen. The production of antibodies can only be induced by exposure to and infection by a pathogen, after an incubation period in which the adaptive immune system recognizes the pathogen and sets in motion the molecular machinery to build specifically modified antibody molecules. The adaptive immune system can generate other responses besides antibodies, and the term *antigen* is used for any substance that can produce an adaptive immune response. The adaptive immune system inserts and retains a molecular memory in its section of the DNA, which can provide *protective immunity* against repeated infection by a specific pathogen.

| Fig. 11.1 | *Structure of an antibody: the antigen-binding fragment binds to a specific antigen with a high degree of selectivity.* |

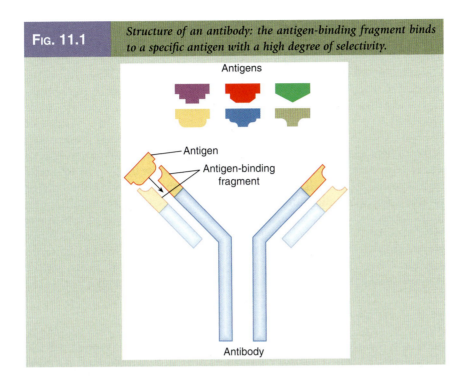

This is the basis for *vaccines*, substances to which the adaptive immune system can be exposed in order to generate antibodies without creating an infection (**Fig. 11.1**).

Biomolecular Specificity of Antibodies. The field of immunology is relatively young in terms of the amount to be learned at the molecular level, but an enormous amount of information has been laboriously elucidated on cellular recognition through complex signaling pathways. This research has discovered the role of the structure of the cell membrane and how sensors imbedded in the membrane interact with the expression of genes by regulating DNA in the cell nucleus to produce antibodies and stimulate the generation of phagocytes in response to infection. In this section we will give some examples of the antibody nano-machinery involved in the immune processes.

Protein–peptide complexes in the immune system are an important example of cellular signaling, trafficking, and targeting mechanisms in the body that are mediated by interactions between proteins and peptides. This sample covers only the barest introduction to the full richness of the immune system, but it is illustrative to anyone asking how assemblages of molecules can collectively recognize self and nonself, attack invaders, learn and remember threats, and organize perfectly matched tools to eliminate them. In effect the immune system displays a kind of molecular machine intelligence that is intimately connected with the genetic molecular machinery at the heart of life.

Antibodies in the Adaptive Immune System. Antibodies and antigens are produced by the protein-synthesizing machinery of the cell in a special region of the DNA genome called the *major histocompatibility complex* (MHC). This large genomic region exists only in vertebrates, as a very highly evolved and intricate defense mechanism. The proteins expressed by the DNA in the MHC are positioned in the cell membrane as signaling markers on the cell surface. The MHC proteins are templates that hold fragmented pieces of antigens on the cell surface, where they can be sensed and interrogated by other cells, especially white blood cells.

Mammalian cells constantly process endogenous proteins and present their peptide fragments on the cell membrane attached to MHC proteins. As long as the proteins being processed are endogenous, the white blood cells recognize the displaying cell as "self," and all is well. But if the cell is processing foreign protein expressed by infection from a bacterium, virus, or cancerous disruption of the molecular machinery, then the proteins presented on its surface by the MHC template will be recognized by white blood cells as "nonself." White blood cells are constantly engulfing foreign bodies (bacteria, viruses, particles, macromolecules), breaking them apart using lysozymes, and displaying the fragments using their own particular version of the MHC proteins.

11.1.4   *White Blood Cells and Antibodies*

In order to survive in a competitive world, all organisms have evolved defense mechanisms to protect against competitors, predators, and pathogens. These defenses first evolved on the nanoscale: they are so basic an aspect of life that they evolved before life diverged into plants and animals at the single-cell level. Thus all cells have a heritage of molecular defense mechanisms to suppress competitors, deter predators, fight pathogens, and respond to pathogen virulence factors. In multicellular organisms, these molecular defenses must interplay with cellular signaling pathways that orchestrate cooperation between the organism's own cells. And in multicellular organisms, there is the opportunity of some cells to specialize in fighting invaders and toxins that threaten the whole organism—these are the white blood cells.

The innate and adaptive immune systems work together in vertebrates through a coordinated mechanism at the cellular and molecular level, involving white blood cells and biomolecular-selective protein antigens and antibodies.

White Blood Cells.   The number and appearance of white blood cells under the microscope, their separation from the much more numerous red blood cells by centrifugation, and their importance to the practical diagnosis of infection play a large and important part in the practice of medicine and study of immunology. The specificity of stains was discovered by trial and error in the early days of microscopy. Early biochemists and medical researchers speculated that if a chemical stain would selectively attach itself to a pathogen, there might be substances which could attach to pathogens and selectively kill or disable them without harming other cells. This suggested the concept of a "magic bullet" that would seek out and destroy disease germs, even long before the molecular mechanisms of antigens were understood, or before the concept of germ-destroying nanobots was proposed.

In humans and vertebrates the *leukocytes* or white blood cells play the major role in both innate and adaptive immune response. A number of different types of leukocytes are produced by special stem cells in various organs of the immune system such as the bone marrow and spleen. They are released into the blood at a rate of billions per day. The number and type of white blood cells circulating in the body varies in response to infection. Under the microscope they can be distinguished and classified according to their staining properties as agranulocytes and granulocytes; the latter have granules that show up with staining under the microscope. The granulated structures are due to specialized lysosomes associated with the digestion of engulfed particles, so that the staining differences correspond to different cell structures and functions.

Innate leukocytes include the macrophages, dendritic cells, and the granulocytes: neutrophils, basophils, and eosinophils. The macrophages, dendritic cells, and neutrophils are phagocytic, and engulf and digest pathogens. Other innate nonphagocytes are mast cells and natural killer cells. The nonphagocytes disable pathogens by contact or by attachment to mark them for attack by the adaptive immune system.

Molecular signaling pathways control production of leukocytes and direct them to find, identify, and fight sources of infection or antigens. Each of the innate leukocyte cell types plays a specialized role in the immune response:

1. Macrophages move through the body engulfing pathogens that they encounter and recognize or are summoned to in response to chemical signals. Macrophages migrate within tissues and release enzymes, proteins, and other factors which act on pathogens. Macrophages ingest and destroy worn out and dead cells of their own organisms, in addition to pathogens.

2. Neutrophils are carried through the bloodstream and follow chemical signals by chemotaxis to the site of infections where they concentrate as part of the inflammation response. Dendritic cells are phagocytes active mainly in tissues close to the external environment, such as skin, respiratory passages, and digestive tract. Their name refers to root-like appendages, similar in appearance but unrelated to some nerve cells. Dendritic and macrophage innate cell types play a role in activating the adaptive immune system, signaling the presence of new antigens by presenting them to special adaptive immune cells.

3. Natural killer (NK cells) cells are leukocytes that attack and destroy tumor cells, or cells that have been infected by viruses.

4. Basophils and eosinophils are related to neutrophils. They release chemical signals in response to parasites and play a role in the inflammations associated with allergic reactions.

5. Mast cells are very similar to basophils. In response to injury or antigens their granules are released and play an important part in inflammation.

The white blood cells of the adaptive immune system are special types of leukocytes, called lymphocytes. B cells and T cells are the major types of lymphocytes. B cells are involved in the humoral immune response, whereas T cells are involved in the cell-mediated immune response. When the MHC in precursor white blood cells are presented with an antigen, the genetic mechanisms produce new lines of adaptive white blood cells with new specific receptors for the antigen. This process takes time, so there is an incubation period before immunity takes hold.

11.2   USING ANTIBODIES IN BIOSENSORS: IMMUNOASSAYS

Antibodies have evolved to be molecular detectors playing a vital function in the maintenance of the organism. They possess the properties that are most desirable in a sensor: sensitivity, specificity, and discrimination. Antibodies work at the level of cell membranes and macromolecules, the nanoscale. Thus they are extremely sensitive, capable of detecting and latching onto a single molecule of antigen. Antibodies are highly specific: each one is custom designed to recognize and capture a specific molecular shape, electronic bonding, and pattern. Antibodies are highly discriminating: they can distinguish between antigens that are very similar to each other, and distinguish pathogens that are similar to cells that belong to their own organism (self).

Antibodies can be used in bulk chemistry as analytical indicators. Historically they were employed in serum assays where a precipitate or other visible change indicated the presence of a pathogen. Such tests are called *immunoassays*. With micro- and nanotechnology, antibodies can be harnessed at the molecular level to make very sophisticated and sensitive immunoassays. The design of ways to immobilize antibodies onto microchips and combine them with optical or electrical indicators has become a major area of research and development with the goal of producing analytical tools for research and medical diagnosis. In this section we give a brief introduction to this field and define some key terminologies.

11.2.1   Antibodies in Molecular Recognition Sensors

Use of Antibodies in Molecular Recognition Devices.   Antibodies can be removed from blood and other tissues, and used for vaccines or other purposes. Like DNA and RNA, antibodies can be used as a macromolecular template to recognize and bind to other selected molecules. DNA segments can be fixed onto biochip substrates and used to search for matching complementary segments with sensitivity, specificity, and discrimination, even in complex mixtures of similar molecules. Readout of the biochip can be with optical, electronic, or chemical indicators.

11.2.2   Production of Antibodies

Making Antibodies.   Most antigens have a variety of features that generate more than one antibody. So exposing whole animals to antigens produces a serum vaccine with heterogeneous mixtures of antibodies. This has advantages for fighting pathogens, as it gives more than one way for the body to localize and eliminate the pathogens. But for research and diagnostic purposes a single, specific antibody is required. And for safety and precision of therapy, serum antibodies have been replaced by *monoclonal antibodies* for vaccines and other medical uses [22].

11.2.3   Monoclonal Antibodies

Monoclonal antibodies, derived from a single cloned line of cells, are produced by molecular biotechnology techniques. In 1975 Köhler and Milstein, after many years of work, produced a technique for hybridizing myeloma cells with antibody-secreting cells from mice. The myeloma cancer cells could be grown indefinitely in culture, unlike normal cells, thus providing a way to manufacture antibodies. (Köhler and Milstein were awarded a Nobel Prize for this work.) Monoclonal antibodies produced in culture or in mice can be bonded to radioactive atoms, magnetic particles, fluorescent molecules, or other additives for use against cancer cells.

11.2.4   *Reverse Transcriptase*

Reverse Transcriptase.   Genetic engineering provides a means of producing proteins by modifying the DNA sequence of cells so that they replicate the desired amino acid sequence. It was made possible by the discovery in 1970 that certain viruses use an enzyme—*reverse transcriptase*—to transcribe DNA from RNA, the opposite of the normal direction for nucleotide synthesis. (H. Temin and D. Baltimore received a Nobel Prize in 1975 for this discovery.) The RNA viruses that reproduce using reverse transcriptase are called *retroviruses*, and genetic material produced using *reverse transcriptase* is called retroviral DNA. Retroviral DNA techniques can be used to produce many proteins in culture or in transgenic animals or plants, providing another source for peptides, including antibodies.

11.2.5   *Recombinant DNA*

Transfer of Genes between Organisms.   To produce a bioengineered product, gene sequences that code for the product must be inserted into organisms that can be easily cultured or harvested. Plasmids, small circles of DNA found in bacteria, can introduce operational DNA into bacterial cells. Plasmids used to deliver DNA are called *vectors*. Stitching a sequence of DNA into a plasmid ring is carried out by cut and paste enzymes, the restriction endonucleases and ligating enzymes. DNA altered in this manner is called *recombinant DNA*. Once in the bacteria, the plasmid is reproduced along with the products coded for by the recombinant DNA, so bacterial cultures can be used to produce genetically engineered products like human insulin, growth hormone, and interferon. Recombinant DNA technology can also be applied to plants and animals, producing genetically engineered organisms.

Recombinant DNA technology has advanced and been enhanced by refinements so that it is now used for DNA sequencing and large-scale production of hormones and vaccines. It employs a combination of traditional chemical, microbiological techniques along with molecular biology and genetics. Its chief relevance to nanotechnology is as a source for special macromolecules with specific properties, which can be used in nanomachinery and sensors. In turn, nanotechnology is contributing new abilities to analyze and manipulate DNA and other molecules in the cell.

11.2.6   *Antibodies as Selection Tools for Biosensors*

Antibodies are one of the biomolecules that can be used as highly specific selection and capture tools at the molecular level. Antibodies are produced by the adaptive immune system of mammals in response to all kinds of proteins and macromolecules, making them much more versatile than DNA and RNA. While DNA and RNA are very important for genetic and medical research and diagnosis, antibodies can be adapted for many more diverse purposes. In the next section we look at how antibodies can be attached to micromolecular devices to make extremely sensitive analytical detectors, and how nanoscale phenomena make the devices work.

11.3   CANTILEVERS AS NANO-BIOSENSORS

Sensors are devices which can be used to read and report information on their environment. Sensors typically change state in the presence of a targeted chemical or physical condition such as pressure or temperature.

Biosensors use a biological molecule or structure to effect the function of the sensor. Nano-biosensors are biosensor devices whose function takes advantage of some properties or phenomena unique to the nanoscale of interactions between matter and its environment.

Small size alone may be the chief attribute of a nanosensor, but often when we explore devices that are scaled down into the nano region, we find new behaviors that were not present at the macroscale. These phenomena can be due to changes in surface-to-volume ratios, internal material stress, heat and electrical conductivity, ratio of light wavelengths to size of material, and other phenomena. These behaviors, unique to the nanoscale, are what make nanosensors interesting and especially useful.

Natural biological sensors work at the nanoscale, and are sensitive to very small amounts of material and small changes in their environment. But they generally do not offer us a ready and quick reporting mechanism that would make them useful in an engineered device. Use of biological sensors without nanotechnology usually involves waiting for an incubation period to see the results in terms of a biological outcome, or at best involves a number of separation and amplification steps.

Integrating biomolecules into nanodevices gives us a way to combine the sensitivity and selectivity of biomolecular sensors with the speed and versatility of interfacing and readout that we require of our electronic and mechanical systems. Micro- and nanoscale cantilevers provide a good example of this paradigm.

11.3.1   Sensing Physical Properties

Physical parameters include mass, pressure, temperature, electrical potential, force, acceleration, tension, electric charge, etc., which can be related to biological measurements such as membrane potential, metabolism rate, and concentration of specific species (such as oxygen, hydrogen ions = pH). Physical measurements on the nanoscale can be made by exploiting physical, optical, and/or electronic properties of materials and structures on the scale of a few thousand atoms, the nanoscale. One especially important type of sensor is based on measuring the nanoscale motion of microscopic cantilevers with nanoscale coatings interacting with nanoscale quantities of substances in their environment.

11.3.2 Cantilevers and Selective Binding

Cantilever Sensors. Microscopic cantilevers are widely used tools that have become a mainstay of nanotechnology; they can be etched from silicon and other materials, as described elsewhere in this book. A cantilever or microprobe is the basis for the atomic force microscope, and arrays of cantilevers have been adapted to fabricate microelectronic memory devices and a variety of actuators and resonators [23–26]. Microchip cantilevers can also be used as mirrors, resonators, and capacitors to fabricate electro-optical and electromechanical devices that can be adapted as sensors for biomedical assays [27–29]. We will discuss some of the biomedical applications for this technology in a later section. First, we will look at how the properties of cantilevers on the nanoscale have made it possible to adapt them as sensors that have demonstrated potential for nanomedicine. In recent years this has been one of the most promising directions for a practical application in medical diagnosis of cancer, identification of proteins, and elucidation of molecular signaling pathways in molecular biology.

11.3.3 Active Cantilever Sensors

Active Cantilever Sensing—Vibration Resonance. A tiny cantilever etched from silicon or other substrates will have a characteristic resonant vibration frequency, like the tine of a microscopic tuning fork. The physics of vibrations of bulk materials applies well to submicroscopic cantilevers provided that their composition is uniform, without defects that lead to loss of elasticity. [30,31] A carbon nanotube or fiber is composed on bonded atoms, so it can be used as a cantilever that is more perfectly elastic than a deflecting or resonating structure etched from even the best single-crystal silicon, but nonlinear effects at the nanoscale must be taken into account. [32]

Nanoscale cantilevers can be used to measure density and viscosity of fluids, because their rate of vibration is damped by collisions with molecules in the fluid. The rate of vibration of a cantilever can be monitored electronically by a capacitive circuit design with the cantilever as an element. The deflection of a cantilever can also be measured optically, by changes in the angle of reflection or by interferometry. Cantilevers can be used to detect and determine the mass of molecules that are absorbed or bonded to their surface, since the resonant frequency of oscillation of a vibrating object depends on its mass as well as its elasticity. At the nanoscale, the difference in mass made by the addition of just a few large molecules such as proteins or DNA is large enough in relation to the mass of the cantilever to make it possible to detect minute quantities of biological substances [33].

11.3.4 Passive Cantilever Sensors

Passive Cantilever Sensing—Measuring Deformation. Nanoscale cantilevers can be distorted passively, without vibrating, due to differences in the amount and strength of absorption on opposite sides of the sliver of material. This is perhaps the most striking example of how things are different at the nanoscale with regard to cantilevers. This phenomenon follows the same laws of surface chemistry and physics as for larger-scale bulk materials, but at the nanoscale, the surface forces dominate over the internal properties of the material.

For a cantilever on the order of 500-μm long, 100-μm wide, and 0.5-μm thick, the free-energy differences between a surface covered with absorbed molecules and a clean surface, although small, are enough to force a displacement in the shape of the cantilever which can be detected when one side of a cantilever is coated with an agent that binds with an absorbate. The absorption of the analyte may be highly specific, as between an antibody and an antigen, or may be merely weakly selective. In the latter case, combinations of coatings on groups of cantilevers can be used to obtain patterns for recognition, as in the artificial nose device [34].

11.3.5   *Surface Effects on Nanocantilevers*

Absorption Strength and Surface Crowding: Absorption Affinity.   To envisage the effect of absorbed molecules on the cantilever, consider that as bonds are formed with the absorbate on one specially coated or prepared side of the sliver of material, the absorbed molecules will tend to push their way into the matrix of molecules that make up the surface of the cantilever. This crowding into the surface will tend to distend the surface ever so slightly. The stronger the affinity or attractive force that binds the absorbed molecules to the substrate, the greater the crowding force will be and the more it will tend to push the molecules apart near that surface of the cantilever. If no absorption takes place on the other side, there will be a slight differential expansion on the absorption side, and this expansion will cause a deflection, with a convex curvature on the absorption side, and a concave curvature on the opposite side.

11.3.6   *Steric Effects*

Surface Crowding: Steric Effects.   An additional force for expansion on the absorption side will occur if the absorbed molecules repel each other or push against each other; in that case they will tend to pull apart the molecules in the cantilever to which they are bonded. These two forces are sufficient to produce a measurable deflection if the cantilever is small enough that the effects on the surface overpower the forces holding the internal molecules in their preferred places relative to each other in the bulk material matrix.

Complexity due to Interplay between Affinity and Steric Effects.   Some of the leading researchers in the field of cantilevers have pointed out that understanding the molecular causes of the deformation is critical, because the observed bending depends on both the strength of absorption affinity and the nanomechanical response of the cantilever. These two factors may be convoluted such that a strong binding affinity will not necessarily produce a large surface stress. Hans Peter Lang, Martin Hegner, and Christoph Gerber in Switzerland have shown that steric crowding of the absorbed molecules may be more important than free-energy changes on the surface, although the two factors are interlinked. [34,35] What is certain is that as the size of the cantilever becomes smaller, classical models for describing the interaction with its surroundings begin to break down, and care must be taken in designing a device and interpreting the results. [36,37]

Even with classical elastic behavior of the cantilever, thermal effects may bias the degree of displacement, for example. Thus, in designing a device for practical application in biomedicine, reference cantilevers are placed alongside the elements that are coated to bind with the analyte, so that any purely thermal effects or other environmental influences can be compensated.

11.3.7 *Surface Free Energy at the Nanoscale*

Cantilever motion can be explained by changes in the surface free energy of one surface of the cantilever relative to the opposite side. The strong intermolecular forces associated with specific binding between molecules, such as between an antibody and an antigen, result in much higher free-energy change than for nonspecific bonds such as physical absorption layers. Cantilever deflections, large enough to be useful for sensors, should be a result of specific binding.

An estimate for the magnitude of a cantilever deflection Δh is given by Stoney's equation:

$$\Delta h = 3\sigma(1-v) / E \cdot (L/d)^2 \qquad (11.1)$$

where
  σ is the change in surface free-energy density (or surface stress) due to specific binding
  E is the elastic modulus (or Young's modulus) of the cantilever material
  v is its Poisson ratio
  L and d are the length and the thickness of the cantilever, respectively

Longer and thinner cantilevers will result in larger deflections for the same degree of surface stress. Note that the above formula does not take account of steric crowding forces [35,38,39].

11.4 MICRO- AND NANOSENSORS AND APPLICATIONS

Many cantilever and similar devices for detecting molecules have been described and developed. In principle, a string or wire of nanodimensions could be used in a similar way, provided that a means of inducing and measuring vibrations were devised [35]. A number of books and reviews are available on the subject of nanoscale cantilever molecular sensors [34,35]. Among the many types of cantilever devices used to measure biomolecules, we present a few examples that have particular potential or interest for nanomedicine.

11.4.1 *Biomedical Cantilever Applications*

To make a cantilever sensor with a medical application, one may coat one surface with a reactant such as an antibody specific for an antigen. When the surface is exposed to solutions of the antigen, it becomes coated with molecules of the antigen, producing a small but detectable deformation of the cantilever due to

FIG. 11.2 *Absorption and steric crowding on an antibody coated cantilever.*

Nonadherent
cantilever straight

Adherent
cantilever bends
upward

surface forces. This deformation can be detected by optical reflection. Only for devices of nanoscale proportions are the intermolecular nanomechanics at the surface sufficient to produce a detectable deflection usable as a sensor (**Fig. 11.2**).

11.4.2 Cantilever Sensor for Cancer Screening

An interdisciplinary team of coworkers at University of California Berkeley, University of Southern California, and Oak Ridge National Laboratory developed an application of this technique to detect the prostate cancer antigen (PSA) [38]. Using 600-μm-long and 0.65-μm-thick silicon nitride cantilevers, it was feasible with their technique to detect PSA concentrations of $0.2\,ng\cdot mL^{-1}$. Since this antigen is produced by the body in response to cancer cells, this sensor could be useful in a blood screening test for early detection of cancer in humans. This technique may also lead to devices for high-throughput label-free analysis of protein–protein binding, DNA hybridization, and DNA–protein interactions, as well as drug discovery.

11.4.3 Biotechnology Applications of Cantilevers

In actual medical use, diagnosis for complex diseases such as cancer would require quantitative detection of multiple proteins. Therefore, current issues in research are how to improve sensitivity, specificity, and detection of real-world diagnostic concentrations of antigen in the presence of interferences under clinically relevant conditions. Many designs for using micro- and nanoscale cantilevers have been developed for biomedical detectors.

The Quartz Crystal Microbalance. The quartz crystal microbalance is a special case of the use of nanoscale effects in a cantilever. It is used to measure mass by the change in frequency of a piezoelectric quartz crystal when a small mass such as a virus or a large biomolecule is added. The frequency change is proportional to the added mass, so long as the added mass has elastic properties similar to those of the quartz itself. The mass change is given by the *Sauerbrey equation*:

$$\Delta f = \left[\frac{-2 f_{\text{ro}}}{A \sqrt{\rho_q G_q}} \right] \Delta m_q \tag{11.2}$$

where

Δf is the change in the resonance frequency due to the added mass
f_{ro} is the resonance frequency of the unloaded resonator
ρ_q and G_q are the density and shear modulus of the quartz
A is the surface area of the resonator [40–43].

Arrays of Cantilevers. As we mentioned earlier, arrays of microelectronic cantilevers have been developed for use as semiconductor memory and digital light-processing devices [23–26,44,45]. Digital versions of cantilever arrays can be adapted not only to sensitize patterns for making biochips, in maskless photo-lithography [26–29], but also to read out the status of microchip sensors by reflecting precisely focused laser beams onto sensors in an array. These devices can be used to prepare and interrogate arrays of antibodies, and also with DNA and RNA segments to bind to complementary sequences, making arrays that can select and identify genetic information [46–48]. Microarrays can even use a mixture of DNA and antibody sensors to detect genetic and protein information at the same time.

In the case of the digital micromirror array, each element is not a freely suspended cantilever, but is designed to switch between two states. Thus the mirror elements are not used directly as sensors, but to prepare and interrogate arrays of sensors by directing light to precise addresses in a microarray. The use of cantilevers as digitally controlled mirror arrays has had a large impact on the efficiency of DNA sequencing, perhaps one of the more direct examples of the interplay between nanoscale fabrication technology and molecular biology [49–52].

11.4.4 Surface Acoustic Wave Nanosensors

Another surface interface phenomenon which can be exploited for nanoscale detectors is the *surface acoustic wave* (SAW). Surface acoustic waves are compression vibrations traveling over the surface of an elastic material. SAW waves have a longitudinal and a vertical shear component. This type of wave propagation is distinct from *body waves* which travel through the interior of a material. The amplitude of a surface acoustic wave decays exponentially with depth into the material.

In dielectric materials, surface acoustic waves can be generated by application of electric fields, and detected by the transduction of the waves into electromagnetic fields. This behavior is used in electronic circuit devices. SAW devices are designed to act as filters, oscillators, and transformers in circuits, using one transducer to excite the wave and another to detect it at opposite ends of a strip of piezoelectric material.

SAW waves exhibit strong coupling with adsorbed material on the device surface. This effect allows SAW devices to be used as sensors by measuring changes in frequency and amplitude of the wave in response to contact with external substances.

Coating a SAW device with a specific antigen or other compound that bonds to a biological target molecule turns the device into a sensitive and specific detector that can be used to monitor for known antigens or even for bacteria [53]. SAW devices for biological applications have not yet seen widespread application, but they are an interesting example of the use of a nanoscale phenomenon which may in future have significant applications.

Conventional SAW devices use piezoelectric material. Researchers at Pennsylvania State University have developed SAW devices based on *magnetoelastic* materials. These have the advantage that a coil can be used to detect changes in the frequency of the waves, thus providing for wireless operation [54]. The resonant frequency of a magnetoelastic sensor changes in response to temperature, pressure, ambient flow rate, and liquid viscosity and density, as well as to absorbed mass, making this type of sensor very versatile.

11.4.5 *Electrochemical Nanosensors*

Electrochemical sensors convert chemical reaction energy into electrical potential or current. They are extremely sensitive, but must be in electrical contact with a tissue or fluid in order to be effective. Gases can be detected by electrochemical sensors if a moistened membrane is maintained on the surface of the sensor, in which the gas is absorbed.

Electrochemical probes measure the *electronic potential difference* at the interface between the surface of two materials, usually a solid and a liquid. Every chemical element has an inherent tendency to attract or donate electrons to its surroundings. Any time that two different materials are in contact with one another, an electronic potential exists across the interface. The greater the difference in the chemical tendency to donate or accept electrons, the larger the potential difference. If there are no barriers to chemical reactions at the interface, the reaction proceeds with exchange of electrons until the potential is neutralized and equilibrium is reached.

The difference in potential between two substances can be measured when both are in contact with a conductive medium, an *electrolyte*, through which electrons can be exchanged. In an *electrochemical cell*, two dissimilar electrodes are in contact with an electrolyte. An electronic potential is created by the donation or acceptance of electrons through chemical reactions which take place between the electrodes and the electrolyte.

If the two electrodes are connected by an electronic conducting path in parallel with the path through the electrolyte, a circuit is formed over which electrons can flow. Electronic charge flows over the conductor from the electrode with the greater tendency to accept electrons from the electrolyte towards the electrode which donates electrons in reactions with the electrolyte. This process accelerates the chemical reactions at the electrodes by providing a path for electron transfer that is of lower resistance than the path through the electrolyte. The energy of the chemical reaction is converted into electrical current, which is harnessed in batteries and fuel cells, and used in electrochemical sensors [55].

The potential difference at the interface between electrode and electrolyte is affected by adsorbed substances at the electrode surface. This provides a sensitive means for measuring the nature and concentration of molecules which interact with the surface of an electrode [56,57].

Electrochemical Detection of Free Radicals and Nitric Oxide in Tissues. An example of the usefulness of nanoprobes for electrochemical sensing is the detection of nitric oxide (NO) in tissues. Nitric oxide is a simple free radical that has been found to act as a cell signaling messenger playing a number of roles in the body. One of its most important functions is to signal the muscles in the wall of blood vessels to relax, allowing more blood to circulate. Nitroglycerin's effectiveness in relieving cardiovascular distress is due to the release of nitric oxide. This was one of the clues which led to nitric oxide being identified as a key endothelium-derived relaxing factor. Nitric oxide has been found to play a number of roles in different cell signaling cycles; its absence or excess can lead to tissue damage in heart, kidneys, and muscle.

Researchers at the Oak Ridge National Laboratory have developed an electrochemical probe to track concentrations of NO in tissue, using a functionalized carbon nanotube. Important free radicals and other species which can be monitored with the nanoprobe include NO, H_2O_2, O_2, H_2S, ATP, glucose, and superoxides [58,59]. The nanoprobe uses an electrochemical sensor principle and is based on functionalized carbon nanotubes with a sensor diameter of 100 nm (**Fig. 11.3**).

The advantages of an electrochemical nanoprobe compared with luminescence assays and other spectroscopic methods are

1. It is real-time rather than based on a derivative product.
2. It provides improved spatial resolution.
3. It does not require luciferin or other expensive bioluminescence or fluorescence reagents.

FIG. 11.3   *An electrochemical sensor probe made from a carbon nanofiber coated with silver chloride and a semipermeable membrane, with a 100-nm measurement tip.*

Ag/AgCl   Glass tube

Naflon PFA-selective membrane

100 nm

<~159 μm>

Functionalized carbon nanotube

4. It can be applied to direct measurement of cells in a culture or tissue without use of a luminometer or spectrometer, and requires no shielding from external light sources.

A Carbon Fiber Nanoelectrode for Glucose Detection.   Electrochemistry is a good method for detecting glucose and other simple compounds in blood and tissue. As one example, a glucose micro-biosensor has been developed by researchers in Austria using electrochemical co-deposition of glucose oxidase (GOx) along with MnO_2 as a mediator, onto a single carbon fiber microelectrode with a tip diameter of 100 nm [60]. A thin silver/silver chloride film is applied as an immobilization and interference-free protective layer. This type of micro-biosensor can be used as an amperometric glucose detector probe in blood or tissue. Improvements in glucose detection are important for research in diabetes, obesity, and metabolism.

Significance for Nanobiotechnology.   Nanoscale electrochemical sensors work on the same principles as macroscale sensors; what makes them significant for use in nanobiotechnology is the newly developed ability to fabricate such sensors in nanoscale sizes for detection of very small amounts of substances in complex mixtures typically found in biological systems [61–63]. The ability to use nanoscale electrochemical probes is also due to advances in electronics, which enables the precise measurement of extremely small potentials and currents. Arrays of electrochemical microsensors are being fabricated with both DNA and antibody probes, in order to analyze DNA and proteins on a single platform for proteomics and medical applications [64]. The importance of interfaces, chemical boundaries, and electrical forces in biological systems makes electrochemical measurement an important route to many biological measurements [65,66].

11.5   Optical Nanosensors

In this section we look at some examples of optical sensors, which allow direct measurement of light without depending on cantilevers or other mechanical nanodevices, and some of their biomedical applications. These include *photonic sensors*, which use the interaction of light with molecules in spectroscopic absorption, reflectance, fluorescence, or luminescence; and *plasmon sensors*, which use the interaction of light at a surface layer, waveguide, or nanoscale particle to detect the presence of a molecule or change in state of an absorbate.

11.5.1   *Photonic Nanosensors*

Photonic sensors can report biochemical reaction events at the molecular level by converting chemical energy into light signals. At the macroscale, this is a standard technique for spectroscopic identification. At the nanoscale, it can be used to interrogate very precise and subtle details of the inner workings of cells. This is accomplished by tagging specific parts of the cell's molecular pathway with a molecule that converts energy into light through fluorescence or bioluminescence, which can be detected by a nanoprobe or by nanoscale-capable microscopy techniques [67–72].

Photonic Sensor for Antibody Detection Inside a Single Cell. Tuan Vo-Dinh and colleagues in a research team at the Oak Ridge National Laboratory have developed an optical nanoprobe for measurements inside a single cell [73]. The nanoprobe uses a fluorescent antibody that attaches to a molecule produced when the carcinogen benzo[a]pyrene (BaP) attaches to DNA to form a mutation-causing adduct that can lead to cancer. Detection of this biomarker is significant in monitoring for DNA damage due to BaP exposure and for possible precancer diagnosis. BaP is a polycyclic aromatic hydrocarbon of serious environmental and toxicological interest because of its mutagenic/carcinogenic properties and its widespread presence in the environment [74]. The measurements were performed on model cells from a rat liver cell line which contained the carcinogen metabolite.

The nanoprobes were fabricated by the same technique used for cell patch clamp experiments (described in chapter 14 of *Introduction to Nanoscience*) [75]. A glass tube was heated and drawn out until its cross section at the tip was on the order of 50 nm. A reflective coating of silver was deposited on the outside of the glass probe, whose average diameter with the coating was 250–300 nm. The nanoprobes, which required great care to be made consistently, acted as fiberoptic waveguides which could be inserted through the cell membrane to take laser light into individual cells. The nanoprobes were coated with the fluorescent antibodies specific for the carcinogenic antigen, inserted into individual cells, incubated 5 min to allow antigen–antibody binding, and then removed for fluorescence detection. A concentration of 9.6×10^{-11} M for the carcinogen in the individual cells was determined.

Fiberoptic chemical sensors and biosensors offer important advantages for in situ monitoring applications because of the optical nature of the detection signal. The application of submicron fiberoptic chemical probes has been pioneered by Kopelman and coworkers, who developed probes for monitoring pH [76,77] and nitric oxide [78]. The use of submicron tapered optical fibers has also been demonstrated and used to investigate the spatial resolution that is possible using near-field scanning optical microscopy [79,80].

This example of an antibody-based nanoprobe for measurements of chemicals inside a single cell is being followed by further improvement of the technique for applications in biotechnology, molecular biology research, and medical diagnostics. This technique could eventually be used to develop advanced biosensing systems to study intracellular signaling and gene expression inside single cells.

11.5.2 *Surface Plasmon Nanosensors*

A *surface plasmon* is a charge density wave that is induced by light striking an interface between a thin film and another medium. *Surface plasmon resonance* (SPR) is the coupling of energy from incident light with the charge density in the surface, resulting in energy being transferred into a thin layer on the surface instead of being reflected. Resonance occurs only when the energy of the incident light is of the right frequency and angle of incidence for coupling.

SPR has been introduced in earlier chapters, as the phenomenon which gives rise to intense colors in butterfly wings, diatoms, nanodots, and other materials with thin layers on the nanoscale, corresponding to resonant frequencies with incident light. It is a useful phenomenon for optical detection and measurement where thin films can vary in composition and thickness. The waveform and intensity of the plasmon depend strongly on the thickness of the film and the type of material on which the film is deposited.

SPR is used to study the formation of self-assembled layers and structures of organic molecules formed at the surfaces of electrodes [81]. SPR can also be used to detect biological substances, such as DNA and proteins. A surface can be coated in patterns with different specific antigens or DNA complementary test strands to make an array in which a combination of biological target molecules can be detected (**Fig. 11.4**).

Dual Polarization Interferometry.   Dual polarization interferometry (DPI) is an extension of the application of SPR which divides focused laser beams into two waveguides, one, the "sensing" waveguide, with an exposed surface, and the other to serve as a reference beam. When light that has passed through both waveguides is combined, an interference pattern is created. Measurement of the interference pattern from two different polarizations of light permits the refractive index and the thickness to be determined for an adsorbed layer on the surface of the sensing waveguide. Because the polarization can be switched rapidly, real time measurements can be obtained for chemical reaction or flows taking place on the surface of the sensing waveguide. The DPI technique is used to obtain

FIG. 11.4   *Harnessing molecular recognition with electronics. (a) Treated metal surfaces are modified to anchor ligands—modified biological macromolecules with specific receptor sites for analytes. The optical and/or electronic properties of the ligand-coated surfaces change when analytes are captured. (b) The analyte is captured from the solution onto the active sites.*

structural information on proteins and other molecules adsorbed on the waveguide interface [82,83].

Plasmon Waveguide Resonance. Plasmon waveguide resonance (PWR) spectroscopy is a highly sensitive extension of the surface plasmon technique which is especially suited for characterizing the kinetics, affinities, and conformational changes involved in receptors in cell membranes, without the need for radioactive or other labeling techniques.

PWR spectroscopy utilizes a thin dielectric coating (e.g., silica) on a thin metal film (e.g., silver) deposited onto a prism. The dielectric functions as a waveguide allowing plasmon excitation by polarized light. Plasmon resonance in the waveguide provides narrow resonance and high sensitivity. The use of light with different polarizations to probe immobilized biological films provides detailed structural information on oriented molecular arrays.

Lipid bilayers can be deposited onto the resonator surface with their incorporated membrane proteins, and changes in resonance monitors ligand binding to the proteins. Lipid bilayers containing microdomains whose refractive index anisotropies and thicknesses differ from those of the bulk membrane will display multiple resonances and thus can be mapped.

PWR can measure differences in ligand-binding constants, and protein properties that result from differences in the microenvironment. Thus, this technology provides useful insights into the structural and functional consequences of microdomains in cell membranes [84].

11.5.3 Nanoscale Optical Resonance Grids—Using the Butterfly Wing Effect

At the nanoscale, there are new types of opportunities for photonic effects, due to resonance, as we have seen elsewhere in this volume. Now we consider a new type of nanoscale sensor which works on the same principle as the butterfly wing photonics, taking advantage of the nanoscale effects of guided mode optical surface resonance to detect very small amounts of material trapped by absorption between vanes of a nanoscale optical resonance grid. This and similar nanoscale optical phenomena are leading to many new types of biological sensors. These new systems are capable of sensing at extremely sensitive levels and are suitable for parallel integration for detection of multiple signals.

11.5.4 Guided-Mode Resonance Sensors

Leaky optical resonance modes arise on periodic films when an incident light beam couples to the waveguide-grating layer system. This results in generation of a *guided-mode resonance* (GMR) field response in the spectrum. The resonance effect leads to dramatic redistribution of the diffracted energy and may manifest as sharp reflection and transmission peaks radiating from the structure. Application of this fundamental effect to sensors was first proposed in 1992 [85].

The sensor is based on the high parametric sensitivity inherent in the basic GMR effect [85,86]. Figure 11.5 schematically illustrates a generic GMR sensor and its operation. As an attaching biomolecular layer changes the parameters of the resonance element, including thickness, fill factor, and refractive index, the resonance frequency changes [87]. Thus, incident light is efficiently reflected in a narrow spectral band whose central wavelength (and resonance angle) is highly sensitive to chemical reactions occurring at the surface of the sensor element. A target analyte interacting with a bio-selective layer on the sensor can thus be identified without additional processing or use of foreign tags. Computed examples and experimental results have been presented that illustrate key sensor properties and chart paths for establishing useful sensor technology based on this effect [88].

11.5.5 Applications of Guided Mode Sensors

These photonic resonance concepts enable a new class of highly sensitive and selective biosensors and chemical sensors with applications in medical diagnostics, drug development, environmental monitoring, homeland security, and others.

Guided-Mode Resonance Sensor for Protein Detection.   **Figure 11.5** illustrates the measurement of protein binding to a surface in air environment utilizing a double-layer resonant element illuminated at normal incidence. The clean grating surface is first chemically modified with amine groups by treating with a 3% solution of aminopropyltrimethoxysilane in methanol (**Fig. 11.6 Top**, **Left**). The device is then washed in a solution of bovine serum albumin (BSA, $100\,mg\cdot mL^{-1}$) and a deposited 38 nm thick layer of BSA results in a reflected resonant peak spectral shift of 6.4 nm (**Fig. 11.6 Bottom**, **Left**). It is significant that minimal signal degradation occurs as the biomaterial attaches to the sensor surface with reflectance remaining at ~90% before and after BSA attachment. The biomaterial layer thickness is determined by fitting the measured data with theoretically computed values. The observed shift in resonance wavelength of 6.4 nm is easily detected with economic spectrum analyzers.

FIG. 11.5   *A generic resonant sensor and its response. Interaction of a target analyte with a bioselective layer on the sensor surface yields measurable spectral shifts that directly identify the binding event without additional processing or foreign tags.*

FIG. 11.6

Measured leaky-mode resonance sensor spectral response in air and comparison with theory (Top, Left). The sensor is illuminated by a TE-polarized laser beam and the sensor surface is modified with a silane chemical linker; a 38-nm layer of attached bovine serum albumin (BSA) results in a 6.4-nm peak shift (Bottom, Left). Sensor parameters are $n_C = 1.0$, $n_1 = 1.454$ (SiO_2), $n_2 = 1.975$ (HfO_2), $n_S = 1.454$, $d_1 = 135\,nm$, $f = 0.58$, $d_2 = 208\,nm$, $\Lambda = 446\,nm$, and $\theta = 0°$. A scanning electron micrograph (SEM) of the sensor surface is also shown (Top, Right).

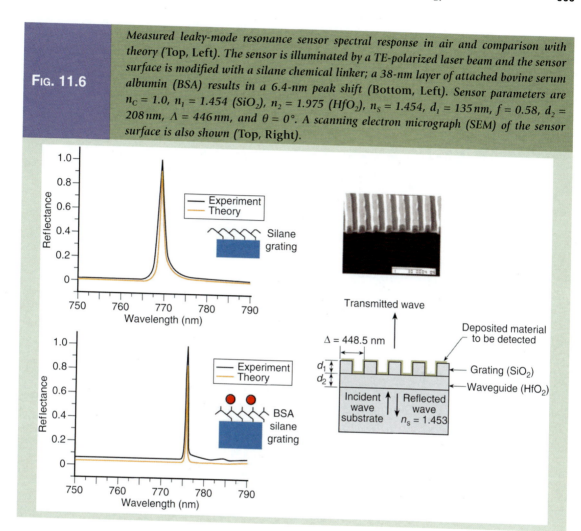

Significantly, polarization diversity permits monitoring of thickness and refractive index variations simultaneously. Occurrence of multiple leaky modes at resonance contributes to sculpting diverse spectral bands that are useful for new sensor types and many other optical elements. **Figure 11.7** shows an example application of a GMR sensor where both polarization states simultaneously monitor the reaction, providing added accuracy and precision in the quantitative determination of the biomolecular response.

Other Applications of Guided Mode Optical Sensors. Nanoscale optical sensors based on the butterfly wing resonance principle are leading to many new types of biological sensors. These new systems are capable of sensing at extremely sensitive levels and are suitable for parallel integration for detection of multiple signals [89–91]. Researchers are using butterfly wings as templates to fabricate optical grids and nanotubes for use in sensors [92,93]. The sensor systems being produced based on the butterfly wing principle dramatically outperform existing

FIG. 11.7 *Demonstration of the use of polarization diversity in GMR biosensing. The plot shows measured streptavidin binding dynamics for a biotin-coated sensor element with varying background represented by 75% milk solution and 25% milk solution.*

nanoengineered photonic sensors. They are being studied to develop new applications for detection of water, alcohols, and chlorinated hydrocarbons for a variety of biological, environmental, and medical applications.

11.6   NANOTECHNOLOGY FOR MANIPULATION OF BIOMOLECULES

In addition to sensing on the nanoscale, biomolecular nanotechnology involves ways of manipulating cells and macromolecules using nanoscale forces. Some of the most important nanoscale manipulation tools are the atomic force microscope and atomic force microprobe, which are discussed elsewhere in this book. Two other techniques for nanoscale manipulation are especially useful for biological studies: dielectrophoresis and optical tweezers. These techniques depend on the electromagnetic field phenomena acting on very small scales.

11.6.1   *Optical Tweezers*

Light consists of oscillating electronic and magnetic fields, a beam of light can exert electromagnetic forces when it interacts with matter. Light also has momentum, so when it is refracted, reflected, or absorbed, it produces a force on an object. On the macroscale, these forces are so small relative to the mass of an object that they are hardly detectable. But on the micro- and nanoscale, these forces become significant. Random scattering of suspended microscopic particles by incident light can be observed with a microscope. When light is highly coherent, as in a laser beam, these forces are aligned and have very directed and controllable effects.

Optical Trapping. A tightly focused laser beam which converges through a pinched focal point creates an optical trap for a charged or polarizable particle small enough to fit inside the focal area of the beam. This trapping effect is only possible on the microscale and smaller, where the electromagnetic forces on the particle are large relative to gravity and inertia, and where the wavelength of the laser light is on the same order as the size of the particle. The density gradient of the electromagnetic fields combines with the momentum transfer to create the trap geometry. This effect was discovered by Askin and coworkers in 1986 [94,95]. Since then it has been applied using laser focusing arrangements called *optical tweezers* to manipulate objects on the micro- and nanoscale. It is especially applicable to biological cells, since living cells are highly polarizable.

Optical Tweezers. Optical tweezers can manipulate particles as small as a single atom. In the biological sciences these instruments have been used to apply forces in the pN-range and to measure displacements in the nm range of objects ranging in size from 10 nm to over 100 mm. Optical tweezer instruments are now available as sophisticated computer controlled systems devices that are able to measure displacements and forces with high precision and accuracy.

Applications of Optical Tweezers. Optical tweezers have been used to trap and manipulate dielectric inorganic particles, viruses, bacteria, living cells, organelles, proteins, and DNA segments. They can be used for confinement and sorting of cells, tracking of movements of bacteria and motile cells, measurement of small forces, and manipulation and rearrangement of cell membranes and structures.

Because visible light forms ideal traps at the micron scale, smaller objects are usually studied under the microscope with visible wavelength lasers by attaching specimens to micron-sized beads, which are then trapped, where the effects of light forces can be studied. Forces required to rearrange DNA or proteins can be measured using this technique. Molecular motors such as kinesin, myosin, and RNA polymerase can be studied to determine the size and continuity of motion steps and forces required to produce movements [96–99].

11.6.2 Dielectrophoresis

Electrophoresis is the phenomenon of movement, or *electrokinetics*, of a particle induced by an electrical field gradient, known since the early 1800s. An electrical field induces a force on a polarizable particle. As in the case of optically induced forces, the electrophoresis forces become significant at the micro- and nanoscale. Electrophoresis is used to separate colloidal particles and biological macromolecules such as proteins and DNA fragments, and has become a highly useful and refined technique in molecular biology and pathology [100–103].

Dielectrophoresis is a more complex phenomenon in which polarizable particles move in response to alternating electromagnetic fields [104]. Complex standing waves and gradients can be created around alternating current electrodes, which can trap and move polarizable particles such as biological cells. Dielectrophoresis has become a refined and useful technique in molecular biology and nanotechnology with the availability of sophisticated electronics and computer analysis to direct the patterns of fields [105]. Dielectrophoresis can be employed as the active principle for submicron scale instruments to (1) separate and isolate specific cells of interest, (2) concentrate and amplify cell concentrations, (3) accurately measure indicators for different cells, and (4) measure a quantitative result based on cell differences. Dielectrophoresis is being designed in microfabricated devices to separate cells for research, diagnosis, and preparation of purified fractions.

In dielectrophoresis an alternating current is applied to a volume containing cells to create a nonuniform temporal and geometric pattern of electromagnetic fields. A dielectrophoretic force acting on the polarized cells causes them to move along the electric field gradient created by the nonuniform electrode geometry. At specific radio frequencies, cells are moved differently depending on their polarizability and geometry. Polarizibility can distinguish genetically similar but distinct cells, which will have different movement responses at given frequencies.

Dielectrophoresis can be used like the more conventional methods of cell concentration and separation, such as centrifugation, filtration, fluorescence-activated cell sorting in flow cytometry, or optical tweezers. Dielectrophoresis does not require the chemical binding of labels or tags to the cell; it is based on the inherent polarizability of the cell's molecular contents. Thus it can operate directly on native, unlabeled cells, without the expense of labeling, or development of labels and tags. A single dielectrophoretic instrument can isolate and analyze a wide range of particle types (cells, bacteria, viruses, DNA, and proteins) using the same basic procedure.

Dielectrophoresis does not generally harm cells, so they remain viable for further study, processing, or therapy. Research is being undertaken to find whether, under certain conditions, dielectrophoresis could be capable of selectively destroying cells such as bacteria and cancer cells [106].

Development is also being done to refine dielectrophoresis separation and analysis techniques to analyze intact cells to provide detailed selective information on their DNA makeup, and to electrically lyse cells to break out their contents for processing and analysis.

11.6.3   *Some Dielectrophoresis Applications*

Separation of Cancer Cells from Blood.   Dielectrophoresis can be used to separate cancer cells from normal cells, based on differences in their polarizability. A measure of the dielectric properties of cells is the plasma membrane capacitance, which has been found to differ greatly between normal and cancer cells. For example, MDA-231 human breast cancer cells have been found to have a mean plasma membrane specific capacitance of $26\,mF\cdot m^{-2}$, more than double the value ($11\,mF\cdot m^{-2}$) observed for resting T lymphocytes.

Researchers at the M. D. Anderson Cancer Center have used a dielectrophoresis affinity column to separate several different cancerous cell types from blood. The affinity column consists of a series of thin, flat chambers with microelectrode arrays on the bottom wall, through which the blood flows. Dielectrophoresis forces generated by the application of alternating current fields to the electrodes influence the rate of elution of cells from the chamber. Dielectric properties were measured for the various cell types in the mixtures to be separated, and theoretical modeling was used to design the optimal flow column and applied field parameters. Various ratios of cancerous to normal cells at different concentrations separated in test runs in the column. The results included a 100% efficiency for purging MDA-231 cells from blood at a ratio of about 1 cancer cell to 100,000 normal cells. Cell viability was not compromised, and separation rates were at least 103 cells·s⁻¹ [107]. These results could lead to a diagnostic test for very early detection of cancer.

Concentrating Bacteria for Water Analysis. One of the problems that arise with nanoscale detection and analysis is finding the molecule or cell of interest when it is hidden in a large volume of material. It does little good to have a sensor that is capable of detecting a single molecule if the sensor is unlikely to ever encounter the molecule of interest in the midst of billions of others. Hence ways of concentrating or replicating the molecule or cell to be detected are sought.

The traditional way of amplifying the bacteria count so that an assessment of bacterial contamination can be made is to allow a small sample of bacteria to grow for hours or days in a culture medium, until the colony, held in place by a gel, is sizeable enough to be seen and evaluated. DNA can be amplified by a cyclical reaction technique, the *polymerase chain reaction* (PCR), by which a single strand of DNA is allowed to replicate repeatedly in a medium of base components and replication enzymes, through temperature cycles that cause the DNA chains to replicate and separate, doubling the amount of DNA on each cycle [108]. Centrifugation, filtration, evaporation, and other concentration techniques are not practical for detecting small numbers of bacteria in large volumes of water for a number of reasons.

Because it can selectively and efficiently separate different types of cells, dielectrophoresis can be used to concentrate cells into a small volume for efficiency of counting and analysis.

Insulator-Based Dielectrophoresis for Concentration of Bacteria. Researchers at Sandia National Laboratories have developed a novel form of dielectrophoresis, called *insulator-based dielectrophoresis* (iDEP), to selectively—and very quickly—concentrate live pathogenic bacteria from large water samples. iDEP can enable detection of small amounts of biological material in large sample volumes, without bacterial cell culturing. This significantly lowers costs, speeds analysis, and reduces risks of contamination [109].

iDEP could be used in medical diagnostic tests where a few anomalous cells result from diseases, such as cancer, sickle cell anemia, and leukemia. It could also be useful for protein isolation and concentration, and mass spectrometry sample preparation for proteomics and drug discovery.

Conventional dielectrophoretic sorters have electrodes within a sampling chamber and use the nonuniform electric field adjacent to electrodes to induce dielectrophoretic motion of cells. Dielectrophoretic electrodes typically require precision microfabrication, produce bubbles, and electrolyze products that can harm device operation, and can damage cells with strong field gradients.

In contrast, iDEP uses electrodes located on the outside of the sample chamber. Current from the electrodes conducts through a particle-bearing liquid into the device where patterned walls or insulating obstacles produce the required nonuniform electric field. This arrangement eliminates many of the disadvantages of conventional devices: insulating structures can be replicated economically, produce no electrolytic effect, and can be contoured to be gentle on cells.

Integration of Dielectrophoresis into Highly Parallel Devices for High Throughput. Dielectrophoresis is well suited for integration into highly parallel devices for rapid screening and analysis of multiple types of cells or macromolecules in complex mixtures [110].

Dielectrophoresis Summary. Dielectrophoresis makes use of the effects of small amplitude electromagnetic forces interacting with very small particles. At the scale of microsized particles and cells, the nanoscale geometric landscapes produce effects that are not intuitive or familiar to us from experience with macroscale objects and forces. Understanding and using these nanoscale phenomena to design new types of dielectrophoresis devices requires mastery of the descriptive equations and modeling techniques for predicting the behavior of alternating electromagnetic fields, which is beyond the scope of this introductory chapter. Further aspects of electromagnetics, physical chemistry, biochemistry, molecular and cell biology, and other disciplines are found in other chapters of this book and in the references.

11.6.4 *Micro- and Nanofluidics*

For large bodies of liquid, shape and flow characteristics can be understood with negligible consideration for electric charge and surface tension effects. As the volume of a stream or pool of liquid decreases, surface tension and evaporation become increasingly important factors in the behavior of the liquid, leading to the formation of droplets. At the microscale and below, electric charge becomes a very important force in the behavior of the droplets. These phenomena lead to the behaviors of aerosols and clouds, and have been exploited in ink-jet printers and similar technologies. They have also been exploited in biological and chemical applications, a few of which we will explore here.

Ink-Jet Printing. Ink-jet printing ejects ink under pressure through very small capillary nozzles in a series of droplets. In some versions of the technology the droplets are electrically charged, and can be accelerated and focused by electric fields. A similar technique is used in biology to inject minute amounts of material into cells. This procedure is called ballistic injection, or variously biolistic, particle bombardment, or gene-gun injection. It is used to transfer genetic DNA material into plants and animal cells.

Ballistic Injection. To transfer genes into cells, plasmid DNA encoding the gene of interest is coated onto 1- to 3-μm-sized gold or tungsten microbeads, and these particles are then accelerated by one of several motive force techniques to penetrate cell membranes. Ballistic DNA injection has been successfully used to transfer genes to a wide variety of mammalian cell lines in culture, and in vivo to the epidermis, muscle, liver, or other organs which can be exposed surgically. Ballistic DNA injection can deliver precise DNA dosages, but genes delivered by this method tend to be expressed only transiently and considerable cell damage may occur at the discharge site [111].

Ink-Jet Coating to Functionalize Cantilever Sensors.   In an earlier section we saw how a nanomechanical cantilever could be functionalized to serve as a chemical or biochemical sensor. Ink-jet printing can be used to rapidly and accurately direct controlled deposition of functional layers onto cantilevers for sensor fabrication. Alkanethiols can be deposited onto gold-coated cantilever surfaces to make sensors for pH or ion concentrations in liquids. Single-stranded DNA oligomers can be bonded onto gold surfaces by thiol linkages to make detectors for gene fragments. Chemical gas detectors can be made by printing thin layers of selected polymers from dilute solutions onto cantilevers. The ink-jet method is noncontact, and is simpler, faster, and more accurate than applying coatings with microcapillaries or pipettes; it is more easily scalable to large arrays and can coat arbitrary patterns [112].

Electrospray Ionization for Mass Spectrometry Ion Sources.   An ink-jet type system is used in mass spectrometry as a way to produce gas phase ions of large molecules such as proteins and DNA fragments, which otherwise tend to break up if volatized and ionized by electron bombardment or photolysis. This technique, along with matrix-assisted laser desorption from surfaces, makes mass spectrometry a useful technique for the analysis of large biomolecules. In electrospray ionization, a liquid is forced from a very small charged metal capillary opening. The liquid contains a dilute solution of the substance to be analyzed in the mass spectrometer, in a volatile solvent mixture. The analyte molecules are in solution in the ionic state, either as anions or cations. The liquid exits the capillary and forms an aerosol mist of droplets about 10 μm across. As the solvent evaporates, the analyte molecules are forced closer together, repel each other, and break up the droplets in a process called coulombic fission. Driven by repulsive coulombic forces between the charged molecules, the process repeats itself until the analyte is free of solvent, leaving isolated ions. The ions are separated in the mass analyzer stage of the spectrometer [113,114]. This is a very important application of nanoscale science which was recognized with a Nobel Prize in Chemistry to John Bennett Fenn in 2002.

Other Applications for Capillary Jet Injection.   Ink-jet-like spray techniques are used to create a variety of unique particles and structures, taking advantage of intermolecular forces that become significant in microdroplets of mixed molecules as solvents evaporate and charges interact [115]. Ink-jet techniques are used to produce microencapsulation and nanoencapsulation of particles, as micelles and similar structures are formed in droplets composed of mixtures of a drug and an encapsulating polymer or other coating [116]. Ink-jet and electrospray techniques can be used to spray coatings with controlled nanosurface structures for biocompatible microdevices and materials [117]. Microfluidic spray droplets are used in surgery and dentistry, to coat patterns of functionalization in all kinds of biochips, and many other biotechnology applications [118].

11.6.5   *Biochips, Labs on Chips, and Integrated Systems*

Any of the sensor modalities surveyed in this section can be used to create arrays with multiple sensors, thus yielding very powerful screening and analysis tools. In diagnostic and chemical analysis applications in the clinic or laboratory, an array with a large number of sensors will be preferred, each primed to detect a specific analyte of interest. The various types of "biochips" that have been developed are based on printing a pattern of sensors in an array on a chip which can be interrogated for the presence or absence of the analyte to be detected [119].

Biochips.   The engineering of biochips represents an integration of molecular biology and biochemistry with microelectronics and digital control. Demanding engineering challenges continue to exist in making such arrays smaller, more efficient, and more reliable, and exploration of the disciplines and technologies that go into the making of biochips is beyond the scope of this chapter. The future application of this technology will be involved with higher degrees of integration with electronics and digital wireless communications. We will describe a few examples by way of introduction.

Labs on Chips.   Microfluidics can play a role in conjunction with nanosensors to create a "lab on a chip" in which extremely small samples are delivered to the sensing area for analysis by a nanosensor, an array of sensors, or a series of sensors. In some cases micro- and nanodot technologies are used to deliver samples, using injection techniques similar to those used for ink-jet printers, precision microelectronics soldering systems, and nanospray ion sources used in mass spectrometry. Many biochips rely on wired connections to the sensors to monitor the result of interaction with the analyte, but optical or wireless methods will be preferred where possible [120].

Integration of Cells with Biochips.   One example of how microfluidics supports the use of nanosensors in cell research is seen in the use of a microfluidic device that allows neurons to be grown at low density. Brain researchers seek to analyze the signals between neurons with the goal of finding new treatments for brain damage and disease. Elucidation of neuron signaling is made difficult because neurons die without communication with networks of other neurons. Thus they can only be kept alive in vitro in dense cultures of many neurons, making it difficult to distinguish the signaling patterns.

A research group at the University of Illinois at Urbana-Champaign has developed a method of culturing neurons at low density, using a microfluidic chamber. The group has optimized conditions in microfluidic devices so that neurons can survive for around 11 days as opposed to 3 or 4. They used a material commonly employed in microfluidics but by heating the chip and washing it with buffer have made the nanoliter channels a suitable environment to grow neurons. This work was reported in a journal titled *Lab on a Chip*, which as the name implies is a good source for further information on this subject. [121]

Carbon Nanofibers for Inserting DNA into Cells.   Carbon nanofibers have been coated with plasmid DNA and inserted into cells, where the DNA was viable in expressing RNA and proteins. This represents a radically new way of modifying the genetic behavior of cells in culture. If it were able to be applied in the human body, it could lead to "gene therapy on a stick," with DNA plasmids containing genes for expressing insulin, growth hormone, or neurotransmitters, for example, could be inserted into cells to repair deficiencies or alter organ and brain functions [122].

In experiments cells have survived for more than three weeks with the plasmid coated carbon nanotubes continuing to express their DNA. If the DNA coating is only physically absorbed onto the surface of the carbon nanofibers, it is transferred into the cell and is passed on to daughter cells when the cells divide. But if the DNA is chemically bonded to the nanofiber, it is not transferred into the cell's endoplasm domain and thus is not inherited by progeny cells upon cell division. This could provide a way of controlling genes that turn on cell growth, in order to repair bone or nerves or grow new skin or tendons, but without unlimited growth that would lead to tumors or cancer.

This development is just one example of cellular engineering on the molecular level using probes that control biomolecular reactions with far more precision than could be possible without nanoscale probes. The possibilities of these technologies have only just begun to be realized.

11.7 SUMMARY

Biomolecular nanotechnology is a rapidly emerging field, with many more aspects than can be covered in a short introductory study. The fields of biomolecular motors, DNA computation engines, artificial muscles and organs, implantable medical devices, and many others have only been mentioned or not covered at all, and will be left to future chapters and sections. Many aspects of what we have called biomolecular nanotechnology overlap strongly with medical nanotechnology, which is the subject of the next chapter.

In selecting topics we have tried to provide continuity with the material covered in *Introduction to Nanoscience*, by illustrating how natural nanomaterials such as the *Morpho* butterfly and nanoscale biomolecular systems such as the dielectric potential of cellular membranes can be harnessed in technical applications. The examples selected are intended to give a broad overview of the field, and the references provided will point the reader to many additional applications and techniques.

Acknowledgments

We wish to acknowledge the significant contributions from the authors who graciously gave permission for use of their work and illustrations, citation in the references, especially Resonant Sensors Incorporated, for giving Robert Magnusson's and Debra Wawro's permissions for contributions and use of figures and data.

References

1. D. Springham, G. Springham, V. Moses, and R. E. Cape, *Biotechnology: The science and the business*, Taylor & Francis, Boca Raton, FL (1999).
2. P. Rabinow, *Making PCR: A story of biotechnology*, University of Chicago Press, Chicago, IL (1996).
3. C. Ratledge and B. Kristiansen, *Basic biotechnology*, Cambridge University Press, Cambridge, UK (2006).

4. D. S. Goodsell, *Bionanotechnology: Lessons from nature*, Wiley-Liss, Inc., Hoboken, NJ (2004).
5. M. Koehler and S. Diekmann, Biomolecular nanotechnology, *Reviews in Molecular Biotechnology*, 82, 1–2 (2001).
6. N. H. Malsch ed., *Biomedical nanotechnology*, CRC Press, Boca Raton, FL (2005).
7. R. R. H. Coombs and D. W. Robinson, *Nanotechnology in medicine and the biosciences*, CRC Press, Boca Raton, FL (1996).
8. T. Vo-Dinh, ed., *Nanotechnology in biology and medicine: Methods, devices, and applications*, CRC Press, Boca Raton, FL (2007).
9. J. D. Bronzino, ed., *Biomedical engineering handbook*, IEEE Press/CRC Press, Boca Raton, FL (2005).
10. J. Moore and G. Zouridakis, *Biomedical technology and devices handbook*, CRC Press, Boca Raton, FL (2004).
11. A. Guiseppi-Elie and T. Vo-Dinh eds., *Biochips handbook*, CRC Press, Boca Raton, FL (2007).
12. M. J. Heller and A. Guttman, *Integrated microfabricated biodevices*, Marcel Dekker, New York (2002).
13. J. S. Albala and I. Humphery-Smith, *Protein arrays, biochips and proteomics*, Marcel Dekker, New York (2003).
14. M. Schena, *Microarray analysis*, John Wiley & Sons, Inc., New York (2002).
15. K. Mitchelson, ed., *New high throughput technologies for DNA sequencing and genomics 2*, Elsevier, Amsterdam, the Netherlands (2007).
16. C. M. Niemeyer and C. A. Mirkin, *Nanobiotechnology*, Wiley-VCH, Weinheim, Germany (2004).
17. J. Benyus, *Biomimicry: Innovation inspired by nature*, William Morrow and Co., New York (1997).
18. G. Virella, *Medical immunology*, 7th ed., Informa Healthcare, New York (2007).
19. C. A. Janeway Jr., P. Travers, M. Walport, and M. J. Shlomchik, *Immunobiology*, 5th ed., Garland Publishing, Inc., New York (2001).
20. L. Du Pasquier and G. W. Litman, *Origin and evolution of the vertebrate immune system*, Springer Verlag, Berlin, Germany (2000).
21. R. Latorre and J. C. Sáez, eds., *From ion channels to cell-to-cell conversations*, Springer Verlag, Berlin, Germany (1997).
22. J. Haurum and S. Bregenholt, Recombinant polyclonal antibodies: Therapeutic antibody technologies come full circle, *IDrugs*, 8, 404–409 (2005).
23. H. G. Craighead, Nanoelectromechanical systems, *Science*, 290, 1532–1535 (2000).
24. C. A. Savran et al., Micromechanical detection of proteins using aptamer-based receptor molecules, *Analytical Chemistry*, 76, 3194–3198 (2004).
25. R. Berger et al., Micromechanics: A toolbox for femtoscale science, *Microelectronic Engineering*, 35, 373–379 (1997).
26. P. Vettiger, J. Brugger, M. Despont, U. Drechsler, U. Dürig, W. Häberle, M. Lutwyche, H. Rothuizen, R. Stutz, R. Widmer, and G. Binnig, Ultrahigh density, high-data-rate NEMS-based AFM data storage system, *Microelectronic Engineering*, 46, 11–17 (1999).
27. S. Singh-Gasson, R. D. Green, Y. Yue, C. Nelson, F. Blattner, M. R. Sussman, and F. Cerrina, Maskless fabrication of light-directed oligonucleotide microarrays using a digital micromirror array, *Nature Biotechnology*, 17, 974–978 (1999).
28. R. McKendry, J. Zhang, Y. Arntz, T. Strunz, M. Hegner, H. P. Lang, M. K. Baller, U. Certa, E. Meyer, H.-J. Guntherodt, and C. Gerber, Multiple label-free biodetection and quantitative DNA-binding assays on a nanomechanical cantilever array, *The Proceedings of the National Academy of Science USA*, 99, 9783–9788 (2002).
29. Y. Arntz, J. D. Seelig, H. P. Lang, J. Zhang, P. Hunziker, J. P. Ramseyer, E. Meyer, M. Hegner, and C. Gerber, Label-free protein assay based on a nanomechanical cantilever array, *Nanotechnology*, 14, 86–90 (2003).

30. I. G. Main, *Vibrations and waves in physics*, Cambridge University Press, Cambridge, UK (1993).

31. A. P. French, *Vibrations and waves*, W. W. Norton, New York (1971).

32. S. I. Lee, S. W. Howell, A. Raman, R. Reifenberger, C. V. Nguyen, and M. Meyyappan, Nonlinear tapping dynamics of multi-walled carbon nanotube tipped atomic force microcantilevers, *Nanotechnology*, 15, 416–421 (2004).

33. H. P. Lang, M. Hegner, and C. Gerber, Cantilever array sensors, *Materials Today*, 8, 30–36 (2005).

34. J. Gardner and P. N. Bartlett, eds., *Sensors and sensory systems for an electronic nose*, Kluwer Academic Publishers, Dordrecht, the Netherlands (1992).

35. J. Fritz, M. K. Baller, H. P. Lang, H. Rothuizen, P. Vettiger, E. Meyer, H. -J. Güntherodt, Ch. Gerber, and J. K. Gimzewski, Translating biomolecular recognition into nano-mechanics, *Science*, 288, 316–318 (2000).

36. A. K. Gupta, P. R. Nair, D. Akin, M. R. Ladisch, S. Broyles, M. A. Alam, and R. Bashir, Anomalous resonance in a nanomechanical biosensor, *The Proceedings of the National Academy of Science USA*, 103, 13362–13367 (2006).

37. R. Kamalian, Y. Zhang, and A. M. Agogino, *Microfabrication and characterization of evolutionary MEMS resonators*, Proceedings of the 7th Symposium on Micro- and Nano-Mechatronics for Information-based Society, IEEE Robotics and Automation Society, pp. 109–114 (2005).

38. G. Wu, R. H. Datar, K. M. Hansen, T. Thundat, R. J. Cote, and A. Majumdar, Bioassay of prostate-specific antigen (PSA) using microcantilevers, *Nature Biotechnology*, 19, 856–860 (2001).

39. G. Wu, H. Ji, K. Hansen, T. Thundat, R. Datar, R. Cote, M. F. Hagan, A. K. Chakraborty, and A. Majumdar, Origin of nanomechanical cantilever motion generated from biomolecular interactions, *The Proceedings of the National Academy of Science USA*, 98, 1560–1564 (2001).

40. P. E. Sheehan and L. J. Whitman, Detection limits for nanoscale biosensors, *Nano Letters*, 5, 803–807 (2005).

41. G. Sauerbrey, Verwendung von Schwingquarzen zur Wägung dünner Schichten und zur Mikrowägung, *Zeit Physics*, 155, 206–222 (1959).

42. K. K. Kanazawa and J. G. Gordon, Frequency of a quartz microbalance in contact with liquid, *Analytical Chemistry*, 57, 1770–1771 (1985).

43. B. D. Vogt, E. K. Lin, W. Wu, and C. C. White, Effect of film thickness on the validity of the Sauerbrey equation for hydrated polyelectrolyte films, *Journal of Physical Chemistry B*, 108, 12685–12690 (2004).

44. D. R. Collins, J. B. Sampsell, L. J. Hornbeck, J. M. Florence, P. A. Penz, and M. T. Gately, Deformable mirror device spatial light modulators and their applicability to optical neural networks, *Applied Optics*, 28, 4900–4904 (1989).

45. L. J. Hornbeck, Digital light processing and MEMS: An overview, Advanced Applications of Lasers in Materials Processing/Broadband Optical Networks/Smart Pixels/Optical MEMs and Their Applications: IEEE/LEOS 1996, 7–8 (1996).

46. G. H. McGall, A. D. Barone, M. Diggelmann, S. P. A. Fodor, E. Gentalen, and N. Ngo, The efficiency of light-directed synthesis of DNA arrays on glass substrates, *Journal of American Chemical Society*, 119, 5081–5090 (1997)

47. A. C. Pease, D. Solas, E. J. Sullivan, M. T. Cronin, C. P. Holmes, and S. P. A. Fodor, Light-generated oligonucleotide arrays for rapid DNA sequence analysis, *The Proceedings of the National Academy of Science USA*, 91, 5022–5026 (1994).

48. G. McGall, J. Labadie, P. Brock, G. Wallraff, T. Nguyen, and W. Hinsberg, Light-directed synthesis of high-density oligonucleotide arrays using semiconductor photoresists, *The Proceedings of the National Academy Science USA*, 93, 13555–13560 (1996).

49. E. Maier, S. Meier-Ewert, D. Bancroft, and H. Lehrach, Automating array technologies for gene expression profiling, *Drug Discovery Today*, 2, 315–324 (1997).

50. K. M. O'Brien, J. Wren, V. K. Dave, D. Bai, R. D. Anderson, S. Rayner, G. A. Evans, A. E. Dabiri, and H. R. Garner, ASTRAL, A hyperspectral imaging DNA sequencer, *Review of Scientific Instruments*, 69, 2141–2146 (1998).

51. H. R. Garner, R. P. Balog, and K. L. Luebke, The evolution of custom microarray manufacture, *IEEE Engineering in Medicine and Biology Magazine*, 21, 123–125 (2002).

52. S. E. Lyshevski, *Nano and molecular electronics handbook, series: Nano- and microscience, engineering, technology and medicine volume: 9*, CRC Press, Boca Raton, FL (2007).

53. D. Branch and S. M. Brozik, Low-level detection of a *Bacillus anthracis* simulant using love-wave biosensors on 36° YX LiTaO$_3$, *Biosensors and Bioelectronics*, 19, 849–859 (2004).

54. M. K. Jain, Q. Cai, and C. A. Grimes, A wireless micro-sensor for simultaneous measurement of pH, temperature, and pressure, *Smart Materials and Structures*, 10, 347–353 (2001).

55. V. S. Bagotsky, *Fundamentals of electrochemistry*, 2nd ed., Wiley, New York (2006).

56. J. Wang, *Analytical electrochemistry*, 3rd ed., Wiley-VCH, Hoboken, NJ (2006).

57. H. B. Girault, *Analytical and physical electrochemistry*, Dekker, New York (2004).

58. X. Zhang, L. Cardosa, M. Broderick, H. Fein, and J. Lin, An integrated nitric oxide sensor based on carbon fiber coated with selective membranes, *Electroanalysis*, 12, 1113–1117 (2000).

59. X. Zhang, Y. Kislyak, J. Lin, A. Dickson, L. Cardosa, M. Broderick, and H. Fein, Nanometer size electrode for nitric oxide and S-nitrosothiols measurement, *Electrochemistry Communications*, 4, 11–16 (2002).

60. S. B. Hocevar, B. Ogorevc, K. Schachl, and K. Kalcher, Glucose microbiosensor based on MnO$_2$ and glucose oxidase modified carbon fiber microelectrode, *Electroanalysis*, 16, 1711–1716 (2004).

61. W. Lorenz and W. Plieth, *Electrochemical nanotechnology: In situ local probe techniques at electrochemical interfaces*, Wiley-VCH, Weinheim, Germany (1998).

62. R.-I. Stefan, J. F. van Staden, and H. Y. Aboul-Enein, *Electrochemical sensors in bioanalysis*, Dekker, New York (2001).

63. A. Brajter-Toth and J. Q. Chambers, *Electroanalytical methods for biological materials*, Dekker, New York (2002).

64. J. C. Harper, R. Polsky, D. R. Wheeler, S. M. Dirk, S. M. Brozik, Selective immobilization of DNA and antibody probes on electrode arrays: Simultaneous electrochemical detection of DNA and protein on a single platform, *Langmuir*, 23, 8285–8287 (2007).

65. J. W. Schultze, T. Osaka, and M. Datta, *Electrochemical microsystem technologies (New trends in electrochemical technology, Vol. 2)*, Taylor and Francis, New York (2002).

66. E. Palecek, F. Scheller, and J. Wang, *Electrochemistry of nucleic acids and proteins: Towards electrochemical sensors for genomics and proteomics*, Elsevier, Boston, MA (2005).

67. O. S. Wolfbeis, *Fiber optic chemical sensors and biosensors, Vol 1*, CRC Press, Boca Raton, FL (1991).

68. G. Boisde and A. Harmer, *Chemical and biochemical sensing with optical fibers and waveguides*, Artech House, Boston, MA (1996).

69. S. Donati, *Electro-optical instrumentation: Sensing and measuring with lasers*, Prentice Hall, Upper Saddle River, NJ (2004).

70. G. Orellana, *Frontiers in chemical sensors: Novel principles & techniques, series on chemical sensors & biosensors, Vol. 3*, Springer Verlag, Berlin, Germany (2005).

71. R. M. Nakamura, Y. Kasahara, and G. A. Rechnitz, eds., *Immunochemical assays and biosensor technology*, American Society for Microbiology, Washington, DC (1992).

72. T. Vo-Dinh, M. J. Sepaniak, G. D. Griffin, J. P. Alarie, Immunosensors: Principles and applications, *Immunomethods*, 3, 85–92 (1993).

73. T. Vo-Dinh, J.-P. Alarie, B. M. Cullum, and G. D. Griffin, Antibody-based nano-probe for measurement of a fluorescent analyte in a single cell, *Nature Biotechnology*, 18, 764–767 (2000).

74. T. Vo-Dinh, ed., *Chemical analysis of polycyclic aromatic compounds*, Wiley, New York (1989).

75. B. Hille, *Ion channels of excitable membranes*, 3rd ed., Sinauer, Sunderland, MA (2001).

76. W. Tan, Z.-Y. Shi, and R. Kopelman, Development of submicron chemical fiber optic sensors, *Analytical Chemistry*, 64, 2985–2990 (1992).

77. T. Tan, Z.-Y. Shi, S. Smith, D. Birnbaum, and R. Kopelman, Submicrometer intra-cellular chemical optical fiber sensors, *Science*, 258, 778–781 (1992).

78. S. L. R. Barker, Y. D. Zhao, M. A. Marletta, and R. Kopelman, Cellular applications of a sensitive and selective fiber optic nitric oxide biosensor based on a dye-labeled heme domain of soluble guanylate cyclase, *Analytical Chemistry*, 71, 2071–2075 (1999).

79. E. Betzig, J. K. Trautman, T. D. Harris, J. S. Weiner, and R. L. Kostelak, Breaking the diffraction barrier: Optical microscopy on a nanometric scale, *Science*, 251, 1468–1470 (1991).

80. V. Deckert, D. Zeisel, R. Zenobi, and T. Vo-Dinh, Near-field surface-enhanced Raman imaging of dye-labeled DNA with 100-nm resolution, *Analytical Chemistry*, 70, 2646–2650 (1998).

81. L. L. Norman and A. Badia, Electrochemical surface plasmon resonance investiga-tion of dodecyl sulfate adsorption to electroactive self-assembled monolayers via ion-pairing interactions, *Langmuir ASAP*, 10, 1021 (2007).

82. G. H Cross, A. Reeves, S. Brand, M. J. Swann, L. L Peel, N. J. Freeman, and J. R Lu, The metrics of surface adsorbed small molecules on the Young's fringe dual-slab waveguide interferometer, *Journal of Physics D: Applied Physics*, 37, 74–80 (2004).

83. N. J. Feeman, L. L. Peel, M. J. Swann, G. H. Cross, A. Reeves, S. Reeves, and J. R. Lu, Real time, high resolution studies of protein adsorption and structure at the solid–liquid interface using dual polarization interferometry, *Journal of Physics: Condensed Matter*, 16, S2493–S2496 (2004).

84. G. Tollin, Z. Salamon, S. Cowell, and V. J. Hruby, Plasmon-waveguide resonance spectroscopy: A new tool for investigating signal transduction by G-protein coupled receptors, *Life Sciences*, 73, 3307–3311 (2003).

85. R. Magnusson and S. S. Wang, New principle for optical filters, *Applied Physics Letters*, 61, 1022–1024 (1992).

86. S. S. Wang and R. Magnusson, Theory and applications of guided-mode resonance filters, *Applied Optics*, 32, 2606–2613 (1993).

87. D. Wawro, S. Tibuleac, R. Magnusson, and H. Liu, Optical fiber endface biosensor based on resonances in dielectric waveguide gratings, *Biomedical, Diagnostic, Guidance, and Surgical-Assist Systems II, Proceedings of SPIE*, 3911, 86–94 (2000).

88. R. Magnusson, Y. Ding, K. J. Lee, P. S. Priambodo, and D. Wawro, Characteristics of resonant leaky mode biosensors, *Nanosensing: Materials and Devices II, Proceedings of SPIE*, 6008, 60080U pp. 1–10 (2005).

89. P. Alivisatos, The use of nanocrystals in biological detection, *Nature Biotechnology*, 22, 47–52 (2004).

90. A. R. Parker and H. E. Townley, Biomimetics of photonic nanostructures, *Nature Nanotechnology*, 2, 347–353 (2007).

91. R. A. Potyrailo, H. Ghiradella, A. Vertiatchikh, K. Dovidenko, J. R. Cournoyer, and E. Olson, Morpho butterfly wing scales demonstrate highly selective vapour response, *Nature Photonics*, 1, 123–128 (2007).

92. J. Huang, X. Wang, and Z. L. Wang, Controlled replication of butterfly wings for achieving tunable photonic properties, *Nano Letters*, 6, 2325–2331 (2006).

93. W. Zhang, D. Zhang, T. Fan, J. Ding, Q. Guo, and H. Ogawa, Fabrication of ZnO microtubes with adjustable nanopores on the walls by the templating of butterfly wing scales, *Nanotechnology*, 17, 840–844 (2006).

94. A. Ashkin, J.M. Dziedzic, J. E. Bjorkholm, and S. Chu, Observation of a single-beam gradient force optical trap for dielectric particles, *Optics Letters*, 11, 288–290 (1986).

95. A. Ashkin, History of optical trapping and manipulation of small-neutral particle, atoms, and molecules, *IEEE Journal of Selected Topics in Quantum Electronics*, 6, 841–856 (2000).

96. A. Pralle, M. Prummer, E. L. Florin, E. H. Stelzer, and J. K. Horber, Three-dimensional high-resolution particle tracking for optical tweezers by forward scattered light, *Microscopy Research and Technique*, 44, 378–386 (1999).

97. Y. Ishii, A. Ishijima, and T. Yanagida, Single molecule nanomanipulation of biomolecules, *Trends in Biotechnology*, 19, 211–216 (2001).

98. S. C. Kuo, Using optics to measure biological forces and mechanics, *Traffic*, 2, 757–763 (2001).

99. M. J. Lang, C. L. Asbury, J. W. Shaevitz, and S. M. Block, An automated two-dimensional optical force clamp for single molecule studies, *Biophysical Journal*, 83, 491–501 (2002).

100. R. Westermeier, *Electrophoresis in practice: A guide to methods and applications of DNA and protein separations*, Wiley VCH, Weinheim, Germany (2005).

101. M. G. Khaledi, ed., *High-performance capillary electrophoresis: Theory, techniques, and applications*, Wiley, New York (1998).

102. K. D. Altria, *Capillary electrophoresis guidebook: Principles, operation, and applications*, Humana Press, Totowa, NJ (1996).

103. J. P. Landers, ed., *Handbook of capillary electrophoresis*. 2nd ed., CRC Press, Boca Raton, FL (1996).

104. W. A. Goddard, *Handbook of nanoscience, engineering, and technology*, CRC Press, Boca Raton, FL (2003).

105. T. B. Jones, *Electromechanics of particles*, Cambridge University Press, New York (2005).

106. J. Vykoukal and P. R. C. Gascoyne, Particle separation by dielectrophoresis, *Electrophoresis*, 23, 1973–1983 (2002).

107. P. R. C. Gascoyne, X.-B. Wang, Y. Huang, and F. F. Becker, Dielectrophoretic separation of cancer cells from blood, *IEEE Transactions on Industry Applications*, 33, 670–678 (1997).

108. J. M. S. Bartlett and D. Stirling, *A short history of the polymerase chain reaction*, in PCR protocols, 2nd ed., Methods in molecular biology, Vol. 226, Humana Press, Totowa, NJ (2003).

109. B. H. Lapizco-Encinas, B. A. Simmons, E. B. Cummings, and Y. Fintschenko, Dielectrophoretic concentration and separation of live and dead bacteria in an array of insulators, *Analytical Chemistry*, 76, 1571–1579 (2004).

110. E. Cummings, Streaming dielectrophoresis for continuous-flow microfluidic devices, *IEEE Engineering in Medicine and Biology Magazine*, 22, 75–84 (2003).

111. T. M. Klein, R. Arentzen, P. A. Lewis, and S. Fitzpatrick-McElligott, Transformation of microbes, plants and animals by particle bombardment, *Nature Biotechnology*, 10, 286–291 (1992).

112. A. Bietsch, J. Zhang, M. Hegner, H. P. Lang, and C. Gerber, Rapid functionalization of cantilever array sensors by inkjet printing, *Nanotechnology*, 15, 873–880 (2004).

113. J. B. Fenn, M. Mann, C. K. Meng, S. F. Wong, and C. M. Whitehouse, Electrospray ionization for mass spectrometry of large biomolecules, *Science*, 246, 64–71 (1989).

114. P. Kebarle, A brief overview of the present status of the mechanisms involved in electrospray mass spectrometry, *Journal of Mass Spectrometry*, 35, 804–817 (2000).

115. K.-H. Roh, D. C. Martin, and J. Lahann, Biphasic Janus particles with nanoscale Anisotropy, *Nature Materials*, 4, 750–763 (2005).
116. I. G. Loscertales et al., Micro/nano encapsulation via electrified coaxial liquid jets, *Science*, 295, 1695–1698 (2002).
117. C. Berkland, W. Daniel, D. W. Pack, and K. Kim, Controlling surface nano-structure using flow-limited field-injection electrostatic spraying (FFESS) of poly(D,L-lactide-co-glycolide), *Biomaterials*, 25, 5649–5658 (2004).
118. V. Farkas, L. Daniel, R. C. Leif, and D. V. Nicolau, Imaging, manipulation, and analysis of biomolecules, cells, and tissues, *Proceedings of SPIE*, 6441, 64410Z (2007).
119. P. C. H. Li, *Microfluidic lab-on-a-chip for chemical and biological analysis and discovery*, CRC Press, Boca Raton, FL (2005).
120. J. Melin and S. R. Quake, Microfluidic large-scale integration: The evolution of design rules for biological automation, *Annual Review of Biophysics and Biomolecular Structure*, 36, 213–31 (2007).
121. L. J. Millet, M. E. Stewart, J. V. Sweedler, R. G. Nuzzo, and M. U. Gillette, Microfluidic devices for culturing primary mammalian neurons at low densities, *Lab Chip*, 7, 987 (2007).
122. T. E. McKnight et al., Intracellular integration of synthetic nanostructures with viable cells for controlled biochemical manipulation, *Nanotechnology*, 14, 531–556 (2003).

Problems

11.1 Why is immunology important in nanotechnology? What are some of the ways that the immune system can be used to make sensors?

11.2 What other biological molecular systems can be used to make sensors? Which is the most versatile?

11.3 How would you expect the vibrational frequencies of nanocantilevers to change with increasing (1) mass; (2) dimensions: a. thickness, b. length, c. width; (3) density; (4) modulus of elasticity; (5) bond strength of chemical constituents; (6) amount of material absorbed on the surface? Justify your explanation in terms of physical laws.

11.4 How would you rank the expected vibrational frequencies of nanocantilevers with the same physical dimensions to compare if made from the following substances: (1) silicon, (2) steel, (3) silica glass, (4) quartz, (5) carbon nanotube, (6) copper? Justify your ranking (Hint: look up the physical properties.)

11.5 On the same basis as the preceding examples, calculate the changed vibrational frequency if a single molecule of a protein with a mass of (1) 700 Da, (2) 1200 Da, or (3) 3000 Da is attached to the tip of the cantilever.

11.6 Design a simple circuit to stimulate and monitor resonant vibration of the preceding cantilevers.

11.7 For a cantilever of dimensions $500 \times 100 \times 0.5\,\mu m$ made of single-crystal silicon with an antibody of 14,000 Da attached to its tip, which binds to an antigen of 43,000 Da, calculate the rate of vibration with and without the attached antigen.

11.8 Use Stoney's equation to determine the relative deflections produced for a silicon cantilever of length 500 μm and thickness 0.5 μm versus a silicon nitride cantilever of length 600 μm and thickness 0.65 μm.

11.9 Perhaps the pinnacle of molecular recognition is illustrated by the immune globulins—for example, the antibodies. These classes of proteins are not only able to recognize and neutralize antigenic materials that have invaded the body, but are also able to do so in a dynamic, versatile way. Antigens are considered to be divalent; antibodies are considered to be polyvalent. Explain what this means and relate your understanding to the overall solubility of antibody–antigen complexes?

11.10 What is the piezoelectric effect? Why does a layer of absorbed molecules change the surface acoustic resonance of an electrically excited vibrating circuit element?

11.11 What are the relative merits of optical and electrochemical (potentiometric) nanosensors?

11.12 How do optical tweezers work? How does dielectrophoresis differ from the optical tweezers effect, and how it is similar?

BIOMIMETICS

Nature works for maximum achievement at minimum effort. We have much to learn.

GEORGE JERONIMIDIS
Director, Centre for Biomimetics at the University of Reading, United Kingdom

*C*hapter 12

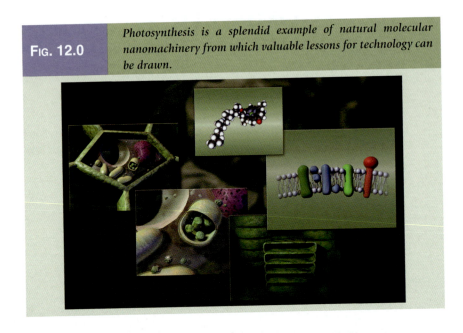

FIG. 12.0   *Photosynthesis is a splendid example of natural molecular nanomachinery from which valuable lessons for technology can be drawn.*

THREADS

We introduced nanobiotechnology in *chapter 11*. Nanobiotechnology is the application of nanoscale biological materials, structures, and processes to synthetic goals. Biomimetics is a related topic that deserves special attention—biomimetics is the process of copying nature into the design of artificial devices.

Four chapters (Nanobiotechnology, Biomimetics, Medical Nanotechnology, and Environmental Nanotechnology) are rooted strongly in biology. Nature is the supreme wizard of nanotechnology—so why not copy its materials, structures, and processes? If "imitation is the sincerest form of flattery" then let's flatter nature to the best of our abilities. Following this chapter and the one on bionanomedicine, we wrap up the book with environmental aspects of nanotechnology—an area with growing applications of nanoscience and nanotechnology.

12.0   THE BIO SCIENCES AND TECHNOLOGIES

In this chapter, we examine the important role that biological science plays in nanotechnology development. A list of the biologically inspired influences that have been transmuted into human technology would be very long [1–14]. We give examples, starting with simple nanoparticles and proceeding through surfaces, membranes, and more complex structures and nanomachinery. Before we explore biologically inspired nanotechnologies, we will give some definitions that clarify the background and aims of biomimetics in general.

12.0.1   *Biomimetics, Bioengineering, and Other Bioengineering Fields*

The use of biological models in other fields is described by many terms—biomimetics, bionics, biomechanics, biophysics, bioengineering, etc., terms that we will define to avoid confusion. Historically many aspects of bio-inspired engineering and technology originated separately, but their overlap and convergence is being accelerated by nanotechnology.

Biomimetics. *Biomimetics* (from the Greek *bios* meaning "life, course or way of living" + *mimetikos* "imitative") has roots in the beginnings of civilization. Biomimetics is science or technology that copies or imitates nature to find new methods and applications. *The Encyclopedia of Biomaterials and Biomedical Engineering* gives the following definition [15].

> Biomimetics is the study of how Nature, building atom by atom, through eons of time, developed manufacturing methods, materials, structures, and intelligence. These studies are inspiring engineering and design of manmade miniature objects.... It is in the nanoworld that Nature is way ahead of human engineering as it has learned to work with much smaller, more versatile building blocks and master the self-assembly of those building blocks.... We believe that, just as in the macroworld, blindly mimicking Nature will be unproductive, but studying Nature and taking hints as to how to build nanostructures from the bottom-up … will be productive.

The term biomimetics was coined in the 1950s by the American engineer and biophysicist Otto Herbert Schmitt. His investigations into the nervous system of the squid led to the development of the "Schmitt trigger," a device used so often in modern electronic devices that its biomimetic origins are often forgotten [16]. Advances in understanding of how nature works on the nanoscale led to new opportunities for biomimetic nanotechnology. Many significant applications are emerging, with huge room for growth [17].

Bionics. Bionics (from the Greek *bios* meaning life + the suffix *-ic* "like") is the application of biology to engineering, especially when used to enhance human capabilities directly, as in spacesuits or prosthetics. Bionics usually refers to augmentation of human performance. Bionics is the *transfer of technology* between living organisms and synthetic devices created by humans. The iron lung (P. Drinker, L.A. Shaw in 1928), the artificial heart (R. Jarvik in 1982), prostheses, cochlear ear implants (G. von Békésy in 1973), electronically stimulated limbic systems (K.D. Wise, D.J. Edell), artificial eyes (R. Birge in the 1990s), and electronic retinas (C. Mead of Cal Tech) are just a few examples of bionic applications to the human condition. The term bionics was coined by J.E. Steele in 1958, a scientist at Wright-Patterson Air Force Base in Dayton, Ohio who worked in aerospace medical systems [11,18].

Biomechanics. *Biomechanics* is the study of the mechanics of living things [19–24]. Biomechanics historically deals with the macroscale, and it is still heavily associated with functional anatomy, physical therapy, and sports medicine. However, biomechanics involves materials and phenomena that begin at the molecular level and extend up the biological hierarchy. *Biodynamics* and kinematics are subsets of this discipline that relate to movement [7]. *Biomaterials* and nanomaterials, and their properties, are important factors in biomechanical research and analysis. Metabolic and energetic factors are also important to biomechanics. Biomechanics was the one of the earliest of the bio-disciplines. Books describing the motion of animals were written by Giovanni Alfonso Borelli of Italy during the Renaissance—*De Motu Animalium I* and *De Motu Animaliam II* [25]. Leonardo da Vinci further advanced biomechanics with studies of anatomy and the design of machines based on biological actions. In the nineteenth century, understanding of biodynamics was advanced by photographic studies of animal and human motion, pioneered by Musgrave and others.

Bioengineering.   This field is the application of engineering principles and practices to medicine and biology—increasingly vested in nanotechnology. It overlaps biomedical engineering (engineering applied to medicine) and genetic engineering (bioengineering that uses biotechnology to manipulate and make use of genetic coding) [26–28]. The National Institutes of Health defines bioengineering as follows [29]:

> Bioengineering is rooted in physics, mathematics, chemistry, biology, and the life sciences. It is the application of a systematic, quantitative, and integrative way of thinking about and approaching the solutions of problems important to biology, medical research, clinical proactive, and population studies. The NIH Bioengineering Consortium agreed on the following definition for bioengineering research on biology, medicine, behavior, or health recognizing that no definition could completely eliminate overlap with other research disciplines or preclude variations in interpretation by different individuals and organizations.
>
> Bioengineering integrates physical, chemical, or mathematical sciences and engineering principles for the study of biology, medicine, behavior, or health. It advances fundamental concepts, creates knowledge for the molecular to the organ systems levels, and develops innovative biologics, materials, processes, implants, devices, and informatics approaches for the prevention, diagnosis, and treatment of disease, for patient rehabilitation, and for improving health.

The NIH is placing increasing emphasis on nanoscience and nanomaterials in its bioengineering initiatives with the emergence of *biomedical nanotechnology* and the application of nanotechnology to the design and development of biomedical devices, for analysis, diagnosis, and medical therapeutics [30,31].

Biophysics.   Biophysics is an interdisciplinary field that applies the theories and methods of physics and related sciences to biology. Biophysics is interdisciplinary: it overlaps with biomechanics and bioengineering [32–34]. It tends to be oriented towards basic science, and is very involved with nanoscale phenomena in biology.

Biotechnology and Bio-Nanotechnology.   The term biotechnology was coined to describe genetic and protein engineering technologies. These use molecular biology, cell biology, and biochemistry for practical applications such as genetic modification of organisms for the production of drugs and development of DNA assays. As nanotechnology has been increasingly applied, terms such as *bio-nanotechnology* and *nanobiotechnology* have emerged and perhaps converged [35–38]. Bio-nanotechnology refers to nanoscale phenomena such as protein and organelle structure and function, and supramolecular mechanisms in proteins and cellular systems [39]. *Nanobiotechnology* focuses on the application of nanotechnologies such as nano-cantilevers, quantum dots, and nanomechanical structures to applications such as biosensors and biochips for diagnostics and

DNA screening. Nanobiotechnology was discussed extensively in chapter 11, and additional examples are given in this chapter [40,41].

Biomimicry. The term *biomimicry* implies copying nature. Originally it was used to describe convergent adaptations or behaviors found in nature (one organism copying another). An example is the viceroy butterfly, with similar coloration to the monarch butterfly (whose taste is repellent to birds). In recent times biomimicry has been used to describe a general approach to engineering and economics with emphasis on sustainability and environmental compatibility [3]. According to the Biomimicry Institute [42]

> Biomimicry (from the Greek bios, "life" + mimesis, "imitate") is a new science that studies nature's best ideas and then imitates these designs and processes to solve human problems.

Biognosis is a related term, used to refer to systems of knowledge and practice based on nature and natural principles.

Biomimetic Nanotechnology: An Interdisciplinary Field. Biomimetics and nanotechnology are both interdisciplinary, new, and rapidly growing. For the purpose of this chapter, we define *biomimetic nanotechnology* as the design of materials and devices based on nanoscale biological structure and function. The purpose of biomimetics is to pattern useful materials and devices after biological blueprints. As stated by Claus Mattheck at the Forschungszentrum Karlsruhe in Germany [43]:

> Modern biomimetics is a systematic approach for researchers who know that new developments and insights can only be achieved in transdisciplinary collaborations. If this is true for examples of co-working between physicists or chemists and engineers, why not networking with biologists?

Table 12.1 summarizes the frequency of the bio-terms discussed above and their relation to nano.

TABLE 12.1	*Google Hits of Bio + Nano Overlap and Place*			
	Nanotechnology **(14,400,000)**	**Nanoscience** **(810,000)**	**Nanomaterials** **(1,350,000)**	**Cumul. score**
Biotechnology	520,000	215,000	466,000	4
(49,300,000)	2	1	1	
Bioengineering	1,280,000	39,600	31,500	5
(3,330,000)	1	2	4	
Biomimetics	310,000	14,100	140,000	10
(1,130,000)	3	5	2	
Biophysics	203,000	38,500	139,000	11
(4,540,000)	4	3	3	
Biomechanics	72,000	16,200	17,000	14
(3,370,000)	5	4	5	
Bionics	21,000	2,400	3,300	18
(908,000)	6	6	6	

The clear winner of the nanotechnology sweepstakes is *biotechnology*—not surprisingly because biotechnology is clearly a technological discipline and a very broad one at that. Bioengineering is next with an average score of five—another fairly broad category. *Biomimetics* has more specificity and it is placed third overall with a high overlap with nanomaterials. Biophysics and biomechanics have been around for quite a while. *Bionics* rounds out the table. Bionics, based on the findings of this table, is not as nano-driven as the others. Interestingly biotechnology, biomimetics, and biophysics appear to have the strongest nanomaterial component. Clearly, the "field" of nanotechnology by itself is a powerful force in today's science and technology—at least when considering the number of Google hits.

12.0.2 Biomimetics as an Emerging Science and Engineering Discipline

Nature has many solutions to problems that intrigue scientists and engineers. How does bone adapt to growth, concentrating material resources where they are needed? How does the anisotropic structure of bird feathers contribute to its toughness? How can an apple hold its firmness and shape when it consists of 97% water?

The Biomimetic Dialogue. Biological and synthetic designs are dramatically different. Nature does not utilize steel, flat surfaces, or sharp corners—nature uses proteins, curves, and adaptable shapes. As our knowledge of biology at the cellular and subcellular level increases dramatically, biomimetics is becoming a major player in nanobiotechnology. Nanotechnology enables scientists to understand the mechanisms that allow basilisk lizards to walk on water, penguins and sharks to reduce drag during swimming, and insects to fly and hover. The nanostructure of the lotus plant leaf is the key to understanding how it stays clean in the muddy environment that it inhabits. Swarm intelligence and artificial neural networks are another outgrowth of biologically inspired thinking [44].

12.0.3 Biomimetic Systems

The study of life on the nanoscale fosters appreciation for the self-organizing and evolutionary properties of biosystems: how they process and store energy, materials, and information, with hierarchical organization from the macro to the nanoscale [45,46].

Macroscopic Biomimetics. Velcro is a familiar biomimetic product which originated with an observation from nature—the clinging mechanism of *cockleburs* to dog hair and to fabric. In 1948, a Swiss engineer named George de Mestral, inspired by the microscopic hook-like structures in cockleburs, designed the first Velcro fastener (from the French *velour* and *crochet*)—a simple two-component device with hooks on one side and soft loops on the other. The result—Velcro Industries N.V. became a multimillion dollar industry [8].

Another macroscopic example of biomimetics is architecture based on termite mounds—structures that are able to maintain constant temperature and humidity regardless of ambient conditions that vary from near 0°C to over 40°C. A high-rise office complex in Zimbabwe called the Eastgate Centre is modeled after the internal structure of the termite mound, with energy consumption 10% that of a conventional building [9,47].

Microscopic Biomimetics.   Nature provides us with numerous models of hierarchically structured materials. Bones, teeth, shells, skeletal components, and other materials are biocomposites with micro- and nanostructure composed of lipids, proteins, and polysaccharides and inorganic mineral materials such as hydroxyapatite, calcium carbonate, and silica. In the microscopic structure of natural materials we see a hierarchy from the bottom up, starting with the basic building blocks of biology—proteins, lipids, carbohydrates, and the nucleic acids. Complex architectures are then assembled that contain different types of building blocks held together by combinations of different bonding forces [45].

Molecular Biomimetics.   The design of molecules is the domain of synthetic chemistry rather than nanotechnology. In earlier chapters we saw examples of how small molecules regulate nanoscale machinery in the cell, as key players in neurotransmission, hormonal regulation, and immunity. Because small molecules are so important in the regulation of cellular nanomachinery, it is worth examining some instances of biomimetic design of molecules where the structure and activity of the small molecule produces profound supramolecular effects. This is the most common mode of action for molecules used as drugs and pesticides. In such cases, molecular biomimetics has implications for the nano- and macroscale, as we shall see.

Industrialization brought the production of tremendous quantities of synthetic detergents and pesticides, which had no natural breakdown pathways and thus begin to build up in people, animals, and the environment, along with their toxic derivatives. One of the goals of biomimetic chemistry, or molecular biomimetics, is to find and create molecules that are closer to natural products, so that their fate in the environment and food chain is to break down without harmful effects. Another goal is to find more effective and lower cost drugs by looking for elegant molecules in natural systems as models.

12.0.4   *The Nano Perspective*

Most biomimetics has occurred on the macroscale to microscale. Now, a confluence of biomimetics with nanotechnology is in full force—the mimicking of nature at the molecular level. Organisms have exploited nanotechnology for nearly 4 billion years—DNA, RNA, proteins, and inorganic nanomaterials all contribute to the structures in living cells that perform a variety of functions at the nanoscale. Nature's nanotechnology is optimized and efficient and has undergone, via evolutionary pressures, the most rigorous product development and testing in the most demanding laboratory over the longest periods of time. It is no wonder then that we scientists seek to duplicate nature's marvelous materials and devices in order to create our own nanomaterials and machines from the bottom up.

So, what clues does nature provide to help us with fabrication of our biomimetic devices? According to P. Ball from his book *Made to Measure* [10], biomimetic construction (nanochemistry) should follow these guidelines (from nature):

- Use of composite materials that are made of alternating layers of aragonite ($CaCO_3$) and biopolymer rather than monolithic materials. Examples include bone, nacre (e.g., abalone shell, mother of pearl)—in which the composite outperforms either component separately
- Liquid phase manufacturing
- Materials based on carbon (and not silicon or metals)
- Parallel processing for high throughput
- Hierarchical organization for high strength-to-density ratios and multifunctional performance
- Self-repair
- Soft, flexible materials (not hard stiff materials with flat surfaces, sharp corners)
- Self-assembly (molecular recognition) and self-replication (intelligence) via weak intermolecular forces
- Template synthesis (genetic replication)
- Compartmentalization: tissues $\rightarrow$ cells $\rightarrow$ organelles $\rightarrow$ nanostructured biological materials

Biomimetic systems arising from the collaboration of physicists, chemists, molecular and cellular biologists, and bioengineers are expected to have numerous applications in bioengineering, pharmacology, and medicine. We will start our exploration of biomimetics with a look at design of molecules, and then proceed to the nanoscale.

12.1   BIOMIMETIC DESIGN OF MOLECULES

Biomimetic principles of design can be applied on any size scale from the macroscopic down to the molecular level. Molecular biomimetics is the design of novel molecules based on structures and functions of natural products from plants and animals. Molecular biomimetics may use the chemical structure of a natural chromophore to produce improved dyes and pigments, or use other properties, as in artificial sweeteners or synthetic flavor compounds. Even using formic acid, the active ingredient in ant venom, as a starting point for making Formica polymer could be considered a type of biomimicry. But deeper biomimetics involves understanding how simple molecules produce large biological effects, and designing synthetic routes to achieve similar effects for drugs.

On the chemical level, synthetic drugs and insecticides act by blocking or duplicating the stereochemistry and functional activity of natural hormones, antibodies, and other biochemical substances. Plants, bacteria, and marine creatures, with their interactions in biodiverse ecosystems, offer a natural laboratory for discovery of biological interactions. Bacteria and archaea, plants and sea creatures, and other primitive life forms have interacted and developed over billions of years of evolution, to produce an enormous library of biochemical and nanomachinery processes, which can be studied and harnessed for applications to produce subtle and advanced medicines, foods, materials, catalysts, pesticides, and energy and growth regulators.

Centers of biodiversity such as tropical rainforests and coral reefs represent especially rich storehouses of biochemical information and materials that are only beginning to be tapped, even as they are disappearing due to ignorance of their value. Sources of biodiversity are not limited to tropical rainforests and reefs: boreal and alpine environments, and semidesert regions that have remained undisturbed by glaciation for millions of years, such as southwest Australia, also hold unique highly developed communities of plants and animals. Temperate forests, plains, savannahs, and wetlands also possess abundant networks of species with their own finely developed diverse biochemistry.

Learning from these natural biochemical networks is a more efficient way of discovering biochemical nanomachinery pathways than attempting to deduce them from basic chemistry and molecular biology, simply because the number and complexity of all the biological possibilities is so vast. The study of natural compounds synthesized by plants and animals yields valuable insights into substances and mechanisms that act on cellular pathways. Once these effects are discovered in nature, biochemical and molecular biology can then elucidate their mechanisms, and can guide the design of drugs with similar effects.

Ethnobotany, the study of the lore for traditional uses of plants in ancient indigenous cultures, has been found to accelerate the discovery of new drugs from plants, animals, and marine life, because cultural traditions can represent the tested results of thousands of years of trial and error experience. Plants produce large numbers of allelopathic chemicals which act as attractants, repellents, growth inhibitors, and poisons on competitors, pests, and predators. Knowledge of these compounds and how they work can improve the odds in the search for new and more effective drugs and pesticides. Currently, it is estimated that fewer than 5 in every 10,000 compounds investigated in the effort to develop new pharmaceuticals results in an effective approved new medication.

12.1.1 Design and Discovery of Drugs

The drug discovery process can be top down or bottom up. The top-down approach observes an effect produced by a substance and then works to establish the mechanism and chemical basis of the action. This traditional approach relies on evaluating compounds from traditional remedies or from exhaustive screening of classes of chemicals.

The bottom-up approach is more targeted, based on knowledge of chemical signaling pathways obtained from biochemical and molecular biology research: drugs are designed and synthesized to precisely match receptors, to block or enhance metabolic and signaling paths that control functions in cells, to regulate the homeostasis of the body, or to enhance the performance of the body's natural immune response mechanisms. When design moves on to production, biotechnology is used to manufacture drugs in quantity by redirecting the natural DNA–RNA molecular synthesis process in yeasts, bacteria, or genetically modified plants and animals. Modern drug discovery employs both top-down and bottom-up approaches.

The same processes apply to the discovery and production of pesticides and plant growth regulators as to pharmaceuticals. The traditional screening of compounds for pesticide activity is now supplemented by targeted design of compounds aimed at blocking metabolic pathways. New knowledge of how receptors and cytoplasmic reticulum structure operate in cellular processes is leading to drugs that act on the macromolecular and nanostructural levels.

As we shall see in this section, the chemicals produced naturally by animals and plants to regulate their internal and external interactions can yield a wealth of information for the discovery of new drugs, pesticides, growth regulators, and other useful molecules. Learning the modes and mechanisms by which these molecules act can reveal surprising aspects of the internal nanomachinery of cells which we might otherwise have never suspected or discovered in millions of years—but then, that is how long it took plants and insects to develop them.

12.1.2 *Targeting with Magic Bullets*

The path to synthesis of new drugs historically developed from use of simple inorganic poisons to highly sophisticated targeting based on knowledge of the nanoscale functioning of cells. Early medicines relied heavily on toxic metals such as mercury and arsenic to kill pathogens in the body—the cure offered by a quacksalver (from quicksilver → mercury + salve → healing potion or lotion) could be worse than the disease.

In 1908, Paul Ehrlich, a brilliant bacteriologist who developed new staining methods for classifying bacteria, set up a systematic search for improved drug molecules. Stains had been used in microscopy for nearly 100 years, first simply to enhance microscopic images, and later to identify microorganisms. Certain dyes were highly specific in binding to different types of bacteria. Ehrlich theorized that by combining aniline dye stains with arsenic he could produce a "magic bullet" that would bind specifically to bacteria in preference to human cells, delivering the poison to the pathogen with minimal harm to the body.

Previously, in 1859, the French chemist Antoine Béchamp had reacted aniline and arsenic acid to produce an organic preparation of arsenic which was hoped to be less toxic than the inorganic acid. The new compound, named atoxyl, though still highly poisonous, was indeed less toxic than free arsenic alone in the treatment of skin conditions. In 1905, the British physicians H.W. Thomas and A. Breinl discovered that atoxyl was active against the spiral trypanosome microorganism which causes sleeping sickness. Ehrlich teamed with the organic chemist Alfred Bertheim who synthesized new derivatives of the atoxyl molecule that were then systematically tested in Ehrlich's laboratory by his colleague Sahachiro Hata (**Fig. 12.1**).

Hata went through hundreds of trials of different aniline arsenic derivatives in an attempt to find one that was more specifically toxic to disease causing organisms than to humans; eventually the 606th compound, azobenzene, was effective against the syphilis pathogenic spirochete *Treponema pallidum* (**Fig. 12.2**). The new compound, although crude and still toxic, was a great improvement over previous drugs. Named salvarsan, or "Ehrlich 606," it was hailed as a magic bullet against the feared disease. It marked the first instance of targeted chemotherapy design, and set the pattern for drug research for most of the next century: systematic evaluation of derivatives of a lead compound by teams of bacteriologists, chemists, and clinicians. Ehrlich received the Nobel Prize for Medicine together with Ilya Ilyich Mechnikov in 1908.

The work of Ehrlich, Bertheim, and Hata was a great advance over random trials of chemicals for drug effectiveness, but at the same time other, remarkably

FIG. 12.1

Paul Ehrlich, originator of the "magic bullet" concept, with his colleague Sahachiro Hata, who carried out the testing that discovered the effect against the syphilis microbe.

FIG. 12.2

The syphilis pathogenic spirochete Treponema pallidum.

effective drugs were being inspired by natural biomimetic phenomena other than staining dyes: compounds from natural and traditional remedies led to more effective and safer medicines [48,49].

Many natural plant compounds are found to act specifically against devastating disease organisms, with very low toxicity to humans. Quinine was isolated from the bark of a tropical tree as a remedy for malaria. When its widespread use for a century or more led to resistant strains of pathogens, it was another natural product from traditional medicine, artemisin, which provided an alternative. In the next section we look at a natural plant substance which acts not by inhibiting disease organisms but by regulating the metabolism of our own bodies.

12.1.3 Aspirin: Signaling Pathways Revealed by the Willow

A remarkable example of a drug derived from plants is aspirin, one of the first drugs to be designed by a deliberate attempt to imitate the molecular structure and function of a natural product—the first biomimetic drug. Aspirin was developed as a safer version of willow bark extract, a traditional remedy for aches and fever which had the drawback of releasing strong acids into the stomach, producing digestive distress and even ulcers with continued use.

Willow bark has been used since ancient times for pain relief in arthritis and other inflammations, and to reduce fever. In the nineteenth century the active compound was isolated from willow bark, and named salicylic acid (from Latin *salix* = willow) (**Fig. 12.3**). Salicylic acid is a derivative of a more complex compound found in many related trees, the β-glycoside salicin (**Fig. 12.4**). Salicin breaks down into D-glucose (**Fig. 12.5a**) and salicylic acid (**Fig. 12.5b**), a ubiquitous plant hormone which acts as a signaling agent in some very basic and ancient biochemical pathways involving nonspecific immune responses [50].

The carboxylic group in the salicylic acid extracted from willow bark is made more acidic by the presence of a phenol radical, leading to its harmful effects on the stomach lining. In 1897, Felix Hoffmann, working at the German chemical company Bayer, developed a synthetic method for modifying salicylic acid, substituting the phenol to make it less harmful to the stomach, creating aspirin (**Fig. 12.6**). (Anecdotally, Hoffman supposedly was seeking a remedy to aid his father, who suffered from stomach ulcers from taking salicylic acid for arthritis over a long period.)

Aspirin's recognized therapeutic uses, as well as its phenomenal success as an over the counter remedy, generated significant profits, and much research effort was expended seeking improvements and alternatives. It was realized that research efforts into the mechanism of aspirin's many remarkable effects would open the door to a major class of new drugs; but for many years the mechanism of action of aspirin was not fully understood. Eventually, aspirin was found to inhibit the *arachidonic acid pathway* that leads to the synthesis of *eicosanoids*, potent mediators of pain and inflammation. In 1971, researchers led by the British pharmacologist John Robert Vane, at the Royal College of Surgeons in London, showed that aspirin suppresses the production of the prostaglandins and thromboxanes involved in inflammation [51]. This discovery led to major

FIG. 12.3   *The willow tree, source of salicin.*

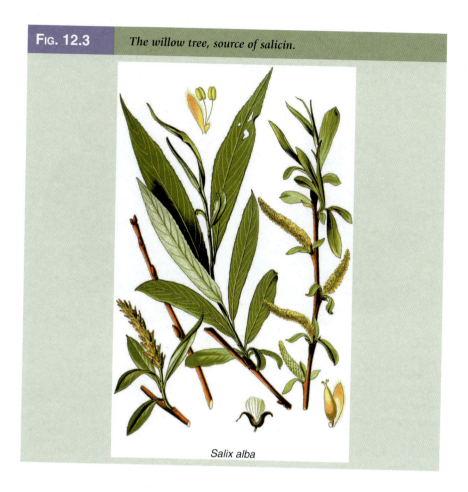

Salix alba

FIG. 12.4   *Salicin molecule. (a) Three-dimensional model and (b) chemical bond structure.*

(a)   (b)

FIG. 12.5	*Salicin component molecular structures. (a) Glucose and (b) salicylic acid.*

(a)   (b)

breakthroughs in understanding molecular signaling pathways between cells. The Nobel Prize for Medicine for 1982 was awarded to Professors Bengt Samuelsson, John Vane, and Sune Bergstrom for this discovery.

Aspirin was the model for a new class of pharmaceutical agents known as nonsteroidal antiinflammatory drugs (NSAIDs). Many but not all NSAIDs are derivatives of salicylates; all have similar effects—most act by nonselective inhibition of the enzyme cyclooxygenase, needed to synthesize prostaglandin and thromboxane. *Prostaglandins* are local (*paracrine*) hormones whose diverse effects include transmission of pain information to the brain, modulation of the hypothalamic thermostat, and regulating inflammation. Thromboxanes are involved in aggregation of platelets that form blood clots. Aspirin can irreversibly block the formation of thromboxane A2 in platelets, producing an inhibitory effect on platelet aggregation. This is the mechanism of aspirin's anticoagulant effects used to reduce the incidence and severity of heart attacks. A side-effect is a general reduction in the ability of the blood to clot, which may result in excessive bleeding with the use of aspirin [52].

FIG. 12.6	*Aspirin: Acetylsalicylic acid molecule chemical structure.*

FIG. 12.7   *Action of aspirin on the structure of COX-2 (prostaglandin H synthase). The COX-2 molecule is a dimer; the blue and green halves are identical. In each monomer, the active serine site has been acetylated by aspirin, inactivating it.*

Source: Image courtesy of Jeff Dahl.

Aspirin suppresses the production of prostaglandins and thromboxanes by its *irreversible* inactivation of the cyclooxygenase (COX) enzyme, which is required for prostaglandin and thromboxane synthesis (**Fig. 12.7**). Aspirin acts as an acetylating agent, covalently bonding an acetyl group to a serine residue in the active site of the COX enzyme. This *irreversible* mode of action is different from other NSAIDs (such as diclofenac and ibuprofen), which are *reversible* inhibitors. Recently, *reversible* blocking of COX-2 by synthetic NSAID drugs has been found to lead to harmful effects, showing that we still have some surprising biomimetic lessons to learn from the subtle mode of action of the natural product.

In addition to inactivation of prostaglandin and thromboxane production, aspirin has two additional modes of action, contributing further to its strong analgesic, antipyretic, and antiinflammatory effects. Aspirin buffers and transports protons across membranes involved in energy release by ATP in the mitochondria, where it uncouples oxidative phosphorylation and disrupts energy release. As a weak acid, aspirin can carry protons, diffusing from the inner cell membrane space into the mitochondrial matrix, where it ionizes to release protons.

Aspirin stimulates the formation of NO radicals that enable the body's white blood cells (leukocytes) to fight infections more effectively. Dr. Derek W. Gilroy was awarded Bayer's International Aspirin Award in 2005 for his research revealing the effects of aspirin on NO production [53].

The study of aspirin has contributed to our understanding of inflammation and how it has evolved as a protective response to insult or injury, a primordial response that eliminates or neutralizes foreign organisms and materials. Salicylic acid and its derivatives have been found to modulate signaling through a number of transcription factor complexes that play central roles in many biological processes, including inflammation. Inflammation involves many mechanisms that protect us against tissue injury and promote the restoration of tissue after damage: our well-being and survival depend upon its efficiency and carefully balanced control [54,55].

There are many other examples of drugs acting on human regulatory pathways, derived from compounds found naturally occurring in plants: *digitalis* acts specifically on heart muscle; *coumarins* act on blood clotting mechanisms. Today more than 120 important and widely used medicines derive directly from plant precursors, and many others derive from fungi and marine organisms [56].

In the next section we look at an example of a molecule whose action would have been even more difficult to predict without its existence in nature; it works not by modulating a signaling pathway, but by blocking a fundamental nanoscale process involved in cell replication.

12.1.4   *Taxol: Novel Drug Actions on the Nanolevel*

Use of a biomimetic approach can be especially useful in discovering nanoscale effects, providing insights into interactions between macromolecules involved in cellular metabolism. One example of a nanoscale drug discovered by examining natural plant metabolites is taxol. Taxol was discovered by screening extracts from yew trees (**Fig. 12.8**). The yew is important in European folk traditions

FIG. 12.8   *Pacific yew tree, source of taxol.*

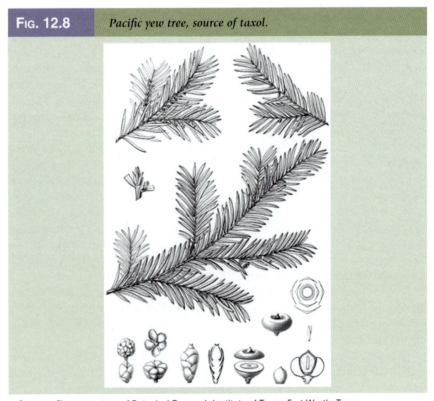

Source: Figure courtesy of Botanical Research Institute of Texas, Fort Worth, Texas.

dating back to the Druids and earlier. Yew berries are poisonous, and extracts of yew leaves and bark were used in Celtic medicines and ceremonies.

From the 1960s the United States National Institutes of Health (NIH) sponsored research, screening thousands of natural product extracts for potential use as cancer drugs. By the 1980s tests of extracts from the bark of the Pacific yew tree yielded promising results on cancer-prone mice strains. Further tests found that a compound in the Pacific yew extract, taxol, was especially effective in halting the growth of certain types of human breast cancer which were resistant to other treatments. The structure and mechanism of taxol were elucidated by further research sponsored by the NIH and pharmaceutical companies (**Fig. 12.9**). It was found that taxol disrupted the functioning of the microtubules involved in guiding chromosomes as they separated in mitosis during cell division. Rapidly growing tumor cells were preferentially affected compared to normal cells, as they were blocked from reproducing by the action of taxol. Taxol, also known by the generic name paclitaxel, has been found effective in treatment of lung, ovarian, breast, head and neck cancers, and advanced stages of Kaposi's sarcoma, and is used for the prevention of *restenosis* in patients with blocked blood vessels.

Research work developed ways to synthesize taxol and related compounds using precursors and biotechnology without the need to harvest large amounts of raw materials from the trees. Thus the value of the trees was not material, but informational. Taxol is an especially interesting drug from the nano perspective, because it acts by absorbing on to specific sites in the microtubule structures that act as scaffolding and guideways in the separation of chromosomes during cell division. Taxol stabilizes these structures, making them rigid and unable to function—it "gums up the works." It acts in the domain between the chemical and the mechanical—the nanoworld.

FIG. 12.9   *Taxol molecule chemical structure.*

Modern biomimetic science relies on natural models like the yew tree to find examples of specific biological activity at the macromolecular level that would otherwise be unexpected. The action of taxol is at the nanoscale level of microtubules formed in mitosis, thus it lies between the domains of biochemistry and cellular biology. Taxol-derived drugs are examples of discovery of drug design and action by observations from nature. Taxol is also a cautionary tale of the costs of ignoring the importance of preserving the irreplaceable information contained in ancient biodiverse ecological systems [57,58].

12.1.5   *Pyrethrum: Learning from the Daisy*

The pyrethrum daisy is a flowering plant of the chrysanthemum family, native to the Caucasus mountains (**Fig. 12.10**). Use of the flowers as an insecticide and insect repellent dates back to antiquity; the dried flowers have long been prized and traded across the world. The plant produces natural organic compounds called *pyrethrins* with potent insecticidal activity. The pyrethrins are secreted in the seed cases in varying concentrations in different varieties. One species, *Chrysanthemum cinerariaefolium*, is grown commercially as a source of the insecticides.

The chemical structure of pyrethrins was first published by Hermann Staudinger and Lavoslav Ružička in 1924 (**Fig. 12.11**) [59]. The different variations of natural pyrethrins are all structurally related esters with a cyclopropane core [60,61]. Natural pyrethrins are viscous lipophilic liquids. They are nonpersistent, biodegradable, and break down relatively quickly on exposure to light or oxygen. Pyrethrins are neurotoxins that attack the nervous systems of all insects, being especially effective against the muscles of insects such as mosquitoes,

FIG. 12.10   **Pyrethrum chrysanthemum** *daisy, source of pyrethrin.*

Source: Figure courtesy of Botanical Research Institute of Texas, Fort Worth, Texas.

FIG. 12.11	*Pyrethrin chemical structure.*

fleas, and houseflies. Low concentrations can have an insect repellent effect. Because of their powerful "knock-down" effect, insects are slow to develop into resistant strains. They are harmful to fish, but have very low toxicity to mammals and birds. In humans they can irritate eyes, skin, and respiratory systems but are not toxic to the human nervous system or metabolic pathways. They are considered to be amongst the safest insecticides for use around food [62].

The chemical structure of pyrethrins has served as a model for a variety of synthetic insecticides called pyrethroids such as permethrin and cypermethrin, which share their specificity, low mammalian toxicity, and nonpersistence in the environment, while being even more effective against insects than the natural substances. Thus the natural molecule has served as a source of information for the design of effective insecticides, not only for their action on insect muscle, but also to show how to design molecules that do not build up in the food chain and environment. Pyrethroids act on muscle, which is itself a very interesting example of biological nanomachinery. Studies of how pyrethrins work to disrupt muscle function can shed light on mechanisms of paralysis caused by diseases such as tetanus, polio, and muscular dystrophy. In a later section we will examine the structure of muscle and how it works, and how it serves as a model for some biomimetic engines.

Other Biomimetic Pesticides.   Many other plants produce pesticides that have revealed new chemical paths for fighting crop and human pests. The Mexican marigold produces thiols, which are effective against nematodes that attack plant roots in valuable crops. Other plants produce allelopathic growth inhibitors against competing plants. Natural plant compounds can be used as models for safe, specific, and effective weed killers that do not build up in the environment, causing unintended harm to beneficial insects and animals, and creating chemical resistance. For these reasons plant scientists and agronomists are looking for natural and biomimetic substances as alternatives to costly petrochemical-based synthetics.

Biomimetic Methods of Molecular Synthesis.   Natural processes can serve as models as well as natural materials. Organisms make complex molecules and link them together in polymers and nanostructures, working at ambient temperatures without high energy expenditure or large amounts of wasted by-products. The enzymes and catalytic proteins that make this possible are being understood and harnessed by biomimetic synthetic methods being developed by chemists and engineers. Chemo-enzymatic synthetic methods are a growing area of biomimetic chemistry for drugs, macromolecules, polymers, and biofuels [63–68]. We will see further examples of this theme as we survey other biomimetic materials and how they are made.

The Importance of Molecular Biomimetics in the Discovery of Signaling Pathways and Mechanisms.   It is extraordinarily difficult to predict biological regulatory pathways from observations based purely on biochemistry, molecular biology, or cell biology alone, partly because many of these effects depend upon the nanoscale—they are based on structures and configurations, and interactions of chemistries, macromolecular structures, and nanomachinery involved in cell biology. Observation of interactions involving compounds produced by the organisms with which we share our natural environment can reveal effects, which when investigated using the tools of nanotechnology and the scientific methods of biochemical, cellular, and molecular biology can be unraveled down to their nanoscale mechanisms.

Areas for Further Study.   As a hint of the interesting rewards of further study in this field, we give just two examples. Many models for useful macromolecules come from marine organisms. For example diazonamide A, a model for a family of anticancer drugs with a unique type of architecture and activity, was isolated from the marine acidian *Diazona angulata*, a rare invertebrate found in the Philippine sea [69].

Other sources for molecular biomimicry come from insects, spiders, and mites. In the earlier discussion of aspirin, we mentioned that aspirin acts upon the important arachidonic acid pathway. You may have wondered how this pathway got its name, which comes from the Greek word for spider. Arachidonic acids are found in the venom of many spiders, ticks, and other arachnid arthropods. These animals use the toxic versions of these acids to trigger violent inflammation responses in the victims of their bites, to help break down and digest their meals.

In a particularly interesting instance of biomimicry, researchers have isolated a highly effective class of macromolecular peptide pesticides that work very selectively against disease-carrying ticks, which have become resistant to other more conventional pesticides. The peptides come from a web-weaving spider which preys on other arachnids [70].

It is particularly satisfying to be able to turn the nanomolecular tools of these voracious bloodsuckers against such troublesome pests of their own family, thanks to their own indiscriminate predatory adaptations. It is more significant to consider this as an instructive example of our interconnectedness with the web of life—and sources for biomimetics.

12.2   BIOMIMETIC NANOMATERIALS

In this section, we will discuss how natural materials inspire advanced nanomaterials with unprecedented performance characteristics: high strength, light weight, self-cleaning, and unique optical and electronic behaviors. We will survey some of these emerging materials, their properties, and their applications. We will also look for emergent themes and design principles that natural materials demonstrate [71–77].

Smart materials mimicking biological functions are increasingly closer to nature's technology. These include tough artificial shell and bone, artificial muscle made with electroactive polymers, nanoscale molecular actuators, smart membranes, nanoengineered encapsulation, and nanoadhesives. All of these nanotechnologies incorporate insights into how nature works. Their development is helping us cure diseases, save energy and water, and improve efficiency of food and materials production. By understanding nature on the nanoscale, we will be able to live in our environment more intelligently, sustainably, and elegantly.

12.2.1 *Biomimetic Mineral Nanoparticles*

The control of size and morphology of microcrystals is an area in which biological organisms excel, but which has until recently been beyond the capabilities of human engineers and scientists. Living organisms not only replicate organic structures such as DNA and cell membranes but are also capable of building inorganic structures with precise control of form and properties. As we saw in the previous book, these include shells, teeth, bone, and silica structures.

Control of Crystal Growth by Mimicking Natural Biological Processes. One of the pioneers of early studies which explained precise crystal structures of mineral micro- and nanoparticles grown by biological organisms is Professor Stephen Mann, who observes that [78]

> Inorganic building blocks play a fascinating and crucial role in self-organised assembly for many biological structures including bone, teeth and shell. Biomineralisation uses a limited number of solid inorganic materials such as calcium carbonate, silica and iron oxides to form new materials that bear no relation to underlying structures. Biological systems have developed an exquisite control of inorganic processes. There are species of magnetotactic bacteria that produce nanoscale magnets in their cells. Not just depositing iron oxide but producing crystals that are both the perfect size and shape and aligned in a chain.

Biomimetic Synthesis of Novel Inorganic Crystalline Structures. Mann's group has explored the use of protein-based micellular structures to grow bioinorganic nanocomposites with magnetic and quantum resonance properties, surfactant molecule assemblies as templates for inorganic crystal growth, and microemulsions to create minireaction chambers in micelles. They use biological microstructures to synthesize inorganic complexes, adapting bacterial filaments to fabricate ordered silica macrostructures and tobacco mosaic virus to make inorganic nanotubes. They also recently developed a new process using inorganic nanoparticles to make magnetic spider silk.

Nanoparticle Architectures. How to direct and control the self-assembly of nanoparticles is a fundamental question in nanotechnology. The success, growth, and application of nanotechnology depend upon our ability to manipulate nanoscale objects. A group led by Nicholas A. Kotov is examining three critical questions in the growth of nanoparticle structures:

1. What are the methods of organization of nanocolloids in more complex structures?
2. What kind of structures do we need for different applications?
3. What are the new properties appearing in nanocolloid superstructures?

Professor Kotov at the University of Michigan studies how to organize nanoparticles into a useful variety of larger and more complex systems, making new materials utilizing biomimetic models. His team developed a technique to make nanocrystals in a fluid assemble into free-floating sheets the same way some structures form in living organisms. This fluid process lends itself to automated production that builds materials one nanoscale layer after another. A robotic arm applies nanolayers onto substrates such as glass or silicon wafers, alternating compositions from different liquid source materials, to make composite nanoengineered plastics [79,80].

12.2.2   *Shell as a Biomodel*

In a previous volume, we looked at the natural structure of shells, based on layered, hierarchical nanostructures of mineral crystals, polymers, and proteins. Shells combine proteins and mineral crystals in a nanoscale architecture that produces high strength and toughness, using very little energy and wasting no material in the process. These natural materials and processes are being used as models to design new generations of strong and resilient materials.

Abalone Shell as a Model for Armor.   Researchers led by Marc A. Meyers at the University of California, San Diego (UCSD), are pursuing a number of biomimetic projects, using the shell of the abalone as a model. Meyer describes his team's basic research on new materials in biomimetic terms:

> We have turned to nature because millions of years of evolution and natural selection have given rise in many animals to some very sturdy materials with surprising mechanical properties. In our search for a new generation of armors, we have exhausted the conventional possibilities, so we've turned to biology-inspired, or biomimetic, structures.

Mollusk shells, bird beaks, and other natural biocomposites are based on a hierarchy of structures from the molecular level to the macroscale. At the nanoscale, shell is made of thousands of layers of calcium carbonate "tiles," about 10-μm across and 0.5-μm thick. The layered stacks of thin tiles refract light to yield the characteristic luster of the mother of pearl. The shell nacre's nanostructure of interlocking calcium carbonate tiles and shock-absorbing protein adhesive give it the ability to absorb heavy blows without breaking (**Fig. 12.12**).

The main constituent of the abalone shell is calcium carbonate. Only about 3% of the shell is made up of organic components, but the fracture resistance of the nanocomposite nacre is about 3000 times higher than for the pure carbonate mineral. A key to the strength of shell is a positively charged protein adhesive that binds to the negatively charged top and bottom surfaces of the calcium

FIG. 12.12

Abalone shell tiles: (a) Schematic representation of stacked layers of aragonite tiles. (b) Arrangement of tiles on inner surface of 10-mm shell; back-scattered image (SEM). (c) Schematic drawing of stacking of abalone tiles and their separation under tension. Abalone shell composite is some 3000 times tougher than the calcium carbonate crystals from which it is made. When an impact force is applied, the organic layer deforms slightly. This absorbs some of the shock; the rest causes the tiles to slide until frictional forces oppose the movement. The aragonite tiles can fracture, but the fracture is limited to the individual tiles. As each layer is offset laterally from the other, cracks have a difficult time propagating from one tile layer to the next. If a crack should spread it generally has a difficult time passing through the organic, elastic layer to an adjacent mesolayer.

Source: Image adapted from A. Lin and M. A. Meyers, *Materials Science and Engineering A*, 390, 27–41 (2005).

carbonate tiles. The glue holds layers of tiles firmly together, but is soft enough to permit the layers to slip apart, absorbing the energy of a heavy blow in the process.

Meyers discovered previously unknown details about the nanoscale assembly of the nacre building blocks that help to explain their resilience and provide a guide to the design of stronger biomimetic materials.

> Contrary to what others have thought, the tiles abutting each other in each layer are not glued on their sides, rather they are only glued on the top and bottom, which is why adjacent tiles can separate from one another and slide when a strong force is applied.

The elastic adhesive and free edges in the shell's interior allow it to yield to impacts without breaking, unlike conventional laminated materials (**Figs. 12.13** and **12.14**).

FIG. 12.13 *Micron-scale structure of abalone shell: The many lamella of aragonite tiles form larger mesolayers some 300-μm thick. Each mesolayer is separated by a 20-μm layer of organic material that is primarily composed of proteins, glycoproteins, and beta-chitin (a polymer of glucose). As in bone, the proteins play an important role in the deposition and arrangement of the aragonite, and the organic layer acts as a glue to hold the mesolayers together.*

Source: Electron micrograph image from A. Lin and M. A. Meyers, *Materials Science and Engineering A*, 390, 27–41 (2005). With permission.

Abalones fill fissures within their shells that form due to impacts; they deposit "growth bands" of organic material alternating with mineral deposition. Meyer's group inserted glass slides beneath the mantle of abalone grown in a laboratory aquarium. When they withdrew the slides at various time intervals they were able to observe the pearly growth layers on the slide with a transmission electron microscope. These growth layers showed that the abalone mantle seeded calcium carbonate crystal precipitates at intervals separated by about 10 μm. Tiles began to form from the seed locations, growing 0.5-μm thick and spreading slowly outward. The tiles formed a hexagonal shape as individual tiles in each layer gradually grew to abut a neighboring tile. Microscopic imaging showed the growth surface of the shells with a Christmas-tree appearance as abalones add layers of tile faster than each layer is filled in. The UCSD group is developing a mathematical description of the growth process which will serve as a systematic design tool for biomimetic materials [81].

Other Biomimetic Nanocomposite Materials Modeled on Shell. Meyer's group is not alone in using natural shell structures as biomimetic models for strong materials. At the Particle Technology Laboratory of the Eidgenössische Technische Hochschule (ETH) in Zurich, researchers led by Ludwig J. Gauckler are pursuing bioinspired design principles, making strong nanocomposites through the

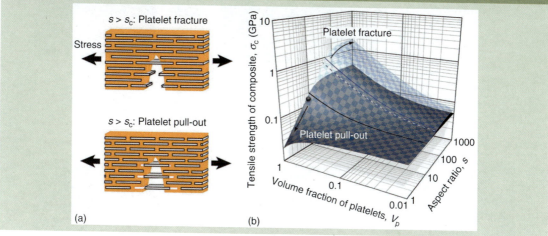

FIG. 12.14 *(a) Fracture mechanisms and (b) estimated tensile strength of platelet-reinforced composites. The aspect ratio of the platelets (s) determines whether the composite fails under the platelet fracture mode (s > s_c) or platelet pull-out mode [s < s_c(a)]. The tensile strength of polymer matrix composites reinforced with $CaCO_3$ and Al_2O_3 platelets is estimated in (b) (black and blue surfaces, respectively) on the basis of a shear-lag mechanical model. The limit between the platelet fracture and pull-out modes is indicated by the black and blue full lines for $CaCO_3$ and Al_2O_3 platelets, respectively [82].*

Source: Image from L. J. Bonderer, *Science*, 319, 1069–1073 (2008). Reprinted with permission from AAAS.

bottom-up colloidal assembly of sub-micrometer-thick ceramic platelets within a ductile polymer matrix [82] (**Figs. 12.14** and **12.15**). In 2007, a research group led by Hansma in California reported how bioinspired optimized adhesives combined with carbon nanotubes or graphene sheets can yield strong, lightweight, damage-resistant nanocomposite materials [83,84]. A common feature of biological nanocomposites is that only a small amount of adhesive is needed—excess adhesive actually weakens the material. The optimal amount of adhesive is just enough to transfer the load to the strong elements.

Mimicking the Self-Organizing Power of Proteins to Make Biomimetic Nanocomposites. At the University of Michigan, Professor Kotov is applying automated production technology to make biomimetic materials using alternating layers of clay nanoparticles and adhesive [79,80]. It takes 300 layers of adhesive polymer alternating with clay nanosheets to create a piece of this material as thick as a piece of plastic wrap. The glue-like polymer used in this nanomaterial is polyvinyl alcohol, which forms cooperative hydrogen bonds with the clay nanosheets. This "nanoglue" gives rise to what Kotov called "the Velcro effect." When hydrogen bonds are broken, they can reform easily in a new place. The dense hydrogen bonds allow for effective load transfer between the nanoparticle and polymer layers.

Biomimetic Materials Based on Wood Nanostructure. The microfibril structure of wood inspired one of the earliest successful biomimetic applications based on micro- and nanostructure of natural materials. R. Gordon, C.R. Chaplin, and

FIG. 12.15   *Bottom-up colloidal assembly of multilayered hybrid films. Surface-modified platelets are assembled at the air–water interface to produce a highly oriented layer of platelets after ultrasonication. The 2D-assembled platelets are transferred to a flat substrate and afterwards covered with a polymer layer by conventional spin coating.*

Source: Image from L. J. Bonderer, *Science*, 319, 1069–1073 (2008). Reprinted with permission from AAAS.

G. Jeronimidis at Reading University patented a bio-inspired composite structural panel with high strength and toughness based on ultrastructural features in wood [85–87]. The orientation of the fibers in this biomimetic composite is based on angles found in microfibrils of wood tracheids. Reading is among a number of centers around the world that are investigating the micro- and nanostructure of wood, feathers, and other materials for biomimetic applications [88–96].

Modern lightweight composite structures are manufactured with so-called gradient textile techniques. As in nature, every single fiber strand is exactly laid within the structure in the direction necessary to neutralize outer forces so that no unnecessary fibers or weight are incorporated. Manufacturing of these ultralight composites was made possible by the development of adequate "finite element" computing methods for the calculation of forces in curved and irregular shapes. Biomimetic manufacturing technologies based on plant models are the special focus for physicists, biologists, and engineers led by Claus Mattheck at the Forschungszentrum Karlsruhe in Germany, where they are studying the biological design rules that govern the development of optimized shapes in growing plant structures [43,90].

12.2.3 Nanoengineering Bone

Crustaceans and other shell makers create their exoskeletons primarily as protection from external threats; so the shell structure is an excellent adaptation to protect against crushing and cracking forces applied from the outside. Vertebrates rely on bony endoskeletons which must perform a much more varied and complex set of functions; bones and teeth have to bear complex loads of moving bodies, provide a protective cage for vital organs, anchor tendons and muscles, and act as joints, fulcrums, and levers. Because of this functional complexity, the nano- and microstructure of bone are more complex than the carbonate nanostructure of shells. Different parts of the same skeleton must have different directional, compression, and tension strengths to prevent fracture and distortion under stresses, loads, impact, and fatigue.

Because of its medical importance, research into the nanostructure of bone has focused mainly on artificial bone for replacement, scaffolding for bone augmentation and growth, and interfaces between bone and implants. With the rapid growth of biomimetics, the structure of bone is serving as a model for materials development. Efforts have been made both to mimic the bony material itself as well as to mimic the process by which bone forms, but such efforts are still in their early stages compared to the biomimetics of shell-like materials [78].

Bone is an interesting natural material in its own right, due to its combination of mechanical properties. Its combination of strength and resilience with light weight is the result of its nanocomposite structure composed of mineral and hydrated organic macromolecules (see chapter 13, *Introduction to Nanoscience*). Bone consists of layers of mineral crystals (hydroxylapatite) interspersed with protein and biopolymers that add resilience to the compression-bearing strength of the mineral component **(Fig. 12.16)**. Teeth have additional strength from enamel and fluorides **(Fig. 12.17)**.

Biomimetic Composites for Artificial Bone. One approach to promoting the growth of bone for healing is to provide scaffold material into which bone-forming cells can migrate, leading to fusion of new bone growth with the scaffold. Early versions of scaffolding involved coatings of hydroxylapatite applied to fractures. More recently sol-gel nanocoating techniques have been shown to improve bone grafting. Bone forms by mineralization of precursor cartilage tissues. Successful growth, maintenance, and healing of bone depend on each portion of the structure matching the types of forces to which it is subjected in the body. Much work is being done to understand how precursor cells are programmed and influenced to promote mixtures of mineralization and organic fibers that optimally match the compressing and tension loads that must be borne by the skeleton, and how this can be mimicked and encouraged by scaffolds, growth-promoting substances, and electromagnetic stimulation.

F.Z. Cui and coworkers in Beijing, China have developed a bone scaffold by biomimetic synthesis of nano-hydroxyapatite and collagen assembled into mineralized fibrils [97]. This material shows some features of natural bone in composition and hierarchical microstructure, with three-dimensional porous scaffold materials that mimic the microstructure of cancellous bone. In cell culture and animal tests, bone-forming osteoblast cells from rats adhered, spread, and proliferated throughout the pores of the scaffold material within a week. This scaffold composite has promise for the clinical repair of bone defects. This and similar work being done elsewhere shows how biomimetic structures can integrate with natural tissues [98,99].

FIG. 12.16

Bone resists tension, torsion, and compressive forces due to its nanostructure, made of hydroxyapatite mineral and collagen protein fibers. (a) The hydroxyapatite is arranged in concentric layers (called lamella) within a cylindrical, functional unit called an osteon. The apatite is arranged in small crystalline plates about $60 \times 30 \times 8$ nm in size found within and around collagen fibrils (bundles of collagen filaments.) The plates overlap and are aligned parallel to the long axis of the fibril. The plates within each fibril have the same angular orientation; adjacent fibrils are oriented at different angles to each other. Compression deforms the collagen, allowing the plates to slide towards each other until they start to touch. This initial movement absorbs some of the applied force and prevents the bone from breaking. As the plates touch, fictional forces take over and the plates can no longer slide. What happens when the forces are too great for the crystalline plates? They fracture, but because the plates are so small large cracks cannot form. Because the plates of adjacent fibrils are oriented at different angles small cracks cannot propagate easily from one plate to the next. (b) Micrograph of osteoblath reveals layered bone structure.

Source: Image courtesy of Max Planck Institute of Colloids and Interfaces. With permission.

12.2.4 *Sponge Fiber Photonics*

In the previous sections we looked at natural nanomaterials whose main functions are structural and load bearing; their nanostructures are optimized for interaction with forces, tensions, loads, and pressures. In this section we look at interactions with light—in photonic nanomaterials—and the biomimetic inventions that they are inspiring. The natural optical fibers found in the sea sponge *Euplectella* are tougher than current man-made glass fibers, and contain a unique mixture of dopants that enhance their optical properties. Fiber-optic engineers are studying the nanostructure of these natural silica fibers to learn how to improve the engineering of fiber-optic cables.

Euplectella is commonly called the "glass sponge." It is also known as the "Venus flower-basket" or "wedding basket" because its silica skeleton forms an intricate cage, which often houses a pair of mating shrimp. The skeleton of the

FIG. 12.17

Teeth are adapted for withstanding pressure and wear. Extreme examples are the teeth of coral-eating fish. The enameloid outer layer of parrot fish teeth consists of fluoroapatite crystals and collagen fibers, where the latter only comprises about 3% of the enameloid. The high mineral content makes a very hard surface, withstanding both compressive and shearing forces. The orientation of the fluoroapatite crystals determines the hardness of the enameloid. The crystals, 100 nm in diameter and several microns in length, are bundled together to form large fibrils. At places where the strongest shearing forces are present (where the grinding action takes place) the fibrils are perpendicular to the surface. This presents a rather small cross-sectional area to the shearing forces and a lower risk for crystal fracture. Should a fracture occur, propagation of the fracture is limited because of the small crystal size. (a) Collagen fibers, (b) mineral plates.

sponge is supported by a network of amorphous hydrated silica fibers, called spicules, with especially long and clear fibers growing from the base of the animal. These silica fibers resemble optical glass fibers used as waveguides in communication networks, of similar size and with a highly refractive core surrounded by a lower refractive index cladding. Researchers found that the sponge fibers carried light like single-mode or low-mode waveguide fibers (**Fig. 12.18**).

The natural spicules are tougher than man-made glass fibers. They can be bent without breaking, whereas telecommunications fibers are brittle and have to be protected. The natural fibers are composed of nanoscale layers of silica separated by organic molecules that form a crack-arresting elastic buffer. The fibers contain silica spheres between 50 and 200 nm in diameter, assembled together between many organic layers to yield a 100-μm fiber.

The natural fibers grow by self-assembly at the ambient temperatures, compared to the high energy and hard-to-control glass furnaces from which optical fibers are drawn. Low-temperature self assembly would allow engineers to dope glass fibers with precise concentrations of alkaline earth ions such as sodium, calcium, and magnesium in definite layers to selectively control the refractive index of the glass for improved optical transmission performance.

FIG. 12.18   *Micro- and nanostructure of a silica fiber from a glass sponge.*

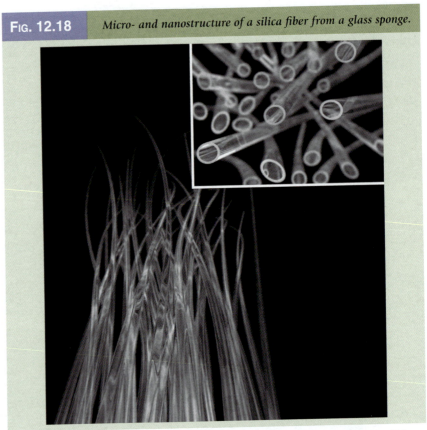

Source: Image courtesy of Jeff Christiansen, Halff Associates, Visual Science and Technology, Dallas, Texas. With permission.

Further biological studies to see how fibers in different sponges are adapted to depth, temperature, and available light could suggest ways to optimize the optics of man-made fibers. The sponge fiber has stimulated thinking about new ways to design and construct optical fibers for communications, and should "shed light on low-temperature, biologically inspired processes that could give rise to better fiber-optical materials and networks" [100].

Biomimetic Silica Nanofabrication Processes. A research group headed by Dr. Dan Morse at the University of California, Santa Barbara, is investigating the biomolecular mechanisms that direct the nanofabrication of silica in living organisms such as sponges, diatoms, and grasses. Mimicking these living organisms could lead to ways to direct the synthesis of photovoltaic and semiconductor nanocrystals of titanium dioxide, gallium oxide, and other materials. Applying natural synthetic methods to materials with which nature has never built structures before is a way to expand the limits of optical, electronic, and mechanical performance. Using natural examples could lead to production of semiconductors and photovoltaic materials in more energy efficient and environmentally benign ways.

Morse summed up the advantages of using natural templates: "Sponges are abundant right here off-shore and they provide a uniquely tractable model system that opens the paths to the discovery of the molecular mechanism that governs biological synthesis from silicon. This sponge produces copious quantities of fiberglass needles made from silicon and oxygen." Silicon, one of the most abundant elements on earth, is the basis for semiconductor technologies for computers, telecommunications devices, fiber optics, and other high-tech applications.

Morse and his group discovered that the silicon fiber of the sponge contains a protein filament that controls the synthesis of the spicules. They cloned and sequenced the DNA of the gene that codes for the protein filament, to determine how it acts as a catalyst to seed growth of silica. This was the first discovery of a protein catalyst that could control the growth of a rock-hard silica biomaterial. The protein actively promotes silicon deposition from low concentrations in sea water while simultaneously serving as a template to guide the nanostructure of the growing glass fiber [101–103].

Morse extended this work to develop synthetic mimics of the natural protein, and show that the same type of catalytic activities can be applied to the nanostructured synthesis of oxides of metals such as titanium and gallium with valuable photovoltaic and semiconductor properties. Low temperature, biomimetic catalytic fabrication of valuable materials used in electronics, photonics, and energy conversion devices could eventually replace high-energy, high-temperature processes that require vacuums, caustic chemicals, hazardous waste products, difficult control regimes, and consequent low yields and lost efficiencies.

Other Photonic Biomimetic Nanomaterials.   Sponges and diatoms are not the only natural models for photonic nanomaterials, nor is silicon the only natural element on which such materials are built. In chapter 13, *Introduction to Nanoscience*, we saw the photonic properties of opals and butterfly wings, and in chapter 11 of this book, we see how the photonic nanostructure of the butterfly wing can be adapted to make chemical nanosensors.

Insect chitins, feathers, diatom shells, and many other natural substances have photonic nanostructures. Reflectors, diffraction gratings, and two- and three-dimensional photonic crystals have been found in nature, including some designs not encountered previously in physics [104–107]. Some optical biomimetic engineering methods make direct analogues of the reflectors and antireflectors found in nature [108]. However, recent nanotechnology ventures beyond merely mimicking in the laboratory what happens in nature, leading to thriving new areas of research.

12.2.5   The Lesson of the Lotus—Nanocontrol of Surfaces

The lotus has been a symbol of purity for thousands of years because its leaves and petals shed contamination. Electron microscopy enabled researchers to resolve the mechanism by which the lotus and other plants and animals repelled water and dirt. It took decades to understand the full details of the effect, through in-depth studies of extreme water repellency (superhydrophobicity), at centers like the Nees Institute for Biodiversity of Plants at the University of Bonn by Boris F. Striffler, Zdenek Cerman, and Wilhelm Barthlott and others [109–110].

The "Lotus Effect" is now understood to be associated with very high contact angles for water on surfaces, based on low-energy hydrophobic surfaces and surface micro- and nanostructures that decrease contact and increase effective hydrophobicity [111–118].

Other Plant and Insect Surfaces.   Variations of the lotus effect have been found in many other plants and animals. Superhydrophobicity based on nanofibers has been found in gecko feet, water bugs, caterpillars, spiders, spider silk, wool, feathers, and leaves of many types [119–124]. Nanostructured superhydrophobic surfaces of many aquatic and semiaquatic plants and animals allow them to retain a layer of air under water. Besides keeping them clean and dry, the trapped air layer aids buoyancy, reduces drag, and allows for respiration [125–127].

Control of Surface Interactions by Nanosurfaces.   Studies of the shapes of natural surfaces on the nanolevel—*nanosurfaces*—have led to understanding the interactions of liquids with surfaces [128,129]. On rough surfaces, depending on the interplay between surface roughness and surface chemistry, different modes of wetting can occur. Surface wetting phenomena are classified into three categories:

1. Drops suspended on top of hydrophobic roughness features, with air trapped underneath ("fakir" drops). This situation is frequently referred to as "Cassie" or "composite" wetting because the liquid rests on a composite surface composed of solid and air.
2. Wenzel wetting, where drops are impaled on roughness features.
3. Superwetting, where drops are sucked into the surface structure, ultimately spreading to a contact angle of $0°$.

The category where drops rest on top of the rough nanofeatures is the Lotus leaf case; there the wetting contact angle is reduced, allowing drops to roll around easily in response to very small forces. When the surface tilts, the drops roll off, tending to take along particles lying on the surface, and leaving the surface dry and clean. The other extreme, superwetting, is found in the surfaces of insects that absorb moisture from the air in desert climates [130].

Lotus Effect Nanotechnology.   Active research continues into enhanced biomimetic superhydrophobic and superhydrophilic materials. Studies reveal how nanoscale geometry, chemistry, and topology affect surface behavior [131–134]. Active switching between superhydrophilicity and superhydrophobicity of nanosurfaces can be effected by photoswitching, pH, temperature, and other methods [135–140]. The effect of plasma etching and interactions with surfactants on nanosurfaces has also been studied [141,142]. A range of fabrication methods in different kinds of materials has been developed [143–155].

Self-Cleaning Surfaces.   The Lotus Effect (trademarked as "Lotus-Effect") has generated many successful, eco-friendly products such as self-cleaning paints, and fabrics that require less washing, detergents, and dry cleaning than conventional surfaces [156–159].

Transportation and Lighting. Another application is for windows and surfaces in vehicles and buildings—for prevention of icing, lubricating, and slowing corrosion. Automobile makers are investing in lotus research for self-cleaning surfaces and antifogging windows and lights. Compound insect eyes have been mimicked for water repellent, nonreflective, antiglare properties [160–165]. Nanoengineered surfaces mimic the ability of aquatic and semiaquatic plants and animals to retain a layer of air under water that can reduce drag on ships [166].

Biomedical Applications. Surface properties are vital in processes such as blood clotting and cell adhesion. Biomimetic nanosurfaces have been designed as anticlotting surfaces for medical implants, wetting of pharmaceutical and food powders, and providing cell-friendly surfaces for tissue culture. Nanosurface structure is a factor in biocompatibility of materials, and is crucial for attachment of cells to growth scaffolding, implants, and culture vessels [167–170].

Technological Applications. Nanosurface control is employed by chemical analysis, for mixing and laminar flow in microfluidic systems, for high-speed computer disk drive heads, and to protect surfaces of nano-manipulator instrumentation [171–174]. Biomimetic nanosurfaces are being directed to the development of papers and printing techniques for publications, signs, displays, and paper money for both improved wettability and self-cleaning applications [175].

Significance for Soil and Agriculture. Hydrophobicity of nanosurfaces is important for the efficiency of photosynthesis, plant metabolism, and the storage of fruits and vegetables. Research has shown how this effect increases the efficiency of photosynthesis, leaf respiration, and protection from pathogens. It has also been studied for the application of pesticides and herbicides to leaves and insect cuticles [176,177].

One of the most important areas of nanosurface research is revealing factors that affect hydrophobicity of soils and productivity of agriculture, especially with increased droughts and global temperatures. Understanding absorption of water by soil and plants, reducing runoff and surface evaporation, and erosion, reducing irrigation requirements and improving productivity of lands, could be among the most important outgrowths gained from the lotus leaf [178–184].

Technical and Economic Significance. Commercial products modeled on lotus leaf superhydrophobicity continue to grow in importance. The Lotus effect has become one of the best examples of how basic research in biological nanoscience produces practical applications through biomimetics. It shows that biomimetics is not simply a matter of imitating nature but of understanding the principles involved and applying them in an intelligent scientific manner.

Biomimetic Synthesis and Surface Modification of Natural Materials. Using natural substances like insect chitin and wood as starting material, chemists, materials scientists, and nanoengineers are developing new materials with improved properties. In many cases surface properties are modified to create superhydrophobicity or hydrophilicity for applications such as water-resistant fabrics, paints, or coatings. Other new materials are based on synthesis of biomimetic versions of natural materials with properties not found in nature [185–194].

Nanoscale interaction of solid surfaces with layers of molecules in liquids and gases surrounding them has inspired biomimetic materials with unprecedented properties and capabilities. They may even lead to technologies like "Spiderman" suits with advanced camouflage, self-cleaning, and adaptable superadhesive materials [195].

12.2.6 *Gecko Glue and Other Biomimetic Nanoadhesives*

Nanostructures can prevent adhesion of water and other substances, or nanostructures can create strong adhesion effects, as we saw in the gecko foot. In this section we will review how natural surfaces like the gecko foot provide adhesion without glues, and how those same surfaces can be manipulated to turn adhesion on and off as needed—it all depends on the nanostructure.

Geckos' toes are covered with millions of tiny hairs called setae; each seta branches out into billions of nanoscale spatulae. By studying geckos, spiders, flies, and other animals, researchers have learned how their fibrillar, self-cleaning feet can control adhesion by conforming to the nanoscale contour of surfaces, without using glue or leaving residues following detachment. This form of adhesion is due to forces that become important at the nanoscale; when van der Waals and capillary forces act at millions of gecko hairs in close contact to a surface, they add up to a bond that is a thousand times stronger than the force geckos need to hang onto a wall. The gecko can adhere strongly without having to press down on the surface, and it is able to detach and reattach rapidly without having to exert force [196,197].

Biomimetic Gecko Adhesives.   In 2002, an interdisciplinary research team found that the network of gecko setae forms intermolecular bonds with surfaces by means of van der Waals forces [198]. The team later experimentally demonstrated the mechanism for gecko adhesion with synthesized biomimetic gecko hair tips. Researchers have continued developing artificial biomimetic adhesives based on the gecko effect, and we give some examples of their work here.

Gecko Adhesive Using Hard Polymer Microfibers.   Researchers at the University of California, Berkeley, have developed adhesive gecko foot surfaces for use with climbing robots. They fabricated patches of microfiber arrays with 42 million polypropylene microfibers per cm^2. The patches can support $9\,N \cdot cm^{-2}$, with preloading of just $0.1\,N \cdot cm^{-2}$—a patch $2\,cm^2$ can support a 400 g load. Like the natural gecko foot, the Berkeley patches do not adhere when pressed down, but only when they slide over a surface, producing a shear force. The nanofiber adhesive surface is "smart" in that greater load increases the contact area and causes more fibers to engage, leading to greater adhesion strength [199,200].

Professor Kellar Autumn of Lewis and Clark College (Oregon) has identified seven key benchmark properties for gecko adhesives [201]. The Berkeley polypropylene gecko adhesive was able to demonstrate five of these seven properties in a single material:

1. Anisotropic directional attachment—the microfibers do not attach by being pressed into the surface, and instead require a sliding motion parallel to the surface for the fibers to bend and attach.

2. High pulloff force to preload ratio—a preload of less than 0.1 N is sufficient to engage the fibers, and after the preload is removed, the patch can sustain a shear load of 4 N.
3. Low detachment force—the polypropylene gecko adhesive is directional and shows adhesion in the direction parallel to the surface, and the patch can be easily removed with a force of less than 0.001 N.
4. Anti-self-matting—the polypropylene fibers do not stick to one another, and do not clump even with 100s of loading/unloading cycles.
5. Nonsticky default state—the polymer used in the synthetic adhesive, polypropylene, is almost as hard as the keratin used by natural gecko, and the surface of the patch is nonsticky.

Achievement of Keller Autumn's gecko properties six and seven continues to be the goal of further research efforts:

6. Topography independence (sticking to rough surfaces) and material independence (adhesion by van der Waals rather than chemical forces)
7. Self-cleaning

Eventually researchers hope to build arrays incorporating the necessary features to approach or surpass the adhesion achieved by geckos, which is about $10 \, \text{N} \cdot \text{cm}^{-2}$.

Effects of Geometry and Materials on Adhesive Fiber Arrays. Low detachment force, self-cleaning, and nonsticky default states are best provided by hard polymers, rather than the soft polymers typically used in pressure sensitive adhesives. Researchers at Carnegie Mellon University (CMU), using the gecko and spider as models, have developed microfibers made of polyurethane. The adhesion strength is dependent on the material properties as well as the tip shape and fiber size [202–204]

Theoretical and experimental work on the adhesive mechanisms of geckos, flies, spiders, and other animals has also been actively pursued at the Max Planck Institute in Germany. Researchers there have produced the first artificial gecko adhesives with some degree of self-cleaning ability [205,206]. The work at Berkeley and elsewhere shows that hard plastic microfibers adhere when bent over by sliding forces at the surface, but are not sticky when pressed down normal to the surface. The adhesive force increases with sliding distance, which provides a natural automatic braking effect as the gecko runs over the surface. The Berkeley group found that their synthetic patches became stronger the more they were used [207].

Polymer and Carbon Nanotube Composite Tape. Researchers at the Rensselaer Polytechnic Institute and the University of Akron created biomimetic "gecko tape" using polymer surfaces covered with carbon nanotube hairs, which can stick and unstick on a variety of surfaces, including Teflon. By imitating the nanopatterned microtubes on the gecko foot, and varying the shape, pattern, and compliance of the nanofibers, the team fabricated adhesive tapes from carbon nanotubes with high performance [208].

Geckel Nanoadhesives—Combining Wet and Dry Adhesive Strategies. Dr. Philip Messersmith and other researchers at Northwestern University have merged two opposite adhesion strategies found in nature, combining gecko-inspired dry adhesive spatulae with mussel-inspired glue to obtain a nanostructured surface that works better than either alone. The researchers call the hybrid material a geckel nanoadhesive.

Messersmith has done research sponsored by the NIH/National Institute of Dental and Craniofacial Research on the underwater adhesion of mussels. He coated each synthetic gecko-inspired microfiber with a synthetic adhesive protein inspired by the mussel. This double biomimicry opens a potentially superior route to the design of temporary adhesive materials.

The unusual protein compound used to coat the artificial setae mimics the reversible bonding action of a mussel adhesive protein that Messersmith's group previously isolated and studied over several years. The wet adhesive force of each pillar increased nearly 15 times when coated with the mussel mimetic. The dry adhesive force of the pillars also improved when coated with the compound [209].

Gecko Adhesive for Medical Applications.   Dr. Robert Langer at the Harvard-MIT Division of Health Sciences and Technology, and Jeff Karp at MIT and Brigham and Women's Hospital and Harvard Medical School, used gecko nano-adhesives coated with a thin layer of glue to help bandages stick in wet environments, such as heart and lung tissue. Their group included team members working at MIT, Draper Laboratory, Massachusetts General Hospital, and the University of Basel, Switzerland, fabricating materials and performing animal experiments.

The tape is designed for medical and surgery applications. The material can be biodegradable, to dissolve over time without having to be removed. The tape can fold or roll up to be inserted through a small incision, and then unfurled for application in minimally invasive or natural orifice transluminal surgery procedures where suturing is particularly difficult. There are significant design challenges for adhesives used in medical applications. For use in the body, adhesives must hold fast in a wet environment and be constructed of biocompatible materials: they must not cause inflammation; they must be biodegradable, meaning they decompose over time without producing toxins; and they must be elastic, to conform and stretch with tissue. To meet these requirements, the MIT team developed a new "biorubber" polymer. Pillars and holes were etched using micro-patterning technology adapted from computer chip production tools to create different hill and valley profiles at nanoscale dimensions. The stickiest profile determined by tests on pigs was one with pillars spaced just wide enough to grip and interlock with the underlying tissue.

A very thin layer of sugar-based glue was coated onto the profile, which created a strong bond even on a wet surface. In this application, low lift-off force is not required, since the biodegradable bandages are left in place. The effects of patterning with and without glue were tested on intestinal tissue from pigs and in living rats. Nanopatterned adhesive bonds were twice as strong as unpatterned adhesives, and coating nanopatterned adhesive with glue increased adhesive strength more than 100%.

The biorubber can be infused with drugs which are released as the polymer degrades. The elasticity and degradation rate of the biorubber are tunable, as is the pillared landscape, so that the adhesives can be customized to have the right elasticity, resilience, and grip for different medical applications. Despite the differences between the biorubber medical adhesive and the gecko foot, Karp said that his team was inspired by the gecko to create a patterned interface to enhance the surface area of contact and thus the overall strength of adhesion. Nanostructures inspired by nature can be adapted to create new types of materials to fit the complex needs of new applications [210].

There are many other research groups around the world developing adhesive systems based on the gecko foot model, using many different materials and biomimetic designs. Novel gecko-inspired dry adhesives have been produced by a silicon dioxide based MEMS process at the University of California, Santa Barbara [211]. A nanorod dry adhesive has been fabricated at the University of Manchester in the United Kingdom [212]. We leave as a useful exercise for the student the project of coming up with other novel designs and comparing them with those in the literature.

Biomimetic Tribology—Coming Unglued with Nanoscale Surface Control. Precise control of surfaces at the nanoscale can produce repellent, absorbent, adhesive, and lubricating effects. Study of natural systems has revealed that friction between surfaces can be viewed as a continuous phenomenon ranging from adhesion at one end to lubrication at the opposite extreme. Thus tribology—the science of lubrication and friction—can be a beneficiary of biomimetics. Biotribologists gather information about biological surfaces in relative motion, their friction, adhesion, lubrication, and wear, and apply this knowledge to technological innovation. Biological systems excel at optimizing interactions between matter at the micro- and nanometer scale. The miniaturization of devices such as hard-disk drives, MEMS, and biosensors increases the opportunities for applications of nanoscale tribological phenomena [213–219].

Living organisms have many tribological challenges on the nanoscale. Moving surfaces contact each other in joints, muscles, and blood vessels. Examples of moving micromechanical systems can be found in diatoms with hinges and interlocking devices on the micrometer-scale and below. The immune system produces molecules that can switch states from lubricating to adhesive; these glycoproteins control the movement of white blood cells as they move between endothelial cells and adhere to foreign particles. As we saw earlier, protein macromolecules play key roles in absorbing stress in strong self-healing adhesives in bone and shells.

Wear-Resistant Joints and Self-Healing Adhesives in Diatoms. Diatoms are a type of algae with nanostructured amorphous silica surfaces (see chapter 13, *Introduction to Nanoscience*). Diatoms have evolved interconnected junctions and self-healing adhesives that prevent wear between rigid surfaces in relative motion. Their silica cell coverings grow efficiently at ambient conditions to produce an amazing variety of sizes, shapes, and nanostructured patterns, many of which exhibit photonic resonance with visible light [220,221].

Diatoms serve as model organisms for nanotribological biomimetics and as templates for novel three-dimensional microelectromechanical systems. Some diatom species have evolved strong, self-healing underwater adhesives; others have elastic linkages between the halves of their silica coverings. Some species secrete viscous mucilage which binds colonies together while protecting the silica shells from wear as they rub against each other. Hinges and interlocking devices in diatoms are very stable and can be seen under the microscope preserved intact in fossil deposits that are millions of years old. The lubrication mechanisms used by diatoms have been studied as models for nanolubricants to reduce stiction for MEMS [222].

Some diatoms are free floating, but others have evolved adhesives for stable and strong attachment in water. Some diatom species found in Antarctic seas synthesize special proteins that bind to ice, and also prevent recrystallization of water. These may be templates for biomimetic cryoprotective substances [223]. Diatom adhesives show opportunities to tailor new synthetic adhesives, lubricants, and protective coatings for specific applications such as de-icing.

Immuno-adhesives as Models for Biomimetic Switchable Adhesives.   As we saw in the previous chapter (chapter 11, this book), the body's immune system includes white blood cells which flow through vessels and enter the extracellular matrix in tissues, sticking to and engulfing foreign particles. The immune system employs exquisitely sophisticated adhesives to deploy antigens to their targets, including the adhesive portions of the specific antibody macromolecules targeted to bind to recognized antigens.

These adhesives are highly selective—they use molecular and nanopattern templates to distinguish between self and nonself, allowing free flow in the first case and adhering tightly in the other. Their adhesive function can also be turned on by signaling substances produced by parts of the immune system that detect invaders and initiate the process of inflammation. Understanding how these nanoadhesives work is an important goal for biomimetic nanoengineering, both for the development of drugs and the design of artificial switchable adhesives for technological applications [224]. In addition, this is an important research area for medicine, for the prevention and treatment of immune-related disorders, the management of infection and healing, and in pathology and biotechnology applications.

Adhesive Processes in the Immune Response.   Before we discuss adaptive adhesion molecules and how they function in the immune system, let us review the steps that take place in the nonspecific immune response—the *inflammatory response*. The inflammatory response begins when tissue damage or metabolic disruption triggers mast cells to release histamines stored in granules into the neighboring tissue. The histamines stimulate dilation of capillaries, allowing plasma and leukocytes to penetrate into surrounding tissue (The characteristic swelling and redness results). Damaged or infected cells release signaling molecules which attract phagocytes, which adhere to and engulf dead cells, bacteria, and foreign particles. At the same time, activation of antibodies or other triggers may initiate a cascade of complement proteins which attract phagocytes and penetrate the membranes of damaged cells, leading to lysis and acceleration of the inflammation process. The tissue returns to its normal state as histamine and complement protein release ceases [225].

Rolling and Sticking Behavior of White Blood Cells.   When the inflammatory response is observed through the microscope, granulocytes and other leukocytes can be seen to adhere to and release from capillaries and small venules in the inflamed area. Leukocytes are observed to roll slowly along the endothelium, tethering and detaching as they move along in the direction of blood flow. The rolling velocity is typically 10–100 times lower than a nonadherent white blood cell moving next to the vessel wall. Adhesion molecules on the white blood cells and the endothelium regulate their interactions. The molecular mechanism underlying this leukocyte-endothelial interaction is of great interest for understanding switchable and adaptive adhesives [226–228]. The adaptive adhesion of white blood cells onto the endothelium takes place in a cascade of stepwise events [229]. These are classified as:

1. Initial tethering
2. Rolling adhesion (involving probing for signs of inflammation)
3. Firm adhesion
4. Escape from blood vessels into tissue

The Molecular Mechanisms for Adaptive Adhesion.   It is known that phagocyte adhesion is mediated by specific biological macromolecules, which include cell membrane receptors, extracellular ligands, and cytoskeletal components (see chapter 14, *Introduction to Nanoscience*). Interdisciplinary researchers have analyzed adhesion between cells to gain an understanding of the underlying mechanical and molecular properties that govern the processes. Progress has been made towards modeling biochemical and biophysical cellular nanoadhesive processes in terms of quantitative parameters which could be useful in guiding the biomimetic design of tailored synthetic adhesives and lubricants [230,231].

Switching the Permeability of the Endothelium.   One question addressed by this research is how white blood cells interact with the endothelium—the thin layer of cells that line the interior walls of blood vessels. Normally, the endothelium must resist interactions with blood cells, to allow the free flow of blood; but white blood cells may stop at particular sites and pass through the endothelium into the underlying subendothelial matrix—the layers of smooth muscle cells, structural proteins, and fibroblasts that make up the blood vessel wall.

White blood cells may pass through the matrix into surrounding tissue when cells of the endothelium are signaled to relax and open; this retraction of the blood vessel barrier may be induced by signaling compounds in the arachidonic acid pathway, which are produced by platelets as part of the clotting process. This signaling mechanism is important to understand how white blood cells contribute to clotting and build-up of plaques in the arteries, and how certain cancer cells migrate through the blood vessel walls during metastasis [232,233].

Separate Molecular Adhesive Mechanisms for Rolling and Migration.   Researchers have studied white blood cell adhesion to endothelial cells under well-defined flow conditions in a variety of systems. They have been able to distinguish separate mechanisms for the initial adhesion and rolling steps and the firm adhesion and migration of white blood cells through the endothelium [234,235].

Selectin.   Initial adhesion and rolling appear to be controlled by interactions between *selectin* and *carbohydrates* in the cell membranes. Selectins are a special type of lectins, the glycoproteins that bind sugar polymers.

Integrins.   The firm adhesion and white blood cell migration steps are mediated primarily by interactions between *integrins* and *peptides* at inflammation sites. Integrins are integral membrane proteins that bind cells to the intracellular matrix and other cells, as well as acting as signal receptors.

Signaling and Switching of Integrins.   Integrins play important roles in the cell membranes of all animals, and especially in white blood cells. Unlike most transmembrane proteins that form one-way cell membrane receptors, integrins can carry signals in both directions from outside and inside the cell. In the case of inflammation, signals from other receptors on the white blood cell are transmitted to its integrin transmembrane sites. These signals induce the extracellular domains of these sites to undergo conformational movements (changes in their molecular arrangement) that enable calcium ligand binding—switching from a nonadhesive to an adhesive state. In this state integrins rapidly stabilize contacts between white blood cells in the bloodstream and the endothelium at sites of inflammation [236,237].

Adaptive Adhesion and Molecular Properties.   Integrins are the most sophisticated adhesion molecules known, and as such have aroused much interest for nanobiomimicry as well as for their medical importance. Through conformational changes, integrins can mediate both firm and transient types of adhesion. In order to design adaptive nanoadhesives based on integrins, we must first understand the functional properties of these molecules that control the dynamics of adhesion. There is evidence that adhesion depends on physical chemistry in addition to mechanical features such as deformability, morphology, or signaling conformations [238,239]. Possible physicochemical properties that affect dynamic states of adhesion are reaction rate, affinity, mechanical elasticity, kinetic response to stress, and molecular size and length.

Adhesion Dependence on Applied Forces.   The activity of adhesion molecules may also be regulated by the force distribution present in blood vessels. The specifics of molecular adhesion can vary with the local wall shear stress—the force required to produce a certain rate of flow of a viscous liquid. Levels of venous and arterial shear stresses range between 0.1–0.5 Pa and 0.6–4 Pa, respectively. Adhesion of leukocytes to blood vessel walls may depend on shear force in a manner not unlike the adhesion of gecko fibrils to surfaces [240,241].

Integrins as a Biomimetic Model for Nanotools.   The molecular properties that enable integrins to switch their transient adhesion states are of interest for their medical implications [242–246] as well as for technological applications, such as nano-grippers and sensors [247]. The complex synergistic interactions of the large molecules involved in cellular adhesion provide models for biomimetic nanoscale adhesion and manipulation [248–250]. The molecular conformations and forces involved in cellular adhesion have been investigated experimentally as well as theoretically [251–256]. This active and ongoing research provides models and inspirations for sophisticated biomimetic nanobiotechnology that utilizes the full potential of nanoscale mechanical, physical, and chemical forces in an integrated fashion. For example, cellular adhesion molecules have been adapted as the active selective elements for field flow fractionation for a flow cytometry device, giving highly selective cell separation which is aimed at identification and diagnosis of blood disorders [257].

Adaptive Protein Grippers as Antiparasite Weaponry. An example of an adaptive gripper from the immune system as a model for biomimetic drugs comes from the mosquito. An international team discovered this molecular gripper mechanism in research for treatment and prevention of malaria. They unraveled the mechanism for natural resistance shown by some mosquitoes to malarial parasites. This resistance was known to be associated with thioester proteins (TEPs) in the innate immune system of insects, which have some similarities with the complement factor glycoproteins of the mammalian immune system.

TEP1 proteins are activated by a biochemical reaction triggered by the presence of malarial parasites in the mosquito. The reaction cleaves a thioester bond, opening up the protein into an active state that locks covalently onto the parasite's surface, targeting it for elimination. The protein acts in a very specific way as an adaptive adhesive, similar to those we saw earlier in the targeted immune systems associated with white blood cells in vertebrates.

The team used x-ray crystallography to determine TEP1's three-dimensional structure. They found that the genetic differences between mosquitoes that are resistant and those that are susceptible to the parasite mostly manifest in a region of the TEP1 protein they called "the warhead," the portion that grabs the malarial parasite. The senior researcher of the group is Dr. Johann Deisenhofer, who was awarded the 1988 Nobel Prize in chemistry for using x-ray crystallography to describe the structure of a protein involved in photosynthesis.

Another lead researcher in the group is postdoctoral fellow, Dr. Richard Baxter, who described the role of TEP1 as a scout that finds the enemy, then plants a homing signal, and calls in the air strike. Other members of the research team were at UT Southwestern and the Institut de Biologie Moléculaire et Cellulaire in Strasbourg, France. The French group previously determined that the gene for TEP1 occurs in two forms, or alleles. One is found in mosquitoes that are resistant to malarial infection. Detailed analysis of the x-ray crystallography structure showed that the differences cluster around the warhead area, reinforcing the theory that it is a key element of the binding to the malaria parasites, and suggested a gripping type of mechanism similar to that of complement proteins in the immune system of higher organisms (**Fig. 12.19**). Understanding how mosquitoes fight off malarial parasites with macromolecular protein weapons could provide a model for effective drug and control strategies [258].

Tough Underwater Adhesives. In an earlier section we discussed the nanostructure of shell, with emphasis on the mineral platelets and how they were linked by elastic glue. In the immune system we have seen the complex macromolecular mechanisms involved in natural adhesion. In the shell these types of molecules are used to hold mineral building blocks, and in the immune system they are adapted to bind to foreign bodies. The macromolecular glues in shells are adapted for toughness—to allow the shell to absorb and redistribute shock forces. This is a very useful property for engineering, if its mechanism can be understood and imitated [259].

FIG. 12.19 *(a) Thioester protein structures: TEP and human immune complement factor C3, left to right: TEP1r, human C3 inactive state, and human C3b activated state with the reaction center exposed. The thioester is shown as spheres and the remaining protein domains as ribbon cartoons. The thioester-containing domain (TED) is in green, the complement C1r/ C1s, Uegf, Bmp1 domain (CUB) is in navy blue, and the macroglobulin-8 domain (MG8) is in yellow. The anaphylotoxin domain (ANA), present in complement factor but not TEP1, is in red. (b) Thioester proteins as grippers: The thioesters and the domains that are active in gripping the antigens are shown in color for clarity, with the remaining parts of the protein shown as grey tubing.*

FIG. 12.19 (CONTD.)

(c) Thioester protein genetics: a portion of the sequence alignment for alleles of the TEP1 gene from the two laboratory strains of mosquito that are susceptible (S) and refractory (R) to infection by the malarial parasite Plasmodium berghei. Differences between the two alleles are concentrated in the thioester domain and surroundings, and on the right of the figure the pink spheres locating these differences illustrate that they cluster in space about the domains most important for the proteins' function.

(c)

Source: Images courtesy of Dr. Richard Baxter and Professor Johann Deisenhofer, UT Southwestern Medical Center at Dallas. Images produced with PyMOL (http://www.pymol.org) for the proteins and ESPript (http://espript.ibcp.fr) for the sequence alignment.

The binding proteins in shells and strong underwater glues produced by mollusks and diatoms share two important features in their nanoscale structure and mechanism of reacting to stress. These are *self-healing* and *sacrificial bonds*.

Self-Healing Adhesives. The natural organic adhesive in nanocomposite nacre is composed of large, long chain molecules with sacrificial bonds and hidden lengths (unfoldable modules). Sacrificial bonds are weak bonds (such as Coulomb, van der Waals, or hydrogen bonds) between molecular segments. The sacrificial bonds allow the molecule to be reversibly stretched by their breaking and rebonding, dissipating large amounts of energy, and thereby preventing the covalent bonds making up the backbone of the molecular chain from breaking.

Sacrificial Bond Energetics on the Nanoscale Makes Natural Adhesives Self-Healing. Sacrificial bonds and hidden length in the folded conformations of long-chain natural adhesives provide an energy-absorbing mechanism that protects the molecule from being torn apart. This reversible, molecular-scale energy-dissipation greatly increases the fracture toughness of biomaterials. The sacrificial bonds themselves are relatively weak (van der Waals, electrostatic, Coulomb, hydrogen bonds) compared to the covalent bonds linking the backbone of the polymer chain. But breaking the multiple sacrificial bonds also involves the energy needed to reduce entropy and increase enthalpy as molecular segments are stretched after being released from their folded attachments. This energy is on the order of $100\,eV$—large compared to the energy needed to break the polymer backbone, on the order of a few electron volts. In many but not all cases, the breaking of sacrificial bonds is reversible, adding a "self-healing" property to the material [260].

Identifying and Measuring Sacrificial Bonds with Nanotechnology Tools. Historically, the understanding of hidden bonds began with studies of stretching wool fibers. Single-molecule force spectroscopy using an atomic force microscope (AFM) has been the modern nanotechnology tool by which sacrificial bonding has been investigated and quantified. In natural polymers such as glycoproteins, the force versus distance curves can be very complicated. Researchers closely analyze *atomic force microscopy* curves to obtain information about the molecules and their bonding to other components of natural composites.

Sacrificial Bonds Revealed by Sawtooth Force Patterns. AFM investigation of proteins and other biopolymers reveals their folding and bond-breaking behavior under applied stress. When individual strands of polymer fibers are stretched using AFM, the force plot exhibits a repetitive "sawtooth" response, rising and falling as the material is extended. Each tooth in the saw represents an unfolding of the long, complex molecular chain. This pattern is characteristic of material with sacrificial bonds connecting modular structures. This provides direct evidence of the effects of the hydrogen bonds, ionic attractions, covalent bonds, and steric rearrangements into less favorable energetic conformations. The ability of the bonds to re-form when the force is relaxed can be determined by repeating the stretch cycle. The force plots show the action mechanism of the multimodular nanostructure that gives natural adhesives self-healing behavior.

Researchers have observed this sacrificial bond behavior in protein-based fibers from many sources including wool and adhesives from mussels, diatoms, and algae. Dugdale and coworkers at the University of Melbourne, Australia, showed that single adhesive nanofibers produced by certain diatoms to attach to surfaces have the signature fingerprint of modular proteins: their force–extension curves have regular sawtooth patterns [261].

Amyloid Fibrils in Natural Adhesives. Researchers Anika Mostaert and Suzanne Jarvis at Trinity College Dublin studied AFM curves of algal adhesives and found highly ordered amyloid structures in the fibrils. Amyloids are normally associated with neurodegenerative diseases such as Alzheimer's where they are a primary component of plaques in brain tissue.

But the amyloid sheets found by the Dublin researchers provide sacrificial bonds in natural adhesives. They found evidence for similar material in parasitic worms. By exploring the differences between types of amyloid proteins found in different organisms, they seek to find potential templates for the design of biomimetic adhesives. Their work may also shed light on the origins, causes, and treatments for neurodegenerative diseases [262–264].

Future Opportunities for Biomimetic Nanocomposites. Synthetic adhesives and lubricants still have a long way to go to match the performance of many natural systems. Using the lessons learned from nature and armed with the tools of nanotechnology, biotribologists and other biomimeticists draw upon natural designs that have been optimized for millions of years to realize adaptive, self-healing and environmentally friendly lubricants and adhesives.

In this section we have looked at surfaces and their interactions. Next we add one more dimension to our exploration of biomimetic materials: biological membranes.

12.2.7 *Biomimetic Membranes and Nanocapsules*

Through the cell membrane with its gates and chemically selective lipid and protein domains, the cell senses and communicates via chemical signals with its complex biochemical matrix. We will see some biomimetic ways of making artificial lipid bilayers for practical applications and some of the artificial membrane structures that have been made to mimic useful functions such as selective permeability and molecular sensing.

Biomimetic Membranes.   The bilipid membrane of cells has served as a biomimetic model for decades. Engineers duplicate the function and structure of natural membranes in more durable and versatile materials to control molecular transport and fabricate sensitive and selective sensors. There are a number of ways to create biomimetic membranes. Lipid vesicles (artificial liposomes) can be formed by agitation of phospholipids in water, using sonication (high-frequency sound waves) or rapid mechanical mixing. Planar-supported bilayers can also be formed from phospholipids and other bipolar macromolecules. The classic method was developed by Langmuir and Blodget in their studies of the surface forces of polar organic molecules in water. (**Fig. 12.20**).

There are many other ways of producing membranes with nanoscale organization. As we saw in section 12.2.2, membranes of proteins and inorganic crystals

FIG. 12.20   *A Langmuir–Blodget instrument for formation of lipid bilayers from thin layers of lipids on the surface of water. (a) Macromolecular monolayer on liquid surface is entrained as support is withdrawn, (b) coated wafer is withdrawn from apparatus.*

Source: Image courtesy of KSV Instruments SA. With permission.

are being formed in biomimetic processes to make analogues of shell and other biocomposites.

Artificial Membrane Mazes. Artificial lipid bilayers are very useful for investigating material transport and signaling mechanisms in natural cells, but making practical sensors requires a more durable material with long-term stability in many types of environment. Synthetic membranes have been fabricated using nanopores in aluminum, glass, silicon, polyester, and polycarbonate polymers. A simple way of producing a material with extremely small pores is by repeated drawing, bundling, and re-stretching of glass capillaries containing an etchable core, then removing the cores. Such artificial membranes, however formed, can be used to mimic some of the material transport functions of the natural cell membrane.

Forests of vertically aligned carbon nanofibers can control molecular transport in a manner analogous to natural cell membranes. Dense arrays of carbon nanofibers are used as membranes integrated within fluidic structures. Size-dependent transport, perpendicular to the orientation of the fibers, can be controlled based on the wall-to-wall spacing of the individual fibers. When biomolecules, such as DNA and proteins, are attached to the sides of the nanofibers, such membranes can function as sensors, using the resulting combination of size fractionation and chemical specificity to select analytes. Integrating electrically addressable fibers adds another dimension to controlled transport as well as electrochemically based sensing. The biomimetic modification of carbon nanofiber structures is useful for performing chemical separations and for mimicking the properties of natural membranes. [265–270]

Mimicking Natural Encapsulation. We have progressively covered nanostructures and how they interact with forces, optical materials and light, nanoengineered surfaces and their interaction with liquids and nanoparticles, and membranes and their interactions with molecules. Now we will look at a special case of membranes: nanoencapsulation.

Drug Delivery by Nano-Dumplings. Biomimicry has been used to inspire encapsulation in self-assembled structures similar to liposomes, globular proteins, and other nanostructures in cells. A team led by Dr. Karen Wooley at Washington University Saint Louis, Missouri, developed a novel class of synthetic polymer nanocapsules, similar to globular proteins, ranging in size from 10 to 100 nm. Dr. Tomasz Kowalewski, a member of the team who used AFM to obtain images of the nanoparticles, suggested that they be called "knedels"—after a type of Polish dumpling.

Shell-crosslinked knedels (or SCKs) are spheres in which a hydrophobic core is surrounded by a hydrophilic shell that resembles the structure of lipoproteins. With the right conditions, the particles form by self-assembly: amphiphilic block copolymers (polymer chains with separate water-soluble and water-insoluble segments) nucleate to form a micelle made up of from tens to several hundred individual polymer chains. The hydrophobic cores of the resulting micelles are shielded from the surrounding aqueous environment by the solubilized outer shell.

These "nano-dumplings" can be made in different sizes and chemical compositions for drug delivery, gene therapy, and immunology. Their stability, release rates of encapsulated contents, and attraction to different types of external molecules can be manipulated by varying the polymers in the shells. Possible applications range from removing hydrophobic contaminants from aqueous solutions to use as recording materials [271].

Nanoscale Bioreactors. In a demonstration of a multifunctional, smart nanoscale drug delivery system, researchers at the University of Basel have created a drug-loaded nanocontainer that targets specific cells and releases its payload in response to a specific physiological signal. Smart nanocontainers could become anticancer drug delivery vehicles that target tumors and release their contents only when they receive a tumor-specific biochemical signal [272].

Bacteria and Viruses as Templates for Nanocapsules. Diatoms, bacterial cell walls, and other natural capsules are used as templates for nanoencapsulation—these natural structures can be adapted for active drug delivery [273]. This biomimetic technology takes the natural mechanisms used by microbes to invade the body and adapts them for research and medical purposes. For example, the dendrimer plug from a bioengineered virus was adapted as a drug delivery capsule. Researchers developed viral coat nanoparticles that incorporate receptors in their outer shells acting on biological effectors inside cells. The receptor and effector together act to detect a specific biochemical signal that then affects the nanocontainer and its contents. The effects can include drug release or the generation of a diagnostic signal [274].

The viral nanocontainers are loaded with an enzyme that converts substrate molecules into a light-emitting fluorescent form. The substrate molecules are specifically transported by bacterial membrane pore proteins engineered into the viral nanocontainer, where they react to produce light that can be seen using fluorescence microscopy, demonstrating that the delivery capsule is working. This method could insert an enzyme that converts an inactive drug into its active form for release only inside a diseased cell.

12.2.8 Some Other Biomimetic Materials

Nature is rich in examples of nanostructured materials. There are so many new biomimetic nanotechnologies from which we can only select a few examples in this chapter. The reader will want to pursue further reading about many other interesting and informative examples of biomimetic nanotechnology. We close this section with a few more interesting examples.

Biomimetic Water Harvesting. Certain plants and insects possess the ability to capture water from fog. The desert cockroach (*Arenivaga investigata*) can harvest water from apparently dry air [130]. The Namibian desert dwelling beetle Stenocara has bumps on its wing scales with superhydrophilic nanostructured surfaces. The peaks of the bumps are glassy-smooth and hydrophilic; the slopes of the bumps and troughs in between are covered with hydrophobic wax. As droplets accumulate in size, they roll from the tops into the waxy channels to a spot on the beetle's back that supplies its mouth [275].

QinetiQ, Ltd. of the United Kingdom developed sheets that capture water vapors from cooling towers and industrial condensers based on the beetle wing design. They capture 10 times more water than the preexisting ethno-biomimetic technology—the fog catching nets used by inhabitants of remote mountain towns in Peru and Chile (inspired by the capture of fog by cloud forest plants). The sheets are being tested for water collection on tents and roofs in arid regions.

Biomimetic Food and Nutrition Science.   In the introduction to this chapter, we raised questions about how apples store fluids. The nanostructured gels in which nutrients are stored and protected in fruits, as well as how nanoencapsulation affects biological activity of their antioxidative macromolecules, are studied by agricultural scientists and nutritionists [276,277]. Food scientists are exploring the nanoscale structure of fruits and how they are affected by freezing and drying to seek improved hardiness and preservation of freshness [278].

Biomimetic lessons also come from feathers, hair, wool, skin, claws, beaks, eggshells, teeth, corals, sponges, crustacean and insect chitins, spider and silk-worm silks, insect eyes, pollens, seeds, grass stems, wood, celluloses from plants and bacteria, bacterial cell walls, and many others [279,280].

12.3   BIOMIMETIC NANOENGINEERING

Up to now we have looked at biomimetic materials with interesting properties due to their nanostructures. Now we will look at more active nanoscale devices. There is a wealth of natural sources from which to draw inspiration for nano-scale devices. These include muscle that exerts force, viral mimetic batteries for high power density, photosynthetics to capture and store light energy in chemical form, rotors for cilia, gates for ionic transport, and DNA itself, whose storage and processing of information has been adapted in biomimetic molecular computing engines.

12.3.1   *Artificial Muscles*

Muscle is a natural engine that uses chemical energy to contract nanostructures, thereby exerting force which animals use for movement and mechanical work. Muscle generates macroscale forces by a hierarchical organization of millions of nano-actuators acting in parallel. Muscle produces undulatory motion (as in a slug or mollusk foot) or is harnessed to shells and bones to form levers for more complex movements.

Nanostructure of Natural Muscle.   Muscle exhibits a hierarchical organization when viewed microscopically at progressively greater magnification. Typical muscle is made up of fiber bundles connected to bony joints via ligaments and tendons. The active portion of the muscle is composed of fibers called fascicles. Each fascicle is composed of smaller fibers, consisting of individual cells. Each cell contains bundles of myofibrils, the active engine of muscle—bundles of contractile protein filaments aligned in parallel with the ends overlapping.

The filaments are composed of two types of protein: actin and myosin. When seen at high magnification, the overlapping portions of actin and myosin chains form visible bands, or striations. The overlapping nanostructures making up the striations are called sacromeres. A sacromere consists of a parallel bundle of

| **FIG. 12.21** | *Schematic diagram of muscle structure.* |

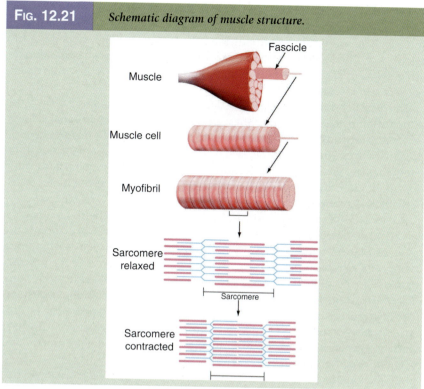

myosin filaments interlaced on each end with actin filaments. The actin filaments are linked together so that their long chains extend between the fibrin filaments like fingers on a hand (**Fig. 12.21**) [281–287].

Action of Natural Muscle.   When a muscle is stimulated by a nerve impulse, the actin filaments move across fibrin filaments, producing a lengthwise contraction in the sacromere. The energy for the contraction comes from the chemical bonds of ATP, driving a reaction in which the myosin and actin bond to form actomyosin, shortening the chain. This process can be repeated cyclically to result in a significant shortening of the entire bundle.

Nanolevel biomolecular investigations have revealed the details of the contraction mechanisms and energetics, in which voltage-gated calcium channels play an important role. The interactions of thin and thick filament proteins of the contractile apparatus are driven by troponin and calcium binding [288–291].

Natural muscles can repeat millions of work cycles before tiring, in part because the work load is shared and distributed over many fibers, some working while others recover. Natural muscles can contract by more than 20% or their length, at rates exceeding 50% per second. The efficiency of conversion of glucose to energy is nearly 40%, more than double the energy efficiency of artificial mechanical engines. Thus natural muscle has been the starting model for many efforts to duplicate some or all of these qualities by artificial means.

Biomimetic Artificial Muscle.   Biomimetic versions of nanoscale-based actuators have been designed to mimic the smooth and efficient action of natural muscle. Some early efforts used polymers that contracted and expanded when treated with different chemicals, but more modern versions are based on *electroactive polymers*, which expand or contract when subjected to electromagnetic fields or currents [292–297].

One type of artificial muscle is based on *electronically conducting polymers* such as polyaniline and polypyrrole. These polymers undergo dimensional changes when dopant ions are inserted into the polymer lattice by electrochemical potential forces, similar to those used in a battery. With these polymers, carbon nanotubes can greatly increase the surface area available for interaction with a dopant solution, thus increasing the effective speed, strength, and volume change of the artificial muscle actuation.

A type of artificial muscle which acts like a capacitor, rather than a battery, can be made based on *dielectric elastomers*. In these polymers, actuation is the result of *Maxwell stress*, which results from the attraction between opposite charges and repulsion between like charges, on different layers of the polymer matrix. Silicone rubber polymers are one class of material used to fabricate dielectric elastomers. Artificial muscles based on dielectric elastomers can generate higher actuator strains than those based on conducting polymers. A company, Artificial Muscle Inc., has been formed to produce elastomeric actuators, which can generate strains of 120%, stresses of 3.2 MPa, and a peak strain rate of 34,000% per second for 12% strain.

Another type of artificial muscle uses the volume change of a polymer electrolyte caused by electrostatic repulsion when ions are absorbed. One version, called the *ionic polymer/metal composite actuator*, is manufactured by Environmental Robots Inc. These artificial muscles amplify low strains using the cantilever effect; they consist of two metal-nanoparticle electrodes filled with and separated by layers of a solid electrolyte. The actuators act as supercapacitors when an applied interelectrode potential injects electronic charge into the high-surface-area electrodes. This charge draws solvated ions to migrate between the electrodes. The volume of the solvated ions causes one electrode to expand relative to the other, thereby bending a cantilever actuator strip. Actuation can be produced by electrolysis, which causes a local pH change and transport of hydrated ions and water into the nanostructure.

Actuation rates and efficiencies for ionic systems are generally lower than for dielectric polymer systems. But a robot based on this system made artificial muscle history in 2005 as the best arm wrestling robot in competition with a human athlete during the first human–robot arm wrestling competition in San Diego, California. This marked a milestone in what Dr. Y. Bar-Cohen of the NASA Jet Propulsion Laboratory calls the "grand challenge" for robotics [296].

Other artificial muscles are based on shape-memory alloys, which can generate strains of up to 8%, but require conversion of electrical to thermal energy, followed by conversion of the latter to mechanical energy, with heat as a by-product. Actuators that use electrochemically generated gases confined in carbon nanotubes have produced strains of 300%, but with low energy conversion efficiency and cycle life.

The Nanotech Institute of the University of Texas at Dallas, led by Dr. R. Baugham, is making low-voltage, low-strain actuators that use electrochemical charge injection into nanostructured electrodes, resulting in electrostatically driven electrode expansion. One device type, carbon nanotube artificial muscles, can generate 100 times the stress of natural muscle and provides comparable actuation rates (20% per second), but the actuator strain is at best 2%. Baugham has suggested improving performance by replacing the carbon nanotubes by nanostructured elastomeric conducting polymers that would have comparable surface areas but would be more easily deformable (by a factor of 100).

Baugham recently reviewed approaches that could radically improve ionic polymer/metal composite actuators by replacing cantilever actuators that operate by bending with actuators that operate in tension. This more biomimetic design could be made by separating the opposing electrodes by a liquid electrolyte, which would provide the solvated ions for actuation [294].

The capacitance of current ionic polymer/metal composite actuators is less than one-tenth that of other supercapacitors. Increasing capacitance by increasing electrode surface area could increase actuator strains, as long as the electrode did not restrict movement in the actuator stroke direction. Baugham proposed that highly twisted carbon nanotube yarns, with their high electrical conductivities and high surface area, would be ideal for this type of electrode.

New directions for improved artificial muscle more closely mimic natural muscle: more elastic material is closer to muscle protein, liquid electrolyte is closer to intercellular plasma, action by tension instead of bending, and the nanoscale structure of the surfaces that provides the actuation force—these design directions are all moving closer to the sacromere structure. Thus we see that biomimetics continues to be a productive approach for design of nanoactuators.

12.3.2 *Viral Energy Storage*

A key theme in biomimetic nanotechnology is the exploitation of high surface areas created by nanoscale structures. This is true for the lotus and gecko effects, and for artificial muscles. This effect is exploited for energy storage in a very direct biomimetic manner by researchers at MIT, who patterned battery electrodes with self-assembling nanostructures grown by genetically engineered viruses [298,299].

Using genetic engineering to alter a few viral genes, the MIT team made the viruses bond to conductive metal nanoparticles of gold and cobalt oxide. They altered nucleotide sequences in the viral DNA which directed the outer coat proteins to add an amino acid that binds to cobalt ions, which react with water to form cobalt oxide, an advanced battery material with higher storage capacity than the carbon-based materials now used in lithium-ion batteries.

To improve the conductivity of the electrodes, the genetic code was further modified to express an additional strand of virus proteins that bind to gold. The viruses then assembled into nanowires with both cobalt oxide and gold particles. The virus used to make the wire, the M-13 bacteriophage, is about 6 nm in diameter and 880 nm in length.

The conductive viral nanoparticles were layered between oppositely charged polymers to form thin, flexible sheets. To make electrodes, support sheets are dipped into a solution of engineered viruses. The viruses, with hydrophobic and hydrophilic ends, assemble into a uniform layer. The coated sheet is then dipped into a solution containing metal ions. The viruses arrange the ions into an ordered crystal structure which is ideal for high-density batteries.

Used as electrodes in thin lithium-ion batteries, the high surface area sheets improve performance in smaller volume. By increasing surface area and eliminating inert supporting material, energy density is raised by a factor of three. Equally important, viral nanowires are grown at normal room temperature, leading to an energy-efficient manufacturing process.

Exploiting the abilities of microbes to recognize the correct molecules and assemble them where they belong, the MIT team is able to precisely control the nanostructure of electronic materials. The team led by Dr. A. M. Belcher is working towards building faster, cheaper, and environmentally friendly transistors, batteries, solar cells, diagnostic materials for detecting cancer, and semiconductor devices [299].

> My dream is to have a DNA sequence that codes for the synthesis of materials, and then out of a beaker to pull out a device. And I think this is a big step along that path.
>
> **Dr. A. M. Belcher**
> *MIT [299]*

12.3.3 *Photosynthesis*

Photosynthesis, the conversion of light into stored chemical energy in plants, is one of the crowning achievements of nature's biological nanotechnology [300,301]. Its fundamental importance to life on earth has been recognized by the award of 10 Nobel Prizes for unraveling its mechanisms. One of these Nobel laureates, P. D. Bower, used the term: "a splendid molecular machine," to describe ATP synthase, a key component of the reaction chain [302]. Increased understanding of this process, and advances in chemical synthesis and nanotechnology have made it possible to create biomimetic devices and semibiological hybrids capable of many of the functions of natural photosynthesis [303–306].

In this section, we describe the basics of the natural photosynthetic process and some biomimetic systems. To begin, photosynthesis requires light-harvesting antennas, linked to reaction centers that convert photoexcitation energy to chemical potential in the form of long-lived electrochemical charge separation. With these two steps, engineers can create molecular-level optoelectronic devices with a variety of uses, but natural photosynthesis goes further—it converts electrochemical energy into chemical bonds where the energy can be stored for use as fuel or to build biological polymer materials. This step involves light-driven proton pumps embedded in the lipid membranes of cells, which build up electrochemical potential to drive the synthesis of adenosine triphosphate (ATP), the natural energy storage and transport substance common to all bacterial, plant, and animal cells [307–314].

Photosynthesis and the Development of Natural Photocells. Photosynthesis (from the Greek *photos* "light" and *syntheses* "put together, combine") is the conversion of solar energy into chemical energy (in the form of carbohydrates) from raw materials CO_2 and H_2O, with O_2 as a by-product. Photosynthesis—the ultimate achievement of natural bio-nanotechnology—harnesses photochemical and biochemical energy to fuel life's functions (**Fig. 12.22**).

Photosynthesis Overview. Sometime over three billion years ago, blue-green algae declared their energy independence by developing the capability to split water into hydrogen and oxygen. The process is called photosynthesis. Perhaps one of the most ambitious directives of biomimetics is to duplicate nature's grand method of energy conversion—to design the next generation of energy production on a photosynthetic model.

Chemical reactions associated with photosynthesis are shown below. The creation of carbohydrates:

$$6CO_{2(g)} + 12H_2O_{(aq)} + h\nu \rightarrow C_6H_{12}O_{6(aq)} + 6O_{2(g)} + 6H_2O_{(aq)} \qquad (12.1)$$

FIG. 12.22	*Photosynthesis in plant leaf cells.*

(continued)

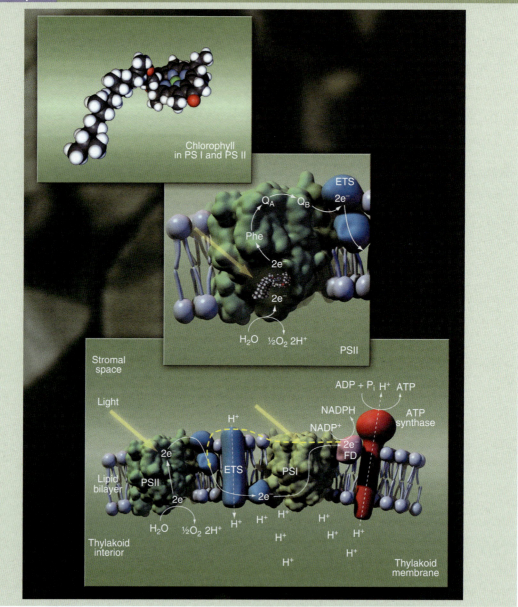

Source: Image courtesy of A. Rao.

The production of oxygen from water:

$$2H_2O + 2NADP^+ + 2ADP + 2P_i + h\nu \rightarrow 2NADPH + 2H^+ + 2ATP + O_2 \quad (12.2)$$

where P_i is inorganic phosphorous. Carbon fixation is given by

$$3CO_2 + 9ATP + 6NADPH + 6H^+ \rightarrow C_3H_6O_3 \sim PO_4 + 9ADP + 8P_i$$
$$+ 6NADP^+ + 3H_2O$$
(12.3)

For this to occur, ATP provides an oxidizer stronger than oxygen. The process of respiration proceeds in the reverse direction.

Plant pigments like chlorophyll, xanthophylls, and carotenes absorb strongly in most of the visible region of the solar spectrum. When the pigments absorb light, photons raise some of their electrons from the ground state to a higher energy. The excited electron can transfer its energy in several ways: (1) it can reemit the energy at the same wavelength (reflectance or emission) or at a longer wavelength (fluorescence), (2) the energy can be dissipated as heat, or (3) the energy can be used to fuel biochemical processes (e.g., photosynthesis). If the process were confined to one pigment molecule (chlorophyll), then the energy would dissipate as the electron regained its ground state. However, an antennae complex consists of several pigment molecules, which allows the excited electron energy to be delocalized from its original site, making it available for charge transfer reactions which can drive the process further.

Significant photocatalytic products produced (in the thylakoids discussed below) include O_2, NADPH + H^+, and ATP—the last two of which provide for the reduction of CO_2 in the surrounding stroma of the thylakoids [315]. The reduction process is modulated by the nanomaterial *ferredoxin*, an iron–sulfur containing protein nanomaterial. Overall, water is oxidized and oxygen released by the plant. Most of the chemical functions are accomplished by five large protein complexes: photosystem I (with bound antennae), photosystem II (with bound antennae), light-harvesting complex II, cytochrome b_6f, and ATP synthase [316].

The photoreaction center consists of ligated chlorophylls that are oxidized to a cation radical following transfer of exciton energy from the antenna pigments—electron transfer (charge separation) reactions occur on the order of picoseconds (corresponding to molecular vibrations). Charge separation induces the creation of a large redox potential between the oxidized chlorophyll a and potential acceptors—the largest known in biology: ~1.5–1.8 V for photosystems I and II, respectively. Electron transfer reactions that follow are energetically downhill. The highly oxidizing chlorophyll reaction centers are prevented from reacting with highly reducing acceptor species—a tribute to ingenious molecular nanodesign. In photosystem I, Fe–S clusters, soluble ferrodoxin and flavoproteins catalyze the reduction of $NADP^+$. In photosystem II, a tetranuclear Mn cluster serves the four-electron oxidation of H_2O to oxygen. The redox reactions are catalyzed by metalloproteins containing either iron or Mn.

The Leaf. Surface nanostructure is a prime driver in biological function. The structure of the leaf is geared to maximize the photosynthesis process—the phenotypical expression of billions of years of development. **Figure 12.22** gives an illustration of the nano and microstructures that make a leaf what it is—a lean photosynthetic nanomachine.

The Chloroplast. Chloroplasts are semiautonomous micron-sized structures that house many kinds of structural and functional nanomaterials. The chloroplast in higher plants is an elongated vesicle filled with an aqueous matrix material called *stroma*. The chloroplast is surrounded by two enveloping membranes.

Thylakoid Membranes. Thylakoid membranes, the chlorophyll-sequestering laminar systems contained within the chloroplast, are stacked pancake-like vesicles (**Fig. 12.22**) found in photosynthetic prokaryotes and eukaryotes. The thylakoid vesicles contain most of the proteins required for the light reactions of photosynthesis and are made primarily of lipids with membrane-embedded proteins. All of the light-harvesting and energy-transducing aspects of photosynthesis take place in the thylakoid membranes.

The thylakoids are ca. 500 nm in size and are stacked to form structures called *grana*. Approximately every other thylakoid possesses an appendage that extends into the *stroma* (the interstitial spaces in chloroplasts) to form a three-dimensional interconnected network. These extensions are known as *stroma lamellae*. The physically contiguous membrane encloses an aqueous phases known as the *thylakoid lumen*. The thylakoid membrane is in a unique class compared to membrane bilayers of other organelles and the cell plasma membrane [315].

Chlorophyll. Chlorophyll pigments are tethered to thylakoid protein components of photosystem I and photosystem II. Chlorophylls trap solar energy and convert it into usable chemical energy to conduct oxidation–reduction reactions. Trapped energy is stabilized by transfer to other chlorophylls, pigments, and secondary redox reactions in protein complexes within the thylakoid membrane [315]. *Chlorophyll* a absorbs energy from the violet-blue, orange red, and some from green-yellow-orange wavelengths of visible light. Chlorophyll a reflects green light (hence the color of most leaves.)

Oxidation and Reduction Processes. Chlorophyll acts as a light antenna in which photons are absorbed, exciting electrons to higher energy states. The energized electrons drive a series of photochemical electron transfer reactions involving *quinones*, which carry the energy in proton bonds to a reaction center where it is stored in the bonds of the ATP molecule, which reacts further to form *nicotine adenine dinucleotide phosphate* ($NADP^+$). In the final step of this chemical process, hydrogen is taken from water to form NADPH, releasing oxygen as a by-product. The NADPH stores the energy until it is used in the next step to energize the formation of carbon–carbon bonds, consuming carbon dioxide in the process (**Fig. 12.23**). The end products are carbohydrates (The general formula of carbohydrates is $[CH_2O]_n$.) Thus the overall process consumes water and carbon dioxide, two greenhouse gases, produces fixed carbon which is the food base for all animal life, and releases oxygen into the atmosphere [300–302].

Photosynthesis and Artificial Nano. Photosystem I (PSI) complex has been isolated from spinach and used to power electronic devices—to fabricate the first biomimetic solid-state photosynthetic solar cell [317]. Creating the interface is nontrivial because it is quite a leap from a biological system that requires water and salts to function—both materials not desired in solid-state electronic devices. Researchers at MIT extracted PSI (10–20 nm in size) from chloroplasts and stabilized them with surfactants in a solid-state device [318]. The PSI systems are relatively large; the tightly coupled antennae complexes contain 175 chlorophyll molecules per PSI. The efficiency of charge separation is close to 100% when in its natural state and the terminal electron-accepting moieties are stable with a low electrical potential (<0.6 V) [319].

The device consists of a bottom transparent layer coated with a conducting material (a thin layer of gold that assists in the self-assembly of the PSI units). A semiconducting layer is evaporated on the biological materials to prevent electrical shorts and then another conducting layer is applied. About 12% of the photonic energy in incident light is converted into electrical charge. Higher levels of charge efficiency (~20%) could be attained by fabricating multiple layers of these PSI sandwiches [317,318].

FIG. 12.23

(a) Schematic of the photosynthesis reaction center in plant leaf showing the essential membrane-bound elements that comprise the light reactions of photosynthesis. Light is absorbed by protein-bound pigments within a light-harvesting complex (green); the energy is transferred to a reaction center (yellow) in which the energy is used to separate charge across the membrane, leading to reduction of a membrane-soluble quinine Q_B; the quinone migrates to a second protein reaction center (red) that couples its reoxidation ($Q_o \Leftrightarrow Q_i \Leftrightarrow Q$) to transfer of protons across the membrane; the backflow of protons across the membrane drives ATP synthesis at another transmembrane protein element (purple). (b) Structure of the photosynthesis reaction center: the light activated reaction center from Rhodopseudomonas viridis *was the first of the integral membrane proteins involved in photosynthesis whose structure was determined to atomic resolution (by Deisenhofer, Epp, Miki, Huber and Michel in 1984).*

(a)

(b)

(continued)

FIG. 12.23 **(CONTD.)**	*(c) Laue diffraction pattern of the bacterial reaction center of* R. viridis. *Resolution at detector edge 2.9 Å, temporal resolution 2 ms.*

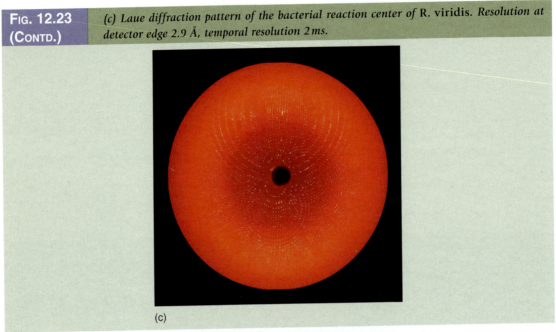

(c)

Source: Images courtesy of Dr. Richard Baxter, UT Southwestern Medical Center, Dallas. With permission.

The Grätzel Cell.   Michael Grätzel of the Swiss Federal Institute of Technology (EPFL) developed the dye-sensitive solar cell in 1991 (known as the *Grätzel cell*) [320,321]. The mechanism of the Grätzel cell is similar to several aspects of photosynthesis in plants—an organic dye (like chlorophyll in plants) captures photons and transfers energy to electrons. In the Grätzel cell, tethered dye molecules absorb photons and transfer energy to titanium dioxide semiconductor nanoparticles [322,323]. The basic concepts of the Grätzel cell are shown in **Figure 12.24**. The schematic flow of electrons is shown in **Figure 12.25**.

Traditional solar cells, like those described earlier, behave like transistors in which the silicon semiconductor materials provide both the "n" and "p" components. The silicon absorbs light and is responsible for charge separation. These materials, as a consequence, must be very pure to prevent recombination of electrons and holes at defects. In Grätzel cells, this dilemma is overcome by differentiation between absorbers and charge separator materials [324–330].

In the photoexcitation process, electron–hole pairs (excitons) are created in the p/n-junction of semiconductors. In semiconductors like TiO_2, energy from the ultraviolet region of the solar spectrum (<400 nm) is required in order to overcome the bandgap energy (~3.2 eV for TiO_2). One disadvantage to pure semiconductors like silicon and titanium oxide is that only photons with the required energy ($>E_g$) or more are able to induce photoexcitation—much of the energy is lost as heat if energies $>E_g$ are absorbed. The p-layer needs to be fairly thick for the process to occur with a high chance of success—a condition that also promotes the chance of electron–hole recombination. In other words, silicon acts both as the source of excitons and as the site of the potential barrier of charge separation. Dye-sensitized solar cells resolve this problem. Dye molecules

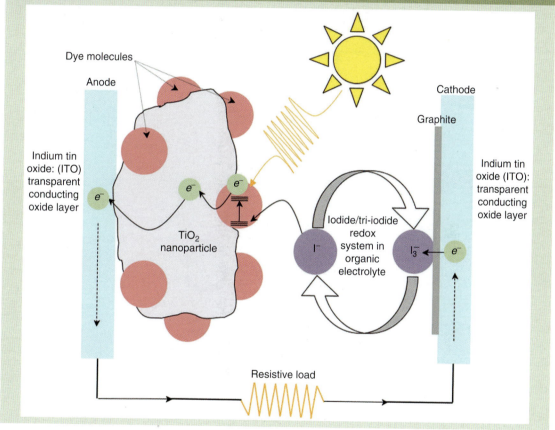

FIG. 12.24

An electronic flow scheme of a Grätzel cell is depicted. A dye molecule tethered to a TiO₂ nanoparticle traps solar energy similar to the action of chlorophyll—photoexcitation of electrons. The excited electron of the dye molecule is in close proximity to the TiO₂ semiconductor. By the process of charge transfer, excited electrons enter the conduction band of the TiO₂ and accumulate at the anode and flow in an external circuit. Electrons from the counter electrode are injected into the iodide/tri-iodide redox cycle from which iodide (the reduced form) relinquishes its electron by transferring it to the dye molecule. In this way, the electronic state of the dye molecule is back to square one and ready for another photoexcitation episode.

absorb lower energy regions of the spectrum. Since the solar spectrum consists of ca. 46% visible light, use of dye molecules allows the cell to capture more energy, as opposed to semiconductors like TiO_2 that absorb ultraviolet radiation (ca. 3% of the solar spectrum).

In a dye-sensitized solar cell, the function of the semiconductor (e.g., the titanium oxide nanoparticles) is to serve as a charge carrier for transfer of electrons from the dye molecules. Visible light absorption is accomplished by a thin unimolecular layer of dye molecules tethered to the semiconductor surface by organic ligands (**Fig. 12.26**).

The efficiency of the Grätzel cell improved to 12.3% by 2004 [331]. Modifications of the concept developed by Michaël Grätzel continue to this day—especially

FIG. 12.25   *A schematic version of the Grätzel cell depicting electronic states of the tethered dye molecule and n-type TiO$_2$ semiconductor conduction and valence bands is depicted. The voltage, against a standard calomel electrode, ranges from +2.5 V to −0.9 V (bottom-to-top). An electron in a transition metal complex excited by a photon transfers the excited electron to the conduction band of the semiconductor by the process of injection. The maximum voltage this device is able to deliver is the difference between the potential of the I⁻/I$_3$ redox couple and the Fermi level of the semiconductor.*

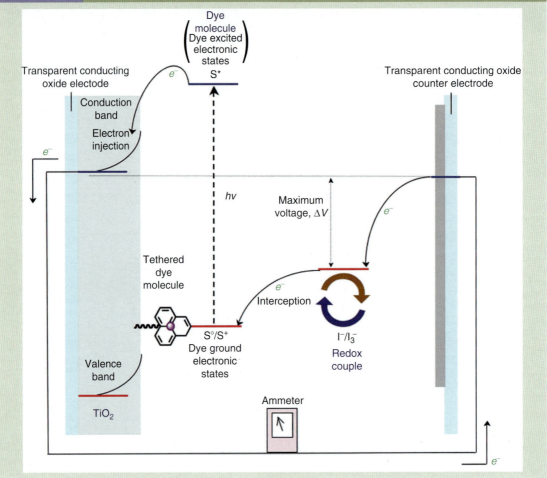

in the electrolyte and dye materials. For example, K.G.U. Wijayantha et al. of the University of Bath in the United Kingdom replaced the organic dye molecules with CdS quantum dots [332–334]—a nanotechnological adaptation of the Grätzel cell (**Fig. 12.27**). CdS quantum dots were self-assembled on the surface of dispersed nanocrystalline TiO$_2$ by tethering with organic ligands as bifunctional linking molecules—3-mercaptopropionic acid HO$_2$C–CH$_2$–CH$_2$–SH (MPA). As with the organic dye depicted in **Figure 12.26** above, the carboxylic acid groups serve as links between the TiO$_2$ and the quantum dots. 3-MPA, a material used often in biosensors, promotes electron transfer reactions between cytochrome *c* and gold surfaces and is known as a promoter molecule [335].

Fig. 12.26

A more realistic rendition of the Grätzel cell is depicted. All dye-decorated semiconductors are connected to the transparent conducting anode. The semiconductor particles form a compact layer with very high surface area, which is a few microns in thickness. The thin-layer ($5–10\,\mu g \cdot cm^{-2}$) Pt mirror also serves to catalyze the cathodic reduction of tri-iodide to iodide. The electrolyte I^-/I_3^- mediator is soluble in a nitrile organic solvent. Once all components are in place, the cell is sealed from the ambient environment. More advanced cells utilize a dry solid-state hole-transporting polymer called PVK (poly(N-vinyl-carbazole). Efficiency of Grätzel polymer glass cells is on the order of 5% while that of the liquid cell is on the order of 10%.

$Ru[4,4\text{-}bis(\text{carboxyl})\text{-bpy}]_2\text{-}(NCS)_2$ is a typical dye molecule with a transition metal center. Ruthenium bipyridyl complexes are commonly used as dye molecules in electron transfer reactions.

Transparent conducting electrode

TiO$_2$ buffer layer (1 μm)

Dye-decorated semiconductor nanoparticles

Redox I^-/I_3^- electrolyte mediator

7 μm

1 μm

Platinum mirror

Glass-reinforced conducting counter electrode

Future Directions for Electronic and Energy Transfer in Supramolecular and Biomolecular Systems. Biomimetic photosynthesis, photoelectricity production, and photonic production of hydrogen offer possible alternatives to conventional sources of fuel and electricity. Research in biomimetic photonics has the goal of providing low-cost renewable energy and fuels without polluting or global warming by-products. Highly focused and skilled research groups around the world are pursuing this goal by various routes. One route would use plants, bacteria, and other cells, enhancing their photosynthetic productivity by genetic engineering and hybrid biomimetics [336–338]. Other approaches include splitting water into hydrogen and oxygen by biomimetic photocatalytic systems.

FIG. 12.27 *Generic rendition of a TiO₂ nanoparticle decorated with CdS quantum dots. The tethering ligand is 3-mercaptopropionic acid. Use of quantum dots allows for nearly 100% capture of the solar spectrum. Although organic dyes function well in the visible range, they are expensive and are relatively unstable over the long term. The use of quantum dots pose several advantages: (1) they are easy (and relatively inexpensive) to fabricate (e.g., spin casting), (2) their optical properties are size dependent, therefore tunable, (3) they are relatively robust, and (4) they are able to absorb light over the IR, near-IR, visible, and UV range of the solar spectrum.*

Others utilize biomimetic methods to create novel and efficient photovoltaic and photoelectric cells to generate electric power directly. No doubt a combination of all of these approaches will eventually find practical application [339–340].

12.3.4 *Sensors Based on Biomimetic Moieties*

The sensitive interactions of natural biological nanostructures with molecules give them great potential for biomimetic sensors. We saw examples in the butterfly wing photonic sensors in the previous chapter. And the adhesive molecules discussed earlier in this chapter offer many possibilities for use as selective biosensors [341]. We will discuss two additional examples that are illustrative in their biomimicry and principle of operation. These are gas sensors using the high surface area of diatom shells, and biomimetic nanoscale temperature sensors.

Silicon Diatom Model for Gas Sensors. A nitric oxide sensor based on the nanopatterned diatoms shells was designed by Sandhage et al. [342]. They started by making pure silicon replicas of the silica (silicon dioxide) diatom frustules. Unlike insulating silica, silicon is a semiconductor which can carry electrical current. The result was micro- and nanostructured silicon shells with high surface area ($>500\,m^2 \cdot g^{-1}$) which readily absorb gases, with resulting rapid changes in electrical impedance. The impedance changes can be measured by resistance to the flow of electrical current or by wireless coupling in a radiofrequency field.

When exposed to gaseous nitric oxide in concentrations as low as 1 ppm, the sensitivity and response speed of the diatom-patterned nanosensor are much greater than for conventional porous silicon NO sensors. The applied bias voltage can be as low as 100 mV—considerably smaller than that needed in other devices. Also, the silicon frustules luminesce strongly in UV light, which could provide a route to other types of sensor design.

Nanoscale Biomimetic Temperature Sensors. Lee and Kotov have reviewed the design of thermal sensors on the nanoscale [343]. This type of measurement is a good paradigm for how everything changes at the nanoscale. Temperature is straightforward in the macroscale. But it ceases to be simple when we leave large-scale molecular statistics for the nanoscale, where random variations of molecular motion and energy are not averaged out so smoothly. This is evident when we consider Brownian motion.

Temperature measurement at the nanoscale poses challenges, where we need to measure heat with high spatial resolution. Nano- and biotechnology require precise thermometry down to the nanoscale regime if temperature of nanodevices is to be calibrated. The development of a nanoscale thermometer is not merely a matter of size—it requires materials with novel physical properties, because all physicochemical and thermodynamic properties are drastically altered at the nanoscale. Progress on nanoscale thermal sensors will require use of molecular and biological moieties, as well as nanoscale superstructures, such as nanosprings and cantilevers.

Biological systems have a number of temperature-dependent molecules and processes which can be adapted for nanoscale thermal measurement. Use of thermo-transformable responsive entities borrowed from biology can lead to high spatial resolution and enhanced biocompatibility because of the moieties' reduced size and direct applicability to biomedical or clinical sensing and imaging. (For example, sensing very small cancer tumors, which have a higher local temperature in the body.)

Examples are temperature-dependent changes in double-stranded DNA structure from B- to Z-DNA, which has been investigated as a possible molecular nanothermometer. Differences in the electronic properties of the two structures and the charge-transfer process from fluorescent probes result in marked changes in optical emission. Certain messenger RNAs (mRNAs) change conformation with temperature. Areas of RNA chains undergo temperature-dependent conformal changes that can be monitored by ultraviolet and nuclear magnetic resonance spectroscopy.

12.3.5 *Biomimetic Molecular Nanoengines*

The design of macromolecular engines is an active area of nanotechnology, and biology is a very rich source of models and ideas for such designs. We briefly mention two topics in the area to call attention to its importance, although it is an advanced topic that is beyond the scope of this introductory chapter on biomimetic nanotechnology.

Biologically Inspired Nanoengines. Much work has been done on design of nanostructures using natural molecular motors such as the rotators in the flagella of bacteria, spermazoa, and similar cellular biomotive engines [344–350]. It is interesting to note that when mankind invented the wheel, no one could have known that nature had already developed highly sophisticated rotary engines. There was a natural precedent that no one would be able to see for thousands of years.

Another very interesting and semibiomimetic area is the adaptation of the information processing and storage properties of the genetic code machinery to design molecular computing engines. This falls somewhat outside what is usually considered biomimetic, but can be considered to be inspired by nature [351–354].

DNA for Parallel Processing. Leonard Adelman proposed using DNA to solve complex mathematical problems in 1994. He mapped the traveling salesman problem (a difficult to solve mathematical formulation also known as the Hamiltonian path problem) whose solution is a path through a set of points from start to end, going through all points once and only once. This problem becomes combinatorically difficult as the number of points increases.

Adelman represented each city by a unique DNA nucleotide code sequence; replication created new sequence combinations. DNA as a computing engine maps naturally to combinatoric problems where parallel processing can create many solutions at the same time for comparison. Most conventional computing architectures have to solve problems one step at a time (linear processing). DNA has the added benefits of being a cheap, energy-efficient, and extremely small for high information-holding density. The main disadvantages are setting up the program and the random errors that can occur in biological systems.

Molecular Penrose Tiling. Molecular tiles can be designed with capillary interactions that correspond to simple logic operations—AND, XOR, etc., and allowed to self-assemble to determine whether they organize into assemblies according to mathematical and logical rules, as in Penrose tilings. Interesting results have been obtained by researchers, including Adleman and Paul Rothemund. Rothemund designed experiments to see whether one could make tiles that obey the Penrose tile matching rules and to see how well they could make a Penrose tiling. There has been much interest in whether Penrose-type tilings can self-assemble with few errors. Experiments have shown that one could make sufficiently complex matching rules using capillary force interactions, and that the energetics of hydrophilic and hydrophobic bonds on the tile elements could skew the distribution of structures [353] (**Fig. 12.28**).

Neural Networks and Swarm Computing. The coordinated firing patterns of neurons inspired abstract mathematical analysis which found deep connections with quantum state phenomena, leading to the development of neural network algorithms for parallel computing and pattern recognition. This field underwent intense development from the 1970s through the present, and has moved into many practical applications [355,356]. The cooperative, independent but coordinated movements of flocks of birds and swarms of ants have inspired studies which found an underlying relationship to distributed computing models and

| **FIG. 12.28** | *Penrose tile patterns produced by molecular computing elements.* |

(a)   (b)

Source: Image from P. W. K. Rothemund, *Proceedings of the National Academy of Sciences USA*, 97, 984–989 (2000). With permission.

to statistical mechanics. This has led to the development of swarm computing algorithms for parallel processing which are finding application in many combinatoric and exponentially difficult problems in graphics, searches, and other areas, including nanotechnology [357].

DNA for Diagnosis.   DNA computing is most naturally applied to biomedical applications, where its ability to recognize, match, and replicate trillions of combinations can be used for diagnosing and treating cancers, virus infections, and other disorders that involve genetic expression. Since the object of the tests is also DNA, there is no input/output interface problem. As a demonstration, researchers at the NSF National Center for Chemical Bonding built an automaton using DNA encoding simple logic to play a programmed game of Tic Tac Toe, always obtaining a win or tie, as any computer does. The automaton is called MAYA (Molecular Array of YES and AND logic gates) and its second generation (MAYA-II) has more than 100 DNA logic circuits. This technology could be used in the future to develop instruments that can simultaneously diagnose and treat cancer, diabetes, or other diseases.

DNA Nanotube Computer Circuits.   DNA can serve as a template for digital and quantum computing [358–360]. IBM has used self-replicating and self-organizing properties of DNA to arrange carbon nanotube into nanowire circuitry [361]. Much DNA computing tends to leave the biomimetic domain once results are pursued— just as early airplanes ceased to resemble birds. This perhaps reflects our lack of real understanding, not of DNA and biomolecules, but of how the brain really works on a much deeper level than is addressed by nanotechnology.

TRIZ—Biomimetic Problem Solving Methodologies. Julian Vincent at the University of Bath has developed a methodology for solving engineering design problems which attempts to emulate the decision-making processes used by nature—perhaps the ultimate application of biomimetics. Vincent and his colleagues adopted a formalism called the "theory of inventive problem solving." Known by its Russian acronym Triz, it was developed by Genrich Altshuller as a way to systemize engineering and economic decision making. It is similar to operational analysis methods, and to design methodologies such as the Boothroyd–Dewhurst design for manufacturing method. Vincent put biology into the matrix to highlight the striking contrasts between the technical solutions favored in engineering and those found in nature [362].

Data is loaded into the matrix for design solutions used in engineering and those found in nature, using variables such as composition, structure, spatial arrangement, time, energy and information, and then the contradictions and agreements are computed. The conclusions are striking: there was only a 12% degree of similarity between the two approaches, suggesting that nature usually has quite different problem-solving strategies from those of engineers. The matrix revealed that engineering solutions varied with size scale. For manipulation at small scales, from molecular to micro, our engineering methods nearly always rely on high use of energy—reflecting that we use heat to drive random recombinations of molecules in chemical- and energy-generating processes.

At larger scales, structure, which involves directed manipulation of matter, becomes gradually more significant than energy for engineers. But in biology, the proportions of each solution class stay more or less constant at all sizes, from nanometers to kilometers. Crucially, this implies that biology manipulates information, with the saving of energy and material, at the nanoscale. This reflects that we can apply information more effectively at large scales where we are more familiar with visualizing the manipulation of the problem elements. But now nanotechnology, especially with the direction provided by insights from nature, is giving engineers the ability to manipulate matter intelligently on the nanoscale.

Vincent is planning to make a systematic version of Triz biomimetic design methodology available to engineers, designers, and planners on the Internet. In time it may become a widely used design tool, leading indirectly to a global infusion of biomimetic technology.

12.4 CONCLUSION

> The new technology we build in the future should be recyclable and sustainable, reliable and energy efficient. By elucidating the delicate and intricate assembly of living organisms, it will be possible to create new materials and systems.
> Markus Milwich et al. Competence Network Biomimetics [187].

By applying the principles learned from nature, biomimetics contributes to realizing "smart," dynamic, complex, environmentally friendly, self-healing, and multifunctional artificial structures, machines, lubricants, and adhesives. By illustrating how mechanical, physical, chemical, and entropic nanoscale forces can work in concert to provide strength, resilience, adaptivity, self-healing, and functionality, biomimetics plays a key role in the development of a mature and integrated nanotechnology that like natural systems, applies the appropriate tools to execute nanoscale tasks.

> The laws of biomimicry are obvious and straightforward–Nature runs on sunlight, uses only the energy it needs, fits form and function, recycles everything, rewards cooperation, banks on diversity, demands local expertise, curbs excesses from within and taps the power limits J. M. Benyus, Biomimicry: Innovation Inspired by Nature [3].

Life itself is still a miracle to us. Organisms are complex open systems through which energy and materials flow on a dynamic homeostatic trajectory far from thermodynamic equilibrium. Engineers, scientists, economists, and managers can learn by understanding and generalizing natural approaches to the challenges of life. Human engineers will need to be able to apply creativity and disciplined thought to solve the many new problems that we encounter and make for ourselves in the built environment of an industrialized world. Understanding natural methods is a powerful starting point, but biomimetics implies taking the natural examples and improving and expanding upon them in new materials and combinations.

References

1. Y. Bar-Cohen, ed., *Biomimetics: Biologically inspired technologies*, CRC Press, Boca Raton, FL (2005).
2. Y. Bar-Cohen, Biomimetics—using nature to inspire human innovation, *Bioinspiration and Biomimetics*, 1, P1–P12 (2006).
3. J. M. Benyus, *Biomimicry: Innovation inspired by nature*, HarperCollins, New York (1998).
4. J. F. Vincent et al., Biomimetics: Its practice and theory. *Journal of Royal Society Interface*, 3, 471–482 (2006).
5. T. Speck and C. Neinhuis, Bionik, biomimetik, *Naturwissenschaftliche Rundschau*, 57, 177–191 (2004).
6. S. Vogel, *Cats' paws and catapults–mechanical worlds of nature and people*, Norton, New York (1998).
7. G. Jeronimidis, Biodynamics, *Architectural Design*, 74, 90–96 (2004).
8. R. Hooper, Ideas stolen right from nature, *Wired Magazine*, www.wired.com/science/discoveries/news/2004/11/65642 (2004).
9. B. Poole, Biomimetics: Borrowing from biology, on *Trendy Science Blogsite*, http://trendyscience.blogspot.com/2007/08/biomimetics-borrowing-from-biology.html (2007).
10. P. Ball, *Made to measure*, Princeton University Press, New York (1997).
11. M. H. Dickinson, Bionics: Biological insight into mechanical design, *Proceedings of the National Academy of Science USA*, 96, 14208–14209 (1999).
12. Centre for Biomimetics, University of Reading, www.rdg.ac.uk/biomim/home.htm (2007).
13. BIONIS: The Biomimetics Network for Industrial Sustainability, Newsletter Number 23, Link at: www.biomimetics.org.uk November (2005).
14. What is Biomimetics? University of Reading, http://www.rdg.ac.uk/biomimetics/about.htm (2007).
15. J. Kikuchi, A. Ikeda, and M. Hashizume, Biomimetic Materials, In *Encyclopedia of biomaterials and biomedical engineering*, G. L. Bowlin and G. Wnek, eds., Informa/Routledge, New York, 96–102 (2004).

16. E. R. Ritneour, *Dr. Otto H. Schmitt*, The Schmitt Charitable Foundation, www. otto-schmitt.org (2008).

17. E. J. Lerner, Biomimetic nanotechnology, *The Industrial Physicist*, Aug-Sep, 18–21 (2004).

18. Bionics, U.S. History Encyclopedia, www.answers.com/topic/bionics?cat = health (2008).

19. Y. C. Fung, *Biomechanics: Mechanical properties of living tissue*, 2nd ed., Springer, Berlin, Germany (1993).

20. C. R. Ethier and C. A. Simmons, *Introductory biomechanics: From cells to organisms*, Cambridge University Press, Cambridge, UK (2007).

21. J. D. Humphrey, Continuum biomechanics of soft biological tissues, *Proceedings of the Royal Society London A: Mathematical, Physical, and Engineering Science*, 459, 3–46 (2002).

22. J. D. Humphrey and S. DeLange, *An Introduction to biomechanics: Solids and fluids, analysis and design*, Springer, Berlin, Germany (2004).

23. J. D. Humphrey, *Cardiovascular solid mechanics: Cells, tissues, and organs*, Springer, Berlin, Germany (2002).

24. S. Vogel, *Comparative biomechanics: Life's physical world*, Princeton University Press, Princeton, NJ (2003).

25. A. Thurston, Giovanni Borelli and the study of human movement: An historical review, *Australian and New Zealand Journal of Surgery*, 69, 276–288 (1999).

26. J. D. Bronzino, ed., *Biomedical engineering handbook*, CRC Press, Boca Raton, FL (2007).

27. S. A. Berger, W. Goldsmith, and E. R. Lewis, *Introduction to bioengineering*, Oxford University Press, Oxford, UK (1996).

28. Y. C. Fung, *Introduction to bioengineering*, World Scientific Publishing, Singapore, Malaysia (2001).

29. NIH Working Definition of Bioengineering, National Institutes of Health, 1997, www.becon.nih.gov/bioengineering_definition.htm (2008).

30. N. H. Malsch, ed., *Biomedical nanotechnology*, CRC Press, Boca Raton, FL (2007).

31. M. P. Hughes, *Nanoelectromechanics in engineering and biology*, CRC Press, Boca Raton, FL (2002).

32. R. M. J. Cotterill, *Biophysics: An introduction*, John Wiley and Sons, New York (2002).

33. J. A. Tuszynski and M. Kurzynski, *Introduction to molecular biophysics*, CRC Press, Boca Raton, FL (2003).

34. R. Glaser, *Biophysics*, 1st ed., Springer, Berlin, Germany (2004).

35. D. Springham, G. Springham, V. Moses, and R. E. Cape, *Biotechnology: The science and the business*, Taylor & Francis, London (1999).

36. K. L. Lerner, B. W. Lerner, and B. Wilmoth, eds., *Biotechnology: Changing life through science*, Gale Cengage Learning, Andover, UK (2007).

37. M. El-Mansi, A. L. Demain, and C. F. Bryce, eds., *Fermentation microbiology and biotechnology*, CRC Press, Boca Raton, FL (2006).

38. K. Fukui and T. Ushiki, *Chromosome nanoscience and technology*, CRC Press, Boca Raton, FL (2007).

39. D. S. Goodsell, *Bionanotechnology: Lessons from nature*, Wiley-Liss, New York (2004).

40. O. Shoseyov and I. Levy, eds., *Nanobiotechnology: Bioinspired devices and materials of the future*, Springer Verlag, Berlin, Germany (2007).

41. C. M. Niemeyer and C. A. Mirkin, eds., *Nanobiotechnology: Concepts, applications and perspectives*, Wiley-VCH, Weinheim, Germany (2004).

42. The Biomimicry Institute, Missoula, Montana, USA, website: http://www. biomimicryinstitute.org/ (2007).

43. C. Mattheck, *Design in nature—learning from trees*, Springer Verlag, Heidelberg, Germany (1998).

44. S. Olariu and A. Y. Zomaya, eds., *Handbook of bioinspired algorithms and applications*, CRC Press, Boca Raton, FL (2005).

45. R. Lipowsky, Biomimetic systems and transport systems, *European Whitebook of fundamental research in materials science*, pp. 78–82, Max Planck Institut für Metallforschung, Stuttgart, Germany (2001).

46. G. E. Wnek and G. L. Bowlin, eds., Biomimetics, In *Encyclopedia of biomaterials and biomedical engineering*, Marcel Dekker, New York, 1009–1016 (2004).

47. R. Webb, Offices that breathe naturally, *New Scientist*, 1929, 38 (1994).

48. K. Brown, *The Pox: The life and near death of a very social disease*, Sutton, Stroud, UK (2006).

49. B. Witkop, Paul Ehrlich and his magic bullets—Revisited, *Proceedings of the American Philosophical Society*, 143, 540–557 (1999).

50. C. C. Mann and M. L. Plummer, *The aspirin wars: Money, medicine, and 100 years of rampant competition*, Harvard Business School Press, Boston, MA (1991).

51. J. R. Vane and R. M. Butting, The mechanism of action of aspirin, *Thrombosis Research*, 110, 255–258 (2003).

52. H. Tohgi, et al., Effects of low-to-high doses of aspirin on platelet aggregability and metabolites of thromboxane A2 and prostacyclin, *Stroke*, 23, 1400–1403 (1992).

53. T. Morris, M. Stables, and D. W. Gilroy, New perspectives on aspirin and the endogenous control of acute inflammatory resolution, *Scientific World Journal*, 6, 1048–1065 (2006).

54. T. Lawrence and D. W. Gilroy, Chronic inflammation: A failure of resolution? *International Journal of Experimental Pathology*, 88, 85–94 (2007).

55. M. Paul-Clark, et al., 15-epi-lipoxin A 4-mediated induction of nitric oxide explains how aspirin inhibits acute inflammation, *Journal of Experimental Medicine*, 200, 69–78 (2004).

56. C. R. McCurdy and S. S. Scully, Analgesic substances derived from natural products (natureceuticals), *Life Sciences*, 78, 476–484 (2005).

57. J. Goodman and V. Walsh, *The story of Taxol: Nature and politics in the pursuit of an anti-cancer drug*, Cambridge University Press, Cambridge, UK (2001).

58. Y. Ji, J.-N. Bi, B. Yan, and X.-D. Zhu, Taxol-producing fungi: A new approach to industrial production of taxol, *Chinese Journal of Biotechnology*, 22, 1–6 (2006).

59. J. E. Casida, ed., *Pyrethrum: The natural insecticide*, Academic Press, New York (1973).

60. T. J. Class, et al., Pyrethroid metabolism: Microsomal oxidase metabolites of (S)-bioallethrin and the six natural pyrethrins, *Journal of Agriculture and Food Chemistry*, 38, 529–537 (1990).

61. I. T. Baldwin, M. J. Karb, and P. Callahan, Foliar and floral pyrethrins of *Chrysanthemum cinerariaefolium* are not induced by leaf damage, *Journal of Chemical Ecology*, 19, 2081–2087 (1993).

62. G. Vettorazzi, *International regulatory aspects for pesticide chemicals: Toxicity profiles*, Vol. 1, CRC Press, Boca Raton, FL (1979).

63. J. S. Dordick, Enzymatic and chemoenzymatic approaches to polymer synthesis and modification, *Annals of the New York Academy of Sciences*, 672, 352–362 (1992).

64. K. M. Koeller and C.-H. Wong, Complex carbohydrate synthesis tools for glycobiologists: enzyme-based approach and programmable one-pot strategies, *Glycobiology*, 10, 1157–1159 (2000).

65. J. W. Peeters, et al., Chemoenzymatic synthesis of branched polymers, *Macromolecular Rapid Communications*, 26, 684–689 (2005).

66. H. Yu, et al., One-pot three-enzyme chemoenzymatic approach to the synthesis of sialosides containing natural and non-natural functionalities, *Nature Protocols*, 1, 2485–2492 (2006).

67. N. Anand, et al., Stereoselective chemoenzymatic process for the preparation of optically enriched phenylglycidates as precursors of taxol side chain, US Patent

7060471, Issued June 13, (Council of Scientific and Industrial Research, India) (2006).

68. T. Tanaka, et al., A novel glycosyl donor for chemo-enzymatic oligosaccharide synthesis: 4,6-dimethoxy-1,3,5-triazin-2-yl glycoside, *Chemical Communications*, DOI: 10.1039/b801090k (2008).

69. E. Wilson, Total synthesis surprise: Scientists revise structure of coveted anticancer marine natural product, *Chemical and Engineering News*, 79, 11 (2001).

70. A. K. Mukherjee, B. L. Sollod, S. K. Wikel, and G. F. King, Orally active acaricidal peptide toxins from spider venom, *Toxicon*, 47, 182–187 (2006).

71. R. S. Greco, F. B. Prinz, and R. L. Smith, *Nanoscale technology in biological systems*, CRC Press, Boca Raton, FL (2004).

72. M. Stoneham, How soft materials control harder ones: Routes to bioorganization, *Reports on Progress in Physics*, 70, 1055 (2007).

73. N. A. Kotov, *Nanoparticle superstructures - Nanoparticle assemblies and superstructures*, CRC Press, Boca Raton, FL (2005).

74. N. Yui, *Supramolecular design for biological applications*, CRC Press, Boca Raton, FL (2002).

75. J. B. Park and J. D. Bronzino, eds., *Nano- and microscience, engineering, technology and medicine, Volume: 4: Biomaterials: Principles and applications*, CRC Press, Boca Raton, FL (2002).

76. M. J. Schulz, A. D. Kelkar, and M. J. Sundaresan, *Nanoengineering of structural, functional and smart materials*, CRC Press, Boca Raton, FL (2005).

77. J. A. Schwarz, C. I. Contescu, and K. Putyera, eds., *Dekker Encyclopedia of nanoscience and nanotechnology*, (5 Vols), Marcel Dekker, Inc., New York (2004).

78. D. Green, et al., The potential of biomimesis in bone tissue engineering: lessons from the design and synthesis of invertebrate skeletons, *Bone*, 30, 810–815 (2002).

79. P. Podsiadlo, et al., Ultrastrong and stiff layered polymer nanocomposites, *Science*, 318, 80–83 (2007).

80. B. S. Shim, et al., Nanostructured thin films made by dewetting method of layer-by-layer assembly, *Nano Letters*, 7, 3266–3273 (2007).

81. A. Lin and M. A. Meyers, Growth and structure in abalone shell, *Materials Science and Engineering A*, 390, 27–4 (2005).

82. L. J. Bonderer, A. R. Studart, and L. J. Gauckler, Bioinspired design and assembly of platelet reinforced polymer films, *Science*, 319, 1069–1073 (2008).

83. A. K. Geim and K. S. Novosolov, The rise of Graphene, *Nature Materials*, 6, 183–191 (2007).

84. P. K. Hansma, P. J. Turner, and R. S. Ruoff, Optimized adhesives for strong, light-weight, damage-resistant, nanocomposite materials: New insights from natural materials, *Nanotechnology*, 18, 044026 (2007).

85. R. Gordon, C. R. Chaplin, and G. Jeronimidis, Composite material, US Patent 4409274, WestVaco Corp. (1983).

86. G. Jeronimidis, Wood, one of nature's challenging composites, In *The mechanical properties of biological materials*, J. F. V. Vincent and J. D. Currey, eds., Vol. 34, pp. 169–182, Symposia of the Society for Experimental Biology, Cambridge University Press, Cambridge, UK (1980).

87. J. R. Barnett and V. A. Bonham, Cellulose microfibril angle in the cell wall of wood fibres, *Biological Reviews*, 79, 461–472 (2004).

88. I. Burgert, et al., Structure-function-relationships of four compression wood types—Micromechanical properties at the tissue and fiber level, *Trees-Structure and Function*, 18, 480–485 (2004).

89. I. Burgert, N. Gierlinger, and T. Zimmermann, Properties of chemically and mechanically isolated fibres of spruce (*Picea abies* [L.] Karst.). Part 1. Structural and chemical characterization. *Holzforschung* 59, 240–246 (2005).

90. C. Mattheck, *Trees: The mechanical design*, Springer Verlag, Heidelberg, Germany (1996).

91. K. J. Niklasm, *Plant biomechanics. An engineering approach to plant form and function*, University of Chicago Press, Chicago, IL (1992).

92. A. Reiterer, et al., Experimental evidence for a mechanical function of the cellulose microfibril angle in wood cell walls, *Philosophical Magazine A*, 79, 2173–2184 (1999).

93. I. Burgert, Exploring the micromechanical design of plant cell walls, *American Journal of Botany*, 93, 1391–1401 (2006).

94. S. H. Li, et al., Biomimicry of bamboo bast fiber with engineering composite materials, Biomolecular and biomimetic materials: Materials Research Society Fall Meeting Symposium. S3, 125–130 (1995)

95. J. F. V. Vincent, Ideas from skins, *Interdisciplinary Science Reviews*, 24, 52–57 (1999).

96. R. H. C. Bonser, L. Saker, and G. Jeronimidis, Toughness anisotropy in feather keratin, *Journal of Material Science*, 39, 2895–2896 (2004).

97. S. S. Liao, et al., Hierarchically biomimetic bone scaffold materials: Nano-HA/ collagen/ PLA composite, *Journal of Biomedical Materials Research B: Applied Biomaterials*, 69B: 158–165 (2004).

98. G. Heness and B. Ben-Nissan, Innovative Bioceramics, *Materials Forum*, 27, 3–21 (2004).

99. D. Vashishth, K. E. Tanner, and W. Bonfield, Experimental validation of a microcracking-based toughening mechanism for cortical bone, *Journal of Biomechanics*, 36, 121–124 (2003).

100. V. C. Sundar, et al., Fibre-optical features of a glass sponge, *Nature*, 424, 899–900 (2003).

101. M. M. Murr and D. E. Morse, Fractal intermediates in the self-assembly of silicatein filaments, *Proceedings of the National Academy of Sciences*, 102, 11657–11662 (2005).

102. G. Fu, et al., $CaCO_3$ biomineralization: Acidic 8-kDa proteins isolated from aragonitic abalone shell nacre can specifically modify calcite crystal morphology, *Biomacromolecules*, 6, 1289–1298 (2005).

103. G. Fu, et al., Acceleration of calcite kinetics by abalone nacre proteins, *Advanced Materials*, 17, 2678–2683 (2005).

104. A. R. Parker, 515 Million years of structural colour. *Journal Optics A: Pure and Applied Optics*, 2, R15–28 (2000).

105. A. R. Parker, et al., Aphrodite's iridescence, *Nature*, 409, 36–37 (2001).

106. A. R. Parker, et al., An opal analogue discovered in a weevil, *Nature*, 426, 786–787 (2003).

107. A. R. Parker, Z. Hegedus, and R. A. Watts, Solar-absorber type antireflector on the eye of an Eocene fly (45 Ma). *Proceedings of the Royal Society of London B: Biological Sciences*, 265, 811–815 (1998).

108. A. R. Parker and H. E. Townley, Biomimetics of photonic nanostructures, *Nature Nanotechnology*, 2, 347–353 (2007).

109. W. Barthlott and C. Neinhuis, Purity of the sacred lotus or escape from contamination in biological surfaces, *Planta*, 202, 1–7 (1997).

110. A. Solga, et al., The dream of staying clean: Lotus and biomimetic surfaces, *Bioinspiration and Biomimetics*, 2, S126–S134 (2007).

111. M. Callies and D. Quéré, On water repellency, *Soft Matter*, 1, 55–61 (2005).

112. D. Quéré, Non-sticking drops, *Reports on Progress in Physics*, 68, 2495 (2005).

113. J. Bico, C. Marzolin, D. Quere, Pearl drops, *Europhysics Letters*, 47, 220 (1997).

114. X.-M. Li, D. Reinhoudt, and M. Crego-Calama, What do we need for a superhydrophobic surface? A review on the recent progress in the preparation of superhydrophobic surfaces, *Chemical Society Reviews*, 36, 1350 (2007).

115. Y.-J. Sheng, S. Jiang, and H.-K. Tsao, Effects of geometrical characteristics of surface roughness on droplet wetting, *Journal of Chemical Physics*, 127, 234704 (2007).

116. H. Zhang, R. N. Lamb, and D. J. Cookson, Nanowetting of rough superhydrophobic surfaces, *Applied Physics Letters*, 91, 254106 (2007).

117. R. D. Narhe and D. A. Beysens, Water condensation on a super-hydrophobic spike surface, *Europhysics Letters*, 75, 98 (2006).

118. J. Hyvaluoma, et al., Droplets on inclined rough surfaces, *European Physics Journal E*, 23, 289–293 (2007).

119. K. Autumn and W. Hansen, Ultrahydrophobicity indicates a non-adhesive default state in gecko setae, *Journal of Comparative Physiology A*, 192, 1205–1212 (2006).

120. W. R. Hansen and K. Autumn, Evidence for self-cleaning in gecko setae, *Proceedings of the National Academy of Sciences*, 102, 385–389 (2005).

121. G. S. Bakken, et al., It's just ducky to be clean: The water repellency and water penetration resistance of swimming mallard *Anas platyrhynchos* ducklings, *Journal of Avian Biology*, 37, 561 (2006).

122. Y. Fang, et al., Hydrophobicity mechanism of non-smooth pattern on surface of butterfly wing, *Chinese Science Bulletin*, 52, 711 (2007).

123. Y. Zheng, X. Gao, and L. Jiang, Directional adhesion of superhydrophobic butterfly wings, *Soft Matter*, 3, 178 (2007).

124. H. I. Hima, et al., Novel carbon nanostructures of caterpillar-like fibers and interwoven spheres with excellent surface super-hydrophobicity produced by chemical vapor deposition, *Journal of Material Chemistry*, 18, 1245 (2008).

125. X. Gao, and L. Jiang, Water-repellent legs of water striders, *Nature* 432, 36 (2004).

126. W. H. Thorpe, Plastron Respiration in Aquatic Insects, *Biological Reviews*, 25, 344 (1950).

127. P. J. P. Goodwyn, D. Voigt, and K. Fujisaki, Skating and diving: Changes in functional morphology of the setal and microtrichial cover during ontogenesis in *Aquarius paludum fabricius* (Heteroptera, Gerridae), *Journal of Morphology*, DOI: 10.1002/jmor.10619 (2008).

128. N. J. Shirtcliffe, et al., Plastron properties of a superhydrophobic surface, *Applied Physics Letters*, 89, 104106 (2006).

129. C. Dorrer and J. Ruhe, Wetting of silicon nanograss: From superhydrophilic to superhydrophobic surfaces, *Advanced Materials*, 20, 159 (2008).

130. M. J. O'Donnell, Hydrophilic cuticle - the basis for water vapour absorption by the desert burrowing cockroach, *Arenivaga investigate*, *Journal of Experimental Biology*, 99, 43–60 (1982).

131. Y. B. Gerbig, et al., Effect of nanoscale topography and chemical composition of surfaces on their microfrictional behavior, *Tribology Letters*, 21, 161 (2006).

132. X. Zhang, et al., Effect of pattern topology on the self-cleaning properties of textured surfaces, *Journal of Chemical Physics*, 127, 014703 (2007).

133. B. D'Urso, J. T. Simpson, and M. Kalyanaraman, Emergence of superhydrophobic behavior on vertically aligned nanocone arrays, *Applied Physics Letters*, 90, 044102 (2007).

134. N. Zhao, et al., A novel ultra-hydrophobic surface: Statically non-wetting but dynamically non-sliding, *Advanced Functional Materials*, 17, 2739–2745 (2007).

135. N. Verplanck, Y. Coffinier, V. Thomy, and R. Boukherroub, Wettability Switching Techniques on Superhydrophobic Surfaces, *Nanoscale Research Letters*, 2, 577 (2007).

136. P. Roach, N. J. Shirtcliffe, and M. I. Newton, Progress in Superhydrophobic Surface Development, *Soft Matter*, 4, 224–240 (2008).

137. X. Yu, et al., Reversible pH-responsive surface: From superhydrophobicity to superhydrophilicity, *Advanced Materials*, 17, 1289 (2005).

138. A. Tuteja, et al., Designing superoleophobic surfaces, *Science*, 318, 1618 (2007).

139. F. DiBenedetto, et al., Photoswitchable organic nanofibers, *Advanced Materials*, 20, 314 (2008).

140. W. Sun, et al., Reversible switching on superhydrophobic TiO_2 nano-strawberry films fabricated at low temperature, *Chemical Communications*, 2008, 603–605 (2008).

141. M. Morra, E. Occhiello, and F. Garbassi, Surface characterization of plasma-treated PTFE, *Surface and Interface Analysis*, 16, 412 (1990).

142. M. Ferrari, et al., Surfactant adsorption at superhydrophobic surfaces, *Applied Physics Letters*, 89, 053104 (2006).

143. A. Nygard, et al., A Simple approach to micro-patterned surfaces by breath figures with internal structure using thermoresponsive amphiphilic block copolymers, *Australian Journal of Chemistry*, 58, 595–599 (2005).

144. G. R. J. Artus, et al., Silicone nanofilaments and their application as superhydro-phobic coatings, *Advanced Materials*, 18, 2758 (2006).

145. D. Kim, et al., Superhydrophobic nano-wire entanglement structures, *Journal of Micromechanics and Microengineering*, 16, 2593–2597 (2006).

146. P. van der Wal, and U. Steiner, Super-hydrophobic surfaces made from Teflon, *Soft Matter*, 3, 426 (2007).

147. T. Ishizaki, et al., Fabrication and characterization of ultra-water-repellent alumi-na-silica composite films, *Journal of Physics D: Applied Physics*, 40, 192 (2007).

148. M. O. Gallyamov, et al., Formation of superhydrophobic surfaces by the deposi-tion of coatings from supercritical carbon dioxide, *Colloid Journal*, 69, 411–424 (2007).

149. I. A. Larmour, et al., Remarkably simple fabrication of superhydrophobic surfaces using electroless galvanic deposition, *Angewandte Chemie*, 46, 1710 (2007).

150. M. Qu, et al., Fabrication of superhydrophobic surfaces on engineering materials by a solution-immersion process, *Advanced Functional Materials*, 17, 593 (2007).

151. J. Li, et al., Carbon nanofibers "spot-welded" to carbon felt: A mechanically sta-ble, bulk mimic of Lotus leaves, *Advanced Materials*, 20, 420 (2008).

152. T. Mizukoshi, et al., Control over wettability of textured surfaces by electrospray deposition, *Journal of Applied Polymer Science*, 103, 3811 (2007).

153. M. Motornov, et al., Superhydrophobic surfaces generated from water-borne dispersions of hierarchically assembled nanoparticles coated with a reversibly switchable shell, *Advanced Materials*, 20, 200 (2008).

154. X.-J. Huang, et al., A one-step route to a perfectly ordered wafer-scale microbowl array for size-dependent superhydrophobicity, *Small*, 4, 211 (2008).

155. Y. Liu, et al., Superhydrophobic behavior on transparency and conductivity con-trollable ZnO/Zn films, *Journal of Applied Physics*, 103, 056104 (2008).

156. J. E. Ruckman, Water vapour transfer in waterproof breathable fabrics: Part 3: under rainy and windy conditions, *International Journal of Clothing Science Technology*, 9, 141 (1997).

157. I. P. Parkin and R. G. Palgrave, Self-cleaning coatings, *Journal of Materials Chemistry*, 15, 1689–1695 (2005).

158. J.-T. Yeh, C.-L. Chen, and K.-S. Huang, Preparation and application of fluorocarbon polymer/SiO_2 hybrid materials, part 2: Water and oil repellent processing for cot-ton fabrics by sol-gel method, *Journal of Applied Polymer Science*, 103, 3019 (2007).

159. T. Wang, X. Hu, and S. Dong, A general route to transform normal hydrophilic cloths into superhydrophobic surfaces, *Chemical Communications*, 2007, 1849–1851 (2007).

160. H. Saito, K. Takai, and G. Yamauchi, Water- and ice-repellent coatings, *Surface Coatings International*, 80, 168 (1997).

161. A. Marmur, Super-hydrophobicity fundamentals: Implications to biofouling prevention, *Biofouling*, 22, 107 (2006).

162. Y. T. Cheng, et al., Effects of micro- and nano-structures on the self-cleaning behaviour of lotus leaves, *Nanotechnology*, 17, 1359–1362 (2006).

163. Y. C. Chang, et al., Design and fabrication of a nanostructured surface combining antireflective and enhanced-hydrophobic effects, *Nanotechnology*, 18, 285303 (2007).

164. X. Gao, et al., The dry-style antifogging properties of Mosquito compound eyes and artificial analogues prepared by soft lithography, *Advanced Materials*, 19, 2213–2217 (2007).

165. J. A. Howarter and J. P. Youngblood, Self-cleaning and next generation anti-fog surfaces and coatings, *Macromolecular Rapid Communications*, DOI: 10.1002/marc.200700733 (2008).

166. Y. Zhang, S. Sundararajan, Superhydrophobic engineering surfaces with tunable air-trapping ability, *Journal of Micromechanics and Microengineering*, 18, 035024 (2008).

167. H. Schott, Contact angles and wettability of human skin, *Journal of Pharmaceutical Science*, 60, 1893–1895 (1971).

168. W.-C. Liao and J. L. Zatz, Critical surface tensions of pharmaceutical solids, *Journal of Pharmaceutical Science*, 68, 488494 (1979).

169. M. O. Riehle, Biocompatibility: Nanomaterials for cell- and tissue engineering, *Nanobiotechnology*, 1, 308–309 (2005).

170. T. Sun, et al., No platelet can adhere - Largely improved blood compatibility on nanostructured superhydrophobic surfaces, *Small*, 1, 959 (2005).

171. G. McHale, N. J. Shirtcliffe, and M. I. Newton, Super-hydrophobic and super-wetting surfaces: Analytical potential? *The Analyst*, 129, 284–287 (2004).

172. J. Ou, G. R. Moss, and J. P. Rothstein, Enhanced mixing in laminar flows using ultrahydrophobic surfaces, *Physical Review E*, 76, 016304 (2007).

173. K. Fukuzawa, et al., Conformation and motion of monolayer lubricant molecule on magnetic disks, *IEEE Transactions on Magnetics*, 41, 3034 (2005).

174. K. Gjerde, et al., On the suitability of carbon nanotube forests as non-stick surfaces for nanomanipulation, *Soft Matter*, 4, 392 (2008).

175. C.-M. Tøg, et al., Influence of surface structure on wetting of coated offset papers, *Holzforschung*, 61, 516 (2007).

176. P. J. Holloway, Surface factors affecting the wetting of leaves, *Pesticide Science*, 1, 156 (1970).

177. A. N. Round, et al., The influence of water on the nanomechanical behavior of the plant biopolyester cutin as studied by AFM and solid-state NMR, *Biophysical Journal*, 79, 2761–2767 (2000).

178. J. Poulenard, et al., Water repellency of volcanic ash soils from Ecuadorian paramo: Effect of water content and characteristics of hydrophobic organic matter, *European Journal of Soil Science*, 55, 487–496 (2004).

179. R. W. McDowell, The effectiveness of coal fly-ash to decrease phosphorus loss from grassland soils, *Australian Journal Soil Research*, 43, 853–860 (2005).

180. G. McHale, M. I. Newton, and N. J. Shirtcliffe, Water-repellent soil and its relationship to granularity, surface roughness and hydrophobicity: A materials science view, *European Journal of Soil Science*, 56, 445–452 (2005).

181. N. J. Shirtcliffe, et al., Critical conditions for the wetting of soils, *Applied Physics Letters*, 89, 094101 (2006).

182. G. McHale, et al., Implications of ideas on super-hydrophobicity for water repellent soil, *Hydrological Processes*, 21, 2229–2238 (2007).

183. D. A. L. Leelamanie and J. Karube, Effects of organic compounds, water content and clay on the water repellency of a model sandy soil, *Soil Science and Plant Nutrition*, 53, 711–719 (2007).

184. F. Bartoli, A. J. Poulenard, and B. E. Schouller, Influence of allophane and organic matter contents on surface properties of Andosols, *European Journal of Soil Science*, 58, 450–464 (2007).

185. B. Chen and J. Fan, Microstructures of chafer cuticle and biomimetic design, *Journal of Computer-Aided Materials Design*, 11, 1573–4900 (2004).

186. T. G. Rials and W. G. Glasser, Engineering plastics from lignin. XIII. Effect of lignin structure on polyurethane network formation, *Holzforschung*, 40, 353–360 (2006).

187. M. Milwich, et al., Biomimetics and technical textiles: Solving engineering problems with the help of nature's wisdom, *American Journal of Botany*, 93, 1455–1465 (2006).

188. J. Gravitis, Nano level structures in wood cell wall composites, *Cellulose Chemistry and Technology*, 40, 291–298 (2006).

189. J. Cao, R. Wijaya, and F. Leroy, Unzipping the cuticle of the human hair shaft to obtain micron/nano keratin filaments, *Biopolymers*, 83, 614–618 (2006).

190. N. Kohli, et al., Direct transfer of preformed patterned bio-nanocomposite films on polyelectrolyte multilayer templates, *Macromolecular Bioscience*, 7, 789–797 (2007).

191. T. T. Teeri, H. Brumer 3rd, G. Daniel, and P. Gatenholm, Biomimetic engineering of cellulose-based materials, *Trends in Biotechnology*, 25, 299–306 (2007).

192. Y. Kaneko, S. Matsuda, and J. Kadokawa, Chemoenzymatic syntheses of amylose-grafted Chitin and Chitosan, *Biomacromolecules*, 8, 3959–3964 (2007).

193. Technical Research Centre of Finland,. Water Repellent Wood Fiber Products Developed. ScienceDaily 28 January 2008, website at: http://www.sciencedaily. com-/releases/2008/01/080123163554.htm (2008)

194. Z. Lin, S. Renneckar, and D. P Hindman, Nanocomposite-based lignocellulosic fibers 1: Thermal, *Cellulose*, 15, 333–346 (2008).

195. N. M Pugno, Towards a Spiderman suit: Large invisible cables and self-cleaning releasable superadhesive materials, *Journal of Physics: Condensed Matter*, 19, 395001 (2007).

196. K. Autumn and N. Gravish, Gecko adhesion: Evolutionary nanotechnology, *Philosophical Transactions of the Royal Society of London A: Mathematical, Physical, and Engineering Science*, 366, 1575–1590 (2008).

197. E. Arzt, Biological and artificial attachment devices: Lessons for materials scientists from flies and geckos, *Materials Science and Engineering C: Biomimetic and Supramolecular System*, 26, 1245–1250 (2006).

198. K. Autumn, M. Sitti, Y. A. Liang, A. M. Peattie, and W. R. Hansen, Evidence for van der Waals adhesion in gecko setae, *Proceedings of National Academy of Sciences*, 99, 12252–12256 (2002).

199. J. Lee, C. Majidi, B. Schubert, and R. Fearing, Sliding induced adhesion of stiff polymer microfiber arrays: 1. Macroscale behaviour, *Journal of the Royal Society Interface*, (10.1098/rsif.2007.1308) (2008).

200. B. Schubert, et al., Sliding induced adhesion of stiff polymer microfiber arrays: 2. Microscale behaviour, *Journal of the Royal Society Interface*, (10.1098/rsif.2007. 1309) (2008).

201. K. Autumn, Gecko adhesion: Structure, function, and applications, *MRS Bulletin*, 32, 473–478 (2007).

202. S. Kim, et al., Effect of soft backing layer thickness on adhesion of single-level elastomer fiber arrays, *Applied Physics Letters*, 91, 161905–161907 (2007).

203. S. Kim, B. Aksak, and M. Sitti, Enhanced friction of polymer microfiber adhesives with spatulate tips, *Applied Physics Letters*, 91, 221913–221915 (2007).

204. M. Murphy, et al., Adhesion and anisotropic friction enhancement of angled heterogeneous micro-fiber arrays with spherical and spatula tips, *Journal of Adhesion Science and Techology*, 21, 1281–1296 (2007).

205. E. Arzt, S. Gorb, and R. Spolenak, From micro to nano contacts in biological attachment devices, *Proceedings of National Academy of Science USA*, 100, 10603–10606 (2003).

206. R. Spolenak, S. Gorb, and E. Arzt, Adhesion design maps for bio-inspired attachment systems, *Acta Biomaterialia*, 1, 5–13 (2005).

207. C. Majidi, et al., High friction from a stiff polymer using microfiber arrays, *Physics Review Letters*, 97, 076103 (2006).

208. L. Ge, et al., Carbon nanotube-based synthetic gecko tapes, *Proceedings of the National Academy of Sciences*, 104, 10792–10795 (2007).

209. H. Lee, B. P. Lee, and P. B. Messersmith, A reversible wet/dry adhesive inspired by mussels and geckos, *Nature*, 448, 338–341 (2007).

210. A. Mahdavi, et al., A biodegradable and biocompatible gecko-inspired tissue adhesive, *Proceedings of the National Academy of Sciences*, 105, 2307–2312 (2008).

211. A. K. Geim, et al., Microfabricated adhesive mimicking gecko foot-hair, *Nature Materials*, 2, 461–463 (2003).

212. M. T. Northen and K. L. Turner, A batch fabricated biomimetic dry adhesive, *Nanotechnology*, 16, 1159–1166 (2005).

213. I. C. Gebeshuber, Biotribology inspires new technologies, *Nano Today*, 2, 30–37 (2007).

214. B. Bhushan, ed., *Handbook of micro/nanotribology*, CRC Press, Boca Raton, FL (1999).

215. B. Bhushan, ed., *Modern tribology handbook, Vol. 1 - Principles of tribology, Section II*, CRC Press, Boca Raton, FL (2001).

216. B. Bhushan, Tribology: Friction, wear, and lubrication, In *The engineering handbook*, R. C. Dorf, ed., p. 210, CRC Press, Boca Raton, FL (2000).

217. M. Scherge and S. Gorb, *Biological micro- and nanotribology – Nature's solutions*, Springer Verlag, Berlin, Germany (2001).

218. D. Qur, et al., Slippy and sticky microtextured solids, *Nanotechnology*, 14, 1109–1112 (2003).

219. P. K. Hansma, et al., Optimized adhesives for strong, lightweight, damage-resistant, nanocomposite materials: New insights from natural materials, *Nanotechnology*, 18, 044026 (2007).

220. I. C. Gebeshuber, et al., Diatom bionanotribology—Biological surfaces in relative motion: Their design, friction, adhesion, lubrication and wear, *Journal of Nanoscience and Nanotechnology*, 5, 79–87 (2005).

221. R. Gordon, ed., A special issue on diatom nanotechnology, *Journal of Nanoscience Nanotechnology*, 5, 1–4 (2005).

222. Y. X. Zhuang and A. Menon, On the stiction of MEMS materials, *Tribology Letters*, 19, 111 (2005).

223. J. A. Raymond and C. A. Knight, Ice binding, recrystallization inhibition, and cryoprotective properties of ice-active substances associated with Antarctic sea ice diatoms, *Cryobiology*, 46, 174–181 (2003).

224. C. E. Orsello, et al., Molecular properties in cell adhesion: A physical and engineering perspective, *Trends Biotechnology*, 19, 310–316 (2001).

225. R. L. Thurmond, et al., The role of histamine H1 and H4 receptors in allergic inflammation: The search for new antihistamines, *Nature Reviews Drug Discovery*, 7, 41–53 (2008).

226. O. Abbassi, et al., Canine neutrophil margination mediated by lectin adhesion molecule-1 in vitro, *Journal of Immunology*, 147, 2107–2115 (1991).

227. D. A. Jones, et al., P-selectin mediates neutrophil rolling on histamine-stimulated endothelial cells, *Biophysical Journal*, 65, 1560–1569 (1993).

228. K. Ley, et al., Lectin-like cell adhesion molecule 1 mediates leukocyte rolling in mesenteric venules in vivo, *Blood*, 77, 2553–2555 (1991).

229. T. A. Springer, Traffic signals for lymphocyte recirculation and leukocyte emigration: The multistep paradigm, *Cell*, 76, 301–314 (1994).

230. B. R. Alevriadou, et al., Real-time analysis of shear-dependent thrombus formation and its blockade by inhibitors of von Willebrand factor binding to platelets, *Blood*, 81, 1263–1276 (1993).

231. M. B. Lawrence, et al., Effect of venous shear stress on CD18-mediated neutrophil adhesion to cultured endothelium, *Blood*, 75, 227–237 (1990).

232. H.-W. Denker, Molecular approaches to cell-cell adhesion: From leukocyte extravasation to embryo implantation, *Cells Tissues Organs*, 172, 150–151 (2002).

233. K. V. Honn, et al., Enhanced tumor cell adhesion to the subendothelial matrix resulting from 12(S)-HETE-induced endothelial cell retraction, *FASEB Journal*, 3, 2285–2293 (1989).

234. R. P. McEver, K. L. Moore, and R. D. Cummings, Leukocyte trafficking mediated by selectin-carbohydrate interactions, *Journal of Biological Chemistry*, 270, 11025–11028 (1995).

235. C. V. Carman and T. A. Springer, Integrin avidity regulation: Are changes in affinity and conformation underemphasized? *Current Opinion in Cell Biology*, 15, 547–556 (2003).

236. H. Ait-Oufella, E. Maury, B. Guidet, and G. Offenstadt, The endothelium: A new organ (L'endothélium: un nouvel organe), *Reanimation*, 17, 126–136 (2008).

237. J. G. Lock, B. Wehrle-Haller, and S. Strömblad, Cell-matrix adhesion complexes: Master control machinery of cell migration, *Seminars in Cancer Biology*, 18, 65–76 (2008).

238. D. K. Brunk, D. J. Goetz, and D. A. Hammer, Sialyl Lewis(x)/E-selectin-mediated rolling in a cell-free system, *Biophysical Journal*, 71, 2902–2907 (1996).

239. D. K. Brunk and D. A. Hammer, Quantifying rolling adhesion with a cell-free assay: E-selectin and its carbohydrate ligands, *Biophysical Journal*, 72, 2820–2833 (1997).

240. D. F. J. Tees and D. J. Goetz, Leukocyte adhesion: An exquisite balance of hydrodynamic and molecular forces, *News in Physiological Sciences*, 18, 186–190 (2003).

241. S. Reboux, G. Richardson, and O. E. Jensen, Bond tilting and sliding friction in a model of cell adhesion, *Proceedings of the Royal Society of London A: Mathematical, Physical and Engineering Science*, 464, 447–467 (2008).

242. M. Shimaoka and T. A. Springer, Therapeutic antagonists and conformational regulation of integrin function, *Nature Reviews Drug Discovery*, 2, 703–716 (2003).

243. S. Huveneers, H. Truong, and E. H. J. Danen, Integrins: Signaling, disease, and therapy, *International Journal of Radiation Biology*, 83, 743–751 (2007).

244. S. Choi, et al., Small molecule inhibitors of integrin α2β1, *Journal of Medicinal Chemistry*, 50, 5457–5462 (2007).

245. F. G. Giancotti, Targeting integrin β4 for cancer and anti-angiogenic therapy, *Trends in Pharmacological Sciences*, 28, 506–511 (2007).

246. C. Coisne, et al., Therapeutic targeting of leukocyte trafficking across the blood-brain barrier, *Inflammation and Allergy - Drug Targets*, 6, 210–222 (2007).

247. A. R. Aricescu and E. Y. Jones, Immunoglobulin superfamily cell adhesion molecules: Zippers and signals, *Current Opinion in Cell Biology*, 19, 543–550 (2007).

248. R. P. McEver, Adhesive interactions of leukocytes, platelets, and the vessel wall during hemostasis and inflammation, *Thrombosis and Haemostasis*, 86, 746–756 (2001).

249. D. Vestweber, Adhesion and signaling molecules controlling the transmigration of leukocytes through endothelium, *Immunological Reviews*, 218, 178–196 (2007).

250. M. R. Morgan, et al., Synergistic control of cell adhesion by integrins and syndecans, *Nature Reviews Molecular Cell Biology*, 8, 957–969 (2007).

251. R. P. McEver and R. D. Cummings, Cell adhesion in vascular biology. Role of PSGL-1 binding to selectins in leukocyte recruitment, *Journal of Clinical Investigation*, 100, 485–491 (1997).

252. T. A. Springer and J.-H. Wang, The three-dimensional structure of integrins and their ligands, and conformational regulation of cell adhesion, *Advances in Protein Chemistry*, 68, 29–63 (2004).

253. S. Zhuang, et al., Multiple α subunits of integrin are involved in cell-mediated responses of the Manduca immune system, *Developmental and Comparative Immunology*, 32, 365–379 (2008).

254. J. Takagi, Structural basis for ligand recognition by integrins, *Current Opinion in Cell Biology*, 19, 557–564 (2007).

255. A. S. Popel and R. N. Pittman, Mechanics and transport in microcirculation, In *The biomedical engineering handbook*, 2nd ed., J. D. Bronzino, ed., pp. 31–101, CRC Press, Boca Raton, FL (2000).

256. C. W. Patrick, et al., Fluid shear stress effects on cellular function, In *The biomedical engineering handbook*, 2nd ed., J. D. Bronzino, ed., pp. 114–201, CRC Press, Boca Raton, FL (2000).

257. J. Li and W. Zhong, A two-dimensional suspension array system by coupling field flow fractionation to flow cytometry, *Journal of Chromatography A*, 1183, 143–149 (2008).

258. R. H. G. Baxter, et al., Structural basis for conserved complement factor-like function in the antimalarial protein TEP1, *Proceedings of the National Academy of Sciences*, 104, 11615–11620 (2007).

259. B. N. J. Persson, On the mechanism of adhesion in biological systems, *Journal of Chemical Physics*, 118, 7614–7621 (2003).

260. G. E. Fantner, et al., Sacrificial bonds and hidden length: Unraveling molecular mesostructures in tough materials, *Biophysical Journal*, 90, 1411–1418 (2006).

261. T. M. Dugdale, et al., Single adhesive nanofibers from a live diatom have the signature fingerprint of modular proteins, *Biophysical Journal*, 89, 4252–4260 (2005).

262. A. S. Mostaert and S. P. Jarvis, Beneficial characteristics of mechanically functional amyloid fibrils evolutionarily preserved in natural adhesives, *Nanotechnology*, 18, 044010 (2007).

263. A. S. Mostaert, et al., Nanoscale mechanical characterisation of amyloid fibrils discovered in a natural adhesive, *Journal of Biological Physics*, 32, 2887–2893 (2006).

264. A. S. Mostaert, T. Fukuma, and S. P. Jarvis, Explanation for the mechanical strength of amyloid fibrils, *Tribology Letters*, 22, 233–237 (2006).

265. J. H. Fendler, *Membrane-mimetic approach to advanced materials*, (*Advances in polymer science*), Vol. 113, Springer Verlag, Berlin (1994).

266. C. R. Martin, Nanomaterials: A membrane based synthetic approach, *Science*, 266, 1961–66 (1994).

267. X.-Y. Zhang, et al., Synthesis of ordered single crystal silicon nanowire arrays, *Advanced Materials*, 13, 1238–1241 (2001).

268. M. A. Guillorn, et al., Individually addressable vertically aligned carbon nanofiber-based electrochemical probes, *Journal of Applied Physics*, 91, 3824–2828 (2002).

269. M. J. Doktycz, et al., Nanofiber Structures as Mimics for Cellular Membranes, *Nanotechnology*, 3, 420–423 (2003).

270. T. A. Desai, et al., Microfabricated immunoisolating biocapsules, *Biotechnology and Bioengineering*, 57, 118–120 (1998).

271. K. Senior, "Nano-dumpling" with drug delivery potential, *Molecular Medicine Today*, 4, 321 (1998)

272. P. Broz, et al., Toward intelligent nanosize bioreactors: A pH-switchable, channel-equipped, functional polymer nanocontainer, *Nano Letters*, 6, 2349–2353 (2006).

273. D. Akin, et al., Bacteria-mediated delivery of nanoparticles and cargo into cells, *Nature Nanotechnology*, 2, 441–449 (2006).

274. E. R. Ballister, et al., Nanotubes from biomimetically bioengineered viruses for drug delivery: In vitro self-assembly of tailorable nanotubes from a simple protein building block, *Proceedings of the National Academy of Sciences*, 105, 3733–3738 (2008).

275. A. R. Parker and C. R. Lawrence, Water capture by a desert beetle, *Nature*, 414, 33–34 (2001).

276. M. L. Fishman, P. H. Cooke, and D. R. Coffin, Nano structure of native pectin sugar acid gels visualized by atomic force microscopy, *Biomacromolecules*, 5, 334–341 (2004).

277. A. Hentschel, S. Gramdorf, R. H. Müller, and T. Kurz, Beta-Carotene-loaded nanostructured lipid carriers, *Journal of Food Science*, 73, N1–6 (2008).

278. J. Fava, et al., Structure and nanostructure of the outer tangential epidermal cell wall in *Vaccinium corymbosum* L. (Blueberry) fruits by blanching, freezing-thawing and ultrasound, *Food Science and Technology International*, 12, 241–251 (2006).

279. R. H. J. Hannink and A. J. Hill, eds., *Nanostructure control of materials*, Woodhead Publishing Limited, Abington, Cambridge, UK (2006).

280. J. Dyck, The evolution of feathers, *Zoologica Scripta*, 14, 137 (1985).

281. J. A. Tuszynski, *Molecular and cellular biophysics*, CRC Press, Boca Raton, FL (2007).

282. D. Jones, J. Round, and A. de Haan, *Skeletal muscle: From molecules to movement*, Elsevier, New York (2004).

283. R. Bartlett, *Introduction to sports biomechanics*, Taylor and Francis, London (1996).

284. M. M. Dewey, et al., Structure of limulus striated muscle: The contractile apparatus at various sarcomere lengths, *The Journal of Cell Biology*, 58, 574–593 (1974).

285. B. M. Millman, The filament lattice of striated muscle, *Physiological Reviews*, 78, 359–391 (1998).

286. A. M. Herrera, et al., Sarcomeres' of smooth muscle: Functional characteristics and ultrastructural evidence, *Journal of Cell Science*, 118, 2381–2392 (2005).

287. M. Reconditi, et al., Structure-function relation of the Myosin motor in striated muscle, *Annals of the New York Academy of Sciences*, 1047, 232–247 (2005).

288. O. M. Hernandez, et al., Plasticity in skeletal, cardiac, and smooth muscle, Invited review: Pathophysiology of cardiac muscle contraction and relaxation as a result of alterations in thin filament regulation, *Journal of Applied Physiology*, 90, 1125–1136 (2001).

289. J. Arikkath and K. P. Campbell, Auxillary subunits: Essential components of the voltage-gated calcium channel complex, *Current Opinion in Neurobiology*, 13, 298–307 (2003).

290. M. Kang and K. P. Campbell, The gamma subunit of voltage-activated calcium channels, Mini Review, *Journal Biological Chemistry*, 278, 21315–21318 (2003).

291. D. Michele and K. P. Campbell, Cardiomyopathy in muscular dystrophies, In *Molecular mechanisms for cardiac hypertrophy and failure*, R. A. Walsh, ed., pp. 541–567, Taylor & Francis, London (2005).

292. M. Shahinpoor, et al., *Artificial muscles: Applications of advanced polymeric nanocomposites*, Taylor & Francis, London (2007).

293. W. Yim, J. Lee, and K. J. Kim, An artificial muscle actuator for biomimetic underwater propulsors, *Bioinspiration and Biomimetics*, 2, S31–S41 (2007).

294. R. H. Baughman, Playing Nature's game with artificial muscles, *Science*, 308, 63–65 (2005).

295. J. Ayers, J. L. Davis, and A. Rudolph, eds., *Neurotechnology for biomimetic robots. Based on a conference held in May 2000*, MIT Press, Cambridge, MA (2002).

296. Y. Bar-Cohen, ed., Electroactive polymer (EAP) actuators as artificial muscles - Reality, potential and challenges, In *Electroactive polymer actuators and devices conference: Smart structures and materials symposium*, JPL, San Diego, CA (2005).

297. M. Shahinpoor, Ionic polymer–conductor composites as biomimetic sensors, robotic actuators and artificial muscles—a review, *Electrochimica Acta*, 48, 2343–2353 (2003).

298. P. J. Yoo, et al., Spontaneous assembly of viruses on multilayered polymer surfaces, *Nature Materials*, 5, 234–240 (2006).

299. C. Mao, et al., Virus-based toolkit for the directed synthesis of magnetic and semiconducting nanowires, *Science*, 303, 213–217 (2004).

300. D. W. Lawlor, *Photosynthesis*, Routledge, Andover, UK (2004).

301. M. Pessarakli, ed., *Handbook of photosynthesis*, 2nd ed., CRC Press, Boca Raton, FL (2005).

302. P. D. Boyer, The ATP synthase—a splendid molecular machine, *Annual Review Of Biochemistry*, 66, 717–49 (1997).

303. R. Govindjee, J. T. Beatty, H. Gest, and J. F. Allen, eds., *Discoveries in photosynthesis, advances in photosynthesis and respiration*, Vol. 20, Springer, Berlin, Germany (2006) (Reprinted from *Photosynthesis Research*, 73, 76, 80).

304. B. R. Selman and S. Selman-Reimer, eds., *Energy coupling in photosynthesis*, Elsevier, New York (1981).

305. S. Tanaka and R. A. Marcus, Electron transfer model for the electric field effect on quantum yield of charge separation in bacterial photosynthetic reaction centers, *Journal of Physical Chemistry B*, 101, 5031 (1997).

306. J. Deisenhofer, et al., Structure of the protein subunits in the photosynthetic reaction centre of *Rhodopseudomonas viridis* at 3 Å resolution, *Nature*, 318, 618–624 (1985).

307. J. J. Katz and M. R. Wasielewski, Biomimetic approaches to artificial photosynthesis, *Biotechnology And Bioengineering Symposium*, 8, 423–452 (1978).

308. D. Gust and T. A. Moore, Mimicking photosynthesis, *Science*, 244, 35–41 (1989).

309. D. Gust, T. A. Moore, and A. L. Moore, Molecular mimicry of photosynthetic energy and electron transfer, *Accounts of Chemical Research*, 34, 40–48 (2001).

310. L. Hammarstrom, et al., A biomimetic approach to artificial photosynthesis: Ru(II)-polypyridine photo-sensitisers linked to tyrosine and manganese electron donors, *Spectrochimica Acta Part A: Molecular and Biomolecular Spectroscopy*, 57, 2145–2160 (2001).

311. L. Hammarström, L. Sun, B. Åkermark, and S. Styring, A biomimetic approach to artificial photosynthesis: Ru(II)–polypyridine photo-sensitisers linked to tyrosine and manganese electron donors, *Spectrochimica Acta Part A: Molecular and Biomolecular Spectroscopy*, 57, 2145–2160 (2001).

312. D. A. LaVan and J. N. Cha, Approaches for biological and biomimetic energy conversion, *Proceedings of National Academy of Science USA*, 103, 5251–5255 (2006).

313. M. A. Baldo, Photosynthetic photovoltaic cells, Final Report US DTI ADA469444 (2006).

314. A. F. Collings and C. Critchley, eds., *Artificial photosynthesis: From basic biology to industrial application*, Wiley-VCH Verlag, Weinheim, Germany (2005).

315. D. R. Ort and C. F. Yocum, eds., Electron transfer and energy transduction in photosynthesis: An overview, chap. 1, In *Oxygenic photosynthesis: The light reactions*,

advances in photosynthesis, Vol. 2, Kluwer Academic Publishers, the Netherlands, 1–9 (1996).

316. L. A. Staehelin and G. W. M. van der Staay, Structure, composition, functional organization and dynamic properties of thylakoid membranes, chap. 2, In *Oxygenic photosynthesis: The light reactions, advances in photosynthesis*, Vol. 2, D. R. Ort and C. F. Yocum, eds, Kluwer Academic Publishers, the Netherlands, 11–30 (1996).

317. P. J. Kiley, et al., Self-assembling peptide detergents stabilize isolated photosystem I on a dry surface for an extended time, *PloS Biology*, 3, 230–237 (2005).

318. R. Das, et al., Integration of photosynthetic protein molecular complexes in solid-state electronic devices, *Nano Letters*, 4, 1079–1083 (2005).

319. B. D. Bruce, M. A. Baldo, and S. Zhang, Integration of photosynthetic complexes into novel biomolecular electronic devices, 2005 NSF Nanoscale Science and Engineering Grantees Conference, December 12–15, Arlington, VA (2005).

320. B. O'Regan and M. Grätzel, A low-cost, high efficiency solar cell based on dyre-sensitized colloidal TiO_2 films, *Nature*, 353, 737–740 (1991).

321. M. Grätzel, Applied physics: Solar cells to dye for, *Nature*, 421, 586–587 (2003).

322. P. Wang, et al., A stable quasi-solid-state dye-sensitized solar cell with an amphiphilic ruthenium sensitizer and polymer gel electrolyte, *Nature Materials*, 2, 402–407 (2003).

323. D. Di Censo, et al., Synthesis, characterization, and DFT/TD-DFT calculations of highly phosphorescent blue light-emitting anionic Iridium complexes, *Inorganic Chemistry*, 47, 980–989 (2008).

324. U. Bach, et al., Solid-state dye-sensitized mesoporous TiO_2 solar cells with high photon-to-electron conversion efficiencies, *Nature*, 395, 583–585 (1998).

325. M. K. Nazeeruddin and M. Graetzel, Transition metal complexes for photovoltaic and light emitting applications, *Structure and Bonding*, 123, 113–175 (2007).

326. J.-H. Yum, et al., Efficient co-sensitization of nanocrystalline TiO_2 films by organic sensitizers, *Chemical Communications*, 44, 4680–4682 (2007).

327. C. Lee, et al., Phenomenally high molar extinction coefficient sensitizer with donor-acceptor, *Inorganic Chemistry* 47, INOCAJ ISSN:0020–1669. AN 2007:997170 (2008).

328. M. Grätzel, A high molar extinction coefficient charge transfer sensitizer and its application in dye-sensitized solar cell, *Journal of Photochemistry and Photobiology A: Chemistry*, 185, 331–337 (2007).

329. H. Choi, et al., A highly efficient and thermally stable organic sensitizers for solvent free electrolyte based dye-sensitized solar cells, *Angewandte Chemie*, 47, 327–330 (2008).

330. R. Buscaino, et al., A mass spectrometric analysis of sensitizer solution used for dye-sensitized solar cell, *Inorganica Chimica Acta*, 361, 798–805 (2008).

331. J.-J. Lagref, et al., Artificial photosynthesis based on dye-sensitized nanocrystalline TiO_2 Solar Cells, *Inorganica Chimica Acta*, 361, 735–745 (2008).

332. L. M. Peter, et al., Photosensitization of nanocrystalline TiO_2 by self-assembled layers of CdS quantum dots, *Chemical Communications*, 10, 1030–1031 (2002).

333. K. G. U. Wijayantha, et al., Fabrication of CdS quantum dot sensitized solar cells via a pressing route, *Solar Energy Materials and Solar Cells*, 83, 363–369 (2004).

334. L. M. Peter, et al., Transport and Interfacial transfer of electrons in dye-sensitized nanocrystalline solar cells, *Journal of Electroanalytical Chemistry*, 127, 524–525 (2002).

335. I. L. Medintz, et al., Quantum dot bioconjugates for imaging, labelling and sensing, *Nature Materials*, 4, 435–446 (2005).

336. T. Kuritz, et al., Molecular photovoltaics and the photoactivation of mammalian cells, *IEEE Transactions on Nanobioscience*, 4, 196–200 (2005).

337. B. R. Evans, et al., Enhanced photocatalytic hydrogen evolution by covalent attachment of Plastocyanin to Photosystem I, *Nano Letters*, 10, 1815–1819 (2004).

338. H. M. O'Neill and E. Greenbaum, Spectroscopy and photochemistry of Spinach photosystem I entrapped and stabilized in a hybrid organosilicate glass, *Chemistry of Materials*, 17, 2654–2661 (2005).

339. University of Leiden, Harnessing Solar Energy for the Production of Clean Fuels, EC ESF Task Force White Paper, at http://www.ssnmr.leidenuniv.nl/content_docs/cleansolarfuels.pdf (25 March 2008).

340. X.-G. Zhu, A. R. Portis, and S. P. Long, Would transformation of C3 crop plants with foreign Rubisco increase productivity? A computational analysis extrapolating from kinetic properties to canopy photosynthesis, *Plant Cell and Environment*, 27, 155–165 (2004).

341. H. S. Sakhalkar, et al., Leukocyte-inspired biodegradable particles that selectively and avidly adhere to inflamed endothelium in vitro and in vivo, *Proceedings of National Academy of Science USA*, 100, 15895–15900 (2003).

342. Z. Bao, M. R. Weatherspoon, S. Shian, C. Ye, P. D. Graham, S. M. Allan, G. Ahmad, M. B. Dickerson, B. C. Church, Z. Kang, H. W. Abernathy III, C. J. Summers, M. Liu, and K. H. Sandhage, Chemical reduction of three-dimensional silica micro-assemblies into microporous silicon replicas, *Nature*, 446, 172–175 (2007).

343. J. Lee and N. A. Kotov, Thermometer design at the nanoscale, *Nano Today*, 2, 48–51 (2007).

344. C. J. Brokaw, Molecular mechanism for oscillation in flagella and muscle, *Proceedings of the National Academy of Science USA*, 72, 3102–3106 (1975).

345. L. M. Godsel and D. M. Engman, Flagellar protein localization mediated by a calcium–myristoyl/palmitoyl switch mechanism, *The EMBO Journal*, 18, 2057–2065 (1999).

346. G. M. Whitesides, The Once and Future Nanomachine: Biology outmatches futurists' most elaborate fantasies for molecular robots, Nanotechnology web site at: http://www.mtmi.vu.lt/pfk/funkc_dariniai/nanostructures/nano_robots.htm (2008) (Keynote paper, Symposium on Functional Combinations in Solid States, Finland, 2002).

347. K. Namba, Revealing the mystery of the bacterial flagellum - A self-assembling nano-machine with fine switching capability, *Japan Nanonet Bulletin*, 11, 5th Feb. (2004).

348. M. Manghi, X. Schlagberger, and R. R. Netz, Propulsion with a rotating elastic nanorod, *Physics Review Letters*, 96, 068101 (2006).

349. G. Jensen, A nanoengine for gliding motility, *Molecular Microbiology*, 63, 4–6 (2007).

350. Z. Wang, Synergic mechanism and fabrication target for bipedal nanomotors, *Proceedings of the National Academy of Sciences*, 104, 17921–17926 (2007).

351. M. R. Diehl, K. Zhang, H. J. Lee, and D. A. Tirrell, Engineering cooperativity in biomotor protein assemblies, *Science*, 311, 1468–1471 (2006).

352. L. M. Adleman, Computing with DNA, *Scientific American*, Aug. 54–61 (1998).

353. P. W. K. Rothemund, Using lateral capillary forces to compute by self-assembly, *Proceedings of the National Academy of Science USA*, 97, 984–989 (2000).

354. M. Amos, *Theoretical and experimental DNA computation*, Springer, New York (2005).

355. L. Medskar and L. C. Jain, eds., *Recurrent neural networks (CRC Press International Series on Computational Intelligence)*, CRC Press, Boca Raton, FL (1999).

356. L. C. Jain and V. R. Vemuri, Industrial applications of neural networks, CRC Press, Boca Raton, FL (1998).

357. J. B. Waldner, *Nanocomputers and swarm intelligence*, ISTE, London (2007).

358. J. Macdonald, et al., Medium scale integration of molecular logic gates in an automaton, *Nano Letters*, 6, 2598–2603 (2006).

359. E. Braun and K. Keren, From DNA to transistors, *Advances in Physics*, 53, 441–496 (2004),

360. K. Keren, et al., DNA-templated carbon nanotube field-effect transistor, *Science*, 302, 1380–1382 (2003).

361. C. M. Lieber, The incredible shrinking circuit, In *The rise of nanotechnology*, Scientific American, New York, 285, 58–64 (2001).

362. J. F. V. Vincent, et al., Putting biology into TRIZ: A database of biological effects, *Creativity and Innovation Management*, 14, 66–72 (2005).

Problems

12.1 Which of the following technological developments were most likely based on observations of biological analogs? (a) Airplane, (b) wheel, (c) axe, (d) steam engine, (e) artificial intelligence, and (f) jet engine.

12.2 Why do you think that biomimetics and bionics have become increasingly more important (and achievable)?

12.3 Do some research and determine how many kinds of Velcro fasteners are on the market. Are there any that approach the nanoscale with regard to working components?

12.4 Research, define, and draw the hierarchical structure of rope. Is its structure related to any biological structure(s) that you know?

12.5 Explain the meaning of "hidden structure" and "sacrificial bonds." How do they contribute to strength of materials? Draw a diagram illustrating sacrificial bonds.

12.6 What percentage of glue in relation to mineral bricks is found in shells? Compare the structure of shells to the structures of different types of brick walls. Would the wall be stronger if less mortar is used? If more?

12.7 What is the function of steel bars used in concrete? What are some analogous structural features in nanofabrications?

12.8 What is meant by "rolling" in white blood cells? What are the steps in the process? What types of materials mediate each step?

12.9 Why can discovery of an adhesive molecule in a mosquito help prevent malaria in humans?

12.10 What are some of the practical commercial applications learned from the lotus leaf?

12.11 Do humans use more energy to make materials and power their engines than animals and plants? Explain the reason for your answer.

12.12 What is the difference between biomimetics and hunting and gathering natural materials?

MEDICAL NANOTECHNOLOGY

While some may dream of nanorobots circulating in the blood, the immediate applications in medicine will occur at the interfaces among …nanotechnology, micro-electronics, microelectromechanical systems (MEMS) and microopti-calelectro-mechanical systems (MOEMS). … The bounty will not be realized until those trained in these new paradigms begin to … address basic medical and scientific questions.

D.A. LaVan and R. Langer, MIT,
NSF Symposium 2001

*C*hapter 13

FIG. 13.0

What if doctors could reach out and correct diseases at the macromolecular level? Nanotechnology provides tools to help bring that dream closer to reality.

THREADS

Chapter 13 presents an introduction to nanotechnology as applied to medicine. This is a rapidly developing area, with many impacts. This chapter emphasizes device and materials nanotechnology rather than molecular and protein engineering. The impact of nanotechnology on treatment is emphasized: keeping in view that since this is about medicine, an integrated approach focused on the patient rather than the technology will be taken. This introduction focuses on some examples of current applications which are active areas of research and development in biomedicine, with an overview of bio-nanoengineering developments that have promise for the practice of medicine or which point to new directions for research and development. Several areas are discussed which are currently more microtechnology than nanotechnology; in these cases the path to feasible benefits through application of nanotechnologies that are in development is evident. Hopefully this will serve to motivate engineering and biomedical students to

become a participant in making those benefits a reality.

As the practice of medicine is an applied art as much as a science, this chapter is placed as part of the book on nanotechnology rather than nanoscience. Biomedical science draws heavily upon biomolecular nanoscience, chemistry, physics, and engineering, which are covered elsewhere in the books on nanoscience and nanotechnology. This chapter is intended to be useful to those in medicine as an introduction to nanotechnology and its relevance to their field. Emphasis is therefore placed on current and near-term applications and developments rather than future long-term possibilities. This introductory survey should help clinicians understand the potential of nanotechnology-based innovations for those in their care. In keeping with evidence-based and peer-mentoring approaches in medicine, this chapter includes extensive references to medical and bioscience sources for further examination and study.

13.0 INTRODUCTION TO MEDICAL NANOTECHNOLOGY

13.0.1 *Definitions: Medicine and Medical Nanoscience*

Medicine is the knowledge and practice of maintaining and restoring health. Health is the state of a person, an organism, or an organ in which its systems are able to perform their functions without failure in the face of external threats and internal complications. Living systems are constantly meeting challenges such as stress; injury; malformation in development; genetic errors; invasions by viral, bacterial, and parasitic agents; cancer; degeneration; and challenging normal life events such as pregnancy and delivery; puberty and menopause; aging; and death. It is the goal of medicine to support, maintain, and restore productive functioning of life while minimizing suffering and doing no harm. The roots of medicine lie in our empathy for our fellow creatures, starting with our fellow human beings. Medicine is essentially a human art which is supported by observation, evidence, recording and passing on of knowledge and experience, training, and standards. "There is no one division of medicine by which we know and another by which we act" [1]. Medical science is inseparable from medical practice, as the ultimate significant observation is the outcome for a patient.

Medical science is the application of scientific methods to the study of living systems with the goal of improving medical practice. Medical science is based on any scientific or technological discipline that can contribute knowledge and techniques that advance the practice and understanding of medicine. These have historically included anatomy, physiology, chemistry, physics, engineering, and other disciplines. The development of medical science is inextricably involved with the other sciences. The student who aspires to work in medical practice or research must be prepared with a solid base in multiple disciplines relevant to human health. Increasingly, these disciplines will include aspects of nanoscience and nanotechnology as applications to medicine emerge.

Health and Molecular Signaling. Modern understanding of health is based on the concept of regulation of metabolism by a complex network of molecular-based communication mechanisms known as cell signaling that governs basic cellular activities and coordinates cell actions. Cells in the body perform their life cycle functions in part by genetic programming, but also by responding to molecular signals generated within the cell and received through receptors on the cell membranes. These networks respond to, are controlled by, and can be disrupted by processes that take place on the electrical, molecular, macromolecular, and supramolecular scales. The latter are the domain of nanotechnology, where current advances are offering applications for medicine.

Homeostasis. Healthy organisms tend to maintain homeostasis, from the Greek words meaning "like or same" and "still or static." Homeostasis is defined as the stable state controlled by a system of feedback: the system reacts to changes sensed in its state and/or environment to counter influences that tend to desta-bilize it or divert its development from the normal path. For example, body temperature, blood pressure, levels of carbon dioxide in the lungs and tissues, the osmotic pressure within cells are homeostatically regulated. Living organisms and systems as a whole are not static: they undergo growth, development, and death in their normal life cycle. Health must be considered as a dynamic rather than a static process, by which a healthy cell or organism responds appropriately to environmental and developmental challenges. Medical science advances the understanding of how these responses are regulated through a complex network of molecular and supramolecular interactions.

Medical science draws upon engineering concepts and methods to create its own unique models for understanding biological networks as not only chemical but also physical and structural—as complex machinery with subtle control systems acting through specific detailed interactions at the macromolecular and nanoscale level. This approach underlies medical nanoscience, a perspective which gives us a framework to model, understand, and intervene in living processes at the level of supramolecular machinery with selectivity and precision.

13.0.2 *Historical Origins: Medical Breakthroughs*

Historically, medical science originates with empirical observations of outcomes of surgical and pharmaceutical applications. Prior to development of written records and a system of critical evaluation of outcomes, primeval societies had oral traditions regarding herbs and other remedies. Archeological evidence for primitive surgical operations has been found. The development of written records accelerated the communication of medical remedies and the development of a long-term body of experience and outcomes. Stone Age sites in Baluchistan have revealed evidence of dental drilling with flint tools in an early farming culture [2]. Some of the oldest known medical writings are Egyptian papyri from 1600 to 1700 B.C., which record surgical cases and outcomes, and document-established practices which may predate the papyri by several thousand years. [3]

Early civilizations held healers in high esteem; their reputations became legendary and they were elevated to god-like status. In ancient Egypt, Imhotep (meaning "one who comes in peace"), was a historical figure who rose from common beginnings to become the vizier to the Pharaoh Djoser (reigned 2630–2611 B.C.). In ancient cultures, the roles of doctor, priest, scribe, sage, poet, astrologer, architect, vizier, and chief minister were all intertwined. There was a degree of interaction between the Egyptian culture and the civilizations of the Tigris and Euphrates valley, such as the Persians. After his death Imhotep (or Greek: Imouthes) was elevated to demigod status and became a cult figure for healing and a focal point for collections of writings and medical practices in temples and courts in many cultures [4,5,6]. He became identified with the Greek god of healing, Aslepius, and the Greek and Arabian civilizations continued the medical heritage of the Egyptians and Asians.

The most famous Greek doctor was Hippocrates, who was also elevated to cult status, with an attributed genealogy descending from Aslepius. Modern Western science-based medicine reveres Hippocrates as a physician who applied philosophical enquiry to medicine and encoded ethical standards for practice. The admonition in the Hippocratic oath: "First, do no harm," is still the foundation of medical practice.

In the Indus civilization, the Ayurvedic tradition of medicine (meaning "life knowledge") is documented by the works of Agnivesha, whose writings were later revised by Charaka, about 300 B.C. Charaka held that health is not predetermined and life may be prolonged by human effort. This was a major advance in outlook for prescientific societies. He defined the goals of medicine as to cure the diseases of the sick, protect the healthy, and to prolong life. Medical students in the Charaka tradition were given a code of practice with parallels to the Hippocratic oath, including honesty, devotion to learning, sober living, and respect for patient confidentiality. The Indus tradition was very systematic with division of medicine into categories such as toxicology, obstetrics, etc., and a rigorous qualifying exam for practitioners.

In China, medical knowledge was documented in the classical period in texts with parallels to the other ancient traditions, basing illness on a balance of the primary elements and astrological influences. A unique aspect of Chinese medicine was the practice of acupuncture, which is systematically treated in the early treatises, with highly developed anatomical descriptions.

In large part because of the care to avoid doing harm, the practice of medicine has traditionally been experience based and very conservative when adopting new practices and theories. The germ theory of disease was famously resisted by medical practitioners in the nineteenth century until the evidence gained from actual practice became overwhelming, notably due to the work of Louis Pasteur, Claude Bernard, and Robert Koch [7]. At least as much harm has been done in the history of medicine by adherence to theories as doctrines as by rational experimentation with new techniques. That said, the goal of minimizing harm is best served by a skeptical approach to new developments.

Medical Breakthroughs.   In 2000, the *New England Journal of Medicine* reviewed major medical progresses of the past 500 years [8], presenting a number of breakthrough medical milestones in chronological order. Besides such major advances in practice such as anesthesia and rigorous statistical clinical trials, the scientific breakthroughs included a series that trends toward understanding of life science on a broadly smaller scale, from the gross to the sub-microscopic:

Elucidation of human anatomy and physiology, beginning in the sixteenth century
> Accurate anatomies, circulation, pulse, blood pressure, electrical nerve stimulus, control of muscle by nerves

Discovery of cells and their substructures, from the sixteenth to the twentieth centuries
> Microscope, discovery of bacteria and protozoa, complex inner structures of plant and animal tissues, plant cells, animal cells, cell division, nucleus, etc., electron microscope, isolation of mitochondria

Development of biochemistry, from the seventeenth to the twentieth centuries
> Concept of active "fermet" replaces idea of passive "humours," discovery of oxygen, other gases, development of organic and physical chemistry, enzyme chemistry, pathways of metabolism, Krebs cycle, role of calcium, sodium, potassium, magnesium, etc., chlorophyll, hemoglobin, hormones, neurotransmitters, cell signaling

Discovery of the relation of microbes to disease beginning in the nineteenth century

Displacement of spontaneous generation theory of microbial life by continuous inheritance principle, association of microbes with diseases, discovery of viruses, pasteurization, vaccination with weakened microbes and viruses, isolation of bacteria in pure culture, antiseptics, recognition of importance of sanitation and sterilization, disease vectors

Elucidation of inheritance and genetics beginning in the nineteenth century

Mendelian laws of inheritance, inheritance of errors of metabolism, genes, chromosomes, mapping of genes on chromosomes, specification of enzymes by genes, identification of DNA as genetic material, base pairing rules in DNA, isolation and x-ray diffraction of DNA molecules, identification of double helix structure of DNA, role of RNA in transcription of proteins coded in DNA, messenger RNA, methodology for decoding sequences of bases in DNA, discovery of reverse transcriptase which converts RNA into DNA, polymerase chain reaction method for DNA amplification, establishment of a relationship between a molecular mutation and a specific disease, "molecular disease" concept

Knowledge of the immune system

Discovery of first known antibodies (in diphtheria antitoxin), identification of role of phagocytes engulfing foreign bodies, cellular theory of immunity, major advances in vaccines, killed virus vaccines, first vaccine produced by DNA technology (hepatitis B)

Medical imaging and biomarkers for diagnosis and research

Discovery of x-rays, use to image bone and hard tissue structure, radioactive tracers for imaging, ultrasound, development of contrast agents; use of power of electronic controls and digital computation to generate detailed multilayer and dimensional images of organs and vessel structure: magnetic resonance imaging, computed tomography, Doppler ultrasound, functional magnetic resonance imaging, image enhancement agents, radiotracers for metabolic and cellular pathways, photochromic markers, bioluminescent markers of molecular activity within cells; use of medical imaging to guide surgery and minimally invasive procedures, use of antibodies labeled with photonic and magnetic targets to identify cancer cells

Discovery and development of antimicrobial agents

First antibiotic compound: Salvarsan, discovery of antibiotic activity of dyes, Prontosil for strep infections, sulfa drugs, penicillin, discovery of antibiotics in soil organisms, development of streptomycin, understanding the ability of microbes to develop resistance to antibiotics

Development of molecular pharmacotherapy

Concept of chemotherapy, extension of pharmaceutic application from microbes to cancer cells, removal or deactivation of hormone-producing glands (ovaries, testes, etc.) to treat cancer of organs regulated by hormones (breast cancer, prostate cancer, etc.), treatment of lymphomas by nitrogen mustard, methotrexate for leukemia, cis-platinum for cancer, beta-blockers; design of drugs for specific molecular targets, explanation of genetic-based variability of drug response

When we review the above catalog of medical advances, there is a trend from general treatment of whole organs and bodies to specific focus on detailed molecular mechanisms in both diagnosis and treatment. Over the past 500 years advances in chemistry, physics, and electrical engineering went hand in hand with advances in medicine to apply therapy more precisely and efficiently. Advances in science and technology have led to new concepts and tools, such as chemical reactions and chemical analysis and synthesis, which have been applied in the life sciences with resulting advances in medical understanding and practice.

In many cases, studies and observations from medicine and the life sciences have led to new scientific breakthroughs in chemistry and physics. The discovery of oxygen by Priestley [9], which led to the replacement of the phlogiston theory of combustion, was based on his experiments with fermentation, respiration, and the ability of plants to restore the life-supporting properties of air in confined spaces. Priestley's interactions with people like Benjamin Franklin accelerated the advance of medical and scientific knowledge, part of a newly emerging network of organizations which spurred and supported science, sometimes in the face of political and religious opposition. The activities of these early societies, supported by the wealth and protection of enlightened patrons, led to the meetings between Priestley and Lavoisier, who completed the conceptual development of the theory of combustion, a fundamental breakthrough in the understanding of matter and energy [10,11].

In an equally important breakthrough, the discovery of electrical stimulation of movement in frog legs by Galvani led to understanding of electric potential and the development of electric batteries by Volta, as well as understanding of how nerve impulses are transmitted. The story of Galvani and Volta and their disagreements provides an instructive example of tensions produced between the medical and technological outlooks [12].

In the twentieth century, the elucidation of *electronic charge transfer* in the process of photosynthesis and energetics of cellular metabolism has led to breakthroughs in photoelectric engineering and solar power generation. The study of human speech and hearing pathologies and development had impacts on computer speech synthesis and recognition. There is an inseparable relationship between advances in science and technology and advances in medicine. New ways of understanding, interpreting, and investigating the world based on nanotechnology will create new understanding of health and disease, and will generate new technical tools for study and application to medical practice.

13.0.3   *Medical Nanoscience: Roots in Medical Science*

The field of nanoscience is essentially a new perspective on science and technology, with new possibilities opened by vastly more powerful tools for examination and manipulation of matter on the sub-micron scale. This new paradigm is rapidly opening possibilities in medical research and practice. The impact of nanoscience on medicine parallels those made by microscopy, chemistry, physics, electronics, and computing which led to new theories of disease, more effective approaches to treatment, and powerful new imaging and surgical tools.

Early medical diagnosis and treatment was based on what could be deduced from appearance and feel of the body, pain and sensitivity, fever, examination of bodily fluids, external appearance of symptoms, and course of the disease. For example, early physicians diagnosed diabetes by the sweetness of urine. With the development of chemistry and the microscope, diagnoses could be made on the basis of the appearance of cells and composition of bodily fluids; for example, the relative number of different types of blood cells (the blood count) became a diagnostic yardstick; fever was quantified by use of the thermometer; clinical chemistry analyzed the pH and other chemical composition of the body fluids; microscopic examination showed the presence of identifiable microbes in sputum. With the development of x-rays and ultrasound, the anatomy of the body could be examined and the presence of fractures and foreign bodies could be diagnosed accurately.

Without advanced tools to directly observe molecules at the nanoscale, these discoveries were made by a combination of careful observation and experiment, insightful hypotheses, and skillful deduction. For example, viruses were first postulated by their ability to convey disease after passing through nanopore filters, whose extremely small pore sizes blocked the passage of previously known cellular disease agents. This early example of nanotechnology used the ceramic filter developed by Charles Chamberland. Viral structure was later elucidated with the electron microscope and x-ray diffraction. Clinical chemistry, histology, and molecular biology produced more subtle and precise windows into biology, with determination of multiple antibody assays and/or genotypes conducted at the patient's bedside or in the clinic.

Drugs are now designed to target specific metabolic pathways and cell membrane receptors or cell processes. For example, the taxols such as paclitaxel disrupt the tubulin process involved in the configuration of microtubules essential in cell division (mitosis), thus preferentially interfering with the growth of cancer cells [13].

13.0.4 Future Possibilities for Medical Nanotechnology: Nanomedicine

In this section, we take a quick look at some of the more far-reaching proposals and research areas for nanomedicine. The spirit of these proposals is summed up in the following quotation from the U.S. National Institutes of Health (NIH) Roadmap for Medical Research [14]:

> What if doctors could search out and destroy the very first cancer cells that would otherwise have caused a tumor to develop in the body? What if a broken part of a cell could be removed and replaced with a miniature biological machine? What if pumps the size of molecules could be implanted to deliver life-saving medicines precisely when and where they are needed? These scenarios may sound unbelievable, but they are the long-term goals of the NIH Roadmap's Nanomedicine initiative that we anticipate will yield medical benefits as early as 10 years from now.

The NIH defines nanomedicine as the highly specific application of nanotechnology to medical intervention at the molecular scale for curing disease or repairing damaged tissues, such as bone, muscle, or nerve. The nanometer size scale—about 100 nm or less—is the scale on which biological molecules and structures inside living cells operate.

Medical science has powerful tools to examine the parts of cells in detail down to the molecular level. It is the goal of nanomedicine to understand further how intracellular structures operate, in order to build "nano" structures or "nano" machines that are compatible with living tissues and can safely operate inside the body. The ultimate goal is to design diagnostic tools and engineer structures for highly specific and precise treatments of disease and repair of tissues.

Nanomedicine Research Programs.　Nanoscience is being developed in conjunction with advanced medical science for further precision in diagnosis and treatment. Multidisciplinary biomedical scientific teams including biologists, physicians, mathematicians, engineers, and computer scientists are working to gather information about the physical properties of intracellular structures upon which biology's molecular machines are built. New emphasis is being given to moving medical science from the laboratory to the bedside and the community.

As researchers gain knowledge of the interactions between molecules and larger structures, patterns will emerge, and we will have a greater understanding of the intricate operations of processes and networks inside living cells. Mapping these networks and understanding how they change over time will, in turn, enable researchers to use this information to correct biological defects in unhealthy cells [15]. New tools that will work at the nanoscale and allow scientists to build synthetic biological devices, such as nanosensors to scan for the presence of infectious agents, or metabolic imbalances that could spell trouble for the body, and miniature devices to destroy infectious agents or fix the "broken" parts in cells.

The NIH Nanomedicine Roadmap.　The NIH has developed a series of roadmaps planning research to develop new tools to intervene at the nanoscale or molecular level [16]. These include the *National Technology Centers for Networks and Pathways*, a network of centers which will create new tools to describe the dynamics of protein interactions. The centers will develop instruments, methods, and reagents for quantitative measurements at subcellular resolution and very short timescales. In addition, the NIH is creating *Nanomedicine Development Centers* (see boxes) that will focus on the engineering of new tools for medical interventions and diagnosis at the nanoscale or molecular level. The Nanomedicine Centers' research will support the development of synthetic biological devices, such as miniature, implantable pumps for drug delivery and implantable or mobile sensors to scan for signs of infectious agents, metabolic imbalances, or other biomarkers to detect disease.

Emerging medical nanoscience is targeted to impact the following areas:

1. New medical materials for cell growth scaffolding and tissue repair
2. Enhancement of diagnosis and imaging
3. Enhancement of drug delivery
4. Understanding and control over biomolecular mechanisms
5. Discovery of properties and medical effects of smaller units of life and nanoparticles

Areas of medical care that will benefit from the above nanoscience advances are

1. Plastic surgery and wound healing using nanogels and nanoengineered scaffolding materials
2. Repair of cut nerves using nanofabricated growth channels with cell growth coating patterns
3. Improved healing and fusion of bone fractures using nanopatterned porous implants
4. Selective image enhancement of diseased cells with antibody-coated nanoparticles
5. Reduction of MRI interaction with surgical and sensor probes by nanoengineering coatings
6. Targeted drug delivery with surface-modified and -coated nanoparticles
7. Drug delivery across the blood–brain barrier with "smart" nanoparticles
8. Custom-designed molecular enzymes to selectively switch cell functions on or off
9. Custom-designed phage-like molecular machines to deliver drugs or kill cancer cells
10. Custom-designed molecular enzyme machines to diagnose and repair subcellular structures
11. A suite of molecular enzyme machines to selectively initiate cell death in cancer cells
12. Artificial molecular agents to engulf and deactivate prions, attack viruses, and digest refractory plaques

Center for Nucleoprotein Machines. For example, the NIH-funded National Nanomedicine Center for Nucleoprotein Machines based at Georgia Tech, in collaboration with Emory University and the Medical College of Georgia, will take a biomedical engineering design approach to the repair of DNA, focusing on a model nanomachine that carries out nonhomologous end joining (NHEJ) of DNA double-strand breaks. This and other DNA repair machines have relatively simple structures (<20 components) and significant biological and clinical relevance, and thus are promising as feasible models for nanoscience engineering approaches. DNA repair is vitally important to human health, as both normal metabolic activities and environmental factors can cause DNA damage, resulting in as many as 100,000 individual molecular lesions per cell per day. If allowed to accumulate without repair, these lesions interfere with gene transcription and replication, leading to premature aging,

THE FIRST EIGHT NIH NANOMEDICINE CENTERS AND THEIR RESEARCH AREAS

Summary of the work at the Nanomedicine Development Centers funded to date (2006) through the National Eye Institute and their goals for diseases and medical research. Each of the centers is highly multi-disciplinary and involves multiple institutions such as schools of biomedical science, medicine, engineering, and hospitals. For a guide and links to further information see the NIH Web site. Source: http://nihroadmap.nih.gov/nanomedicine/fundedresearch.asp (accessed May 2007).

1. **Nanomedicine Center for Mechanical Biology**—Columbia University leading partnership of six institutions
   Research on the roles of force, rigidity, and form in regulating cell functions signaling pathways and gene expression, and their influence on diseases such as cancers, immune disorders, genetic malformations, and neuropathies, using tools of nanotechnology and modern cell biology.

2. **UCSF/UCB Center for Engineering Cellular Control Systems**—University of California, San Francisco, and University of California, Berkley
   Work on engineering "grand challenges" to develop modified cells or cell-like molecular assemblies that perform intelligently guided precision therapeutic functions, such as tissue repair or "search and deliver" treatment of microscopic tumors or cardiovascular lesions, by focus on reengineering cell guidance, cell force generation, and cell motility systems.

3. **National Center for Design of Biomimetic Nanoconductors**—University of Illinois, Urbana-Champaign
   Research to design synthetic arrays of ion transport channels based on biological ion channels and other ion transport proteins, inserted in arrays of pores on substrates to study cell signaling, energy transport, and generation of osmotic pressures and flows, in order to gain insight into biological processes and disease targeting, and to develop practical applications such as biosensors, osmotic pumps, and power generation.

4. **Center for Protein Folding Machinery**—Baylor College of Medicine with Stanford University
   An interdisciplinary program to define the basic chemical and physical principles used by molecular chaperones in the folding of proteins, in order to engineer protein machines to assist the folding of any protein of interest, and to develop strategies to alleviative or prevent protein misfolding associated with human diseases.

5. **Nanomedicine Development Center for the Optical Control of Biological Function**—University of California Lawrence Berkeley National Laboratory
   Developing methods for rapidly turning select proteins in cells on and off with light, developing chemical and molecular toolkits for integration of optical control into proteins, viral delivery of photo-switchable proteins into cells, and light delivery systems to address these nanodevices in vivo, with the aim of treating retinal and cardiac pathologies by gaining optical control over the signaling and enzymatic activity of cells.

6. **The Center for Systemic Control of Cyto-Networks**—University of California, Los Angeles
   The goal for this center is to use engineering principles to develop global system control methods to investigate and manipulate the complex cell signaling network governing homeostasis of cells, in order to control and correct perturbations in the network by invading organisms, accumulation of pathologic substances, and uncontrolled cell growth that are the hallmarks of most morbid and mortal illness, especially conditions like cancers, infectious diseases, and stem cell related disorders.

7. **Nanomotor Drug Delivery Center**—Purdue University
   Creation of biocompatible membranes and arrays with embedded phi29 in vitro viral packaging biomotors for DNA insertion applications in medicine, by reverse engineering the phi29 motor; incorporating the active nanomotor into lipid bilayers; and developing active nanomotor arrays that enable drug delivery and diagnostics.

8. **Nanomedicine Center for Nucleoprotein Machines**—Georgia Tech, with Emory University and the Medical College of Georgia
   Using nanotechnology and biomolecular approaches, elucidate the structure–function relationships within and among DNA repair nanomachines, for precise modification of DNA and RNA, leading to therapeutic strategies for a wide range of diseases.

apoptosis, or unregulated cell division. The nucleoprotein machine engineering approach is

1. Develop protein tags and fluorescence probes including quantum dot bioconjugates for nanomachine targeting
2. Decipher structure–function relationship for the NHEJ reaction
3. Characterize the dynamics of nanomachine assembly and disassembly in the repair process
4. Determine the dimensions and structure of repair foci at high resolution in fixed cells
5. Establish engineering design principles for DNA double-strand break repair

The Georgia Nucleoprotein Machines center will complement the other NDCs that focus on filaments, membranes, and protein-folding enzymes. The probes, tools, and methodologies developed in these centers will be useful as tools for biological and disease research, and may ultimately provide genetic cures for common human diseases based on the ability to manipulate the somatic human genome using nanomedicine.

To match the Nanomedicine Initiative, many of the NIH Institutes that focus on specific medical specializations and diseases are supporting programs to find areas of application for the tools being developed at the nanomedicine centers. The goal is to improve diagnosis and treatment with nanotechnology-based techniques and materials in the areas of cancer, radiology, and others. For example, the National Cancer Institute (NCI) has a nanotechnology plan that is distinct from, but complementary to, the NIH Nanomedicine Roadmap. The NCI plan

> focuses on using knowledge from basic research discoveries and translating that into clinical oncology applications. The endpoints of this effort will be technology platforms in the context of diagnostics and therapeutics. [17]

The goals of nanomedicine are extrapolations of advances that have been made in the past by application of new science and technology to medicine. Not all will be successful in their application. Many will be controversial. This is not new in the history of medicine.

13.0.5 *Putting Medical Nanoscience into Practice: Medical Nanotechnology*

In this chapter we explore how nanoengineered materials and devices are being applied in clinical medicine in many areas: radiology, oncology, endocrinology, neurology, orthopedics, cardiology, otology, ophthalmology, emergency care, obstetrics and gynecology, gastroenterology, surgery, and others. (We discuss nanotechnology applications to laboratory medicine in chapter 11. These include biomolecular nanotechnology and nanodevices such as microfluidic biochips, diagnostic lab-on-a-chip devices, nanodroplet dispensers, cell manipulation and separation chips and micromanipulators, and DNA/RNA nanotechnology, with applications in pathology, diagnostics, cytology, genetics, forensic medicine, etc.)

Clinical nanotechnology applications are being applied to enhance or enable new diagnostic and therapeutic methods, including

1. Enhancement agents for medical imaging
2. Finding and destroying cancer cells
3. Delivering drugs deep into tumors (cancer)
4. Delivering insulin through novel routes (diabetes)
5. Delivering drugs through the blood–brain barrier (Alzheimer's, Parkinson's, etc.)
6. Guiding and stimulating nerve regeneration (spinal cord injury, paralysis)
7. Improving neural stimulation (cardiac pacemakers, neuroprosthetics)
8. Noninvasive, sensitive detection of nerve activity (ECG, brain–machine prosthetics)
9. Less invasive hearing and vision prosthetics (hearing and vision loss)
10. Improved remote medical monitoring (preventive, postoperative, recuperative, etc.)
11. Wearable and minimally invasive wireless physiological sensors (GI, Ob-Gyn, etc.)
12. New tissue scaffolds and artificial tissues and organs (surgery, wound care, etc.)
13. Advanced minimally invasive and effective surgical tools and techniques

In the following sections of this chapter, we give examples of the above applications of nanotechnology to medicine that are currently being developed and put into practice, along with discussion of some microscale technologies that are leading to future, improved, nanoscale techniques. Ultimately these families of technologies will merge across the micro-, nano-, and molecular scales into an integrated medical science supporting more powerful and effective medical practice, made more accessible and cost-effective by the availability of tools that operate on the scale of the machinery of the cell.

13.1 NANOPARTICLES AND NANOENCAPSULATION FOR MEDICAL APPLICATIONS

Nanoparticles made of metal, carbon nanotubes, polymers, or other materials can be used in a variety of medical applications, especially when combined with antigen-specific coatings or functional groups on their surfaces. In this section we look at some examples.

The therapeutic and diagnostic usefulness of inorganic nanoparticles depend on their size and physical properties in addition to their chemical composition. They must be small enough to circulate through the bloodstream and tissues without becoming lodged in capillaries or other microanatomies. But they must be larger than atomic size in order to lend enhancement to images or separation techniques.

13.1.1   Nanoparticles for Medical Imaging

Nanoparticles have been used as contrast and image enhancement agents for x-ray and computed tomography imaging. Conventional image contrast agents are molecules such as iodinated benzoic acid derivatives, but these have risk factors and side effects associated with intravenous iodine injection. These chemicals are typically of low molecular weight, and they clear from the human body rapidly, making it difficult to target these agents to disease sites. Iodinated molecules have been encapsulated into liposomes to make a nanoscale particle, but the stability and concentration of agent delivered to the imaged site by this means is low [18].

Experimental image enhancement agents containing gadolinium (Gd) and radioisotopes have been developed for CT imaging, based on dendritic conjugation compounds [19]. However, in such systems, only a relatively small number of gadolinium atoms may be delivered to/in the vicinity of the target tissues. Both approaches deliver, at most, a couple of hundred heavy atoms (i.e., iodine or gadolinium). Another approach is to encapsulate solid nanoparticles of iodine compounds (sized from 200- to 400-nm diameter) in a polymer coating [20], but this involves risk of iodine toxicity should the coating break down.

In order to enhance an x-ray image, an agent must deliver a detectable number of heavy atoms into the imaged tissue without toxic effects. Nanoparticles of elemental heavy metals have the highest density (number of heavy metal atoms/volume), but they must be biologically inert and stable. Nanoparticles of inert metals such as gold are not very cost-effective. To overcome these issues, researchers at General Electric developed nanoparticles made of heavy metal compounds encapsulated in gold shells [21].

Gold-coated nanoparticles can be made by vapor or electrodeposition onto nanoparticles; gold can also be deposited into nanoscale mold templates of silicon, carbon, alumina, or other material with nanosize pores or wells. In the case of silicon, the wells can be created artificially in a silicon wafer using nanofabrication techniques. In the case of carbon nanotubes or alumina the templates are made by controlling the synthesis or electrodeposition of the material.

Organic compounds with sulfide (–S–H) groups (thiols) can be used to coat gold particles with uniform organic monolayers. The sulfide group attaches to the metal surface leaving the organic portion of the molecules exposed as an organic layer. By functionalizing proteins with thiol groups, gold particles can be coated with selectively binding antigens, antibodies, or target compounds for receptors on the surfaces of cells. By targeting receptors unique to certain types of cancer cells, gold nanoparticles can be made to enhance an x-ray image to increase the ability to detect the cancer cells by many orders of magnitude.

Metal and silicon nanoparticles can be used to enhance MRI. Silicon particles fabricated into shapes and coated with conductive layers can have enhanced magnetic resonance interactions with an imaging field. Such specially fabricated nanoparticles are being developed and evaluated at Johns Hopkins and Chicago [22,23].

Coatings and RF filters made from nanoengineered particles can reduce image artifacts and enhance the visibility of many biomedical devices, both implantable and interventional, that today are difficult to image due to eddy currents and other problems that interfere with MRI fields. Special coatings can also improve the ability to image guidewires and devices used in many surgical procedures.

Thin-film nanomagnetic particle coatings have been developed that can shield conductive wires and surgical instruments from radiofrequency (RF)-induced fields in MRI instruments. The high magnetic fields used for MRI, plus the RF signals, normally prevent use of conductors inside the field space. This is a problem for pacemaker wires and other devices. But specially engineered coatings based on nanoparticles have been able to shield such devices, allowing their unimpeded use with MRI.

These magnetic nanoparticle coatings can enable devices contraindicated for MRI due to safety concerns—devices such as pacemakers, defibrillators, neuro-stimulators, guidewires, endoscopes, etc.,—to be used in the MRI. Safety problems usually involve device heating and, in some cases, induced voltages that can cause very rapid heartbeats. An MRI safe pacemaker and ECG lead using an RF filter developed at Johns Hopkins has been successfully tested on pacemaker leads and licensed to Biophan Technologies Inc. for pursuit of applications and Federal Drug Administration (FDA) approvals for clinical use. Biophan also have a license from Nanoset LLC of Rochester, for medical rights to thin-film nanomagnetic particle coatings that provide a magnetic shield without electrical conductivity, and exclusive rights to a carbon composite material developed at the University of Buffalo. These developments will open the way to increased and simplified use or MRI for guiding surgical procedures.

Nanoshell particles with optical resonances in the infrared have been functionalized and used to enhance imaging of cancer cells. Metal nanoshells are composite spherical nanoparticles consisting of a dielectric core covered by a thin metallic shell, which is typically gold. By varying the relative dimensions of the core and the shell, the optical resonance of nanosize particles can be tuned from the near-UV to the mid-infrared. Work on this type of nanoshell for cancer treatment is being carried out by research groups at Rice and Arizona universities [24].

Nanoshells can destroy attached cells by absorbing infrared light at a frequency that is not absorbed by tissue. The plasmon resonance absorption heats the particles and destroys cells selectively bound to the nanoshell particle [25,26]. The nanoshells consist of a dielectric core and a gold shell, whose core–shell ratio determines their optically resonant frequency. Because these nanoparticles show intense absorption, light scattering, and emission properties in the "water window" of the near infrared (800–1300 nm), they are optimally suited for bioimaging and biosensing applications. In addition, the Rice group has developed rare earth nanoemitters, which are brightly luminescent rare earth ions incorporated into a silica nanoparticle matrix.

13.1.2 Nanoparticles for Targeting Cancer Cells

In this section, we look at a few of the many ways in which nanoparticles are being used to devise new therapies for cancer. In this rapidly advancing field, we will take some examples that show the directions and possibilities among the many new applications [27].

Ferromagnetic micro- and nanoparticles can be functionalized with antibodies, allowing cancer cells to be separated out of tissue samples such as blood and concentrated manyfold for diagnostic analysis. This is an important promising technique because cancer cells are released into the bloodstream in large numbers by microscopic tumors too small to be detectable by imaging modalities. If the circulating cancer cells can be concentrated and detected from a blood test, it would provide a means for early detection of cancer, with greatly improved prognosis for treatment versus detection after the tumor has grown to a size detectable by imaging. This technique is being developed and evaluated in a number of research centers and is being introduced into therapeutic use by more than one medical device company [28–32].

In principle, separation could be based on mass or charge rather than magnetic susceptibility, with centrifugation; or use of nanoparticles with electrically polarizable or charged functionalities with electrophoretic separations. In practice, separation by strong permanent magnets is simpler. In some cases, separation and concentration are enhanced by a combination of techniques, as in magnetic separation followed by concentration from suspensions by centrifugation. A system has been developed which uses 20-nm diameter luminescent/magnetic nanocomposite particles composed of superparamagnetic particle cores coated with CdSe/ZnS quantum dot shells, for ease of quantitative measurement of separated cells [33].

Nanomagnetic particles or ferrofluids can be used for magnetically controlled drug targeting. This technology is based on binding established anticancer drugs with ferrofluids that concentrate the drug in the area of interest (tumor site) by means of magnetic fields. Then, the drug desorbs from the ferrofluid and acts against the tumor. Nanoparticles are one option along with magnetic liquids for magnetically controlled anticancer chemotherapy [34,35]. Magnetic particle separations can also be used to separate cancer cells from bone marrow and other tissues, and for the isolation, identification, and genetic analysis of specific DNA sequences [36,37].

13.1.3 Nanoencapsulation for Drug Delivery to Tumors

In the previous section we described how nanoparticles can be coated with selective compounds to adhere to cancer cells for imaging and for delivering killing blows of energy. In a similar manner, nanoparticles can be filled with absorbed or encapsulated drugs and targeted onto cancer cells or disease agents. The most straightforward way of using nanoparticles to attack cancer is to embed them into tumors. This approach has advantages over simply circulating the drug through the body, because it enables release of the concentrated drug on the site of the tumor, with minimal effect on healthy cells [38].

Drug-laden nanoparticles can be injected into tumors with minimally invasive procedures. Their effect can be enhanced by using drugs that are further activated by radiation that can be directed onto the tumor. Drugs can be infused into the tumor which may be very potent, but difficult to deliver selectively through the circulatory system because of toxicity, insolubility, or reaction with enzymes or other compounds [39].

Nanoparticle Delivery of β-Lapachome. An example of a drug that can be enhanced by nanoparticle delivery is the promising compound *β-lapachone*, an *o*-naphthoquinone found in the bark of the South American lapacho tree.

β-Lapachone is known to induce cytotoxic effects in a wide variety of malignant human cell types including colon, lung, prostate, breast, pancreatic, ovarian, and bone cancers, as well as some blood cancers and retinoblastoma [40].

Dr. David Boothman and his colleagues at the University of Texas Southwestern Medical Center found that β-lapachone interacts with an enzyme called NQO1 (NAD(P)H:quinone oxidoreductase), which is present at high levels in certain types of solid cancer tumors. In tumors, the compound is metabolized by NQO1 and produces cell death but does not initiate apoptosis in noncancerous tissues, since they normally do not express this enzyme. In the tumor cells, β-lapachone induces a novel apoptotic pathway dependent on NQO1, which reduces β-lapachone to an unstable hydroquinone that rapidly undergoes a two-step oxidation back to the parent compound, perpetuating a self-sustaining redox cycle. A deficiency or inhibition of NQO1, such as is the case in normal cells, protects them from the effects of β-lapachone—but when β-lapachone interacts with NQO1 in the tumor cell, the cell kills itself [41].

Thus, β-lapachone has great potential for the treatment of specific cancers with elevated NQO1 levels. (e.g., breast, non-small cell lung, pancreatic, colon, and prostate cancers). Dr. Boothman's team is developing β-lapachone mono(arylimino) prodrug derivatives, specifically a derivative converted in a tumor-specific manner (i.e., in the acidic local environment of the tumor tissue), in order to reduce normal tissue toxicity while eliciting tumor-selective cell killing by NQO1 bioactivation [42].

In order to ensure delivery of β-lapachone and its derivatives to the local environment of the tumor without losing or diluting them in the body, one could simply inject them into the tumor. But experiments showed that the drug is carried away by the blood circulation relatively rapidly, before it has time to fully react with a large number of tumor cells. Dr. Boothman and his group are developing a variety of polymer implants that can be placed in the tumors to slowly release the anticancer drug in an effective manner. The implants include nanoscale polymer plugs molded from nanocells derived from a number of natural nanomaterials [43].

13.1.4 Nanoencapsulation for Penetration of the Blood–Brain Barrier

Delivery of drugs across the blood–brain barrier is another area where nanotechnology gives important new routes of access. The existence of the blood–brain barrier was discovered in the nineteenth century when Paul Ehrlich and his student Edwin Goldman found that dyes used to stain tissues would not pass between the central nervous system and the other tissues of the body. In the central nervous system, the epithelial cells lining the walls of blood vessels overlap in tight junctions, unlike those in the rest of the body. This closes off easy transport of large molecules (greater than molecular weight around 500 Da) between the blood and the brain. This helps protect the sensitive and vital central nervous system from disturbance by chemicals and pathogens (for example, viruses) that are tolerated by the more robust tissues of the body.

The tight epithelial barrier is highly lipophilic. Small lipophilic molecules can dissolve through the lipid bilayer and penetrate the barrier. Other molecules needed for the brain to function make use of specific natural transport mechanisms in the cell membranes (See chapter 14, *Introduction to Nanoscience*). Small polar molecules, such as glucose and amino acids, and larger proteins, like insulin and transferrin, are transported through the blood–brain chemical traffic by "gatekeeping" processes. Each of the required small molecules has its own transporter protein that carries it through the cell membranes—this process is called carrier-mediated transport. For proteins, specific cell membrane receptors bind the large molecules and pull them across the barrier in a mechanism called receptor-mediated transcytosis. In addition, some ionic proteins (e.g., cationic albumin) bind to and penetrate the blood–brain barrier using electrostatic interactions, in a process called absorptive-mediated transcytosis [44].

Many of the mechanisms that mediate transport across the blood–brain barrier are unknown; elucidating them is an active area of research in genomics, proteomics, and molecular biology. In the meantime, pharmaceutical research is seeking to exploit the pathways that are known. Some success has been made in modifying drugs, linking them to molecules that have transporter proteins, and thus hitching a ride across the barrier. For example, nipecotic acid, which has potential for treating Parkinson's disease, has been conjugated to ascorbic acid, which has access to ascorbate transporters, and has been delivered across the blood–brain barrier in rats, while the unconjugated nipecotic acid is barred. Other pathways have also been exploited using these "Trojan horse" and "chimeric peptide" techniques [45].

Potential drugs for treating Alzheimer's disease, Huntington's disease, stroke, and brain cancers often have molecular weights from 10,000 or 100,000 da or even greater. Many new high molecular weight peptides are being identified with potential use for central nervous system therapy. It has been estimated that up to 98% of potential drugs for the brain are not usable because of the blood–brain barrier, but this area has until recently been underdeveloped in the neurosciences. One reason is the difficulty of identifying each transport mediator path and synthesizing a chimeric or Trojan horse version of each drug to fit it.

Nanotechnology is beginning to offer a possible alternative for transport through the blood–brain barrier that is more generally applicable to a wide range of drugs. Drugs can be encapsulated in biodegradable polymers to make artificial liposomes; the coatings contain active sites to which antibodies can be attached. The antibodies are recognized by the brain-capillary receptors, which mediate their passage through the blood–brain barrier. Once inside the central nervous system, the liposomes release their contents.

These biodegradable polymeric nanoparticles, with appropriate surface modifications that can deliver drugs of interest through the blood–brain barrier, are being formulated with various physicochemical properties. Different surfactant concentrations, stabilizers, and amyloid-affinity agents are being evaluated to determine how they influence the transport mechanism.

Recently, the radiolabeled Cu^{2+} or Fe^{3+} metal chelator clioquinol, which has a high affinity for amyloid plaques which are a factor in neurodegenerative disease, has been encapsulated within small, spherical, lipophilic drug carriers capable of crossing the blood–brain barrier [46]. This and similar nanoencapsulation formulations have the potential to deliver many drugs to the central nervous system, opening new possibilities for therapy.

13.1.5 Nanoparticles and Nanoencapsulation for Insulin Delivery

One of the requirements for effectiveness of a therapeutic drug is delivery—getting the right amount in the right form to the right place. Proteins make up the nanomachinery of cells, and therefore proteins or peptides would make the ideal drug for many diseases, but proteins can be broken down and modified by many enzymes in the body. Thus, most successful pharmaceuticals are not peptides; many are small molecules, such as aspirin, which act to regulate complex biological networks. Some drugs, such as pacitoxol, are larger molecules (but still small compared to proteins), which inhibit or disrupt a precise part of the cellular machinery.

Insulin. One of the first peptides to be a successful therapeutic agent was insulin, which was discovered and applied to treat diabetes early in the twentieth century [47–51]. This represented an enormous breakthrough for one of the oldest documented diseases. The main cause of type I (insulin-dependent) diabetes mellitus is degeneration of insulin-producing cells in the islets of Langerhans, located in the pancreas. Named after the German pathologist Paul Langerhans, who discovered them in 1869, the islets are clusters of specialized cells which produce a number of hormones, including insulin. The islets contain five types of cells: alpha cells that make glucagon, which raises the level of glucose in the blood; beta cells that make insulin, which lowers blood sugar and is needed by cells to metabolize it; delta cells that make somatostatin, which inhibits the release of numerous other hormones in the body; and PP cells and D1 cells, about which little is known.

Insulin used to control diabetes has to be injected because like other peptides it is broken down into amino acids by digestive enzymes if taken orally. Soon after its discovery, researchers began to seek alternatives to subcutaneous injection with hypodermic needles. Besides the inconvenience and discomfort of injection, there are risks of improper dosage and rates of release. Insulin is released in controlled amounts by the pancreas in response to changes in blood sugar and other network stimuli, which are difficult to simulate by injections. And accidental injection of insulin directly into the bloodstream results in insulin shock, a dangerous and potentially fatal condition of hypoglycemia.

One approach to the insulin delivery problem is to synthesize or find compounds that have similar activity to insulin, but would survive modification in the digestive tract, passing into the bloodstream in an active form [52]. Other approaches seek injectable forms of insulin that have a more controlled release, and/or which are suitable for alternative forms of delivery, such as inhalation or intravenous or transcutaneous administration by micropumps. Progress has been slow on most fronts.

Nanoencapsulation of Insulin. A widely explored path is to encapsulate insulin in a protective coating that would allow its release after passing through the digestive system. A variant of this technique that has been extensively developed is to encapsulate insulin for injection into the soft tissues, in order to achieve a gradual, controlled rate of release into the bloodstream for a basal level of insulin, which can be supplemented as needed by injections. Many methods of encapsulation for injection have been tried, including liposomes, which can be administered intravenously [53].

The encapsulation of insulin and other drugs into liposomes, microcapsules, and nanocapsules is an example of practical nanoengineering to which much effort has been devoted over many years with the goal of seeking improved treatments for diabetes. A related approach is to infuse insulin into porous or absorbent polymer particles for gradual release [54]. The greatest benefits would come from a noninjected delivery mechanism, so most effort has been directed towards usable oral insulin formulations [55–59].

In a typical method for preparing nanoparticles of insulin or other peptide, the peptide is dissolved in an aqueous solution; then a nonsolvent such as a low molecular weight (C1 to C6) alcohol is stirred in with the aqueous solution. The alcohol absorbs up to 100% of its weight of water, causing the peptide to precipitate out of solution, with particles having diameters in the range of about 100–200 nm. If the mixture contains a suitable polymer, the particles are spontaneously coated as they precipitate; this process is called phase inversion nanoencapsulation.

Zinc insulin is a slowly released form of insulin that has been encapsulated in various polyester and polyanhydride nanosphere formulations using phase inversion nanoencapsulation. The encapsulated insulin maintains its biological activity and is released from the nanospheres over a span of hours. Some formulations have been shown to be active orally. These formulations typically have about 10% of the efficacy of intraperitoneally delivered zinc insulin, but they are able to control plasma glucose levels when faced with a simultaneously administered glucose challenge. The key properties that make such formulations promising for oral administration are size of dosage, release kinetics, bioadhesiveness, and ability to traverse the gastrointestinal epithelium.

As colloidal and nanofabrication techniques have advanced, more sophisticated forms of encapsulation have been developed and tested, using new types of polymer and inorganic coatings and matrices. With the advent of genetic engineering and biotechnology on a large scale, numerous bioactive peptides besides insulin are available in large quantities. Administering these substances by the oral route remains a formidable challenge due to their insufficient stability in the gastrointestinal tract and their poor absorption pattern. This has given new impetus to investigating new approaches to improve their oral bioavailability. The use of polymeric microparticles and nanoparticles is an actively pursued concept. Encapsulating or incorporating peptides in particles should at least protect these substances against degradation and, in some cases, also enhance their absorption.

Chitosan Nanoencapsulation.   Nanoencapsulation with polymeric nanoparticles is offering new drug delivery routes for therapy and diagnosis of a number of diseases. An example of one promising encapsulation starting material is chitosan [60], a derivative of the chitin polysaccharide which we encountered in chapter 13, *Introduction to Nanoscience*. Extensive research is taking place on many formulations for encapsulating insulin, including detailed models of absorption and desorption rates [61].

A typical method developed for encapsulating nanoparticles with chitosan is the polyionic coacervation fabrication process, used for protein encapsulation and subsequent release. This process has been systematically manipulated and studied in a number of laboratories in efforts to obtain predictable effectiveness [62,63]. Bovine serum albumin (BSA) is widely used as a model protein, which is encapsulated using the polyanion tripolyphosphate (TPP) as the coacervation cross-link agent to form chitosan–BSA–TPP nanoparticles.

The BSA-loaded chitosan–TPP nanoparticles are characterized for particle size, morphology, zeta potential (colloidal electrokinetics), BSA encapsulation efficiency, and subsequent release kinetics. These properties have been found to be dependent on chitosan molecular weight, chitosan concentration, BSA loading concentration, and chitosan/TPP mass ratio. Protein-loaded nanoparticles can be prepared under varying conditions in the size range of 200–580 nm, with a high positive zeta potential. An advantage of chitosan over some other encapsulation materials is that later stage particle degradation and disintegration does not yield a substantial follow-on release, as the remaining protein molecules, with adaptable 3-D conformation, seem to be tightly bound and entangled with the cationic chitosan chains.

The polyionic coacervation process for fabricating protein-loaded chitosan nanoparticles offers simple preparation conditions and a useful range for manipulation of physiochemical properties of the nanoparticles (e.g., size and surface charge). A weakness of chitosan nanoparticle encapsulation is typically with difficulties in controlling initial burst effects which can release large quantities of protein molecules [64].

Insulin Pumps. Another approach to diabetes therapy is the use of external and implantable pumps to deliver insulin in response to fluctuations in blood sugar. An early external full size pump is the artificial beta cell (patented in 1979), which regulates blood glucose concentration by continuously analyzing blood from the patient and deriving a computer output signal to drive an insulin infusion pump. Advances in electronics and integration led to wearable pumps with small reservoirs of insulin formulation, but there are many problems with any type of artificial insulin pump. The pump must react to changes in blood sugar, so it must have an accurate and timely measurement device to drive its feedback loop. Any error or delay in reading or delivery is a potentially serious problem. And completely implanting a pump leaves the problem of how to maintain the supply of insulin. A challenging problem for insulin pumps and any system implanted into the body is the buildup of plaques and bacterial infections. Following a period in which further miniaturization and integration of insulin pumps looked promising; attention is now turning to other methods. Nevertheless, there will be a role for pumps in many therapeutic and evaluation situations; research to solve the challenges is continuing on silicon micropumps and minimally invasive microneedle designs [65–67].

Nanoparticles for Inhalation Therapy. Shortly after the discovery of insulin, efforts began to investigate the possibility of delivery by inhalation. The surface area of the interior of the lungs in an adult human is roughly the size of a tennis court, so if insulin could be delivered to the capillaries and alveoli of the lung, there is high potential for direct absorption into the bloodstream, by-passing the problems of digestion and absorption in the gastrointestinal tract and avoiding the problems of injection [68,69]. The fact that no successful inhalation therapeutics for insulin have emerged since the first work is testimony to the challenges of understanding the processes by which micro- and nanoparticles are processed by the cilia and alveoli in the lungs.

Much work has been done on pulmonary diseases, airborne bacteria and virus infections, smoking, and air pollution, which provides insights that may be useful in formulating nanoparticles for inhalation therapies. Nanotechnology has much to contribute to solving lung diseases as well as finding new effective inhalation drug delivery methods. Inhalation routes to drug delivery are an important and growing area or biomedical research in many areas besides diabetes, and are likely to be an area where nanotechnology will make a large impact [70].

13.1.6 *Nanoencapsulation for Protection of Implants from the Immune System*

Another area where nanotechnology is showing promise is encapsulation of living cells for implantation. One option for treatment of diabetes is transplantation of healthy pancreas beta cells to the patient, but rejection of the foreign cells by the host immune system is a major problem. For several decades medical researchers have tried various attempts to encapsulate or shield the transplanted tissue with a barrier that would protect it from immune attack. Only in recent years, with advances in nanotechnology for fabricating nanostructured porous biocompatible materials, has this approach been brought closer to feasibility.

Nanoscale capsules are being fabricated to contain living cells. Pores in the sides of the nanoparticle cages allow small molecules such as nutrients, oxygen, and carbon dioxide to pass through, but can be sized to keep out antibodies and protect the enclosed cells from attack by macrophages. Assemblies of encapsulated cells, enclosed in silica gel, silicon, and other materials, are being used to make bioartificial organs, which are being tested for effectiveness in various types of tissue implants ranging from pancreatic beta cells to bone marrow [71].

In one of the most striking examples of the transfer of silicon-based nanoengineering to biomedical use, a group at Johns Hopkins have developed self-assembling silicon nanocubes which can be used for cell encapsulation. Living cells have been successfully enclosed in the containers as they fold from their flat silicon lithography state into closed cubes with windows in their walls for circulation [72].

Most work on transplanting encapsulated cells has been in the area of insulin-producing pancreatic beta cells. After years of having experiments that result in the implants being smothered by plaques and invaded by the immune system of the host, promising results are beginning to appear, using new nanoengineered encapsulation materials and techniques [73].

In a recent example of work in this field, researchers in Germany achieved the first successful transplantation of functioning microencapsulated islets of Langerhans. They used a novel alginate-based microencapsulation formulation to implant human islets into immunocompetent diabetic mice [74]. This and perhaps other approaches, which must be subjected to many rigorous trials in animals and humans, may soon lead to a treatment for diabetes that restores normal blood sugar regulation.

13.2 GUIDING AND STIMULATING TISSUE FUNCTION AND GROWTH

Besides encapsulating cells and tissues, nanostructures can be used to guide and stimulate the growth of cells, serving as scaffolding for growing new tissues. Tissue scaffolding, or tissue engineering, is an emerging technique in surgery and wound healing, which is being given new options and opportunities by the development of new nanomaterials and nanostructures. Tissue scaffold materials must be biocompatible; in some cases biodegradability is desirable.

Scaffolding should have good properties for cell adhesion and binding to connective tissue and, if appropriate, bone material. Depending on the application, porosity allowing cells and cell extensions to penetrate the material is needed. Major areas for application of tissue scaffolding are treatment of burns for regrowth of skin, guiding and stimulating regrowth of bone after surgery or injury, reconstruction and growth of ligaments and connective tissues, and guiding regrowth of nerves. This is a large field with many materials being developed and applied to many tissue types [75,76]. We will discuss some of the structural and surgical uses for tissue scaffolding in a later section, but first we will look at the very important special case of systems for support of nerve growth and nerve activity.

13.2.1 Nanoguides for Neural Growth and Repair

The technology of silicon microchips has been in use for several decades to make structures to guide cell growth on the microscale. The technique of growing cells on micropatterned and nanopatterned glass or silicon substrates has become important for evaluating nerve growth [77,78]. Nanotechnology is now available to fabricate more detailed and finer structures for use in sensor functionalization and cell growth patterning, by fabricating micro-channel cell growth guides with specific degrees of surface roughness and coatings nano-bioengineered to promote cell adhesion [79] (**Fig. 13.1**).

Guiding and Monitoring Nerve Growth. Micropatterned plates have been developed for use as a template for evaluating the growth of neurons in the presence of growth stimulators and inhibitors. This type of growth guide is used in evaluating the effects of different neuronal growth stimulants and inhibitors in the laboratory, with the goal of understanding how to promote regeneration of severed or damaged nerves [80,81]. Fabricated cell growth surfaces with patterns and coatings, and printing equipment to apply cells, adhesion agents, and reagents to plates is now readily available from a number of sources [82]. This type of work is relevant to treatment of spinal cord injuries and other forms of paralysis.

Technology for neural growth guidance is now being taken from the research and diagnostic laboratory into the clinical laboratory, where experiments are

FIG. 13.1	*A template for molding experimental nerve growth guides.*

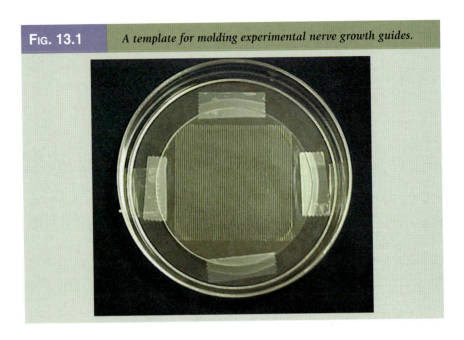

being conducted on promoting repair and growth of nerves, including the spinal cord. After injury, axonal regeneration occurs across short gaps in the peripheral nervous system, but regeneration across larger gaps remains a challenge. Cellular channels during development and after peripheral nerve injury have been shown to provide guidance cues to growing axons. In the following section we look at some examples of technologies aimed at bridging the nerve regeneration gap, which may one day be applied to restore mobility to patients with spinal nerve column damage.

Promoting Nerve Repair.   A number of research groups are obtaining promising results in animal studies using various types of micro- and nanoporous guides for nerve regeneration. In the complex biological system of nerve repair, early physiological intervention to minimize the spread of injury will always be the first line of defense in treating nerve damage. Devices and techniques based on nanotechnology may eventually contribute to such intervention.

Basic research in prevention and treatment of permanent nerve injury, especially of the spinal cord, include (1) reduction of edema and free-radical production, (2) rescue of neural tissue at risk of dying in secondary processes such as abnormally high extracellular glutamate concentrations, (3) control of inflammation, (4) rescue of neuronal/glial populations at risk of apoptosis, (5) repair of demyelination and conduction deficits, (6) promotion of neurite growth through improved extracellular environment, (7) cell growth and replacement therapies, (8) transplantation approaches, (9) gene therapy to activate expression of growth factors, (10) rehabilitation to retrain and relearn motor tasks, (11) restoration of lost function by electrical stimulation, and (12) relief of chronic pain syndromes [83,84].

Nanotechnology will impact many of these areas, but it must be integrated into the entire therapeutic regime. Possible benefits from nanotechnology capabilities will be in (1) rapid, efficient, and minimally invasive surgical repair, (2) improved automation of rehabilitation through sensors and smart materials to provide feedback, (3) improved electrical stimulation devices for prostheses and pain relief, and (4) nanoengineered microdevices to promote cell growth.

Certain cell, molecular, and bioengineering strategies for repairing the injured spinal cord are showing encouraging results (either alone or in combination) in animal models. The most promising route is application of nanoengineered nerve growth guidance matrices in combination with (1) seeding of neuronal support cells such as glial and Schwann cells, and (2) molecular coatings and growth factors embedded into the matrix material.

To understand why this nanoengineering approach is being pursued so actively, consider the status of the alternatives. Transplantation of stem cells faces many difficulties and is in an early stage of investigation. Gene therapy is another potentially promising approach. Research using cell cultures and gene-altered mice has shown that switching on just two genes can induce considerable regeneration of damaged nerve fibers in the spinal cord, suggesting that genetic therapy or drugs that activate perhaps only a handful of genes might be enough to induce regeneration of spinal cords in humans. This result was surprising but promising because of the large number of different genes known to be involved in nerve growth.

In experiments at Duke University, using cell cultures and gene-altered mice, researchers have found that switching on just two genes can induce considerable regeneration of damaged nerve fibers in the spinal cord. Inserting the genes that expressed the two nerve growth regulatory proteins GAP-43 and CAP-23 increased spinal cord regeneration by as much as 60 times in transgenic mice as compared to controls [85].

Further research has been stimulated by these discoveries, but several obstacles remain to using gene therapy for nerve regeneration. Transgenic animals used in the experiments expressed the axon growth promoter genes throughout life, whereas normally they turn off after development of the spinal cord is complete. For the process to be used therapeutically, there would have to be some means of turning the genes on after an accident (and turning off other genes that suppress neuron growth) in a safe and rapid procedure. Other questions that would have to be answered would be how long the genes need to be expressed to get an effect, and what would be the side effects (such as neural cancer) of leaving them turned on in adults, and how to turn them off at the right time. These are challenging research problems that require much new effort in molecular and cell biology, genomics, and proteomics.

In the meantime, rapid progress is being made in developing implantable devices for neural growth promotion and support, some of which may be useful in delivering future gene therapies. The size scale for fabricating nerve growth conduits is in micrometers, but features such as patterning on the inside of the growth channels, porosity, and cross-linking of polymer structures involve nanoscale materials engineering. This is a challenging and productive area with many multidisciplinary teams actively pursuing the goal of nerve repair with promising results. Here are some examples of the different materials and designs for neural conduits.

Polylactide Foams. A research team at the University of Liege in Belgium has made macroporous polylactide foams and assessed the ability of dorsal root ganglion (DRG) derived neurons to survive and adhere in vitro [86]. (Polylactides are biodegradable, aliphatic polyester thermoplastic polymers made from lactic acids in one of several chiral forms.) The foams were fabricated using a thermally induced polymer–solvent phase separation. Two types of pore structures were obtained, oriented or interconnected pores. The foams were coated with polyvinyl alcohol to improve the wettability for cell culture. Microscopic observations of the cells seeded onto the polymer foams showed that the interconnected pore networks were more favorable to cell attachment than the anisotropic ones.

The Liege group investigated the capacity of the highly oriented foams to support in vivo peripheral nerve regeneration in rats. A sciatic nerve gap of 5-mm length was bridged with a polymer implant showing macrotubes of 100-μm diameter. At 4 weeks postoperatively, the polymer implant was still present and well integrated anatomically. An abundant cell migration was observed at the outer surface of the polymer implant, but not within the macrotubes. This dense cellular microenvironment was found to be favorable for axogenesis.

Polylactide Filaments. A research project conducted jointly between the University of Texas at Arlington, the University of Texas Southwestern Medical Center, and the University of Kentucky is producing nerve guidance channels made from laminin-coated poly(L-lactide) filaments to induce directional axonal growth and to enhance the rate of axonal growth after injury [87]. Dorsal root ganglia grown on these filaments in vitro extend longitudinally oriented neurites in a manner similar to native peripheral nerves. The extent of neurite growth is significantly higher on laminin-coated filaments compared with uncoated and poly-L-lysine-coated filaments. Schwann cells were found to grow on all types of filaments, and were associated with greater neurite growth.

To improve regeneration across extended nerve defects, the team fabricated wet-spun microfilaments of different fiber densities, with the capability for drug release to support cellular migration and guide axonal growth across a lesion. In bundles that were not loaded with drug release, after 10 weeks, nerve cable formation increased significantly in the filament bundled groups when compared to empty-tube controls. At lower packing densities, the number of myelinated axons was more than twice that of controls or the highest packing density. In a consecutive experiment, PLLA bundles with lower filament-packing density were examined for nerve repair across 1.4- and 1.8-cm gaps. After 10 weeks, the number of successful regenerated nerves receiving filaments was more than twice that of controls. These initial results demonstrate that PLLA microfilaments enhance nerve repair and regeneration across large nerve defects, even in the absence of drug release. Ongoing studies are examining nerve regeneration using microfilaments designed to release neurotrophins or cyclic AMP.

Polylactide Tubules. A group at the University of Iowa has developed biodegradable conduits that provide a combination of physical, chemical, and biological cues at the cellular level to facilitate peripheral nerve regeneration [88]. The conduit consists of a porous poly(D,L-lactic acid) tubular support structure with a micropatterned inner lumen. Schwann cells were pre-seeded into the lumen to provide additional trophic support.

In evaluation experiments, tubular conduits with micropatterned inner lumens seeded with Schwann cells were compared with three types of conduits used as controls: M (conduits with micropatterned inner lumens without pre-seeded Schwann cells), NS (conduits without micropatterned inner lumens pre-seeded with Schwann cells), and N (conduits without micropatterned inner lumens, without pre-seeded Schwann cells).

The conduits were implanted in rats with 1-cm sciatic nerve transections and the regeneration and functional recovery were compared in the four different cases. The number or size of regenerated axons did not vary significantly among the different conduits. The time of recovery and the sciatic function index, however, were significantly enhanced using the MS conduits, based on qualitative observations as well as quantitative measurements using walking track analysis. This and other experiments indicated that the micropatterning and the Schwann cells provide a combination of physical, chemical, and biological guidance cues for regenerating axons at the cellular level. The patterned and seeded conduits performed significantly better than conventional biodegradable conduits.

Biosynthetic Nerve Implants. Another type of conduit for promoting nerve regeneration, the biomimetic biosynthetic nerve implant (BNI), was developed by a group based at the Texas Scottish Rite Hospital for Children [89]. The BNI is a hydrogel-based, transparent, multichannel matrix designed as a 3-D substrate for nerve repair. Polymer scaffold casting devices were designed for reproducible fabrication of grafts containing several microconduits. A number of different polymers were evaluated for making the grafts, including cellulose, hydroxymethyl cellulose, hydroxyethyl cellulose, carboxymethyl cellulose, carboxymethyl chitosan, poly-2-hydroxyethyl-meth-acrylate, poly(R-3-hydroxybutyric acid-co-(R)-3-hydroxyvaleric acid)-diol (PHB), collagen, gelatin, glycinin, both neat and as mixtures.

The grafts have been tested in vivo using a sciatic nerve animal model for repair of the adult hemitransected spinal cord. At 16 weeks postinjury of the sciatic nerve, empty tubes formed a single regenerated nerve cable. In contrast, animals that received the multiluminal BNI showed multiple nerve cables within the available microchannels, better resembling the multifascicular anatomy and ultrastructure of the normal nerve. In the injured spinal cord, the BNI loaded with genetically engineered Schwann cells were able to demonstrate survival of the grafted cells with robust axonal regeneration through the implant up to 45 days after repair.

Carbon Nanotubes Enhance Cell Adhesion Surfaces. In a related series of experiments, the Texas Scottish Rite group tested electrodeposited, photolithographic, and micromachined gold microelectrodes for nerve cell stimulation [90]. The gold microprobe interface was modified by addition of conductive polymers and carbon nanotubes. It was observed that the addition of carbon nanotubes favors the formation of nodules and increases the surface roughness. Also, electrochemical impedance spectroscopy revealed that conductive polymer composites lower the impedance of gold microelectrodes by three orders of magnitude. The carbon nanotube/polymer composite coated electrodes maintain intimate contact with axons, enabling high-quality nerve spike signals and electrical stimulation of neurons.

Carbon Nanotube Sheets as Neuron Growth Support.   In cooperation with the University of Texas Southwestern Medical Center and the University of Texas at Dallas NanoTech Institute, the Texas Scottish Rite group used sheets and yarns made from multiwalled carbon nanotubes to support the long-term growth of a variety of cell types ranging from skin fibroblasts and Schwann cells to postnatal cortical and cerebellar neurons [91]. The study found that the carbon nanotube sheets stimulate fibroblast cell migration compared to plastic and glass culture substrates, entice neuronal growth to the level of those achieved on polyornithine-coated glass, and can be used for directed cellular growth. The carbon nanotube yarns were recently developed at the NanoTech Institute [92]. These findings have positive implications for the use of this type of material in applications such as nerve growth channels, as well as for tissue engineering, wound healing, neurostimulation, and biosensors.

Biocompatibility Issues with Carbon Nanotubes.   The published studies on biocompatibility of carbon nanotube materials are contradictory [93]. A number of recent studies found that neural cells adhere to multiwall carbon nanotubes [94,95]. Studies with cardiac cells found some short-term effects attributed to physical rather than chemical interactions, but no long-term toxicity [96]. Studies on toxicity in the lung have produced some ambiguous results; these appear to be related to absorption of other nanoparticles on the carbon nanotubes [97,98]. Biocompatibility issues for nonspecific protein absorption onto single-walled carbon nanotubes can be circumvented by co-adsorption of a surfactant and poly(ethylene glycol), whereas specific binding is achieved by co-functionalization of nanotubes with biotin and protein-resistant polymers, thus making the nanotubes essentially inert to most forms of interaction with biological systems, and providing a substrate which can be used as a base for preparing protein-specific molecular recognition systems [99].

The Texas Scottish Rite studies used multiwalled carbon nanotubes produced with a minimal residual content of catalytic transition materials to obtain good cell growth. The sheets were found to stimulate fibroblast cell migration, and neuronal growth was enticed to the level achieved on polyornithine-coated glass, which is the standard used for directed cellular growth.

Natural Material Scaffolds from Agarose and Laminin.   One way to avoid issues of biocompatibility is to use a well-characterized natural material and coat it with laminin (for a description of laminin, see chapter 14, *Introduction to Nanoscience*). Researchers at the Cell and Tissue Engineering Laboratory of Case Western Reserve University made hydrogels from agarose and loaded the gel structure with laminin and nerve growth factors to create a three-dimensional scaffold for neurite growth [100]. The agarose hydrogel scaffolds were engineered to stimulate and guide neuronal process extension in three dimensions in vitro. The extracellular matrix protein laminin was covalently bound to agarose hydrogel using the bifunctional cross-linking reagent 1,1'-carbonyldiimidazole. Compared to unmodified agarose gels, laminin-modified gels significantly enhanced neurite extension in chick dorsal root ganglia cells. The Case Western team used inhibitors to study which types of receptors on the surfaces of the ganglia cells were active in the adhesion and growth process on the laminin. They also embedded nerve growth factors into the hydrogels. The resulting trophic factor gradients stimulated directional neurite extension. As a result of this and similar research, agarose hydrogel scaffolds may find application as biosynthetic 3-D bridges that promote regeneration across severed nerve gaps.

Natural Material Scaffolds with Collagen. Another natural connective material used for nanoengineered nerve growth guides is collagen (see chapter 13, *Introduction to Nanoscience*). For this purpose collagen polymer can be cross-lined chemically or with microwave radiation. Collagen polymers thus made can incorporate peptides to promote nerve growth. A number of research groups around the world have fabricated collagen tube and fiber microdevices with nanoengineered substructures and have demonstrated their ability to support nerve regeneration.

The Institute for Frontier Medical Science, at Kyoto University in Japan, made tubeless grafts with 2000 collagen filaments in each, to bridge 20-mm defects in rat sciatic nerve. Effective growth of myelinated axons was observed in the collagen filament nerve guides [101,102].

At the Bio-Organic and Neurochemistry Laboratory of the Central Leather Research Institute in Chennai, India, researchers fabricated multilayered collagen tubes by a lamellar evaporation technique and successfully used 14-mm tubules for regeneration of 10-mm nerve gaps in a rat model. Fourier transform infrared spectra of the collagen films showed that the native triple helicity of collagen was unaltered during the multilayered preparation process. Several different means of inducing cross-linking in the fibers were studied, including treatment with glutaraldehyde and microwave radiation [103,104].

Scanning electron microscopy of cross-linked tubes showed porous, fibrillar structures of collagen filaments in the matrices. Microscopic histology analysis showed that the tubule surfaces provide for good adherence and proliferation for the sprouting axons from the cut proximal nerve stumps. Among the two types of cross-linking, the microwave-irradiated collagen conduits results in ample myelinated axons compared with the GTA group, where more unmyelinated axons were observed. Solute diffusion studies on the tubes indicated that they are highly porous to a wide range of molecular sizes during regeneration.

Functional evaluations of the regenerated nerves were performed by measuring the sciatic functional index (SFI), nerve conduction velocity (NCV), and electromyography (EMG). The conduction velocity and recovery index improved significantly after 5 months, reaching the normal values in the autograft and microwave-induced cross-linked collagen groups compared to glutaraldehyde and uncross-linked collagen tubes.

Studies were conducted with nerve growth promoting peptides incorporated into the collagen matrix. Immunofluorescence studies demonstrated the staining of S100 proteins in the peripherally located cells indicating the proliferation of Schwann cells in the early days of regeneration. The staining pattern of integrin-αV was observed mostly in the perineural regions in close proximity to the peptide-incorporated collagen tubes. Evaluation of the sciatic functional index and conduction velocity at 90 days postoperatively showed regeneration of lesioned nerves with the peptide incorporated collagen implants [105,106].

Extensive evaluations were carried out for different cross-linking methods, including microwave, glutaraldehyde, di-tertiary butyl peroxide, and dimethyl suberimidate. The physical properties of collagen-based biomaterials are profoundly influenced by the method and extent of cross-linking. Cross-linking density, swelling ratio, thermomechanical properties, stress–strain characteristics, and resistance to collagenase digestion were determined to evaluate the physical properties of cross-linked matrices. The spatial orientation of amino acid side chain residues on collagen plays an important role in determining the cross-linking density and consequent physical properties of the collagen matrix. The microwave cross-linked matrices gave the best result for nerve regeneration [107].

Summary of Progress in Nerve Regeneration.   Nerve regeneration research being conducted around the world offers possible solutions for the need to sacrifice a healthy nerve to make a graft, and for the shortage of graft material available, for the repair of severed nerves [108]. This work is progressing towards clinical treatments for the repair of spinal cord injuries and cures for paralysis. More than 50 clinical trials are in progress worldwide on various treatments for spinal cord injury. Consequently, in this millennium, unlike in the last, no spinal cord injury patient will have to hear "nothing can be done" [83]. Nanoengineering of tissues will have played a significant part in making such cures possible.

13.2.2   *Neuronal Stimulation and Monitoring*

The cardiac pacemaker is one of the best-known and most widely used neuroprostheses. Since 2001, the number implanted in the United States has approached 200,000 per year. Other pacemakers are not implanted permanently but are used during cardiac catheterization and other procedures [109–112]. Other types of electronic stimulators include cardiac defibrillators, cochlear implants, bone growth stimulators, and neural stimulators for the deep brain and spinal cord, and vagus, sacral, and other nerve stimulators [113].

Nanotechnology advances these devices with improved battery technologies, biocompatible materials and surface treatments for enclosures and leads, electrode miniaturization and efficiency improvements, and smaller-sized integrated circuits for control and power, while speed and processing capabilities increase. These improvements made possible the accessible, inexpensive emergency cardiac defibrillators in public places and on transportation such as aircraft, with control systems safe to be used by nonspecialists.

These indirect benefits, like battery life and power, will not be covered in this chapter, although they are an important result of nanotechnology innovations. The design and use of each of the above types of stimulator involves specialist knowledge in cardiology, neurology, and other medical fields. In this chapter we have selected applications in nerve repair, stimulation, and neuroprosthetics as examples which illustrate challenges and opportunities for nanotechnology. Advances discussed, such as improvements in electrodes, nanoengineered biocompatible materials, stimulation methods, etc., are applicable to cardiac pacemakers and other types of implantable neuroprosthesis which use electrodes.

An area in which nanotechnology is likely to benefit cardiac pacemakers directly is in improvements in electronic leads and electrode biocompatibility and durability. The leads have been a weak link for cardiac pacemakers while size, power, and battery life have improved, thus requiring longer life service for the wires and electrodes. Cardiac pacemaker leads must be biocompatible with tissue, but unlike surgical implants, the internal portion near the heart must not promote tissue adhesion and growth which would make them difficult to remove. In addition they must heal to skin and surface tissues to seal to prevent becoming a channel for infection. They must resist the growth of bacterial and fungal biofilms. These are challenges worthy to be taken up by nanoengineering of materials. Bioengineering must also be advanced to understand long-term factors affecting cell growth and adhesion to leads.

13.2.3 Neurostimulation for Pain and Nervous Disorders

Nanofabrication is increasing the resolution and capabilities of neurostimulation devices. Neurostimulation is used medically for cardiac pacemaking, deep brain stimulation to control tremors in Parkinson's disease, management of chronic pain, stimulation of tissue healing, prevention and reversal of nerve degeneration, and other conditions and therapies, including chronic neuropathy, diabetic neuroarthropathy, and cardiomyoplasty [114].

Integrated micro- and nanoscale devices make it possible to apply much finer resolution with many more electrode stimulation points, which can be dynamically programmed. Every advance in computer and signal processing power resulting from electronics nanoengineering contributes directly to the power and sophistication of programmable medical devices such as neurostimulation systems. These advantages may appear in new generations of cochlear implants, cardiac pacemakers, deep brain stimulation, and in new types of devices [115].

An important aspect of the development of cortical prostheses is the enhancement of suitable implantable microelectrode arrays for chronic neural recording. A promising approach to this function is the use of implantable silicon-substrate micromachined probes. Work on these probes has improved their reliability and signal quality. In rodent models the probes provide high-quality spike recordings over extended periods of time lasting up to 127 days. More than 90% of the probe sites consistently record spike activity with signal-to-noise ratios sufficient for amplitudes and waveform-based discrimination. Histological analysis of the tissue surrounding the probes generally indicated the development of a stable interface sufficient for sustained electrical contact [116]. Surface treatments, new electronic circuit materials, and other advances contributed by nanotechnology will result in continual improvements in making such probes resistant to the challenging environment of implantation. In addition to the general improvements in size, power, and mobility made possible by nanotechnology, we have seen in a previous section how nanoengineering of the surfaces of electrodes, including use of carbon fiber nanotubes, is contributing to improved interfaces between neurons and electrostimulation devices [90].

Electrical stimulation of neural tissue by surgically implanted neuroelectronic devices is now an approved modern therapy. Reduction of the size and power requirements with integrated microelectronic devices makes it feasible in many cases to energize an implanted device by RF electromagnetic transmission of power, eliminating wires and batteries [117–120]. Implanted neurostimulation devices such as pacemakers are already available that receive power from RF energy, thus eliminating transcutaneous wires that are a source of infection and complications [121,122]. The question of RF interferences becomes an important design consideration for such devices (see section 13.2.6).

Improvements in energy storage through nanoengineered supercapacitors and hypercapacitors, aerocapacitors, and conductive polymers [123], coupled with lower power requirements for nanoengineered electronics, allow room for great improvements in size and capability of embedded devices, thus allowing very small implantable devices to perform electrostimulation in selected points of the nervous, sensory, and neuromuscular systems. Such devices may make it practical for increased use of implanted electrostimulation for bone and tissue grafts, and to stimulate function in the endocrine system and other organs.

13.2.4 *Neuroprosthetics*

Neural interfaces to nano- and microelectronic devices open new opportunities to design more powerful neurostimulators for prosthetics. Broadly defined, a neural prosthesis is a device implanted to restore a lost or altered neural function [124,125]. The Greek word *"prosthesis"* originally refers to the addition of a syllable to the beginning of a word. In classical medicine it means an artificial replacement for a missing part of the body [126]. The term "prosthetics" denotes the medical art of providing prostheses to improve the life of patients.

Assistive Devices.   Neuroprostheses are distinct from "assist devices" such as heart ventricular assist pumps, which do not interface to the voluntary nervous system [127]. Neuroprostheses are also distinguished from assistive devices which translate or amplify movement, for example, systems which enable paralyzed persons to control computers or devices by eye, tongue, or small muscle movement [128]. Nanotechnology is a key driver advancing the state of the art for assistive devices, and most dramatically for neuroprosthetics. As we have seen, nanotechnology is improving the electrodes that interface to nerves, and it is providing smaller and more powerful sensors, actuators, and distributed control systems to make prosthetics more natural and effective.

SCALABLE, DISTRIBUTED NETWORKS OF NANOELEMENTS

Nanotechnology will have its greatest impact on medicine with assemblies of cooperating interconnected networks of computing, communicating, and sensing nanoprocessors driving assemblies of modular interworking nanoactuators to make up a micro- or macrodevice like a powerful but subtle motor neuroprosthesis.

Types of Neural Prostheses.   Neural prostheses are of two types—motor and sensory. Sensory neuroprostheses are devices that translate external stimuli such as sound or light into signals that are interfaced to the brain either directly or via neural pathways, to restore lost or damaged perception ability. Glasses and external hearing aids are prostheses, but a sensory neuroprosthesis is an active device that delivers electrical stimulus to the nervous system, such as a cochlear implant or artificial retina.

Motor Neuroprostheses.   Motor neuroprosthetic devices take signals from the brain or motor nerve pathway and convert that information into control of an actuator device to execute the user's intentions. Motor neuroprosthetics work in one of two ways, either by (1) translating motor nerve impulses to electrical stimulation that excites or inhibits neuromuscular paths to paretic or paralyzed organs and limbs (functional electrical stimulation), or (2) picking up electricity generated by the brain or nerves and interpreting it to control prostheses or assistive devices (device control). In both cases, nerve signals can be interfaced to the neuroprosthesis by recording electrical impulses externally through the surface of the skin (myoelectric control) or through implanted electrodes.

Functional stimulation enables neural command signals to control muscle movement where native motor nerve function is paretic or impaired. Device control collects and maps nerve impulses to control electronic or electromechanical aids—actuated braces, artificial limbs or hands, synthetic speech generators, devices to allow control of bowel or bladder sphincter function, etc. Nerve signals interfaced to an active neuroprosthesis, or brain–computer interface, can be used to drive assistive technologies such as computer-based communication programs, environmental controls, and assistive robots [129,130].

For a motor neuroprosthesis to restore function, it must be integrated with the human owner's nervous system. The signals picked up by the electrodes from the nerves or brain must be translated into smooth and controlled actuation of the prosthesis [131]. This involves some combination of learning and adapting by the user and some sophisticated electromechanical control strategies to decode signals from either (1) the remaining peripheral nerves, or (2) the brain, and translate them into actions [132].

The design and implementation of a device capable of complex motor tasks—such as grasping, manipulation, and walking with smooth gait, coordinating movement with vision and balance, responding with appropriate force and velocity—all require that the prosthesis have a sophisticated distributed network of control, actuation, and feedback [133]. Limb and hand prostheses also need a high degree of fidelity to mechanical and dynamic properties of the natural limb. Otherwise the task of learning to use the device will be difficult [134,135]. Therefore to be effective, the nanotechnical design of nanoactuators and nanosensors must be fully integrated with the control systems from design to implementation, including the very special ergonomic interface between patient and device [136].

There is a high degree of overlap between prosthetics, robotics, virtual reality, design of space suits, and other human augmentation technologies [137–145]. These highly multidisciplinary fields present many opportunities for application of nanotechnologies in materials as well as electronics. Nanotechnology is already enabling more natural prostheses by providing (1) smaller (and more affordable) sensors, processor elements, and the wiring and interconnections to network them into a distributed control systems [146–151], (2) smaller, more powerful, efficient, and responsive scalable actuators whose mode of movement is natural and smooth because it is based on molecular forces, similar to those in natural muscle [152,153], and (3) engineered materials that match the strength/weight ratios, elasticity/rigidity, and mechanical energy storage characteristics of key components of natural extremities [154–157].

Nanomaterials such as conductive polymers and carbon nanotubes offer routes to nanosensors [158], and nanoscale-distributed computing elements [159], as well as to modular, lightweight, and strong artificial muscles that can be ganged in parallel to match force requirements [160–163]. Nanoscale magnetometers, accelerometers, pressure sensors, and gyroscopic devices will be able to more precisely detect even minute movements and angle changes; these will support the design of internal device rotation mechanisms that feature smooth, accurate movement as well as accurate transmittance of control and feedback information for human operation.

Nanotechnology will have its greatest impact on medicine not with tiny mobile robots, but with assemblies of cooperating interconnected networks of computing, communicating, and sensing nanoprocessors, driving assemblies of modular interworking nanoactuators to make up a micro- or macrodevice like a powerful but subtle motor neuroprosthesis. To see the current state of the art in prosthetics for limbs, and how integrated nanotechnology can be harnessed by distributed control to enable their design, consider some examples of artificial limbs.

Leg, Knee, and Foot Prostheses.   Loss of a foot or leg is one of the most common amputations, due to war, civilian encounters with land mines, accidents, and complications of diabetes. Before the 1990s artificial legs, knees, and feet were difficult to use, and required more energy for movement than normal walking [164].

Without a control and feedback system that reproduces the natural interaction of the limb with the body of the wearer and the external environment, amputees walking with a leg prosthesis consume more energy than a nonamputee at comparable walking velocities. The elastic tendons of the foot and leg store and release energy with each step, and artificial limbs have been designed to reproduce this spring mechanism, with improved results [165–167].

Designs with distributed microprocessor control in a complex prosthesis have given improved usability [168], but until recently, processors, sensors, and actuators have been too large, power-hungry, and expensive to offer a practical solution. The most challenging function to reproduce was the proprioception or kinesthetic sense that provided virtually unconscious and autonomous feedback to the body, enabling us to keep our balance through all types of movement and terrain. This requires low power, rugged, fast sensors, processors, and actuators that have only recently become available through advances in micro- and nanotechnology.

Advanced strong and lightweight materials, including smart materials with built-in sensors, are also necessary to achieve the full potential of the electromechanical control system. As these prerequisites have become available, being able to build and test prototypes has led to better understanding of how the limb interacts with its physical environment as well as with the body and the nervous system.

Laboratories and companies around the world working on the leading edge of nanotechnology and control have now produced highly advanced artificial feet, knees, and legs that are giving life-like restoration of function to amputees [169–171]. A walk through a few step cycles with a leading commercial prosthesis shows how it uses highly miniaturized but powerful accelerometers, processors, and actuators with an optimized control strategy that reproduces essential features of natural gait and posture. A prosthesis like the Proprio Foot [172], for example, might use adaptive neural network algorithms to learn from the user's movements and stride to optimize its response [173]. Such a system would use fuzzy logic feedback algorithms to ensure smooth reactions and avoid over-shooting responses [174]. Distributed parallel processing enables the system to plan the positioning of parts of the limb for the next movement in real time while executing a step. Force sensors and accelerometers provide feedback for adjustment to slope, speed, and sudden off-balance movements by making thousands of measurements per second.

Distributed and embedded artificial intelligence and electrophysical feedback control the autonomous functions involved in walking that we do not have to think about—like keeping the leg positioned relative to the center of gravity of the body, rotating the ankle, and controlling the angle of the foot to the ground, with compensation for forward speed, slope, and other factors. At the same time, the prosthesis interfaces to the user's nervous system to respond to motor commands and execute voluntary movements. A requirement for success of the system is that it should interface, interact, and adapt to the forces and patterns of movement from the physical environment and the rest of the body, as well as to the nerve impulses, just like a natural limb would do. Thus we see that our definition of a motor neuroprosthesis was somewhat limited: it has to interact appropriately with its entire environment, not just with the nervous system.

TWO IMPORTANT CONCEPTS FOR DESIGN OF MOTOR PROSTHESES

Proprioception. The sum of all the tactile information constantly being fed to your brain that tells you where you are and what you're doing.

Kinesthesia. The ability of the brain and body to precisely and reproducibly know where limbs and fingers, etc., are positioned, how fast they are moving, and how much force is being applied. Kinesthetic memory enables musicians and athletes to achieve high performance.

Hand Prostheses.   The hand has one of the highest densities of motor nerves and sensory nerve endings of any part of the human body, serving its highly developed tactile and sensory capabilities with 22 degrees of freedom in its four fingers and thumb. Prostheses to replace a lost hand are one of the oldest forms of prosthetics: the surgeon Ambroise Paré designed an anthropomorphic hand for wounded soldiers in the sixteenth century. Modern motor neuroprosthetics provide considerable functionality, thanks to advances in microelectronics, microactuators, and robotics [175].

Like the lower extremity, the hand interacts with its environment as well as with the nervous system, but the hand is much more intimately and intricately connected with the brain in the performance of voluntary planned and executed tasks. A number of prosthetic hands have been developed with embedded controls [176–179]. Much has been learned about hand movement from development of robotic hands for industry and space, as well as prosthetic use [180–182]. As in the case of the artificial foot, the most recent and advanced designs use distributed control and sensors with local feedback rather than requiring the user to decide and communicate how much force is exerted by each finger at a given time [183].

Artificial hands with advanced capabilities can be used for robotics or human augmentation aids as well as prosthetics [184–186]. Devices for robotics and prosthetics have been developed with 20 degrees of freedom, using shape memory alloy actuators [187,188]. Some advanced designs implement the complex degrees of freedom of the fingers and thumb and employ sensitive microphones to detect vibration when a grasped object is slipping [189–191].

Ongoing progress towards more natural artificial hands is making use of MEMS accelerometers, as well as force and pressure sensors (and hence more nanotechnology). Designs have been proposed for smart artificial skin with embedded nanosensors, which will give more options for designers of prostheses.

Prosthetics, like other medicine, is human oriented, seeking to adapt assistive devices to the unique needs of each patient, within inevitable constraints of cost, technical feasibility, and rehabilitative learning [192]. With the advance of nanotechnology making devices adaptable as well as affordable, this ideal becomes more realizable. The range of options for fitting a pirate hook was much narrower than with a modern nanosensor and processor-enhanced prosthesis.

For example, when a patient has some functioning tendons leading to a missing hand, these can be harnessed to give a more natural interface to the prosthesis. The loss of motor function suffered by many patients is not complete. For these patients who still possess residual functions, modular, more naturally controllable systems for supporting these functions are needed rather than complex systems to replace them [193].

Neuromotor devices that are one-way, sending signals collected from the brain side of the gap to the muscles, leave the patient relying primarily on visual feedback for control (or auditory feedback in the case of voice prostheses) [194]. Such control can be learned with therapy [195,196]. Indeed, the brain is very plastic; patients are able to learn to activate and use prostheses via a number of different neural pathways, regardless of the nature of the function [197].

An ideal prosthetic hand would replicate sensory–motor capabilities of the amputated limb. The special feedback that the body, and especially the hand, provides for such tasks is referred to as the haptic sense. Haptics involves primarily the sense of touch and perception of resistance to applied force, with some elements of kinesthetics. Haptics is important in the skilled performance of dexterous tasks such as surgery.

A number of research centers are working on biomechatronic hands which aim to approach natural dexterity, speed, and strength with normal weight and appearance. A challenge is to minimize power requirements, operate noiselessly, and be resistant to water, oils, food, and be easily cleanable. The goal is to have an intelligent, adaptable self-programming control system that makes it easy for the user to develop skill. Some current designs have excellent dexterity, but are lacking in some of the other requirements.

This is a very active and exciting area for application of new nanotechnology. In the case of actuators, miniature electronic motors or pneumatic muscles have yet to be replaced by nanoengineered artificial muscles in an integrated design for an artificial hand. Lifelike, durable, soil-resistant skin coverings have yet to be developed and evaluated (perhaps using some of the soil- and water-shedding features learned from the lotus petal) (see chapter 12, *Introduction to Nanoscience*).

NANOTECHNOLOGY IS BRINGING THE BIONIC MAN CLOSER TO REALITY—THE BRAIN–MACHINE INTERFACE

"Abstract systems that allow a brain to control a computer are inching ever closer to reality—but their most important applications may be different from those envisaged by science fiction."

Is this the bionic man? *Nature* (2006) [198]

The Brain–Machine Interface. The control of physical objects by the power of thought alone has long captured the human imagination. Using our thoughts to control a computer or robot used to be the realm of science fiction writers. Now with the aid of new technology and years of study of neural activity in the brain, the control of machines and computers by the brain is becoming a reality. Systems are being made in which patterns of neuronal firing in the brain can be translated into electronic controls to support communication, mobility, and independence for paralyzed people [198–200]. Nanoengineered electronic and magnetic detection devices are helping brain–machine interfaces reach a level of speed and responsiveness that will make a brain interface a usable prosthesis that does not require long and difficult concentration and training for the user [201].

Motor and Sensory Interfaces. To fully interface with the brain, a neuroprosthesis must not only receive signals from the brain but must also return sensory information for feedback. This feedback is visual, auditory, kinesthetic, and haptic. In later sections we will discuss sensory neuroprostheses, but first we will examine aspects of motor control.

Promising Breakthroughs in Brain–Machine Prostheses. Many research groups have been working for decades to integrate sensors, computers, and knowledge of the patterns of nerve firing with which the brain controls movement, in order to help people who have brain or spinal cord damage to communicate and interact with the outside world. The goal is to use prostheses to replace or restore lost motor functions in paralyzed humans by routing movement-related signals from the brain, around damaged parts of the nervous system, to external effectors.

Much remains to be done before neuroprostheses that can respond to the brain become a clinical reality, but in experiments, paralyzed patients with electrodes implanted in the part of the brain that controls movement, the cerebral motor cortex, have been able to control computers and televisions, open e-mail, and move objects using a robotic arm [202]. Although control using the system was rather slow, it did not require complete focused attention to operate: the patient was able to engage in conversation while doing tasks with the neuroprosthesis. Experiments using implants in monkeys that were not paralyzed have demonstrated the potential for faster response times, with techniques to speed up the brain–machine interface [203].

Challenges to Be Overcome. It is possible to control prosthetic devices by extracellular recordings from the cortical neurons in the brain, where movement control originates [204,205]. Extensive preclinical experimentation with implanted neurosensor electrodes is laying the foundation for design of such systems; the pattern of neuron firing has to be at least approximated in order to map signals onto movement in the prosthesis. The development of smaller and faster neuro-electrodes is important for future success in patients.

To make a brain–machine interface requires the estimation of a mapping from neural spike trains collected in motor cortex areas onto the kinematics of a normal limb—but fortunately, this mapping does not have to be absolutely accurate—otherwise the task would be impossible [206]. Imagine that someone ripped the dashboard and steering column out of a car, or the cockpit out of an airplane, and handed it to you and asked you to reconnect all of the wires and controls back together so that it worked again. That would be an extremely simple task compared to mapping the connections between the brain and muscles after a spinal cord injury. That a brain–machine neuroprostheses can work is as much a tribute to the adaptive and regenerative power of the brain and nervous system as it is to the years of patient experimentation and intricate deciphering by neuroscientists. If the brain is presented with an interface that is at all workable and predictable, the cortex can adapt in most cases to learn how to associate patterns with movement [207].

There are purely clinical factors and obstacles for success of a brain neuro-prosthesis. Firing patterns and maps from brain to motor effector neurons must be measured in intact animals and human subjects, to establish a baseline for translation from brain to machine. Movement signals in the motor cortex of the brain may not persist after nerve paths to a limb are cut off. ("Use it or lose it" holds true in the brain.)

But given the clinical obstacles, early experiments give indications that workable prostheses may be possible. There are nanoengineering challenges to be solved in both materials and control system architectures. The systems for recognizing and interpreting brain signals to movement must be robust and adaptable enough to deal with changes in pattern presented after injury, during learning, and due to individual differences. Current experimental systems typically sample ensembles of 100–200 cortical neurons. Advances in nanotechnology will allow denser and larger arrays to be implemented. The neural interfaces must be compatible with nerve cells and their environment, so that the transmission of signals does not fade over time due to buildup of plaques on the electrodes or withering away of the neurons in contact with the surfaces. In the opinion of many scientists, "most of these difficulties are now engineering challenges, rather than problems of principle" [198].

Design of Control Algorithms.   Much work by neuroscientists has been done to tell us what part of the cortex controls which limbs, and how control spreads out down the neural networks [208,209]. If a control system could simply be "hardwired" to the brain to talk to a prosthesis, the paradigm would be to interface electrodes to the appropriate part of the brain's motor cortex or motor pathway, and analyze the relationship between the cortical activity and measured arm or hand movements; this relationship would then map cortical activity to similar prosthetic arm movements. However, the pathway to the brain is not so simple, and there is the clinical problem that measured limb movements are not feasible for amputees or patients with physical mobility limitations.

The firing patterns of the nerves controlling natural neuromuscular systems are very complex, and are modulated by chemical neurotransmitters and inhibitors, so a purely electronic interface can never be completely natural. The motor signals spread out, and sensory feedback returns, over a network of dendritic paths in patterns that can be mapped experimentally to the musculature and sensory surfaces of the body [210,211]. The areas of the cortex that control particular motor movements set up predictable patterns prior to initiating motor commands [212]. For a neuroprosthesis we merely want to provide the brain a tool to work with, rather than try to duplicate its operation.

We know from neuroscience that, to a large degree, the brain is adaptable and can learn new control tasks by response to feedback presented to it. If we can present the cortex cells with a coding system of responses that activate the prostheses, then the brain can learn to exercise control [213–215]. Hopefully, this learning can be automatic, implemented by feedback loops within the cortex that are similar to natural control mechanisms for the original limb. Otherwise control of the prosthesis will require extra effort and conscious attention to execute. Thus getting the control coding right is as important as the weight, power, and agility of the robotics in the prosthesis. Fortunately, progress is being made in designing workable control schemes that enable the brain to successfully map intent onto movements of the prostheses. And the brain has shown itself to be remarkably adaptable in learning to control devices by neural stimulation.

Adaptive Coding and Training.   Recent research has shown that adaptive control strategies shorten the learning time and increase the effectiveness of brain–machine interfaces. A number of coding and training methods have been used to interface such neuroprostheses [216]. These coding techniques compute statistical scores to match patterns in the cortex with movements of the limb or prosthesis. Adaptive methods can automatically improve their performance with practice, by iterative computation that optimizes the correlation between nerve firing patterns and movements with practice.

Systematic investigation of several linear (Wiener filter, LMS adaptive filters, gamma filter, and subspace Wiener filters) and nonlinear models (time-delay neural network and local linear switching models) are applied to experiments in monkeys performing motor tasks (reaching for food and target hitting), in order to map cortical function onto motor movements [217]. Other correlation models that have been used include Bayesian graphical networks, Bayesian quadratic, Fisher linear and hidden Markov model classifiers [218] as well as neural networks, Kalman filters [219], and Kalman filters with smoothing [220], and other statistical component analysis techniques [221,222]. The timing and synchronization of firing between groups of neurons have been analyzed as a means of predicting cortex output based on coincidence and synchronization of firings [223].

Distributed Control Networks of Nanocomputers.   Because the brain is more universal than any specialized branch of the neural pathways connecting it to the extremities, and is the starting point for voluntary control signals, one might assume that the neuroprosthetic interface could be controlled largely and directly by the brain, by-passing the functions of the spinal cord, ganglia, motor nerves, etc. But distribution of function, especially motor functions, serves to prevent overload of the brain; much preprocessing of stimuli takes place along the pathways.

In the case of the leg, a great many of the control feedback loops are automatic, and can be implemented entirely within the limb. But for the hand or for speech, a complex mixture of voluntary initiation and automatic cascades of control must take place, with the option for voluntary control to override the automatic function at any point playing a much greater role. To match the learning and adaptive power of neural networks that control complex tasks like manipulation or speech, the control systems must be adaptive [224].

Analysis of such control systems reveals that they are best implemented with distributed, communicating networks of processing elements [225–227]. Thus a network of nanofabricated small processors with communication links, sensors, and actuators will be an optimal platform for implementing agile and responsive control systems for prosthetics (as well as robotics) which can adapt to external conditions and to the stimuli from the user's brain. Such a network also has the advantage of requiring less energy for the same computation capability of one or two fast processors for controlling prostheses.

Noninvasive Interfaces for Brain–Machine Interfaces. Noninvasive or minimally invasive methods of communicating between brain and prosthesis would be highly desirable—eliminating surgery, electrodes, and wires. Some obvious minimally invasive techniques, such as using finger, toe, or eye movements, have the disadvantage of requiring a high degree of the patient's attention, even if muscle movement is possible [228]. One way to avoid the disadvantages of wires through the skull and skin would be to communicate with an implanted electrode via wireless RF signals—but this raises many technical problems and still requires an implant.

In principle, it is also possible to control computer cursors through noninvasive electrodes monitoring the brain from outside the head. The electrical potential of neurons collectively firing is detectable on the scalp. Its recording is termed the *electroencephalogram* (EEG), whereas the pattern of signals measured by electrodes on the surface of the cerebral cortex is called the *electrocortiogram* (ECoG). The EEG shows an average of many neurons firing in a broad region of cortex, filtered through the skull and scalp, but a number of patterns can be detected nevertheless. Considerations for control programming are similar to those for ECoG, but mapping is more arbitrary [229,230]. The lower interface resolution of the EEG makes it difficult to provide a wide range of subtle and distinct movement controls, and requires long training periods. EEG-based systems have generally been too slow for controlling rapid and complex sequences of movements, so this type of interface remains a less desirable option.

Some direct comparison studies have been conducted between EEG and ECoG, in order to measure the relative information-processing capacity that can be achieved using brain–computer interfaces with the two interface modes. Methods of translating brain computer interface data between noninvasive EEG and invasive ECoG are sought with the goal of further improving EEG control [231,232].

Researchers have used asynchronous EEG analysis and machine learning techniques to implement brain control of advanced robots with impressive results. Using an EEG-based brain–machine interface that recognized three mental states, they achieved mental control comparable to manual control on the same task, with a performance ratio of 0.74 [233].

EEG readings have been combined with other physiological measurements in multimodal monitoring to give an enhanced picture of the status and activity in the brain. Other data integrated with EEG include hemodynamics, functional MRI, and others [234].

Magnetic Neural Stimulation and Monitoring.   Advances in micro- and nano-technology are on the cusp of giving an alternative to low-performance EEG interfaces and invasive electrode implants. With advances in materials and electronic nanosensors, it is becoming feasible to use magnetic fields to stimulate and monitor the brain. Magnetic stimulation is noninvasive; focused magnetic fields can stimulate nerves deep within the body and brain without implanted electrodes or shocks through the skin [235].

Basic Principles.   The firings of neurons and the travel of ion currents along axon membranes generate magnetic fields. A steady current induces a static magnetic field, and any change in current creates a change in the magnetic field. Thus, magnetism is a second-order effect of the movement of an electrical charge: the magnetic field is proportional to the velocity of the charge and the change in magnetic field is proportional to the acceleration or deceleration of the charge. This indirectness makes magnetic fields more difficult to use, but it also creates the possibility of noninvasive communication with nerves, without implanted electrodes or painful transcutaneous shocks.

High-intensity changes or pulses in strong magnetic fields can induce electrical fields in the nervous system, which can exceed the firing threshold for neurons (*the action potential*). The induced electrical currents are proportional to the rate of change of magnetic field (dB/dt). Thus magnetic stimulation can induce electrical currents in the neuron cell membranes like those induced by implanted electrodes, but without physical contact [236] (**Fig. 13.2**).

| **FIG. 13.2** | *The brain can be stimulated noninvasively by magnetic fields. Nanosensors make it possible to detect magnetic fields produced by nerve activity.* |

Magnetic Stimulation.   The induction of nerve firing by application of strong focused magnetic fields is a new medical technique used to stimulate motor nerves in the limbs (*functional stimulation*) or neurons in the brain (*transcranial magnetic stimulation*). Magnetic stimulation requires strong magnetic fields, which must vary or pulse in order to generate an electric field—generation of an electrical field requires movement of an electrical charge relative to a magnetic field. Electrical stimulation of nerves could be obtained with a simple voltaic jar or electrostatic spark, but magnetic stimulation requires equipment that can generate short, intense, pulsed and focused magnetic fields of very high strength, about 0.5 T at the surface of the cortex, which typically requires a 2 T magnetic coil outside of the body.

To localize the stimulation, arrays of magnetic coils are positioned outside of the head to focus the combined fields inside the brain. Relatively low-frequency magnetic fields can be used (typically a few kilohertz) which are not absorbed by the skull or spinocerebral fluid. The electromagnets are controlled so that the strength of the magnetic pulse exceeds the excitation threshold only at the focal point. The resolution that can be achieved is less than with electrodes, but the noninvasive advantage is spurring efforts to increase the precision of the magnetic focus [237].

Development of Medical Applications.   Electric and magnetic fields are complementary: electrical currents generate magnetic fields, and changes in magnetic fields induce electrical fields. The relationship between electricity and magnetism was demonstrated experimentally by Faraday in 1831 and explained theoretically by Maxwell in the latter half of the nineteenth century. Applications of the phenomena were developed by Edison, Marconi, and many others since, forming the basis of our electronic economy.

Electrical stimulation of nerves was known since the days of Galvin and Volta, but the idea of magnetic stimulation had to wait for Faraday and Maxwell to demonstrate the relationship between electricity and magnetism. There were a number of early attempts to stimulate the brain and motor nerves using magnetic fields, but the technical challenges are formidable for either stimulation or monitoring with magnetism.

The first successful development of technology for magnetic stimulation was conducted at the University of Sheffield in England starting in 1976. In 1982 the Sheffield team demonstrated supermaximal stimulation (simultaneous firing of all of the motor nerves in a nerve bundle) followed in 1985 by transcranial stimulation [238]. The repetitive stimulator (rTMS), which can generate up to 30 pulses per second, became available in the 1990s. Since then the field has expanded rapidly, with several companies producing clinical equipment and obtaining regulatory approvals for experimental and some clinical uses. One use of magnetic stimulation is to temporarily shut down portions of the neural network. This allows connections to be mapped, and is being studied for treating conditions such as Parkinson's disease [239,240].

Magnetic Monitoring.   The magnetic fields produced by nerve currents are weak but can be detected with sensitive magnetic field detectors (*magnetometers*) [241–243]. Extremely sensitive magnetometers are needed to detect the fields produced within the brain. Fortunately, a great deal of development has been invested toward high-performance magnetic sensors for many applications such as computer disk drives, oil exploration, and security detection. The current state-of-the-art *magnetoencephalography* (MEG) can map brain activity on a one millimeter grid or less. With powerful signal processing and statistical analysis, MEG images can be co-registered with MRI scans with good accuracy. Because magnetic fields are induced perpendicular to the direction of current flow, MEG gives orthogonal information to EEG in terms of the types of neural tissue and direction of nerve impulses that are revealed.

The first generation of MEG equipment was typically bulky, requiring shielded rooms, high power consumption, cryogenic cooling of detectors, and significant processing times to deconvolute data from relatively few sensors. Thus MEG has until recently been limited to research and highly specialized diagnostic applications for life-threatening conditions [244]. This is a rapidly developing area which is being accelerated by applying nanofabrication to existing types of magnetic sensors and entirely new concepts made possible by nanotechnology. The results are 1000-fold improvements in sensitivity and reductions in size and power requirements by factors of 10 to 100.

Devices for Magnetic Stimulation and Monitoring. The improvements in performance necessary to make noninvasive magnetic communication with the brain a practical reality are already being delivered by nanotechnology. For stimulation, the impacts of nanotechnology are largely indirect; nanofabrication of interstitial compounds and alloys is producing better high-temperature superconducting materials to reduce the size and cryogenic environmental constraints for high-performance superconducting magnets, and nanoparticle thin films are being used to fabricate magnetic shielding materials.

Historically three classes of magnetic sensors have been developed: mechanical, electronic, and quantum. Researchers are re-examining magnetic sensing to find opportunities for enhancement based on phenomena that appear at the nanoscale, with very small masses and volumes. Many older types of magnetic measurement devices can be sub-miniaturized with nanofabrication, but nanotechnology is also making new designs possible based on previously inaccessible physical phenomena. Both types of development are producing concrete results, and applications are expanding.

Mechanical magnetic sensors include geometric magnetometers, where the sensor is moved or deformed by interaction with the magnetic field, and resonance sensors, whose vibration rate is influenced by field forces [245]. Electronic sensors include Hall effect sensors [246], which measure the resistance to flow of electrons caused by their deflection in a magnetic field; and magnetoresistive, giant magnetoresistive, and colossal magnetoresistive sensors, based on thin film conduction effects (the 2007 Nobel Prize in Physics was awarded to the discoverers of giant magnetoresistance, Albert Fert and Peter Grünberg) [247–250], and flux-gate devices, which compare the difference in current required to magnetize a coil in two directions. Some sensor designs utilize more than one physical effect in the same device for enhanced performance. Quantum sensors include the *superconducting quantum interference device* (SQUID), based on *Josephson junction* currents—the magnetically sensitive tunneling of electrons through a thin insulating barrier separating two superconductors [251].

The SQUID is the highest sensitivity magnetometer commercially available. Magnetic scanning systems approved for mapping neural activity are based on SQUID sensors. Although the first generation was bulky, it has been used successfully for brain and cardiac imaging. A second generation design has been optimized with highly sophisticated software and good engineering to increase resolution and reduce the size and weight of the cooling and shielding systems. Applications include diagnostic imaging for neonatal brain assessment, liver susceptometry, and gastric ischemia, and for difficult to diagnose and serious conditions [252,253].

Advances in Magnetic Sensor Design.   A number of new nanoscale magnetometer designs are being developed which approach or exceed the sensitivity of SQUID, without cryogenic cooling, and with less power consumption, lower cost, and smaller size. One, the *optical atomic magnetometer*, developed by the U.S. National Institute of Standards and Technology with the University of Colorado and Sandia Laboratories, is based on the interaction of laser light with atoms oriented in a magnetic field in the gas phase [254]. Workers at Princeton University and at the University of California, Berkeley, and elsewhere are also developing optical atomic magnetometers, and improvements in performance continue to be published [255,256]. Another promising new magnetometer is a nanoscale cantilever design developed at Lucent Technologies' Bell Labs [257]. These and other designs may open new possibilities for magnetic medical imaging.

Optical Atomic Magnetometers.   The NIST optical atomic magnetometer measures the change in alignment when atoms with a magnetic spin moment interact with a beam of laser light. In the absence of an external magnetic field, the atoms will align with the laser beam's crossed electric and magnetic fields. Any perturbation by a magnetic field will disorient the alignment with the beam, reducing the amount of light transmitted through the gas. Magnetic shielding is used to make the detector selective and directional. The fabrication of a cell containing the gas, a small solid-state laser, and a detector for the transmitted light can be scaled down to microchip form, with nanoscale geometries, to make an extremely small, sensitive, and economical sensor element.

At NIST a prototype, millimeter-scale microfabricated rubidium vapor cell with a low-power laser, was able to detect the heartbeat of a rat. In Berkeley researchers used the atomic magnetometer to detect magnetic particles flowing through water. Princeton researchers using a high-sensitivity atomic magnetometer based on potassium vapor performed MEG experiments [258]. Physicists at the University of Wisconsin and elsewhere are refining optical atomic magnetometer designs to reduce noise for biomedical applications [259].

The millimeter-scale prototype at NIST contains about 100 billion atoms of rubidium gas in a vial the size of a grain of rice. The change in spin alignment was easily detectable, and scalable down to much smaller sizes. The atomic magnetometer is about 1000 times more sensitive than previous devices of a similar size. With sensitivity below 70 femtotesla (fT) per root Hertz, it is comparable to, or even exceeds SQUID sensors. It can be made much smaller than a SQUID, and operates at much higher temperatures, at around 150°C.

Currently the complete NIST device is a few millimeters on each side. Developers predict that with the small size and high performance such sensors could lead to magnetocardiograms that provide similar information to an electrocardiogram (ECG), without requiring electrodes on the patient's body, even from outside clothing. The current versions of the atomic detector are sensitive enough to detect alpha waves from the human brain, which produce magnetic fields of about 1000 fT just outside the skull.

To pick up the full range of magnetic fields emanating from the human head, the atomic optical devices would need to be more sensitive—down to 10 fT or less, which is projected to be feasible. The thermal magnetic noise level generated by the human brain is on the order of 0.1 fT. A sensitivity of 0.2 fT is projected for the Princeton potassium-based atomic magnetometer, if supercooled shielding were used to reduce the noise level at the detector. This would enable imaging of individual cortical modules in the brain, which have a size of 0.1–0.2 mm. This could provide an alternative to MRI and PET imaging, without injection of contrast enhancement agents or tracers.

Previous atomic magnetometers and SQUIDs are larger and require much more power than the gas laser design. Even with the laser and heating components, the new devices use relatively low power and can be extremely small. Thus, they could be used in high-resolution arrays of distributed sensors. The small size allows the sensor to get close to the heart or brain for magnetic measurements. Developers project that the sensors could even be used to make portable MEG helmets for brain–machine interfaces. They could also be used to identify markers for specific chemicals by measuring nuclear quadrupole resonance of excited atoms, opening further possibilities for monitoring and research.

Nanoscale Electromechanical Resonator Magnetometers. Mechanical magnetometers can be made using nanoscale cantilevers or bridges, coated or implanted with magnetic materials to harness the sensitivity of nanoscale resonance vibrations. Fundamental breakthroughs in nanotechnology made in the past few years by Bell Labs and the New Jersey Nanotechnology Center (NJNC) have led to a new nanomechanical magnetometer design with performance that is potentially 100- to 1,000-times greater than existing commercial devices, at extremely low cost based on silicon lithographic fabrication [257].

The new Bell Labs MEMS magnetometers employ a silicon resonator carrying an electric circuit. Oscillation of the resonator in a magnetic field generates a current around a closed loop circuit damped by a resistor. Variations in magnetic field strength alter the amplitude and frequency of the resonator. This mechanical sensitivity can be used to measure the magnetic field by coupling the mechanical motion of a silicon bar or paddle to the ambient magnetic field.

In order to give a sensitive measure of a magnetic field, this nonlinear resonator must have negligible internal damping—a high Q-factor. Nanoscale crystalline oscillators made from quartz or silicon can be made with much higher Q numbers than all but superconducting electronic oscillators.

Magnetometers that use electronic detection (Hall, magnetoresistance, or flux-gate devices) have sensitivity limited by their electronic Q-factor, which depends on the resistance to electrons traveling through the metal in the circuit;

it is difficult to reduce this factor (and increase sensitivity) without resorting to superconducting materials (which is why SQUIDs remain the ultimate purely electronic detector). A tuning fork resonator made from single-crystal silicon (with less internal friction than that of the hardest metal) will vibrate almost a thousand times longer than the best room temperature electronic oscillators.

Researchers are working to optimize resonator magnetometer designs to achieve substantial improvements in sensitivity by modifying the microscale geometries with nanoscale features. Nanoscale mechanical resonators with mechanical Q-factors approaching 10,000 or more at room temperature can be made from semiconductor-grade silicon and similar single-crystal materials. This is a huge improvement over electronic detectors, without cryogenic cooling for superconductivity. Electromechanical resonator magnetometers should be up to 100- to 1,000-times more sensitive than existing commercial devices.

In the meantime, improved designs using nanotechnology continue to be applied to optical atomic magnetometers, as well as to devices based on the Hall effect, magnetoresistance, and SQUID [260–263]. Advances in signal processing are being applied to provide capacity to extract, analyze, and efficiently present magnetic sensing data for medical use [264].

A challenge in the design and application of MEG nanosensors is that the forces measured and signals generated by nanodevices can be many orders of magnitude smaller than the environmental magnetic noise. This is a general problem for all nanosensors, so it is illustrative to see how this significant metrological challenge is solved in the case of magnetic brain sensing. The environmental noise can be attenuated by a combination of shielding, primary sensor geometry, and synthetic analytical methods (signal processing). One of the most successful applications of computer power for noise cancellation is the use of synthetic higher-order gradiometers [265].

The MEG signals measured on the scalp surface must be interpreted and converted into information about the distribution of currents within the brain. This task is complicated by the fact that such inversion is nonunique. Additional mathematical simplifications, constraints, or assumptions must be employed to obtain useful source images, along with sophisticated signal extraction algorithms [266]. Beam-forming techniques such as synthetic aperture magnetometry beamformers can also reduce extraneous signals and focus the detector into the body [267].

Future Opportunities in Medical Magnetic Sensing. Imaging systems based on the newest sensors have not yet been built, but they will undoubtedly bring wider use of MEG, perhaps taking the portable noninvasive, efficient brain–machine interface closer to reality. In the meantime, other techniques for brain stimulation are being explored, such as stimulation of neural tissue by light [268,269], and vibrotactile or acoustic stimulation [270]. And the possibilities of feedback with MEG and other modalities for brain–machine interfaces are being explored [271]. So there are entirely new paths that the development of neuroprostheses could take. Whatever they are, nanotechnology will play a key role in implementing them (**Fig. 13.3**).

Sensory Neuroprosthetics: Haptics. Haptics refers to the sense of touch, and more generally to the sense of pressure and force feedback from the body to the

FIG. 13.3   *Nerve cells can be stimulated by light, even if they are not photo-receptors, if the light pulse depolarizes the membrane potential (experimental setup for cells in culture).*

brain [272]. This brings us to the area of sensory neuroprosthetics, of which haptic sensors are a special case, since they are usually an integral part of motor prostheses. We discussed the importance of haptics and kinesthetics in the sections above on control and feedback of motor prostheses, and for brain–machine interfaces in general.

Haptics, along with vision, hearing, and balance, is an essential component of the stream of feedback that the nervous system sends to be brain [273]. A deficit in the haptic sense is usually the result of loss of the sensory nerve endings due to injury or amputation, but in rare and more difficult cases it can be due to brain injury. Haptics is a diffuse sense, so there is no one prosthetic device that satisfies the brain's requirement. It has been less well understood than more obvious senses, even taste and smell, which are more localized and less intuitive.

Because tactile feedback is so important to the fine motor control necessary for dexterous and delicate tasks, as well as athletic performance, it is becoming an important research area. Nanotechnology-based haptic sensors are having an impact on robotics, design of spacesuits, and in medicine—not only for neuroprosthetics, but in the design of robotics-assisted surgery systems [274,275]. It is extremely important in design of prosthetic strategies and in rehabilitation. Surgical connection of sensory as well as motor nerves to a prosthesis can give dramatically improved results [276,277]. So-called smart materials and embedded nanosensors will provide many options for implementing haptics in medical devices and surgical tools, and will assist in improving the quality of telemedicine [124,129,147].

Cognitive prosthetics.   The concept of a *cognitive prosthesis*, a system developed to support and augment the cognitive abilities of its user, has only been made possible recently with the advances in computing and interface technologies [278]. To the extent that cognitive prosthetics includes augmentative and alternative communication for people with impaired communication, and virtual reality systems, there is not a sharp dividing line between the term and the functions provided by other sensory–motor augmentation modes such as implants to relieve seizures, hearing prostheses, and brain–computer interfaces in general [124,279]. However the concept does raise the possibilities of powerful augmentation of human capabilities, as well as treatment for mental deficits, both small and profound [280]. The idea raises many psychological, social, ethical, and medical concerns, as well as research issues.

Medical, Social, and Ethical Issues.   Typically neuroprosthetics are resorted to only after pharmacologic and neurosurgical options have been exhausted. Bioengineers and other designers of systems to augment human capabilities are cognizant that their role should be to assist the body's adaptation and compensation for a deficit, rather than replace any remaining function. Systems that surpass natural capabilities can be intimately interfaced to the human body, in "bionic human" or cyborg scenarios [281,282]. Nanotechnology is making such capabilities more feasible and affordable, obliging us to confront the social, medical, and ethical consequences [283].

Future Directions for Brain–Machine Interfaces.   The progress that has been made is remarkable, but many obstacles must still be overcome. Without invasive implants, current experimental brain–machine prostheses require the patient to be tethered to bulky equipment, which needs tuning and maintaining by a team of technicians. Prototype implants have wires that penetrate the skull and skin, with the risk of infection. Wireless signal transmission for brain implants is still in the future, along with wearable magnetic brain–machine interfaces.

A more difficult obstacle is that the performance of microelectrodes recording from neurons tends to fall off over time; better engineering of interfaces using nanoengineered materials is needed to improve biocompatibility and allow lower stimulation potentials. Even if implants are supplanted by noninvasive magnetic communication, there is still research to be done in neurocognition and how to interface learning systems.

Although patterns of control for hands and arms have been mapped, individual differences are not fully explored—some experimental patients are able to control prostheses much more easily than others. Concerted efforts are developing adaptive learning systems that require less effort from the patient by embedding neural networks and adaptive filters into the control system of the prosthesis, with impressive results.

Paths and mechanisms for feedback is another area where more research is needed—to succeed in duplicating or restoring limb function instead of merely controlling machines, researchers have to work out how the body tells the brain where its limbs are positioned in space—proprioception. Better pressure and vibration sensors, accelerometers, actuators, and force sensors are needed in order to develop improved artificial proprioception, haptics, and kinesthetics.

All of these pieces need to be integrated seamlessly in intelligent control systems that respond precisely and adapt over time to changes in their environment. Nanotechnology is playing an important role in sensors, actuators, communications, and computing elements to make the brain–machine interface a clinical reality. Most of the difficulties for motor neuroprosthetics are now engineering challenges, rather than problems of principle. They will be solved by closely knit interdisciplinary teams that include doctors, engineers, rehabilitation specialists, and patients.

13.2.5 *Neuroprosthetics for the Ear*

Restoring sensory pathways is as important for neural prosthetics as restoring motor function. Sight and hearing are valuable not only because of their role in performing tasks, but because they are the bridges for social communication. Hearing loss is the most common form of sensory impairment. Electronic aids for hearing have a long history intertwined with the telecommunications inventions that shape the modern world. Alexander Graham Bell was an audiologist; the telephone was a by-product of his interest in making an electronic hearing aid [284–286].

The study of how the human ear distinguishes sounds is important in design and optimization of large-scale telephone networks sending usable speech over long distances. Voice compression, recognition, and synthesis are modeled largely on an understanding of how the ear and brain process speech. The study of how hearing works—how information is encoded and decoded in sounds by the brain—has been essential in developing technology, and in turn has helped to develop aids to ameliorate hearing disorders [287]. Research to optimize telecommunications led to signal processing technology that made advanced hearing aids possible. Artificial stereocilia MEMS, modeling the resonators in the cochlea, are being fabricated in nanotechnology laboratories in order to understand the mechanisms of hearing [288–294].

Cochlear implants for hearing disorders are one of the most mature and best established areas of any electronic neuroprostheses. Nanotechnology has enhanced microelectronics, batteries, and micromechanical transducers in cochlear implant devices and thus has contributed significantly to the quality of life of persons with hearing impairment. Patients with cochlear implants benefit from improved understanding of speech in noise, sound quality, and localization of sounds compared to patients with acoustic hearing aids, without the ear canal occlusion, acoustic feedback, and inconvenience and cosmesis of external devices [295–297].

Cochlear prostheses are implanted in the middle ear, where they stimulate the ossicles electromechanically, rather than acoustically, through either electromagnetic or piezoelectric transducers. Sound signals transduced by an externally worn microphone are sent to the implant by wireless transmission. Cochlear implants achieve an average threshold improvement of 10 dB from 500 to 4000 Hz. At 6000 Hz the gain is about 20 dB compared with conventional fitted acoustic hearing aids [298].

Another type of hearing neuroprosthesis, the auditory brainstem implant, uses electrodes placed over the cochlear nucleus or inserted into the brainstem. This direct interface is capable of restoring some residual hearing in many patients who have lost both hearing nerves [299,300].

Because the organ to which the prosthesis connects the brain is inside the skull, neuroprostheses for hearing can bypass many of the problems faced in getting signals from the organ and its prosthesis to the relevant part of the brain. Cochlear implants (like cardiac pacemaker or spinal stimulators or deep brain stimulators) can be self-contained, with power transmitted through the skull by electromagnetic induction. Thus the main problems are long-term biocompatibility (build up of biofilms and plaques) and refinement of the signal processing to improve performance by attempting to match the natural function of the ear. Challenges include recognition of pitch for understanding and enjoyment of music, and preventing bacterial and fungal infections. The latter is an area that has prospects of being improved through nanomaterials [301–305].

Neuroprosthesis for Balance.   Besides the sensory organ for hearing, the inner ear contains the vestibular arches, filled with fluid and cilia which can sense microscopic inertial fluid flows caused by head and body movements. Disruption of this function can cause severe dizziness, loss of balance, inability to walk or even sit upright, and sensations of sea-sickness or air-sickness. Some research is being pursued to develop an implantable MEMS neuroprostheses that could restore or compensate for loss or disturbance of this sensory organ [306–310].

Neuroprosthesis for Tinnitus.   Tinnitus is a condition which results in a constant sensation of sound, regardless of whether an audial stimulus is present. It is difficult to treat, and can be very troubling. Most treatment seeks to modulate the patient's response, rather that treating the tinnitus itself. One alternative development is an implantable device to deliver electrical stimulation in the middle ear, close to the cochlea; the goal is to turn off the nerve pathway, similar to stimulators for relief of pain and tremor [311].

Anatomy of the Ear.   In order to see the relevance of nanotechnology to cochlear implants, consider the neuroanatomy of the inner ear and how it is stimulated.

Mechanoacoustical pressure is delivered to the membrane that seals off the inner ear, the cochlea, by an extremely delicate linkage from the eardrum to the malleus, incus, and stapes (hammer, anvil, and stirrup) bones. Sound is collected by the eardrum, causing the stapes to vibrate against the membrane which separates the cochlea from the outer ear, transmitting pressure waves to fluids in the interior. This linkage is a powerful mechanical amplifier—the human ear can detect motions of the eardrum on the order of a picometer—smaller than the diameter of an atom.

The cochlea is a hollow tapering helix supported by a bony spiral shelf, the *osseous spiral lamina*, which winds around a central core, the *modiolus*. The cochlea's spiral cone geometry, like a French horn or conch shell, acts as a mechanical acoustical transform to select for different vibration frequencies along its interior. The interior of the cochlea is separated into two fluid-filled chambers (the *scala vestibuli* and *scala tympani* or upper and lower ducts) by a thin sac, called the *cochlear duct*, filled with gelatinous material. The large end of the spiral is sealed from the outer ear by two membranes, the oval and round windows, on either side of the cochlear duct. The duct separates the two chambers all the way up the spiral to its apex, where there is a small opening between them. The sensory hair cells are inside the cochlear duct adjacent to a thin layer of tissue (the *tectorial membrane*). Each hair cell has a group of stereocilia projecting into the viscous gelatin, which resonate with sound [312,313].

The chambers on either side of the cochlear duct are filled with an electrolyte solution which conducts sound from the oval window, to which the stapes is attached, into the scala vestibuli. When pressure waves travel through the upper side of the cochlear duct to the apex of the spiral, and down the lower side, the duct is compressed by the fluid, causing movements of the cell hairs. The shearing movement of the hair cells opens potassium ion channels in hair cell membranes, depolarizing the cells and initiating an electrical signal. (See chapter 14, *Introduction to Nanoscience*, for a review of potassium ion channels in nerve cells.)

The mechanical stress is transmitted to the ion channels in hair cells via tiny filaments that connect neighboring hairs in a bundle. **Figure 13.4** shows a schematic of the hair cells in the ear. The hairs are about 500 nm in diameter and tip links are on the order of 2 nm in diameter.

> Each hair cell is a micromachine that is constructed from hundreds of much smaller components (e.g. ion channels and tip links). But hair cells do not operate in isolation. Hair cells are part of a larger system: the inner ear.
>
> Thus the inner ear represents VERY LARGE SCALE INTEGRATION of hundreds of thousands of biological microelectromechanical devices.
>
> **A. J. Aranyosi**
> *MIT Micromechanics Group*

The stereocilia are linked in a complex network, with an inner and outer layer. The cochlea contains about 28,000 hair cells in humans. The absolute number of cells does not directly relate to auditory acuity; cats have 39,000, bats, rats, and dolphins about 15,000. In the human, the interior linear extent of the cochlea is about 3.5 cm. So the density of sense cells is about 800 per millimeter (spaced at about 1.25 μm along the cochlea).

Behind the hair cells is a layer of neurons, the spiral ganglion, which contains four or five times more cells than the sensory cell layers. The network of auditory neurons in the spiral ganglion lies close to the interior or modiolar wall of the cochlear duct. The spiral ganglion neurons control the selection and organization of stimuli that are sent to the auditory cortex in the brain.

The hair cells are extremely sensitive to acoustical vibrations, and vulnerable to damage. Up to one-third of the sensory hair cells typically die with age and damage by intense or chronic sound overload, chemicals, and disease. Unlike disorders in the mechanical acoustical path, nerve damage cannot be compensated by external devices.

Design of Cochlear Implants.   Cochlear implants serve to bypass damaged inner ear hair cells and transfer auditory information to the brain. A cochlear implant takes sound from an external microphone and converts it into electrical impulses to stimulate the cochlea. The microphone and the signal processing module are worn outside the head, and the stimulator is implanted behind the ear, with wires inserted into the cochlea. Stimulation is usually applied to the ganglion layer, which is accessed by inserting electrodes into the scala tympani, the lower cochlear chamber next to the modiolus where it can be placed close to the spiral ganglion.

Fig. 13.4 *Hair cells in the cochlear duct have stereocilia (microvilli) linked to potassium ion channels. Movements induced in the stereocilia by mechanical stimulus open the ion channels to depolarize the cell, initiating nerve firing.*

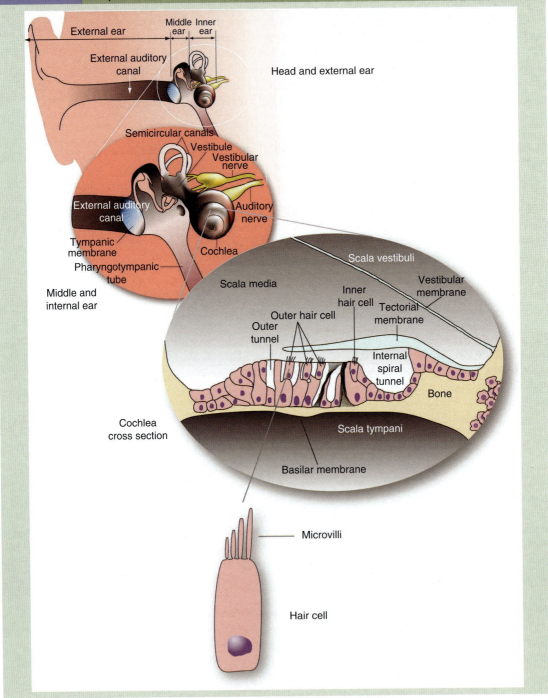

Coding of sound and mapping stimulation onto the cochlea are guided by neurocognitive research in how the cochlea receives and processes sound. Early cochlear implants used a single electrode and encoded sounds by converting sound frequencies into electronic pulse frequencies, which are perceived as sound when applied to the ganglion. This encoding is oversimplified and does not reproduce fine details of sound perception. Newer cochlear implants map different frequencies spatially along the cochlea, from lower to higher frequencies. This improved mapping is closer to the natural perception of sound by the ganglion, and is the coding used by most current cochlear implant devices.

Current devices have between 16–100 electrodes. While it may never be necessary to approximate the number of discrete stimulation points on the spiral ganglion that are presented by the 20,000 or more hair cells, clearly there is plenty of room at the bottom for delivery of sound coded impulses with higher resolution along the cochlear duct.

Researchers and developers are working to overcome a number of challenges in what is called the *electroneural bottleneck*. Currently, the voltage required to stimulate the neurons in the spiral ganglion is not localized—it stimulates a relatively large area. The electrodes cannot be placed very close together because it would result in cross-talk. The electrodes cannot be too conforming or embedded in the cochlea because the device may have to be removed without damage to the tissues. The electrodes are currently hand-made assemblies; integrated device electrode assemblies have been proposed and used experimentally in the lab, but performances in electrode resistance, durability, reliability, and biocompatibility are still not sufficient for clinical use. Currently the signal-processing power available to fit into a low-power wearable package is not sufficient to process received sound into many more channels for discrete delivery, and a large number of channels would present a wiring challenge.

Work has been done on designs using nested wiring with electrode contacts that can be de-insulated by laser, which may resolve some obstacles. Electrode coatings and treatment with platinum, iridium, gold, platinum black, and alloys have been studied to lower resistance and reduce corrosion. Shape metal alloys have been proposed to produce better conformance to the modiolus and spiral ganglion. Designs with quadrupolar electrodes, with contact points on various places in the modiolus have been proposed, which might help to focus the excitation area. Coating of electrodes with brain-derived growth factors has been studied experimentally in an effort to reduce long-term atrophy of spiral ganglion cells [314–316].

Long-term wear of cochlear implants has given rise to biofilm contamination and infection; the ear is particularly vulnerable to infection since it is open to the mouth and throat.

Nanotechnology may in future offer solutions to these challenges with better materials, circuits, signal-processing arrays, and stimulation methods. Application of nanotechnology may even come up with convenient, low-cost nanoengineered smart acoustical materials that could filter out damaging frequencies and noise levels when inserted or injected into the ear as an expanding foam, while allowing the wearer to hear normal sounds in comfort, thus preventing hearing loss.

13.2.6   *Vision Prosthetics*

Loss of sight has profound psychological as well as social and physical consequences; some 30% of the sensory input to the brain comes from the eye [317,318]. The sense of sight involves highly parallel processing of image data from light focused onto a surface. Vision involves the process by which the eye and the optical nerves gather light, extract image data by sampling areas on the projection, detect features in the image, and send the information to the brain for further processing, recognition, and analysis.

The brain is very plastic, especially with respect to the pathways to the visual cortex; with a prosthesis that maps pixels onto the surface of a suitable area of skin, with dense nerve receptors, it is possible for the brain to map the signals to the visual cortex so that the patient learns to visualize from the haptic inputs. The adaptability of the brain gives a good prognosis for the development of a number of visual prostheses types, whether mapping light images to haptic nerve endings or stimulating some level in the layers of processing that lead from the eye to the brain, all the way up to direct stimulation of the visual cortex.

Any of these routes is a formidable undertaking from both technical and neurological viewpoints. If the ear is the foot of sensory prosthetics, then the eye is the hand. Both are difficult, but hearing is a mapping onto a one-dimensional sensory cell space—the linear array of cilia—whereas vision is a mapping onto a two-dimensional space of rods and cones in the retina. Additional dimensionality for both hearing and sight is added by parallelism in time—sensation from millions of cells is simultaneous over the sensing space rather than serial. One of the jobs of the layers of neurons behind the retina is to organize a sampling scheme for transmitting images to the brain. Vision is not totally asynchronous—somewhere on the path to the brain the incoming data is organized into scans, which is why the visual cortex can be satisfied to generate images from strobed motion picture and television screens, so long as their output rate exceeds the sampling rate of the brain's optical system. The latter is not a simple scanning process, but in practice a frame rate of about 30 frames per second is perceived as continuous [319–321].

The Retina.   The retina is the point at which light is converted into neural impulses, which are processed into images by networked layers of neurons carrying information to the visual cortex. The retina samples images with photoreceptor cells—rods and cones—whose overall size is on the order of microns. Cones are color sensitive. Rods have a sharper acuity than cones, but do not discriminate colors [322,323].

The human retina contains approximately 120 million rod and 1 million cone cells. The densely packed cones in the center of the retina where vision is most acute, the fovea, have a center to center spacing of about 2.5 μm. Cone density in the fovea is between 100,000 and 300,000/mm², but rods and cones are both present in surrounding periphery of the retina. Rods are absent in the fovea, and are packed at a density of 80,000–100,000/mm² in the periphery.

Fovea.   The area of the fovea, where rods are absent, has a radius of 200–300 μm; the central part of the fovea where cones are packed most densely is only about

$50 \times 50 \, \mu m$. The total number of cones in the fovea is approximately 200,000. The total number of cones in the entire retina is approximately 6,400,000. The total number of rods in the retina is 110,000,000–125,000,000.

The peak rod density is located in a ring around the fovea between 1.5 and 5 mm from the center, where the rod density is between 100,000 and 160,000 rods/mm^2. In terms of the radius of the field of vision, the peak is between 5 and 18° from the center. The rod-free area of the fovea is only 1 or 2° of the visual field.

The photoreceptors communicate with ganglion neurons located in a tissue layer 150–300 μm behind the retina, and separated from it by a layer of support cells, the retinal pigment epithelium. There are 1 or 2 cones and about 20 rods for each ganglion cell. A network of interconnected neurons process the information from more than six million receptors in the retina down to where it is carried by approximately one million axons in the optic nerve to the brain.

By-Passing the Retina.   Retinal degeneration or detachment is one of the main causes of vision loss. Therefore most efforts to develop a visual neuroprosthesis have attempted to stimulate the ganglia cell layer behind the retina, as the simplest strategy to interface to the visual signal processing that is in place in the optical nerve path. Some prostheses have been designed and tested that stimulate the visual cortex directly, producing a low resolution pattern of visual sensation, and some have mapped digitized imaging onto touch sensors in the back or other skin areas, in a kind of transposed Braille that delivers images rather than encoded letters.

Artificial Retinas.   Most recent research in visual neuroprosthetics has been focused on artificial retina replacements or bypasses converting visual information into patterns of electrical stimulation onto inner retinal neurons. More than a dozen projects around the world are aimed at using advances in nanotechnology and high-density integrated microelectronics to develop an implant analogous to the cochlear implant for the ear, to restore lost vision. Like the cochlear implant, an eye implant is not anticipated to fully restore all lost function. Research devices are expected to provide enough visual perception of contours, outlines, and shades of light to allow a blind person to move freely in unfamiliar environments (**Fig. 13.5**).

Treatment of diseases such as retinitis pigmentosa is focused on growing knowledge of affected biochemical pathways, development of animal models, and possible gene therapy, especially for genetically defined subsets of patients, based on newly identified genes [324]. As is the case with many other parts of the nervous system, vision, its development, and its degeneration are controlled by a large number of different genes, and this approach is still in the early stages of development. Another possible approach being investigated is transplantation of cells to the retina [234]. Visual neuroprostheses will probably still be needed, even when other therapies become available, because of the large number and diversity of causes of vision loss.

In the remainder of this section, we look at some projects developing prostheses that can be implanted in the visual cortex, around the optic nerve, or in the eye. These approaches have shown promise for useful perception to patients with visual impairments [325–330].

Fig. 13.5 *Retinal prostheses can be implanted in the eye to electronically stimulate the optical nerves, bypassing a damaged retina.*

A large-scale project has been undertaken for a number of years by the U.S. Department of Energy National Laboratories and the National Science Foundation, with a team that includes a number of universities, institutes, and private industry. The artificial retina device bypasses nonfunctioning retinal cells to transmit signals directly to the optic nerve. The device consists of a tiny camera and microprocessor mounted in eyeglasses, a receiver implanted behind the ear, and an electrode-studded array that is tacked to the retina. Power is provided by a wireless battery pack worn on the belt.

A microprocessor converts the camera image and transmits information to an implanted wireless receiver. The receiver sends the signals through a tiny cable to the electrode array to generate stimulating pulses. The pulses are perceived as patterns of light and dark spots corresponding to the electrodes stimulated. Patients have learned to interpret the visual patterns produced, enabling them to detect when lights are on or off, describe an object's motion, count individual items, and locate objects. To evaluate the long-term effects of the retinal implant, five devices have been approved for home use (**Fig. 13.6**).

The DOE project has produced three successively more sophisticated models, which are being tested and evaluated. Surgical time was reduced from the 6 h required for the first model to 2 h for the second version, which has 60 electrodes. The third model will have more than 200 electrodes, and will use more advanced materials than previous ones. A special coating, only a few microns thick, will replace the bulky sealed package used in previous models. The new model will use flexible conductive materials for the electrode array so that it will conform to the shape of the inner eye. The latest model will be many times smaller than earlier models. Engineering goals include enhancing the resolution with more electrodes, and decreasing the size of the device and complexity of the surgical procedure [331–335].

FIG. 13.6 *An artificial retina system consists of an external camera with a signal processor which transmits encoded image stimuli to an electrostimulator array in the eye, which can trigger nerve firings behind the retina that can produce visual perceptions. A cochlear implant works much the same way with an external microphone and a stimulator next to the nerves in the ear. Experiments are underway to stimulate the visual nerve network with chemical neurotransmitters using microfluidic devices with nanoscale membranes.*

Other retinal prostheses systems have been developed including subretinal versions, and versions with high pixel density, in Australia, Europe, Japan, and elsewhere including international teams [336–341].

Researchers face numerous challenges in developing retinal prosthetic devices that are effective, safe, and durable enough to last for the lifetime of the individual. Materials must be biocompatible with delicate eye tissue, yet able to withstand corrosion. The device must remain fixed to a precise area of the retina and not compress or pull the tissue. The apparatus also needs to deliver enough power to stimulate electrodes, without generating excess heat to damage the remaining functional nerve tissue. Image processing needs to be performed in real time, so there is no delay in interpreting an object in view. In addition, effective surgical approaches are critically important to ensure a successful implant.

A number of interdisciplinary research teams are looking at the effect of implant surgery on the retina, the sensitivity of the retina to electronic charge, spatial resolution in relation to stimulation, the best patterns and locations for nerve stimulation, and evaluating the learning and adaptation of patients with early versions of the devices [342–367].

The possibility of electrode or tissue damage limits excitation schemes to those that may be employed with electrodes that have relatively low charge densities. The excitation thresholds that have been required to achieve vision have been found to be relatively high. This may result in part from poor apposition between neurons and the stimulating electrodes and is confounded by the effects of the photoreceptor loss, which initiates other pathology in the surviving retinal tissue. The combination of these and other factors imposes a restriction on the pixel density that can be used for devices that actively deliver electrical stimulation to the retina. The resultant use of devices with relatively low pixel densities presumably will limit the degree of visual resolution that can be obtained with these devices. Further increases in pixel density, and therefore increased visual acuity, will necessitate either improved electrode–tissue biocompatibility or lower stimulation thresholds. To meet this challenge, innovations in materials and devices are needed, as well as experiments on the factors and functional parameters relevant to the designing of implants, such as thresholds and electrical point spread functions [342].

Nanotechnology offers a potential way to avoid the obstacles with electrical stimulation, by using nanofabricated microfluidics to inject chemical neurotransmitter stimulants into the subretinal ganglion. Researchers at Stanford University have developed a prototype test system to study the possibility, by characterizing the stimulus produced with a microfluidics system fabricated on a 500-nm thick silicon nitride membrane, with a single 5- or 10-μm aperture overlaying a microfluidic delivery channel in a silicone elastomer. Controlled excitation based on picoliter amounts of neurotransmitter delivered was obtained in rat neurons grown in culture on the surface of the apparatus. In the experiments the stimulation radius was as small as 10 μm, comparable to what has been achieved in electrical stimulation experiments with a micrometer-size electrode. This experimental study shows how future neurostimulation systems, scaled up with arrays of delivery channels, might be possible with advances in integrated nanofabrication and cell engineering [368].

13.3 SUMMARY

Most of the medical devices which we covered in this chapter are on a borderline between micro- and nanotechnology, but there is a clear trend towards more subtle nanoscale mechanisms. The same is true for other types of devices which we did not have the scope to cover, such as robotic surgery, and implantable sensors for blood pressure, gastric reflux, and other monitors [369–371].

Nanomedicine will have finally come of age when design of devices has fully merged with design of drugs—when the devices that are used for medicine may be made of protein and DNA and macromolecules as easily as silicon or polymers, all designed to fit their work in a natural way in cells and biological systems—and when the drugs and pharmaceuticals are atomic- and molecular-scale biochemical, genomic, and proteomic devices, designed with a full understanding of how they will fit into and do their work in cells and organisms—and when sensors and nanosurgical tools no longer have a clear distinction from genetically engineered bacteria and macrophages. This is the direction that the science and technology are going, and it will be reached only when a much deeper and fuller understanding of the fundamentals of biology are understood by engineers, and engineering by biomedical researchers, and the principles of physics, chemistry, electronics, and systems organization understood by both.

Even if that day comes, it will not bring a health utopia. There will always still be many challenges, as life is continually adapting and evolving. Solving one metabolic pathway or gene expression disorder, or finding a cure for one disease, will not eliminate the eternal work of survival and adaptation. New disorders and new organisms will arise to take advantage of any possible pathway that is open to feed energy, sustenance, and propagation. Living systems will always need to be changing to maintain their dynamic existence, which is like a flame—a structure maintained by a complex of dynamic chemical reactions as fuel and oxidants pass through. Though each organism may appear to us as a solid, stable body, we are dynamic equilibrium structures, apt to be disintegrated by a sufficient disturbance. The wonder of life is how stable and resilient it is, how all the systems work together to compensate for disturbances and counter multifold attacks, and how such complexity is all able to work together smoothly and beautifully in a harmonious whole. So the conclusion is that there will always be plenty of work to be done by nanoscientists and nanoengineers, so study diligently and apply yourself intelligently.

Conclusion.   We do not have to wait for the future to find ways to use nanotechnology. To advance sensors in performance, size, cost, and power, some designs are moving into new nanoscale territory, such as the interaction between gas phase atomic magnetic resonance and laser optics, while others go down in the nanoscale to the mechanical principles of cantilevers. At the nanoscale, the electromechanical rules yield different results from the micro- or atomic scale, so re-examination of the tuning fork magnetometer or the gas phase atomic clock at the nanoscale gives new opportunities.

What is so fascinating and exciting about the nanoscience field at this time is that it is eminently probable that an even better design for some important medical device or process, improved by orders of magnitude, simpler, easier to build and assemble into arrays and interface into imaging systems, will be invented by someone who is now just now learning about nanoscience and nanotechnology—perhaps even someone who is studying this book.

Acknowledgments

The author wishes to thank the many colleagues who provided material, comments, and review of the drafts. Much material has been drawn from experiences in projects at the University of Texas Southwestern Medical Center, the University of Texas at Dallas, the University of Texas at Arlington, and the Texas Scottish Rite Hospital for Children. Any errors remain his own. Special thanks to the staff of the UT Southwestern Library.

References

1. H. J. Cook, What stays constant at the heart of medicine, *British Medical Journal*, 333, 1281–1282 (2006).
2. A. Coppa, L. Bondioli, A. Cucina, D. W. Frayer, C. Jarrige, J. -F. Jarrige, G. Quivron, M. Rossi, M. Vidale, and R. Macchiarelli, Palaeontology: Early Neolithic tradition of dentistry, *Nature*, 440, 755–756 (2006).
3. B. Morris, Surgery on Papyrus, *Student British Medical Journal*, 12, 309–348 (2004).

4. M. Kennedy, *A brief history of disease, science, and medicine*, Asklepiad Press, Mission Viejo, CA (2004).

5. I. Shaw, *The Oxford history of ancient Egypt*, Oxford University Press, Oxford, UK (2000).

6. J. F. Nunn, *Ancient Egyptian medicine*, University of Oklahoma Press, Norman, OK (1996).

7. R. Porter, *The greatest benefit to mankind: A medical history of humanity from antiquity to the present*, HarperCollins, New York (1997).

8. The Editors, Looking back on the millennium in medicine, *New England Journal of Medicine*, 342, 42–49 (2000).

9. K. Conley, *Joseph Priestley and the discovery of oxygen*, Mitchell Lane Publication Inc., Hockessin, DE (2005).

10. American Chemical Society, National Historical Chemical Landmarks website: http://acswebcontent.acs.org/landmarks/landmarks/priestley/discovery_O2.html (2008).

11. B. Jaffe, *Crucibles: The story of chemistry from ancient alchemy to nuclear fission*, 4th ed., Dover Publications, Mineola, NY (1976).

12. M. Pera, trans. by J. Mandelbaum, *The ambiguous Frog: The Galvani-Volta controversy on animal electricity*, Princeton University Press, Princeton, NJ (1992).

13. H. Xiao, P. Verdier-Pinard, N. Fernandez-Fuentes, B. Burd, R. Angeletti, A. Fiser, S. B. Horwitz, and G. A. Orr, Insights into the mechanism of microtubule stabilization by Taxol, *Proceedings of the National Academy of Science USA*, 103, 10166–10173 (2006).

14. J.-M. Lehn, Perspectives in supramolecular chemistry—From molecular recognition: Towards molecular information processing and self-organization, *Angewandte Chemie*, 29, 1304–1319 (2003).

15. NIH Office of Portfolio Analysis and Strategic Initiatives, website at: http://nihroadmap.nih.gov/nanomedicine/ (2008).

16. NIH Roadmap webpage at http://nihroadmap.nih.gov/initiatives.asp (2008).

17. NCI Nanotechnology Plan webpage at http://www.cancer.gov/researchfunding/NIHRoadmapFAQs (2008).

18. J. U. Leike, A. Sachse, and K. Rupp, Characterization of continuously extruded iopromide-carrying liposomes for computed tomography blood-pool imaging, *Investigative Radiology*, 36, 303–308 (2001).

19. A. T. Yordanov, et al., Novel iodinated dendritic nanoparticles for computed tomography (CT) imaging, *Nano Letters*, 2, 595–599 (2002).

20. J. R. Swanson, H. W. Bosch, K. J. Illig, D. M. Marcera, and R. L. Mueller, Process of preparing x-ray contrast compositions containing nanoparticles, US Patent 5543133, NanoSystems L.L.C., Collegeville, PA (1996).

21. P. J. Bonitatebus Jr., O. H. E. Axelsson, O. H. Erik, A. M. Kulkarni, B. C. Bales, D. J. Walter, A. S. Torres, and C. Treynor, Nanoparticle-based imaging agents for x-ray/computed tomography, US Patent 20070098641, General Electric Company, Fairfield, CT (2007).

22. J. W. M. Bulte and D. L. Kraitchman, Iron oxide MR contrast agents for molecular and cellular imaging, *NMR in Biomedicine*, 17, 484–499 (2004).

23. S. Eroglu, B. Gimi, L. Leoni, B. Roman, G. Friedman, T. Desai, and R. L. Magin, NMR imaging of biocapsules for monitoring the performance of cell and tissue implants, *IEEE-EMB Special Topic Conference on Microtechnology in Medical Biology, 2nd Annual International*, pp. 193–198 (2002).

24. C. Loo, A. Lin, L. Hirsch, M.-H. Lee, E. Chang, J. West, N. Halas, and R. Drezek, Gold nanoshell bioconjugates for molecular imaging in living cells, *Optics Letters*, 30, 1012–1014 (2005).

25. C. Loo, A. Lin, L. Hirsch, M.-H. Lee, J. Barton, N. Halas, J. West, and R. Drezek, Nanoshell-enabled photonics-based imaging and therapy of cancer, *Technology In Cancer Research and Treatment*, 3, 33–40 (2004).

26. C. Loo, A. Lowery, N. Halas, J. West, and R. Drezek, Immunotargeted nanoshells for integrated cancer imaging and therapy, *Nano Letters*, 5, 709–711 (2005).

27. M. M. Amiji, *Nanotechnology for cancer therapy*, CRC Press, Boca Raton, FL (2006).

28. I. Safarik and M. Safarik, Use of magnetic techniques for the isolation of cells, *Journal of Chromatography B*, 722, 33–53 (1999).

29. M. Lewin, N. Carlesso, C.-H. Tung, X.-W. Tang, D. Cory, D. T. Scadden, and R. Weissleder, Tat peptide-derivatized magnetic nanoparticles allow in vivo tracking and recovery of progenitor cells, *Nature Biotechnology*, 18, 410–414 (2000).

30. Z. M. Saiyed, S. D. Telang, and C. N. Ramchand, Application of magnetic techniques in the field of drug discovery and biomedicine, *Biomagnetic Research and Technology*, 1, 1–2 (2003).

31. B. Molnara, F. Siposb, O. Galamb, and Z. Tulassay, Molecular detection of circulating cancer cells: Role in diagnosis, prognosis and follow-up of colon cancer patients, *Digestive Diseases*, 21, 320–325 (2003).

32. Q. A. Pankhurst, J. Connolly, S. K. Jones, and J. Dobson, Applications of magnetic nanoparticles in biomedicine, *Journal of Physics D: Applied Physics*, 36, R167–R181 (2003).

33. D. Wang, J. He, N. Rosenzweig, and Z. Rosenzweig, Fe_2O_3 beads-CdSe/ZnS quantum dots core-shell nanocomposite particles for cell separation, *Nano Letters*, 4, 409–413 (2004).

34. A. S. Lübbe, C. Alexiou, and C. Bergemann, Clinical applications of magnetic drug targeting, *Journal of Surgical Research*, 95, 200–206 (2001).

35. U. Häfeli, et al., *Scientific and clinical applications of magnetic carriers*, Springer Verlag, Berlin, Germany (1997).

36. M. Shinkai and A. Ito, Functional magnetic particles for medical application, *Advances in Biochemical Engineering/Biotechnology*, 91, 191–220 (2004).

37. B.-I. Haukanes and C. Kvam, Application of magnetic beads in bioassays, *Biotechnology*, 11, 60–63 (1993).

38. C. G. Thanos and D. F. Emerich, The pinpoint promise of nanoparticle-based drug delivery and molecular diagnosis, *Biomolecular Engineering*, 23, 171–184 (2006).

39. K. K. Jain, Targeted drug delivery for cancer, *Technology in Cancer Research and Treatment*, 4, 311–454 (2005).

40. P. Merritt and L. A. Snyder, Pharmacology of β-Lapachone and Lapachol, *Cyberbotanica*, URL: http://biotech.icmb.utexas.edu/botany/beta.html, BioTech Resources, Austin, TX (1997).

41. D. A. Boothman and A. B. Pardee, Inhibition of Radiation-Induced Neoplastic Transformation by β-lapachone, *Proceedings of the National Academy of Science*, 86, 4963–4967 (1989).

42. K. E. Reinicke, E. A. Bey, M. S. Bentle, J. J. Pink, S. T. Ingalls, C. L. Hoppel, R. I. Misico, G. M. Arzac, G. Burton, W. G. Bornmann, D. Sutton, J. Gao, and D. A. Boothman, Development of β-Lapachone prodrugs for therapy against human cancer cells with elevated NAD(P)H:Quinone oxidoreductase 1 levels, *Clinical Cancer Research*, 11, 3055–3064 (2005).

43. D. Sutton, S. Wang, N. Nasongkla, J. Gao, and E. E. Dormidontova, Doxorubicin and β-lapachone release and interaction with micellar core materials: Experiment and modeling, *Experimental Biology and Medicine*, 232, 1090–1099 (2007).

44. D. Filmore, Breaching the blood-brain barrier, *Modern Drug Discovery*, 5, 22–27 (2002).

45. K. S. Soppimath, T. M. Aminabhavi, A. R. Kulkarni, and W. E. Rudzinski, Biodegradable polymeric nanoparticles as drug delivery devices, *Journal of Controlled Release*, 70, 1–20 (2001).

46. C. Roney, P. Kulkarni, V. Arora, P. Antich, F. Bonte, A. Wu, N. N. Mallikarjuana, S. Manohar, H. F. Liang, A. R. Kulkarni, H. W. Sung, M. Sairam, and T. M. Aminabhavi,

Targeted nanoparticles for drug delivery through the blood–brain barrier for Alzheimer's disease, *Journal of Controlled Release*, 108, 193–214 (2005).

47. J. Rafuse, Seventy-five years later, insulin remains Canada's major medical-research coup, *Canadian Medical Association Journal*, 155, 1306–1308 (1996).

48. M. Bliss, *The discovery of Insulin*, McClelland and Stewart, Toronto (1982).

49. R. D. Simoni, R. L. Hill, and M. Vaughan, The discovery of Insulin: The work of Frederick Banting and Charles Best, *Journal of Biological Chemistry*, 277, 26 (2002).

50. L. Rosenfeld, Insulin: Discovery and controversy, *Clinical Chemistry*, 48, 2270–2288 (2002).

51. L. Rosenfeld, Margaret and Charley: The personal story of Dr. Charles Best, the co-discoverer of Insulin, *Journal of the American Medical Association*, 291, 1903–1904 (2004).

52. B. Zhang, G. Salituro, D. Szalkowski, Z. Li, Y. Zhang, I. Royo, D. Vilella, M. T. Díez, F. Pelaez, C. Ruby, R. L. Kendall, X. Mao, P. Griffin, J. Calaycay, J. R. Zierath, J. V. Heck, R. G. Smith, and D. E. Moller, Discovery of a small molecule Insulin mimetic with antidiabetic activity in mice, *Science*, 284, 974–977 (1999).

53. R. S. Spangler, Insulin administration via liposomes, *Diabetes Care*, 13, 911–922 (1990).

54. S. Furtado, D. Abramson, L. Simhkay, D. Wobbekind, and E. Mathiowitz, Subcutaneous delivery of insulin loaded poly(fumaric-co-sebacic anhydride) microspheres to type 1 diabetic rats, *European Journal of Pharmaceutics and Biopharmaceutics*, 63, 229–236 (2006).

55. E. Allémann, J.-C. Leroux, and R. Gurny, Polymeric nano- and microparticles for the oral delivery of peptides and peptidomimetics, *Advanced Drug Delivery Reviews*, 43, 171–189 (1998).

56. M. Aboubakar, F. Puisieux, P. Couvreur, M. Deyme, and C. Vauthier, Study of the mechanism of insulin encapsulation in poly(isobutylcyanoacrylate) nanocapsules obtained by interfacial polymerization, *Journal of Biomedical Materials Research*, 47, 568–576 (1999).

57. S. Watnasirichaikul, N. M. Davies, T. Rades, and I. G. Tucker, Preparation of biodegradable Insulin nanocapsules from biocompatible microemulsions, *Pharmaceutical Research*, 17, 684–689 (2000).

58. G. P. Carino, J. S. Jacob, and E. Mathiowitz, Nanosphere based oral insulin delivery, *Journal of Controlled Release*, 65, 261–269 (2000).

59. J. S. Jacob, Y. S. Jong, D. T. Abramson, E. Mathiowitz, C. A. Santos, M. J. Bassett, and S. Furtardo, Nanoparticulate therapeutic biologically active agents, US Patent 20050181059, Spherics, Inc. (2004).

60. S. Salmon and S. M. Hudson, Crystal morphology, biosynthesis, and physical assembly of cellulose, chitin, and chitosan, *Polymer Reviews*, 37, 199–276 (1997).

61. S. A. Agnihotri, N. N. Mallikarjuna, and T. M. Aminabhavi, Recent advances on chitosan-based micro- and nanoparticles in drug delivery, *Journal of Controlled Release*, 100, 5–28 (2004).

62. Y. Zheng, Y. Wu, W. Yang, C. Wang, S. Fu, and X. Shen, Preparation, characterization, and drug release in vitro of chitosan-glycyrrhetic acid nanoparticles, *Journal of Pharmaceutical Sciences*, 95, 181–191 (2005).

63. Q. Gan and T. Wang, Chitosan nanoparticle as protein delivery carrier—Systematic examination of fabrication conditions for efficient loading and release, *Colloids and Surfaces B: Biointerfaces*, 59, 24–34 (2007).

64. C. Pinto Reis, R. Neufeld, A. Ribeiro, and F. Veiga, Nanoencapsulation II. Biomedical applications and current status of peptide and protein nanoparticulate delivery systems, *Nanomedicine: Nanotechnology, Biology, and Medicine*, 2, 53–65 (2006).

65. E. Renard, G. Costalat, and J. Bringer, From external to implantable insulin pump, can we close the loop? *Diabetes and Metabolism*, 28, S19–S25 (2002).

66. J. D. Zahn, Y.-C. Hsieh, and M. Yang, Components of an integrated microfluidic device for continuous glucose monitoring with responsive insulin delivery, *Diabetes Technology and Therapeutics*, 7, 536–545 (2005).

67. S. L. Tao and T. A. Desai, Microfabrication of multilayer, asymmetric, polymeric devices for drug delivery, *Advanced Materials*, 17, 1625 (2005).

68. Y. Y. Huang and C.H. Wang, Pulmonary delivery of insulin by liposomal carriers, *Journal of Controlled Release*, 113, 9–14 (2006).

69. G. T. McMahon and R. A. Arky, Inhaled insulin for diabetes mellitus, *New England Journal of Medicine*, 356, 497–502 (2007).

70. Z. T. Bloomgarden, Insulin treatment and type 1 diabetes topics, *Diabetes Care*, 29, 936–944 (2006).

71. E. J. A. Pope, K. Braun, and C. M. Peterson, Bioartificial organs I: Silica gel encapsulated pancreatic islets for the treatment of Diabetes Mellitus, *Journal of Sol-Gel Science and Technology*, 8, 635–639 (1997).

72. B. Gimi, T. Leong, Z. Gu, M. Yang, D. Artemov, Z. M. Bhujwalla, and D. H. Gracias, Self-assembled three dimensional radio frequency (RF) shielded containers for cell encapsulation, *Biomedical Microdevices*, 7, 341–345 (2005).

73. P. de Vos, A. F. Hamel, and K. Tatarkiewicz, Considerations for successful transplantation of encapsulated pancreatic islets, *Diabetologia*, 45, 159–173 (2002).

74. S. Schneider, P. J. Feilen, F. Brunnenmeier, T. Minnemann, H. Zimmermann, U. Zimmermann, and M. M. Weber, Long-term graft function of adult rat and human islets encapsulated in novel alginate-based microcapsules after transplantation in immunocompetent diabetic mice, *Diabetes*, 54, 687–693 (2005).

75. W. T. Godbey and A. Atala, In vitro systems for tissue engineering, *Annals of the New York Academy of Sciences*, 961, 10–26 (2002).

76. L. G. Griffith and G. Naughton, Tissue engineering—Current challenges and expanding opportunities, *Science*, 295, 1009–1014 (2002).

77. G. W. Gross, W. Wen, and J. Lin, Transparent indium-tin oxide patterns for extracellular, multisite recording in neuronal culture, *Journal of Neuroscience Methods*, 15, 243–252 (1985).

78. U. Egert, B. Schlosshauer, S. Fennrich, W. Nisch, M. Fejtl, T. Knott, T. Muller, and H. Hammerle, A novel organotypic long-term culture of the rat hippocampus on substrate-integrated multielectrode arrays, *Brain Research Protocols*, 2, 229–242 (1998).

79. M. Bani-Yaghoub, R. Tremblay, R. Voicu, G. Mealing, R. Monette, C. Py, K. Faid, and M. Sikorska, Neurogenesis and neuronal communication on micropatterned neurochips, *Biotechnology and Bioengineering*, 92, 336–345 (2005).

80. Editors' Choice, Ephrins: From axon guidance to neurite inhibitor to viral receptor, *Science STKE*, 2005 tw281(2005).

81. M. D. Benson, M. I. Romero, M. E. Lush, Q. R. Lu, M. Henkemeyer, and L. R. Parada, Ephrin-B3 is a myelin-based inhibitor of neurite outgrowth, *Proceedings of the National Academy of Sciences*, 102, 10694–10699 (2005).

82. Application Note 111, *Versatile biomolecular printing on a variety of surface types*, BioForce Nanosciences, Inc., Ames, IA 50010 USA (2007).

83. C. E. Hulsebosch, Recent advances in pathophysiology and treatment of spinal cord injury, *Advances in Physiology Education*, 26, 238–255 (2002).

84. M. E. Schwab, Repairing the injured spinal cord, *Science*, 295, 1029–1031 (2002).

85. H. M. Bomze, K. R. Bulsara, B. J. Iskandar, P. Caroni, and J. H. P. Skene, Spinal axon regeneration evoked by replacing two growth cone proteins in adult neurons, *Nature Neuroscience*, 4, 38–43 (2001).

86. V. Maquet, D. Martin, B. Malgrange, R. Franzen, J. Schoenen, G. Moonen, and R. Jérôme, Peripheral nerve regeneration using bioresorbable macroporous polylactide scaffolds, *Journal of Biomedical Materials Research*, 52, 639–651 (2000).

87. T.-T. B. Ngo, P. J. Waggoner, A. A. Romero, K. D. Nelson, R. C. Eberhart, and G. M. Smith, Poly(ʟ-lactide) microfilaments enhance peripheral nerve regeneration across extended nerve lesions, *Journal of Neuroscience Research*, 72, 227–238 (2003).

88. G. E. Rutkowski, C. A Miller, S. Jeftinija, and S. K. Mallapragada, Synergistic effects of micropatterned biodegradable conduits and Schwann cells on sciatic nerve regeneration, *Journal of Neural Engineering*, 1, 151–157 (2004).

89. M. Romero-Ortega and P. Galvan-Garcia, A biomimetic synthetic nerve implant, US Patent 20070100358, Texas Scottish Rite Hospital for Children (2007).

90. T. Kmecko, G. Hughes, L. Cauller, J.-B. Lee, and M. Romero-Ortega, Nanocomposites for neural interfaces, In *Electrobiological interfaces on soft substrates*, J. P. Conde, B. Morrison III, and S. P. Lacour, eds., Materials Research Society Symposium Proceedings, 926E, 926-CC04-06 (2006).

91. P. Galvan-Garcia, E. W. Keefer, F. Yang, M. Zhang, S. Fang, A. A. Zakhidov, R. H. Baughman, and M. I. Romero, Robust cell migration and neuronal growth on pristine carbon nanotube sheets and yarns, *Journal of Biomaterials Science, Polymer Edition*, 18, 1245–1261 (2007).

92. M. Zhang, K. R. Atkinson, and R. H. Baughman, Multifunctional carbon nanotube yarns by downsizing an ancient technology, *Science*, 306, 1358–1361 (2004).

93. S. K. Smart, A. I. Cassady, G. Q. Lua, and D. J. Martin, The biocompatibility of carbon nanotubes, *Carbon*, 44, 1034–1047 (2006).

94. J. Chłopek, B. Czajkowska, B. Szaraniec, E. Frackowiak, K. Szostak, and F. Béguin, In vitro studies of carbon nanotubes biocompatibility, *Carbon*, 44, 1106–1111 (2006).

95. M. A. Correa-Duarte, N. Wagner, J. Rojas-Chapana, C. Morscheck, M. Thie, and M. Giersig, Fabrication and biocompatibility of carbon nanotube-based 3D networks as scaffolds for cell seeding and growth, *Nano Letters*, 4, 2233–2236 (2004).

96. S. Garibaldi, C. Brunelli, V. Bavastrello, G. Ghigliotti, and C. Nicolini, Carbon nanotube biocompatibility with cardiac muscle cells, *Nanotechnology*, 17, 391–397 (2006).

97. A. Magrez, S. Kasas, V. Salicio, N. Pasquier, J. W. Seo, M. Celio, S. Catsicas, B. Schwaller, and L. Forró, Cellular toxicity of carbon-based nanomaterials, *Nano Letters*, 6, 1121–1125 (2006).

98. J. M. Wörle-Knirsch, K. Pulskamp, and H. F. Krug, Oops they did it again! Carbon nanotubes hoax scientists in viability assays, *Nano Letters*, 6, 1261–1268 (2006).

99. M. Shim, N. W. S. Kam, R. J. Chen, Y. Li, and H. Dai, Functionalization of carbon nanotubes for biocompatibility and biomolecular recognition, *Nano Letters*, 2, 285–288 (2002).

100. X. Yu, G. P. Dillon, and R. V. Bellamkonda, A laminin and nerve growth factor-laden three-dimensional scaffold for enhanced neurite extension, *Tissue Engineering*, 5, 291–304 (1999) doi:10.1089/ten.1999.5.291.

101. S. Yoshii and M. Oka, Peripheral nerve regeneration along collagen filaments, *Brain Research*, 888, 158–162 (2001).

102. S. Yoshii, M. Oka, M. Shima, A. Taniguchi, and M. Akagi, 30 mm regeneration of rat sciatic nerve along collagen filaments, *Brain Research*, 949, 202–208 (2002).

103. M. R. Ahmed and R. Jayakumar, Peripheral nerve regeneration in RGD peptide incorporated collagen tubes, *Brain Research*, 993, 208–216 (2003).

104. M. R. Ahmed, U. Venkateshwarlu, and R. Jayakumar, Multilayered peptide incorporated collagen tubules for peripheral nerve repair, *Biomaterials*, 25, 2585–2594 (2004).

105. V. Charulatha and A. Rajaram, Crosslinking density and resorption of dimethyl suberimidate-treated collagen, *Journal of Biomedical Materials Research*, 36, 478–486 (1997).

106. V. Charulatha and A. Rajaram, Influence of different crosslinking treatments on the physical properties of collagen membranes, *Biomaterials*, 24, 759–767 (2003).

107. M. R. Ahmed, S. Vairamuthu, M. Shafiuzama, S. H. Basha, and R. Jayakumar, Microwave irradiated collagen tubes as a better matrix for peripheral nerve regeneration, *Brain Research*, 1046, 55–67 (2005).

108. S. Yoshii, M. Oka, N. Ikeda, M. Akagi, Y. Matsusue, and T. Nakamura, Bridging a peripheral nerve defect using collagen filaments, *Journal of Hand Surgery*, 26A, 52–59 (2001).

109. K. A. Ellenbogen and M. A. Wood, *Cardiac pacing and ICDs*, 4th ed., Blackwell, Malden, MA (2005).

110. R. E. Klabunde, *Cardiovascular physiology concepts*, 4th ed., Lippincott Williams & Wilkins, Philadelphia, PA (2005).

111. D. E. Mohrman, *Cardiovascular physiology*, McGraw-Hill, New York (2002).

112. R. M. Berne and M. N. Levy, *Cardiovascular physiology*, 8th ed., Mosby, St. Louis, MO (2001).

113. P. J. Rosch and M. S. Markov, *Bioelectromagnetic medicine*, Informa Health Care, London (2004).

114. K. Mullett, State of the art in neurostimulation, *Pacing and Clinical Electrophysiology*, 10, 162–175 (1987).

115. A. L. Benabid, G. Deuschl, A. E. Lang, K. E. Lyons, and A. R. Rezai, Deep brain stimulation for Parkinson's disease, *Movement Disorders*, 21(Suppl. 14), S168–S170 (2006).

116. R. J. Vetter, J. C. Williams, J. F. Hetke, F. A. Nunamaker, and D. R. Kipke, Chronic neural recording using silicon-substrate microelectrode arrays implanted in cerebral cortex, *IEEE Transactions on Biomedical Engineering*, 51, 896–904 (2004).

117. N. M. Neihart and R. R. Harrison, Micropower circuits for bidirectional wireless telemetry in neural recording applications, *IEEE Transactions on Biomedical Engineering*, 52, 1950–1959 (2005).

118. F. Silveira and D. Flandre, *Low power analog CMOS for cardiac pacemakers design and optimization in bulk and SOI technologies*, Springer Verlag, Berlin, Germany (2004).

119. M. Ghovanloo and G. Lazzi, *Transcutaneous magnetic coupling of power and data, in Wiley Encyclopedia of biomedical engineering*, John Wiley & Sons, New York (2006).

120. C. M. John and V. John, Device for neuromuscular peripheral body stimulation and electrical stimulation (ES) for wound healing using RF energy harvesting, US Patent 2000161216, San Francisco, CA (2006).

121. M. Glikson and P. Friedman, The implantable cardioverter defibrillator, *Lancet*, 357, 1107–1117 (2001).

122. R. Allen, *Medtronic sets the pace with implantable electronics*, Electronic Design Online ID #5951 (2003).

123. R. Kötz and M. Carlen, Principles and applications of electrochemical capacitors, *Electrochimica Acta*, 45, 2483–2498 (2000).

124. W. E. Finn and P. G. LoPresti, *Handbook of neuroprosthetic methods*, CRC Press, Boca Raton, FL (2002).

125. K. W. Horch and G. S Dhillon, eds., *Neuroprosthetics, theory and practice*, World Scientific Publishing, Singapore (2004).

126. P. Rebelo and M. Van Walstijn, Designing acoustic thresholds, In *Les journées du design sonore*, IRCAM, Paris (2004).

127. D. H. Delgado, V. Rao, H. J. Ross, S. Verma, and N. G. Smedira, Mechanical circulatory assistance: State of art, *Circulation*, 106, 2046–2050 (2002).

128. H. H. Hu, P. Jia, T. Lu, and K. Yuan, Head gesture recognition for hands-free control of an intelligent wheelchair, *Industrial Robot: An International Journal*, 34, 60–68 (2007).

129. H.-N. Teodorescu and L. C. Jain, *Intelligent systems and technologies in rehabilitation engineering*, CRC Press, Boca Raton, FL (2001).

130. D. Taylor, Neural control of assistive technology, In *Wiley Encyclopedia of biomedical engineering*, John Wiley & Sons, Inc., New York (2006).

131. C. M. Light, P. H. Chappell, B. Hudgins, and K. Engelhart, Intelligent multifunction myoelectric control of hand prosthesis, *Journal of Medical Engineering and Technology*, 26, 139–146 (2002).

132. G. M. Friehs, V. A. Zerris, C. L. Ojakangas, M. R. Fellows, and J. P. Donoghue, Brain–machine and brain–computer interfaces, *Stroke*, 35, 2702–2705 (2004).

133. J. D. Weingarten, G. Lopes, R. E. Groff, M. Buehler, and D. E. Koditschek, Automated gait adaptation for legged robots, IEEE International Conference on Robotics Automation (ICRA), New Orleans, LA (2004).

134. R. Rupp and H. J. Gerner, Neuroprosthetics of the upper extremity: Clinical application in spinal cord injury and future perspectives, *Biomedizinische Technik, (Berlin)*, 49, 93–98 (2004).

135. R. Gailey, Rehabilitation of a traumatic lower limb amputee, *Physiotherapy Research International*, 3, 239–243 (2006).

136. M. Lowe, A. King, E. Lovett, and T. Papakostas, Flexible tactile sensor technology: Bringing haptics to life, *Sensor Reviews*, 24, 33–36 (2004).

137. H. Atmani, F. Merienne, D. Fofi, and P. Trouilloud, Computer aided surgery system for shoulder prosthesis placement, *Computer Aided Surgery*, 12, 60–70 (2007).

138. M. Ciocarlie, C. Goldfeder, and P. Allen, Dimensionality reduction for hand-independent dexterous robotic grasping, presentation, IROS: IEEE/RSJ International Conference on Intelligent Robots and Systems, San Diego, CA (2007).

139. P. S. Blaer and P. K. Allen, Data Acquisition and View Planning for 3-D Modeling Tasks, presentation, IROS: IEEE/RSJ International Conference on Intelligent Robots and Systems, San Diego, CA (2007).

140. R. D. Howe and Y. Matsuoka, Robotics for surgery, *Annual Review of Biomedical Engineering*, 1999, 1, 211–240 (1999).

141. S. J. Weghorst, K. S. Morgan, and H. B. Sieburg, *Medicine meets virtual reality: Health care in the information age*, IOS Press, Incorporated, Fairfax, VA (1996).

142. J. D. Westwood, *Medicine meets virtual reality 2001: Outer space, inner space, virtual space (Studies in health technology and informatics)*, IOS Press, Incorporated, Fairfax, VA (2001).

143. B. Beckwith, Medicine meets virtual reality 2001 (Review), *Clinical Chemistry*, 47, 2190 (2001).

144. S. Canright, ed., *Amy Ross, space suit designer*, NASA Education Home, STS-118, Resources for Educators, webpage at http://www.nasa.gov/audience/foreducators/stseducation/stories/Amy_Ross_Profile.html (2007).

145. P. Danaher, K. Tanaka, and A. R. Hargens, Mechanical counter-pressure vs. gas-pressurized spacesuit gloves: Grip and sensitivity, *Aviation, Space, and Environmental Medicine*, 76, 381–384 (2005).

146. S. E. Lyshevski, *Nano- and micro-electromechanical systems: Fundamentals of nano- and microengineering*, CRC Press, Boca Raton, FL (2005).

147. R. Zurawski, *Embedded systems handbook*, CRC Press, Boca Raton, FL (2006).

148. M. Rieth, *Nano-engineering in science and technology: An introduction to the world of nano-design*, World Scientific Publishing, Singapore (2003).

149. M. Gad-el-Hak, *The MEMS handbook*, CRC Press, Boca Raton, FL (2002).

150. B. G. Lipták, *Instrument engineers' handbook: Process control and optimization*, CRC Press, Boca Raton, FL (2005).

151. D. Hristu-Varsakelis, W. S. Levine, R. Alur, K.-E. Arzen, J. Baillieul, and T. A. Henzinger, *Handbook of networked and embedded control systems*, Birkhäuser, Boston, MA (2005).

152. M. Shahinpoor, K. J. Kim, and M. Mojarrad, *Artificial muscles: Applications of advanced polymeric nanocomposites*, Taylor & Francis, New York (2007).

153. Y. Osada and D. E. De Rossi, *Polymer sensors and actuators (Macromolecular systems—Materials approach)*, Springer Verlag, Berlin, Germany (1999).

154. Y. Gogotsi, *Nanomaterials handbook*, CRC Press, Boca Raton, FL (2006).

155. H.-J. Fecht and Y. Champion, *Nano-architectured and nanostructured materials: Fabrication, control and properties*, Wiley-VCH, Weinheim, Germany (2006).

156. H. S. Nalwa, *Handbook of nanostructured biomaterials and their applications in nanobiotechnology*, (2 Vols), American Scientific Publishers, Stevenson Ranch, CA (2005).

157. B. D. Ratner, A. Hoffman, F. Schoen, and J. Lemons, *Biomaterials science: An introduction to materials in medicine*, 2nd ed., Academic Press, Burlington, MA (2004).

158. R. W. Bogue, Nanotechnology: What are the prospects for sensors? *Sensor Reviews*, 24, 253–260 (2004).

159. W. Lu and C. M. Lieber, Nanoelectronics from the bottom-up, *Nature Materials*, 6, 841–850 (2007).

160. K. Bullis, Ultrastrong carbon-nanotube muscles: Artificial muscles made from carbon nanotubes are 100 times stronger than human muscles, *MIT Technology Review*, Dec. 8, 1–2 (2006).

161. University of Texas at Dallas, Nano technologists demonstrate artificial muscles powered by highly energetic fuels, *Science Daily*, March 17, webpage at: http://www.sciencedaily.com-/releases/2006/03/060317110801.htm (2006).

162. Y. Bar-Cohen, Electroactive polymers as artificial muscles: Reality, potential and challenges, EAP Actuators and Devices Conference: Smart Structures and Materials Symposium, San Diego, CA, March 6, NASA Jet Propulsion Laboratory, Pasadena, CA (2005).

163. NASA/Jet Propulsion Laboratory, Scientists "muscle" Sci-Fi into reality, *Science Daily* June 11, webpage at: http://www.sciencedaily.com-/releases/2002/06/020611071940.htm (2002).

164. D. H. Nielsen, D. G. Shurr, J. C. Golden, and K. Meier, Comparison of energy cost and gait efficiency during ambulation in below-knee amputees using different prosthetic feet—A preliminary report, *Journal of Prosthetics and Orthotics*, 1, 24–31 (1989).

165. S. E. Irby, K. R. Kaufman, R. W. Wirta, and D. H. Sutherland, Optimization and application of a wrap-spring clutch to a dynamic knee-ankle-foot orthosis, *IEEE Transactions on Rehabilitation Engineering*, 7, 130–134 (1999).

166. B. S. Farber and J. S. Jacobson, An above-knee prosthesis with a system of energy recovery: A technical note, *Journal of Rehabilitation Research and Development*, 32, 337–348 (1995).

167. J. Dewen, W. D. Pilkey, J. Zhang, and W. A. Gruver, Analytical evaluation of an energy-storing foot prosthesis, *System Man Cybernetics: IEEE International Conference on Intelligent Systems 21st Century*, Vol. 1, Waseda University International Conference Center, Tokyo, Japan, pp. 513–517 (1995).

168. K. R. Kaufman, B. K. Iverson, D. J. Padgett, R. H. Brey, J. A. Levine, and M. J. Joyner, Paired outcome assessment of a microprocessor controlled knee versus a mechanical prosthesis, *Gait and Posture*, 24, S57–S59 (2006).

169. G. Chamberlain, Artificial foot and knee designed jointly by U.S., Russian labs, *Design News*, September (1999).

170. Sandia National Laboratories, American and Russian nuclear labs work with prosthetics company to develop artificial knee for landmine victims, *Science Daily*, August 12, On website at http://www.sciencedaily.com-/releases/1999/08/990812081041.htm (1999).

171. *Ossur Proprio Foot*, Ossur hf, 110 Reykjavik, Iceland, web page at: http://bionics.ossur.com/Home/INTRODUCTION/LAWS-OF-BIONICS (2008).

172. The New York Times, *Electronic brain in an artificial foot*, October 3, web animation at:http://www.nytimes.com/packages/html/science/20061003_FOOT_GRAPHIC/index.html (2008).

173. P. Chen, Y. Hasegawa, and M. Yamashita, Grasping control of robot hand using fuzzy neural network, *Advances in Neural Networks*, 1178–1187 (2006).

174. J. M. Winters, Commentary: A case for soft neurofuzzy control interfaces for humans with disabilities, In *Biomechanics and neural control of movement*, J. M. Winters, and P. E. Crago, eds., chap. 39, pp. 548–550, Springer Verlag, New York (2000).

175. G. Lundborg, Tomorrow's artificial hand, *Scandinavian Journal of Plastic and Reconstructive Surgery and Hand Surgery*, 34, 97–100 (2000).

176. J. Yang, E. P. Pitarch, K. Abdel-Malek, A. Patrick, and L. Lindkvist, A multi-fingered hand prosthesis, *Mechanism and Machine Theory*, 39, 555–581 (2004).

177. M. C. Carrozza, B. Massa, S. Micera, R. Lazzarini, M. Zecca, and P. Dario, The development of a novel prosthetic hand-ongoing research and preliminary results, *IEEE/ASME Transactions on Mechatronics*, 7, 108–114 (2002).

178. P. J. Kyberd, C. Light, P. H. Chappell, J. M. Nightingale, D. Whatley, and M. Evans, The design of anthropomorphic prosthetic hands: A study of the Southampton hand, *Robotica*, 19, 593–600 (2001).

179. P. Kyberd, The intelligent hand, *IEE Review*, 46, 31–35 (2000).

180. F. L. Lewis, D. M. Dawson, and C. T. Abdallah, *Robot manipulator control: Theory and practice*, 2nd ed., CRC Press, Boca Raton, FL (2003).

181. T. R. Kurfess, *Robotics and automation handbook*, CRC Press, Boca Raton, FL (2004).

182. M. C. Carrozza, G. Cappiello, L. Beccai, F. Zaccone, S. Micera, and P. Dario, Design methods for innovative hand prostheses, *IEMBS04: 26th Annual International Conference of the IEEE Engineering Medicine and Biology Society*, San Francisco, Vol. 6, pp. 4345–4348 (2004).

183. P. H. Chappell, A. Cranny, D. P. J. Cotton, N. M. White, and S. P. Beeby, Sensory motor systems of artificial and natural hands, *International Journal of Surgery*, 6, doi:10.1016/j.ijsu.2006.06.028, 436–440 (2008).

184. M. C. Carrozza, B. Massa, S. Micera, M. Zecca, and P. Dario, A "wearable" artificial hand for prosthetics and humanoid robotics applications, *Proceedings of the 2001 IEEE –RAS International Conference on Humanoid Robots*, 22–24 (2001).

185. M. C. Carrozza, F. Vecchi, S. Roccella, L. Barboni, E. Cavallaro, S. Micera, and P. Dario, The ADAH project: An astronaut dexterous artificial hand to restore the manipulation abilities of the astronaut, 7th ESA Workshop on Advanced Space Technologies for Robotics and Automation "ASTRA 2002" ESTEC, Noordwijk, the Netherlands, November 19–21 (2002).

186. K. J. De Laurentis, Design of a rapidly fabricated, smart material actuated robotic manipulator for space applications, 55th International Astronautical Congress, International Astronautical Federation, International Academy of Astronautics, International Institute of Air and Space Law IAC-04-I.4.11, Vancouver, Canada (2004).

187. K. J. De Laurentis and C. Mavroidis, Actuators for artificial limbs, *Technology and Health Care*, 10, 91–106 (2002).

188. W. Craelius, R. L. Abboudi, N. A. Newby, and J. Flint, Control of a mutli-finger prosthetic hand, *IEEE Transactions on Rehabilitation Engineering*, 7, 121–129 (1999).

189. M. Harris, P. Kyberd, and W. S. Harwin, Design and development of a dextrous manipulator, *Transactions of the Institute of Measurement and Control*, 27, 137–152 (2005).

190. A. Poulton, P. J. Kyberd, and D. Gow, Progress of a modular prosthetic arm, In *Universal access and assistive technology*, S. Keates, P. Langdon, J. Clarkson, and P. Robinson, eds., pp. 193–200, Springer Verlag, Berlin, Germany (2002).

191. P. J. Kyberd, N. Mustapha, F. Carnegie, and P. H. Chappell, A clinical experience with a hierarchically controlled myoelectric hand prosthesis with vibro-tactile feedback, *Prosthetics and Orthotics International*, 17, 56–64 (1993).

192. J. M. Winters and M. F. Story, eds., *Medical instrumentation: Accessibility and usability considerations*, CRC Press, Boca Raton, FL (2006).

193. R. Rupp and H. J. Gerner, *Neuroprosthetics of the upper extremity-clinical application in spinal cord injury and future perspectives*, Biomedizinische Technik, (Berlin), 49, 93–98 (2004).

194. F. W. J. Cody, ed., *Neural control of skilled human movement*, Portland Press Ltd., Colchester, UK (1995).

195. M. Kuttiva, G. Burdea, J. Flint, and W. Craelius, Manipulation practice for upper-limb amputees using virtual reality, *Presence: Teleoperators and Virtual Environments* 14, 175–182 (2005).

196. J. M. Winters and P. E. Crago, eds., *Biomechanics and neural control of movement*, Springer Verlag, New York (2000).

197. R. K. Shields, Muscular, skeletal, and neural adaptations following spinal cord injury, *Journal of Orthopaedic and Sports Physical Therapy*, 32, 65–74 (2002).

198. Editorial, Is this the bionic man? *Nature*, 442, 164–171 (2006).

199. W. Craelius, The bionic man: Restoring mobility, *Science*, 295, 1018–1021 (2002).

200. S. H. Scott, Neuroscience: Converting thoughts into action, *Nature*, 442, 164–171 (2006).

201. T. M. Vaughan and J. R. Wolpaw, The third international meeting on brain-computer interface technology: Making a difference, *IEEE Transactions on Neural Systems and Rehabilitation Engineering*, 14, 126–127 (2006).

202. L. R. Hochberg, M. D. Serruya, G. M. Friehs, J. A. Mukand, M. Saleh, A. H. Caplan, A. Branner, D. Chen, R. D. Penn, and J. P. Donoghue, Neuronal ensemble control of prosthetic devices by a human with tetraplegia, *Nature*, 442, 164–171 (2006).

203. G. Santhanam, S. I. Ryu, B. M. Yu, A. Afshar, and K. V. Shenoy, A high-performance brain–computer interface, *Nature*, 442, 195–198 (2006).

204. M. A. Lebedev and M. A. L. Nicolelis, Brain–machine interfaces: past, present and future, *Trends in Neurosciences*, 29, 536–546 (2006).

205. P. D. Cheney, J. Hill-Karrer, A. Belhaj-Saïf, B. J. McKiernan, M. C. Park, and J. K. Marcario, Cortical motor areas and their properties: implications for neuro-prosthetics, *Progress in Brain Research*, 1, 135–160 (2000).

206. V. Brezina, I. V. Orekhova, and K. R. Weiss, The neuromuscular transform: The dynamic, nonlinear link between motor neuron firing patterns and muscle contraction in rhythmic behaviors, *Journal of Neurophysiology*, 83, 207–231 (2000).

207. M. A. Lebedev, J. M. Carmena, J. E. O'Doherty, M. Zacksenhouse, C. S. Henriquez, J. C. Principe, and M. A. L. Nicolelis, Cortical ensemble adaptation to represent actuators controlled by a brain machine interface, *Journal of Neuroscience*, 25, 4681–4693 (2005).

208. W. O. Friesen and R. J. Wyman, Analysis of Drosophila motor neuron activity patterns with neural analogs, *Journal of Biological Cybernetics*, 38, 41–50 (1980).

209. A. E. Lindsay, K. A. Lindsay, and J. R. Rosenberg, New concepts in compartmental modeling, *Computing and Visualization Science*, 10, 79–98 (2007).

210. K. A. Lindsay, J. M. Ogden, and J. R. Rosenberg, Dendritic subunits determined by dendritic morphology, *Neural Computation*, 13, 2465–2476 (2001).

211. D. J. Weber, R. B. Stein, D. G. Everaert, and A. Prochazka, Decoding sensory feedback from firing rates of afferent ensembles recorded in Cat dorsal root ganglia in normal locomotion, *IEEE Transactions on Neural Systems and Rehabilitation Engineering*, 14, 240–245 (2006).

212. J. Wessberg, C. R. Stambaugh, J. D. Kralik, P. D. Beck, M. Laubach, J. K. Chapin, J. Kim, S. J. Biggs, M. A. Srinivasan, and M. A. Nicolelis, Real-time prediction of hand trajectory by ensembles of cortical neurons in primates, *Nature*, 16, 361–365 (2000).

213. J. P. Donoghue, Connecting cortex to machines: Recent advances in brain interfaces, *Nature Neuroscience*, 5, 1085–1088 (2002).

214. S. G. Mason, Z. Bozorgzadeh, G. E. Birch, The LF-ASD brain-computer interface: On-line identification of imagined finger flexions in subjects with spinal cord injuries, *Proceedings of the ASSETS 2000*, ACM Press, Washington, USA, 108–109 (2000).

215. J. M. Carmena, M. A. Lebedev, R. E. Crist, J. E., O'Doherty, D. M. Santucci, D. F. Dimitrov, P. G. Patil, C. S. Henriquez, and M. A. L. Nicolelis, Learning to control a brain-machine interface for reaching and grasping by primates, *PLoS Biology*, 1, 193–208 (2003).

216. S.-P. Kim, J. C. Sanchez, Y. N. Rao, D. Erdogmus, J. M. Carmena, M. A. Lebedev, M. A. L. Nicolelis, and J. C. Principe, A comparison of optimal MIMO linear and nonlinear models for brain–machine interfaces, *Journal of Neural Engineering*, 3, 145–161 (2006).

217. S. Rezaei, K. Tavakolian, A. M. Nasrabadi, and S. K. Setarehdan, Different classification techniques considering brain computer interface applications, *Journal of Neural Engineering*, 3, 139–144 (2006).

218. M. Bogdan, M. Schröder, and W. Rosenstiel, Artificial neural net based signal processing for interaction with peripheral nervous system, *Proceedings of the 1st International IEEE EMBS Conference on Neural Engineering 2003*, pp. 134–137 (2003).

219. G. J. Gage, J. K. Otto, K. A. Ludwig, and D. R. Kipke, Co-adaptive Kalman filtering in a naive rat cortical control task, *IEMBS04: 26th Annual International Conference of IEEE Engineering, Medical and Biological Society*, Vol. 6, pp. 4367–4370 (2004).

220. W. Wu, A. Shaikhouni, J. R. Donoghue, and M. J. Black, Closed-loop neural control of cursor motion using a Kalman filter, *IEMBS-4: 26th Annual International Conference of IEEE Engineering, Medical and Biological Society*, Vol. 6, pp. 4126–4129 (2004).

221. R. Tomioka and K. Aihara, Classifying matrices with a spectral regularization, *Proceedings of 24th International Conference on Machine Learning*, ICML07, pp. 895–902, ACM Press, Washington, USA (2007).

222. N. J. Hill, M. Schröder, T. N. Lal, and B. Schölkopf, Comparative evaluation of Independent Components Analysis algorithms for isolating target-relevant information in brain-signal classification, *Brain-Computer Interface Technology*, 3, 95 (2005).

223. H. Parikh, G. Gage, T. C. Marzullo, D. Kipke, Real-time detection of unitary events for cortical control, *EMBS05: 27th Annual International Conference of IEEE Engineering, Medical and Biological Society*, Shanghai, China, pp. 2122–2125 (2005).

224. M. C. Lovett, C. D. Schunn, C. Lebiere, and P. Munro, *Sixth International Conference on Cognitive Modeling: ICCM-2004*, Psychology Press, Routledge, London (2004).

225. M. R. Jane, D. Marini, and A. De Gloria, *Transputer Applications and Systems '94: Proceedings of the 1994 World Transputer Congress*, Cernobbio, Italy, IOS Press, Amsterdam, the Netherlands (1994).

226. R. R. Burridge, A. A. Rizzi, and D. E. Koditschek, Sequential composition of dynamically dexterous robot behaviors, *International Journal of Robotics Research*, 18, 534–555 (1999).

227. G. A. D. Lopes and D. E. Koditschek, Visual servoing for nonholonomically constrained three degree of freedom kinematic systems, *International Journal of Robotics Research*, 26, 715–736 (2007).

228. J. R. Wolpaw, N. Birbaumer, D. J. McFarland, G. Pfurtscheller, and T. M. Vaughan, Brain-computer interfaces for communication and control, *Clinical Neurophysiology*, 113, 767–791 (2002).

229. M. Schröder, T. N. Lal, T. Hinterberger, M. Bogdan, N. J. Hill, N. Birbaumer, W. Rosenstiel, and B. Schölkopf, Robust EEG channel selection across subjects for brain computer interfaces, *EURASIP Journal on Applied Signal Processing, Special Issue: Trends in Brain Computer Interfaces*, 19, 3103–3112 (2005).

230. M. Schröder, M. Bogdan, W. Rosenstiel, T. Hinterberger, and N. Birbaumer, Automated EEG feature selection for brain computer interfaces, *Proceedings of the 1st International IEEE EMBS Conference on Neural Engineering, 2003*, pp. 626–629 (2003).

231. J. A. Wilson, E. A. Felton, P. C. Garell, G. Schalk, and J. C. Williams, ECoG factors underlying multimodal control of a brain–computer interface, *IEEE Transactions on Neural Systems and Rehabilitation Engineering*, Vol. 14, pp. 246–249 (2006).

232. E. A. Felton, J. A. Wilson, J. C. Williams, and P. C. Garell, Electrocorticographically controlled brain–computer interfaces using motor and sensory imagery in patients with temporary subdural electrode implants: Report of four cases, *Journal of Neurosurgery*, 106, 495–500 (2007).

233. D. M. Taylor, S. I. Tillery, and A. B. Schwartz, Direct cortical control of 3D neuroprosthetic devices, *Science*, 296, 1829–1832 (2002).

234. M. S. Hämäläinen, Progress and challenges in multimodal data fusion, *CARS 2007—Proceedings of the International Congress*, 1300, 15–18 (2007).

235. A. Pascual-Leone, N. Davey, J. C. Rothwell, E. M. Wassermann, and B. K. Puri, *Handbook of transcranial magnetic stimulation*, CRC Press, Boca Raton, FL (2002).

236. V. Walsh and A. Pascual-Leone, *Transcranial magnetic stimulation*, MIT Press, Cambridge, MA (2003).

237. P. B. Fitzgerald, S. Fountain, and Z. J. Daskalakis, A comprehensive review of the effects of rTMS on motor cortical excitability and inhibition, *Clinical Neurophysiology*, 117, 2584–2596 (2006).

238. A. T. Barker, R. Jalinous, and I. L. Freeston, Non-invasive magnetic stimulation of human motor cortex, *Lancet*, 1(8437), 1106–1107 (1985).

239. T. Kujirai, Corticocortical inhibition of the motor cortex, *Journal of Physiology*, 471, 501–509 (1993).

240. A. Pascual-Leone, D. Bartres-Faz, and J. P. Keenan, Transcranial magnetic stimulation: studying the brain-behaviour relationship by induction of virtual lesions, *Philosophical Transactions of the Royal Society of London B: Biological Science*, 354, 1229–1238 (1999).

241. S. Tumanski, *Thin film magnotoresistive sensors*, CRC Press, Boca Raton, FL (2001).

242. J. R. Brauer, *Magnetic actuators and sensors*, IEEE/Wiley Interscience, New York (2006).

243. S. L. Mouaziz, *Micro and nano tools for magnetic field imaging—Series in microsystems*, P. A. Besse, J. Brugger, M. Gijs, R. S. Popovic, and P. Renaud, eds., Vol. 21, Hartung-Gorre Verlag Konstanz, Germany (2007).

244. S. H. Allos, D. J. Staton, L. A. Bradshaw, S. Halter, J. P. Wikswo, Jr., and W. O. Richards, Superconducting quantum interference device magnetometer for diagnosis of Ischemia caused by mesenteric venous thrombosis, *World Journal of Surgery*, 21, 173–178 (1997).

245. D. King, A resonant MEMS magnetometer, *IEE Seminar and Exhibition on MEMS Sensor Technology, 2005*, pp. 1–12 (2005).

246. C. Schott, F. Burger, H. Blanchard, and L. Chiesi, Modern integrated silicon Hall sensors, *Sensor Reviews*, 18, 252–257 (1998).

247. A. P. Ramirez, Colossal magnetoresistance, *Journal of Physics: Condensed Matter*, 9, 8171–8199 (1997).

248. E. Paperno and B.-Z. Kaplan, Sub-nano-tesla in-plane vector magnetometer employing single magnetoresistor, *Nineteenth Convention on Electrical and Electronics Engineering in Israel*, 1996, pp. 188–191 (1996).

249. S. R. Brankovic, X. M. Yang, T. J. Klemmer, and M. Seigler, Electrodeposition of 2.4T Co37Fe63 alloys at nanoscale for magnetic recoding application, *IEEE Transactions on Magnetics*, 42, 132 (2006).

250. S. A. Solin, T. Thio, D. R. Hines, and J. J. Heremans, Enhanced room-temperature geometric magnetoresistance in inhomogeneous narrow-gap semiconductors, *Science*, 289, 1530–1532 (2000).

251. J. Clarke, Principles and applications of SQUIDs, *Proceedings of IEEE*, 77, 1208–1223 (1989).

252. Y. Okada, K. Pratt, C. Atwood, A. Mascarenas, R. Reineman, J. Nurminen, and D. Paulson, BabySQUID: A mobile, high-resolution multichannel magnetoen-cephalography system for neonatal brain assessment, *Review of Scientific Instruments*, 77, 024301 (2006).

253. L. A. Bradshaw, A. Irinia, J. A. Sims, M. R. Gallucci, R. L. Palmer, and W. O. Richards, Biomagnetic characterization of spatiotemporal parameters of the gastric slow wave, *Neurogastroenterology and Motility*, 18, 619–631 (2006).

254. V. Shah, S. Knappe, P. D. D. Schwindt, and J. Kitching, Subpicotesla atomic magnetometry with a microfabricated vapour cell, *Nature Photonics*, 1, 649–652 (2007).

255. S. Xu, M. H. Donaldson, A. Pinesb, S. M. Rochester, D. Budkerc, and V. V. Yashchuk, Application of atomic magnetometry in magnetic particle detection, *Applied Physics Letters*, 89, 224105 (2006).

256. I. K. Kominis, T. W. Kornack, J. C. Allred, and M. V. Romalis, A subfemtotesla multichannel atomic magnetometer, *Nature*, 422, 596–599 (2003).

257. D. S. Greywall, Sensitive magnetometer incorporating a high-Q nonlinear mechanical resonator, *Measurement Science Technology*, 16, 2473–2482 (2005).

258. H. Xia, A. B.-A. Baranga, D. Hoffman, and M. V. Romalis, Magnetoencephalography with an atomic magnetometer, *Applied Physics Letters*, 89, 211104 (2006).

259. Z. Li, R. T. Wakai, and T. G. Walker, Parametric modulation of an atomic magne-tometer, *Applied Physics Letters*, 89, 134105 (2006).

260. L. Balcells, E. Calvo, and J. Fontcuberta, Room-temperature anisotropic magne-toresistive sensor based on manganese perovskite thick films, *Journal of Magnetism and Magnetic Materials*, 242–245, 1166–1168 (2002).

261. M. M. Raja, R. J. Gambino, S. Sampath, and R. Greenlaw, Thermal sprayed thick-film anisotropic magnetoresistive sensors, *IEEE Transactions on Magnetics*, 40, 2685–2687 (2004).

262. G. B. Donaldson, SQUIDs—ultimate magnetic sensors, *Physica Status Solidi*, 2, 1463–1467 (2005).

263. A. Candini, G. C. Gazzadi, A. di Bona, M. Affronte, D. Ercolani, G. Biasio, and L. Sorba, Hall nano-probes fabricated by focused ion beam, *Nanotechnology*, 17, 2105–2109 (2006).

264. J. Vrba and S. E. Robinson, Signal Processing in magnetoencephalography, *Methods*, 25, 249–271 (2001).

265. A. A. Fife, J. Vrba, S. E. Robinson, G. Anderson, K. Betts, M. B. Burbank, D. Cheyne, E. Cheung, S. Govorkov, G. Haid, V. Haid, C. Hunter, P. R. Kubik, S. Lee,

J. McKay, E. Reichl, C. Schroyen, I. Sekachev, P. Spear, B. Taylor, M. Tillotson, and W. Sutherling, Synthetic gradiometer systems for MEG, *IEEE Transactions on Applied Superconductivity*, 9, 4063–4068 (1999).

266. J. Vrba and S. E. Robinson, Linearly constrained minimum variance beamformers, synthetic aperture magnetometry, and MUSIC in MEG applications, *Signals, Systems and Computers, 34th Asilomar Conference 2000*, Vol. 1, pp. 313–317 (2000).

267. R. Frostig, ed., *In vivo optical imaging of brain function (Methods and new frontiers in neuroscience)*, CRC Press, Boca Raton, FL (2002).

268. J. Wells, C. Kao, K. Mariappan, J. Albea, E. D. Jansen, P. Konrad, and A. Mahadevan-Jansen, Optical stimulation of neural tissue in vivo, *Optics Letters*, 30, 504–506 (2005).

269. J. Wells, C. Kao, P. Konrad, T. Milner, J. Kim, A. Mahadevan-Jansen, and E. D. Jansen, Biophysical mechanisms of transient optical stimulation of peripheral nerve, *Biophysical Journal*, 93, 2567–2580 (2007).

270. G. Caetano and V. Jousmäki, Evidence of vibrotactile input to human auditory cortex, *Neuro Image*, 29, 15–28 (2005).

271. T. N. Lal, M. Schröder, J. Hill, H. Preissl, T. Hinterberger, J. Mellinger, M. Bogdan, W. Rosenstiel, T. Hofmann, N. Birbaumer, and B. Schölkopf, A Brain Computer Interface with Online Feedback based on Magnetoencephalography, In *Proceedings of the 22nd International Conference on Machine Learning*, L. De Raedt and S. Wrobel, eds., pp. 465–472, ACM Press, Washington, USA (2005).

272. D. Hecht and M. Reiner, Field dependency and the sense of object-presence in haptic virtual environments, *Cyber Psychology and Behavior*, 10, 243–251 (2007).

273. C. R. Wagner and R. D. Howe, Mechanisms of performance enhancement with force feedback, *First Joint Eurohaptics Conference and Symposium on Haptic Interfaces for Virtual Environment and Teleoperator Systems WHC05*, pp. 21–29 (2005).

274. A. M. Okamura, Methods for haptic feedback in teleoperated robot-assisted surgery, *Industrial Robot: An International Journal*, 31, 499–508 (2004).

275. G. A. Calvert, C. Spence, and B. E. Stein, *The handbook of multisensory processes*, MIT Press, Cambridge, MA (2004).

276. L. Hochberg and D. Taylor, Intuitive prosthetic limb control, *Lancet*, 369, 345–346 (2007).

277. T. A. Kuiken, L. A. Miller, R. D. Lipschutz, B. A. Lock, K. Stubblefield, P. D. Marasco, P. Zhou, and G. A. Dumanian, Targeted reinnervation for enhanced prosthetic arm function in a woman with a proximal amputation: A case study, *Lancet*, 369, 371–380 (2007).

278. J. Arnott, N. Alm, and A. Waller, Cognitive prostheses: communication, rehabilitation and beyond, *IEEE International Conference on Systems Man and Cybernetics IEEE SMC 1999*, Vol. 6, pp. 346–351 (1999).

279. W. Barfield and T. Caudell, *Fundamentals of wearable computers and augmented reality*, Lawrence Erlbaum Associates/CRC Press, Boca Raton, FL (2001).

280. S. K. Rosahl, Neuroprosthetics and neuroenhancement: Can we draw a line? *Virtual Mentor*, 9, 132–139 (2007).

281. M. E. Clynes and N. S. Kline, Cyborgs and space, *Astronautics*, September, 26–27 (1960).

282. R. Clarke, Human-artefact hybridisation: Forms and consequences, *Ars Electronica 2005 Symposium on Hybrid - Living in Paradox*, Linz, Austria, September 2–3 (2005).

283. F. Jotterand, Nanomedicine: How it could reshape clinical practice? *Nanomedicine* 2, 401–405 (2007).

284. US NIDCD: National Institute on Deafness and Other Communication Disorders, *Cochlear Implants*, NIH Publication No. 00-4798, Website at: http://www.nidcd.nih.gov/ (2007).

285. US NIDCD: National Institute on Deafness and Other Communication Disorders, Statistics about hearing disorders, ear infections, and deafness, Website at: http://www.nidcd.nih.gov/health/statistics/hearing.asp (2008).

286. R. D. Kent, *The MIT Encyclopedia of communication disorders*, MIT Press, Cambridge, MA (2004).

287. D. E. Ingber, Cellular mechanotransduction: Putting all the pieces together again, *FASEB Journal*, 20, 811–827 (2006).

288. D. I. Margolin, *Cognitive neuropsychology in clinical practice*, Oxford University Press, New York (1992).

289. P. R. Cook, *Music, cognition, and computerized sound: An introduction to psychoacoustics*, MIT Press, Cambridge, MA (2001).

290. V. K. Madisetti and D. Williams, *The digital signal processing handbook*, CRC Press, Boca Raton, FL (1997).

291. M. Kahrs and K. Brandenburg, *Applications of digital signal processing to audio and acoustics*, Springer Verlag, Berlin, Germany (1998).

292. B. Wilson, Digital signal processing applications for hearing accessibility, *IEEE Signal Processing Magazine*, 20, 14–18 (2003).

293. J. Tierny, M. A. Zissman, and D. K. Eddington, Digital signal processing applications in cochlear-implant research, *Lincoln Laboratory Journal*, 7, 31–62 (1994).

294. G. J. M. Krijnen, M. Dijkstra, J. J. van Baar, S. S. Shankar, W. J. Kuipers, R. J. H. de Boer, D. Altpeter, T. S. J. Lammerink, and R. Wiegerink, MEMS based hair flow-sensors as model systems for acoustic perception studies, *Nanotechnology*, 17, S84–S89 (2006).

295. G. Clark, *Cochlear implants: Fundamentals and applications*, Springer Verlag, Berlin (2003).

296. US FDA: Food and Drug Administration, *Cochlear Implants*, website at: http://www.fda.gov/cdrh/cochlear/index.html (2008).

297. M. Valente, H. Hosford-Dunn, and R. J. Roeser, *Audiology: Treatment*, Thieme Medical Publishers, New York (2000).

298. J. W. Hall, *New handbook for auditory evoked responses*, Allyn & Bacon, Boston, MA (2006).

299. G. Miller, *Sensory organ replacement and repair*, Morgan & Claypool Publishers, San Rafael, CA (2006).

300. A. R. Moller, *Cochlear and brainstem implants (Advances in otorhinolaryngology)*, Karger, Basel, SZ (2006).

301. K. S. Pawlowski, Anatomy and physiology of the Cochlea, In *Ototoxicity*, P. S. Roland and J. A. Rutka, eds., BC Decker Inc, London (2004).

302. K. S. Pawlowski, D. Wawro, and P. S. Roland, Bacterial Biofilm Formation on a Human Cochlear Implant, *Otology and Neurotology*, 26, 972–975 (2005).

303. T. A. Johnson, K. A. Loeffler, R. A. Burne, C. N. Jolly, and P. J. Antonelli, Biofilm formation in cochlear implants with cochlear drug delivery channels in an in vitro model, *Otolaryngology–Head and Neck Surgery*, 136, 577–582 (2007).

304. R. Cristobal, C. E. Edmiston Jr., C. L. Runge-Samuelson, H. A. Owen, J. B. Firszt, and P. A. Wackym, Fungal biofilm formation on cochlear implant hardware after antibiotic-induced fungal overgrowth within the middle ear, *The Pediatric Infectious Disease Journal*, 23, 774–777 (2004).

305. B. Gold and N. Morgan, *Speech and audio signal processing: Processing and perception of speech and music*, John Wiley & Sons, New York (1999).

306. G. P. Jacobson, C. W. Newman, and J. M. Kartush, *Handbook of balance function testing*, Springer Verlag, Berlin, Germany (1997).

307. J. M. Goldberg and C. Fernandez, Vestibular mechanisms, *Annual Review of Physiology*, 37, 129–162 (1975).

308. K. W. Lindsay, T. D. Roberts, and J. R. Rosenberg, Asymmetric tonic labyrinth reflexes and their interaction with neck reflexes in the decerebrate cat, *Journal of Physiology*, 261, 583–601 (1976).

309. M. F. Reschke, J. M. Krnavek, J. T. Somers, and G. Ford, A brief history of space flight with a comprehensive compendium of vestibular and sensorimotor research conducted across the various flight programs, NASA/ SP–2007–560, National Center for AeroSpace Information, Hanover, MD (2007).

310. A. M. Shkel and F.-G. Zeng, An electronic prosthesis mimicking the dynamic vestibular function, *Audiology and Neurotology*, 11, 113 (2006).

311. A. A. Maltan and T. K. Whitehurst, Stimulation using a microstimulator to treat tinnitus, US Patent 20070021804, Advanced Bionics Corporation, Valencia, CA (2007).

312. S. A. Gelfand, *Essentials of audiology*, 2nd ed., Thieme, New York (2001).

313. V. P. Eroschenko and M. S. H. di Fiore, *Di Fiore's atlas of histology*, 10th ed., Lippincott Williams & Wilkins, Philadelphia, PA (2004).

314. S. S. Corbett, III, J. W. Swanson, J. Martyniuk, T. A. Clary, F. A. Spelman, B. Clopton, A. H. Voie, and C. N. Jolly, Multi-electrode cochlear implant and method of manufacturing the same, US Patent 5630839, PI Medical Corporation (Portland, OR); University of Washington, Seattle, WA (1997).

315. C.-P. Richter and S. Ho, Cochlear implant including a modiolar return electrode, US Patent 7194314, Northwestern University (2007).

316. D. Rejalia, V. A. Leec, K. A. Abrashkina, N. Humayuna, D. L. Swiderskia, and Y. Raphaela, Cochlear implants and ex vivo BDNF gene therapy protect spiral ganglion neurons, *Hearing Research*, 228, 180–187 (2007).

317. D. Seybold, The psychosocial impact of acquired vision loss, *Vision 2005—Proceedings of the International Congress*, 1282, 298–301 (2005).

318. S. H. Schwartz, *Visual perception*, 3rd ed., McGraw-Hill, New York (2004).

319. J. S. Glaser, *Neuro-ophthalmology*, 3rd ed., Lippincott Williams & Wilkins, Philadelphia, PA (1999).

320. P. L. Kaufman and A. Alm, *Adler's Physiology of the eye*, 10th ed., Harcourt/Mosby, London, UK (2002).

321. J. L. Smith, *Neuro-ophthalmology*, Vol. 4, C. V. Mosby Co., St. Louis, MO (1968).

322. B. Cassin and M. L. Rubin, eds., *Dictionary of eye terminology*, 5th ed., Triad Publications, Gainesville, FL (2006).

323. M. E. Brezinski, *Optical coherence tomography: Principles and applications*, Academic Press, Boston, MA (2006).

324. D. T. Hartong, E. L. Berson, and T. P. Dryja, Retinitis pigmentosa, *Lancet*, 386, 1795–1809 (2006).

325. T. Suzuki, M. Akimoto, H. Imai, Y. Ueda, M. Mandai, N. Yoshimura, A. Swaroop, and M. Takahashi, Chondroitinase ABC treatment enhances synaptogenesis between transplant and host neurons in a mouse model of retinal degeneration, *Cell Transplantation*, 16, 493–503 (2007).

326. W. Roush, Envisioning an artificial retina, *Science*, 268, 637–638 (1995).

327. R. R. Lakhanpal, D. Yanai, Douglas, J. D. Weiland, G. Y. Fujii, S. Caffey, R. J. Greenberg, E. de Juan, Jr., and M. S. Humayun, Advances in the development of visual prostheses, *Current Opinion in Ophthalmology*, 14, 122–127 (2003).

328. J. D. Weiland, W. T. Liu, and M. S. Humayun, Retinal prosthesis, *Annual Review of Biomedical Engineering*, 7, 361–401 (2005).

329. P. Hossain, I. W. Seetho, A. C. Browning, and W. M. Amoaku, Science, medicine, and the future—Artificial means for restoring vision, *British Medical Journal*, 330, 30–33 (2005).

330. M. Javaheri, D. S. Hahn, R. R. Lakhanpal, J. D. Weiland, and M. S. Humayun, Retinal prostheses for the blind, *Annals in Academy of Medicine Singapore*, 35, 137–144 (2006).

331. G. Dagnelie, Visual prosthetics 2006: Assessment and expectations, *Expert Reviews in Medical Devices*, 3, 315–326 (2006).

332. D. E. Casey, ed., Envisioning sight for the blind: The DOE artificial retina project, *Artificial Retina News*, 1, 1–3 (2006).

333. J. D. Weiland and M. S. Humayun, A biomimetic retinal stimulating array, *IEEE Engineering in Medicine and Biology Magazine*, 24, 14–21 (2005).

334. J. D. Weiland and M. S. Humayun, Intraocular retinal prosthesis—Big steps to sight restoration, *IEEE Engineering in Medicine and Biology Magazine*, 25, 60–66 (2006).

335. C.-Y. Wu, F. Cheng, C.-T. Chiang, and P.-K. Lin, A low-power implantable Pseudo-BJT-based silicon retina with solar cells for artificial retinal prostheses, *Proceedings of 2004 International Symposium. on Circuits and Systems*, ISCAS 04. IV, pp. 37–40 (2004).

336. G. J. Suaning and N. H. Lovell, CMOS neurostimulation system with 100 electrodes and radio frequency telemetry, Conference of IEEE EMBS (Vic), Melbourne (1999).

337. D. C. Rodger and Y.-C. Tai, Microelectronic packaging for retinal prostheses, *IEEE Engineering in Medicine and Biology Magazine*, 24, 52–57 (2005).

338. F. Paillet, D. Mercier, and T. M. Bernard, Second generation programmable artificial retina, *Proceedings of the 12th Annual IEEE International ASIC/SOC Conference 1999 XII*, pp. 304–309 (1999).

339. F. Gekeler and E. Zrenner, Status of the subretinal implant project. An overview, *Ophthalmologe*, 102, 941–52 (2005).

340. E. Funatsu, Y. Nitta, Y. Miyake, T. Toyoda, J. Ohta, and K. Kyuma, An artificial retina chip with current-mode focal plane image processing functions, *IEEE Transactions on Electron Devices*, 44, 1777–1782 (1997).

341. J. Ohta, et al., Silicon LSI-based smart stimulators for retinal prosthesis—A flexible and extendable microchip-based stimulator, *IEEE Engineering in Medicine and Biology Magazine*, 25, 47–59 (2006).

342. A. Stett, A. Mai, and T. Herrmann, Retinal charge sensitivity and spatial discrimination obtainable by subretinal implants: Key lessons learned from isolated chicken retina, *Journal of Neural Engineering*, 4, S7–S16 (2007).

343. T. Schanze, H. G. Sachs, C. Wiesenack, U. Brunner, H. Sailer, Implantation and testing of subretinal film electrodes in domestic pigs, *Experimental Eye Research*, 82, 332–340 (2006).

344. M. T. Pardue, M. J. Phillips, H. Yin, B. D. Sippy, S. Webb-Wood, A. Y. Chow, and S. L. Ball, Neuroprotective effect of subretinal implants in the RCS rat, *Investigative Ophthalmology and Visual Science*, 46, 674–682 (2005).

345. M. T. Pardue, M. J. Phillips, B. Hanzlicek, H. Yin, A. Y. Chow, and S. L. Ball, Neuroprotection of photoreceptors in the RCS rat after implantation of a subretinal implant in the superior or inferior retina, *Retinal Degenerative Diseases*, 572, 321–326 (2006).

346. M. T. Pardue, S. L. Ball, M. J. Phillips, A. E. Faulkner, T. A. Walker, and A. Y. Chow, Status of the feline retina 5 years after subretinal implantation, *Journal of Rehabilitation Research and Development*, 43, 723–732 (2006).

347. A. P. Fornos, J. Sommerhalder, B. Rappaz, A. B. Safran, and M. Pelizzone, Simulation of artificial vision, III: Do the spatial or temporal characteristics of stimulus pixelization really matter? *Investigative Ophthalmology and Visual Science*, 46, 3906–3912 (2005).

348. S. I. Fried, H. A. Hsueh, and F. S. Werblin, A method for generating precise temporal patterns of retinal spiking using prosthetic stimulation, *Journal of Neurophysics*, 95, 970–978 (2006).

349. L. Johnson, et al., Impedance-based retinal contact imaging as an aid for the placement of high resolution epiretinal prostheses, *Journal of Neural Engineering*, 4, S17–S23 (2007).

350. E. Margalit and W. B. Thoreson, Inner retinal mechanisms engaged by retinal electrical stimulation, *Investigative Ophthalmology and Visual Science*, 47, 2606–2612 (2006).

351. T. Yagi and Y. Hayashida, Artificial retina implantation, *Nippon Rinsho*, 57, 1208–1215 (1999).

352. H. A. Hassan, S. R. Montezuma, and J. F. Rizzo, In vivo electrical stimulation of rabbit retina: Effect of stimulus duration and electrical field orientation, *Experimental Eye Research*, 83, 247–254 (2006).

353. W. H. Dobelle, Artificial vision for the blind by connecting a television camera to the visual cortex: State of the art, *American Society for Artificial Internal Organs Journal*, 46, 3–9 (2000).

354. W. H. Dobelle, J. Turkel, D. C. Henderson, and J. R. Evans, Mapping the representation of the visual field by electrical stimulation of human visual cortex, *American Journal of Ophthalmology*, 88, 727–735 (1979).

355. J. F. Doorish, A wireless photovoltaic Mini epi-Retinal Prosthesis (MeRP) 1: Concept and design, *Journal of Modern Optics*, 53, 1267–1285 (2006).

356. A. Y. Chow, V. Y. Chow, K. H. Packo, J. S. Pollack, G. A. Peyman, and R. Schuchard, The artificial silicon retina microchip for the treatment of vision loss from *Retinitis pigmentosa*, *Archives of Ophthalmology*, 122, 460–469 (2004).

357. K. Hungar, M. Gortz, E. Slavcheva, G. Spanier, W. C. Weidig, and W. Mokwa, Production processes for a flexible retina implant (Eurosensors XVIII, Session C6.6), *Sensors and Actuators A-Physical*, 123, 172–178 (2005).

358. P. Walter, Z. F. Kisvárday, M. Görtz, N. Alteheld, G. Rossler, T. Stieglitz, and U. T. Eysel, Cortical activation via an implanted wireless retinal prosthesis, *Investigative Ophthalmology and Visual Science*, 46, 1780–1785 (2005).

359. J. Ohta, T. Tokuda, K. Kagawa, S. Sugitani, A. Taniyama, A. Uehara, Y. Terasawa, K. Nakauchi, T. Fujikado, and Y. Tano, Laboratory investigation of microelectronics-based stimulators for large-scale suprachoroidal transretinal stimulation (STS), *Journal of Neural Engineering*, 4, S85–S91 (2007).

360. J. S. Pezaris and R. C. Reid, Demonstration of artificial visual percepts generated through thalamic microstimulation, *Proceedings of the National Academy of Science USA*, 104, 7670–7675 (2007).

361. R. J. Jensen, O. R. Ziv, and J. F. Rizzo, Thresholds for activation of rabbit retinal ganglion cells with relatively large, extracellular microelectrodes, *Investigative Ophthalmology and Visual Science*, 46, 1486–1496 (2005).

362. L. B. Merabet, J. F. Rizzo, A. D. Amedi, D. C. Somers, and A. Pascual-Leone, What blindness can tell us about seeing again: Merging neuroplasticity and neuroprostheses, *Nature Reviews Neuroscience*, 6, 71–77 (2005).

363. F. Duret, M. E. Brelen, V. Lambert, B. Gerard, J. Delbeke, and C. Veraart, Object localization, discrimination, and grasping with the optic nerve visual prosthesis, *Restorative Neurology and Neuroscience*, 24, 31–40 (2006).

364. D. W. Chun, J. S. Heier, M. B. Raizman, and B. Michael, Visual prosthetic device for bilateral end-stage macular degeneration, *Expert Review of Medical Devices*, 6, 657–665 (2005).

365. H. L. Hudson, S. S. Lane, J. S. Heier, R. D. Stulting, L. Singerman, P. R Lichter, P. Sternberg, and D. F. Chang, Implantable miniature telescope for the treatment of visual acuity loss resulting from end-stage age-related macular degeneration: 1-year results, *Ophthalmology*, 113, 1987–2001 (2006).

366. J. O. Winter, S. F. Cogan, and J. F. Rizzo, Retinal prostheses: Current challenges and future outlook, *Journal of Biomaterials Science Polymer Edition*, 18, 1031–1055 (2007).

367. M. C. Peterman, D. M. Bloom, C. Lee, S. F. Bent, M. F. Marmor, M. S. Blumenkranz, and H. A. Fishman, Localized Neurotransmitter Release for Use in a Prototype Retinal Interface, *Investigative Ophthalmology and Visual Science*, 44, 3144–3149 (2003).

368. W. H. Ko, A review of implantable sensors, *Pacing and Clinical Electrophysiology*, 6, 482–487 (1983).

369. R. Bogue, MEMS sensors: Past, present and future, *Sensor Review*, 27, 7–13 (2007).

370. M. Norris, Design considerations for wireless implants, *Design News*, May 14 (2007).

371. B. Sarmentoa, A. Ribeirob, F. Veiga, and D. Ferreira, Development and characterization of new insulin containing polysaccharide nanoparticles, *Colloids and Surfaces B: Biointerfaces*, 53, 193–202 (2006).

Problems

13.1 Is nanotechnology relevant to medicine? How does medicine depend on technology in general? What are the considerations medical researchers and caregivers take when evaluating a new technology for patient care?

13.2 What are the nine major medical breakthroughs identified in the text? What three, if any, do you consider the most important? (Discuss the reasons for your choices.)

13.3 What role does the NIH consider nanotechnology will play in medical research and healthcare, and why?

13.4 What is the earliest application of nanotechnology in medicine of which you are aware?

13.5 What are the typical size ranges of (a) human cells, (b) bacteria, (c) viruses, and (d) DNA?

13.6 How are nanoparticles used to enhance medical imaging? What advantages do they have over other image contrast agents? Can you think of any potential disadvantages? (Discuss for specific types of contrast agents and nanoparticles.)

13.7 What radiation frequencies are used with nanoparticles to destroy cancer cells? How can the nanoparticles be made selective for cancer cells?

13.8 What are three ways that nanotechnology can be used to treat diabetes?

13.9 How can nanoencapsulation be used to deliver drugs through the blood–brain barrier?

13.10 How can nanotechnology be used to stimulate nerve growth? What are the factors needed for successful application of this technique?

13.11 Is nanotechnology applicable to (a) motor prosthetics, (b) sensory prosthetics, (c) neither, and (d) both? Explain.

13.12 What are some potential improvements that nanotechnology presents for (a) neurostimulation for pain, (b) hearing aids, and (c) prosthetics for the blind? What are some challenges faced in improving these types of devices and where is nanotechnology relevant?

13.13 What are the factors that limit the density of placement of electrodes in a cochlear implant?

13.14 What are the factors that limit the density of stimulation electrodes in a retinal implant?

ENVIRONMENTAL NANOTECHNOLOGY

Environmentally friendly cars will soon cease to be an option … they will become a necessity.

FUJIO CHO
President of Toyota Motors, 2004

THREADS

We save this chapter on environmental nanotechnology for the last—the one to leave the student, hopefully, with an important perspective of things to come and, unequivocally, to place value on the place that we all live. We need to know how nanotechnology will impact our environment, our health and our safety, and our society in general. There are numerous positive attributes to nanotechnology in this regard. The environmental footprint of nano is expected to be three orders of magnitude less than that observed for today's technology, the microtechnology. There are also potentially negative effects of this wondrous technology—as you might imagine, ones that are able to tap into the very fiber of nature itself. Which nanomaterials are environmentally neutral? Which ones are beneficial? Which ones are potentially dangerous to human health?

The nanotechnology we have presented and discussed so far included topics as varied as nanometrology, nanomanufacturing, optical-electromagnetic devices, thin films and interfaces, catalysis, polymers and fibers, and those associated with bionanotechnology and nanomedicine—truly a vast and diverse assembly of materials, devices, and applications. Once manufactured, used or disposed, what happens to the nanomaterials from which they originated? Because nanomaterials possess remarkable properties, how are we to know that these remarkable properties keep on expressing themselves once we are through with the technology? Will these remarkable properties work against us and against the environment? Or will the balance between the positive and the negative be in favor of the positive? We certainly hope the latter is emphasized in all possible scenarios.

We limit our discussion to the nanotechnology of environmental devices and materials (and natural nanotechnological devices and materials) that are designed to measure, mitigate, or provide power—the sensors, filters, practices, and other commercial and industrial (and natural) applications. We need to know more how our materials, by-products, and devices affect the environment, and thereby us. What do we do and how do we do it? Although nanotechnology will help us become better informed, better equipped, and better defended, unfortunately, the answer to this question transcends solutions rendered to them through nanotechnology.

14.0   The Environment (and Technology)

Nanotechnology is expected to play a critical role in environmental-related issues—from sensing, monitoring, mitigation, and power to perhaps generating its own brand of pollution, contamination, and infection. The stages upon which all these actors play are the traditional ones that we are all familiar—the air, the soil, and the water—and of course, the workplace and home and the human body itself. Pollutants, contaminants, and pathogens come in a variety of forms and are released from an incredible number of diverse sources—ranging from trash, industrial wastes and spills, seepage, combustion to more subtle expressions of human activity like radiation (e.g., cell phones), noise, congestion, recycled air, and daily consumer activities.

Nanotechnology is expected to impact all industrial sectors—from aerospace to energy and from construction to medicine. Nanotechnology's impact on the environment is also expected to be significant—its vector remains to be determined. There is a need for sensors that have better detection limits, versatility, and durability. For example, since 1992, there are over two million storage tanks under surveillance by the Environmental Protection Agency (EPA) that sequester toxic and/or volatile materials—and what of storage tanks that are not included in the monitor registry [1,2]? With an adequate arsenal of micro- and nanosensors, we should be able to detect low-level leaks from unregistered tanks and determine their source before any serious environmental damage ensues.

Where can nanomaterials and devices take us? Use of nanomaterials in instruments, devices, equipment, and other products and the energy required for their manufacture and operation both have direct and indirect impact on our environment—in both cases, the less the better. In addition, detection and mitigation of low-level pollutants can be accomplished with better nanoscale catalysts, inline and remote detectors, and nanochemical reactors. Facilitative and affordable field analysis technologies have and will continue to spring from advances in nanotechnology. For example, surface-enhanced Raman fiber-optic probes have been around for quite some time [3]. The surface-enhanced resonance Raman scattering (SERRS) method is capable of attomole level sensitivity [4]—no big deal anymore since single-molecule detection has already been achieved with this technique [5]. SERS is without question a nanotechnology-enhanced analytical tool that relies on nanosized silver or gold particles to generate a gigantic signal.

Current environmental procedures rely on a significant investment in time, cost, and energy—including sample acquisition, preparation, and laboratory analysis [6]. Better, more accurate methods able to scan larger sample sizes with higher throughput and greater sensitivity need to be developed. The methods and equipment must be robust, reliable, and reusable with the capability for remote sensing and operation in real time—all with enhanced facility and cost-effectiveness [6]. Quite the wish list! We present a broad spectrum of environmental issues and technology in this chapter. We divide the chapter into three major categories: water and soil, air, and energy. With this chapter, we round out *Fundamentals of Nanotechnology*, and *Introduction to Nanoscience and Nanotechnology*, the combined volumes.

14.0.1 *Background*

Agriculture, villages, and cities have brought on pollution and resource exploitation—upsetting Earth's natural balance, like never before, in order to provide food, shelter, and energy for burgeoning populations. Environmental insults have occurred since *Homo sapiens'* opposable thumb and frontal lobe started working together. For example, evidence of the use of fire to clear forests was found at an archeological dig in Tanzania dating back 60,000 years [7]—an early example of anthropomorphic alteration of the natural environment. As technology developed and populations continued to explode, the concomitant levels of pollution and resource exploitation increased proportionally—air pollution from dust, wood, smoke, tanneries, and animal manure; and water pollution from large cities in the form of sewage. Specific toxins found in lead-sweetened wine ("sugar of lead," lead acetate) and lead piping resulted in poisoning among the Romans [7]. It is a well known that ancient Babylonians, Minoans, Phoenicians, and Romans stripped local forests of timber to supply their cities.

Although the technology of the age was not sophisticated, the devastation was widespread nonetheless. This is not to say that all this early form of environmental insults was conducted without some consciousness. The Sumerians for example passed laws as early as ca. 2700 B.C. to protect remaining forests [7]. From the human health viewpoint, acid mists in copper mines were known to endanger the lungs of miners—so observed the Greek physician Galen in ca. 200 B.C. [8]. In first century Rome, workers in zinc smelting operations used animal bladders as respirators—so observed Pliny the Elder [8]. Plutarch recommended that only criminal slaves should work in the lead and mercury mines—an early societal issue associated with a technology. Yes, to the doubters out there, anthropomorphic generated activities do impact the environment—and yes, as global populations continue to swell, these impacts are manifesting themselves on a global scale.

The effects of fossil fuel exploration, drilling, mining, and combustion are well known. In thirteenth century England, the combustion of coal fouled the city—forcing "royal personages" to flee to cities with cleaner air [9]. The combustion of coal and other fossil fuels, exacerbated by population increases worldwide, has contributed most to pollution and other forms of environmental anxiety and aesthetics—simply because with the advent of the "Oil Age," humankind was able to alter the energy balance by tapping energy resources manufactured in the distant past. Tapping ancient resources enabled us to produce more food than we could have otherwise thereby shifting the natural equilibrium of things [10]. The subsequent usurping of the world's food supply, greater than 40% of NPP (net primary production) [11], resulted in our current dilemma—overpopulation, insatiable energy demand, and rampant resource exploitation and waste.

We have certainly upset the natural balance of things. People will comment readily that "the smog is really bad today"; "the water tastes terrible"; or "look at all the trash everywhere." Most people will agree that all pollution is bad, ugly, and unhealthy and some believe in science and unfortunately, some don't—but that's what makes our world go around. What is real and tangible is that nanotechnology is here to stay. Nanotechnology is ready to enhance and enable preexisting technologies and to develop new ones that will help the environment. Both the aware and the unaware would benefit and/or pay the price—perhaps perceptibly or more likely, perhaps imperceptibly.

14.0.2   *Traditional Methods of Detecting Environmental Contaminants*

Many traditional methods exist to detect environmental pollutants, contaminants, and pathogens. Since most of us are quite familiar with most of those methods, much detail and discussion is not allocated to them in this section. Analytical instruments and procedures require the following: (1) rapid response, (2) portability, (3) measurements in real time, (4) multiparameter capability, (5) simple design, (6) low detection limits, (7) high throughput, and (8) large working range [12]. Quite the wish list! How can nanotechnology help achieve these goals? A brief catalog of traditional laboratory analytical techniques is provided in **Table 14.1**. In the third column of the table, improvements due to

TABLE 14.1	*Traditional Laboratory Analytical Techniques Used to Measure Environmental Pollutants*		
Technique	**Acronym**	**Samples**	**Nano-enablement and nano-uses**
Gas chromatography	GC	Organic pollutants, halogenated organic compounds	Thinner capillary columns filled with nanomaterial supports; enhanced detection systems
GC/mass spectrometry	GC/MS	Organic samples	Microfluidic samples containing nanogram quantities of material [13]. In situ analysis of carbon nanotubes by MS
High-performance liquid chromatography	HPLC	Persistent pesticides, herbicides, and polychlorinated biphenyls (PCBs)	Postcolumn derivatization; nanospray sample introduction (emitters), nanoflow separation (eliminating dead volume); HPLC-chip/MS systems, biocompatible HPLC-chips [14]; and micro-nano HPLC columns (75 μm diameter—ca. 4000x increase in sensitivity) [15]
Fourier transform infrared spectroscopy	FTIR		Live cell FTIR [16]
Atomic absorption/emission spectroscopy	AAS/AES	Heavy metals	Nanospray enhancements
Inductively coupled plasma spectroscopy	ICP	Heavy metals	ICP/MS with ppt to ppq (parts per quintillion) level of detection, nanoparticle introduction systems; ICP-MS to analyze gold nanoparticles [17]
Ion and ion-exclusion chromatography	IC and IEC	Waterborne cations and anions, carbon nanotubes	Ion exchange chromatography to separate single-walled carbon nanotubes (SWNTs) based on electrical properties [18]
Electrode methods		Heavy metal ion pollutants	Separation based on redox potentials + point-contact spectroscopy and conductance; in situ nano-contact sensor for heavy metal detection—15 pairs of nanoelectrodes on Si chip [19];
Fluorescence	FS	Pathogens	Nanophotonic light sources for fluorescence spectroscopy and cellular imaging [20]

nanotechnology of various aspects of the technique and some nano-applications are provided—whether for the detector, the mechanism that delivers the sample or other associated features of the technique.

14.0.3 Types of Environmental Sensors

For any environmental assessment or risk analysis, it is good to have data upon which to base conclusions (obviously!). In order to acquire data, detectors, sensors, collectors and analysis are required. Generic classes of sensors are listed and described in **Table 14.2**. Natural sensors are able to detect single molecules (e.g., the pheromone bombykol by the *Bombyx* moth). The Defense Advanced research Projects Agency (DARPA) wants to create a "dog's nose" to detect land mines with greater sensitivity (e.g., for amyl acetate, 1–10 ppb level of detection is possible by a dog's nose).

TABLE 14.2	*General Kinds of Sensors Used in Environmental Analysis*	
Category of sensor	**Basic mode of operation**	**Examples of nano-applications**
Biosensors	Biosensors consist of a working enzyme, nucleotide, or biomolecule tethered to a detector surface. The working enzyme must show a strong affinity for a type of molecule or pathogen Biosensors using lanthanide oxides have large Stokes shift, sharp emission spectra, long luminescence lifetimes, and photostability [21]	Acetylcholinesterase (for organophosphates) [6]; Luciferase (for Hg^{2+} detection) [6] GABA (γ-aminobutyric acid for pyrethroid and bicyclophosphate detection) [6] Antibodies (that bind with insecticides, herbicides, or microbes) [6] Immunosensors and DNA sensors on the surface of magnetic La- or Eu-doped Gd_2O_3 phosphor nanoparticles to measure pesticides in water, toxins in food, and MTBE (methyl *t*-butyl ether) remediating bacterial DNA in soil [21]
Electrochemical sensors	Detection of changes in electrical resistance induced by contact with substrate Electric current passed between or on electrodes that interact with chemicals Electrodes can be coated with nanometer layers of substrate-selective materials	Analytical improvements in detection limits, fabrication, and remote communication [22]. Ultrasensitive electrochemical sensors using carbon nanotubes—able to detect trace concentrations of Pb in the ppb range—best yet for an electrochemical sensor [23]
Mass sensors	Measure change in mass associated with substrate reaction or sample collection	Cantilever sensor arrays on a chip that rapidly detect proteins; measurement of mass loading effects, stress, and charge simultaneously; integration with microfluidic channels capable of picoliter volumes and mass detection down to a few attograms [24] Quartz crystal microbalance (QCM) and surface acoustic wave (SAW) are refined nano-processes that have nearly achieved their theoretical limits [25]
Optical sensors	Measures change in visible light (energy or flux) following chemical or physical reaction of substrate	Semiconductor photocatalysis of dye molecules SERS detection of adsorbed species
Thermal sensors	Measures temperature changes in gaseous, aqueous, or solid-state material or environment; collect sensitive data with high accuracy from many kinds of environment	Monitoring of global warming Systems-on-a-chip containing nanosensors and actuators equipped for remote sensing

Biosensors.   One of the first biosensors was the canary in the mine—a small creature with a high metabolic rate sensitive to toxic gases. Biosensors abound in nature and nearly any function can be traced to some form of biosensing. From the perspective of *Homo sapiens*, Leland C. Clark, Jr. is considered to be the father of the synthetic biosensor [26]. In the 1954, Clark invented the *Clark electrode*, an electrochemical sensor that is capable of measuring oxygen in blood [27]. Biosensor history began in 1962 with the glucose sensor [26–28]. The glucose sensor is derived from the Clark electrode [26–28]. The electrochemical reduction of oxygen occurs on a catalytic platinum surface according to

$$O_2 + 2e^- + 2H_2O \rightarrow H_2O_2 + 2OH^- \qquad (14.1)$$

Clark encased the working electrode (and the counter electrode) in nonconductive polyethylene—a material with limited permeability to oxygen. Later in the 1960s, Clark experimented with trapping an enzyme that reacted to oxygen against an electrode with a dialysis membrane. He believed that by monitoring enzymatic activity the concentration of oxygen could be determined. Thus, the first chemobiosensor was created—equipped with the first enzyme transducer electrode [27,28]. The trapped enzyme was glucose oxidase (GOD). A decrease in measured oxygen (or an increase in H_2O_2) was proportional to the glucose concentration in solution. Thus, the first glucose sensor was based on the amperometric detection of hydrogen peroxide [27,28].

A biosensor combines a biological moiety with a physicochemical sensor to detect a targeted analyte [29]. The analyte may be biological, organic, or inorganic. The substrate-sensitive component may be an enzyme, a nucleic acid, an antibody, a cell receptor, or other biologically based materials. The biosensing element is linked to a transducer that is able to translate physical or chemical changes based on optical, electrical, electrochemical, piezoelectric, thermal, or other parameters. When an appropriate substrate interacts with the biological component of the sensor, a signal is generated that is converted by a *transducer* into a signal that is easy to measure, for example, electrical resistance, current, or voltage change. In the case of the first glucose sensor, the measured material was hydrogen peroxide, the result of an electrochemical transformation of oxygen and water at a Pt electrode. According to Martin Chaplin of the London South Bank University Faculty of Engineering, Science, and the Built Environment, a successful biosensor must have the following [30]:

- The enzyme (or biocatalyst) must be highly specific, be stable during storage and repetitive analytical applications.
- The analytical reaction should be independent to the best degree possible of physiochemical conditions such as stirring, *p*H, and temperature.
- The response of the biosensor should be accurate, precise, and reproducible over the analytical range without dilution or concentration.
- The sensor must be small, sterilizable, and compatible with in vivo samples and not be prone to fouling or proteolysis.
- The biosensor should be relatively inexpensive, portable, and user friendly (and easy to fabricate?).
- A preexisting market should be in place for the biosensor.

There are many kinds of biosensors—calorimetric, potentiometric, amperometric, optical, piezoelectric, and immunobiosensors—just about all the kinds of sensor types mentioned earlier except with an added biological twist.

Biosensors are typically based on the phenomenon of molecular recognition—recognition between an enzyme and a substrate for example. Many biosensors, therefore, are enzyme based. For example, an oxidoreductase-type enzyme (the sensor), anchored to a detector system electrode (the transducer), is placed in a natural water stream. An environmental toxin that happens along is oxidized in the active pocket of the enzyme. The substrate is oxidized yielding two electrons that are able to reduce nearby FAD (flavine adenine dinucleotide, a redox coenzyme, or prosthetic group) to $FADH_2$ (**Fig. 14.1**). The electrode subsequently oxidizes the $FADH_2$ back to FAD thereby releasing two electrons and a signal is

Fig. 14.1   Basic mechanism of an oxidoreductase enzyme-based biosensor is depicted. A substrate makes contact with the active pocket of the oxidoreductase. The substrate is oxidized and two electrons are generated that transform FAD into FADH$_2$. The FADH$_2$ is oxidized back to FAD by a mediator yielding the two electrons. The electrons are scooped up by an amperometric or potentiometric transducer (an electrode) and converted into a signal for processing.

generated. The size of the current ideally is proportional to the concentration of the analyte.

Aptamer-based sensors are a new class of ultrasensitive biosensors called "smart materials" that employ oligonucleotides like double- or single-stranded DNA and single-stranded RNA or peptides to bind molecular targets such as proteins or metabolites with high affinity and specificity. Aptamers (from the Latin *aptus* "to fit") are engineered through iterations of in vitro selection (called SELEX, systematic evolution of ligands by exponential enrichment) [32]. Aptamers, often referred to as DNAzymes, are segments of DNA (or RNA) that possess enzymatic function, and can be coupled with fluorophores or gold nanoparticles to selectively detect targeted analytes [31]. Nucleic acid aptamers, for example, have been shown to inhibit the replication of HIV-I [33].

The range of substrates that aptamers can selectively bind is enormous—well on the order of antibodies. However, advantages of aptamers over antibodies are multifold: (1) they can be engineered in laboratories by chemical synthesis methods (do not require biological hosts), (2) are easily tagged or cross-linked with thiols, (3) have desirable storage properties and longevity while immobilized, and (4) engender little or no immune response [32].

Juewen Liu and Yi Lu of the Department of Chemistry at the University of Illinois at Urbana-Champaign recently developed aptamers linked to nanomaterials that are sensitive to cocaine, adenosine, and K+ [31]. In one system, two types of DNA-functionalized 13-nm diameter gold nanoparticles were produced: one functionalized with the 3'-end thiol and the other with the 5'-end thiol (**Fig. 14.2**). Already a built-in templating component is installed in the system as the 3' or 5' ends of the nucleotides direct further modification by self-assembly. The two types of precursors were then linked to an adenosine aptamer and a cocaine aptamer, respectively, that self-assembled into an agglomerated form (the color of the aggregate is blue). In the presence of adenosine, the adenosine aptamer adjusted its structure to accommodate two adenosine molecules. In the presence of cocaine, the appropriate aptamer adjusted its structure to

FIG. 14.2

An example of an aptamer sensor that simultaneously targets cocaine and adenosine. The presence of adenosine and cocaine are necessary in order to disperse the gel nanoparticle aggregation. This is evidenced by a change in color from blue (the aggregate phase) to red (the dispersed phase). Different permutations of adenosine–cocaine systems can be generated: a highly cooperative system is depicted in the figure, systems with three kinds of links to gold nanoparticles, and lastly, systems with no cooperativity (respond individually to bind the substrates). This aptamer scheme is quite an ingenious tribute to nanotechnology—depending on the aptamer, the nanoparticles, and the way they are linked.

Source: J. Liu and Y. Lu, *Advanced Materials*, 18, 1667–1671 (2006). With permission.

accommodate one molecule of cocaine. Complete dispersion of the aggregate occurred only when both cocaine and adenosine were present—resulting in a red color indicative of dispersed gold colloids [31].

Due to design generalities among the aptamers, sensors that respond to any combination of the three can be developed, a process called controlled cooperativity. According to J. Liu et al., the gold nanoparticles can be replaced with other nanomaterials such as quantum dots, nanotubes, or polymers that would open the door to a diverse range of applications [31]—quite an amazing display of nanoengineering!

Potentiometric biosensors are able to transfer information generated from a biological reaction into an electronic signal. Ion-selective electrodes resemble a modified pH probe. They consist of an outer immobilized enzyme encased in a semipermeable membrane around an active glass membrane that encloses a pH probe (an internal Ag/AgCl electrode in dilute HCl). An electric potential is formed between the probe and an external reference electrode. The probe detects changes in the hydrogen ion concentration, whether by absorption or generation of H^+, due to catalytic activity of the enzyme. There are three generic kinds of ion-selective electrodes that are exploited in ion-selective biosensors—glass electrodes for cations (with hydrated glass membrane) based on cation competition; glass pH electrode coated with gas-permeable membrane selective for CO_2, NH_3, or H_2S or other gases; and solid-state electrodes that utilize specific ion conductors [30]. Each kind can be enhanced with nanotechnology.

Optical biosensors invoke one of two mechanisms for the purposes of detection. In the first, a change in light absorption following a chemical reaction is quantified, and in the second, light output generated by a luminescent process is measured. An example of an optical biosensor is one that incorporates luciferase as the immobilized enzyme to detect the presence of bacteria. ATP from the bacteria is reacted with D-luciferin and O_2 to produce yellow light (ca. 560 nm) in high quantum yield with sensitivity less than 10^{-12} M ATP [30].

Calorimetric biosensors rely on heat generated during a catalytic reaction as the analytical parameter. Many enzyme-modulated reactions are exothermic. Therefore, the heat produced in a reaction is an indication of the analyte concentration. Temperature changes are determined by thermistors at the ingress and egress ports of an immobilized enzyme array. Heats of reaction (ΔH_{rxn}) for several enzyme-catalyzed reactions are known—for substrate–enzyme couples such as cholesterol/cholesterol oxidase ($53 \, \text{kJ} \cdot \text{mol}^{-1}$), glucose/glucose oxidase ($80 \, \text{kJ} \cdot \text{mol}^{-1}$), urea/urease ($61 \, \text{kJ} \cdot \text{mol}^{-1}$), hydrogen peroxide/catalase ($100 \, \text{kJ} \cdot \text{mol}^{-1}$), esters/chymotrypsin ($4$–$16 \, \text{kJ} \cdot \text{mol}^{-1}$), and starch/amylase ($8 \, \text{kJ} \cdot \text{mol}^{-1}$) [30].

Piezoelectric biosensors obey the simple form of the Sauerbrey equation in which the change in frequency of a piezoelectric crystal is dependent on the adsorbed mass: $\Delta f = K f^2 \Delta m \cdot A^{-1}$ where f is the natural resonant frequency of the crystal, Δm (usually in terms of g) is the change in mass of adsorbed material, K is a constant relating to the crystal parameters, and Δf (in Hz) is the change in the crystal frequency due to the adsorbed mass. A quartz crystal microbalance (QCM) and simple electronics are all that are needed to detect adsorbed materials. A formaldehyde biosensor, for example, is fitted with immobilized formaldehyde dehydrogenase coating on a quartz crystal. QCMs are able to detect nanolayers of adsorbed material.

Immunosensors employ an antibody–antigen couple as the basis of detection. Biosensors are able to incorporate enzyme-linked immunosorbent assays (ELISA) within the sensing mechanism [30]. In ELISA, an antibody specific for a target antigen is immobilized on the surface of a transducer. Then, a mixture of a known concentration of antigen–enzyme complex plus an unknown amount of antigen is flowed over the transducer—a competition-based detection scenario. After time for equilibration, the antigen and antigen–enzyme conjugates become distributed between two states: either bound or free depending on their relative concentration. The unbound material is rinsed away and the level of antigen–enzyme complex that is bound is determined by the rate of the enzymatic reaction (or the transducer signal) [30].

Inorganic Electrochemical Sensors. Adsorption of gases and subsequent hetero-geneous catalytic activity (oxidation or reduction) at the surface of semicon-ducting metal oxides results in concomitant changes in the electrical resistance of the sensor material, for example, a signal that is characteristic of a specific gas–surface interaction [6]. The magnitude of the change in electrical resistance is dependent on the type of gas, it's partial pressure, and ambient conditions such as temperature. For nanocrystalline sensor materials, crystallite size is also an important consideration [6]. Selectivity in gas sensors is problematic and a number of strategies have been applied to help with this issue.

Tethered DNA electrochemical sensors were fabricated by M. Curreli et al. to produce functionalized In_2O_3 nanowire electrodes for the purposes of biosens-ing, identification, and quantification of infectious agents [34]. Considering the properties of nanomaterials (e.g., high surface-to-volume ratio), low detection limits can be achieved due to miniscule changes in the conductivity of the nano-sensor. Recognition of the analyte is afforded by analyte-specific molecular recognition groups. Indium oxide nanowires provide an alternative to silicon nanowires and carbon nanotubes. One advantage over silicon nanowires is that In_2O_3 has no oxide layer—therefore it can act as a gas sensor with ppb level detection [34]. Immobilization of DNA-based sensors on the nanowire is accomplished by tethering tailored single-stranded DNA receptors to In_2O_3 nanowires with HQ-PA [4-(1,4-dihydrobenzene)butyl phosphonic acid] linking groups [34].

HQ-PA is an electrochemically active hydroquinone that readily undergoes redox reactions at low potentials [35]. The oxidized form (**Fig. 14.3**) is able to react with many kinds of functional groups that can be spliced into biomole-cules. Phosphonic acids are known to interact with In_2O_3. The reversible redox behavior of HQ-PA on indium–tin oxide (ITO) and In_2O_3 is similar. Cyclic voltammetry studies conducted on ITO showed that the oxidation wave was centered at +330 mV and the reduction wave at –200 mV [35].

Mass Sensors. MEMs and NEMs sensors that rely on microcantilever transduc-ers are also used to detect the presence of chemicals, pathogens, explosives, and ionic species. Once again a tried and true chemical procedure is applied—the functionalization of a surface in order to bind target molecules or entities. In this case, it is the surface of the silicon (or silicon oxide, silicon nitride) micro-cantilever, and a small and sensitive force or mass detector that is chemically modified. Analytical signals are generated following molecular adsorption of

Fɪɢ. 14.3

A functionalized electrochemical cell made of In$_2$O$_3$ nanowires and tethered, DNA-functionalized hydroquinone electroactive groups is shown. In$_2$O$_3$ nanowires and electrodes were formed by photolithography and metal deposition on SiO$_2$/Si to form an electrode array. A self-assembled monolayer of HQ-PA was applied to the indium oxide nanowires. Thiol-terminated C$_{12}$ hydrocarbons (dodecane-1-thiol) formed a self-assembled monolayer on the gold electrodes to prevent thiol-terminated DNA from interacting with the gold electrodes. The DNA sequence is generic and shown only for image clarity and is not provided by the authors of the paper.

Source: M. Curreli, C. Li, Y. Sun, B. Lei, M. A. Gundersen, M. E. Thompson, and C. Zhou, *Journal of American Chemical Society*, 127, 6922–6923 (2005). With permission.

target molecules by monitoring bending and changes in resonant frequency or the position of the cantilever [6]. Several techniques such as optical beam deflection (the principle behind AFM) (**Fig. 14.4**), piezoresistivity or electricity, interferometry, capacitance, and electron tunneling are some of the most applied [36]. For example, forces and surface stresses in the micronewton to nanonewton range allow for quantification of the concentration of adsorbed analyte [37].

An array of silicon cantilevers is coated with sensitive substrate-specific layers. Each cantilever is capable of sensing a unique pollutant. Physical or chemical processes induced by adducts are transduced into mechanical motion due to changes in surfaces stresses like bending. Surface stress leads to bending (static mode) and mass increase leads to lower resonance frequency of the cantilever (dynamic mode). Surface stress is not necessarily related to mass but rather to molecular forces such as steric hindrance, electrostatic repulsion, or conformational

FIG. 14.4

Mass sensor based on cantilever detection is displayed. Cantilevers are typically 10–500 μm in length and as thin as a few microns or less. The sensitivity of a cantilever depends on its "spring constant": the lower the constant, the higher the sensitivity. Laser light is shined on a specific area of the surface and the reflected beam is captured by a position-sensitive photodetector. Polymeric materials coating cantilever surfaces serve to differentiate among VOCs in air. At the lower left corner of the image, a schematic of a cantilever array is depicted. Each cantilever is uniquely derivatized in order to identify and bind a specific substrate.

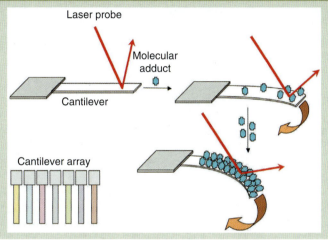

changes within the functionalized layer. Therefore, two complementary sets of data can be acquired from a single measurement [6].

Cantilever sensors have been developed to detect VOCs, TNT, heavy metal ions and anions like chromate, Salmonella bacteria, viruses, fungal spores, and pesticides. Calixarene-crown ether complexes on cantilevers were employed to detect Cs^+ in the range of 10^{-7} to 10^{-11} M [38]. In this case, a selective cesium ion sensor based on an ion-selective SAM-coated microcantilever was capable of detecting low concentrations of Cs^+ in the presence of high concentrations of K^+ and Na^+ cations. A thin gold layer was applied to the surface of a microcantilever and a subsequent thiol layer as well (**Fig. 14.5**).

The sensitivity of this nanotechnologically enhanced electrode is several orders of magnitude better than the best available ion-selective electrode (ISE) [38].

Optical Sensors. Optical sensing is usually accomplished with surface plasmon resonance spectroscopy, interferometry, and luminescent spectroscopy. One class of optical sensing is based on the photoluminescent properties of nano-crystalline porous silicon. The photoluminescent response of this material depends on the chemical nature of its surface [6]. Adduct metal ions and organic materials are able to quench the photoluminescence of porous silicon.

14.0.4 *Introduction to Environmental Mitigation*

We are all quite familiar with many kinds of environmental pollutants and pathogens and therefore will not list them. There are thousands of kinds of hazardous

FIG. 14.5 *Functionalized cantilever specific for cesium cations is shown. The cesium recognition agent is 1,3-alternate-25,27-bis(11-mercapto-1-undecanoxy)-26,28-calix[4]benzocrown-6 (highlighted in black). This molecular recognition system was able to discriminate cesium ions in high concentrations of sodium and potassium cations.*

Source: H.-F. Ji, R. Dabestani, G. M. Brown, and P. F. Britt, *Chemical Communication*, 6, 457–458 (2000). With permission.

substances, and they exist as particulates, organics, inorganics, and biological materials—in water (including groundwater), soil (and sediments), and air (indoor and outdoor)—and as toxins in our bodies. Volatile organic compounds form the biggest class of hazardous pollutants. Inorganic elements such as heavy metals and radionuclides are also prevalent. Polyaromatic hydrocarbons (PAHs), halogenated pesticides, organophosphates, nitrosamines, and other organic chemicals have drawn much attention over the years.

There are three generalized types of mitigation: (Type I) *post de facto* treatment of pollutants released into the environment; (Type II) treatment of pollutants before whole-scale release into the environment—a.k.a., "end-of-pipe" management and cleanup; and (Type III) elimination of pollutant from a process by substitution with another process or by alteration of further the process—a.k.a., prevention. Post de facto treatments are (were?) the most common. Treating the pollutant before release obviously reduces the amount of further post de facto remedial applications in which the pollutant may achieve a highly widespread and dispersed state. The best option is, of course, to remove the pollutant from the process in the first place. This can be accomplished by substituting with alternative sources, modifying processes in such a way as to eliminate the detrimental material or by releasing less material. Selected examples of mitigation, and remedial or preventive processes are listed in **Table 14.3**.

TABLE 14.3	*Environmental Remediation, Mitigation, or Preventative Processes Using Nanotechnology*
Pollutant	**Nano-mitigation**
Automobile exhaust gases	Supported (on alumina) precious metal (Pt-Pd, Au) nanocatalysts of precious metals to remove CO, NO_x, and HCs [39,40] NO_x reduction by Rh supported on alumina or vanadium oxides
Mercury	Coal burning plant and industrial wastewater. High surface area nanoporous silica-based ceramic material with tailored pore size coated with attached monolayer specific for attracting Hg. 99.9% Hg in simulations captured [43,44]
Asbestos	Stopped using asbestos
Volatile organic compounds	5–100 nm MnO nanoparticle catalysts—removal to ppb levels [46] Catalytic destruction of indoor VOCs by nanoparticle scrubbers at room temperature Photocatalytic destruction of VOCs in air scrubbers Manufacture of nanotechnology-based paints without VOCs but also to act as an antimicrobial and antifungal surface Ethylene oxidation by Ag on support
Heavy metals	Granular microscale Fe > 50 μm replaced by nanoscale Fe and Fe/Pd that are more reactive in reduction of inorganics [48]
Halogenated organic compounds	Complete and rapid dechlorination of aqueous species by nanoscale Fe and Fe–Pd (99.9% Fe) [41]
Dense nonaqueous phase liquids (DNAPLs)	Emulsified zero-valent iron (EZI) nanoparticles act like DNAPL in water (sinks to bottom in organic phase), then attract pollutant into hydrophobic sphere and reduces the chlorinated hydrocarbon [42] Removal of lindane (γ-hexachloro-cyclohexane) by FeS nanoparticles
Perchlorate	Complete reduction of perchlorate in contaminated groundwater to chloride by Fe nanoparticles
Pathogens	Viruses, bacteria, endotoxins, and DNA filtered by nanofilters that remove 200-nm particles with 99.9% efficiency [45]
Nuclear waste	Copper canisters have been used for nuclear cleanup in the past. Now "super-sponges" made of inorganic–organic SAMM (self-assembled monolayer on mesoporous support) with channel diameter 20–200 nm can reduce concentrations to 1 ppb [47]
Water treatment	Capacitive deionization technology (CDT) based on carbon aerogels—to remove chloride, nitrate, silica at the positive electrode, and calcium, magnesium, and sodium at the negative electrode [49]. No chemicals are required

Application of Catalysts. Nanotechnology has been applied for many years to reduce the amount of toxins produced by the combustion of gasoline. Rhodium, platinum, and palladium nanosized catalysts have been utilized in catalytic converters for decades—but they are very expensive! The use of transition nanoparticle metal carbides and oxycarbides may provide a lower cost alternative [6]. However, numerous nanotechnology-enhanced contaminant mitigation techniques have been developed recently. Bio- and genetic engineering have provided recombinant DNA to create artificial protein polymers (tunable biopolymers) coupled with molecular recognition capability to target the removal of specific toxic wastes [6,50]. Nanoparticle materials ranging from 1 to 10 nm in size offer routes to chemical remediation of many potential environmental pollutants. The key to this success of the new generation of nanocatalysts is the incredible surface to volume ratio, enhanced surface activity of such small particles, and in many cases, modified surfaces that have the ability to recognize targeted pollutant materials.

Catalytic Gold.   Gold, not known as a catalyst, is quite a good one when reduced in size [51]. As a matter of fact, research and development of gold catalysts has increased such that from 1991 to 2001 over one-third of the patents awarded for pollution control using catalysts have involved gold nanoparticles [52,53]. Gold catalysts on clay particles (magnesium silicate hydrate) coupled with ozone destroyed odors [54]. Gold nanoparticles were very effective in controlling trichloroethylene (TCE) levels, a groundwater contaminant. Michael Wong of Rice University found that Au–Pt nanoparticle co-catalysts broke down TCE 100 times faster than did catalysts made of the traditional material, palladium, and more economically [55]. Detection systems using gold nanoparticles developed at the Indian Institute of Technology found that nanogold was able to attract and then bind pesticide residues in flowing water [55]. Coal-fired power plants are one of the primary sources of anthropogenic mercury. Since gold has a strong affinity for mercury, a pilot study showed that gold catalytic systems were effective in oxidizing mercury in power plant effluent streams from the Lower Colorado River Authority Fayette Power Plant [55].

Nanoparticulate gold on oxide supports, in particular, showed versatility in many kinds of pollution control studies. For example, supported nanogold was able to remove CO from room air at ambient conditions and from fuel cell hydrogen feed gases [52]. The oxidation of methane, propane, and the removal of nitrogen oxides from the atmosphere have also been accomplished at relatively low temperatures with gold catalysts (as opposed to traditional catalyst methods) [56]. More recently, Nanostellar Inc. (a leader in nanoengineered catalyst design and manufacture) announced that gold oxidation catalysts reduced diesel hydrocarbon emissions by as much as 40% over commercially available materials in tests [39]. Since there are over 14 million light-duty and 2 million heavy-duty diesel vehicles produced annually worldwide, the environmental impact provided by such newly developed nanocatalyst systems is expected to be, and in no uncertain terms, enormous [39].

Photocatalysts.   Photocatalytic-driven reactions are gaining widespread acceptance in purifying indoor air, among other applications. Nanosized semiconductor materials such as TiO_2 and ZnO are capable of interfacial charge transfer to organic pollutants situated on the nanoparticle surface. When TiO_2 nanoparticles are irradiated with ultraviolet radiation (<400 nm), electron–hole pairs are created (**Fig. 14.6**). Recombination of surface charges is reduced by the presence of contaminant species. Both electrons and holes are able to create reactive species that can destroy toxic species at the surface. The surface reactions of both are shown in **Figure 14.6**. Holes are capable of creating reactive species that can destroy detrimental organic species (e.g., halogenated hydrocarbons) and transform them into CO_2, H_2O, and HCl while electrons are able to reduce harmful chemicals by unleashing O^{2-} species that accomplish similar reactions.

If noble metals like gold and platinum are chemisorbed to the TiO_2 or ZnO, photocatalytic activity is accelerated due to enhanced electron attraction. The presence of metal also helps to keep electrons and holes from recombining in the semiconductor and thereby increase the efficiency of the photocatalyst [6]. Many methods have been developed to overcome the dependence on UV radiation (less than 3% of the solar spectrum). These include doping the

FIG. 14.6

Schematic illustration of the photocatalytic effect of nanoparticulate titanium dioxide. Energetic light in the ultraviolet regime of solar radiation generates an electron–hole pair in TiO₂. Electrons and holes either recombine on the surface (neutralizing the photocatalytic action) or react with surface-adsorbed species to create reactive species that are able to oxidize organic compounds or kill bacteria. If chlorinated organics are degraded, mineral acids like HCl can form. By this way, pollutants are converted into nontoxic or environmentally benign chemicals without input of artificial sources of energy. One major disadvantage of TiO₂ and other semiconductors is that UV light is required—a wavelength range that comprises only a few percent of the solar spectrum.

relatively wide-bandgap semiconductor with manganese or other metals and semiconductors to produce bandgap states with lower energy or to attach organic dye molecules that are able to transfer excited electrons to the photocatalyst. In this way, the doped or otherwise altered oxide makes possible the absorption of visible light, e.g., light that comprises just less than 50% of the solar spectrum—more effective than UV radiation that contributes only about three percent of the solar light energy reaching the Earth's surface.

There are many kinds of photocatalysts. The most popular is TiO_2. Other photocatalytic oxides include SnO_2, ZnO, Fe_2O_3, and V_2O_5. Many sulfides are also able to act as photocatalysts: FeS, ZnS, MoS_2, CdS, and PbS; and selenides like $CdSe$. Semiconductor system duos utilize TiO_2 with CdS, SnO_2, ZnO, or $CdSe$. ZnO/CdS couples have also been studied. **Figure 14.7** illustrates the products of catalytic oxidation versus direct exposure.

In order for photocatalysts to be effective, however, immobilization onto a solid substrate is recommended. This action not only conserves catalytic material resources (by regeneration) but also leads to higher reaction rates. In particular, polypeptide-cased membranes that sequester nanoscale metals have become more important in recent years [59].

FIG. 14.7

Two modes of pentachlorophenol degradation is depicted. The pathway utilizing the photo-catalyst produces CO_2 and HCl (the process of total mineralization), much preferred over the production of octachlorodibenzo-p-dioxin, a species that is more toxic than its precursor, by direct photolysis [57]. Microbial decomposition of 4-chlorophenol proceeds very slowly— the reaction has a half-life of 500 days [58].

Pentachlorophenol (PCP)

Direct photolysis

Photocatalysis with TiO_2 nanoparticles

Octachlorodibenzo-*p*-dioxin

CO_2 + HCl

14.0.5 National Security and Defense

Security issues involve many aspects of the environment, health, and safety. Water poisoning (e.g., cyanides), radiation exposure (e.g., dirty bombs), explosives (e.g., land mines), nerve gases (e.g., sarin), bioagents (e.g., anthrax), drugs (e.g., cocaine)—and yes, and perhaps nanotechnology-enhanced materials themselves— are some well-known materials known to terrorists. Security programs focus on detection, preparedness, response, and recovery (mitigation). We will focus on detection and mitigation in this section. Toxic materials come in all forms and shapes. Bacterial toxins such as botulin, diphtheria, and anthrax require, in terms of $g \cdot kg^{-1}$ of body weight, 0.001, 0.10, and 0.004–0.02 $g \cdot kg^{-1}$, respectively, to deliver a lethal dose that kills 50% of subjects tested (a.k.a. LD-50) [60]. In this section, we will review several aspects of defensive techniques applied to mitigate potential chemical, biological, and radioactive agents.

Nanosensors. Room temperature CdZnTe sensors have been developed to detect gamma rays emitted by radionuclides from Cs and Co [61]. Gold nanowire sensors able to detect ppb levels of mercury in air or water have been developed [62]. Robust and regenerable sensors have shown superior performance over thin-film devices [62]—most likely due to enhanced surface area. A research team at the Lawrence Livermore National developed compact, low-power piezoresis-tive cantilever-based sensor arrays for chemical detection of many kinds of organic vapors over a wide concentration range [63]. Gas fingerprinting using carbon nanotube transistor arrays has been demonstrated by Frenchmen P. Bondavalli et al. [64]. Carbon nanotube field-effect transistors (CNT-FET) for gas sensing rely on the sensitive response of the Schottky junction between CNTs and numerous kinds of drain/source metal electrodes [64]. The mechanism is based on the change induced in the work function of the contact metal by a specific gas (e.g., sarin), thereby modifying the CNT-FET electrical response [64].

Nanomaterials as Toxic Agents. Nanomaterials themselves can pose formidable hazards to national security. A study by Y. Zhang et al. at the Arizona State University showed that nanomaterials (CdSe quantum dots, hematite, and six commercial metal oxides) in water possessed varied stability and that conventional water treatment methods were not effective in removing them completely [48]. Toxicity studies showed that nanomaterials flatten the microvilli in intestine and can pass easily through epithelial cells [48]. S.L. Harper et al. of Oregon State University conducted in vivo studies (zebra fish embryos) of the biodistribution and toxicity of nanomaterials [65]. Fullerenes and gold nanoparticles were used to investigate the effects of size and surface functionalization, and 11 kinds of nano-metal oxides were used to study the effects of chemical composition. They determined that toxicity was dependent on surface functionalization and chemical composition. Core size and functionalization also influenced the toxicity of gold nanoparticles. Polystyrene and CdSe fluorescent nanomaterials were employed in the investigation of biodistribution [65].

T. Knight et al. of the University of Southern Maine investigated the cytotoxicity of 15-nm diameter gold nanoparticles [66]. They found that significant cytotoxicity was developed in Jurkat human T cells by gold nanoparticles [66]. The gold nanoparticles had differential effects on other kinds of cell types [66]. J. Chen et al. of the Harvard Medical School evaluated the potential toxic effects of metal oxide nanoparticles (Fe_2O_3, CuO, ZnO) and their metal ion counterparts (Fe^{3+}, Cu^{2+}, Zn^{2+}) on human neuroglioma and neuroblastoma cells [67]. They found that CuO caused significant dose-dependent episodes of cellular death while the cytotoxicity of the other Fe_2O_3 and ZnO nanoparticles was marginal when compared to control results [67]. Of the ionic renditions, Cu^{2+} demonstrated a significant effect on cell toxicity [67].

Carbon nanotubes are proving to be useful materials in drug delivery and carbon nanocages filled with metals in MRI applications [68]. However, the potential toxicity of carbon nanotubes has received scrutiny over the past few years. R. Sharma et al. of Florida State University showed that carbon nanotubes (30-nm diameter) penetrated skin tissue quickly (within 2–3 min) and were able to damage the epidermal layer—at the level of the membrane and nucleus [68].

Surface-Enhanced Raman Probes. Land mines can be detected quite well by metal detectors (sensitivity to 0.5 g), but metal detectors respond to anything made of metal—tin cans, old utensils, and automobile parts. Therefore, new generation detectors are required that sense organic constituents of explosives and not specifically the metal components. Surface-enhanced Raman spectroscopy (SERS) analytical tools have been recently developed to detect parts per billion levels of explosives and degraded explosives from land mine vapors emanating from the soil [69]. Many types of explosives contain nitro groups attached to an aromatic moiety (**Fig. 14.8a** and **b**). These include picric acid, TNT (2,4, 6-trinitrotoluene), DNT (2,4-dinitrotoluene, an impurity), and degradation products (by the action of bacteria) 1,3-dinitorbenzene and 4-amino-2,6-dinotrotoluene vapors. All can be detected by SERS [70]. The diagnostic signal is due to the nitrate-bending modes at 820 cm^{-1} and the symmetrical stretching modes at

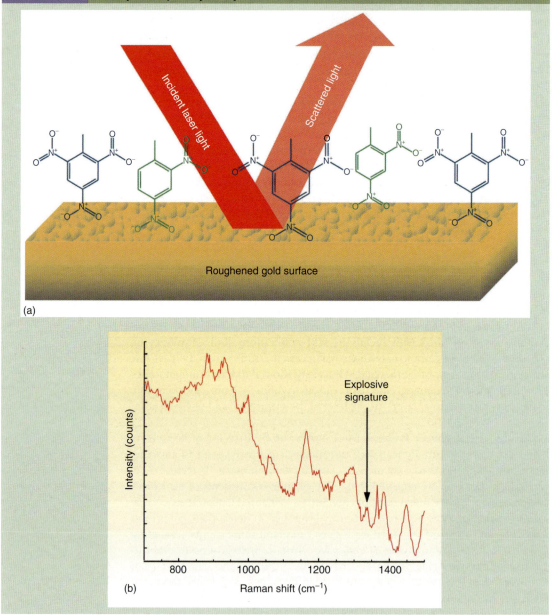

FIG. 14.8 *(a) Schematic representation of 2,4,6-TNT and 2,4-DNT detection by a SERS sensor. The nitrate moiety of TNT and DNT makes contact with the roughened gold substrate. When laser light is shined on the surface (785 nm, in the near-IR range), the symmetrical vibrational modes of the nitrate are modulated by the scattering. (b) The Raman peak at ca. 1384 cm^{-1} is diagnostic for TNT vapors. It can easily be seen above the background signals. Since natural nitrated aromatic compounds are scarce, there is little chance of interference. Therefore, the peak for adsorbed nitrate is diagnostic. Naturally occurring nitrogen dioxides undergo catalytic conversion to NO$_3^-$ at the surface. A sharp peak at 1035 cm^{-1} is characteristic of these species [70–72].*

Source: Images courtesy of Kevin M. Spencer, EIC Laboratories, Inc. With permission.

ca. $1337\,cm^{-1}$. The nitrate signature is strongly enhanced and shows that adsorption does not take place by the aromatic ring. TNT and DNT have been detected down to 1 ppb level in concentration (1 fg or 4.4×10^{-18} mol, the attomole level) by SERS detectors used in the field [71]. Interference from the background in field tests was not significant [72].

The surface-enhanced Raman substrate consists of metallic (gold, sliver, copper) micro- to nanospherical or ellipsoidal particulates (or roughened surfaces with micro- or nanofacets) upon which analyte substrates are adsorbed. SERS is capable of $10^{12}-10^{14}$ enhancement of detection limits [72]. The wavelength of the excitation laser matches the resonance of the surface plasmon of the metal nanoparticles. For the detection of TNT, wavelength in the near-IR (785 nm) provided the probing beam. The metal plasmon is tunable by adjusting the size, shape, and orientation of the metal nano or microparticles with respect to the probing radiation [73]. The SERS surface is oriented towards a commercial fiber-optic probe. A small fan directs the ambient vapor over the substrate [72]. SERS requires no sample preparation and with regard to spectral acquisition, sample evaluation takes place in ca. 30s. SERS is also able to detect many kinds of suspected chemical agents as long as a diagnostic peak exists for that material.

Cyanide found in drinking water is also detectable by SERS. Cyanide has a very strong Raman band at ca. $2200\,cm^{-1}$ and levels down to 2 ppb have been recorded. The detection of cyanide has been shown to be quantitative at ppm levels of concentration [72]. While SERS is perfect for chemical agents, resonant Raman spectroscopy (RRS) has shown utility for detecting biological species [72]. In this case ultraviolet radiation is used to excite biological molecules—especially proteins and nucleotides. Because Raman signal intensity is proportional to v^4 (e.g., the higher the energy of probing radiation, the more intense the Raman signal), nucleotides at concentrations on the order of 500 nM have been successfully detected [72]. This is a new area of national security study and there is still a way to go before a viable system is on the market.

Immunosensor Detection of TNT. Environmental detection and subsequent remediation of hazardous pollutants such as explosives is needed. On the detection front and in addition to SERS methods, portable fluorescence immunoassay biosensors have been developed by the Naval Research Laboratory (NRL) to meet the need [74]. The biosensor is based on a competitive fluoroimmunoassay in which a fluorescent molecule similar in structure to the analyte competes with the analyte for binding sites on antibodies immobilized on the surface of an optical probe [74]. In other words, by this method, fluorescent signal intensity is inversely proportional to the amount of analyte in the sample.

A fiber-optic biosensor (FOB) makes use of molecular recognition and evanescent wave sensing to detect many kinds of analytes. Optical fibers are able to excite fluorescent molecules that are near to the core of the fiber. An evanescent wave, the electric field generated from internally reflected light, has a penetration depth of 125 nm. Fluorescent molecules that enter the evanescent wave (bound by the antibodies on the surface of the fiber-optic probe) are excited and emit light at longer wavelengths (e.g., fluoresce). A portion of the fluorescence is transmitted by the fiber to a detector. Molecules out of range of the evanescent wave are not detected. This is an example of a competition-type immunoassay. A schematic of a fiber-optic detector is shown in **Figure 14.9**.

Source: Image courtesy of Naval Research Laboratory.

A continuous flow immunosensor (CFI) also utilizes immobilized antibodies and fluorescent dye marked signal molecules. Signal molecules similar to the analyte are tagged with a fluorophore (an organic dye). Analyte molecules are detected as they displace the fluorophore, thereby reducing the fluorescent signal. As opposed to the competitive type of immunoassay, the CFI technique is an example of a displacement-type immunoassay [74].

Chemical and Biological Agents and Nanomaterials.   Chemical warfare *nerve agents* include VX (methylphosphonothiolic acid: LD-50 = 15 mg·kg⁻¹ body weight), GB (sarin: LD-50 = 0.01 mg·kg⁻¹ body weight; 26x more deadly than cyanide, 0.5 mg lethal dose for adults), GD (soman), GF (cyclosarin), and GA (tabun). *Blister or vesicant agents* include HD (sulfur mustard yperite), HN (nitrogen mustard), L (lewisite), and CX (phosgene oxamine). *Choking agents* include CG and DP (phosgene and diphosgene), Cl (chlorine), and PS (chloropicrin). *Blister agents* like mustard gas irreversibly alkylate selected amines in proteins, enzymes, and DNA that leads to cell dysfunction and eventual death, depending on the level, of the organism [75]. Most nerve agents are based on phosphonic acid. Nerve gases bind to active centers of phosphatase enzymes such as acetyl-cholinesterase that are necessary for nerve impulse transmission [75]. Selected chemical structures are shown in **Figure 14.10**.

Successful mitigation of chemical agents requires the following: (1) that formation of nontoxic by-products, (2) that decontamination is accomplished quickly, (3) that procedures and materials are applied easily, (4) that detox materials are not corrosive or hurtful to nearby surfaces, (5) that mitigation be accomplished quickly, and (6) that waste products be disposed of safely and efficiently [75]. Mitigation proceeds by physical methods such as adsorption or catalytic reactions that involve hydrolysis or oxidation.

FIG. 14.10 *Chemical structures of various warfare agents.*

Soman (GD)

2,2′-Dichloroethyl sulfide (mustard gas or HD)

Lewisite (L)

Methylphosphonothiolic acid (VX)

Sarin (GB)

Phosgene (CG)

Traditionally, activated carbons adsorb warfare agents quite well although the adsorbate remains in a toxic form. Engineered nanoparticles offer the same surface advantage as do activated carbons but in addition are able to neutralize the toxins via chemical action. VX, GB, DP, and HD have been shown to hydrolyze at room temperature after contact with nanosized Al_2O_3 [76]. The by-products of the alumina-catalyzed reaction are harmless surface-bound nontoxic phosphonates or aluminophosphonates like $Al[OP(O)(CH_3)OR]_3$ [76].

Magnesium oxide nanoparticles possess basic sites that interact with HD by hydrohalogen elimination.

Nerve gases like VX can be neutralized by nanosized MgO to form hydrolysis products. Technological challenges facing catalytic procedures involve keeping the catalyst surface from clogging and aggregation of the catalyst. Aluminas possess a high level of Lewis acid sites that enhance hydrolytic processes. In this case however, the surface of the alumina can be eroded by the chemical agents during reaction. The erosion behavior keeps the surface fresh and active, and this is one of the reasons nanoscale aluminas are expected to contribute favorably to warfare agent mitigation [75].

Biological warfare agents include botulin toxin (LD-50 = 0.001 mg·kg^{-1} body weight), diphtheria toxin (LD-50 = 0.10 mg·kg^{-1} body weight) and anthrax toxin (LD-50 = 0.004–0.02 mg·kg^{-1} body weight) [60]. Traditional methods to address biological toxins include disinfection with chlorine and bleach (chloramine solutions). These materials, however, are quite corrosive and quite toxic themselves. Oil-in-water emulsions containing antitoxins are effective against bacteria but are difficult to dispose and are ineffective against airborne pathogens [75].

One of the best ways to mitigate biological pathogens is by application of nanopowders like MgO, ZnO, and CaO [75]. Nanoparticles have incredibly high adsorption capacity toward potential bactericides like chlorine bromine, or iodine and are able to store them on the surface until required. Halogen-loaded MgO, for example, becomes positively charged in solution—a perfect scenario for attracting bacteria that are negatively charged [77]. Laser confocal microscopy data showed that Br- and Cl-decorated MgO nanoparticles coagulate spontaneously with bacteria to form clumps [77]. In the case of anti-fungal treatment, halogen-coated nanoparticles are able to enter cells through holes developed in the membranes damaged by halogen-treated nanoparticles. Due to the basicity of many nanoparticle surfaces, the outermost layer of base-sensitive proteins of spores is partially or completely removed [75].

In summary, nanoparticles are able to destroy biological agents because they are physically abrasive, have acidic or basic sites, are oppositely charged to bacteria, and are able to support a large amount of oxidizers like chlorine and bromine [75].

Detection of Thermal Neutrons. Another national security risk is posed by dirty bombs—bombs that rely on traditional explosives but are laced with nuclear wastes. The LD-50 of radioactive plutonium is $1 \, mg \cdot kg^{-1}$ body weight [60]. Conventional detectors like the ^3He-tube can attain >80% efficiency but are expensive to maintain and operate (>1000 V, leakage, expensive gas). Solid-state scintillator detectors have difficulty with discrimination, operate under high voltages, and longevity of the photomultiplier tube is questionable. Another kind of detector consists of polycrystalline boron nitride materials and is able to convert neutrons to alpha particles and generate an electrical signal. The problem is that it is not single crystalline. A recently developed detector employs ^{10}B pillars embedded in silicon—P–I–N diode pillars (doped-p/intrinsic undoped layer/doped n-layer) that are 2-μm width and spacing and 50-μm in height [78]. Detector efficiencies on the order of 65% have been reported and better results are expected with smaller features [78]. Advantages over other kinds of detectors are that the pillar detector is three-dimensional and that the interstitial converter material is in very close proximity. Theoretical efficiency can be 70% with the current configuration but better results can be obtained if the device geometry is rendered in the nanoscale [78].

14.0.6 *The Nano Perspective*

There are always positive factors associated with any technology. There are always negative factors. There are always factors that lie somewhere in between in the grey area. Positive impacts of nanotechnology are expected to be numerous and varied. According to T. Masciangoili et al. of the Environmental Protection Agency (EPA, 2004), nanodevices will require less material upfront, during production and as a by-product of manufacturing processes [79]—the case for dematerialization (reduction of the need for raw and manufactured materials), minimization of industrial wastes and effluents, and reduction of toxins entering the environment. Nanotechnology is expected to exert positive societal changes that benefit the environment by enhancement of vehicular transportation systems, urban planning and development, and information management. A significant impact of nanotechnology on the environment is expected to come in the form of green manufacturing, pollution prevention, treatment, mitigation and remediation, and the development of sensors [79].

Pollution can be mitigated a priori by a few general common sense practices: (1) source and resource reduction and conservation, for example, the less material one works with, the less potential there is for pollution; (2) process efficiency, for example, the more efficiency that is built into a process with regard to water use, energy input, raw materials, etc., the less material is wasted or produced that is polluting in nature; (3) effective postprocess pollution mitigation, for example, highly sensitive and efficient "end-of-pipe" means of reducing pollutants before they enter environmental conduits; and (4) substitution of toxic substances (i.e., substances that exert detrimental effects at low levels) with those that are benign to the environment, health, and safety. These factors are relevant for any kind of material regardless of size and any kind of technology regardless of kind. Two of the grand challenges issued by the National Nanotechnology Initiative (NNI, at nano.gov) involve environmental themes: (1) efficient energy and storage, and (2) nanoscale processes for environmental improvement.

The Extreme Importance of Gold and Silver Nanoparticles. According to the American financier Bernard M. Beruch, in the early 1900s

> Gold has worked down from Alexander's time. When something holds good for two thousand years I do not believe it can be so because of prejudice or mistaken history.

Although gold is a coveted commodity worldwide, its value in the nanoworld is ever increasing exponentially. Gold's nanoscale properties have proven to be quite remarkable, especially in its application as a catalyst. David T. Thompson of the World Gold Council [51–53]

> Many of the applications of gold are based on its unique properties, and this uniqueness can be rationalized in terms of the large relativistic contraction of its 6s orbitals resulting in a very small atomic radius compared with that which might be expected from the position of gold in the periodic table...

Although we focus on environmental issues in this chapter, gold's importance in all of nanotechnology cannot be understated. Christopher W. Corti, consultant to the World Council, states that [53]

> The gold industry wants to see increasing demand for gold in industrial and medical applications and has recognised that it must make a commitment to new science and its exploitation if that objective is to be achieved. We also live in an age where protection of the environment and the deleterious impact of global warming are issues of international importance. These two issues are central to the gold mining industry's aim to develop important new industrial applications for gold.

And guess what? Much of this new science is centered on the applications of nanogold. Gold is yet again proving that it is one of the most remarkable (and valuable) of materials—whether in bulk or nanophase! Nanosilver materials have made major environmental impacts in air purification, disinfection, and odor mitigation technology. The importance of nanosilver will be discussed in more detail later on in the text.

Nanotechnology Impact on the Environment. We are beginning to realize how nanotechnology and nanomaterials will be able to mitigate environmental insults and reduce the amount of industrial materials, but what of the nanomaterials themselves? New technologies bring along new benefits and new consequences. Nanotechnology is no different. Research must stay ahead of potential negative environmental impacts of nanotechnology—especially with regard to human health and safety where a few parts per million could exert toxic or even lethal effects. We must understand the transport, transformation, and fate of nanomaterials in various environmental media. We must learn from the impacts of earlier technologies and understand how nanomaterials would be different with regard to environmental mobility, reactivity, and accumulation from earlier technological materials [80].

Nanomaterials are smaller and are therefore able to penetrate regions otherwise forbidden to larger particles and are quite reactive due to enhanced surface-to-volume ratio. Many nanomaterials, for example, are able to penetrate skin. The size and surface (and surface chemistry) of nanomaterials allow for facilitated transport, reactivity, and bio-uptake (e.g., endocytosis, membrane penetration, or transmembrane channels) [80]. Nanomaterials can be hydrophilic or hydrophobic.

Environmental transport of nanomaterials may occur in air, surface water, groundwater, or soils. Gravity, inertial forces, and buoyant forces (like turbulence) affect the dispersion of large particles (>100 nm). The transport of nanoparticles, on the other hand, may occur primarily due to diffusion processes [81]. Nanomaterials, therefore, can mix and disperse quite rapidly and be subjected to the movement of air and water. It also must be kept in mind that smaller particles are subject to forces originating from electrical charge, van der Waals attraction, and diffusion (as stated before). Agglomeration is always a possibility with nanomaterials. Fullerenes like C_{60}, typically insoluble in water, form colloids that remain suspended in water, and therefore are able to persist [82].

14.1 WATER AND SOIL QUALITY, MONITORING, AND MITIGATION

Water covers over 70% of the Earth's surface of which 0.3% is fresh [83]. Seventy percent of the fresh water is used for agricultural purposes. Energy demand contributes to ca. 50% of the cost of desalination. There is nothing more critical to maintain life processes than water. The percent of water in our bodies ranges from 60% to 80% depending on our age. We get our daily requirement, at a level that varies greatly from person to person, from water directly, from foods and from drinks like tea, coffee, and juices. Where do we get our water? Well, we get it from the environment—from rivers, rainwater, snow, groundwater, lakes, and the oceans. All these sources are susceptible to pollution from human activity.

Pollution (from the Latin *pollutionem* "defilement" based on the Greek *lyma* "filth, dirt, disgrace," *lymax* "rubbish, refuse") is the introduction of materials into the environment that cause harm to human health, other living organisms, and the environment itself. Pollution occurs when substances exceed natural levels and have detrimental effects—in the form of aesthetics, and health and sustainability of ecosystems. Water pollution comes in many forms. Pollutants arise from industrial processes, mining, agricultural runoff, domestic sources, hospitals and just about every other anthropogenic source. Pollution also arises from natural sources like volcanic eruptions, floods, windstorms, and biological phenomena. Pollutants take the form of metal cations, particulates, viruses, organic chemicals, radiation sources, or other species that simply do not belong in natural or treated water supplies. Once in water, pollutants are exposed to great environmental mobility, depending on their size, concentration, solubility, and chemical characteristics.

So long as populations were relatively disperse, water pollution was not an issue. With the onset of civilization, pollution soon followed wherever cities were founded. Accumulation of sewage was one of the first water pollution concerns to confront humanity. Some cultures dealt with it efficiently like the Hindus and Israelis (due to adherence to strict religious codes) and the Romans (due to efficient engineering capability). Indus River valley civilizations developed city-wide sanitation and codes of hygiene as early as 2500 B.C. The Romans built *cloaca maxima* (the great sewer) in 500 B.C. Pollution from waterborne heavy metal contaminants were known to the ancient Romans. In 1388, the British Parliament passed an act forbidding the disposal of garbage into ditches and rivers. In 1690, Paris becomes the first city to install an extensive sewer system. Before the advent of carbon nanotube water filters [83,84], humans resorted to less sophisticated methods to cleanse their water.

14.1.1 *Traditional Water Treatment*

Understanding the factors involved in drinking water quality was not well understood by the ancients although there is evidence that water treatment in the form of filtering through charcoal, exposure to sunlight, and boiling and straining were accomplished by ancient Sanskrit cultures 6000 years ago and the ancient Greeks [85]. The driving force behind water treatment in early times was to reduce the level of turbidity. Although the historical record shows that aesthetic issues such as water odor, appearance, and taste have been noted, the notion of water quality took thousands of years to develop [85]. The Egyptians used alum hydrated aluminum potassium sulfate, $KAl(SO_4)_2 \cdot 12H_2O$ to force suspension of solids from solution. The practice of filtration reemerged in the 1700s and 1800s and it was not uncommon to filter water slowly through sand [85]. Fine sand can be 60 μm or less in diameter—not quite a nanomaterial. The link between disease and waterborne microbes was made in the mid-1800s, and it was not until Louis Pasteur linked germs to disease that major sanitation operations were launched. Filtration was used to reduce turbidity and microbial contaminants such as cholera, typhoid, and dysentery [85].

In 1908, chlorine was used as a disinfectant in New Jersey as was ozone in Europe. The U.S. Public Health Service set drinking water standards in 1914. By 1974, the Safe Drinking Water Act was passed by Congress but new kinds of pollutants made the scene by the 1960s: industrial, agricultural, and synthetic chemicals and new treatments that involved aeration, flocculation, and activated carbon adsorption were reinvigorated. More than 800 organic and inorganic chemical species originating from industrial, agricultural, and municipal discharge have been identified in common drinking water [86]. Other estimates raise the level to 1500 or more [86]. Modern techniques designed to remove chemical species, many of them carcinogenic, include aeration, flocculation (coagulation and sedimentation), ion exchange, and activated carbon filtration. Application

FIG. 14.11 *Traditional water treatment process is shown. Nanomaterials have already contributed to such methods. In the future, more solutions are expected to emerge from nanotechnology.*

Water source

Coagulation and mixing
Alum and other chemicals form *floc* (flakes) to attract dirt particulates

Sedimentation
Flocculants settle leaving clear nonturbid water

Filtration
Water filtered through layers of sand, gravel, and charcoal

Disinfection
Chlorine or ozone is added to kill bacteria and microorganisms. Ultraviolet light is also effective

Storage
Water is stored to allow time for effective disinfection

Nanoscale considerations of water treatment

Colloid chemistry plays a major role in the coagulation and mixing phase of water purification. Flocculation is a process by which particles of sub-micron-sized clay aggregate into larger macrosized particles. Flocculation occurs as a result of a chemical reaction between the clay particles (the dirt) and flocculants (e.g., salt water or alum). Flocculants cause colloids, otherwise too small to be filtered, to aggregate. Flocculants usually contain multivalent cations—aluminum sulfate, calcium oxide, iron(II) sulfate, iron(III) chloride, sodium silicate, sodium aluminate, and aluminum chlorohydrate—or can be found in the form of polymers. Under the appropriate pH, temperature, and salinity conditions, positively charged flocculants, interact with negatively charged colloids thereby removing electrostatic barriers to aggregation. Flocculation (reversible aggregation?), coagulation (irreversible aggregation?), and aggregation are terms that are often used interchangeably. Van der Waals and stronger electrostatic forces play major roles in the flocculation process. According to Malvern Instruments, Ltd., suspended colloids in a stable colloidal phase can be removed from water by flocculation, coagulation (aggregation), sedimentation, and finally phase separation—in that order.

Nanomaterials play a role in the filtration process. Sand particles range in size from submicron to several micron dimensions. Sand particles may be coated with a monolayer to enhance separation properties. Charcoal consists of a porous structure of micro-, meso-, and/or macropores. Die to its special surface chemistry, charcoal has a natural affinity for organic materials and is able to remove them from aqueous media. Activated charcoal (or activated carbon) is usually reserved for the next level of water purification—found in the home or laboratory. GACs and PACs (granular and powered activated carbons, respectively) are effective at removing particulates and organics from water. PAC can be removed from water by coagulation, flocculation, and sedimentation. Removal or organics such as *p*-chlorophenol occurs via adsorption and diffusion into pores and onto surfaces of activated carbons mentioned earlier [87].

of activated carbon is the "best broad spectrum control technology available today" [86]. **Figure 14.11** illustrates a generalized scheme of water purification adopted by most municipalities in the United States.

The Clean Water Act. The Clean Water Act (CWA) of 1972 established guidelines to clean up America's waterways. The CWA (and specifically the Federal Water Pollution Control Amendments of 1972) is the primary federal law that regulates the quality of water in the United States. The act identified point sources such as industrial facilities, municipalities, and agriculture as targets for regulation. It set the standards for levels of toxic substances with the goal of eliminating additional water pollution by 1985. Over the years, several modifications to CWA were passed by a number of congressional sessions.

14.1.2 Nanomaterial Contamination in Aqueous Environments

Groundwater is water found beneath the surface of the Earth. Approximately one-third of all freshwater supplies originated from some form of groundwater in the United States. In Asia, the level of use is increased to over 50%, and in Europe the percentage is even higher. Groundwater sources are usually cleaner than surface water sources. Aquifers, the source of water for much of agricultural use, are replenished as part of the hydrologic cycles: rainfall, percolation, or directly by surface water recharging. Seepage from landfills, septic tanks, underground fuel storage tanks, road salts (ice melt), hazardous waste sites (over 200,000 abandoned sites in the U.S. alone), mine runoffs, agricultural runoffs, and poorly managed municipal treatment sites all combine to contaminate groundwaters. What are the contaminants? We are quite familiar with most of them by now in this section: VOCs, SOCs (synthetic organic compounds), HOCs (halogenated organic compounds), heavy metals, inorganic salts (anions and cations), radioactive materials, particulates and biological pathogens like bacteria, protozoans, and viruses, and associated biological toxins.

Natural Nanoparticles in Soil and Water. Natural soils consist of aggregations of particles composed of silicates (sand and silt), aluminosilicates (zeolites and clays), iron oxides, organics, and an assortment of biological materials—a veritable plethora of nanomaterials, aggregates, and conglomerations. Nanophase materials found in the soil are responsible for water percolation, purification, and water content in general. What happens to the natural nanomaterials in soils as we introduce synthetic ones?

Synthetic C_{60} and Carbon-Based Nanoparticles. According to researchers at Purdue University, C_{60} nanoparticles have little impact on the structure and function of the soil microbial community [88]. Naturally occurring oil samples were treated with $1\,\mu g\ C_{60} \cdot g^{-1}$ soil in aqueous suspension or $1000\,\mu g\ C_{60} \cdot g^{-1}$ soil in granular form and incubated for 180 days. Soil respiration byproducts and soil community structure were monitored [88]. Neither the fullerene or the application solvent caused any measurable decrease in the amount of organisms found in the soil. Neither were the levels of oxygen and enzyme activity affected, that is, under laboratory conditions [88]. In real world waterways and soils, exposure pathways may affect toxicity [88].

John D. Fortner et al. at Georgia Tech have compiled significant level of research concerning C_{60} and carbon nanotube effects in soils and water [89–93]. They showed that fullerenes tend to form a stable clump (as rectangular solids—**Fig. 14.12**) with the size of the clump in the range of 25–500 nm (called nano-C_{60}) and recommended further environmental modeling be conducted on not only C_{60} but the C_{60} aggregates as well. The size of the clumps is affected by pH, rate of addition to water, and the level of dissolved salts (that cause C_{60} to form a solid mass that settles on the bottom) [89].

Fortner et al. also found that aggregates inhibited prokaryote growth and decreased aerobic respiration [89]. In subsequent studies with *Escherichia coli* (Gram-negative) and *Bacillus subtilis* (Gram-positive) bacteria, they demonstrated that C_{60} associated strongly with both kinds of bacteria and displayed antimicrobial

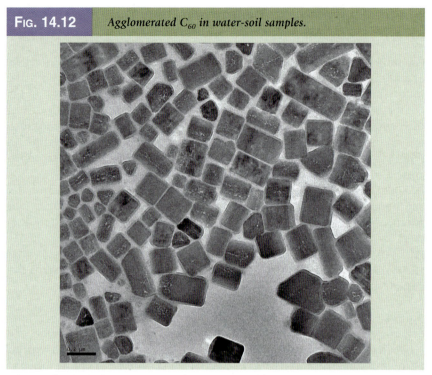

Fig. 14.12   *Agglomerated C_{60} in water-soil samples.*

Source: Reprinted with permission from Georgia Institute of Technology: Professor Joseph Hughes,
Dr. John Fortner, Rice University: Dr. Joshua Falkner and Professor Vicki Colvin.

behavior [90]. As before, higher salt concentration inhibited or eliminated the antimicrobial properties of C_{60} [90].

Aggregation dissolution occurs with the formation of fullerene oxides—to the tune of an average of 29 oxygens per fullerene molecule arranged in repeating hydroxyl-hemiketal moieties [93]. C_{60} in aqueous phase was also shown to produce reactive singlet oxygen in the presence of UV irradiation [91]. These results suggest that such aqueous phase reactivity to a strong oxidizer and ultraviolet radiation found in the environment should be considered when studying the transport, fate, and risk analysis of C_{60} materials [93]. Studies of the behavior of multiwalled carbon nanotubes in the presence of natural organic matter (NOM) revealed that MWNTs are readily dispersed. The individually dispersed MWNTs remained stable for over a month [92]. The authors concluded that the dispersal of carbon-based nanomaterials in natural, aqueous environments occurs better than expected and that this too must be considered when future transport, fate, and risk analyses are conducted [92].

Silver Nanoparticles.   The effects of silver nanoparticles in soils and waterways are relatively unknown although nanosilver is a proven bactericide that is found in more and more consumer products—food packages, odor-resistant socks and clothing, household appliances, and now in wound dressings like Band-Aid [94]. Certain heavy metals in colloidal form and metal salts are known to be toxic to humans and environmentally hazardous to beneficial soil species and organisms such as fish, algae, crustaceans, some plants, fungi, and bacteria [95–99]. Silver nanoparticles are currently used in combination with TiO_2 (nanosilver-titanium dioxide coating, NS-TDC) in an effort to disinfect surfaces at train stations in Hong Kong—escalators, handrails, machines, elevator buttons, etc. [100]. Its manufacturers claim that the NS-TDC spray kills a variety of bacteria, mold, and viruses like the H_1N_1 flu virus [100]. It is not known how much silver is introduced into waterways due to application and erosion of NSTDC materials.

A new kind of energy-efficient washing machine, manufactured by Samsung, was introduced to Swedes in 2005. The machine makes use of silver nanoparticles at low water temperatures [94]. However, the Swedish version of the EPA is concerned that nanosilver will cause damage to water organisms and that it will be costly to remove silver particles from wastewater effluents. The Stockholm Water Authority claims that the washing machines are expected to dump two to three times more silver into waterways than the current level. Another potential problem facing the use of silver nanoparticles is that they may destroy essential bacteria required to treat sewage.

14.1.3 Activated Carbon—A Simple Traditional Nanotechnology

Activated carbon is a nanomaterial—a material with an enormous surface area per gram—on the order of $400–2000 \, m^2 \cdot g^{-1}$. Activated carbon is composed of an "amorphous material" with microcrystalline structure—quasi-graphitic particulates—that consists of highly developed porosity and extended interparticulate surface area [86]. Quasi-graphite consists of stacked graphene sheets cross-linked in a random manner. The irregular arrangement generates interstitial spaces that give rise to a network of polydisperse pores of nonuniform size and shape [86]. The interlayer d-spacing between individual graphene sheets is larger (ca. 0.35 nm) than that of graphite (0.335 nm). ESR analysis revealed that the aromatic structures in activated carbon contain free-radical elements, and hence unpaired electrons at unsaturated edge carbons. The material, although stabilized by the resonance structure typical of graphene materials, renders the material reactive to heteroatoms such as hydrogen, oxygen, nitrogen, and sulfur. Therefore, typical activated carbons contain 88% C, 0.5% H, 0.5% N, 1% S, and ~7% O and the rest inorganic ash residues [86]. E.G. Furuya et al. showed the action of interparticle mass transport of phenol in activated carbons [87].

Activated carbon is manufactured by a two-step process: (1) pyrolysis of carbon-containing materials at 600–900°C in the absence of oxygen and (2) activation by oxidation in CO_2, O_2, or H_2O at 600–1200°C or higher. Pyrolysis of organic materials above 1800°C forms turbostratic graphite. Turbostratic graphite possesses only short-range order on the scale of a few to tens of nanometers. Activation by chemical means also occurs—in acids, bases, or materials like $ZnCl_2$ at 450–900°C. Activation in acidic media produces carbons with affinity for heavy metals but minimally so for chlorinated hydrocarbons—that are preferred if activation takes place in basic media (and heavy metals are not favored). Materials most often adsorbed in water industrial and municipal treatment facilities are free available chlorines (FACs), halogenated organics (HOCs like DDT, endrin, lindane, TCE and others), aromatics (benzene, toluene, dioxins, PCBs, and plasticizers), heavy metals (Pb, Cd, Hg as dissolved ions, colloidal oxides, or carbonates), and taste and odor compounds (T&O) produced by microbes [99].

Nanosized pores and pockets are formed in carbon particles that contribute to its tremendous overall surface area. The pores, spaces, or cavities depending on pretreatment conditions range from less than 2 nm (micropores that comprise ~95% of the surface area) to mesopores with diameter <50 nm and to macropores, in which the pore diameter is greater than 50 nm. Powdered activated carbon (PAC) is ca. 40 μm in diameter. Granulated activated carbon (GAC) is 400–600 μm in size.

Activated carbon, due to the small size of its features, adsorbs organic pollutants by means of van der Waals forces—a characteristic of nanomaterials and due to the random interactions of electron clouds and unsaturated valencies with target contaminants. X-ray studies have shown that heteroatoms are most likely found adsorbed onto edge sites [86]. The overall reactivity and catalytic properties of activated carbons are directly related to their nanostructure (adsorption capacity) and chemical nature (adsorptive species). Active sites, active centers, and attached heteroatoms determine the material's propensity for polar or non-polar adsorption. Regeneration of activated carbon is energy intensive.

Adsorption of contaminants takes place by two well-known processes—physisorption (van der Waals, 10–20 kJ · mol^{-1}) and chemisorption (sharing or exchange of electrons, 40–400 kJ · mol^{-1}). The adsorption profile is described by traditional isotherm formulae: Langmuir, Freundlich, or linear. In drinking water treatment, the most commonly used isotherm application is the Freundlich isotherm—a Type I isotherm in which only surface adsorption is considered and there is no interaction between one adsorbed molecule with another (e.g., no capillary condensation like in BET analysis).

$$\theta = K \cdot (C_{equil})^{n^{-1}} \tag{14.2}$$

where
> K and n are empirical, experimentally determined constants (usually with $0 < n < 1$)
> C_{equil} is the equilibrium concentration of the solute (equal to p_{equil} if in gaseous form at solid–gas interface)
> θ is a measure of surface coverage in terms of

$$\theta = m_{solute} / m_{adsorbant} \tag{14.3}$$

where
> m_{solute} is the mass of solute adsorbed
> $m_{adsorbant}$ is the mass of the adsorbing material

If $n = 1$, then $\theta = K \cdot C_{equil}$; if $n \to \infty$, then $\theta \to 1$ as the limits. Plots of log θ versus log C_{equil} yield a linear relationship. In soil analysis, the values of K and n are dependent on soil properties such as pH, organic matter content, clays, aluminum, and iron oxides.

Adsorption capacities can be determined by preparing identical bottles of powdered activated carbon suspended in water. Different levels of a potential contaminant are added to the bottles and mixed until equilibrium is achieved. The aqueous adsorbent–solute mixtures are filtered and the concentration of remaining unadsorbed contaminant is determined. A plot of log θ versus log C_{equil} is then derived. The average line is the Freundlich isotherm. The Freundlich isotherm for a specific contaminant is then applied to determine the amount of contaminant adsorbed in terms of $mg \cdot g^{-1}$ of activated carbon. By convention,

---

EXAMPLE 14.1   *Freundlich Isotherm Adventures*

Data were collected from different bottles of granular activated carbon (GAC). Varying concentrations of trichloroethylene (TCE) were added to each, C_{equil}. After filtering, the remaining TCE, the unadsorbed concentration was calculated and the adsorbed amount was determined in $mg \cdot g^{-1}$ carbon capacity. The resulting data is shown in **Table 14.4** below. Derive a Freundlich plot (**Fig. 14.13**). Is it economically reasonable to remove TCE from water using GAC adsorbents?

TABLE 14.4   *Isotherm Data*

	1	2	3	4	5	6
C_{equil} TCE $(mg \cdot L^{-1})$	0.0010	0.0030	0.030	0.30	0.50	1.0
log C_{equil}	−3.00	−2.52	−1.52	−0.523	−0.301	0.00
θ, Capacity $(mg \cdot g^{-1})$	0.40	0.75	3.4	15	20	30
log θ	−0.400	−0.125	0.531	1.18	1.30	1.48

FIG. 14.13   *Freundlich isotherm derived from data in* **Table 14.4.**

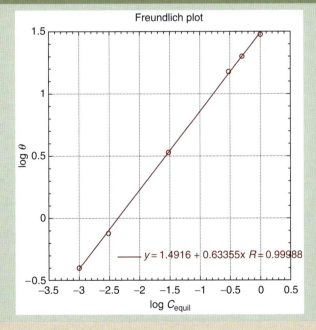

TCE is much better adsorbed than is chloroform. Therefore, TCE can be removed economically from water streams by GAC adsorbents [99].

capacity values $(mg \cdot g^{-1})$ corresponding to $1.0 mg \cdot L^{-1}$ are used to compare adsorptions of different contaminants [101].

In industry practice, chloroform is one of the least absorbable materials by activated carbons $(2.6 mg \cdot g^{-1}$ @ $1.0 mg \cdot L^{-1})$. Therefore, any material with a Freundlich isotherm value less than that of chloroform is considered to be unadsorbable by activated carbon [101]. In industrial practice, activated carbons are used as prefilters in many applications, specifically in reverse osmosis.

14.1.4 *Membranes and Separation Technology*

Water purification, desalination in particular, is accomplished by a few tried and true techniques—*reverse osmosis* (RO), electrodialysis, and distillation. Before we describe reverse osmosis, let us describe the phenomenon of osmosis first. Osmotic pressure is a colligative property of a solution and is a function of the number of solute particles per unit volume. Osmotic pressure, however, is independent of the molecular composition or shape of the solute. Osmotic pressure is often used to determine the molecular weight of macromolecules. Osmosis is the diffusion of water through a cell membrane or semipermeable membrane (only allows water to transport) from a solution of low solute concentration to that of a higher solute concentration [102]. Osmotic pressure is the hydrostatic pressure that is generated by differential concentrations across a semipermeable membrane [102,103]. The process of osmosis is pressure driven. Osmotic pressure Π is defined (in the ideal limiting case) as

$$\Pi = iMRT \tag{14.4}$$

where
  i is the van't Hoff factor (the number of moles of solute in solution)
  M is the molarity (concentration) of the solute
  R is the gas constant equal to $0.082 L \cdot atm \cdot mol^{-1} \cdot K^{-1}$
  T is temperature in Kelvin

In cells, isotonic solutions have $\Pi_{sln} = \Pi_{cell}$; in hypertonic solutions, $\Pi_{sln} > \Pi_{cell}$ (e.g., cells shrink), and in hypotonic solutions, $\Pi_{sln} < \Pi_{cell}$ (cells swell). Another form of this equation is often used

$$\Pi_{osmosis} = \frac{n_{solute}RT}{V} \tag{14.5}$$

The molecular weight of macromolecules like proteins and polymers can be determined from osmotic pressure measurements (osmometry) and applying

$$MW_{solute} = c\frac{RT}{\Pi} \tag{14.6}$$

where
  MW_{solute} is the molecular weight of the solute in $g \cdot mol^{-1}$
  c is the concentration in $g \cdot L^{-1}$.

Reverse Osmosis. Reverse osmosis (RO) is a process by which water is purified (e.g., increasing the concentration of solute on one side) against the natural chemical potential of the solvent/solute system. It requires energy to do so [104]. The nanotechnology involved in RO processes is confined to the chemistry and structure of the semipermeable membrane. Some RO membranes make use of specially coated 20-nm polyamide nodules (with polysulfone as a porous layer support) under high pressure to overcome natural osmotic pressure. Although adapted by industry to treat waters with a wide range of salinity and brackishness, the process is energy intensive and RO membranes have relatively low throughput. Reverse osmosis processes require significant energy, on the order of 1.5–2.5 kW-h of electricity to produce one cubic meter of fresh water. *Electrodialysis* employs an electric potential to force ions through a highly resistive membrane but does not work well with seawater substrates. *Distillation*, the most energy-intensive method, is one we are all quite familiar (1000 kW·AF^{-1} to separate H$_2$O from its salts and 800,000 kW·AF^{-1} just to overcome the heat of vaporization, AF = acre·foot, 326,000 gallons). When measured in terms of MJ·cm^{-3} (not considering distillation which has prohibitively high energy demand), desalinating water with 20,000 ppm salt content requires ~20 MJ·cm^{-3} by electrodialysis and ~15 MJ·cm^{-3} by RO. An optimal level of energy use considered to be is in the neighborhood of ~10 MJ·cm^{-3}.

Mass transport of a solvent through a semipermeable membrane driven by a difference in chemical potential on either side of the membrane is called *osmosis* (from the French *endosmose* "inward passage" on the Greek *osmo* "pushing"). The chemical potential of a solvent decreases with solute but increases with pressure. The process of osmosis strives to create a balance between the two. The solvent flows from the solution with the lower solute concentration (greater μ_{slv}) into the one with the higher solute concentration (lesser μ_{slv}).

$$\mu_{\text{solvent}} = \left(\frac{\partial G}{\partial n_{\text{solvent}}}\right)_{T,P,n_{\text{solute}}} = \mu^o_{\text{solvent}}(T,P) - k_{\text{B}}T\left(\frac{n_{\text{solute}}}{n_{\text{solvent}}}\right) \qquad (14.7)$$

In reverse osmosis, the solvent is forced to flow from the solution with higher solute concentration into the solution with low or no solute concentration. A reverse osmosis configuration is shown in **Figure 14.14**.

RO is a separation process in which solutes are retained and concentrated on one side of a semipermeable membrane while fresh water is forced through the membrane on the other side. Problems associated with RO membranes in addition to energy demand and low throughput include fouling by bacteria and colloidal particles; compaction due to exposure to high pressure that rearranges the polymeric elements; and chemical decomposition, membrane defects and poor sealing, and oxidation and degradation of matrix elements—polyamide by Cl$^-$ and cellulose acetate from *p*H below 3.5 or higher than 7.5. Categories of RO systems depend on pore size and composition: *particle filtration* removes particles >1 μm, *microfiltration* for particles >50 nm, *ultrafiltration* for particles >3 nm, *nanofiltration* for particles >1 nm, and finally *hyperfiltration* for the removal of sub-nanometer particles [104].

FIG. 14.14	*A reverse osmosis cell is depicted above. High pressure is applied to the solute side on the order of 2–17 atm (30–250 psi) for brackish water and from 40–70 atm (600–1000 psi) for seawater. The semipermeable membrane keeps the solute from transiting through the membrane that allows only water to flow.*

RO and Nanoparticles, Proteins, and Colloids. Oncotic pressure is a form of osmotic pressure and is a characteristic of the circulatory system. The walls of capillaries in the circulatory system are semipermeable—permeable to water but not to plasma proteins. Therefore, an osmotic pressure is generated. There is also a tendency for proteins (mostly negatively charged) to attract cations (the

---

EXAMPLE 14.2   *Desalination and Reverse Osmosis*

What pressure is required to desalinate water with sodium chloride concentration of 0.08 M at 25°C?

Solution:
Apply $\pi = iMRT$
Sodium chloride dissociates into sodium cations and chloride anions in solution

$$NaCl \rightarrow Na^+ + Cl^-$$

Therefore, $i = 2$.

Concentration is equal to 0.08 M, $T = 298$ K, and $R = 0.082$ L·atm·mol^{-1}·K^{-1}.

$\pi = iMRT = (2)(0.08\,M)(0.082\,L \cdot atm \cdot mol^{-1} \cdot K^{-1})(298\,K) = 3.9$ atm pressure required to purify water with NaCl concentration of 0.08 M.

Gibbs–Donnan effect) that contributes to the osmotic gradient between the plasma and the interstitial fluids. Because of these two effects, water is drawn from the interstitial regions into the plasma. The resulting pressure is called *colloid oncotic pressure*. The pressure is proportional to the difference in protein concentration between the plasma and the interstitial fluid. The hydrostatic pressure in the capillary causes fluid to leave the plasma (e.g., flow) and the oncotic pressure tends to pull it back into the capillary.

Semipermeable Nanocomposite Membranes.   A new semipermeable nanocomposite membrane developed by Erik M.V. Hoek et al. of UCLA is capable of producing highly purified water by high-pressure reverse osmosis [105]. The membrane is comprised of specially synthesized nanoparticles dispersed throughout a porous polymeric matrix. The membrane is designed to enhance water flow at the expense of contaminants such as dissolved salts, organics, and bacteria. The surfaces of the nanoparticles are rendered hydrophilic. Repelling organic materials and bacteria prevent clogging of the membrane, a problem associated with current technology.

Carbon Nanotube Water Filters.   Researchers at Lawrence Livermore National Laboratory have developed a membrane on a silicon chip that is the size of a quarter that has potential applications in the areas of desalination, demineralization, and purification of water [83,84]. The membrane consists of single- and double-walled carbon nanotubes 1–2-nm in diameter and a silicon chip (**Fig. 14.15**). Flow of contaminants is restricted through the hollow pipeline of the nanotube while the flow of liquids and gases is allowed. Molecular dynamic simulations have predicted that fast flow through carbon nanotubes is likely [83]. According to Olgica Bakajin and Alexandr Noy, membranes containing millions of aligned carbon nanotubes, formed by chemical vapor deposition (CVD) on a silicon chip, have incredibly smooth interior walls [83]. The challenge lies in the fabrication of the membrane—specifically, to fill the gaps between the carbon nanotubes so that seepage is prevented. In order to overcome this obstacle, the nanotubes were coated with silicon nitride. Gap-free membranes were produced. Excess silicon nitride was removed and ends opened by reactive ion etching.

The performance of the membrane showed promise—the membrane was able to block ingress of 2-nm diameter gold nanoparticles while water was allowed through the nanotubes. This is a nano-example of a size-exclusion process. Remarkably, measured gas flow exceeds predictions of a Knudsen diffusion model by ca. 100x [83,84]. And more remarkably, water flow exceeds values calculated by continuum hydrodynamic models by three to four orders of magnitude—comparable to flow rates calculated by molecular dynamic simulations [83,84]. The membranes also outperformed commercially available polycarbonate membranes by several orders of magnitude despite having pore channels an order of magnitude less in diameter [83,84]. Enhanced water flow is explained by few theories: (1) the hydrophilic–hydrophobic relationship between the water molecules and the internal walls of the carbon nanotube results in frictionless flow and (2) confinement-induced ordering that results in the formation of "water wires of extremely long length" [83].

Carbon nanotube filters are expected to reduce the consumption of energy by as much as 75% compared to reverse osmosis [106]. H.G. Park et al. at the Lawrence Livermore National Laboratory also have developed an ion-exclusion method to purify water using carbon nanotubes that should offer high throughput along with cost-effective manufacturing [107]. The principal mechanism of synthesis is similar to the size-exclusion system described above in which open carbon nanotubes were fixed within a silicon nitride matrix except that in this case, the openings of the carbon nanotubes were modified chemically—functionalized with negative chemical moieties that repel anions but allow cations to pass. The permeability of such membranes has been shown to exceed that of a conventional RO membrane by a factor of 100x [107].

Chemically Modified Silica Particles.   Peter Majewski and Chiu Ping Chan of the Ian Wark Research Institute at the University of Southern Australia have developed a water purification system based on specially coated silica nanoparticles, e.g., functionalized self-assembled monolayers, nanometers in thickness. The active nanometer coatings were shown to eliminate biological molecules, viruses, oocysts, proteins, bacteria, and waterborne parasites like *Cryptosporidium parvum* [108]. The pathogenic agents were removed at the *p*H of drinking water by electrostatic attraction and subsequent immobilization between the surface-engineered-silica (SES) and the pathogens. Stirring for 1 h and filtration remove the pathogen–silica complexes from the water [108,109].

Bacteria and Nanofilters. Bioremediation is the process of water purification by bacterial action. Treated water is filtered through a porous membrane with pore size ranging from 10 μm–1 nm [110,111]. N. Hilal et al. reported in 2008 that the new bioremediation membranes can be cleaned without removal from the medium and the waste products used for fuel—a definite improvement over conventional membranes that are easily fouled.

14.1.5 Oil Spills

There are many means of mitigating the effects of an oil spill via nanotechnologies. *Bioremediation methods*, a procedure that follows mechanical skimmers and other physical methods of cleanup, consist of "microbial cultures, enzyme additives, or nutrient additives" that are designed to operate in wetlands or shallow water [112]. The purpose of the additives is to boost the natural nanotechnology of the microbial community to decompose oil materials. *Chemical dispersants* made of surfactants and other supramolecular precursors keep oil from settling onto shores and sensitive habitats by emulsifying oil and rendering it soluble in water. The oil particulates remain afloat in a dispersed state and are made available for accelerated biodegradation [112].

Another method that is gaining acceptance is the use of *specially modified aerogels* decorated with hydrophobic moieties to enhance oil interaction. Such hydrophobic aerogels possess extremely high surface area, a property characteristic of nanomaterials, and are therefore able to absorb sixteen times their weight of oil [113]. The inorganic–organic hybrid aerogels incorporate trifluoromethylpropyl groups into the silicate backbone of the cross-linked aerogel during synthesis.

$$(CH_3O)_4Si + CF_3CH_2CH_2Si(OCH_3)_3 \rightarrow aerogel - CH_2CH_2CF_3 \qquad (14.8)$$

Once the oil has been absorbed, the solid oil-soaked aerogel can be plucked from the surface. Synthesis of aerogels however is not cost-effective and alternatives that utilize charcoal are under consideration [113]. *Magic sand*, another surface-modified silica material, is synthesized from sand coated with an organosilane monolayer—trimethylsilane. In this case, the oil adsorbs to the hydrophobic surface and the sand-monolayer-oil particles sink to the bottom where the oil can be recovered by dredging [114]. Although the sand particle is on the order of microns in size, the monolayer, depending on the silane, is less than a nanometer to a few nanometers in thickness.

The use of nanoparticulates in conjunction with a self-assembled monolayer (SAM) material perhaps shows the most promise. Interface Scientific Corporation developed such a material in 2005 but details about its structure are not available. The developers claim that the nanoparticles absorb 40 times its weight in oil—a level that exceeds any method to date—and the oil can be recovered. The combination of the nanoparticle and SAM produces a hydrophobic environment that excludes all water. The properties can be altered by altering the length and functional groups of the monolayer—and perhaps by the size of the nanoparticle—typical techniques involved in nanochemistry.

14.1.6 *Chemical and Biological Sensors and Detectors*

We have introduced chemical and biological detectors in an earlier section but we reiterate at this time their importance in environmental mitigation strategies—in order to mitigate a problem one needs to know how how what kind and how much of a problem it is. Many kinds of pollutants exist in water supplies [6]: Pathogens that arise from human activity and natural sources; heavy metals like arsenic generated by geological conditions, mining runoff and industrial use; pesticides from aerial spraying and agricultural runoff; algal and bacterial sources of neurotoxins, hepatotoxins, and cytotoxins that originate in waters with high levels of nutrients (due to synthetic chemical fertilizers); nitrates from natural and agricultural sources; fluorides from water supply purification practices and geological sources; and organic compounds and halogenated organic materials from industry and transportation. There are many kinds of waterborne pollution that are toxic to living things—quite the challenge for nanotechnologists developing sensors to help detect them!

14.2 AIR QUALITY, MONITORING, AND MITIGATION

Several factors and sources affect global and local air quality. Many sources are anthropogenic; others are naturally occurring. The U.S. Clean Air Act (CAA) was passed in 1963 followed by the Air Quality Act in 1967, the Clean Air Act Extension of 1970, and the Clean Air Act amendments in 1977 and 1990 [115]. The 1990 revision added statutes concerning emissions trading, acid rain, ozone depletion, and newly classified toxic air pollutants [115]. The CAA sets standards for emission of fine particulates and regulates hazardous chemicals such as volatiles and heavy metal vapors. The EPA's Office of Air and Radiation is currently assessing the impact of engineered nanomaterials and is in the process of determining whether or not nanomaterials should be classified as "potential criteria pollutants" under CAA Section 110, as "potential hazardous air pollutants" under section 112 or pursue other implications [116–118].

The U.S. Toxic Substances Control Act (TSCA) was originally passed in 1976 and is administered by the U.S. EPA. The law authorizes the EPA to secure information on new chemical substances—nanotechnology; however, it is not specifically a chemical substance in the traditional sense. According to the American Bar Association meeting in May 2007, TSCA needs to address the following [116–118]:

- What guidance should EPA provide to nanomaterial manufacturers to determine if a nanomaterial is a new chemical under TSCA?
- What regulatory alternatives are there for nanomaterials that do not qualify as new chemicals?
- Should EPA amend its regulations to address nanomaterials specifically?

- What materials should manufacturers provide to the EPA to help it conduct risk assessments?
- How does TSCA relate to international standards for nanomaterials?

Atmospheric Nanoparticles. Particles found in the atmosphere come in a variety of sizes—ranging from particles as large as 10 μm to 1 nm with the most common sizes detected near 5, 50, and 300 nm [119]. Both natural and synthetic processes are responsible for the production of atmospheric nanoparticles, but anthropogenic sources contribute significantly more to fine particle fractions [120]. Combustion processes produce fly ash. Atmospheric particles impact air quality by reducing visibility, and catalyzing cloud formation and radiation forcing—a process that affects the ratio of absorbed and reflected sunlight. The impact of nanoparticles is both mitigated and acerbated, depending on your local and the prevailing winds, by traveling long distances from one continent to another [119]. Particles are removed from the atmosphere by gravity (via coalescence), diffusion, and rain but residence times can last from minutes to days or longer.

Nanoparticles are capable of reacting chemically or catalyzing chemical reactions while airborne, e.g., the conversion of SO_2 and NO_2 gases into sulfuric and nitric acids, respectively, or depleting ozone by catalyzing the formation of chlorine compounds [119].

14.2.1 Gas Separation: Advanced Membrane Technology

Membrane-based gas transport and separation are accomplished by several kinds of membranes that possess altered chemistry and geometric configuration. Depending on the prevailing pressure (from vacuum to high pressure), mean free path factors determine the different types of mass transport through membranes.

Viscous Flow. Viscous flow occurs when the mean-free path λ_{mpf} of the gas atom or molecule is significantly less than the diameter of the pore channel. Flow patterns are determined by the collisions of the molecules. No effective separation occurs in this scenario and bulk flow of gas proceeds very much like it does for a liquid (turbulent or laminar), relatively unimpeded through pore channels. The pressure domain characteristic of viscous flow is between 10^{-1} and 10^3 torr. In *molecular (Fickian) diffusion*, the mean-free path is much less than 0.01 d_{pore} and follows Fick's law of diffusion and the pressure domain lies between 10^{-6} and 10^{-3} torr. The transport diffusivity relates the macroscopic flux of the gas species to the concentration gradient. This type of diffusion is like Brownian motion in which the movement of the gas molecules is random and not a function of its previous motion.

Knudsen (or Transition) Flow. This is based on the inverse square root ratio of molecular weights of gas species—in this case, the mean-free path of the gas molecules is greater than 1/100th of the diameter of the pore (0.01 d_{pore} < λ_{mpf} < 1.00 d_{pore}). Collisions with the pore wall occur frequently. Knudsen flow dominates when the pore diameter is between ca. 2 and 50 nm. The Knudsen number Kn is the ratio of the mean-free path to the pore diameter

$$Kn = \frac{\lambda_{mfp}}{d_{pore}} \tag{14.9}$$

The transport equation of Knudsen diffusion of species i is given by J_i:

$$J_i = \left(\frac{\varepsilon}{d_{mem}\tau}\right)\left(\frac{D_K}{RT}\right)\Delta P_i \qquad (14.10)$$

where

ε is the porosity (volume pores/volume membrane)
d_{mem} is the thickness of the membrane
D_K is the Knudsen diffusivity
R is the gas constant
T is the absolute temperature
ΔP is the pressure differential across the membrane
τ is the tortuosity

$$D_K = \frac{2}{3}r_{pore}\sqrt{\frac{8RT}{\pi M_i}} \qquad (14.11)$$

where

r_{pore} is the radius of the pore channel
M_i is the molecular weight of the gas in terms of $g \cdot mol^{-1}$

The enrichment factor of the mass transport between two gases is the ratio of the squares of their respective molecular masses.

$$\text{Enrichment}_{ij} = \frac{J_i}{J_j} = \sqrt{\frac{M_i}{M_j}} \qquad (14.12)$$

Ultra-Microporous Molecular Sieving.   This depends on diffusion rates of molecules, for example, with small molecules diffusing quicker than larger ones, but also depends on specific adsorption differences between similar-sized molecules like O_2 and N_2. These membranes may have a tortuous but continuous network of channels.

Solution-Diffusion Separation.   This mode of separation occurs through membranes that are essentially solid (with no continuous passages—nonporous polymeric membranes like PDMS, PS, etc.) and is based on solubility and mobility factors of gas molecules [121]. Passage of gas through such membranes occurs due to thermally agitated motion of polymer chain segments that form transient penetrant-scale gaps—driven by diffusion phenomena [121]. Pressure and concentration gradients provide the driving force with working pressures in the neighborhood of 100 atm. Metallic membranes (Pd and Ag alloys) break down H_2 on one side into H atoms that diffuse through the metal film and reassemble on the other side.

Traditionally Established Industrial Methods.   Gases in purified form are desired for many applications in research and industry [121]. Separation of methane, propane, and other hydrocarbons from H_2S, H_2O, CO_2, and N_2 and other impurities found in natural gas streams and biogas is required in order to tap new energy reserves. Compression of the gases for storage also requires energy investment. Recovered hydrogen from industrial product streams generated by steam reforming and other processes are required in ammonia synthesis and gasification reactions of oil refinery processes. High-purity hydrogen is required to power fuel cells. The effort to separate CO_2 from air has gathered significant momentum over the recent past in order to mitigate the ingress of greenhouse gases into the atmosphere. Traditional gas separation methods are energy intensive. These include cryogenic-distillation methods, amine absorption/regeneration, and pressure swing adsorption (PSA).

PSA is used to remove CO_2 during the commercial synthesis of hydrogen for applications in oil refineries, ammonia production, and to elevate methane content in biogas [122]. For example, a nitrogen purification PSA process passes 95%–99.5% pure N_2 (at 6–8 atm.) over a sorbent material (activated carbon molecular sieve) that removes oxygen and water, the *retentates*, from the input gas stream by adsorption [123]. The purified nitrogen *raffinate* is not adsorbed and the product stream proceeds over the bed to collection (also energy intensive) [123]. If higher purity is required, catalytic processes combine H_2 with O_2 to produce water. Cooling or an additional adsorption process is applied to remove accumulated water [123]. Regeneration occurs by depressurization to atmospheric pressure. Oxygen PSA exploits alumina to remove water vapor. Zeolite molecular sieves are used to remove water, CO_2, and other gases [123]. Application of PSA to capture CO_2 produced by coal-fired plants is currently under consideration [122]. Another version of PSA is adapted to provide carbon dioxide and humidity control to portable life support systems (PLSS) for NASA. A sorbent is coupled to the PSA system and is regenerated by exposure to vacuum in space during extravehicular activity (EVA) [124]. Such microporous membranes offer a nonenergy intensive alternative to gas separation—using the built-in vacuum of space to perform the purging function.

Zeolite Membranes. Zeolites are microporous materials (according to IUPAC convention) comprised of aluminosilicates. Zeolites are natural polymorphic minerals, but many synthetic versions have been developed for experimental and industrial use. Applications of zeolites fall into these major categories— separation via sieving, ion-exchange, size-exclusion etc., and catalysis and tem- plate synthesis. Zeolites are chemically resistant, can tolerate high temperatures and pressures, are able to form into pinhole-free membranes, and the porous structure can be tailor-made to suit end user criteria [125]. Low-cost facilitated synthesis, highly ordered structures, and flexibility in design have made zeolites a leading nanomaterial for membrane separation technology. Pore, or channel, diameter in six and eight-member ring zeolites range from 0.3 to 0.4 nm; 10-member ring zeolites have characteristic pore diameter ranging from 0.5 to 0.6 nm; and larger pore zeolites having 12-member rings have pores 0.7–0.8 nm [125]. Membranes can be supported on alumina or silica substrates, called composite membranes, or fabricated as free-standing contiguous films—limited in size to a few square centimeters. Membrane-based separation, as opposed to conventional techniques such as pressure swing adsorption, should reduce facility and equipment costs and energy consumption and allow for continuous operation [125].

Hydrogen is a clean burning fuel and developmental efforts are underway to power cars, electrical devices, and household appliances with hydrogen. In order to reach these goals, one of the holy grails of renewable fuel development, in addition to efficient production and storage, is to find an efficient way to separate hydrogen from other light gas by-products of industrial processes like steam-reforming and gasification reactions.

Kanna Aoki of the Nanoelectronics and Collaborative Research Center at the University of Tokyo and colleagues prepared a highly hydrophilic A-type zeolite by hydrothermal synthesis on the outer surface of a porous α-alumina tube [126]. The membrane possessed two types of pores: 0.4- to 0.43-nm zeolitic pores and nonzeolitic pores. Molecules such as ethylene were not able to penetrate the zeolitic pore system. Separation was dependent on the hydrophobic nature and molecular size of permeates, and affinity to the pore walls by the permeate gases. Combinations of gases also affected the permeation rates. The presence of water molecules reduced the permeation character of hydrophobic gases [126].

14.2.2   *CO₂ Mitigation*

Established industrial methods of removing CO_2 from combustion flue gases include amine absorption recovery and PSA. Sequestration (capture and storage) or conversion of CO_2 is desired for several reasons, topics upon which we shall not elaborate. Addition of a separation technology to fossil fuel conversion systems would help mitigate anthropogenic sources of CO_2 from reaching the atmosphere. Experimental carbon-based membranes, functionalized mesoporous oxide membranes, and sieving zeolitic membranes are members of a new family of regulated nanostructured materials that have all shown promise in selectively removing CO_2 from complex gas streams.

Y. Ohta et al. utilized dynamic Monte Carlo simulations to estimate the permselectivity of a binary gas mixture consisting of CO_2/N_2 in zeolite-like porous membranes with seven-member ring structure and 0.34-nm pore diameter [127]. The separation factor and permeability of CO_2 were shown to be superior, and simulation methods should prove to be useful in further investigations [127].

Kiochi Yamada, Shingo Kazama, and Katsunori Yogo of the Research Institute for Innovative Technology for the Earth (RITE) in Japan have developed several kinds of gas separation membranes.

T. Koutetsu et al. devised a dendrimer composite membrane on porous substrates for CO_2 separation. Poly(amidoamine), or PAMAN, dendrimers demonstrated superior in situ CO_2/N_2 gas separation [128]. The surfaces of commercially available ultrafiltration membranes were modified to produce a gas-selective layer. Chitosan layer (200 nm) formed the base layer to attach PAMAN dendrimers (total thickness = ca. 300 nm with 20%–40% dendrimer). Separation characteristics were determined under pressure differential of 100 kPa and 40°C. The analyte mixture consisted of 5%-vol. CO_2 and 95-vol% N_2. The selectivity of CO_2 over N_2 was determined to be 400 and the permeance at 40°C was 1.6×10^{-7} m²·s⁻¹·kPa⁻¹ [128].

N. Hiyoshi et al. in 2005 reported that in the aminosilane-modified mesoporous SBA-15 (a silaceous zeolite) membranes resulted in an increase in CO_2 per nitrogen atom via aminosilane adsorption increased as the surface density of the amine was increased [129]. The authors claim that amine pairs onto which CO_2 adsorbed formed alkylammonium carbamates—making a case for a chemisorption mechanism. The adsorption also depended on the structure of the amine [129]. S. Duan et al. in 2006 also developed PANAM dendrimer composite membrane for CO_2 separation [130]. Under similar conditions, Duan et al. achieved better permeance but less selectivity: 230 CO_2/N_2 and 1.6×10^{-7} $m^2 \cdot s^{-1} \cdot kPa^{-1}$, respectively [130]. Kovvali et al. also accomplished work with dendrimers for CO_2 separation [131].

There is great incentive to develop CO_2 capture via innovative membranes with sub-nanoscale materials control [132]. Objectives include separating CO_2 from other gases present in flue gases or synthetic processes as is means of storing CO_2. The strategy is, in general, to develop low-cost polymeric and inorganic membranes for the purposes of separation and storage. Current systems that are not permselective enough require complex cascading structures to obtain pure streams of carbon dioxide.

14.2.3 *Hydrogen Production and Purification*

From the environmental viewpoint, hydrogen is the premier fuel of combustion processes.

$$H_{2(g)} + \frac{1}{2}O_{2(g)} \rightarrow H_2O_{(g)} \tag{14.13}$$

Catalytic reforming produces hydrogen with 40%–92% purity. The effluent is channeled into a PSA treatment system to achieve 99.95% grade hydrogen. Petrochemical off-gases produce 10%–20% pure hydrogen that is funneled into a cryogenic system to achieve 99.95% H_2 that is separated from various hydrocarbon gases. Cryogenic separation is based on the difference in boiling point of H_2 from feedstock impurities—the method is also used to recover CO and other gases. Refrigeration is accomplished by a process based on Joule–Thompson expansion. Water and carbon dioxide are usually removed before application of the cryogenic distillation process. Hydrogen with ~40% purity is generated from petrochemical off-gases, but in this case the effluent is channeled to membrane or PSA purification processes [133].

Bifunctional PSA systems are able to produce H_2 at purity levels 99.95%–99.9999% from 40 to 90 + %H_2 feedstock. An activated carbon layer (a great traditional nanomaterial) removes CO, CH_4, CO_2, H_2O, and H_2S. Molecular sieves (another traditional nanomaterial) remove N_2, Ar, O_2, and the remaining CH_4 and CO [133]. PSA is widely used in H_2 purification for fuel cell applications and is the principal method for hydrogen purification in use today [133]. Operational issues include extensive maintenance due to wear and tear of valves and their inherent complexity. Air Products and Chemicals, Inc. in 2006, produced over 1.75 million tons per year of hydrogen [133].

Polymeric membranes operate on the phenomena of partial pressure difference and the degree of gas–polymer interaction. The rate of diffusion is inversely proportional to the thickness of the membrane. Polymer membrane methods have the following issues: they require high feed pressures (20–200 atm) irreversible membrane damage occurs from impurities, physical defects like cracks that reduce efficiency, and permeability to other gases. Polymeric membranes are therefore not used for purification of hydrogen for fuel cell applications.

14.2.4   *Chemical Sensing and Detection*

We continue again with a section on gas sensing to supplement previous material. Nanotechnology, of course, is expected to play a major role in this area.

Carbon Nanotube–Nanocomposite Gas Sensors.   The detection capability of gas sensors (hydrogen, ammonia, and acetone in experiments) has been recently enhanced with the application of conductive polymers and single-walled carbon nanotubes [134,135]. One type of sensing material is a composite consisting of 1-wt% SWNTs in poly(3,3′-dialkyl-quarterthiophone). According to Sara Vieire et al. at the University of Cambridge, doping with gas at levels of 10 ppb yielded exponential changes in the overall conductivity profile of the nanocomposite sensors. Conducting polymers are good materials to use for sensing due to inherent physical properties such as conductivity, lightness, ease of manufacture, and cost-effectiveness. However, the materials lack specificity in distinguishing among different kinds of gases, for example, lack of selectivity. Semiconducting SWNTs, on the other hand, are highly selective but are expensive to manufacture. Just like for fiberglass, a composite material that outperforms either component by itself, the combination of conductive polymers with carbon nanotubes makes good sense. The device consists of a commercially available silicon-on-insulator complementary metal-oxide semiconductor (SOI-CMOS).

Gas Sensing by Thin Metal/Metal Oxide Films.   Tin oxide (SnO_2) is one of the most prevalent of gas-sensing materials. It is used to detect AsH_3, H_2S, NO, NO_2, H_2, NH_3, CO, CCl_4, C_3H_6, CH_4, O_2, MeOH, EtOH, CO_2, and numerous other hydrocarbon gases. SnO_2 is often supported on a wide variety of materials: alumina, Pt, Pd, Ag, ZrO_2, CuO, and La_2O_3.

The use of nanoparticles has significantly enhanced the performance of gas-sensing devices [136]. H. Ogawa in 1982 showed that enhanced sensitivity was obtained with SnO_2 nanoporous films rather than dense nanoparticulate films of the same material [137]. He explained his results by applying Hall measurements—specifically, that a space–charge region develops in the nanocrystallite when its size is approximately twice the Debye length (the distance over which significant charge separation can occur). This condition results in higher sensitivity to adsorbed gas species [6], and in concert with high, specific surface area of the porous film ($\varepsilon \sim 98\%$ porosity), fast transport of the analyte gas adsorbing and desorbing species is enhanced. This explains why carrier mobility is strongly dependent on the concentration of the substrate in porous films while no such dependence is found for dense thin films [137].

One of the greatest challenges facing gas sensing is specificity (e.g., selectivity)—a challenge where nanotechnology can certainly help.

14.3   ENERGY

The late Nobel laureate Richard E. Smalley toured the country in the early 2000s giving his "Our Energy Challenge" lectures [138]. These lectures summarized the state of humanity and listed the 10 most important challenges that our species will face in the next 50 years! He also made significant mention of nanotechnology and how many of the solutions will arise from that wondrous technology. We will summarize some of his points and discuss the ones relevant to environmental issues.

According to Smalley, humanity's "top ten" problems for the next 50 years are [138]: (1) energy, (2) water, (3) food, (4) environment, (5) poverty, (6) terrorism and war, (7) disease, (8) education, (9) democracy, and (10) population. You may wish to reorder this list to suit your priorities, but most will agree that it does cover some of the most critical issues that we as a species will face (and are already facing) in the coming years. Population is actually at the root of all of those issues and could easily displace energy as the No. 1 issue. For example, the global population is expected to exceed 10 billion by the year 2050 [138]. Energy consumption, however, does not scale linearly with population growth. Although human population has quadrupled over the twentieth century, energy consumption has increased by a factor of 16 [138].

In 2004 (14.5 TW or 220 MBOE·day^{-1}), oil contributed nearly 35% to our total energy resources; coal 25%, gas 20%, fission 5%, biomass 10%, hydroelectric 5%, and solar wind, and geothermal a mere 0.5%. In 2050 (expected 30–60 TW or 450–900 MBOE·day^{-1}), these ratios are expected to undergo a drastic change: oil <5%, coal ~5%, gas >10%, fusion–fission ~15%, biomass >10%, hydroelectric ~5%, and solar, wind, and geothermal ~50%. The acronym MBOE·day^{-1} stands for "millions of barrels of oil equivalents per day." "TW" is of course the terawatt. When reviewing the list of potential energy sources for the future, conservation efficiency, hydroelectric (maxed out), biomass, wind, and wave and tide cannot possibly provide enough energy. Chemical sources such as natural gas and clean coal are faced with extraction and purification costs. Nuclear sources and others such as geothermal, solar terrestrial, solar satellite, and lunar–solar are burdened with high cost, engineering difficulties, and minimal potential or waste disposal issues.

The sun delivers 165,000 TW daily to the Earth—of which we can potentially capture 20 TW by placing six solar farm grids of 10,000 sq. mi. each on the continents [138]. By way of a proposed *Distributed Store-Gen Grid*, energy is transported by wire (rather than rail) on a continental scale that interconnects 100 million local storage and generation sites, electricity can be provided to most of the civilized world. The network includes local storage and generation capability as well as a centralized power grid. Highways in the United States cover about 0.2% of the land area. If we were to cover that amount of area with solar cells that deliver 50% efficiency, we would be able to generate 1 TW of electricity [139].

What can nanotech do to support this master plan? To begin with, the cost of photovoltaics could drop 10-fold due to the use of less material, lowered processing costs, and revolutionary manufacturing breakthroughs. Nano-enhanced materials and devices would be capable of photocatalytic reduction of CO_2 to methanol and direct photoconversion of H_2O to H_2 fuel sources. A likely 10- to 100-fold improvement of batteries, supercapacitors, and flywheels for automotive (plug-in hybrid vehicles) and Store-Gen Grid applications is expected. Power cables made of superconducting or quantum materials would enable worldwide transmission of electricity and replace aluminum and copper in electrical motors.

Carbon nanotube quantum wires would exhibit negligible current loss with one-sixth the weight of copper. Hydrogen storage on carbon nanotubes and other suitable nanomaterials and advanced materials for high-pressure storage of liquid hydrogen will be enabled by new nanomaterials. If the cost of reversible, low temperature fuel cells were dropped 10- to 100-fold, widespread use of portable and reliable energy sources would become reality [138].

14.3.1 *Solar Energy and Nano*

In 1839, 19-year old French scientist Alexandre E. Becquerel discovered the photoelectric effect while illuminating metal electrodes in an electrolyte. English engineer Willoughby Smith in 1873 was the first to observe the phenomenon of photoconductivity while working with a new material with high resistance called selenium (that happened to be photoactive)—a material used to check for flaws in transoceanic cables. In 1877, W.G. Adams and R.E. Day, Willoughby's students, verified that selenium produces a current when exposed to light. C.E. Fritts a few years later in 1883 constructed the first photovoltaic cell consisting of a sandwich of Se between a gold leaf and a brass plate. In 1887, J. Moser described dye-sensitized solar cells. Albert Einstein described the photoelectric effect in 1904. In 1918, Polish scientist J. Czochralski devised a means of fabricating monocrystalline Si and the first Si solar cell was manufactured in 1941. PV cell efficiency started out at 2%. In 1959 and 1960, 9% and 14% efficiencies were achieved, respectively. In July 2007, 42.8% efficiency was achieved with an experimental VHESC (yes, a "very high efficiency solar cell") multijunction by a University of Delaware team. In August 2008, NREL created a reliable solar cell with 40.8% efficiency (www.NREL.gov/news/press/2008/625.html).

There are many kinds of solar cells, all of which have useful environmental potential with regard to efficiency or reduced costs: multijunction concentrators (>40%), silicon cells (single crystal, polycrystalline, and thin film with efficiencies <30%, ~20%, and ~16%, respectively); thin-film technologies such as $Cu(In,Ga)Se_2$ or CIGS (~18%), CdTe (~16%) and amorphous stabilized Si:H (~12%); and emerging PV materials such as dye-cells (~14%) and organic cells (~5%). An overview of solar cell efficiency is presented in **Figure 14.16**.

Nanotechnology offers numerous advantages to solar cell design and manufacture: (1) large surface area collectors, (2) antennae consisting of quantum dots that are able to capture 90% or better of the solar spectrum, (3) tunable solar material interfaces that demonstrate enhanced efficiency, (4) multielectron stimulation, and (5) reduced manufacturing costs due to revolutionary nanoprocesses such as nano-imprint lithography that operate at lower temperatures with higher throughput.

The groundwork for a new solar plant, owned and operated by Nanosolar, Inc., was announced in 2006. The plant is expected to manufacture paper-thin flexible solar materials that are expected to produce over 430 MW of power per year [140]. The solar active component of the thin-film nanotechnology is a copper–indium–diselenide ink that absorbs light and converts it into energy. In December 2006, Boeing-Spectrolab developed a solar cell that is capable of 41% efficiency and earlier in the year, researchers at Lawrence Berkeley National Laboratory developed a cell from a new semiconducting material made of zinc–manganese–tellurium doped with a few atoms of oxygen that demonstrated 45% efficiency [141].

The development of more efficient solar cells from 1975 to the present is depicted. Efficiency greater that 40% has been achieved by the National Renewable Energy Laboratory. The theoretical maximum efficiency of photosynthesis is estimated to be 11%. The efficiency of photosynthesis is limited by the wavelength range (400–700 nm). This range accounts for only 45% of the total solar spectrum available for photosynthesis. Then, there are issues with quantum efficiency per CO_2. Therefore, only 25% of the 45% is actually considered to be photosynthetically active radiation (PAR). Other reasons such as reflection, respiration, and lack of optimal solar radiation levels also serve to reduce efficiency. The comparison with the NREL cell is not direct due to variable circumstances.

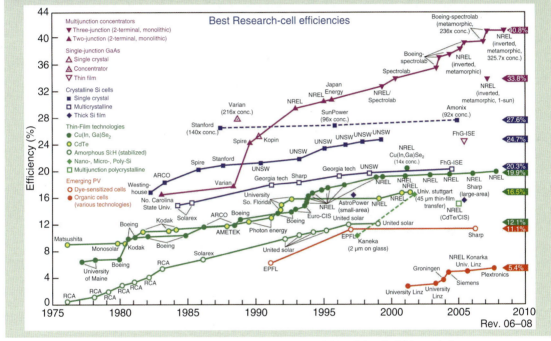

Source: Larry Kazmerski, National Renewable Energy Laboratory (NREL), Golden, Colorado. With permission.

Conventional solar cells are made of an *encapsulant*, a *front contact grid*, an *antireflective coating*, *n*-type-silicon, *p*-type-silicon, and a *back electrical contact* (**Fig. 14.17**).

During our description, think of ways that material components can be enhanced with nanotechnology. The encapsulating material serves to seal the cell and keep the external environment out. It is usually made of glass or clear plastic. The antireflective coating effectively channels photons into the matrix of the solar cell (reduces solar loss). A grid made of conducting material forms close contact with the n-doped silicon as is a metal contact with the *p*-doped silicon.

The ultimate goals of any photoelectric device is to generate free charge carriers from interaction with photons and to separate the charge carriers to generate an electric current before the excited states can decay or recombine [142]. Electron charge carriers exist in the conduction band of the semiconductor—electrons that were promoted due to energy input. The vacancies left by the excited electrons are called holes and are known as virtual charge carriers. The simultaneous process forms *excitons* and the basis for photovoltaic mechanisms. In the

FIG. 14.17

A generic silicon-based solar cell is depicted. The photocell has three major functions: (1) harvest a maximum amount of photons, (2) convert photons into electrical charge, and (3) generate electrical charge from the device.

The glass coating protects the cell from the environment. The antireflective coating reduces solar loss to 5%. The contact grid is electrically conducting and can be metal or a transparent conductor. The n-type silicon is from phosphorous-doped silicon; the p-type from boron-doped silicon to support holes. The back contact is made of metal.

Load

Antireflective coating

Protective glass coating

Conductive grid

p-Type silicon   n-Type silicon

p/n-Junction

broadest sense, excitons are mobile combinations of an electron and a virtual particle called a hole in an excited semiconductor. In the narrower sense, excitons are strongly coupled electrons and holes. Wannier excitons are typical of covalent semiconductors and insulators, are separated by a distance that exceeds the lattice spacing of the solid-state material, and move like unbound particles. Frenkel excitons occur in molecular crystals, are separated by a distance on the order of the lattice spacing, and are usually localized to one site.

Doping alters the electronic properties of semiconductors by introducing excess positive or negative charges. In most bulk silicon solar sells, a thin n-type Si layer is deposited on top of a much thicker p-type Si layer. The *p/n* junction interface establishes a bias that results in recombination of electrons and holes, for example, depleted zone of paired charges. When electromagnetic radiation is applied to the surface, a higher population of charge carriers is induced creating a stronger bias across the interface. Charge carriers then move towards their respective conducting electrode terminals and are made available to do work.

Efficiency is the ratio of input to output energy. In 1961, W. Shockley and H.J. Queisser determined that a conventional PV cell is capable of ~32% efficiency, the Shockley–Queisser limit [143]. Only a small fraction of the solar spectrum matches the bandgap of Si. Impinging photonic energy above or below the bandgap, therefore, will produce heat in the semiconductor. The natural (theoretical) efficiency of commonly used bulk solar materials ranges from ca. 23% to 30%.

There are inherent limits to solar collection efficiency due to a variety of loss mechanisms: (1) spectral range of absorption limitations due to bandgap energy, (2) transformation of surplus light energy into heat, (3) optical losses due to reflection and shadowing of the cell surface, and (4) losses due to electrical resistance and other material factors (contamination, surface effects, crystal defects, etc.) [144]. In bulk semiconductors, one photon creates a single electron–hole pair regardless of photon energy—whether a high-energy photon or a low-energy photon.

Energy conversion efficiency (ECE) η_{ECE} is a parameter used to measure the efficiency of solar cells. *ECE* is the percentage or radiant power (incident irradiance $E_o = W \cdot m^2$) converted to electrical energy.

$$\eta_{ECE} = \frac{P_{max}}{E_o A_{cell}} \tag{14.14}$$

where
P_{max} is the maximum power output of the solar cell in watts (W)
E_o is the incident irradiance
A_{cell} is area of the cell

Quantum efficiency (QE), a subset of the greater category of *quantum yield* (the ratio of any quantum induced phenomena to input photons) should not be confused with *energy conversion efficiency*. Quantum efficiency is the percentage of photons of a select wavelength that produce electron–hole pairs. Good examples of devices that are measured by QE are charge-coupled devices (CCD, QE ≈ 90%) and photographic films (QE ≤ 10%) [145]. QE is often measured as a function of wavelength. If QE is integrated over the whole spectrum, the potential electric current that a certain cell produces can be estimated.

$$\eta_{QE} = \frac{N_e}{N_p} = \frac{\phi_o / h v_o}{\phi_{al} / h v_o} = \frac{\phi_o}{\phi_{al}} \tag{14.15}$$

where
N_e is the number of electrons
N_p is the number of photons
ϕ_o and ϕ_{al} are the incident optical power (radiant flux, in watts or $J \cdot s^{-1}$) and the absorbed optical power, respectively

Optical power is absorbed in the depletion layer. For solar cells, the external quantum efficiency (*EQE*) is an important parameter. *EQE* is the current measured outside the solar cell device and represents efficiency in both the absorption of photons as well as the collection of electrical charge.

$$EQE = \frac{electrons \times s^{-1}}{photons \times s^{-1}} = \left(\frac{i}{e}\right)\left(\frac{h v_o}{\phi_o}\right) \tag{14.16}$$

Quantum Dot Solar Cells.   In cells that contain quantum dots, the number of charge carrier pairs produced depends on the energy of the photon. In 2001, 14% efficiencies were achieved by photocells fabricated from nanoparticles in conducting polymer films [146]. The shape of the quantum dot also is expected to play a role in its operational efficiency [147,148].

In 2006, I. Robel et al. at Notre Dame University fabricated solar cells based on CdSe quantum dots (QDs) tethered with the bifunctional surface linker $HS-R-CO_2H$ onto mesoscopic TiO_2 films [147]. The TiO_2–CdSe QD composite (acting as the photoanode) was able to exhibit a photon-to-charge-carrier efficiency of 12%. The function of the CdSe QDs (ca. 3 nm in diameter) was similar to that of a dye molecule. Following visible excitation, CdSe QDs inject electrons into TiO_2 nanocrystals (ca. 40–50 nm in diameter). The process was verified with femtosecond transient absorption as well as emission quenching experiments. In 2007, the research team found that the rate constant for electron transfer from the thermalized s-state of CdSe increased with decreasing particle size: 7.5-nm diameter particle ($E_g = 1.92$ eV) $\rightarrow 10^7$ s^{-1} and the 2.4-nm diameter particle ($E_g = 2.4$ eV) $\rightarrow 10^{10}$ s^{-1} [148]. The researchers claimed that the significant level of electron loss was due to scattering, charge recombination at the TiO_2/CdSe interface, and internal TiO_2 grain boundaries [147]. If a liberated electron collides with a nearby atom, it loses energy by the creation of atomic vibrations that end up heating the semiconductor.

Due to the size dependence of absorption by quantum dots, the solar cell can be tuned to absorb wavelengths ranging from 650 to 400 nm by decreasing the diameter of the CdSe dot. In this way, a greater range of visible light can be harvested for energy conversion. The thickness of a photocell based on QDs is able to influence the degree of absorption of light energy. Tenfold increase in the thickness, from 0.2 to 2 μm, resulted in significant darkening of the solar cell. This phenomenon is obviously due to the availability of more sites for CdSe binding with concomitant increase in efficiency of the solar cell.

Quantum dots also have the ability to generate multiple charge carriers from a single photon, unlike bulk semiconductors that generate at the most one exciton pair when exposed to an energetic near ultraviolet photon [149–151]. The remainder of the photon energy is converted into heat. QDs are able to generate more electron–hole pairs by a process called impact ionization—a mechanism first proposed by A.J. Nozik of the National Renewable Energy Laboratory in Golden Colorado [152–155]. Impact ionization is a process whereby high-energy charge carriers are able to distribute energy by the creation of more charge carriers. When in refined form, theoretical photon conversion efficiencies greater than 100% are possible [149].

In 2004, researchers at the Los Alamos National Laboratory discovered that three excitons each were generated when 5-nm lead selenide quantum dots were exposed to high-energy blue light [150,151]. In these experiments, in which quantum dots consisting of 1000 atoms were mixed in a liquid and sealed inside a glass sheath, laser pulses over a wide range of energies were directed at the mixture. Recently, 8-nm diameter quantum dots made of lead sulfide were shown to release seven electrons per dot when exposed to ultraviolet radiation [150,151]. This apparent leap in photon conversion efficiency should be able to benefit other types of solar conversions systems such as water splitters that produce hydrogen for fuel cells [150,151].

Organic Solar Cells. The photoactive elements of organic photovoltaics (OPVs) are made of polymers and/or nanocrystalline phases. Most OPVs are called bulk heterojunction devices in which p-type (electron donating) and n-type (electron accepting) organic materials are blended into a polymer matrix [156]. Power efficiencies (η) near 3% have been achieved with conjugated organic polymers like PCBM (phenylbutyric acid methyl ester) with electron-accepting substituted fullerenes [60] and soluble donors like polythiophene or poly(phenylenevinylene) derivatives [157]. Organic heterojunctions are defined by four processes: absorption, exciton diffusion, charge transfer, and charge collection. Charge transfer and collection are highly efficient processes with $\eta_{EXE} \rightarrow$ 100%. The major difference among OPVs is the efficiency of absorption ($\eta_A \rightarrow 50\%$, optical absorption path ~100 nm) and exciton diffusion ($\eta_{ED} \rightarrow 10\%$, diffusion length ~5 nm, HOMO–HOMO, LUMO–LUMO). Since absorption and exciton diffusion are limiting, the overall efficiency of the heterojunction OPV is expected to be around 10%.

C.W. Tang in 1975 conducted some of the first investigations to duplicate nature's ultimate solar cell—the one that exploits the amazing properties of chlorophyll-a in photosynthesis [158]. Tang also created a chlorophyll-a-PV cell in 1975. Monomers of chlorophyll-a were prepared on a chromium electrode by electrodeposition to form a microcrystalline array. The structure has a strong absorption band in the far red range of the spectrum at 740–745 nm [158]. The power conversion efficiency of the sandwich structure (Cr-Chl-a-Hg) was reported to be 0.01%—the highest for an organic PV cell of the day. [158]. Later on in 1986, he developed one of the first organic photocells, called the Tang cell [159]. Via lessons learned from nature, a new generation of solar cell development will most likely be modeled on biological materials like chlorophyll.

Advantages of organic solar materials are multifold: (1) the solar "cell" materials can be applied by painting on roofs, walls, or other surfaces that are targeted for solar collection—a incredibly low-tech solution to a high tech problem, (2) offer great potential for facilitated upscale (high throughput and can cover large areas), (3) are flexible and lightweight, (4) offer an economical solution as the cost-to-efficiency ratio for silicon is teetering towards the former, (5) can be applied on any kind of surface, especially large area flexible plastic surfaces [160], (6) ultra-fast optoelectronic response and charge carrier generation at organic donor–acceptor interface, (7) continuous tunability of optical energy bandgaps via molecular design, synthesis, and processing steps, and (8) integrability into textiles, packaging, and consumer products [161]. Underlying these advantages is the ability to synthesize organic molecules with great flexibility—flexibility that allows for tunability for adoption to specific applications. The major disadvantage is that OPVs promise, at best, only moderate solar conversion efficiencies. Low efficiencies are due to photon loss, exciton loss, and carrier loss.

Photoreactive polymers serve as the basis of OPV technology. In standard solar cells, both the donor and acceptors are basically the same material—doped silicons. On the other hand, in OPVs the donor material may not be anything like the acceptor material. To start, donor and acceptor molecules must be able to absorb light. Many donors contain conjugated molecules with a electron-donating character. Acceptor molecules need to be able to stabilize electrons originating from the donor molecules. Stabilizing molecules include electron-withdrawing groups. Carbon nanotubes and fullerenes, for example, are known for their electron stabilization ability.

C.W. Tang's first cell consisted thin films of copper-phthalocyanine (Cu-PC) donor and perylene tetracarboxylic acid derivative acceptor [159]. Cu-Pc is a porphyrin-like ring that hosts a Cu cation. A power conversion efficiency of 1% was achieved with this cell. In this new concept, bias voltage does not control the charge-separation efficiency, and therefore cells with fill-factor up to 65% can be achieved. In addition, the interface between the two organic materials and not the organic material–electrode contact is the primary factor that determines the PV properties of the cell [159].

In 2006, Klaus Müllen et al. of the Max Planck Institute for Polymer Research in Mainz, Germany developed a bulk heterojunction cell consisting of poly(2, 7-carbazole) (PCz) and perylene tetracarboxydiimide dye (PDI) donor–acceptor pair for organic solar cell applications [156]. PCz, in a highly soluble state, served as the electron donor material and the PDI was the electron acceptor material. The pair demonstrated complementarity within a broad range of the solar spectrum. The reported EQE efficiency was 16% and the power efficiency was 0.6% under solar conditions. The bandgap of PCz is ~3 eV, well into the ultraviolet range of the spectrum. To make this material practical in solar cells, it must be paired with a material that has a lower characteristic bandgap—enter PDI (bandgap ~2 eV). Previously, the Müllen group has shown that PDI in combination with discotic liquid crystalline hexa-*peri*-hexabenzocoronone (HBC) displayed a maximum external quantum efficiency (EQE) of 34% with η = 1.95% at 490 nm [162]. Müllen et al. stated that photoinduced charge transfer occurred between the HBC and the perylene complement in addition to efficient charge transport through vertically segregated perylene and HBC π-systems—all constructed with simple bottom-up nanofabrication solution techniques [162]. The cell is shown in **Figure 14.18**.

Müllen et al. showed that conjugated polymers with high bandgap can be used for solar applications if a proper electron acceptor is selected [156]. Cyclic voltammetry (Ag/Ag⁺, calibrated against ferrocene) of PCz/PDI on ITO revealed that the HOMO of PDI was –5.8 eV, below that of PCz (–5.6 eV). The estimated LUMOs of each, determined by bandgap calculations from absorption spectra, were –2.6 eV for PCz and –3.8 eV for PDI. According to Müllen, PDI are able to form excitons, and hole migration from excited PDI states to PCz is driven by the 0.2 eV difference in HOMO energies. With regard to the corresponding LUMO energies, the 0.8 eV difference in energy is able to drive electron migration from PCz to PDI. The group is working on mixing PCz with other dyes that have broader absorption spectra [156].

Fullerene-modified polymers are increasingly popular for use in OPVs. In such cells, charge separation mechanisms in bulk heterojunction PV cells are based on electron transfer from an absorbing polymer to a fullerene-modified material such as (6,6)-phenyl-C_{61}-butyric acid methyl ester (PCBM). Liu et al. have shown that resonant energy transfer from a red-emitting organic chromophore (Nile red) to PCBM

Tae Wan Kim and his group of the Department of Physics at Hongik University is Seoul, South Korea studied the photovoltaic effects of the Cu-Pc layer thickness—10- to 50-nm [163]. The group studied ITO/Cu-Pc/Al and ITO/CuPc/C_{60}/BCP/Al cells. P. Peumans et al. in 2001 developed an ITO/Cu-Pc/C_{60}/BCP/Al with power conversion efficiency of 3.6% at the AM-1.5 solar spectrum [164]. Excitons generated in OPV semiconductors are characterized by excitons

FIG. 14.18

The PCz-PDI organic solar cell is shown. Thin films of PDI and PCz were prepared by a spin-coating application onto activated (oxygen plasma) ITO surface. A thin layer of silver (100 nm) was evaporated through a mask to form the cathode. Monochromatic light ranging from 300 to 800 nm was supplied with a maximum intensity of 6 W·M⁻² at 600 nm. Solar light at 100 W·m⁻² was supplied by a solar simulator. The PDI films yielded a charge-carrier mobility of 0.2 cm²·V⁻¹·s⁻¹ in its crystalline phase. PCz was synthesized from 2,7-dibromocarbazole by alkylating with 2-decyltetradecylbromide (R = decyltetradecyl) and subsequent polymerization with bis-(1,5-cyclooctadiene)Ni, a.k.a. Ni(COD)₂. Characterization by electrochemistry was conducted with a voltammeter, a three-electrode cell with a working electrode of ITO, a silver quasi-reference electrode (AgQRE calibrated against Fc/Fc⁺ redox couple with E° = −4.8 eV), and a Pt counter electrode. The electrolyte was 0.1. M tetrabutylammonium perchlorate. Independently, PDI showed absorption maxima at 490 and 530 nm, PCz at 395 and 270 nm. When blended, a broad absorption between 300 and 600 nm was displayed—perfect for a solar cell application.

Source: Image courtesy of Klaus Müllen, Max–Planck Institute for Polymer Research in Mainz. With permission.

with higher binding energy than those found in inorganic material counterparts. Therefore, an electric field is applied to help separate the excitons into holes and electrons. The exciton diffusion length is approximately 10 nm in organic PV materials, therefore, according to the authors, the photoactive layer thickness should impact the efficiency of the cell [163].

A 5-mm wide strip of ITO (sheet resistance = 15 $\Omega \cdot sq^{-1}$) was fabricated with selective etching in an HCl:HNO$_3$ (3:1) solution and placed on a glass support.

Cu-Pc was deposited by thermal evaporation to thicknesses ranging from 10- to 50-nm in samples. In other specimens, C_{60} and BCP and then a 150-nm thick layer of aluminum were also deposited via thermal evaporation—quite simple straightforward processes [163]. Under dark conditions, nonlinear current density-voltage behavior was noted as Cu-Pc layers became thicker. Rectifying behavior was exhibited by films thicker than 30 nm indicating that exciton diffusion length is a factor in OPV materials—a trait desirable in solar cells. Therefore, layers of Cu-Pc less than 30 nm in thickness are not suitable for OPV cells. Open-current voltage (VOC, the voltage measured without the presence of current) depends on the Fermi energy difference between the electrodes and the energy levels (HOMO/LUMO) of the organic materials.

The current flow in ITO/Cu-Pc(20 nm)/C_{60}(20 nm)/BCP(15 nm)/Al exhibited more than 100x the current flow measured in single layer cells [163]. The group demonstrated that the efficiencies of heterojunction cells outperformed those of the single-layered Cu-Pc materials. The multilayered cells were more stable and yielded higher efficiencies [163]. The cell chemistry is shown in **Figure 14.19**.

Carbon Nanotube Materials in Solar Cells. In standard TiO_2 cells, electrons undergo a circuitous path, jumping from TiO_2 nanoparticles until they reach the electrode. Many do not achieve this goal, and thereby effectively reduce the efficiency of the cell. Researchers at the University of Notre Dame, P.V. Kamat et al.,

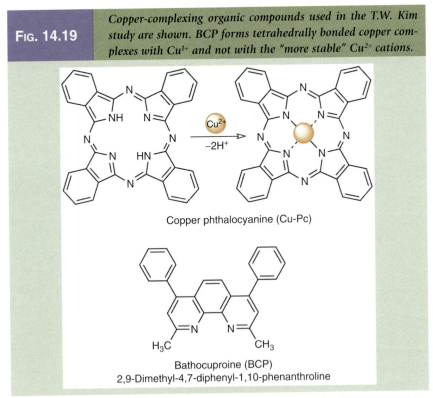

| FIG. 14.19 | *Copper-complexing organic compounds used in the T.W. Kim study are shown. BCP forms tetrahedrally bonded copper complexes with Cu^{1+} and not with the "more stable" Cu^{2+} cations.* |

Copper phthalocyanine (Cu-Pc)

Bathocuproine (BCP)
2,9-Dimethyl-4,7-diphenyl-1,10-phenanthroline

Source: S. W. Hur, T. W. Kim, and J. W. Park, *Journal of Korean Physics Society*, 45, 627–629 (2004). With permission.

have adapted single-walled carbon nanotubes to solar cells [165–169]. P.V. Kamat et al. have developed a cell that utilizes single-walled carbon nanotubes as conduits for electrons between the tethered TiO_2 and the electrode [165]. The SWNT-enhanced cells demonstrated twice the efficiency over cells comprised only of TiO_2 nanoparticles. An ultraviolet light lamp served as the source of the radiation. Carbon nanotubes are known to be good electron acceptors. A shift of approximately 100 mV of SWNT-TiO_2 system when compared to unsupported TiO_2 suggests that Fermi level equilibrium between the two systems was achieved [165].

The efficiency of solar cells using SWNTs decorated with TiO_2 can be enhanced if a one-molecule thick layer of organic dyes is adsorbed onto the semiconductor surface (as before). The dyes are able to expand the range of adsorption by the cell into the visible region of the solar spectrum. Another strategy involves the tethering quantum dots to the ends of the SWNTs. In this case, not only would the cell be able to convert visible range energy into electricity but also the ability to convert high-energy photons into multiple excitons.

Double-Walled Carbon Nanotube Solar Cells.   J. Wei et al. in 2007 showed how double-walled carbon nanotubes (DWNTs) can be applied to solar cell technology [170]. The DWNTs served as both photogeneration sites and electron collection-charge carriers. A semitransparent layer of DWNTs was applied on an *n*-type crystalline Si substrate to create high-density *p–n* heterojunctions in which charge separation takes place by means of electron extraction by the *n*-Si and holes through nanotubes [170]. A power conversion efficiency of >1% was achieved by these cells [170]. The main difference from other cells that use organic PV materials and CNTs is that in those cells the conjugated polymers generate the excitons and the CNTs act as a conduit for electrons. In this cell, the CNTs play a role in charge separation [170].

Ultimate Black Materials—The Super Black Object.   Ideal absorbers absorb all wavelengths without reflection (e.g., like blackbody cavities)—as even graphite and black paint reflect 5%–10% at the air–dielectric interface [171]. The major roadblock to developing such a material has been the inability to reduce the material's index of refraction to unity, so that optical reflection is totally eliminated. [171]. According to the authors, simulations do predict low index of refraction ($1.01 < n < 1.10$) by low-density, vertically aligned, multiwalled carbon nanotubes (VA-CNTs, 8–11 nm in diameter) in an array (e.g., a structure with high aspect ratio). The thickness of the CNT layer was between 10–800 μm and spacing between tubes was ca. 50 nm. These CNTs showed low refractive index, super low reflectance, diffuse reflection, birefringence, and strong absorption in the visible range of light [171]. The optical properties of such a surface is tunable by varying nanotube diameter, spacing, and length—all pretty standard features of nanotechnological engineering. Applications of such materials could impact pyrolytic detectors and solar energy conversion [171].

14.3.2   *Batteries*

The battery is one of the most important inventions. Benjamin Franklin in 1748 called an array of charged glass plates a battery. Luigi Galvani determined that electricity was involved in the mechanism of nerve impulses. Alessandro Volta invented the first real wet-cell battery in 1800—the voltaic pile—and found a means to conduct the charge over some distance. The Daniel cell (a cell that employs two electrolytes) was developed in 1836 by J.F. Daniels. In 1839, the first fuel cell was developed by W.R. Grove by combining oxygen and hydrogen. Frenchman G. Leclanche in 1866 invented the fundamental lead acid battery. The first commercially available dry cell was invented by C. Gassner in 1881. The alkaline storage battery was developed in 1901 by T.A. Edison. 1n 1954, the first solar battery was developed by G. Pearson, C. Fuller, and D. Chapin. Of course, there are many scientists and inventors who have made significant contributions but were left off this list. Nanomaterials should be able to provide battery technology with several advantages that include enhanced surface area, high electrical conductivity, electrical conducting networks, and perhaps even flexibility [172].

Some battery terms are listed and defined in this paragraph. A *primary battery* is not rechargeable. In other words, the electrochemical reaction that powers the cell is not reversible and the battery is usually discarded after it is drained. A dry cell alkaline battery is an example of a primary battery. A *secondary battery* is a battery that can be recharged (its electrochemical reaction is reversible) by passing a current through the battery opposite in direction to that of discharge during use. The lead acid battery found in most cars is an example of a secondary battery. An *ampere-hour* ($A \cdot h$) is equal to a current of one ampere over an hour's time. Battery capacity is measured in terms of $A \cdot h$. Specific battery capacity is measured in terms of $A \cdot h \cdot g^{-1}$ (or more appropriately, $mA \cdot h \cdot g^{-1}$). The *anode* in a battery is the site of electron release; the *cathode*, the site of electron absorption. More electrolyte and more electrode material translate into greater capacity for a battery. Smaller cells made of the same material as a larger cell, therefore, have less capacity although generating the same voltage [173]. The specific capacity of standard AAA through D batteries is ca. $100–140 \, mA \cdot h \cdot g^{-1}$. The standard graphite anode battery is capable of ca. $372 \, mA \cdot h \cdot g^{-1}$ capacity [174].

New enhanced batteries are expected to power automobiles that can operate for 200 miles without the need for recharging—or more modestly, at least be able to run for 24 h without having to plug into a power supply.

Next Generation Batteries.   Desirable characteristics of the next generation of batteries include large discharge capacity at potentials near that of Li, maintenance of high currents, and facile reversability to ensure rechargeability. Silicon actually has one of the highest theoretical capacity because it can store more ions ($\sim 4200 \, mA \cdot h \cdot g^{-1}$), but during Li insertion and extraction, Si-based insertion anodes (materials with highest known intercalation ability and low discharge potential) undergo a 400% change in volume for Li-ion secondary cells. The swelling results in decrepitation, pulverization, and subsequent loss of electrical contact [175]. Si nanowires, rather than bulk materials, improved strain relaxation (without pulverization) and conductivity. Si nanowires were grown directly on the current collector substrates. Improvement was due to good electrical conduction between Si nanowires and the current collector, short Li-insertion distances, high interfacial contact area with electrolyte, good electrical conductivity along the nanowires, and good material durability [176]. Discharge capacities

75% of the maximum were achieved with little diminishment during recycling [176].

High-Capacity Lithium Ion Batteries with Germanium Nanowires.   Maximum lithium capacity germanium–lithium ($Ge_{4.4}Li$) batteries are expected to deliver $1600\,mA\cdot h\cdot g^{-1}$ with 370% volume change. Because the room temperature diffusion of Li in Ge is 400x better than that in Si, Ge is attractive as an anode component in cells that intend to deliver high power rate ratios [175]. C.K. Chan et al. fabricated Ge nanowires by vapor–liquid–solid growth on metallic current collector substrates. An initial discharge capacity of $1140\,mAh\cdot g^{-1}$ was shown to be stable over 20 cycles [174]. Fabrication of Ge nanowire anodes is straightforward. Ge nanowires 50- to 100-nm diameter, 20- to 50-μm long are formed by VLS (vapor–liquid–solid) growth with CVD of GeH_4 from gold catalysts on a metal substrate (the electrical collector, usually a stainless steel foil). The nanowires were single crystalline and required no further processing—a one-step nanotechnology fabrication method [174]. The electrode is shown in **Figure 14.20**.

Carbon Nanotube Based Batteries.   Batteries have been recently fabricated from a random network of carbon nanotubes that serve as charge carriers at the interface with active components [172]. Highly purified SWNTs formed by laser synthesis were applied as anode materials for thin-film lithium ion batteries. The specific surface area and lithium capacity was compared to other traditional carbon materials like graphite, carbon black, and MWNTs [177]. BET-specific surface was found to be $915\,m^2\cdot g^{-1}$, much greater than the other carbonaceous materials mentioned. The discharge capacity of the SWNT-Li ion battery was $1300\,mA\cdot h\cdot g^{-1}$ after 30 charge–discharge cycles with a current density of $20\,\mu A\cdot cm^{-2}$ [177]. Lithium-ion battery anodes made of SWNT paper showed

FIG. 14.20	*Germanium nanowire electrodes on a stainless steel substrate are depicted. The Ge alloys with the SS following an annealing procedure provide good contact with the electrical connector. The purple dots represent gold catalyst nanoparticles.*

Source: C. K. Chan, X. F. Zhang, and Y. Cui, *Nano Letters*, 8, 307–309 (2008). With permission.

lower overall weight due to lack of a binder material and extremely simple preparation [178]. The performance of the paper anode battery was slightly less than the performance of conventional SWNT electrodes [178].

Researchers at Rensselaer Polytechnic Institute (RPI) and MIT have developed a new fully flexible and robust material that eliminates the need for bulky multilayered batteries (e.g., the kind you find in your cell phone). The electrodes, separator, and electrolyte were integrated into a single, flexible nanocomposite unit made of nanoporous cellulose paper embedded with aligned carbon nanotubes and electrolyte [179]. The team grew carbon nanotubes on a Si substrate and plugged the gaps between the tubes with cellulose, thereby making a flexible membrane. Two of these sheets were abutted with the cellulose sides placed inward and saturated in electrolyte to form a supercapacitor material. A 100 g sheet of the paper battery demonstrated 1300 mAh discharge capacity ($130 \, mAh \cdot g^{-1}$) [179]. The use of room temperature ionic liquid (RTIL) electrolytes like 1-butyl, 3-methylimidazolium chloride is able to dissolve cellulose with the assistance of microwave energy and allows for its use as an electrolyte in potential supercapacitor applications [179]. Potential applications of this technology include its use in Li-ion batteries, supercapacitors, and hybrids [179].

SnO₂ Electrodes in Lithium Batteries.   Tin oxide nanoparticles ranging in size from 3 to 8 nm were tested for battery capacity and cyclability. The 3-nm diameter particles showed the best characteristics of the group—ca. $740 \, mAh \cdot g^{-1}$ with negligible capacity fading [180].

14.3.3 *Hydrogen Production and Storage*

Photocatalysis.   Splitting water into hydrogen and oxygen is accomplished regularly and for over 3.5 billion years by photosynthetic organisms. However, how can we accomplish such a fundamental ancient task? In 1972, A. Fujishima and K. Honda achieved the photoelectrolysis of water with a TiO_2 photoanode with a platinum counter electrode. The chemical reactions involved with the photocatalytic splitting of water are given below [181]. The cell of Honda and Fujishima is shown in **Figure 14.21**.

$$TiO_2 + 2h\nu \rightarrow 2e^- + 2h^+ \tag{14.17}$$

$$H_2O + 2h^+ \rightarrow \tfrac{1}{2}O_{2(g)} + 2H^+ \tag{14.18}$$

$$2H^+ + 2e^- \rightarrow H_{2(g)} \tag{14.19}$$

$$H_2O + 2h\nu \rightarrow \tfrac{1}{2}O_{2(g)} + H_{2(g)} \tag{14.20}$$

not including the expression for the photocatalyst in the overall equation and where $h\nu$ = ultraviolet light applied to the photocatalyst.

Other Methods to Manufacture Hydrogen.   There are many chemical and electrochemical means to produce hydrogen. The process of steam methane reforming

FIG. 14.21

The original ultraviolet (<415 nm) photoelectrochemical cell devised by A. Fujishima and K. Honda in 1972 is depicted [181,182]. Water is electrolyzed without the application of an external voltage. The photoanode consisted of an n-type TiO₂ semiconductor material. The cathode was made of platinum. Modern day cells have reached an efficiency of 24% [182]. Charge separation is possible due to the presence of a potential that is provided ultimately from solar energy. In photosynthesis, such a gradient is established by the presence of a 5-nm thick lipid membrane that separates two different aqueous solutions in chloroplasts.

Overall reaction: $2h\nu + H_2O_{(l)} \rightarrow \frac{1}{2}O_2 + H_2$

(SMR) at ca. 800°C is currently the most economic process to synthesize hydrogen. By an ancillary process of chemical shift, carbon monoxide and water react to form more hydrogen [183].

$$CH_4 + H_2O \rightarrow 3H_2 + CO \tag{14.21}$$

$$CO + H_2O \rightarrow H_2 + CO_2 \tag{14.22}$$

The process of partial catalytic oxidation (POX) of methane, natural gas, landfill gas, industrial wastes, solvents, oils, biomass, agricultural wastes, sewage sludge, kerosene, biodiesel, and other hydrocarbon sources proceeds at high temperatures (1100–1350°C) and can achieve ca. 50% conversion efficiency. Gasification occurs with the addition of steam and subsequent desulfurization to produce syngas. Shift reactions and purification results in the production of H₂ and CO₂ [183]. Coal gasification proceeds at 48% efficiency and exploits coal, biomass, and residual oil resources. Pure oxygen is required and the fuel must be in a pulverized state. The operating temperature is problematic: 1100–1300°C [183]. Nuclear-powered H_2 production is capable of driving thermochemical splitting of water but the specter of nuclear wastes and the dropping costs of advanced technologies make the nuclear option less attractive [183].

DOE goals include the development of novel synthetic techniques to achieve >40% efficiency at modest cost, development of materials to generate 10% efficiency solar to stored hydrogen energy, enhance nanocatalyst design to be more efficient, development of carbon-resistant reforming catalysts, better electrolysis catalysts, better photocatalysts, CO-resistant anode materials for fuel cell membranes, better cathode materials with lower overpotentials for fuel cells, and development of hydrogen activation catalysts that depend less on noble metals [183]. With regard to complex hydrides, DOE priorities include development of nanophase structures, transition metal dopants, a new class of hybrid hydrogen storage, and storage of hydrogen at various temperatures and pressures. More will be said about the last item in the next section.

Photosystem II Mimetics by Water Photolysis.   Researchers at Penn State University have recently developed a solar cell that mimics photosynthesis [184]. The prototype is inefficient (~1%) but is able to make hydrogen directly from splitting water. The process mimics natural photosynthesis. The mechanism of the cell is similar to that of the Grätzel cell [185] discussed in chapter 12 except that electrons in dye molecules that have undergone photoexcitation are transferred onto iridium oxide nanoparticles that catalyze the splitting of water into oxygen and hydrogen ions (rather than generate electrons for electricity). The technological challenge is to prevent electrons from recombining with the dye and from hydrogen and oxygen from recombining into their most stable state—water. A catalyst must be found that promotes water splitting but does not promote recombination of hydrogen and oxygen. T.E. Mallouk and W.J. Youngblood recently presented their results at the American Association for the Advancement of Science Annual Meeting, at Boston in February 2008 [184].

Iridium oxide nanoparticles in direct contact with orange-red dye molecules (absorb in blue range, a high energy domain) formed clusters approximately 2 nm in diameter—the catalyst and the dye molecule were approximately the same size. The catalyst was impregnated into a TiO_2 anode. Pt served as the cathode. The electrodes were immersed in separate compartments in a salt solution. Each surface Ir complex is able to convert water into O_2 when receiving an excited electron from a dye molecule. The researchers claim that each surface iridium atom is able to catalyze the water-splitting reaction approximately 50 times per second—a level comparable to the turnover rate exhibited by PSII in green plants.

Tunability can be achieved according to Mallouk by improving the efficiency of the dye, improving the catalyst, and making adjustments to the geometry of the system [184]. The distance between molecules is one parameter that may be adjusted to direct electrons better—lengthening some paths and shortening others [184]. Mallouk et al. also reported about the ability of trivalent rhodium hydroxide nanoparticles on semiconducting layered calcium niobate for catalytic hydrogen evolution [186].

Hydrogen Storage (and a Real Scientific Debate). Hydrogen is the most abundant element in the universe and is the cleanest burning fuel. The combustion of hydrogen is straightforward with no detrimental by-products.

$$H_{2(g)} + \frac{1}{2}O_{2(g)} \rightarrow H_2O \qquad (14.23)$$

What could be better than producing water from a combustion product or use of hydrogen in a fuel cell that also produces water? Hydrogen is easily manufactured from renewable energy sources [187], plentiful, and is lightweight. This is the perfect scenario, right? The only problem with the use of hydrogen is finding a safe, efficient, compact, and economical way to store it [188]. There are two issues facing both compression and liquefaction scenarios—two traditional methods to store hydrogen. First, hydrogen is explosive and will ignite at relatively low concentrations, for example, recall the fates of the Hindenburg airship that used compressed hydrogen or the space shuttle Challenger that used liquid hydrogen. Most hydrogen storage requires "hydrogen boil-off," venting that is most certainly a potential safety hazard. Second, compressing or cooling hydrogen is an energy-intensive process—not desirable from an economic perspective. Development of materials strong enough for use in storage canister walls would also present technological challenges. Quantum Technologies has developed storage canister equipped with composite fiber walls capable of withstanding 10,000 psi.

Other traditional methods of storing hydrogen are by gas-on-solid adsorption (using activated carbon or zeolites) and chemical/metal hydride materials (pure or alloyed metals react with H_2 to store ca. 2–10 wt%) [183].

Scientists are in general agreement that alternative methods of storing hydrogen by materials such as activated carbons and metal hydrides have not lived up to the goals established by the Department of Energy's *FreedomCar Program* roadmap—that on-board mobile hydrogen storage system should provide 6.5 wt% of H_2 in order to be effective as a transportation fuel [189]. The FreedomCar roadmap, by 2010, asks for onboard hydrogen storage systems to achieve $2\,kWh \cdot kg^{-1}$ (6 wt%), $1.5\,kWh \cdot L^{-1}$, and \$4/kWh. By 2015, the bar is raised to $3\,kWh \cdot kg^{-1}$ (9 wt%), $2.7\,kWh \cdot L^{-1}$, and \$2/kWh [189]. On carbon nanotubes or graphite, one hydrogen molecule per carbon atom yields 14 wt% of hydrogen. Other ways to store hydrogen need to be developed if the Freedom (from oil?) Car is to become a reality. Will the solution(s) come through nanotechnology?

Hydrogen Storage with Carbon Nanotubes? In 1997, Dillon et al. at the National Renewable Energy Laboratory (NREL) in Golden, Colorado claimed that single-walled carbon nanotubes were able to store hydrogen gas (at high density) inside the channels of the tubes by a condensation capillary mechanism [190]. The NREL group reported that single-walled carbon nanotubes, found in arc-produced soots, were able to store 8 wt% by weight of hydrogen at room temperature and moderate pressures [190]. The paper generated an incredible level of international interest and a whirlwind of controversy over the next decade (and now beyond). For example, a group at the Max Planck Institute in Germany, led by Michael Hirscher, disputed the claim by reporting that SWNTs adsorb less than 1 wt% H_2. Hirscher et al. reported that the metal alloy contaminants found in the arc-generated soot were responsible for H_2 adsorption reported by Dillon et al. [191].

Most scientists agreed in the years following that a physisorption mechanism mediated by the outside walls (e.g., interstitially between nanotubes that make up a bundle) of the tubes were able to attract H_2 to nanotubes—and that the mechanism of adsorption was not based on capillarity or condensation inside the tubes. Computational methods have since demonstrated that simple physisorption mechanisms cannot account for 6 wt% of hydrogen storage on pristine nanotubes [192]. According to J. Karl Johnson of the University of Pittsburgh, the fact that physisorption by itself is not able to account for the adsorption of hydrogen does not rule out other mechanisms [193]. Johnson showed that only 1–2 wt% hydrogen storage is possible on simulated pristine nanotubes at room temperature and reasonable pressures [188].

The shape of the thermally programmed desorption curve (TPD) also presented a forum for controversy. The curve obtained initially resembled that of a titanium alloy. As it turned out, an ultrasonication system utilized a Ti alloy (Al and V) during a "cutting" process [188]. Cutting is a process by which nanotube caps can be opened with the intent of allowing ingress of gas during exposure to H_2. From further study it was determined that the Ti alloys were able to adsorb 2.5–3.4 wt% H_2 on their own.

Most recently, researchers at the Stanford Synchrotron Radiation Laboratory (SSRL) demonstrated that >7-wt% H_2 storage is possible with saturated (chemically modified with hydrogen) carbon nanotubes [194–196]. Theoretical analyses and total energy density functional theory (DFT) calculations predict that 7.7 wt% storage is possible by chemisorption mechanisms by saturation of the carbon–carbon double bonds of the nanotube (**Fig. 14.22**) [197]. The mechanism of hydrogen adsorption is by the chemisorption process while the contribution from physisorption mechanisms was shown to be negligible at room temperature [194–196]. One likely problem associated with hydrogen storage on saturated carbon nanotube is the potential for slow kinetics during desorption due to the strength of the C–H bond [197].

A. Nikitin et al. claim that chemisorbed hydrogen can be desorbed and recombined at temperatures as low as 200–300°C, not entirely an unreasonable temperature in an automobile system [194–196]. The desorption temperature would be controlled by the desorption kinetics and the nanotube diameter—a characteristic that can be manipulated to tune the strength of the C–H bond—a true nanotechnology in action! The challenge then is to find an economic and efficient means of hydrogenating the SWNTs with molecular hydrogen. The *spillover process* is one such way [198]. H_2 molecules dissociate over the surface of a catalysts deposited on carbon nanotubes. Upon breakdown, hydrogen radicals spill over from the catalyst particle to the carbon nanotube and add across the double bonds to form C–H bonds—one nanotechnology helping out another [194–196]. Y.-W. Lee et al. have shown that Pt nanoparticles inside carbon nanotubes enhance the uptake of hydrogen fivefold [198].

Another aspect of the Stanford technology is that *non-bundled nanotubes* of the *same diameter* are required to provide maximum adsorption. Nanotubes, as we know so well by now, prefer to exist in a bundled state due to strong van der Waals attractions, and to date, no one has been able to fabricate nanotubes

FIG. 14.22

An illustration of the partial hydrogenation of single-wall carbon nanotubes. Hydrogenation proceeds by exposing the nanotubes to atomic hydrogen and the hydrogen radicals add to the double bond of the SWNT. The hydrogenation process is reversible at 500–600°C [199]. Others claim reversibility at lower temperatures [194–196]. Smaller SWNTs are less stable against hydrogenation than larger tubes. Following hydrogenation, SWNTs undergo structural deformations, reduced electrical conductance, and increased semiconductor behavior [199]. The diameter of SWNTs increased from 1.0 to 1.3 nm or from 1.8 to 2.1 nm, depending on the starting material, following hydrogenation.

Single-wall carbon nanotubes

Source: G. Zhang, P. Qi, X. Wang, Y. Liu, D. Mann, X. Li, and H. Dai, *Journal of the American Chemical Society*, 128, 6026–6027 (2006).

with the same diameter, electrical properties, and chirality without extensive postsynthesis separation.

Meanwhile, the investigation into Ti-decorated carbon nanotubes continues [200]. Recent theoretical calculations showed that light transition metal atoms that decorate one- and two-dimensional carbon-based nanostructures (carbon linear chain, graphene and nanotubes) were capable of hydrogen uptake. Interestingly, the researchers claim that even graphene would serve as a potential high-capacity H_2 storage material [200].

In October 2006, the DOE Hydrogen Program decided to pull the plug on research funding to continue work with pure, undoped, single-walled carbon nanotubes for vehicular hydrogen storage. The decision was based on the inability of pure SWNTs to store 6wt% hydrogen at close to room temperature and low pressure—experimental data show that adsorption of 0.6wt% hydrogen—an order of magnitude less than required. At cryogenic temperatures (77°C), only 3–6wt% was achievable at 20-bar pressure. The DOE report, however, affirms that certain areas of carbon nanotube research with metal-doped hybrid materials may warrant further investment by DOE [201]. The NREL group, for example, has achieved 3wt% storage at room temperature and nominal pressure on metal-doped SWNTs [202]. Perhaps the Ti-alloy impurity is promoting a spillover process that feeds hydrogen radicals to the nanotubes. Regardless of current obstacles, it seems likely that hydrogen storage by saturated or metal-doped nanotubes is not too far off in the future.

Storage with Single-Walled Carbon Nanohorns. David B. Geohegan et al. at Oak Ridge National Laboratory have a different approach to hydrogen storage and catalyst-assisted hydrogen storage—by application of single-walled carbon nanohorns (SWCNHs) with tunable pore size [203,204]. The SWCNHs can be therefore optimized, by control of synthesis parameters and postprocess treatment, for hydrogen adsorption and metal-catalyzed spillover. SWCNHs are formed by a high-temperature laser ablation procedure capable of generating $9\,g\cdot h^{-1}$ of nanohorns. The highest measured surface area of SWCNHs is equal to $1892\,m^2\cdot g^{-1}$. Levels of hydrogen storage of 2.5 wt% were achieved with opened-Pt decorated SWCNHs [203,204]. Charging SWCNHs in an electric field is expected to significantly increase the storage capacity of hydrogen. Studies with C_{82} fullerenes revealed that a charge of $6e^-$ was able to increase the capacity of hydrogen storage to 8 wt% [205]. Single metal atoms and organic molecules inside fullerenes are able to generate sufficient electric fields to enhance hydrogen storage [205].

14.3.4 *Fuel Cells*

The fuel cell offers one of the panaceas to the energy challenge posed by Professor Smalley. Fuel cells are best described as a renewable electrochemical battery (or better, a reusable battery). The technology is developed enough so that potential consumer goods are likely on a large scale—products in the form of battery replacement devices, remote power generators, and automobile power. The fuel cell is based on the electrochemical conversion of hydrogen and oxygen to produce water. The result of the electrochemical process results in maximum 1.2 V and $1\,W\cdot cm^{-2}$ of power [206]. Nanostructured catalysts and membranes are integral in the design and operation of a fuel cell. A generic fuel cell is depicted on **Figure 14.23** [206].

Sir William Grove conceived the concept of the fuel cells as early as 1839. Grove thought that by reversing the process of electrolysis, in which hydrogen and oxygen are generated from the electrolysis of water, electricity could be produced. His first fuel cell was called the "gas voltaic battery." Ludwig Mond and Charles Langer coined the term fuel cell later on before the turn of the century. There are many kinds of fuel cells—the polymer exchange membrane (the one we focus on in this section), solid oxide fuel cell (high operating temperatures—generates heat and electricity), the alkaline fuel cell (requires pure oxygen and hydrogen and used in the space program), and the direct methanol fuel cell among others. The efficiency of a fuel cell powered by pure hydrogen can be as high as 80%—when converted into mechanical power, the overall efficiency is still quite high at 64%. The primary reactions of each are shown in **Table 14.5**.

The polymer (or proton) electrolyte membrane (PEM) serves as a conduction medium for protons. Nafion is a sulfonated tetrafluoroethylene copolymer ionomeric membrane invented by Du Pont in the 1960s and developed in the 1970s. Nafion is able to operate at relatively low temperatures—ca. 85°C. Nafion

FIG. 14.23	*A generic rendition of a fuel cell is depicted on the top. Below, a detailed description of the active region is shown.*

Anode
$$H_2 \rightarrow 2H^+ + 2e^-$$

Cathode
$$2e^- + \tfrac{1}{2}O_2 \rightarrow O^{2-}$$

H_2 input

Compression

Coolant channels

Unused H_2 recirculated

O_2 input (from air)

Compression

Heat (85°C)

Air + H_2O output

Gas diffusion layer

Catalyst electrode layer

PEM membrane

Catalyst electrode layer

Gas diffusion layer

H_2

H^+ conduction

O_2 gas

H_2O

Carbon nanoparticles

Pt catalyst

Membrane electrode assembly

| 200 μm | 15 μm | 25 to 170 μm | 15 μm | 200 μm |

Source: Image courtesy of Dr. David Jacobson, National Institute of Standards and Technology. With permission.

TABLE 14.5	**Types of Fuel Cells and Anode/Cathode Reactions**		
Type of fuel cell	**Anode**	**Cathode**	**Ion**
Alkaline fuel cell (AFC)	$H_2 + 2(OH)^- \rightarrow 2H_2O + 2e^-$	$1/2O_2 + H_2O + 2e^- \rightarrow 2(OH)^-$	OH^-
	Operating temperature: 80–100°C; *Efficiency*: 60%; *Electrolyte*: Potassium hydroxide		
Direct methanol fuel cell (DMFC)	$MeOH + H_2O \rightarrow CO_2 + 6H^+ + 6e^-$	$1/2O_2 + 2H^+ + 2e^- \rightarrow H_2O$	H^+
	Operating temperature: 60–90°C; *Efficiency*: 25%		
Molten carbonate fuel cell (MCFC)	$H_2 + CO_3^{2-} \rightarrow CO_2 + H_2O + 2e^-$	$CO_2 + 1/2O_2 + 2e^- \rightarrow CO_3^{2-}$	CO_3^{2-}
	Operating temperature: 600–650°C; *Efficiency*: 45%–60% *Electrolyte*: Carbonate melt		
Phosphoric acid fuel cell (PAFC)	$H_2 \rightarrow 2H^+ + 2e^-$	$1/2O_2 + 2H^+ + 2e^- \rightarrow H_2O$	H^+
	Operating temperature: 200–220°C; *Efficiency*: 40%–45% *Electrolyte*: Phosphoric acid		
Proton exchange membrane fuel cell (PEMFC)	$H_2 \rightarrow 2H^+ + 2e^-$	$1/2O_2 + 2H^+ + 2e^- \rightarrow H_2O$	H^+
	***Operating temperature*: 70–80°C; *Efficiency*: 35%–45% *Electrolyte*: Hydrogen conducting membrane**		
Solid oxide fuel cell (SOFC)	$H_2 + 2O^{2-} \rightarrow H_2O + 2e^-$	$1/2O_2 + 2e^- \rightarrow O^{2-}$	O^{2-}
	Operating temperature: 800–1000°C; *Efficiency*: 50%–65% *Electrolyte*: Solid oxide		

Note: The proton exchange fuel cell system is in bold in the table.

possesses excellent thermal properties like Teflon (up to 190°C), conductive properties, and chemical resistance and durability (does not release fragments). The vehicles for electrical conduction are protons on the sulfonate groups as they hop from one acid site to the next [207]. Only cations (and protons) are allowed to migrate through the porous matrix as anions and electrons are blocked. It is known as a super-acid catalyst because the sulfonic groups attached to the fluorinated (electron withdrawing) backbone and the stabilizing effect of the extended polymer matrix make Nafion a strong proton donor ($pK_a < 0$). The membrane is highly selective and permeable to water and the degree of hydration of the membrane impacts its electrical properties. Nafion is a nanomaterial—consisting of a network of interconnected clusters similar in form to inverted micelles. The pore channels are on the order of 4 nm in diameter although the exact structure is difficult to determine [207]. Nafion is also used in ion-exchange membranes used to produce Cl_2 gas and $NaOH$ during the electrolysis of water, in drying or humidifying applications, and as a super-acid catalyst in the production of fine chemicals [208].

Hydrogen is first circulated through channels made of solid graphite and then passed into a porous mesh layer called the "gas diffusion media" [206]. Once well into the media, the hydrogen gas contacts graphitic nanoparticle supported platinum catalyst and is stripped of electrons. The electrons are swept into a current emanating from the anode. Meanwhile, the remaining protons make their way through the PEM to the cathode and encounter another layer of carbon-supported platinum catalysts. The protons interact with oxygen and incoming electrons from the external circuit to form liquid water that is drained from the system [206].

14.3.5 Solar Heating and Power Generation

Although not as exciting, exotic, or sophisticated as its photovoltaic cousins, solar heating systems, as a general rule, are able to convert 25%–40% of solar radiation into usable heat [144]. For a small hot water heating system (50 L per person at 45°C per day), the recommended dimensions of a solar collector are on the order of 1.2–1.5 m² [144]. The principle of solar heating is straightforward. The process is based on a material that absorbs sunlight energy and releases it in the form of heat whether directly to a water source or to a heat exchange element (heat pump). There are two general types of solar collectors: the *flat plate collector* in which solar energy absorbing coated pipes are enclosed within a glass paneled box, and a solar water heater that employs a *batch collector* that combines the collection and storage systems within one device. Any material that is able to enhance surface area or absorption properties are sure to benefit this aspect of the solar energy applications—enter nanotechnology.

Solar Concentration. The efficiency of solar cells can be increased by external means as well—by the use of concentrating solar collectors that apply Fresnel lenses and parabolic mirrors to concentrate and focus solar energy onto solar cells. Advantages include the use of smaller areas for solar cells and increasing the level of solar flux impinging on the cells. Another advantage is afforded by the installation of solar tracking systems [209].

Concentrating solar energy for heating by an array of reflective mirrors is a means of providing heat to power a generator that is used to create electricity. Such systems require a large amount of space and generous helpings of direct sunlight. Parabolic trough collectors consist of a parabolic mirror with an absorber tube placed at its center, thereby heating the liquid within the tube. The heated effluent is used to boil water that consequently produces steam to run a generator.

Geothermal Energy. Nanotechnology contributions to extracting energy by geothermal mechanisms include development of new materials that exhibit enhanced thermal conductivity or corrosion resistance [210]. Nanoscale materials are expected to better manage geothermal reservoirs, acquire subsurface information, and enhance thermoelectric materials that can detect small thermal gradients. Transport of thermal energy (100–200°C) to within 1 km is practical if geothermal energy sources are to prove useful. Strength enhancement of geothermal drilling materials is also required as higher temperatures are encountered.

14.4 EPILOGUE

We have reached the end of our adventure in nanotechnology with this last chapter—the discussions about the environmental aspects. What we have observed over the course of this book, and perhaps from *Introduction to Nanoscience* as well, is that nanotechnology is causing a paradigm shift with regard to properties, capability, and fabrication and applications. For example, the micro-age was based on lithography. Although lithography is still a viable and thriving top down, not much mention was made of this technique in the latter chapters— rather emphasis was placed on bottom-up self-assembly methods. The shift of fabrication techniques from the engineer to the chemist is in full sway as more amazing and incredible devices are created. Another trend in nanotechnology is also making itself known—the relationship between nature and a new generation of devices. Nature also makes things from the bottom up—from the basic amino acids, carbohydrates, and lipids.

New methods to clean water and store hydrogen, among many other projects, are constantly under development [211–224]. We take No. 1 and 2 from Smalley's list of the grand challenges that face humanity and add a few more examples to take with you: that of the bottom-up assembly of materials that remove heavy metals to purify water (by SAMMS) and the storage of hydrogen with metal organic frameworks (MOFs)—both examples of true chemical synthesis bottom-up nanotechnology.

14.4.1 SAMMS

SAMMS are self-assembled monolayers on mesoporous supports—the combination of organic and inorganic materials to form a device. SAMMS serve as the perfect example of where nanotechnology will take us—better detection, sensing and overall capability, higher throughput, and ultimately more economical and environmental sense.

SAMMS.   Self-assembled monolayers on mesoporous supports (SAMMS) provide a new generation of SAM technology applied to removal of waterborne contaminants—a pure and true nanotechnology! Novel nanocomposite sorbents based on copper ferrocyanide immobilized within a mesoporous ceramic matrix have shown great promise in removing Cs from water streams—complete removal of Cs in the presence of competing metal ions under a variety of conditions for solutions in which the [Cs] = 2 ppm [211–213]. The Pacific Northwest Laboratory group reported loading capacities of more than $1.35\,mmol\ Cs \cdot g^{-1}$ adsorbent [211–213]. The system displayed fast binding kinetics that was due to the rigid pore structure and the extremely high surface area of the ceramic sorbent material. DOE has put into place some goals designed to promote the reduction of vitrified wastes—specifically that just the radionuclide-laden SAMMS need to be vitrified and not the whole volume of all solid wastes [211–213]. Since actinides and other heavy metals form insoluble hydroxides, separation methods usually require alteration of pH to less than 4—a process that doubles the waste volume [211–213]. SAMMS, on the other hand, display solution-solid rations of ca. 100 (and even higher)—a process that actually reduces the amount of vitrified waste by a factor of 50 [211–213]. An image depicting a SAMMS molecular sponge adsorbent is shown in **Figure 14.24**.

SAMMS technology is based on three generations of self-assembled monolayer work. First, the development of surfactant molecules to create micellular templates was accomplished; second, aggregation of silica-coated micelles into

FIG. 14.24

A nanostructured adsorbant SAMMS is depicted. Starting with micelles that served as the original template material, a silaceous gel was generated. Upon high-temperature treatment, the organic component was removed leaving the silica skeleton—a highly porous, uniform, and interconnected system. After activation of the silica surface, monolayers were attached by application of specially modified silanes or other linking groups. Quite a bit of bottom–up nanotechnology is involved in its fabrication and operation. Depending on the functional groups, the size, and configuration of the pore channels, a wide range of selectivity can be achieved.

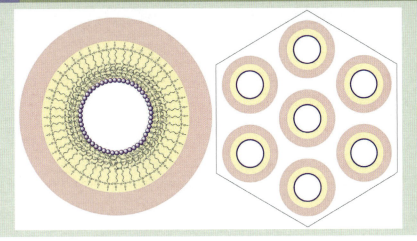

Source: Image courtesy of Pacific Northwest National Laboratory. With permission.

a mesostructured material occurred with high surface area ($\sim 1000\,\mathrm{m^2 \cdot g^{-1}}$); and finally, self-assembly of silanes into an ordered monolayer on the surface of the silicate pore structure with a high concentration of binding groups [211–213]. Many kinds of SAMMS can be prepared, for example, glycinyl-urea (Gly-UR) and salicyclamide (**Fig. 14.25**).

14.4.2   *One More Pass at Hydrogen Storage*

We bring up hydrogen storage one more time because it is one of the most important challenges facing modern nanotechnology. Carbon nanotubes (inconsistency) and metal hydride (cost, low specific uptake, unfavorable kinetics and contamination) storage systems have significant limitations [214]. Metal-organic frameworks (MOFs) provide for a new medium of hydrogen storage [214–217]. Dissociative methods of storing hydrogen (discussed earlier) are hindered by large H–H bond dissociation thresholds. Associated adsorption, the case in which H_2 is adsorbed as a molecule, does not suffer from such an impediment [217]. MOFs are crystalline solids composed of metal clusters and organic ligands. A 7.5wt% level of storage was demonstrated with a type of MOF called MOF-177 at 60 bar and 77 K—conditions that are not particularly practical [217]. Recently, by molecular simulation methods, it was demonstrated that Li-doped MOF (called Li-MOF-C30) showed significantly improved results at much higher H_2 saturation temperatures of $-30°C$ @ 100-bar [217]. The Li-doped material was able to adsorb 6.0wt% H_2. One of the DOE FreedomCar

Various types of SAMMS ligands are depicted. Depending on the end group, a high level of versatility can be achieved to targets specific materials. HOPOs are hydroxy pyridinones. Thiol-derivatized SAMMS are good for cleaning up Pd, Ag, Cd, Pt, Au, Hg, Tl, and Pb; Cu-FC-EDAs are tailor made to extract Cs; CU-EDAs target V, Cr, Mo, Tc, Ge, As, Se, Br, and I (EDA, ethylene diamines form a system of three bidentate ligands to bind Cu^{2+}; HOPOS and Prop-phos work for Th, Pa, U, Np, Pu, and Am mitigation) [211–213].

criteria is to make H_2 storage practical between –30 and 80°C. This material, therefore, shows high compatibility for hydrogen storage.

Highly porous metal-organic frameworks (**Fig. 14.26**) have uniform-sized cavities within a crystalline-metal-organic frame consisting of $[OZn_4]$ conjoined in an octahedral array by $[O_2C–C_6H_4–CO_2]^{2-}$ (a.k.a. 1.4–benzenedicarboxylate or BDC). Other kinds of ligands, for example, the tetrazolates, have four nitrogens at the disposal of metals wishing to become coordinated. The adsorption

Example of MOF cage structures. The yellow sphere represents the scope of the enclosed volume (and not hydrogens). The zinc oxide (blue tetrahedrons and orange dots respectively) are linked with organic ligands. Black dots represent carbon atoms. Very clever— very nano!

Source: N. L. Rosi, J. Eckert, M. Eddaoudi, D. T. Vodak, J. Kim, M. O'Keefde, and O. M. Yaghi, *Science*, 300, 1127–1129 (2003). With permission.

enthalpy of this MOF was as low as $10\,kJ\cdot mol^{-1}$ and an adsorption capability of $6.9\,wt\%$ $(60\,g\cdot L^{-1})$ at $77\,K$ and 90 bar—one of the best demonstrated so far—this according to Professor Jeffery R. Long of the Department of Chemistry at U.C. Berkeley (alchemy.cchem.berkeley.edu/hydrogen.html).

14.4.3 *Concluding Thoughts*

We explored, among many things, the mechanisms, materials, and devices of organic photocells [218], practical applications of bimetallic nanoparticles for groundwater treatment [219], the mechanisms of activated carbon [220], how cadmium chalcogenide quantum dots are anchored onto stable oxide semiconductors to make QD-sensitized solar cells [221,222], how to make SAMMS for specific environmental remediation of heavy metals [223], and fabrication of ordered SAMs onto gold for the purposes of sensing [224]. There is so much more. Sadly, it is impossible to cover all of nanotechnology in a single textbook. The purpose of the book is to expose the student to broad aspects of this wondrous new area and perhaps, hopefully, inspire that student to apply his or her newly acquired knowledge to help protect our environment.

Please keep in mind Smalley's ten grand challenges and how you, perhaps, could be the scientist that discovers ways to mitigate them. As the late Richard Everett Smalley stated "Be a scientist. Save the World!" Best of luck to all of you.

Acknowledgments

We wish to thank with great appreciation the Department of Energy, the National Renewable Energy Laboratory and the National Institute for Standards and Technology for providing information that was vital to the content of this text.

References

1. EPA, Measurement and analysis of Adsistor and Figaro gas sensors used for underground storage tank leak detection, U.S. Environmental Protection Agency (EPA), Report #EPA/600/R-92/219 (1992).
2. EPA, Measurement and analysis of vapor sensors used at underground storage tank sites, U.S. Environmental Protection Agency (EPA), Report #EPA/600/R-95/078 (1995); T. M. Masciangioli, N. Savage, and B. P. Karn, Environmental implications of nanotechnology, Poster, www.environmentalfutures.org/nanotech.htm (2004).
3. K. I. Mullen and K. T. Carron. Surface enhanced Raman spectroscopy with abrasively roughened fiber optic probes, *Analytical Chemistry*, 63, 2196–2199 (1991).
4. J. C. Cook, C. M. P. Cuypers, B. J. Kip, R. J. Meier, and E. Koglin, Use of colloidal SERS for the trace detection of organic species: A survey based on pyridine as the probe molecule, *Journal of Raman Spectroscopy*, 24, 609–619 (2007).
5. C. J. L. Constantino, T. Lemma, P. A. Antunes, and R. Aroca, Single molecular detection of a perylene dye dispersed in a Langmuir–Blodgett fatty acid monolayer using surface-enhanced resonance Raman scattering, *Spectrochimica Acta Part A: Molecular and Biomolecular Spectroscopy*, 58, 403–409 (2002).
6. H. Krug, ed., *Nanotechnology, Volume 2: Environmental aspects*, Wiley-VCH, Weinheim (2008).

7. R. H. Grove, *Green imperialism: Colonial expansion, tropic island edens and the origins of environmentalism*, Cambridge University Press, Cambridge, UK (1995).

8. Ancient Civilizations, *Environmental History Timeline*, www.runet.edu/~wkovarik/ envhist/1ancient.html (accessed 2007).

9. A. Markham, *A brief history of pollution*, St. Martin's, New York (1994).

10. J. Ewing and D. Quinn, *Monkeys are made of chocolate*, PixyJack Press, LLC., Masonville, CO (2005).

11. The flow of energy: Primary production to higher trophic levels, University of Michigan, www.globalchange.umich.edu/globalchange1/current/lectures/kling/ energyflow/energyflow.html (2005).

12. S. Morgan, Nanotechnology applications in water management, Presentation, Nanotechnology Victoria, Nanotechnology_Water_Management_July06.pdf (2006).

13. T. Hatakeyama, D. L. Chen, and R. F. Ismagilov, Microgram-scale testing of reaction conditions in solution using nanoliter plugs in microfluidics with detection by MALDI-MS, *Journal of American Chemical Society*, 128, 2518–2519 (2006).

14. Helping you explore, be novel, be first in nanotechnology, www.agilent.com/find/ nano (2006).

15. Introduction to Micro/Nano-HPLC, www.innosep.com/2.htm (2007).

16. S. K. Sundaram and T. J. Weber, New nano-canary in the nanotoxicology coalmine: The body itself, Pacific Northwest National Laboratory, www.physorg.com/ news10971.html (2006).

17. A. Scheffer, C. Engelhard, M. Sperling, and W. Buscher, ICP-MS as a new tool for the determination of gold nanoparticles in bioanalytical applications, *Analytical and Bioanalytical Chemistry*, 390, 249–252 (2007).

18. X. Li, X. Tu, S. Zaric, K. Welsher, W.-S. Seo, W. Zhao, and H. Dai, Selective synthesis combined with chemical separation of single-walled carbon nanotubes for chirality selection, *Journal of American Chemical Society*, 129, 15770–15771 (2007).

19. N. Tao, A nanocontact sensor for heavy metal ion detection, Proceedings: EPA Nanotechnology and the environment: Applications and implications, STAR (Science to achieve results) Progress Review Workshop, Arlington, VA (2002).

20. O. Hayden and C. K. Payne, Nanophotonic Light Sources for Fluorescence Spectroscopy and Cellular Imaging, *Angewandte Chemie International Edition*, 44, 1395–1398 (2005).

21. D. Dosev, M. Nichkova, M. Liu, B. Guo, G.-Y. Liu, B. D. Hammock, and I. M. Kennedy, Application of luminescent $Eu:Gd_2O_3$ nanoparticles to the visualization of protein micropatterns, *Journal of Biomedical Optics*, 10, 064006–064012 (2005).

22. G. Hanrahan, D. G. Patil, and J. Wang, Electrochemical sensors for environmental monitoring: design, development and applications, *Journal of Environmental Monitoring*, 6, 656–664 (2004).

23. Y. Tu, Y. Lin, W. Yantasee, and Z. Ren, Carbon nanotube based nanoelectrode arrays: fabrication, evaluation, and application in voltammetric analysis, *Electroanalysis*, 17, 79–84 (2005).

24. M. Watari and J. Ransley, Label-free nanomechanical protein detection on sensor arrays, *IRC in nanotechnology*, Cambridge Nanoscience, www.nanoscience.cam. ac.uk/irc/characterisation/22.html (2006).

25. P. Datskos, Chemical sensors based on nanomechanical resonators, *Oak Ridge National Laboratory-fact sheet*, www.ornl.gov/sci/engineering_science_technology/ sms/Hardy%20Fact%20Sheets/Chemical%20Sensors.pdf (viewed 2008).

26. L. C. Clark, R. Wolf, D. Granger, and Z. Taylor, Continuous recording of blood oxygen tensions by polarography. *Journal of Applied Physiology*, 6, 189–193 (1953).

27. L. C. Clark, Jr. and C. Lyons, Electrode systems for continuous monitoring in cardiovascular surgery, *Annals of New York Academy of Science*, 102, 29–45 (1962).

28. L. P. Clark, people.clarkson.edu/~ekatz/scientists/clark_leland.htm (2007).

29. Biosensor, en.wikipedia.org/wiki/Biosensor (2008).

30. M. Chaplin and C. Bucke, *Enzyme technology*, Cambridge University Press, Cambridge, UK (1990).

31. J. Liu and Y. Lu, Smart nanomaterials responsive to multiple chemical stimuli with controllable cooperativity, *Advanced Materials*, 18, 1667–1671 (2006).

32. Aptamer, en.wikipedia.org/wiki/Aptamer (2008).

33. D. M. Held, J. D. Kissel, J. T. Patterson, D. G. Nickens, and D. H. Burke, HIV-1 inactivation by nucleic acid aptamers, *Frontiers in Bioscience*, 11, 89–112 (2006).

34. M. Curreli, C. Li, Y. Sun, B. Lei, M. A. Gundersen, M. E. Thompson, and C. Zhou, Selective functionalization of In_2O_3 nanowire mat devices for biosensing applications, *Journal of American Chemical Society*, 127, 6922–6923 (2005).

35. E. W. L. Chan, M. N. Yousaf, and M. Mrksich, Understanding the role of adsorption in the reaction of cyclopentadiene with an immobilized dienophile, *Journal of Physical Chemistry A*, 104, 9315–9320 (2000).

36. N. V. Lavrik, M. J. Sepaniak, and P. G. Datskos, Cantilever transducers as a platform for chemical and biological sensors, *Review of Scientific Instruments*, 75, 2229–2253 (2004).

37. J. Fritz, M. K. Baller, H. P. Lang, H. Rothuizen, P. Vettiger, E. Meyer, H.-J. Guntherodt, C. Gerber, and J. K. Gimzewski, Translating biomolecular recognition into nanomechanics, *Science*, 288, 316–318 (2000).

38. H.-F. Ji, R. Dabestani, G. M. Brown, and P. F. Britt, A novel self-assembled monolayer (SAM) coated microcantilever for low level caesium detection, *Chemical Communication*, 6, 457–458 (2000).

39. Nanostellar introduces gold in oxidation catalyst that can reduce diesel hydrocarbon emissions by as much as 40 percent more than commercial catalysts, www.nanotech-now.com/news.cgi?story_id= 21935 (2007).

40. Rational catalyst design, Nanostellar, Inc., www.nanostellar.com/technology.htm (2007).

41. C.-B. Wang and W. X. Zhang, Synthesizing nanoscale iron particles for rapid and complete dechlorination of TCE and PCBs, *Environmental Science and Technology*, 31, 2154–2156 (1997).

42. C. A. Dunn, NASA announces 2005 invention of the year: Environmental cleanup technology earns top honors, *NASA technology innovation*, 13 (2006).

43. S. V. Mattigod, G. E. Fryxell, and K. E. Parker, A thiol-functionalized nanoporous silica sorbent for removal of mercury from actual industrial waste, In *Environmental applications of nanomaterials: Synthesis, sorbents and sensors*, G. E. Fryxell and G. Cao, eds, chap. 11, Imperial College Press, London, 275–284 (2007).

44. S. V. Mattigod, G. E. Fryxell, X. Feng, K. E. Parker, and E. M. Pierce, Removal of mercury from aqueous streams of fossil fuel power plants using novel functionalized nanoporous sorbents, In *Coal combustion byproducts and environmental issues*, K. S. Sajwan, I. Twardowska, T. Punshon, A. K. Alva, eds, chap. 10, Springer, New York, 99–104 (2006).

45. Patented electro-static filter cartridges designed for environmental water purification, industrial water clean-up, RO protection and ultra-pure water, Argonide Advanced Filtration Technologies, www.argonide.com (2006).

46. Manganese oxide nanopowder, American Elements Products, www.americanelements.com (2008).

47. Nanoengineered sorbents, *Chemical and Engineering News*, 79, 32–38 (2001).

48. Y. Zhang, B. A. Koeneman, Y. Chen, P. Westerhoff, D. G. Capco, and J. Crittenden, Fate, transport, and toxicity of nanomaterials in drinking water, Nanoscience and Technology Institute, Symposium: Micro/Nano-Technology for National Security Applications, NSTI Nanotech 2007, Santa Clara, May 20–24 (2007).

49. D. Talley, Aerogel process nears market, CDT Systems, Inc., cdtwater.com/press/0401_H2Onews.php (2004).

50. G. Prabhukumar, M. Matsumoto, A. Mulchandani, and W. Chen, Cadmium removal from contaminated soil by tunable biopolymers, *Environmental Science and Technology*, 38, 3148–3152 (2004).

51. G. C. Bond and D. T. Thompson, Catalysis by gold, *Catalysis Review-Science and Engineering*, 41, 319–388 (1999).

52. C. W. Corti, R. J. Holliday, and D. T. Thompson, Developing new industrial applications for gold: Gold nanotechnology, *Gold Bulletin*, 35, 111–118 (2002).

53. C. W. Corti, Going green, *Annual Report 2006*, Rand Refinery Limited, pp. 38–41 (2006).

54. Toyota Chuo Kenkyushu K. K., Japanese Patent, 9150033 (1997).

55. R. Holliday, Evolving industrial uses of gold, *Annual Report 2006*, Rand Refinery Limited, pp. 34–37 (2006).

56. D. Andreeva, V. Idakiev, T. Tabakova, and A. Andreev, Low temperature water-gas shift reactions over Au/Fe_2O_3, *Journal of Catalysis*, 158, 354–355 (1996).

57. G. Mills and M. R. Hoffmann, Photocatalytic degradation of pentachlorophenol on titanium dioxide particles: Identification of intermediates and mechanism of reaction, *Environmental Science and Technology*, 27, 1681–1689 (1993).

58. M. A. Fox and M. T. Dulay, Heterogeneous photocatalysis, *Chemical Reviews*, 93, 341–357 (1993).

59. K. Venkatachalam, V. G. Gavalas, S. Xu, A. C. de Leon, D. Bhattacharyya, and L. G. Bachas, Poly (Amino Acid)-facilitated electrochemical growth of metal nano-particles, *Journal of Nanoscience and Nanotechnology*, 6, 2408–2412 (2006).

60. D. Ratner and M. A. Ratner, *Nanotechnology and homeland defense: New weapons for new wars*; Pearson Educations, Inc., Prentice-Hall, NJ (2004).

61. J. Indusi, Advanced sensors and detectors, Brookhaven National Laboratory, http://www.bnl.gov/homeland/sensors.asp (2006).

62. S. Keebaugh, W. J. Nam, and S. J. Fonash, Manufacturable, highly responsive nano-wire mercury sensors, Nanoscience and Technology Institute, Symposium: Micro/Nano-Technology for National Security Applications, NSTI Nanotech 2007, Santa Clara, May 20–24 (2007).

63. A. Loui, T. V. Ratto, T. S. Wilson, E. V. Mukerjee, Z.-Y Hu, T. A. Sulchek, and B. R. Hart, A compact, low-power cantilever-based sensor array for chemical detection, Nanoscience and Technology Institute, Symposium: Micro/Nano-Technology for National Security Applications, NSTI Nanotech 2007, Santa Clara, May 20–24 (2007).

64. P. Bondavilli, P. Legagneux, and D. Pribat, Gas fingerprinting using carbon nano-tubes transistor arrays, Nanoscience and Technology Institute, Symposium: Micro/Nano-Technology for National Security Applications, NSTI Nanotech 2007, Santa Clara, May 20–24 (2007).

65. S. L. Harper, B. Maddux, J. Hutchison, and R. L. Tanguay, Biodistribution and tox-icity of nanomaterials in vivo: Effects of composition, size, surface functionaliza-tion and route of exposure, Nanoscience and Technology Institute, Symposium: Micro/Nano-Technology for National Security Applications, NSTI Nanotech 2007, Santa Clara, May 20–24 (2007).

66. T. Knight, S. S. Wise, M. D. Mason, J. P. Wise, Sr., and A.-K. Ng, Cell-based assays for cytotoxic and pro-inflammatory effects of gold nanoparticles, Nanoscience and Technology Institute, Symposium: Micro/Nano-Technology for National Security Applications, NSTI Nanotech 2007, Santa Clara, May 20–24 (2007).

67. J. Chen, J. Zhu, H.-H. Cho, K. Cui, F. Li, X. Zhou, J. T. Rogers, S. T. C. Wong, and X. Huang, Differential cytotoxicity of metal oxide nanoparticles, Nanoscience and

Technology Institute, Symposium: Micro/Nano-Technology for National Security Applications, NSTI Nanotech 2007, Santa Clara, May 20–24 (2007).

68. R. Sharma, K. Shetty, R. Liang, and C. J. Chen, Interaction of carbon nanotube material with rat skin by 21 T MRI, Nanoscience and Technology Institute, Symposium: Micro/Nano-Technology for National Security Applications, NSTI Nanotech 2007, Santa Clara, May 20–24 (2007).

69. K. Spencer, Detection of landmines with SERS, Application Summary, EIC Laboratories, Inc. (2004).

70. J. M. Sylvia, J. A. Janni, J. D. Klein, and K. M. Spencer, Surface-enhanced Raman detection of 2,4-dinitrotoluene impurity vapor as a marker to locate landmines, *Analytical Chemistry*, 72, 5834–5840 (2000).

71. N. T. Kawai and K. M. Spencer, Raman Spectroscopy for Homeland Defense Applications, Application Note #18, InPhotonics, Inc. and EIC Laboratories (2004).

72. K. M. Spencer, J. M. Sylvia, J. A. Janni, and J. D. Klein, Advances in land mine detection using surface-enhanced Raman spectroscopy, *Proceedings of SPIE*, Vol. 3710 (2003).

73. G. L. Hornyak, C. J. Patrissi, and C. R. Martin, Surface enhanced Raman simulations of prolate and oblate nanoparticles, *Nanostructured Materials*, 9, 705–710 (1997).

74. Explosives detecting immunosensors, ESTCP Cost and Performance Report, CU-9713, U.S. Department of Defense (2000).

75. P. K. Stoimenov and K. J. Klabunde, Biological and chemical weapon decontamination by nanoparticles, In *Dekker encyclopedia of nanoscience and nanotechnology*, J. A. Schwartz, C. I. Contescu, and K. Putyera, eds., Vol. 1, pp. 241–245, CRC Press, Boca Raton, FL (2004).

76. G. W. Wagner, L. R. Procell, R. J. O'Connor, S. Munavalli, C. L. Carnes, P. N. Kapoor, and K. J. Klabunde, Reactions of VX, GB, GD, and HD with Nanosize Al_2O_3: Formation of aluminophosphonates, *Journal of American Chemical Society*, 123, 1636–1644 (2001).

77. P. K. Stoimenov, R. L. Klinger, G. L. Marchin, and K. J. Klabunde, Metal oxide nanoparticles as bactericidal agents, *Langmuir*, 18, 6679–6686 (2002).

78. R. J. Nikolic, C. L. Cheung, C. E. Reinhardt, and T. F. Wang, Future of semiconductor based thermal neutron detectors, Symposium: Micro/nano technologies for national security applications—explosive detection & radiological defense, Nanoscience and Technology Institute, NSTI Nanotech 2006, Boston, May 7–11 (2006).

79. T. M. Masciangioli, N. Savage, and B. P. Karn, Environmental implications of nanotechnology, Poster, www.environmentalfutures.org/nanotech.htm (2004).

80. V. Colvin, Nanotechnology: Environmental impact, Presentation, Nanotechnology and the environment: Background and resources, www.environmentalfutures.org/nanotech.htm (date unknown).

81. Nanomaterials and the Environment, In *Environmental, health, and safety research needs for engineered nanoscale materials*, chap. 4, The National Nanotechnology Initiative, NSET Subcommittee, Office of Science and Technology Policy, Washington D.C. (2006).

82. G. V. Andrievsky, V. K. Klochkov, E. L. Karyakina, and N. O. Mchedlov-Petrossyan, Studies of aqueous colloidal solutions of fullerene C-60 by electron microscopy, *Chemical Physics Letters*, 300, 392–396 (1999).

83. J. K. Holt, H. G. Park, Y. Wang, M. Stadermann, A. B. Artyukhin, C. P. Grigoropoulos, P. Costas, A. Noy, and O. Bakajin, Fast mass transport through sub-2-nanometer carbon nanotubes, *Science*, 312, 1034–1037 (2006).

84. J. K. Holt, Fast water transport through carbon nanotubes and implications for water treatment, Desalinating Water Cheaply—Exploring Technologies, Australian Academy of Sciences, Canberra, Australia (2006).

85. The history of drinking water, Environmental Protection Agency, Office of Water, Report EPA-816-F-00-006 (2000).

86. R. C. Bansal and M. Goyal, *Activated carbon adsorption*, Taylor & Francis, Boca Raton, FL (2005).

87. E. G. Furuya, H. T. Chang, Y. Miruya, H. Yokomura, S. Tajima, S. Yamashita, and K. E. Knoll, Interparticle mass transport mechanism in activated carbon adsorption of phenols, *Journal of Environmental Engineering*, 122, 909–916 (1996).

88. Z. Tong, M. Bischoff, L. Nies, B. Applegate, and R. F. Turco, Impact of fullerene (C60) on a soil microbial community, *Environmental Science and Technology*, 41, 2985–2991 (2007).

89. J. D. Fortner, D. Y. Lyon, C. M. Sayes, A. M. Boyd, J. C. Falkner, E. M. Hotze, L. B. Alemany, Y. J. Tao, W. Guo, K. D. Ausman, V. L. Colvin, and J. B. Hughes, C_{60} in water: Nanocrystal formation and microbial response, *Environmental Science and Technology*, 37, 3382–3391 (2003).

90. D. Y. Lyon, J. D. Fortner, C. M. Sayes, V. L. Colvin, and J. B. Hughes, Bacterial cell association and antimicrobial activity of a C_{60} water suspension, *Environmental Science and Technology*, 39, 4307–4316 (2005).

91. J. Lee, J. D. Fortner, J. B. Hughes, and J.-H. Kim, Photochemical production of reactive oxygen species by C_{60} in the aqueous phase during UV irradiation, *Environmental Science and Technology*, 41, 179–184 (2007).

92. H. Hyung, J. D. Fortner, J. B. Hughes, and J.-H. Kim, Natural organic matter stabilizes carbon nanotubes in the aqueous phase, *Environmental Toxicology Chemistry*, 24, 2757–2762 (2005).

93. J. D. Fortner, D.-I. Kim, A. M. Boyd, J. C. Faulkner, S. Moran, V. L. Colvin, J. B. Hughes, and J.-H. Kim, Reaction of water-stable C60 aggregates with ozone, *Environmental Science Technology*, 41, 2529–2535 (2007).

94. A. Navrotsky, Nanomaterials in the environment, agriculture and technology (NEAT), *Journal of Nanoparticle Research*, 2, 321–323 (2000).

95. A. Navrotsky, Environmental nanoparticles, In *Encyclopedia of nanoscience and nanotechnology*, Marcel-Dekker, New York, 1147–1156 (2004); Taylor & Francis (2005).

96. R. Senjen, Nanosilver—A threat to soil, water and human health? Friends of the Earth Australia, www.nano.foe.org.au (2007).

97. R. Eisler, A review of silver hazards to plants and animals, In *4th International Conference Proceedings: Transport, fate and effects of silver in the environment*, A. W. Andred and T. W. Bober, eds., pp. 143–144, Madison, Wisconsin (1996).

98. L. J. Albright and E. M. Wilson, Sub-lethal effects of several metallic salt-organic compound combinations upon heterotrophic microflora of a natural water, *Water Research*, 8, 101–105 (1974).

99. L. Braydich-Stolle, S. Hussain, J. J. Schlager, and M. Hofmann, In vitro cytotoxicity of nanoparticles in mammalian germline stem cells, *Toxicological Science*, 88, 412–419 (2005).

100. J. Fraser, Colloidal silver antibacterial liquid sprayed on Hong Kong subways as public health measure, NewsTarget.com, www.newstarget.com/020851.html (2006).

101. W. H. Beauman, Water purification and treatment, Short Course, Everpure, Inc., Westmont, IL (2001).

102. Osmosis, en.wikipedia.org/wiki/Osmosis (2008).

103. Osmotic Pressure, en.wikipedia.org/wiki/Osmotic_pressure (2008).

104. Reverse Osmosis, en.wikipedia.org/wiki/Reverse_osmosis (2008).

105. M. Abraham, Today's seawater is tomorrow's drinking water: UCLA engineers develop nanotech water desal membrane, Henry Samuelli School of Engineering and Applied Sciences, www.engineer.ucla.edu/news/2006/Desal_Membrane.htm (2006).

106. A. Risbud, Cheap drinking water from the ocean, *Technology Review*, MIT, www.technologyreview.com/Nanotech/16977/ (2006).

107. A. Parker, Tiny tubes make the flow go, *Research Highlights*, Lawrence Livermore National Laboratory, S & TR, February (2007).

108. Cheap, clean drinking water purified through nanotechnology, *Science News*, www.sciencedaily.com/releases/2008 (2008).

109. P. J. Majewski and C. P. Chan, Water purification by functionalised self-assembled monolayers on silica particles, *International Journal of Nanotechnology*, 5, 291–298 (2008).

110. Bacteria and Nanofilters: Future of clean water technology, *Science News*, www.sciencedaily.com/releases/2008/02/080222095403.htm (2008).

111. N. Hilal, Bacteria and Nanofilters: Future of clean water technology, *University of Nottingham News*, February 12 (2008).

112. P. S. Zurer, Countering Oils Spills, *Chemical and Engineering News*, 81, 32–33 (2003).

113. J. G. Reynolds, in P.S. Zurer, Countering oils spills, *Chemical and Engineering News*, 81, 32–33 (2003).

114. D. Robson, Magic sand, *ChemMatters Chemistry*, April, 8–9 (1994).

115. Clean air acts, en.wikipedia.org/wiki/Clean_Air_Act (2008).

116. The clean air act and nanotechnology, Nanotechnology Teleconference Series, Section of Environment, Energy and Resources, American Bar Association; www.abanet.org/environ/programs/teleconference/nanotech/session3/ (2007).

117. Nanotechnology and the toxic substances control act, Nanotechnology Teleconference Series, Section of Environment, Energy and Resources, American Bar Association; www.abanet.org/environ/programs/teleconference/nanotech/tsca/ (2007).

118. Nanotechnology and the clean water act, Nanotechnology Teleconference Series, Section of Environment, Energy and Resources, American Bar Association' www.abanet.org/environ/programs/teleconference/nanotech/cwa/ (2007).

119. A. Navrotsky, Nanomaterials in the environment, agriculture and technology (NEAT), *Journal of Nanoparticle Research*, 2, 321–323 (2000).

120. C. Anastacio and S. T. Martin, Atmospheric nanoparticles, In *Nanoparticles and the environment*, J. F. Banfield and A. Navrotsky, eds., Vol. 44, pp. 293–349 Reviews in Mineralogy and Geochemistry; Mineralogical Society of America, Washington D.C. (2001).

121. W. J. Koros, Gas separation, In *Membrane separation systems: Recent developments and future directions*, R. W. Baker, E. L. Cussler, W. Eykamp, W. J. Koros, R. L. Riley, H. Strathmann, eds., William Andrew Publishing, Inc., New York, 349–355 (1991).

122. Pressure Swing Adsorption, en.wikipedia.org/wiki/Pressure_swing_adsorption (2008).

123. Non-cryogenic air separation processes: Nitrogen or oxygen generation using pressure swing adsorption (PSA), Universal Industrial Gases, Inc., www.uigi.com/noncryo.html (2007).

124. G. Alptekin, An advanced rapid recycling CO_2 and H_2O control system for PLSS, NASA SBIR-NNJ05JB96C, TDA Research (2005).

125. W. Xing, W. C. D. da Costa, and Z. F. Yan, Environmental separation and reactions: Zeolite membranes, In *Encyclopedia of nanoscience and nanotechnology*, Vol. 2, Marcel-Dekker, New York (2004); Taylor & Francis (2005).

126. K. Aoki, K. Kusakabe, and S. Morooka, Separation of gases with an A-type Zeolite membrane, *Industrial and Engineering Chemistry Research*, 39, 2245 (2000).

127. Y. Ohtaa, H. Takabab, and S.-I. Nakaoa, A combinatorial dynamic Monte Carlo approach to finding a suitable Zeolite membrane structure for CO_2/N_2 separation, *Microporous and Mesoporous Materials*, 101, 319–323 (2007).

128. T. Kouketsu, S. Duan, T. Kai, S. Kazaman, and K. Yamada, PAMAM dendrimer composite membrane for CO_2 separation: Formation of a chitosan gutter layer, *Journal of Membrane Science*, 287, 51–59 (2007).

129. N. Hiyoshi, K. Yogo, and T. Yashima, Adsorption characteristics of carbon dioxide on organically functionalized SBA-15, *Microporous and Mesoporous Materials*, 84, 357–365 (2005).

130. S. Duan, T. Kouketsu, T. Kai, S. Kazaman, and K. Yamada, Development of PAMAM dendrimer composite membrane for CO_2 separation, *Journal of Membrane Science*, 283, 2–6 (2006).

131. A. S. Kovvali, H. Chen, and K. K. Sirkar, Dendrimer membranes: A CO_2-selective molecular gate, *Journal of American Chemical Society*, 122, 7594–7595 (2000).

132. K. Yamada, S. Kazama, and K. Yogo, Development of innovative gas separation membranes through sub-nanoscale control materials control, Stanford University Global Climate & Energy Project (GCEP), CO_2 Capture, gcep.stanford.edu/research/factsheets/membrane_subnanoscale.html (2005).

133. D. Guro, Fuel cell grade hydrogen purity requirements: Imapct on purification, analysis and cost, USFCC Fuel Cell Seminar, Air Products (2006).

134. B. Dumé, Conductive polymers and CNTs in the mix, Nanotechweb.org, nanotechweb.org/cws/article/tech/32863 (2008).

135. S. M. C. Vieira, P. Beecher, I. Haneef, F. Udrea, W. I. Milne, M. A. G. Namboothiry, D. L. Carroll, J. Park, and S. Maeng, Use of nanocomposites to increase electrical "gain" in chemical sensors, *Applied Physics Letters*, 91, 203111 (2007).

136. M. K. Kennedy, F. E. Kruis, H. Fissan, B. R. Mehta, S. Stappert, and G. Dumpich, Tailored nanoparticle films from monosized tin oxide nanocrystals: Particle synthesis, film formation, and size-dependent gas-sensing properties, *Journal of Applied Physics*, 93, 551–560 (2003).

137. H. Ogawa, M. Nishikawa, and A. Abe, Hall measurement studies and an electrical conduction model of tin oxide ultrafine particle films, *Journal of Applied Physics*, 53, 4448–4455 (1982).

138. R. E. Smalley, Our energy challenge, Presentation, 27th Illinois Junior Science & Humanities Symposium, smalley.rice.edu/smalley.cfm?doc_id= 4862, April 3 (2005).

139. P. D. Persans in "Busy old fool, unruly sun.....," The Energy Challenge, AIP Industrial Physics Forum, blogs.physicstoday.org/industry07/2007/10/busy_old_fool_unruly_sun_1.html, Seattle, WA (2008).

140. B. K. Miller, Largest solar factory coming to bay area, GlobeSt.Direct, www.nanosolar.com/cache/GlobeSt100MM.htm (2006).

141. M. Kanelios, Solar cell breaks efficiency record, CNET News.com (2006).

142. J. Nelson, *The physics of solar cells*, Imperial College Press, London (2003).

143. W. Shockley and H. J. Queisser, Detailed balance limit of efficiency of p-n junction solar cells, *Journal of Applied Physics*, 32, 510–519 (1961).

144. Photovoltaics: Solar electricity and solar cells in theory and practice, *Solar Magazine*, www.solarserver.de/wissen/photovoltaik-e.html (2007).

145. Quantum efficiency, en.wikipedia.org/wiki/Quantum_efficiency (2008).

146. M. Grätzel, Photoelectrochemical cells, *Nature*, 414, 338–344 (2001).

147. I. Robel, R. Subramanian, M. Kuno, and P. V. Kamat, Quantum dot solar cells: Harvesting light energy with CdSe nanocrystals molecularly linked to mesoscopic TiO_2 film, *Journal of American Chemical Society*, 128, 2385–2393 (2006).

148. I. Robel, M. Kuno, and P. V. Kamat, Size-dependent electron injection from excited CdSe quantum dots into TiO$_2$ nanoparticles. *Journal of American Chemical Society*, 129, 4136–4137 (2007).

149. G. Overton, Photovoltaics: Quantum dots promise next generation solar cells, *Laser Focus World*, www.laserfocusworld.com/articles/252486 (2006).

150. P. Weiss, Photon double whammy: Careening electrons may rev up solar cells, *Science News Online*, 165, 259 (2004).

151. P. Weiss, Quantum dot leap: Tapping tiny crystals' inexplicable light-harvesting talent, *Science News Online*, 169, 344 (2006).

152. A. J. Nozik, Exciton multiplication and relaxation dynamics in quantum dots: applications to ultrahigh-efficiency solar photon conversion, *Inorganic Chemistry*, 44, 6893–6899 (2005).

153. R. J. Ellingson, M. C. Beard, J. C. Johnson, P. Yu, O. I. Micic, A. J. Nozik, A. Shabaev, and A. L. Efros, Highly efficient multiple exciton generation in colloidal PbSe and PbS quantum dots, *Nano Letters*, 5, 865–871 (2005).

154. J. E. Murphy, M. C. Beard, A. G. Norman, S. P. Ahrenkiel, J. C. Johnson, P. Yu, O. I. Micic, R. J. Ellingson, and A. J. Nozik, PbTe colloidal nanocrystals: Synthesis, characterization, and multiple exciton generation, *Journal of American Chemical Society*, 128, 3241–3247 (2006).

155. R. J. Ellingson, A. J. Nozik, M. C. Beard, J. Johnson, J. Murphy, K. Knutsen, K. Gerth, J. Luther, M. Hanna, O. Micic, A. Shabaev, and A. Efros, Nanocrystals generating >1 electron per photon may lead to increased solar cell efficiency, *Solar and Alternative Energy*, SPIE, spie.org/x8735.xml (2008).

156. J. Li, F. Dierschke, J. Wu, A.C. Grimsdale, and K. Müllen, Poly(2,7-carbazole) and perylene tetracarboxydiimide: a promising donor/acceptor pair for polymer solar cells, *Journal of Materials Chemistry*, 16, 96–100 (2006).

157. M. M. Wienk, J. M. Kroon, W. J. H. Verhees, J. Knol, J. C. Hummelen, P. A. van Hal, and R. A. J. Janssen, Efficient methano[70]fullerene/MDMO-PPV bulk heterojunction photovoltaic cells, *Angewandte Chemie International Edition*, 42, 3371–3375 (2003).

158. C. W. Tang and A. C. Albrecht, Chlorophyll-*a* photovoltaic cells, *Nature*, 254, 507–509 (1975).

159. C. W. Tang, Two-layer organic photovoltaic cell, *Applied Physics Letters*, 48, 183–185 (1986).

160. S. Mitra and C. Li, Unique nanotube composites constructed for organic solar cells, *Solar and Alternative Energy*, SPIE, spie.org/x8735.xml (2008).

161. S.-S. Sun and N. S. Sariciftci, eds, *Organic photovoltaics: Mechanisms, materials, and devices (Optical engineering)*, CRC Press, Boca Raton, FL (2005).

162. L. Schmidt-Mende, A. Fechtenkötter, K. Müllen, E. Moons, R. H. Friend, and J. D. MacKenzie, Self-organized discotic liquid crystals for high-efficiency organic photovoltaics, *Science*, 293, 1119–1122 (2001).

163. S. W. Hur, T. W. Kim, and J. W. Park, Organic photovoltaic effects depending on CuPc layer thickness, *Journal of Korean Physics Society*, 45, 627–629 (2004).

164. P. Peumans and S. R. Forrest, Very-high-efficiency double-heterostructure copper phthalocyanine/C60 photovoltaic cells, *Applied Physics Letters*, 79, 126–128 (2001).

165. A. Kongkanand, R. M. Domínguez, and P. V. Kamat, Single wall carbon nanotube scaffolds for photoelectrochemical solar cells: Capture and transport of photo-generated electrons, *Nano Letters*, 7, 676–680 (2007).

166. T. Hasobe and P. V. Kamat, Photoelectrochemistry of stacked cup carbon nano-tube films. Tube-Length dependence and charge transfer with excited porphyrin, *Journal of Physical Chemistry C*, 111, 16626–16634 (2007).

167. A. Kongkanand and P. V. Kamat, Electron Storage in single wall carbon nanotubes: Fermi level equilibration in semiconductor–SWCNT suspensions, *ACS Nano*, 1, 13–21, NDRL 4722 (2007).

168. A. Kongkanand and P. V. Kamat, Interactions of single wall carbon nanotubes with methyl viologen radicals: Quantitative estimation of stored electrons, *Journal of Physical Chemistry C*, 111, 9012–9015 (2007).

169. P. V. Kamat, Meeting the clean energy demand: Nanostructure architectures for solar energy conversion, *Journal of Physical Chemistry C*, 111, 2834–2860, NDRL 4697 (2007).

170. J. Wei, Y. Jia, Q. Shu, K. Wang, D. Zhuang, G. Zhang, Z. Wang, J. Luo, A. Cao, and D. Wu, Double-walled carbon nanotube solar cells, *Nano Letters*, 7, 2317–2321 (2007).

171. Z.-P. Yang, L. Ci, J. A. Bur, S.-Y. Lin, and P. M. Ajayan, Experimental observation of an extremely dark material made by a low-density nanotube array, *Nano Letters*, 8, 446–451 (2008).

172. A. Kiebele and G. Grüner, Carbon nanotube based battery architecture, *Applied Physics Letters*, 91, 144104 (2007).

173. Battery (Electricity), en.wikipedia.org/wiki/Battery_(electricity) (2008).

174. C. K. Chan, X. F. Zhang, and Y. Cui, High capacity Li ion battery anodes using Ge nanowires, *Nano Letters*, 8, 307–309 (2008).

175. U. Kasavajjula, C. Wang, and A. J. Appleby, Nano- and bulk-silicon-based insertion anodes for lithium-ion secondary cells, *Journal of Power Sources*, 163, 1003–1039 (2007).

176. C. K. Chan, H. Peng, G. Liu, K. McIlwrath, X. F. Zhang, and Y. Cui, High-performance lithium battery anodes using silicon nanowires, *Nature Nanotechnology*, 3, 31–35 (2007).

177. R. P. Raffaelle, T. Gennett, J. Maranchi, P. Kumta, A. F. Hepp, M. J. hebern, A. C. Dillon, and K. C. Jones, Carbon nanotube anodes for lithium ion batteries, *Materials Research Society Symposium Proceedings*, 706, Z10.5.1–Z10.5.7 (2002).

178. S. H. Ng, J. Wang, Z. P. Guo, J. Chen, G. X. Wang, and H. K. Liu, Single wall carbon nanotube paper as anode for lithium-ion battery, *Electrochimica Acta*, 51, 23028 (2005).

179. V. L. Pushparaj, M. M. Shaijumon, A. Kumar, S. Murugesan, L. Ci, R. Vajtai, R. J. Linhardt, O. Nalamasu, and P. M. Ajayan, Flexible energy storage devices based on nanocomposite paper, *Proceedings of the National Academy Science*, 104, 13574–13577 (2007).

180. C. Kim, M. Noh, M. Choi, J. Cho, and B. Park, Critical size of a nano SnO_2 electrode for Li-secondary battery, *Chemistry of Materials*, 17, 3297–3301 (2005).

181. A. Fujishima and K. Honda, Electrochemical photolysis of water at a semiconductor electrode, *Nature*, 238, 37–38 (1972).

182. I. Okura and M. Kaneko, *Photocatalysis: Science and technology, biological and medical physics Series*, Springer-Verlag, Berlin, Germany (2002).

183. Journey to sustainable energy: The H_2 solution, Presentation, Clean Energy Research Center, University of Southern Florida, cerc.eng.usf.edu/CERC_Overview_Dec04.pdf (2004).

184. A. E. Messer, Solar cell directly splits water for hydrogen, Eureka Alert, www.eurekalert.org/pub_releases/2008-02/ps-scd021108.php (2008).

185. B. O'Regan and M. Grätzel, A low-cost, high-efficiency solar cell based on dye-sensitized colloidal TiO_2 films, *Nature*, 353, 737–740 (1991).

186. H. Hata, Y. Kobayashi, V. Bojan, W. J. Youngblood, and T. E. Mallouk, Direct deposition of trivalent rhodium hydroxide nanoparticles onto a semiconducting layered calcium niobate for photocatalytic hydrogen evolution, *Nano Letters*, 8, 794–799 (2008).

187. L. Becker, Hydrogen storage, Review Article, www.csa.com (2001).

188. R. Dagani, Tempest in a tiny tube, *Chemical Engineering News*, 80, 25–28 (2001).

189. Hydrogen storage technologies roadmap, FreedomCar Fuel Partnership (2005).

190. A. C. Dillon, K. M. Jones, T. A. Bekkedahl, C. H. Kiang, D. S. Bethune, and M. J. Heben, Storage of hydrogen in single-walled carbon nanotubes, *Nature*, 386, 377–379 (1997).

191. M. Hirscher, M. Becher, M. Haluska, U. Dettlaff-Weglikowska, A. Quintel, G. S. Duesberg, Y.-M. Choi, P. Downes, M. Hulman, S. Roth, I. Stepanek, and P. Bernier, Hydrogen storage in sonicated carbon materials, *Applied Physics A*, 72, 129–132 (2001).

192. V. V. Simonyan, P. Diep, and J. K. Johnson, Molecular simulation of hydrogen adsorption in charged single-walled carbon nanotubes, *Journal of Chemical Physics*, 111, 9778–9783 (1999).

193. V. V. Simonyan and J. K. Johnson, Hydrogen storage in carbon nanotubes and graphitic nanofibers, *Journal of Alloys and Compounds*, 330–332, 654–658 (2002).

194. A. Nikitin, X. Li, Z. Zhang, H. Ogasawara, H. Dai, and A. Nilsson, Hydrogen storage in carbon nanotubes through the formation of stable C–H bonds, *Nano Letters*, 8, 162–167 (2008).

195. A. Nikitin, H. Ogasawara, D. Mann, R. Denecke, Z. Zhang, H. Dai, K. Cho, and A. Nilsson, Hydrogenation of Single-Walled Carbon Nanotubes, *Physical Review Letters*, 95, 225507 (2005).

196. A. Nikitin, New carbon nanotube hydrogen storage results surpass Freedom Car requirements, *Nanowerk*,://www.nanowerk.com/spotlight/spotid= 4154.php (2008).

197. T. Furuta, H. Goto, T. Ohashi, Y. Fujiwara, and S. Yip, Theoretical evaluation of hydrogen storage capacity in pure carbon nanostructures, *Journal of Chemical Physics*, 119, 2376–2385 (2003).

198. Y.-W. Lee, R. Bhowmick, and B. M. Clemens, Hydrogen storage in palladium and platinum-doped SWNTs by spill-over mechanism, Materials Research Society 2006 Fall Meeting, Z6.6, Boston, MA (2006).

199. G. Zhang, P. Qi, X. Wang, Y. Lu, D. Mann, X. Li, and H. Dai, Hydrogenation and hydrocarbonation and etching of single-walled carbon nanotubes, *Journal of American Chemical Society*, 128, 6026–6027 (2006).

200. E. Durgun, S. Ciraci, and T. Yildirim, Functionalization of carbon-based nanostructures with light transition-metal atoms for hydrogen storage, *Physical Review B*, 77, 085405–085413 (2008).

201. Go/No-Go decision: Pure, undoped single walled carbon nanotubes for vehicular hydrogen storage, U.S. Department of Energy Hydrogen Program, October (2006).

202. M. J. Heben, A. C. Dillon, K. E. H. Gilbert, P. A. Parilla, T. Gennett, J. L. Alleman, G. L. Hornyak, and K. M. Jones, Assessing the hydrogen storage adsorption capacity of single-wall carbon nanotube/metal composites, In *Hydrogen materials and vacuum systems*, G. R. Myneni and S. Chattopadhyay, eds., pp. 77–89, American Institute of Physics, Conference Proceedings, No. 671 (2003).

203. D. B. Geohegan, H. Hu, A. A. Puretzky. B. Zhao, D. Styers-Barnett, and I. Ivanov, Synthesis and processing of single-walled carbon nanohorns for hydrogen storage and catalyst supports, *FY 2006 Annual Report*, pp. 473–475 (2006).

204. D. B. Geohegan, H. Hu, A. Purtzky, M. Yoon, B. Zhao, and C. M. Rouleau, Single-walled carbon nanohorns for hydrogen storage and catalyst support, DOE Hydrogen Program Review, May 15 (2007).

205. M. Yoon, S. Yang, E. Wang, and Z. Zhang, Charged fullerenes as high capacity hydrogen storage media, *Nano Letters*, 7, 2578 (2007).

206. D. L. Jacobson, PEM Fuel Cells, National Institute for Standards and Testing (NIST), Gaithersburg, MD., physics.nist.gov/MajResFac/NIF/pemFuelCells.html (2006).

207. Nafion, en.wikipedia.org/wiki/Nafion (2007).

208. Nafion: Physical and Chemical Properties, Perma Pure, LLC., www.permapure. com (2006).

209. A. F. Hepp and S. G. Bailey, Inorganic photovoltaic materials and devices: Past, present and future, In *Organic photovoltaics: Mechanisms, materials, and devices (Optical Engineering)*, chap. 2, S.-S. Sun and N. S. Sariciftci, eds., CRC Press, Boca Raton, FL, Chapter 2, pp 19–36 (2005).

210. S. L. Gillett, Nanotechnology: Clean energy and resources for the future, White Paper for Foresight Institute: Preparing for Nanotechnology (2002).

211. G. E. Fryxell, T. S. Zemanina, H. wu, S. Kelly, Y. Lin, K. Kremmer, and K.N. Raymond, Actinide-specific interfacial chemistry of monolayer coated mesoporous ceramics, Final Report, U.S. Department of Energy, Project No. 65370 (2001).

212. Y. Lin, G. E. Fryxell, H. Wu, and M. Engelhard, Selective sorption of cesium using self-assembled monolayers on mesoporous supports, *Environmental Science and Technology*, 35, 3962–3966 (2001).

213. G. E. Fryxell, S. V. Mattigod, K. Parker, and R. Skaggs, Heavy metal sequestration using functional nanoporous materials, U.S. EPA Workshop on Nanotechnology for Site Remediation, es.epa.gov/ncer/publications/workshop/pdf/10_20_05_ nanosummary.pdf, October 20–21 (2005).

214. N. L. Rosi, J. Eckert, M. Eddaoudi, D. T. Vodak, J. Kim, M. O'Keefe, and O. M. Yaghi, Hydrogen storage in microporous metal-organic frameworks, *Science*, 300, 1127–1129 (2003).

215. S. S. Kaye, A. Dailly, O. M. Yaghi, and J. R. Long, Impact and preparation and handling on the hydrogen storage properties of $Zn_4O(1,4,–benzene dicarboxylate)_3(MOF-5)$, *Journal of American Chemical Society*, 129, 14176–14177 (2007).

216. M. Dinca and J. R. Long, High-enthalpy adsorption in cation-exchanged variants of the microporous metal-organic framework $Mn_3[(Mn_4Cl)_3(BTT)_8(CH_3OH)_{10}]_2$, *Journal of American Chemical Society*, 129, 1172–11176 (2007).

217. S. S. Han and W. A. Goddard, III, Lithium-doped metal-organic frameworks for reversible H2 storage at ambient temperature, *Journal of American Chemical Society*, 129, 8422–8423 (2007).

218. J. Perlin, The story of photocells, In *Organic photovoltaics: Mechanisms, materials, and devices (Optical Engineering)*, chap. 1, S.-S. Sun and N. S. Sariciftci, eds., CRC Press, Boca Raton, FL, 3–18 (2005).

219. D. W. Elliott and W.-X. Zhang, Field assessment of nanoscale bimetallic particles for groundwater treatment, *Environmental Science and Technology*, 35, 4922–4926 (2001).

220. D. Engber, How does activated carbon work? www.slate.com/id/2131430/ (2005).

221. H. J. Lee, D.-Y. Kim, J.-S. Yoo, J. Bang, S. Kim, and S.-M. Park, Anchoring Cadmium chlacogenide quantum dots (QDs) onto stable oxide semiconductors for QD sensitized solar cells, *Bulletin of Korean Chemical Society*, 28, 1–6 (2007).

222. Y.-J. Shen and Y.-L. Lee, Assembly of CdS quantum dots onto mesoscopic TiO_2 films for quantum dot-sensitized solar cell applications, *Nanotechnology*, 19, 045602–045608 (2008).

223. R. Dagani, Silica sorbents with a taste for metals: Organic-inorganic nanocomposites selectively take up cesium, mercury, or other pollutants, *Chemical and Engineering News*, 79, 28–30 (2001).

224. T. Sawaguchi, Y. Sato, and F. Mizutani, Ordered structures of self-assembled monolayers of 3-mercaptopropionic acid on Au(111): In situ scanning tunneling microscopy study, *Physical Chemistry Chemical Physics*, 3, 3399–3404 (2001).

Problems

14.1 Reorder the list generated by Richard Smalley to suit your perspective of the 10 most pressing challenges facing our species over the next 50 years.

14.2 What are some advantages of solar cells that employ quantum dots? List some disadvantages.

14.3 Why do you think groundwater would require less treatment than surface water? What kinds of contaminants would you expect to find in groundwater?

14.4 Why do you think natural groundwaters (or spring waters) taste so clean and fresh? Do you believe wetlands play a role in water purification?

14.5 Is single molecule detection important in environmental analysis?

14.6 List some important differences between fuel cells and photovoltaic cells. Which one is an electrolytic cell? Which one a voltaic cell? Which on uses batteries to store energy? Which one was designed to replace batteries?

14.7 What are the challenges affecting the widespread use of photovoltaic cells and that of fuel cells?

14.8 How many other means can you devise to measure the levels of glucose (hint: think of all aspects of the glucose reaction)?

14.9 Why do you think that C_{60}-agglomerated structures look rectangular?

14.10 The equation for osmotic pressure is similar to the gas law $PV = nRT$. Show how this is so.

14.11 What is the osmotic pressure of 10 g of sucrose in 100 mL of solution (MW = 180.16 g·mol^{-1})?

14.12 (a) 2.00 g of an unknown polymer are placed in 100 mL of solution. The measured osmotic pressure Π is 5.0×10^{-3} atm at 25°C. Find its molecular weight (MW). (b) Calculate MW for a 1% solution of a protein (ideal, isoelectric) that gives an osmotic pressure of 55 mm H_2O at 0°C. (Hint: pressure is in terms of mm H_2O and not mm Hg).

14.13 Are 10-nm diameter Au nanoparticles in solution able to cause osmotic pressure?

14.14 Why is a semipermeable membrane necessary for osmosis to occur? What about osmotic pressure? Is a semipermeable membrane necessary for osmotic pressure?

14.15 Calculate the total surface area of an electrode containing Ge nanowires that are 20 nm in diameter and 50 μm in length. The electrode surface area is 25 cm^2 and the wire surface density is 80%.

Index

F

Fano effect, 72
Faraday dark space, 338; *see also* Cathode dark space
Fermi–Dirac distribution, 133
Fermi energy, 133, 152
Fermi level, 222
Ferrimagnets, 207
Ferrofluids, 214–215, 620
Ferromagnetic shape-memory (FSM) alloy, 319
Fiber-optic biosensor (FOB), 703–704
Fiber-optic chemical sensors, 501
Field effect transistors (FET), 127
Field emission display (FED), 190
Finite element analysis (FEM), 96
Fisher–Tropsch synthesis, 398
Flame synthesis, of SWNTs, 444
Fluorescein, *see* Fluorescent isothiocyanate (FITC)
Fluorescent isothiocyanate (FITC), 188
Fluorophores, 187–190
Foresight Institute, 45
Formic acid, thermal decomposition of, 387–388
Fossil fuel exploration, effects of, 686
Fourier filtering, 92
Free-electron lasers (FEL), 94
Free electrons, 143
Functionalized cantilever cesium cations, 696
Functionalized electrochemical cell, 694

G

Gas sensors
    based on biomimetic moieties, 582
    sensing and detection capabilities of, 728
    silicon diatom model for, 582–583
Gas separation
    advanced membrane technology
        Knudsen flow, 723–724
        solution-diffusion separation, 724
        traditionally established industrial methods, 724–725
        viscous flow, 723
        zeolite membranes, 725–726
Giant magnetoresistance (GMR), 209
    biosensing polarization diversity, 506
Glass transition temperature (T_g), 425, 427
Gold
    nanoparticles, 698
    nanoscale properties of, 707
    nanowires, 261–262
    particles, 390
Grain boundary
    hardening, 302
    sliding, 302
Granular activated carbon (GAC), 714–715
Grätzel cells, 578–581
Ground states, 75, 79
Guided-mode resonance sensors, 504

H

Hall effect, 219
Halloysite nanotube clay composites
    formation of, 470
    natural, 471
    as reinforcing elements, 471–472
Hall–Petch relationship, 302–303
Hall plateau number, 88
Hall potential, 84
Hall voltage, 87
Halogenated organic compounds (HOCs), 711
Halogenation, 447
Hamilton's equations of motion, 248
Hard coatings, 370
Hardness testing, 424–425
Harmonic approximation, 247, 251
Heating, ventilation, and air conditioning (HVAC), 38
Heisenberg uncertainty principle, 68–71, 173
H/E values, of Ti–Si–B–C–N coatings, 368
Hexa-peri-hexabenzocoronone (HBC), 736
High-aspect-ratio vertical structures, 276
High-efficiency particulate air (HEPA) filters, 36, 40
High-power pulsed DC magnetron sputtering (HPPMS), 325, 331
High-resolution transmission electron microscopy (HRTEM), 403
HiPco process, 444–445
Homeostasis, 607
Hydrogenation, 447
Hydrophobic aerogels possess, 721
Hyper filtration, subnanometer particles of, 717

I

Immune system, 484
    adaptive, 486–487
        antibodies in, 487–488
    discovery of, 485
    innate, 486
    molecular recognition mechanism, 485
    white blood cells (*see* White blood cells)
Impregnation, 399
Inclusion fraction, 416
Indium–tin oxide (ITO), 693
Induced dipole moment, 179
Inductively coupled plasma (ICP), 402
Injection plasma processing (IPP), 319
Ink-jet-like spray techniques, 511
Innate immune system, 486
Inorganic carbons, 414
Inorganic nanofibers
    boron nitride, 463–464
    silicon carbide, 463
Inorganic–organic hybrid materials, 106–107
Institute of Electrical and Electronics Engineers (IEEE), 36, 99